SECOND EDITION

MANAGING PIG HEALTH

A Reference for the Farm

Michael R. Muirhead
Thomas J. L. Alexander

John Carr
Editor

5m Publishing

Published by 5m Publishing
Benchmark House
8 Smithy Wood Drive
Sheffield
S35 1QN
United Kingdom

ISBN 978-0-9555011-5-9

A CIP catalogue record for this book is available from the British Library.

Note to the Reader

The methods of treatment and control of conditions discussed in the book are guidelines only. Any recommendations given and so used are the responsibility of the producer, and the advice of a veterinarian should be sought in case of doubt. No responsibility is accepted by the authors or publishers for any application of the advice given in this book because each farm and region is different and responses cannot be predicted. Chemical compound names are used throughout. No endorsement is intended nor is any criticism implied of similar products not named.

5m Publishing. would welcome your comments regarding this book. All feedback will be taken into account and hopefully help enhance subsequent editions.
Please write to us at: **5m Publishing**
Benchmark House
8 Smithy Wood Drive
Sheffield, S35 1QN
United Kingdom

Alternatively e-mail us at **books@5mpublishing.com**
You can also contact us via our web sites at:

www.5mpublishing.com or www.thepigsite.com

About the Editor

John Carr PhD, BVSc., DPM, DiplECPHM, MRCVS.

John started working with pigs at the age of 11 and for the next 10 years worked as a general stockperson on pig, poultry and sheep farms. In 1982 he qualified as a Veterinary Surgeon from Liverpool University and for the next 6 years worked in general practice in Liverpool and Dumfries. John returned to Liverpool University and obtained his PhD following studies on cystitis in sows. In 1992 he joined Garth Veterinary Group where he concentrated on improving pig health through stockmanship on farms throughout the world. At the Garth Veterinary Group he became a Diplomate in Pig Medicine and is a Diplomate of the European College of the European College. John is also recognised as a pig specialist by the Royal College of Veterinary Surgeons.

He has taught production medicine at several universities: UK; Liverpool and Royal Veterinary College, London, US; North Carolina and Iowa State and in Western Australia at Murdoch. John runs a consultancy practice with clients in North America, Europe, Asia, Australia and Africa. John specialises in maintaining the health of pigs through promoting excellence in stockmanship and production practices.

John has experienced pig farming throughout the world and in a variety of hot, cold dry and damp climates. He also has a personal interest in pet pigs and was the International Vet of the Year 2000 for pet pigs.

About the Original Authors

Michael R. Muirhead BVM&S, FRCVS, DPM. (1936 – 2003)

Mike Muirhead was the senior partner in the Garth Veterinary Group, a specialist pig practice in the UK. He was a past president of the Pig Veterinary Society, a Fellow of the Royal College of Veterinary Surgeons and a Diplomate in Pig Medicine. For over 30 years he provided consultancy services to a number of breeding and large and small commercial operations both in Europe and North America. He published over 80 papers on management and disease control and lectured extensively to many farming groups. He received the David Black award in 1980, the Bledesloe award from the National Agricultural Society of England in 1985 and the Dalrympe Champney's cup from the British Veterinary Association in 1989. He was a regular contributor to International Pig letter and was also a partner in a large pig unit.

Thomas J. L. Alexander PhD, MVSc., BSc., MRCVS, DPM. (1930 - 2008)

Tom Alexander taught pig husbandry, pig medicine and infectious diseases at the University of Cambridge where he was deputy head of the veterinary school for six years. He was internationally known for his research on vomiting and wasting disease, swine dysentery, streptococcal meningitis and medicated early weaning which was his brainchild. He was one of the first veterinarians world-wide to take a post graduate degree in pig medicine, at Guelph, in Canada in 1960 and followed up with a PhD in enteric viruses of the pig in 1965. He was international consultant veterinarian to the Pig Improvement Company from 1967. He was a partner in a 200 sow pig farm in the 1970s. He was also on the Board of Directors of a large mixed farming business which included 5,000 sows for 15 years, being chairman for six. He was a founding member of the British Pig Veterinarian Society and its president for two years and was the convenor of the organising committee of the first International Pig Veterinary Society Congress in 1969. He was a prime influence on the development of a National Diploma in Pig Medicine for the Royal College of Veterinary Surgeons and chaired its Pig Medicine Board. He regularly give papers and conducted consultancy work in most pig-rearing countries of the world.

How to Get the Best Use Out of This Book.

IMPORTANT – PLEASE READ.

Managing Pig Health has been written with a clear objective – to help you the pig producer or advisor, to understand, identify, manage and treat disease problems that are specific to your farm, with advice as appropriate from your veterinarian.

It should also help to maximise production and profitability because as a practical manual it can be referred to during the working day.

The book is not meant to be a substitute for your veterinarian but to give you a clearer understanding of what should be done to maximise health and increase production and thus be able to interpret and implement veterinary advice more effectively. It also aims to promote communication between the veterinarian and the farmer.

It is designed for easy reference when you encounter a problem or have a query. For example if greasy pig disease is affecting weaners, you will find that the disease is discussed in chapter 9 "Managing Health in the Weaner, Grower and Finishing Periods". If the problem is in sucking pigs the salient points will be found in chapter 8 "Managing Health in the Farrowing and Sucking Period". You will also find it further illustrated under chapter 10 "Skin Conditions".

The Structure of the Book

**IDEALLY, YOU SHOULD HAVE 2 COPIES OF THIS BOOK.
ONE KEPT IN THE FARM LIBRARY AND THE OTHER FOR USE ON/AROUND THE FARM.**

It is essential you familiarise yourself with the layout before you use the book.

Chapter 1 clarifies the terminology used throughout the book and discusses the basic outline of the anatomy and physiology of the pig. For those of you who already have a good biological knowledge this will be elementary but for others it will be the basis for understanding the rest of the book.

Abbreviations are listed at the end of the book.

Chapters 2 and 3 describe the fundamental causes of disease and how infectious organisms are spread and controlled.

The contents and specific problems are outlined at the beginning of each chapter in alphabetical order.

Relevant chapters open with an explanation of the factors that contribute to the management control of the diseases or conditions.

If you are unable to identify a disease refer to the "identifying problems" format in the relevant chapter. These can be used to relate the clinical signs you observe to common causes.

If in doubt consult your veterinarian.

In the appendix a list of medicines is given as a guide to treatments and dose levels*. Space is available by the chemical name for you to add the trade names of these medicines that are available in your country.

The book is extensively cross-referenced and in some cases duplicated – purposely – for ease of access and to give a wide understanding and appreciation of each problem area.

Metric measurements are used throughout the book. Imperial measurements, methods of conversions and quick reference tables are given in the appendix.

MANAGING PIG HEALTH

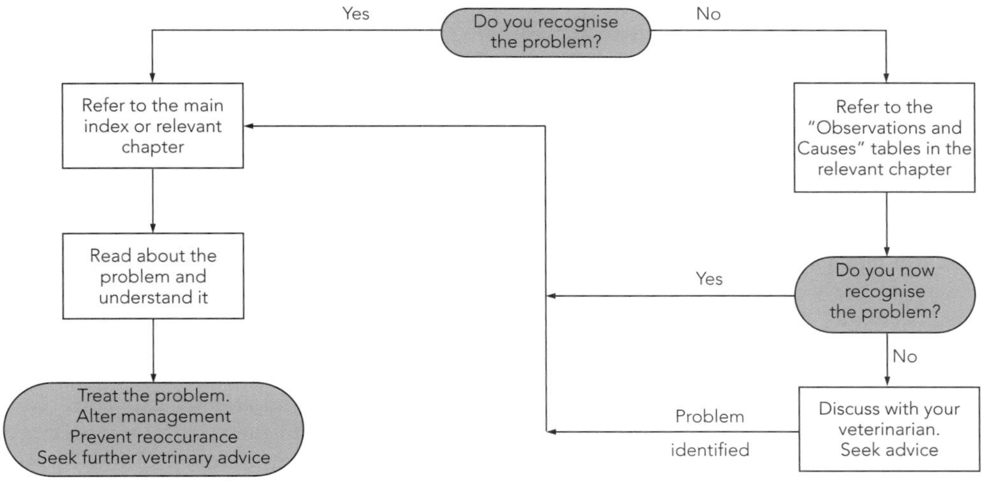

* *The authors are aware that regulations governing medicine usage vary in different countries. In some countries many of the medicines listed may be obtained on prescription only or can be administered only by or under the supervision of a veterinarian. Some products may even be banned. You should, of course, conform to your national regulations. Whatever the regulations, you are advised to consult your veterinarian regularly.*

ACKNOWLEDGEMENTS

John Carr

A further 16 years has passed since Mike and Tom wrote the first edition. We now have the pleasure of presenting the 2nd edition, delayed somewhat due to the untimely passing of Mike and Tom.

The acknowledgements made in the first edition (below) are as relevant today as they were in 1997, especially the efforts of Jim and Pauline who were instrumental (again) in the creation of this new edition.

I must, in addition, extend a personal thanks to all the pig farmers around the world who have taught me about pigs especially Simon Cook. The students at Liverpool University, The Royal Veterinary College, London, North Carolina State University, Iowa State University and Murdoch University for challenging ideas. For the veterinarians in the European Swine and Wine Group; the Garth Veterinary Group, North Tawton Veterinary Group, St David's Farm and Equine Practice and the Howell's Veterinary Group along with the Pig Veterinary Society in the UK; Portec Australia and Noni Mammatt; Hermitage Seaborough and Globinskiy svinokompleks in Ukraine; the Bridge Swine Practice in Korea; the Animal Technology Institute Taiwan; Grand Valley Fortifiers and Nutrition Partners in Canada; for all supplying support. A special thanks to Linda who has put up with me for so many years.

John Carr
August 2013

Mike Muirhead (1936 – 2003)

The production of this book has been my intention for over 20 years following requests by many pig farmers and pig industry personnel for reading references relating to problems on the farm.

Many people in many ways have contributed to the contents and the management health and disease control information it contains.

It would be impossible to refer to everyone by name and it would be unfair to those people left out. Individual references therefore have been purposely omitted. However I would like to thank and acknowledge the contributions that many groups have made over the years including:

Members of the Garth Veterinary Group; Neville Kingston, Paul Thompson, Dr John Carr and associate veterinarians.

The authors of International Pig Letter. The late Al Leman, Professor Richard Penny, Mike Wilson, Gary Dial and Frank Aherne and editors Neal Black and Grant McGinnis.

Members of the American Association of Swine Practitioners.

Pig Improvement Company, National Pig Development Company.

Newsham Hybrid and JSR Healthbred.

Cliff Johnson my farm manager who has helped put into practice many of the new control procedures (or rejected them!).

Mark Enright for his chapter on Health and Safety and Control of Substances Hazardous to Health.

The many pig farmers and veterinarians consulted with over the years.

Tom Alexander who has contributed significantly through his extensive knowledge and in particular with his red pen as editor by making the observations and comments more user friendly and scientifically correct.

This book would not have been completed without the continual devotion to the task by my secretary Pauline Tams and my son James of 5m. Pauline has had to cope with the illegible writings of two [now 3] veterinarians and numerous alterations and corrections.

Finally without the constant encouragement and help from my wife Louise who has endured long hours of silence I would not have finished the project.

No book is ever complete (or without fault) but we hope that this one will provide scope for debate, education, improvements in health and disease control and finally more profitable pig farming.

Michael R. Muirhead
May 1997

CONTENTS

1 AN INTRODUCTION TO THE ANATOMY AND PHYSIOLOGY OF THE PIG

SOME BASIC ANATOMY OF THE PIG

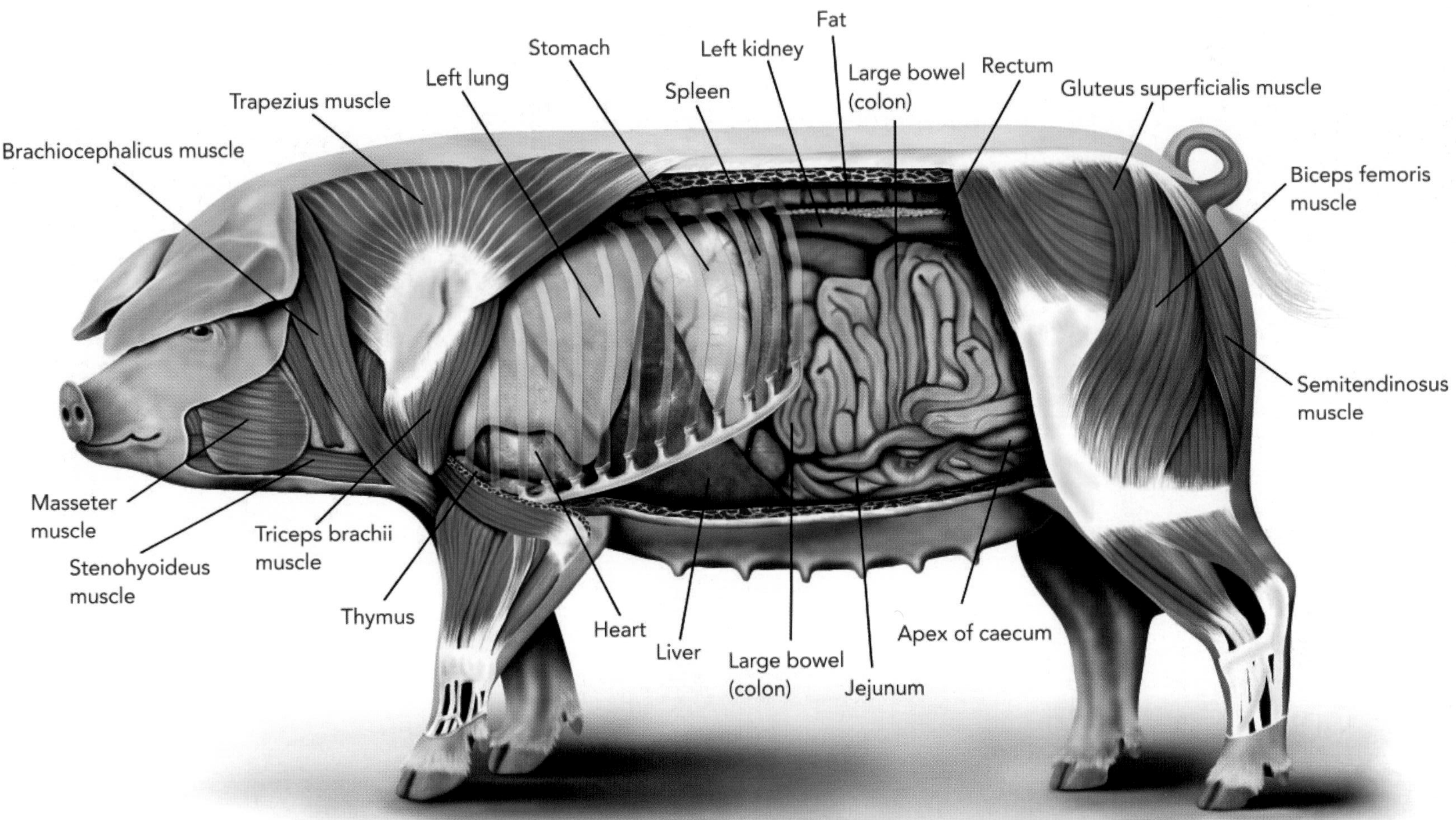

Fig 1-1

1 AN INTRODUCTION TO THE ANATOMY AND PHYSIOLOGY OF THE PIG

Anatomy and Basic Functions of the Various Systems of the Pig

An understanding of disease processes can be difficult if the reader has little scientific background. For those who have not had training in biological subjects this chapter looks at the basic areas of knowledge necessary to appreciate and understand the information given in this book. The basic anatomical layout of the pig is shown in Fig 1-1. All the diseases and conditions mentioned briefly in this chapter are discussed in detail later.

What is the Pig?

The pig has a long history of evolution. All domestic breeds of pig throughout the world, despite their variety of size and shape, belong to one species – *Sus scrofa* – of which the wild variety still exists as the European Wild Boar. Pigs belong to the artiodactyla group of mammals – even-toed animals. In the same group are cattle, deer, camels, hippopotamuses and even the whales.

Anatomy and Organ Layout of the Pig

The anatomy and physiology of the pig can be broadly grouped into the following interrelated systems:

- Circulatory system.
- Digestive system.
- Endocrine system.
- Immune system.
- Integumentary system (the skin).
- Muscular system.
- Nervous system.
- Reproductive system (male and female).
- Respiratory system.
- Sensory systems.
- Skeletal system.
- Urinary system.

Circulatory System

First, study Fig 1-2, then read the following while still referring to the figure.

The circulatory system consists of the heart which is a four chamber suction and pressure pump that moves blood through two separate systems, one to and from the lungs and the other around the body. The blood returns to the heart from the body through a series of veins, which terminate in two large veins called the anterior and posterior vena cava. Blood returns from the lungs through the pulmonary veins. The top two chambers or atria receive the blood from the veins and pass it into the strong muscular bottom chambers called the ventricles. Oxygen-depleted blood from the body enters the right atrium, where it is then pumped into the right ventricle, leaving by two pulmonary arteries that deliver the still un-oxygenated blood to the lungs. Oxygenated blood from the lungs is then returned through the pulmonary veins to the left atrium, where it is pumped to the left ventricle and finally out through the main artery, the aorta, to be transported around the body.

The internal linings of the heart are covered by a smooth, shiny tissue called the endocardium. The various chambers of the heart are separated by valves which control the blood flow. The rate of contraction is known as the pulse rate. This can be felt either at the base of the ear or under the tail and varies from 200 beats per minute in the young piglet to 70 in the adult.

Arteries are the muscular tubes that carry the blood away from the heart. These branch off into smaller arteries like the branches of a tree, eventually becoming very fine arterioles. The arterioles branch further into microscopic tubes called capillaries which exchange fluid through their walls. This enables the cells of the body to receive both oxygen and nutrients and eliminate carbon dioxide. The capillaries then combine to form first small veins, which in turn lead to larger ones. The blood now contains carbon dioxide and reduced levels of oxygen and returns to the heart via the anterior and posterior vena cava to recommence its circulation around the lungs.

There is an important subsidiary circulatory system called the hepatic (i.e. liver) portal system. You will see in Fig 1-2 that two arteries provide oxygen to the stomach and intestines (and also the pancreas and

BLOOD CIRCULATION OF THE PIG

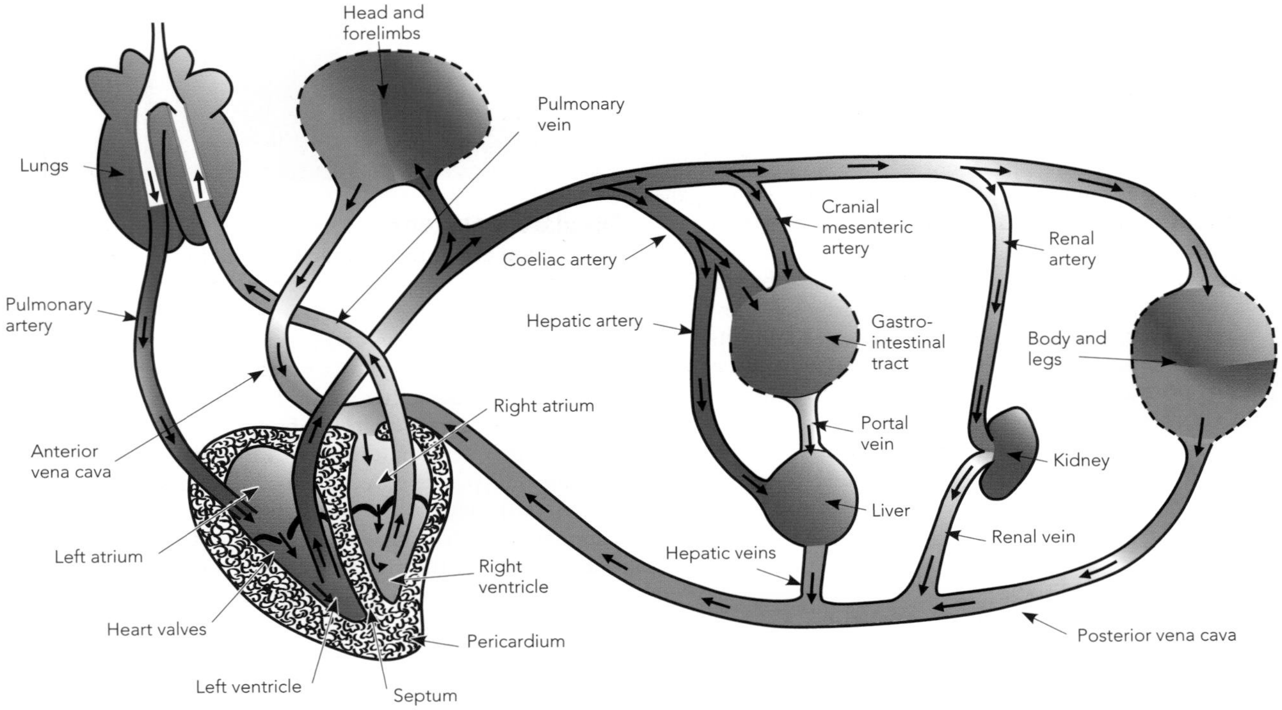

Fig 1-2

spleen). They keep branching until they form capillaries which then join together to form the portal vein which carries the blood to the liver. There the portal vein breaks up into another capillary-type network, where the blood comes into direct contact with the liver cells. The vessels then join together again to form the hepatic veins which discharge the blood into the posterior vena cava. The blood from the intestines carries nutrients from the food eaten and also sometimes harmful substances (toxins). The liver cells are able to modify some of the nutrients for use elsewhere and also to store some. They also detoxify harmful substances. The liver is supplied with oxygen via a separate artery, the hepatic artery.

The blood consists of two main parts, a fluid called plasma, and cells. The blood volume is around 64 ml/kg liveweight (8 % bodyweight). Nutrients such as proteins, sugars and fats are circulated throughout the body in the plasma, and waste products are collected to be detoxified in the liver and excreted via the kidneys. The plasma also carries hormones which are produced in one part of the body and act on another. It also carries antibodies to combat infection. The plasma also supports red blood cells (erythrocytes) which contain the substance haemoglobin whose main function is to transport oxygen around the body and bring back carbon dioxide to be expelled from the lungs. The next largest group in the plasma are the white cells (leucocytes) which are the first line of defence against infectious agents. The leucocytes are composed of five types of cells – lymphocytes, neutrophils, monocytes, eosinophils and basophils. The third type of cells in the plasma are blood platelets. These are really small fragments of cells which are associated with the clotting mechanisms of blood. When blood clots the liquid that remains outside the clot is serum and this contains the antibodies. Measuring the concentration of these antibodies may be used as a useful test for many pathogens. Serum may be used to inject into pigs to provide an immediate source of immunity.

Various other chemicals also circulate in the bloodstream. Measurement of these chemicals can be very useful in the diagnosis of the proper function of various organs.

Effects of disease

If the lungs are damaged by disease such as pneumonia, they cannot oxygenate the blood efficiently; the tissues become starved of oxygen and cannot function properly. When the pig walks or runs its skin may then become blue and it will have difficulty breathing. Chronic pneumonia may also hold back the blood supply causing congestion and heart problems.

If the heart valves become damaged, they will leak, resulting in failure of the pumping action of the heart.

In the presence of many diseases, various parts of the immune system are stimulated, leading to an increase (or decrease) in the various components of the circulation.

Failure of blood to clot and subsequent loss of red cells into the tissues is not uncommon in pigs and occurs in thrombocytopenic purpura – a clotting defect disease – and warfarin poisoning.

Some common diseases/disorders associated with the circulatory system

- Chronic pneumonia.
- Endocarditis.
- Glässer's disease.
- *Mycoplasma suis*.
- Postweaning multisystemic wasting syndrome (PMWS)
- Pneumonia.
- Pulmonary oedema.
- Thrombocytopenic purpura.
- Warfarin poisoning.

Environmental considerations

- Poisoning.

TERMINOLOGY

Albumin – The most abundant protein in the blood.

Anaemia – Any reduction in the number of red cells or in the haemoglobin they contain is described as anaemia and the extent of this is measured by determining either the number of red cells or the level of haemoglobin in the blood. The causes of anaemia include:

- Bowel haemorrhage (proliferative haemorrhagic enteropathy, fungal toxins, acute bowel infection associated with *Escherichia coli (E. coli)* infection of piglets, salmonella infections or swine dysentery).
- Damage to bone marrow.
- *Mycoplasma suis*. This is a blood-borne bacterium that can destroy red blood cells.
- Gastric ulcers and bleeding – or any other cause of haemorrhage.
- Heavy parasite burdens.
- Iron, copper or vitamin deficiencies.

Anoxia – Lack of oxygen. Tissues begin to die after a few minutes.

Antibody – The protective proteins produced in response to antigenic stimulation. They fight infections.

Antigen – The foreign protein contained in viruses, bacteria, fungi or toxins. The body responds by producing an antibody.

Antiserum – Serum containing higher than normal amounts of antibody against a specific antigen. It is used by injection to give an immediate temporary immunity.

Blood count – A laboratory test that determines the numbers of red and white cells and platelets in the blood.

Blood volume – Approximately 8 % of bodyweight, expressed as litres (or 64 ml/kg liveweight).

Blood platelets (thrombocytes) – Cell fragments involved in blood clotting.

Blood poisoning – A common term used to describe large numbers of pathogenic bacteria in the blood.

Capillaries – Very tiny tubes about the diameter of a red cell. These allow water, oxygen and nutrients to diffuse out to the tissues.

Cyanosis – Blueing of the skin and extremities due either to anoxia, toxaemia (toxins in the blood) or septicaemia (pathogenic bacteria in the blood).

Endocardium – The surface tissue lining the inside of the heart. Endocarditis is the end result of the invasion of this tissue by bacteria, in particular *Erysipelothrix rhusiopathiae* (which causes erysipelas) and streptococci. Both organisms often cause growths on the heart valves called valvular endocarditis. This makes the valves leaky and less effective.

Erythrocytes – The red blood cells. In the normal pig there are approximately 7 million per mm^3.

Globulins – The proteins that make up the antibodies. They are called gamma globulins.

Granulocytes – These consist of specialised cells called neutrophils, eosinophils and basophils that engulf and destroy bacteria and viruses.

Haematuria – Blood in the urine often seen in cystitis – inflammation of the bladder.

Haemoglobin – The chemical substance in the red cells that is involved in the transport of oxygen.

Haemoglobinuria – Free haemoglobin in the urine resulting from the breakdown of blood cells.

Haemolysis – The process by which haemoglobin is released from the red cells when the cell envelope is damaged.

Hydropericardium – Excess fluid around the heart. It is often seen in bacterial infections and shock reactions.

Hypoglycaemia – A low level of sugar in the blood. Common in newborn piglets.

Leucocytes – The white blood cells of which there are two types, granulocytes and agranulocytes. The granulocytes contain granules in the cell and depending on how they stain they are called neutrophils, eosinophils and basophils. Neutrophils engulf bacteria (phagocytosis); eosinophils increase in chronic disease, particularly parasitic disease. Basophils produce a substance called histamine during allergic reactions. Agranulocytes consist of monocytes, macrophages and lymphocytes.

Lymph – Excessive tissue fluid drained by the lymphatic system. It is similar to plasma.

Lymphatic system – A drainage system that removes fluids from tissues and the lymph nodes.
Lymph nodes – These act as filters for lymph and are one of the body's first defences against infection.
Lymphocytes – Important cells of the immune system producing immunoglobulins. They are of two types, T and B. The total leucocytes in a normal pig are approximately 15,000 per mm^3 and numbers increase markedly with bacterial infections. However, in some viral diseases their numbers can be significantly reduced.
Macrophages – These take in and usually destroy foreign materials including bacteria and viruses. See Monocytes.
Monocytes – These cells engulf bacteria. When they migrate into tissues they become localised tissue macrophages.
Myocardium – Heart muscle.
Myocarditis – Inflammation of the heart muscle. Any scientific term ending with the term "itis" implies inflammation. Inflammation is the body's response to tissue damage and is associated with swelling, poor circulation, reddening, pressure and pain.
Oedema – Swelling of tissues due to excess fluid. Common in the udder of the newly farrowed sow.
Oxyhaemoglobin – Haemoglobin combined with oxygen. It is the vehicle by which oxygen is carried around the body.
Pericarditis – Inflamation of the pericardium which is the clear sac-like membrane that encloses the heart. Pericarditis occurs as a result of infectious agents which cause respiratory diseases. These include pasteurella, mycoplasma, haemophilus, actinobacillus, streptococci and salmonella bacteria, and viruses such as flu and porcine respiratory reproductive virus.
Plasma – Unclotted blood without the blood cells.
Septicaemia – Pathogenic bacteria in the blood-stream.
Serum – The liquid left after the blood has clotted. It contains large quantities of antibodies which can be used in the laboratory to test for evidence of exposure to diseases or in the field to provide temporary quick protection.
Thrombocyte (blood platelet) – Responsible for blood clotting.
Thrombosis – The formation of a blood clot in an artery or a vein.
Toxaemia – Toxins in the blood-stream.
Spleen – This organ acts as a reservoir for blood.
Vasculitis – Inflammation of either veins or arteries.
Viraemia – Viruses in the blood stream.

Digestive System

The digestive tract can be considered as a tube that starts at the mouth and finishes at the anus (Fig 1-3). In some respects its contents can be considered as outside the body.

The mouth contains teeth to break down the food, and the tongue to sense and taste the food. The pig has incisors (I), canines (C), Premolars (P) and Molars (M) (Fig 1-4). Deciduous (prefix D) are the baby teeth that fall out over the first year. The dentition is described as:
Deciduous: DI 3/3, DC 1/1, DP 3/3, DM 0/0
Permanent: I 3/3, C 1/1, P 4/4, M 3/3

The teeth eruption may be used as a means of determining the age of the pig.

- The first premolar and molar erupt at 5 months.
- Canine and corner incisor erupt at 9 months.
- Central incisor and second molar erupt at 12 months.
- The other premolars erupt at 15 months.
- Last molar and intermediate incisor erupt by 18 months.

ANATOMY OF THE DIGESTIVE SYSTEM

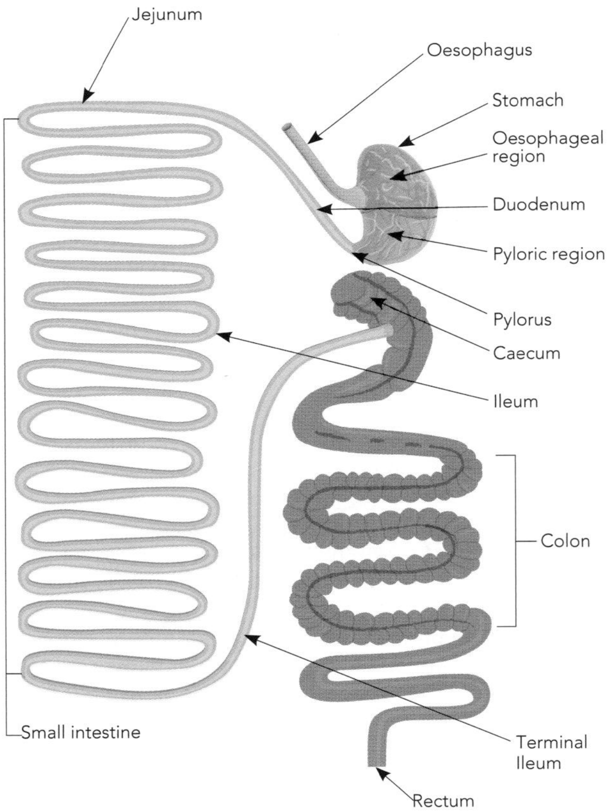

Fig 1-3

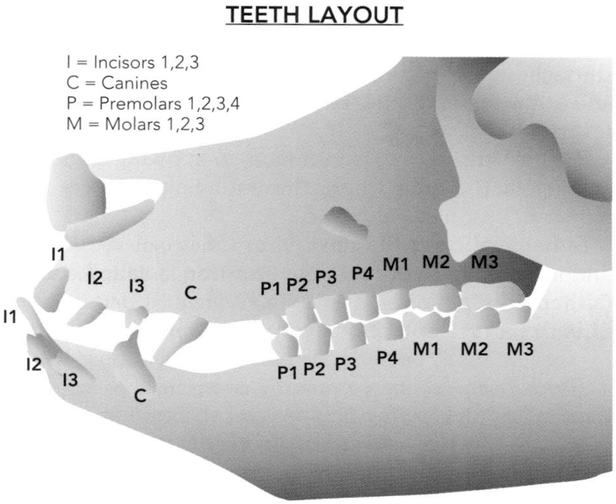

Fig 1-4

The back of the mouth opens into the pharynx which is the common area for the passage of both food and air. A valve or flap of tissue called the soft palate automatically moves to protect the opening into the trachea or windpipe when swallowing. The tonsils of the pig are situated on the surface of the soft palate. The oesophagus is the tube that leads from the pharynx to the stomach, down which food is propelled.

The stomach is a large bag-like structure, which kills many organisms and destroys proteins through the action of enzymes and acids (Fig 1-5). The area where the oesophagus enters the stomach is not protected by mucus, making it vulnerable to ulceration. By the side of the oesophagus entrance is a small pouch, the oesophageal diverticulum. The exit of the stomach is marked by the pylorus, which in the pig has an extra seal called the pyloric torus. This protects the stomach from the alkaline bile salts produced by the liver.

The digestive system of the pig has the ability to convert vegetable and animal materials into highly digestible nutrients. Its anatomy and physiology are similar to that of humans.

The intestine has two distinct parts; the small and the large intestine. The small intestine in cross-section contains millions of finger-like projections called villi (Fig 1-6). These increase the absorptive area enormously and thus the efficiency of the digestive process. The small intestine is divided into three main areas: the duodenum, jejunum and ileum (Fig 1-3).

The large bowel or colon commences with the caecum, the area of the intestinal tract responsible for the digestion of cellulose. The digested food then

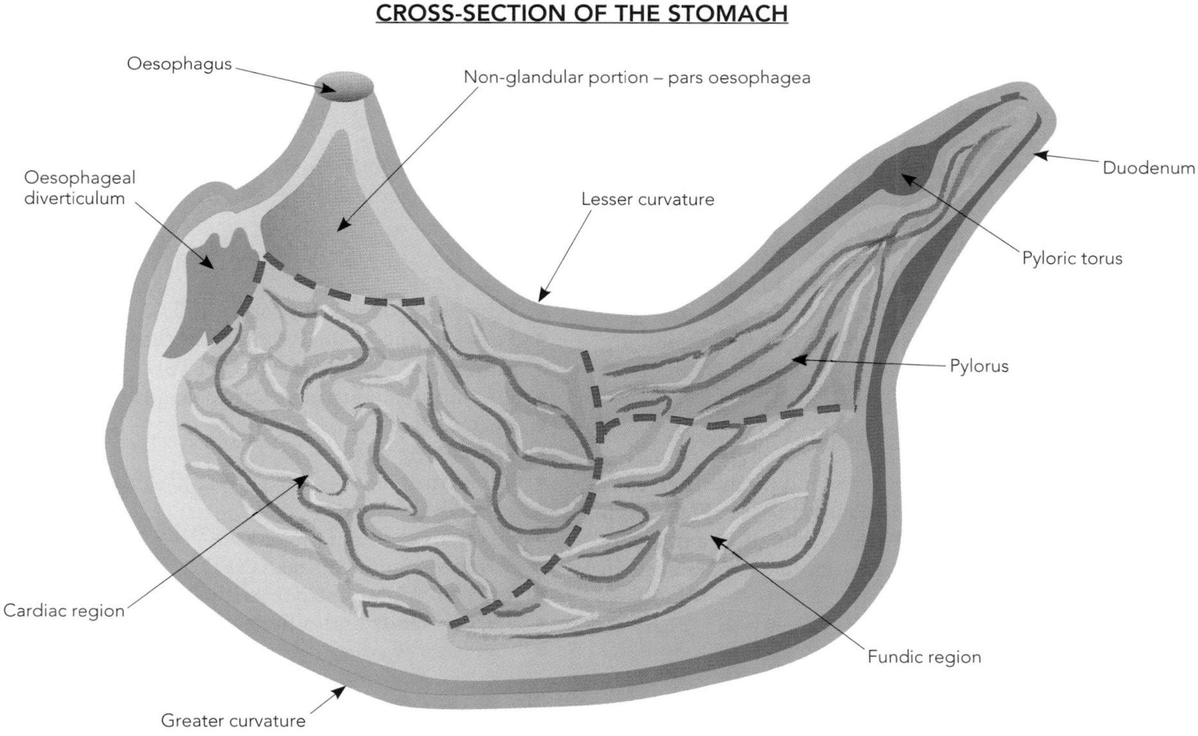

Fig 1-5

CROSS-SECTION OF THE SMALL INTESTINE

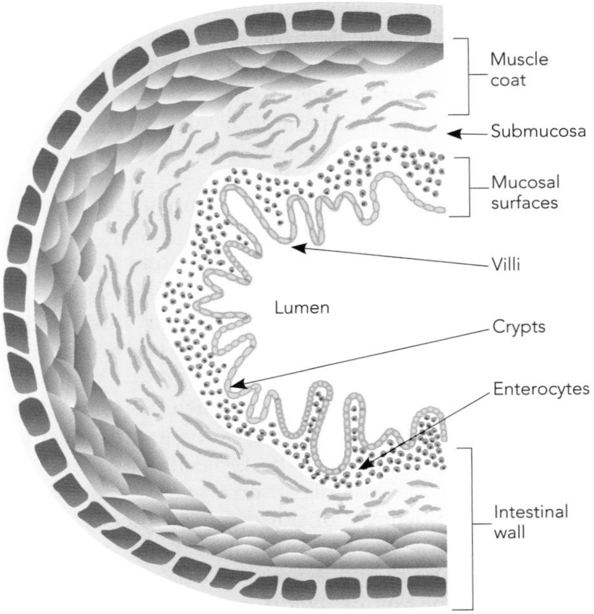

Fig 1-6

passes to the colon whose function is to absorb water from the digesta. In the pig, the colon is spiral-shaped. The digesta finally enters the rectum and is passed out through the anus as faeces.

The liver is a large organ by the right side of the stomach. It functions as a chemical warehouse, producing vital chemicals for the body and detoxifying any unwanted chemicals.

The major disease of the liver is white spot, associated with *Ascaris suum*.

The spleen is a large organ in the abdomen that acts as a reservoir of blood, which the body uses in times of stress and shock. The major diseases of the spleen are associated with Aujeszky's disease, where small spots are visible at post-mortem.

Effects of disease

The main infectious diseases of the mouth are the vesicular ones including foot-and-mouth disease (FMD) and swine vesicular disease (SVD), although occasionally lesions on the skin around the mouth may be seen in Aujeszky's disease/pseudorabies (AD/PRV) and porcine reproductive and respiratory syndrome virus (PRRSV). Infection of both the gums and bones is common following faulty teeth-clipping.

In the stomach the major disease problems are associated with inflammation of its lining called gastritis which may result in vomiting. Vomiting also occurs in systemic disease where the organism has spread throughout the body (in infections such as erysipelas), and from toxins produced by bacteria, or during high fevers.

Gastric ulceration is common in growing pigs occurring in the area where the oesophagus enters the stomach (oesophageal region).

Inflammation of the small intestine is called enteritis and inflammation of the large intestine is called colitis (although sometimes enteritis may mean inflammation of both parts). Enteritis is very common and is caused by specific viral, bacterial or parasitic infections.

Two diseases are commonly seen in the rectums particularly in growing pigs – rectal stricture and rectal prolapse – both of which are discussed in Chapter 9.

Some common diseases associated with the digestive system

- Abdominal catastrophe.
- *Ascaris suum*.
- Bowel oedema.
- Brachyspira colitis.
- Campylobacter.
- *Clostridium difficile*.
- *Clostridium novyi*.
- *Clostridium perfringens*.
- Coccidiosis.
- Colitis.
- Cryptosporidia.
- Dysentery.
- *E. coli*.
- Enteritis.
- Foot-and-mouth disease (FMD).
- Gastric ulcers.
- Gastritis.
- Ileitis.
- Parasites.
- Porcine epidemic diarrhoea (PED).
- Postweaning ill-thrift syndrome/Periweaning failure to thrive syndrome (PFTS).
- Pre and postweaning diarrhoea.
- Rectal prolapse.
- Rectal stricture.
- Rotavirus.
- Salmonellosis.
- Swine dysentery.
- Swine vesicular disease (SVD).
- Transmissible gastroenteritis (TGE).
- *Trichuris suis* and other worms.
- Vomiting wasting disease.

Environmental considerations

- Poisons/toxins

TERMINOLOGY

Ascites – Fluid in the abdomen.

Atrophy – A loss of tissue due to disease or malfunction. Atrophy of the villi in the intestine occurs at weaning time, causing malabsorption.

Bloody gut – A descriptive term applied to haemorrhage in the lower part of the small intestine or the complete digestive tract. The latter is seen where there is complete torsion of the intestines. Ileitis is a common cause (see Chapter 9).

Caecal ileum ligament – A ligament which attaches the caecum to the distal ileum. It can be extremely useful as a tool to find the distal ileum.

Caecum – A blind sac, at the beginning of the large intestine.

Carbohydrates – These consist of two types, crude fibre and soluble carbohydrates. Crude fibre is a mixture of celluloses. Cellulose digestion takes place in the large intestine.

Colitis – Inflammation of the colon or first part of the large bowel. The caecum is often inflamed at the same time (typhlitis). This is a common condition in young growing pigs from 20 to 60 kg (45 to 130 lbs) in weight, caused by nutritional factors and/or infectious agents.

Colon – The spiral part of the large intestine.

Crypts – The bases of the villi.

Duodenum – The first part of the small intestine.

Enteritis – Inflammation of the small intestine. This leads to diarrhoea which is common in sucking pigs, weaners and growers.

Enterocytes – Cells at the base or crypts of the villi in the intestine. They multiply and maintain the length of the villi.

Gall bladder – An organ attached to the liver which produces bile that helps in the digestion and absorption of fats.

Gastric ulcers – Erosions of the mucous lining of the stomach, occurring mainly in the oesophageal region. Very common, and if severe they result in haemorrhage and death.

Gastritis – Inflammation of the stomach lining. Often causes vomiting.

Gingivitis – Inflammation of the gums.

Glossitis – Inflammation of the tongue.

Hepatitis – Inflammation of the liver.

Ileitis – Inflammation of the ileum.

Ileum – The terminal part of the small intestine.

Jejunum – The middle part of the small intestine.

Liver – This organ is the main factory of the body, building new materials and degrading old ones.

Lumen – The open space of the small intestine.

Mucosa – The internal lining of the digestive tract. The cells produce mucus which lubricates the surface and also protects against many pathogenic organisms.

Oesophageal diverticulum – A small pouch to the top of the stomach where some secondary fermentation can take place. In some pigs such as the Babyrousa this is quite well developed. In other artiodactyles such as cattle this develops into the forestomachs vital for the cow's ability to live off grass.

Oesophagus – The muscular tube from the pharynx to the stomach.

Omentum – A reflected net-like membrane from the peritoneum that covers the stomach and intestine.

Pancreas – A gland attached to the duodenum by a tube, which produces digestive enzymes and insulin.

Pars oesophagea – The area of the stomach near the entrance of the oesophagus. A common site for the development of ulcers.

Peritoneum – The smooth, shiny membrane that covers all the surfaces of the abdomen and its contents.

Peritonitis – Inflammation of the peritoneum.

Pharynx – The common passage for food and air at the back of the throat.

Proteins – These are composed of amino acids which contain carbon, hydrogen, oxygen, sulphur, nitrogen and phosphorus. Combinations of different amino acids produce different proteins.

Pyaemia – Invasion of pus-producing organisms throughout the body with small abscess formations.

Pyloric torus – Thickening at the pyloric exit protecting the acid stomach contents from alkaline bile salts.

Pylorus – The last section of the stomach.

Soft palate – The flap of tissue that separates the trachea and the oesophagus. It contains the tonsils.

Salivary glands – There are three of these called the parotid, mandibular and sublingual glands. They secrete saliva into the mouth.

Tonsillitis – Inflammation of the tonsils.

Tonsils – Two patches of lymphatic tissue at the back of the throat on the soft palate.

Villi – Finger-like projections into the lumen of the small intestine (Fig 1-6).

Endocrine System

Endocrines or hormones are the substances produced by various glands, and are carried by blood or other body fluids to influence and control the pig's metabolism. There are eight main glands (Fig 1-7) in the pig which are responsible for controlling a variety of vital functions.

Effects of disease

Generally the disorders associated with the failure of the endocrine glands are not important in the commercial pig. However, when the regulatory and stimulatory mechanisms between the hypothalamus, the anterior pituitary gland and the ovaries fail, anoestrus (not coming on heat) or reproductive malfunctions, including cystic ovaries, result. In the male, testicular function is affected. The hypothalamus stimulates the anterior part of the pituitary gland to release the follicle-stimulating and luteinising hormones (FSH and LH). These in turn act upon the ovaries and the testes to regulate their function (See Chapter 5.)

Some common diseases associated with the endocrine system

- Anoestrus.
- Cystic ovaries.
- Pancreatitis.
- Environmental considerations.
- Stress can result in cystic ovaries.
- Stress can stop the reproductive cycle.
- Mycotoxins – zearalenone can mimic oestrogen.
- Feed deficiencies – iodine – goitre.

TERMINOLOGY

Endocrine glands – These produce chemicals which enter the bloodstream directly without going through a duct.
Follicle stimulating hormone (FSH) – Produced by the anterior pituitary gland. It stimulates the formation of follicles in the ovaries.
Growth hormone – Responsible for promoting growth of most tissues throughout the body. It is produced by the pituitary gland in association with the hypothalamus.
Hypothalamus – An area in the brain responsible for providing both nervous and hormonal control over most other hormone-producing glands.
Luteinising hormone (LH) – Stimulates ovulation and is produced by the pituitary gland.
Oestrogen – The female hormone responsible for all the female sexual characteristics. It is produced by the ovary.
Oxytocin – Produced by the pituitary gland. This stimulates uterine contractions during farrowing and causes milk let-down. It also aids in the movement of sperm and eggs.
Progesterone – The hormone that maintains pregnancy. It is produced by the corpus luteum in the ovary.
Prolactin – This is produced by the pituitary gland and controls milk production.
Prostaglandins – These are produced by the uterus and the placenta and are associated with the initiation of farrowing or abortion.
Testosterone – The male hormone responsible for all the male sexual characteristics. It also controls the development of sperm.

Immune System

The various mechanisms that protect the pig from infectious agents can be considered in seven groups:

1. **Complement system** – This is a non-specific protective mechanism that acts on any foreign cells or viruses that do not possess certain pig proteins on their surface. It consists of a number of chemicals found in the plasma which act together as a cascade to remove or destroy organisms.
2. **Chemical factors** – These include non-specific enzymes (such as lysozyme in saliva) and acids

FIG 1-7: GLANDS AND HORMONE PRODUCTION

Glands	Function
Adrenal	These are attached to the kidney surface. Their hormones control growth, sugar metabolism, kidney function and stress.
Hypothalamus	Found at the base of the brain. The main controlling gland. Its hormones control most body functions and all other glands, together with sexual activity.
Ovaries	Produce the female hormones oestrogen and progesterone.
Pancreas	Located in a fold of the duodenum. Produces insulin to control sugar metabolism.
Parathyroids	Located near the thyroid glands in the neck. Control calcium and phosphorus deposition.
Pituitary	Found at the base of the brain. Hormones control growth, reproduction, lactation and stress.
Testicles	Produce the male hormone testosterone.
Thyroid	Controls metabolism and growth.

which may be found in mucus, saliva and gastric juices. These immobilise or kill pathogens.

3. **Mechanical factors** – These include the skin, mucus, sweat; lining of the nose, mouth, oesophagus, intestine, colon, vagina; flow of urine and passage of faeces.
4. **Macrophage cells** – These are found throughout the body in tissues and in the bloodstream where they are called monocytes. They engulf and digest bacteria. They also have an important role in controlling viral and fungal diseases.
5. **Specific acquired immunity** – This is of two types: that which is activated by cells and called cell-mediated immunity, and antibodies present in the blood called humoral immunity. Cell-mediated immunity arises when T type lymphocytes come into contact with antigens and they are stimulated to produce antibodies. It takes 7 to 14 days for these to develop. Humoral immunity is produced from B lymphocytes which have met the antigen previously and their response is immediate. Some lymphocytes also kill other cells that contain antigens or they may act immediately against antigens.
6. **Immunoglobulins** – Specific antibodies of which there are different types namely immunoglobulins IgG, IgM, IgA and IgE. They are found in blood, in milk and particularly in colostrum. All internal surfaces of the body also contain them.
7. **Normal microbiota** – Protective bacteria which live on and colonise the surfaces of the body and so protect the pig by competitive inhibition.

The normal lymph nodes in the pig are small and can be difficult to find (Fig 1-8). When the lymph nodes are very prominent they may be abnormal. However, many lymph nodes are covered by fat and if this fat disappears the lymph node may appear and may be

THE POSITION OF THE MAJOR LYMPH NODES FROM THE SIDE AND UNDERNEATH (VENTRAL)

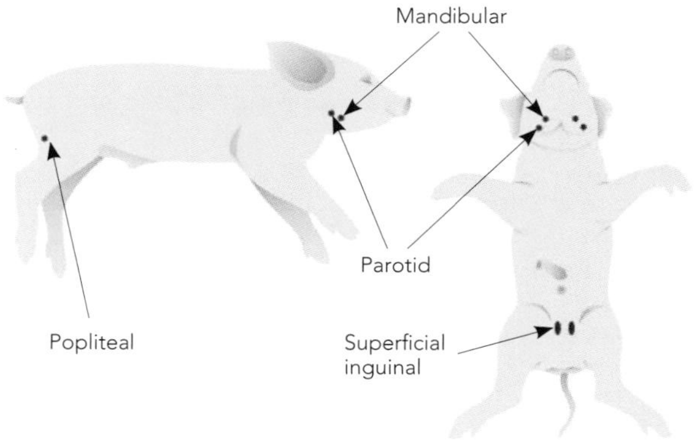

Fig 1-8

misdiagnosed as enlarged. This is a common mistake in cases of postweaning multisystemic wasting syndrome (PMWS).

Infection is considered to have taken place when a virus, bacterium or parasite enters a pig and starts to multiply. If it is a potentially pathogenic organism it may change the normal structure and function of the pig. A study of these changes is called pathology and organisms causing disease are described as pathogens. This is described in more detail in Chapter 3.

Effects of disease

- Poor growth.
- Poor food intake.
- Inability to respond to vaccines properly.
- Inability to fight pathogens.

Some common diseases associated with the immune system

Certain infectious agents can suppress the immune system sufficiently to make the animal more susceptible to other infections.
Examples are:

- Aujeszky's disease/pseudorabies virus (AD/PRV).
- *Mycoplasma hyopneumoniae.*
- *Pasteurella* bacteria.
- Porcine reproductive respiratory syndrome virus (PRRSV).
- Postweaning multisystemic wasting syndrome (PMWS), PCV2.
- Swine influenza virus (SIV).

All of these cause pneumonia. PRRSV is particularly destructive to macrophages in the alveolus of the lung.

Environmental considerations

- Mycotoxins in the feed may reduce the immune response.
- Poor housing conditions may reduce a pig's ability to respond to pathogens.
- Sleeping patterns.

TERMINOLOGY

Adjuvant – A substance added to an inactivated vaccine to make it more effective.
Antibodies – Complex, large proteins (called gamma-globulins) which are produced by specialised cells in response to invading antigens. These stick specifically to the invading antigen, neutralising it or triggering off a destructive reaction.
Antigen – Foreign invading substance (i.e. a substance which is not normally part of the pig's body), usually consisting of protein or part of a protein, which stimulates the body to produce antibodies. Antigens exist on the surfaces of bacteria, viruses and parasites.

Antiserum – Serum with high antibody levels against a specific infection. It has usually been produced experimentally in laboratory animals by injecting the infection into them.
Blood sample – Whole blood sample taken hygienically with a syringe into a bottle or by a pin prick through the skin, absorbing the droplet of blood with blotting paper.
Colostrum – The means through which the piglet acquires antibodies and protective cells from its mother within 12 hours after birth.
Commensal bacteria – Bacteria that live permanently in or on the body without causing disease.
Epithelium – Cellular membrane (e.g. mucous membranes) containing epithelial and other cells.
Humoral immunity – Blood-borne immunity.
Hyperimmune antiserum – The same as antiserum above but emphasising a high titre.
Lymphocytes – Specialised defence cells in lymph nodes, other lymphatic tissue and the blood which produce antibodies or take part in cellular immunity.
Microbiota – The group of normal bacteria living in or on a specific living surface. These can be protective.
Mucous membranes – Cellular membranes (e.g. those lining the gut) which secrete a sticky substance called mucus onto their surfaces.
Mucus – A clear sticky semi-liquid secreted by cells in mucous membranes.
Pathogenic infection – An infectious organism which has the potential to cause disease. This is in contrast to the many organisms that live normally in or on the body which never cause disease and are called commensals.
Phagocytes – Cells of the body whose special task is to engulf bacteria, viruses or parasites in an attempt to destroy them. They are also called macrophages.
Phagocytosis – The process whereby the specialised cells of the body engulf bacteria, viruses or parasites in an attempt to destroy them.
Plasma sample – A whole blood sample taken hygienically with a syringe and mixed with an anti-clotting agent so that it remains liquid. The sample is spun fast in a centrifuge and the red and white blood cells sediment to a firm pellet at the bottom leaving a clear liquid – the plasma.
Serology – Tests done in the laboratory to detect the level of specific antibodies in serum samples ("ology" means study of – so literally serology means "study of serum").
Serum sample – A whole blood sample taken hygienically with a syringe and allowed to clot. The serum is the clear straw-coloured liquid which can be drawn off with a pipette. It contains the antibodies.
Titre – The concentration of a specific antibody in a serum sample. It is expressed as the amount by which the serum has to be diluted before a serological test goes negative.
Virulence – The pathogenicity of an organism. Organisms with a high capability of causing clinical disease are called highly virulent.

Integumentary System (the Skin)

Disorders of the skin are covered in detail in Chapter 10.

Effects of disease

The pig skin is similar to human skin having generally little hair and a subcutaneous fat layer. The major difference is that the pig has very few sweat glands (only a few modified on the wrist (carpus)). When hot, the pig has to cool its skin directly by applying water to the skin and losing heat through evaporation. The pig's skin has good powers of regeneration.

Some common diseases associated with the skin:

- Abscess.
- Areas of skin trauma.
- Dermatitis parakeratosis.
- Epithelium imperfecta.
- Erysipelas.
- Flaking skin.
- Foot-and-mouth disease.
- Greasy pig disease.
- Herniation.
- Mange.
- Pig pox.
- *Pityriasis rosea.*
- Porcine dermatitis and nephropathy syndrome (PDNS).
- Ringworm.
- Swine fever(s).
- Tail biting and other vices.

Environmental considerations

- Poor flooring and protrusions.

TERMINOLOGY

Cyanosis – A blueish discolouration of the skin and/or mucous membranes resulting from inadequate oxygenation of the blood.
Erosion – A gaping wound as a result of skin and flesh being worn away.
Granuloma – A mass of inflamed granulation tissue resulting from chronic infection, inflammation or a foreign body, usually associated with ulcerated infections.
Graze – An abrasion to an area of skin, which is torn or worn away.
Ulcer – A non-healing open sore or wound.

Muscular System

There are three types of muscle in the pig:

1. **Involuntary or smooth muscle** – Found in the digestive and genital systems and the blood vessel walls.
2. **Cardiac muscle** – The heart consists largely of this muscle. It is involuntary.
3. **Voluntary or skeletal muscle** – This is the main muscle mass forming the muscular-skeletal system. These muscles are attached to the surface membrane covering bones called the periosteum (Fig 1-9). Inflammation of this covering is called periostitis.

Effects of disease

Symptoms of muscle injury depend upon which muscle groups are involved. The failure of muscle development may also be due to nerve or bone diseases. The common clinical signs include swelling, pain, wasting or trembling of the muscles. In some diseases there is death of muscle cells. Porcine stress syndrome (PSS) is a generally fatal heritable condition (recessive gene) associated with the sudden onset of prolonged muscle spasms which causes failure of the normal metabolism and the development of acid conditions throughout the body. This has largely been eliminated by investigation of the genome of the breeding stock, through the elimination of carriers of the "Halothane gene".

Some common diseases associated with the muscular system

- Mulberry heart disease.
- Porcine stress syndrome (PSS).

Environmental considerations

- Poor housing conditions.
- Poor nutrition.

TERMINOLOGY

Asymmetric hind quarter syndrome – One hind leg muscle mass appears less than the other. It can arise where poor quality iron injections are given or it may be a congenital condition. It may be part of the porcine stress syndrome (PSS).

Back muscle necrosis – Sudden acute lameness and swellings of the lumbar muscle, often associated with PSS.

Congenital muscle hypertrophy – A breeding defect with excessive muscle formations.

Dark firm dry muscle (DFDM) – Describes the appearance of abnormal muscle at slaughter. Considered part of the PSS condition.

Mulberry heart disease (MHD) – Heart muscle failure associated with unavailability of vitamin E and/or selenium.

Muscle necrosis – Dead muscle tissue. This can arise due to loss of blood supply caused by bacterial

MAJOR SURFACE MUSCLES OF THE PIG

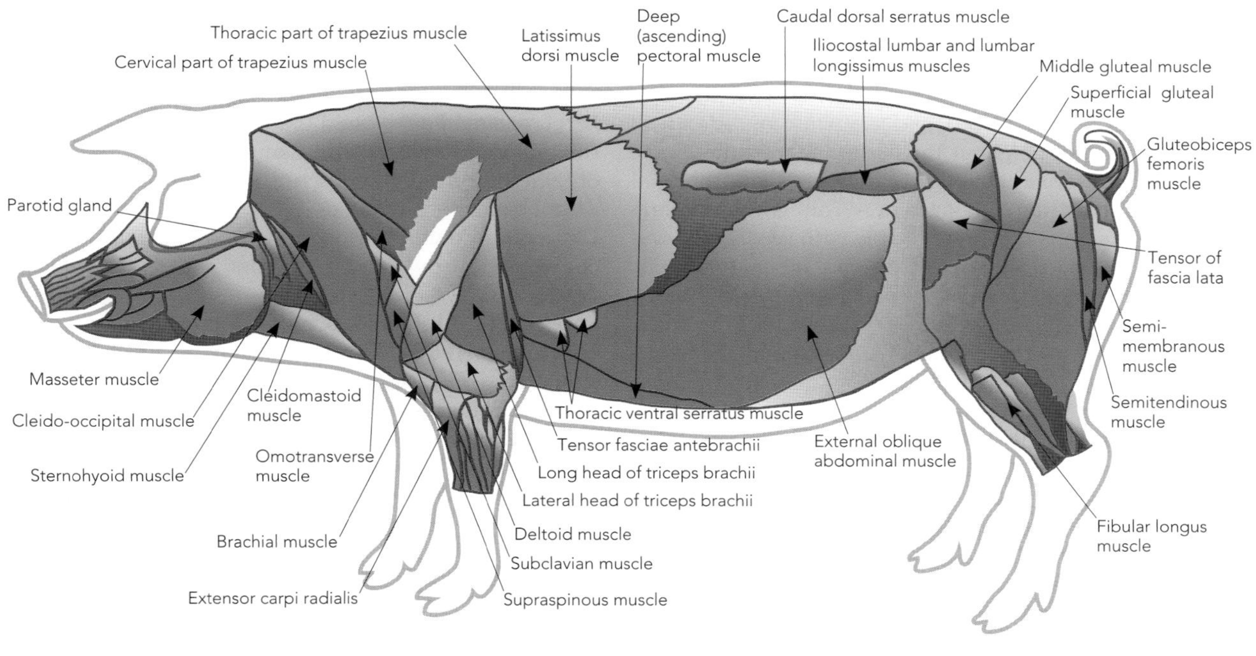

Fig 1-9

thrombosis (bacteria clogging up the blood vessels), physical damage or toxic damage. Iron toxicity, or vitamin E or selenium deficiency are further examples.
Myodegeneration – Loss of function of muscle due to muscle fibres degenerating. Common problems are associated with deficiencies of vitamin E and/or selenium.
Myopathy – This term describes any muscle disease.
Myositis – Inflammation of muscle often caused by trauma or infection.
Pale soft exudative muscle (PSE) – Describes the appearance of abnormal muscle at slaughter. Part of the PSS condition.
Pietrain creeper syndrome – Progressive muscle weakness in pigs from 3 to 12 weeks old. Considered to have a hereditary basis.
Porcine stress syndrome (PSS) – A heritable condition involving defective muscle metabolism.

Nervous System

The nervous system of the pig consists of four basic parts.

1. **The brain** – Part of the central nervous system (CNS) enclosed by the skull. It is covered completely by clear membranes called the meninges.
2. **Spinal cord** – The other part of the CNS. It extends from the brain as a narrowed bore tube, through the spinal canal (in the vertebrae) to the tail. Between each of the vertebra, which make up the spine itself, it sends branches out to different parts of the body. The spinal cord is responsible for transmitting the electrical impulses from the brain to these branches and taking the signals back to the brain.
3. **Peripheral nervous system** – Nerves leave the brain and the spinal cord and transmit the electrical impulses throughout the body. This system is the voluntary one that is under the pig's control.
4. **Autonomic nervous system** – This is the involuntary nervous system of the pig with separate nerves controlling a wide range of involuntary functions. For example, this system partly controls the heartbeat, movement of the muscular walls of the digestive system, the hormonal systems and the excretory systems.

There are a number of important bacterial and viral pathogens that cause clinical nervous signs in the pig. Such signs arise by infection of the brain, the brain covering, the spinal cord or any of the peripheral nerves.

Effects of disease

- Fits/convulsions.
- Head on one side.
- Incoordination.
- Paralysis.
- Trembling/shaking.

Some common diseases associated with nervous signs

- African swine fever (ASF).
- Aujeszky's disease/pseudorabies virus (AD/PRV).
- Classical swine fever/hog cholera (CSF/HC).
- Congenital tremor – caused by an as-yet unidentified virus, swine fever or congenital defects.
- Glässer's disease – *Haemophilus parasuis*.
- Japanese encephalomyelitis virus.
- Middle ear infection.
- Oedema disease (bowel oedema) – *Escherichia coli F18*.
- Porcine stress syndrome (PSS).
- Streptococcal meningitis (SM) – *Streptococcus suis* (various subtypes).
- Teschovirus.
- Tetanus.

Environmental considerations

- Food/mycotoxin poisoning, heavy metal poisoning, iron toxicity.
- Salt poisoning (water deprivation).
- Poor air quality – hydrogen sulphide (H_2S) poisoning.
- Water dripping into the ear of a confined sow can cause severe neurological problems.
- Stray electricity affecting the water or the feeding system can result in bizarre behavioural problems.

TERMINOLOGY

Cerebrospinal fluid – Fluid that circulates around within the brain and spinal cord. Samples of this fluid can be obtained by needle and syringe for laboratory tests to diagnose nervous disease.
Congenital tremor – A condition in newborn piglets characterised by muscle tremors and shaking. (See Chapter 8).
Encephalitis – Inflammation of the brain.
Encephalomyelitis – Inflammation of the brain and spinal cord. Viruses multiplying in the central nervous system primarily cause encephalitis and encephalomyelitis although they may cause meningitis as well.
Meninges – Clear membranes covering the surface of the brain.
Meningitis – Inflammation of the meninges which is extremely painful and often results in dramatic clinical signs. Middle ear infection may be mistaken for meningitis.

THE FEMALE REPRODUCTIVE TRACT

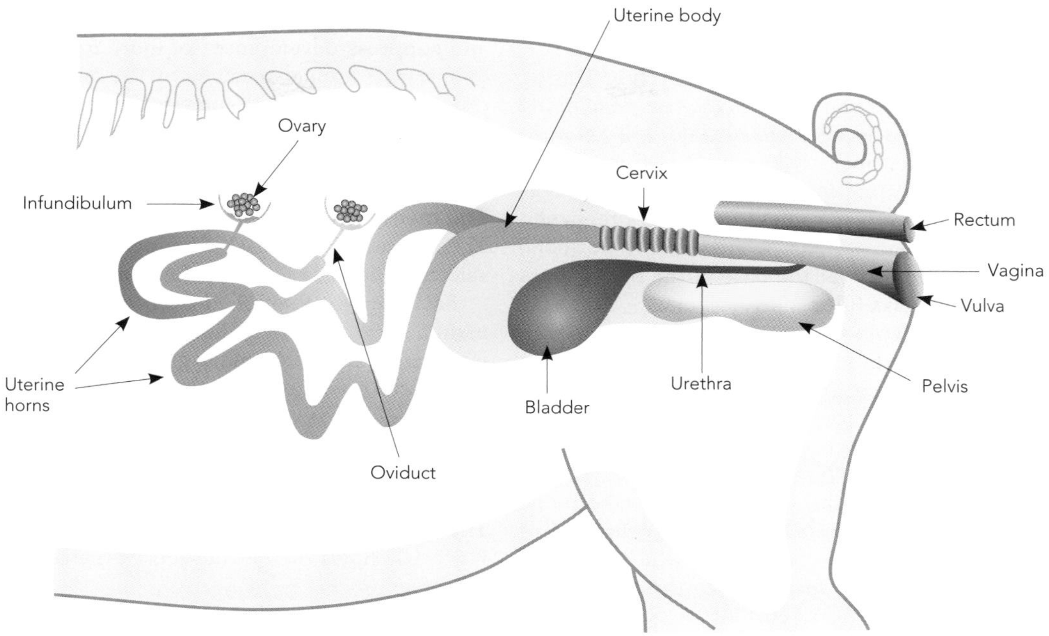

Fig 1-10

Reproductive System (Male and Female)

The terminology of the male and female reproductive systems are outlined below. More detailed information on reproduction is covered in Chapter 5.

The female

The female reproductive tract (Fig 1-10) starts with the vulva lips. When parted, the clitoris may be seen about 3 cm inside the vulva lips. During oestrus the clitoris may be very prominent. The vulva may also present with a small clear mucoid secretion.

Internal to the vulva lips, the beginning of the female reproductive tract is the large and extensible vagina which can be up to 40 cm long. The urethra (the tube from the bladder) enters the floor of the vagina some 15 cm from the vulva. If the AI (artificail insemination) catheter is incorrectly placed it is possible for the catheter to enter this urethra opening.

At the end of the vagina is the cervix. This is easily recognised as a heavily muscled area covered in nobbles. It is between these nobbles that the boar's hooked penis locates during intercourse. Likewise, during artificial insemination, the spiral or the foam tip of the catheter will locate into the cervix. Ejaculation and normal AI will deposit semen into the cervix. The cervix opens into the uterine horn, which is relatively small in the pig. The uterine body divides into two uterine horns. These are 50 cm long when the sow is not pregnant. Gestation of the piglets occurs within the uterine horns.

At the point of farrowing the uterine horns can each be over 2 metres long. Using deep uterine insemination the smaller inner catheter enters the uterine body where semen is deposited. Fig 1-11 shows the general locations of insemination with the different methods.

At the end of each uterine horn is a sharp demarcation into a narrower tube, called the oviduct. Within the oviduct, conception takes place. Its function is to store the semen until conception and protect the fertilised egg for the first 3 days of life. The oviduct ends in a large web structure called the infundibulum. This

SEMEN DEPOSITION

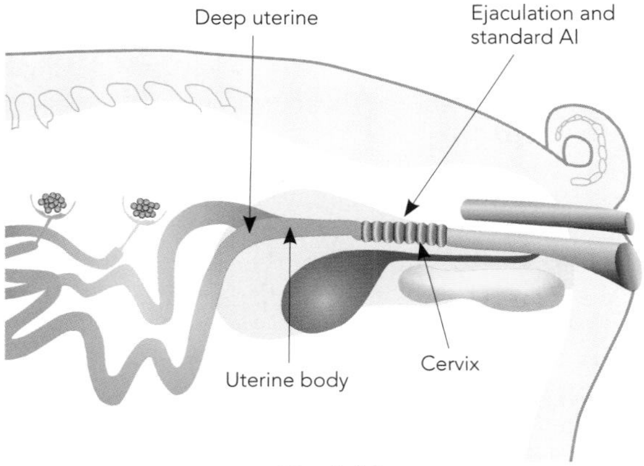

Fig 1-11

acts as a web to catch the eggs when they are released from the ovaries at ovulation. It is interesting that the ovaries are not actually attached to the rest of the female reproductive tract.

The ovaries drive the reproductive cycle. During embryonic development of the female, the developing ovaries produce all the eggs the female will produce during her life. After puberty (around 4 to 6 months of age), the ovaries mature and start producing ripe eggs every 21 days (on average) unless the pig is pregnant.

Under the influence of follicle-stimulating hormone (FSH) from the pituitary (at the base of the brain), eggs develop as small fluid-filled sacs (follicles). Every 21 days some 12 to 16 follicles per ovary start developing and enlarging producing a visible follicle. The developing follicle produces oestrogen. The function of oestrogen is to produce the signs of oestrus, assist female development and encourage protection of the reproductive tract. When the developing follicle reaches 1 cm in size, under the influence of luteinising hormone (LH) produced by the pituitary, ovulation (rupture) of the follicle occurs. This releases the egg which is caught by the infundibulum and passes down the oviduct. A blood clot forms in the ruptured follicle, producing the corpora (body) haemorrhagica. The blood clot is then reorganised into the corpora lutea, which is also about 1 cm in size. The corpora lutea produces progesterone. The function of progesterone is to maintain pregnancy and suppress development of more follicles. If embryonic signals are not received by the ovary on Day 10 of pregnancy, the corpora lutea regresses and a new wave of follicle development starts, resulting in the oestrus cycle every 21 days.

The mammary glands

There should be at least 14 nipples well placed on the selected gilt. See Fig 1-12 for some teat profiles.

Each nipple is supplied from two (sometimes three) mammary glands, the whole unit being referred to as a mammae. Each mammary gland is separate and there is no cistern where milk can be collected (as in a cow). Therefore milk is only available to piglets when the sow releases milk from the gland during suckling under the action of oxytocin.

The male

Fig 1-13 shows the key aspects of the boar's reproductive tract.

This starts with the prepuce. The entrance to the prepuce is protected by short hairs.

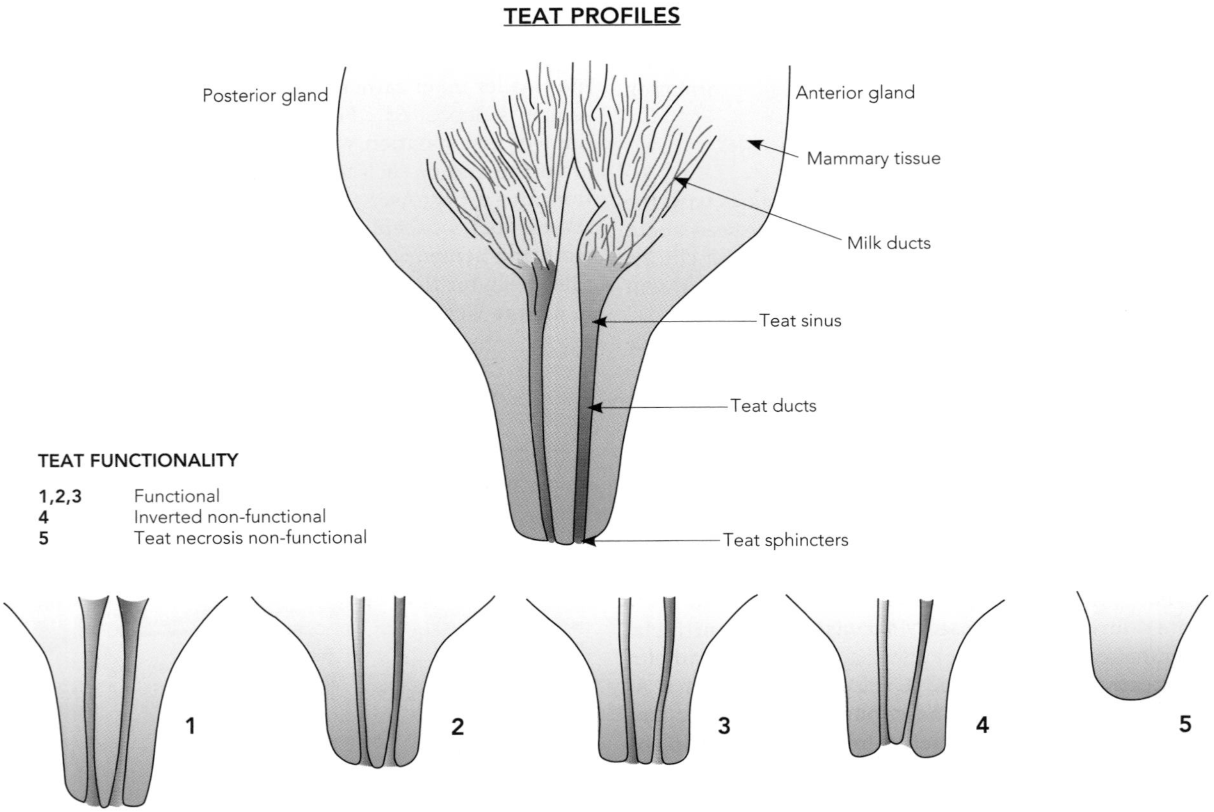

Fig 1-12

THE MALE REPRODUCTIVE TRACT

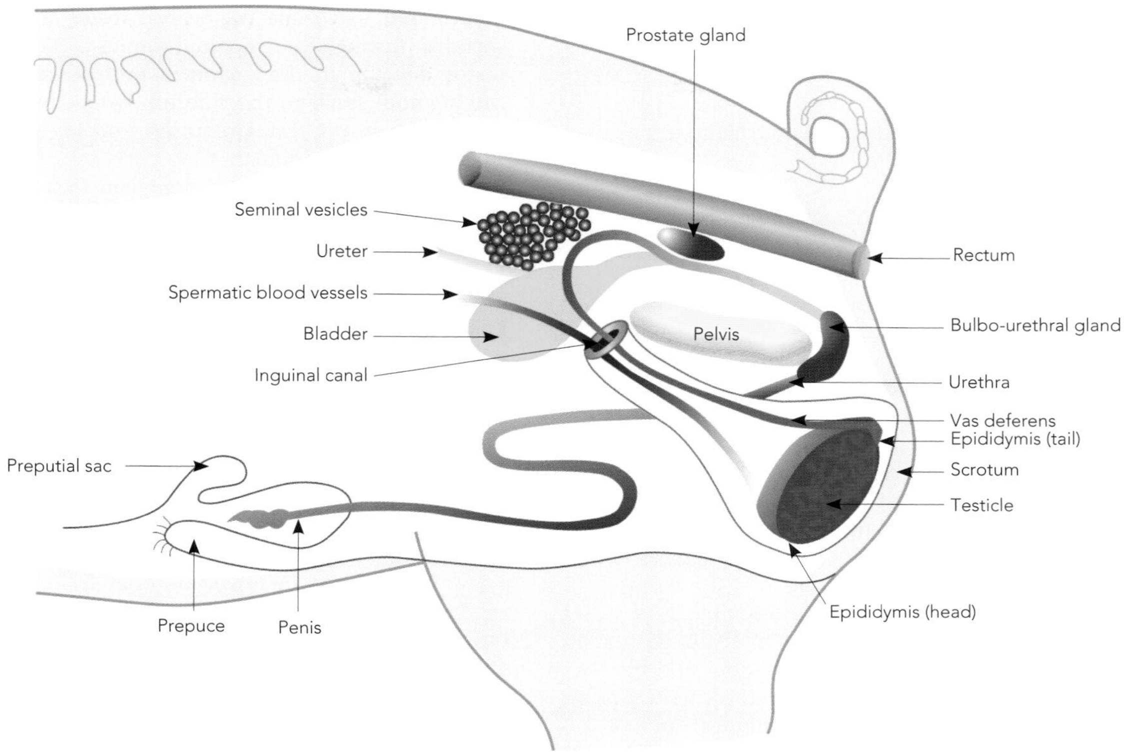

Fig 1-13

Inside the prepuce the penis tip is kept moist. Above the prepuce is a bilobed sac called the preputial sac (diverticulum), which contains foul smelling fluid from urine, sperm and decaying cells. The preputial diverticulum is the normal environment for the bacteria *Actinobaculum suis* – which may cause cystitis and pyelonephritis in the sow.

The penis of the boar is a long extensile tube with a curved end. The curved end locks into the sow's cervix during intercourse. The penis is extended by relaxing the S shaped sigmoid flexure in the urethra.

The urethra is the tube which goes from the end of the penis to the bladder. Down this tube passes urine and semen. Semen is composed of sperm from the testes, fluid from the prostate, sugars (fructose) from the vesicular glands and gel from the bulbo-urethral glands. The prostate, vesicular and bulbo-urethral glands make up the accessory sex glands.

Sperm moves from the testes to the urethra via the vas deferens. This is a tube structure which is removed during a vasectomy or epididectomy.

Sperm is produced in the testes, which are housed within the scrotum, situated at the rear of the boar. The testes are kept cooler than the normal body temperature by being placed 'outside' the main body. One interesting point is that the testes in the boar are actually upside down, with the head of the epididymis (the beginning) at the bottom and the tail of the epididymis (which moves into the vas deferens) at the top. This anatomical feature is utilised when performing an epididectomy (removal of the epididymis tail and some vas deferens).

Sperm are produced in the testes. Sperm are produced at a rate of 6000 per second and take some 3 to 5 weeks to mature before ejaculation. Therefore, infertility can last 6 weeks before healthy new sperm produced after an insult (which could be just a high temperature) are ejaculated.

Effects of disease

- Infertility.
- Loss of production.
- Empty farrowing places.
- Increase in feed costs.
- Time implications – stockpeople breeding more animals.
- Mastitis.
- Colostrum problems.
- Lower weaning weight.

Some common diseases associated with the reproductive system

- 14 to 21 days post-service vulval discharges.
- Abortion in the pig.
- Aujeszky's disease/pseudorabies virus (AD/PRV).
- Brucellosis.
- Common developmental abnormalities.
- Leptospirosis.
- Mastitis.
- Milk production and suckling problems.
- Porcine parvovirus.
- Rectal and vaginal prolapses.
- Stillborn and mummified piglets.
- Tumours of the pig.

Environmental considerations

- Air quality – ammonia reduces the ability to detect oestrus.
- High temperatures can reduce boar fertility and cause abortion in the sow.
- Draughts and chilling.
- Poor flooring can reduce signs of oestrus and cause behavioural standing.
- Food quality – mycotoxin poisoning.
- Lack of adequate water – this results in inadequate cleaning of the bladder and the reproductive tract resulting in cystitis, cervicitis and vaginitis and vulval discharges. In extreme cases the cystitis may develop into a pyelonephritis which is a serious killer of pregnant sows – thus reducing the farrowing rate.

TERMINOLOGY

Abortion – The production of a premature, non-viable litter at 11 to 112 days of gestation.

Agalactia – Failure of milk let-down or shortage of milk or no milk. The udder may be congested, with or without mastitis. In certain conditions, such as mild ergot poisoning, mammary glands fail to develop.

Cervix – The neck of the uterus. Inflammation of the cervix is called cervicitis. Cervicitis is not common in the pig, but erosion of the thick folds occurs in old sows and can cause infertility.

Conceptus – Fertilised ovum and embryo.

Corpora haemorrhagica – When the follicle ruptures to release the egg there is a small amount of haemorrhage. This is the name given to the bloody tissues that remain.

Corpus albucans – After pregnancy or after the animal has been in oestrus the corpus luteum disappears and shrinks to a small white body called the corpus albucans.

Corpus luteum – The corpora haemorrhagica becomes consolidated and forms the corpus luteum. This is the body that produces progesterone, the female hormone that maintains pregnancy.

Cryptorchid – A male pig whose testes have not descended through the inguinal canals. Normally, the testes develop in the abdomen and descend through the inguinal canal to the scrotum before birth. Sperm production in the testes requires a cooler environment than that of the abdomen.

Embryo – The multicellular organism that develops in the uterus from the fertilised egg up to about 20 to 30 days when it becomes a foetus.

Endometritis – Inflammation and infection of the lining of the uterus (the endometrium).

Epididymis – A coiled tube attached to the upper surface of the testicle where the sperm is stored. The sperm leaving it enter the vas deferens. It has a head and a tail. The tail can be cut off (epididectomy) to sterilise the boar (see Chapter 15).

Erythema – Reddening of the skin that is often seen when one or more mammary glands have mastitis.

Farrowing rate (%) – This equals:

$$\frac{\text{No. females farrowed in a batch}}{\text{No. females mated in the same batch}} \times 100$$

It should be remembered that this number is an average number. Half of the time, targeted farrowing rates are not met.

Female – Breeding females including gilts and sows. A gilt becomes a breeding female either from an arbitrary time before mating (e.g. when first brought into the mating area) or, more commonly, when she is first mated. Some pig farmers only include her from the time she farrows but this results in high and less useful indications of herd fertility when farrowing rates and numbers of pigs per female per year are calculated.

Fertilised ovum – The egg as it multiplies and grows to approximately Day 7 post-fertilisation.

Foetus – The developing piglet from when all the major structures are developed, (approximately 30 days) through to farrowing.

Implantation – The attachment of the embryo to the uterine wall by establishment of the placenta, commencing 14 to 17 days post-mating.

Infundibulum – This web-like structure acts to catch the eggs when they are released from the ovaries at ovulation.

Inguinal canal – Gap between the muscles of the abdomen in the groin through which the spermatic cord passes from the abdomen to the testicle.

Inverted nipples – These are shown in Fig 1-12. If the teat sphincter cannot be seen at eye level it is likely that such a teat will remain inverted and will not be functional. This is important to appreciate when selecting or receiving a gilt for breeding. Some inverted nipples will become more normal and be functional when the mammary gland develops, but when selecting gilts you cannot take the chance.

Each teat has two orifices and teat ducts which drain two quite separate mammary glands, front (anterior) and back (posterior).

Irregular return – A return to oestrus more than 25 days after the previous one, with a peak around Day 28 post-mating.

Lactation length – The period from farrowing to weaning in days.

Litters per female per year is calculated by:

$$\frac{\text{No. of farrowings over 12 months}}{\text{Average no. breeding females in the herd}}$$

In large herds this can also be calculated on rolling batch, 3-month and 6-month averages which gives a historical indication of rising or falling fertility, although consideration needs to be given to seasonal infertility.

Mammary oedema – Mammary tissues may contain excess amounts of fluid at farrowing. This fluid can either be under the skin when it can be easily seen and palpated, or deep in the actual tissue itself. Both these conditions can lead to agalactia, mastitis and poor availability of colostrum.

Mammary system – The udder of the sow consists of two parallel rows of 7 or more teats, inter-spaced on each side.

Mastitis – Inflammation of the mammary gland is invariably associated with infection.

Mating – The complete act of copulation involving one or more services.

Mummified pigs – Piglets which died in the uterus and in which the tissues and fluids have been reabsorbed leaving black, shrunken skeletal remains.

Non–productive days (NPD) – These include all the days when the sows and gilts are either not pregnant or suckling. It therefore always includes:

- Entry of the gilt into the herd to point of mating.
- Time from weaning to mating.
- Time from mating to remating if the female is found not to be pregnant and returns to heat.
- Time after a female has been culled until the time it is slaughtered.

NPD is a useful calculation because if it lengthens it may indicate a number of serious problems, including increases in the fail-to-farrow females (not in pig at term – NIP), females dying during pregnancy, and gilts with delayed puberty.

NPD is often abused by producers who deliberately reduce the days by manipulating the gilt introduction or culling programme.

Oestrus (or heat) – The period during which the sow is receptive to the boar (i.e. will stand to be mated). Usually 1–3 days.

Oestrus cycle – The period from one oestrus to another; 18 to 24 days interval is normal.

Orchitis – Inflammation of the testicle. A specific example is infection by *Brucella suis* bacteria. Non-infectious orchitis can arise from trauma to one or both testicles. Occasionally there may be a haemorrhage developing into a haematoma (a pocket of blood).

Ovaries – Two small structures which control the oestrus cycle and from which the follicles are produced and the eggs released.

Oviduct (fallopian tubes) – These are the tubes through which eggs are transported from the ovaries. These are also known as the fallopian tubes.

Oxytocin – A hormone produced by the anterior pituitary gland. Its function is to release milk from the glands and at the same time cause the uterus to contract.

Parity – Used to describe the number of times a female has farrowed, e.g. pregnant gilt = parity 0; gilt farrowed for the first time = parity 1; sow which has had two litters = parity 2.

(N.B. Some people get confused and use the term parity when they mean pregnancy.)

Pigs weaned per sow per year – The number of pigs produced in any 12-month period. In a large herd this is usually calculated as a 3-month and 6-month rolling average of the whole herd.

Preputial diverticulum – This is a sac inside the prepuce, the size of a golf ball, that contains a foul-smelling fluid with a high bacterial content.

Prolactin – A hormone from the pituitary gland involved in the initiation and maintenance of milk production.

Pyometra – Accumulation of pus in the uterus following infection. It is also called pyometritis. This is common when heavy vulval discharges are seen or a retained foetus or placenta are present.

Regular return – A return to oestrus, usually 18 to 24 days after the previous one.

Rig – Synonymous with cryptorchid – A boar in which one or more testicles have not descended into the scrotum.

Salpingitis – Inflammation of the oviducts (fallopian tubes) that carry the eggs from the ovary down towards the uterus.

Scrotum – A sac made of relatively thin pliable skin, which has a muscular inner fibro-elastic layer which contracts in a cold environment and relaxes in a hot environment.

Seminal vesicles – Glands which together with the prostate and bulbo-urethral glands provide fluid and nourishment for the sperm, the fluids being passed out during ejaculation.

Spermatic cord – Fibrous cord, containing the vas deferens and blood vessels, by which the testicles are suspended.
Stillborn pigs – Piglets observed dead behind the sow at birth.
Teat necrosis – Damage to the end of the teat can result in death and sloughing of tissues. This is called necrosis. It is caused by abrasive floor surfaces in the first 18 to 24 hours after birth and can be an important reason for rejecting gilts for breeding.
Testicle – The gland in which the sperm is produced.
Urethritis – Inflammation of the urethra, the tube which carries both sperm and urine down the penis in the boar, or urine from the bladder to the vagina in the sow. Urethritis is uncommon in the boar but can occasionally be caused by small calculi or stones formed in the kidneys. The urethra of the sow is much more likely to become contaminated and infected because its opening is so close to the vulva. Urethritis and cystitis are therefore common in the sow.
Uterus (womb) – Consists of two horns up to 2 metres in length that contain the foetuses.
Vagina – The passageway from the exterior to the cervix. Vaginitis (inflammation) occurs following trauma, infection or multiple matings.
Vas deferens – The muscular tube that at ejaculation propels the sperm from the tail of the epididymis on the testes up through the inguinal canal and into the urethra where it joins just below the neck of the bladder.
Vulva – The vagina opens to the exterior through the fleshy lips of the vulva. Oedema of the vulva (swelling containing fluid) occurs in late pregnancy and trauma and is very common in loose-housed sows. The tissues contain many blood vessels and are prone to haemorrhage. Haemorrhage (haematoma) is also seen in the gilt post-farrowing. Such animals can bleed to death. (See Chapter 15.)

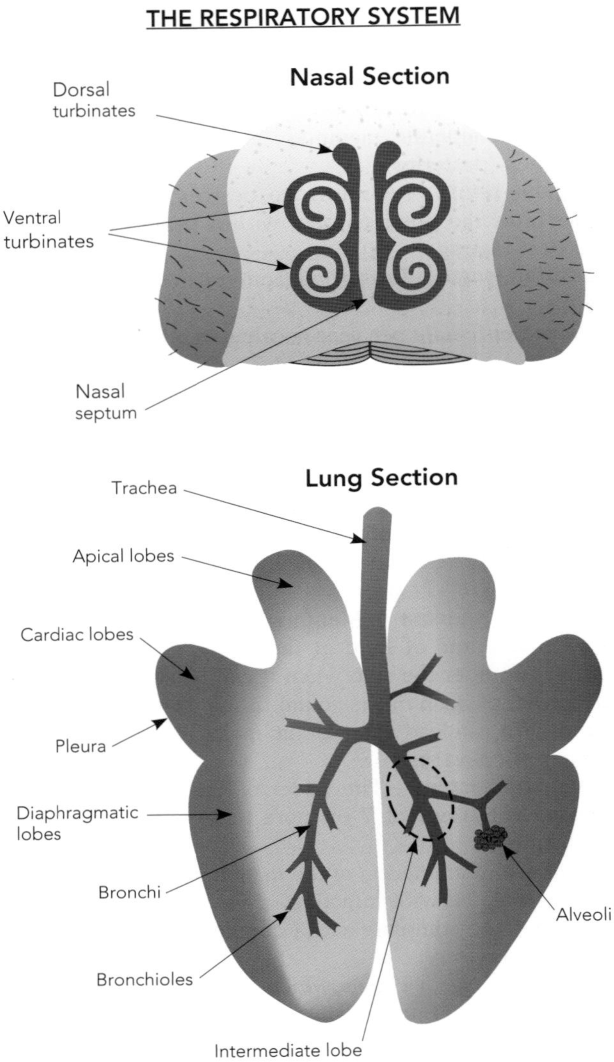

Fig 1-14

Respiratory System

The respiratory system of the pig commences at the nostrils which lead into two nasal passages. These contain the dorsal and ventral turbinate (conchae) bones (Fig 1-14).

The ventral turbinates consist of four thin main bones, two on each side, separated by a cartilaginous septum. You can imagine these as four hair curlers placed inside the nose. The respiratory tract is lined by a smooth membrane called a mucous membrane because it is bathed in a sticky mucus. It is also covered with minute hair-like structures (cilia) which are able to brush the mucus across the surface by their wavy motion. They move the mucus in the nose, bronchial tree and trachea to the throat where it is swallowed. Together the cilia and mucus form the mucociliary escalator. The air breathed in through the nose is warmed by the turbinate bones which, because of their scroll-like shape, cause turbulence. This throws out the larger of the small particles so that they stick to the mucus and are swept to the throat.

The pig also has a third tracheal bronchus on the right. This makes the right apical lobe more susceptible to particles descending into the lungs. The many branches of the bronchi decrease in diameter as they descend into the lung. They branch off the main trunck at right angles. All of this is to increase the speed of the air and make particles present in the air hit the wall of the bronchi. The walls of the bronchi are covered by the sticky mucociliary escalator where particles become trapped. The mucociliary escalator then carries them from the lungs up to the throat. Only the very smallest particles (3 to 1.6 μm) reach the alveoli where the alveolar macrophages engulf and remove them. Particles less than 1.6 μm do not settle within the lung – note viruses are smaller than this – thus viruses need to be "accompanied" into the lung – for example in water droplets or dust particles.

The nasal passages open into the pharynx (throat) which is a common passage for food and air. The food is swallowed down the oesophagus and the air is sucked into the larynx at the back of the throat. The larynx (voice box) controls inspiration and expiration. It opens into the trachea which passes down into the chest where it divides initially into two bronchi.

The bronchi branch into smaller bronchi and continue to branch, gradually reducing in size to become bronchioles which terminate in very tiny air sacs called alveoli. Oxygen is passed from the alveoli into the bloodstream and carbon dioxide is passed out. The lungs are divided into seven lobes, as shown in Fig 1-14.

The lungs are housed within a chest which is lined with the pleura. Infection of the pleura results in pleurisy.

Effects of disease

- Death.
- Respiratory distress.
- Poor feed intake.
- Poor growth rate.
- May be forced to have reduced status within the group.

Some common diseases associated with the respiratory system

- *Actinobacillus pleuropneumoniae.*
- *Actinobacillus suis.*
- *Haemophilus parasuis* (Glässer's disease).
- *Mycoplasma hyopneumoniae.*
- *Mycoplasma hyorhinis.*
- Mulberry heart disease.
- Pneumonic pasteurellosis and streptococci.
- Porcine circovirus 2.
- Porcine reproductive and respiratory syndrome virus.
- Porcine respiratory coronavirus.
- Postweaning multisystemic wasting syndrome.
- Swine influenza virus.

Environmental considerations

- Air quality – ammonia reduces efficiency of mucociliary escalator.
- Dirty floors.
- Dust.
- Hot air temperature.
- Low temperatures and/or draught – chilling.
- Water temperature – especially wet-fed pigs.
- Water flow and availability – reducing efficiency of mucociliary escalator.
- Wet floors.

TERMINOLOGY

Abscess – Area of pneumonia-containing pus where the infection has been sealed off from the remainder of the lung tissue by a fibrous capsule.

Alveolar macrophages – These cells which are located in the alveoli engulf bacteria and viruses. They are destroyed by some viruses e.g. the porcine respiratory reproductive syndrome virus (PRRSV).

Bronchitis – Inflammation of the bronchi or bronchioles in the lung.

Cilia – The small hair-like structures on the surface of cells of the nose and trachea/bronchi. These move the mucociliary escalator.

Consolidating pneumonia – The lung tissue has collapsed and become solid. A common example is Mycoplasma (enzootic) pneumonia (*Mycoplasma hyopneumoniae*) infection which causes inflammation of the anterior lobes of the lungs.

Necrotising pneumonia – Necrosis means death of tissue within the living animal. Necrotising pneumonia occurs where the organism or its toxins kill lung tissue. An abscess may result. A common example is pneumonia caused by *Actinobacillus pleuropneumoniae.*

Pleurisy – Also called pleuritis. The shiny membranes that cover the surface of the lungs and the inside of the chest wall are called the pleura. Infection or inflammation of these surfaces is called pleurisy. This, together with pericarditis, is very common in the pig and accounts for considerable loss through condemnation at slaughter.

Pneumonia – Inflammation in any part of the lung tissue. There are different types of pneumonia.

Progressive atrophic rhinitis – Rhinitis caused by toxigenic (toxin-producing) strains of *Pasteurella multocida*, in which the turbinates lose their tissues (atrophy) irreversibly. This is now called progressive atrophic rhinitis to distinguish it from non-progressive atrophic rhinitis caused by *Bordetella bronchiseptica* (with the addition of other organisms) and/or environmental contaminants, which is less severe and heals when the infection is stopped by the immune response.

Pyaemic pneumonia – Multiple small abscesses scattered through the lungs that have been carried there via the bloodstream. A common example is pyaemia from tail biting. The carcass is condemned at slaughter.

Respiratory rate – This varies from 20 to 40 breaths per minute in piglets and growing pigs and 15 to 20 per minute in sows. More than 40 breaths per minute can be considered to be laboured breathing and can be an important indication of heat stress.

Rhinitis – Describes any form of inflammation to the delicate mucous lining of the nose. Some agents such as dust and gases may cause it but there is no long-term damage to the nose structure. Sneezing always occurs with rhinitis.

Tracheitis – Inflammation of the trachea (windpipe). Influenza may cause a very heavy "barking" cough.

Turbinate bones – Dorsal and ventral. Scrolls of bone inside the nasal passages. They warm and filter air as it passes through the nose.

Sensory Systems

The pig, like the human, experiences sound, sight, smell, taste, heat, cold, pain and balance. The way it responds to these assists us in the recognition of health and disease. For example, pain together with posture will often indicate a specific disease such as fracture of the vertebrae in the spine. Poor balance may be associated with infections of the middle part of the ear, which are common in the young growing pig. Likewise, the stockperson's own senses, particularly sight, smell and touch, are important in assessing whether the pig is healthy or ill and performing to its maximum biological efficiency.

Sight, for example, allows the stockperson to observe the lying patterns, any abnormal excretions, and signs of disease and unevenness of growth. It also helps to appreciate the quality of the environment. Note, the pig only has 2 colour sensors (we have 3) and thus does not see red colours well. This can have an impact on passageway design.

Smell allows the stockperson to detect toxic gases, blocked drains, putrefying tissue and humidity, important points to consider in respiratory disease. He or she may also learn to detect the smell of scour.

Smell and taste are also very important to the pig in helping it to identify unpalatable food.

Effects of disease

- Inability to recognise threats.

Some common diseases associated with sensory systems

- Blue eye disease in Mexico.
- Conjunctivitis.
- Middle ear disease.

Environmental considerations

- Air quality – ammonia in the air may affect the eyes.
- Disinfectant misuse can cause blindness and/or damage by penetrating wounds.
- Mycotoxins in the food may affect ability to taste.

Skeletal System

The general skeletal layout of the pig is shown in Fig 1-15. The structure of a bone and joint are shown in

THE SKELETON AND GENERAL BODY SHAPE OF THE COMMERCIAL PIG

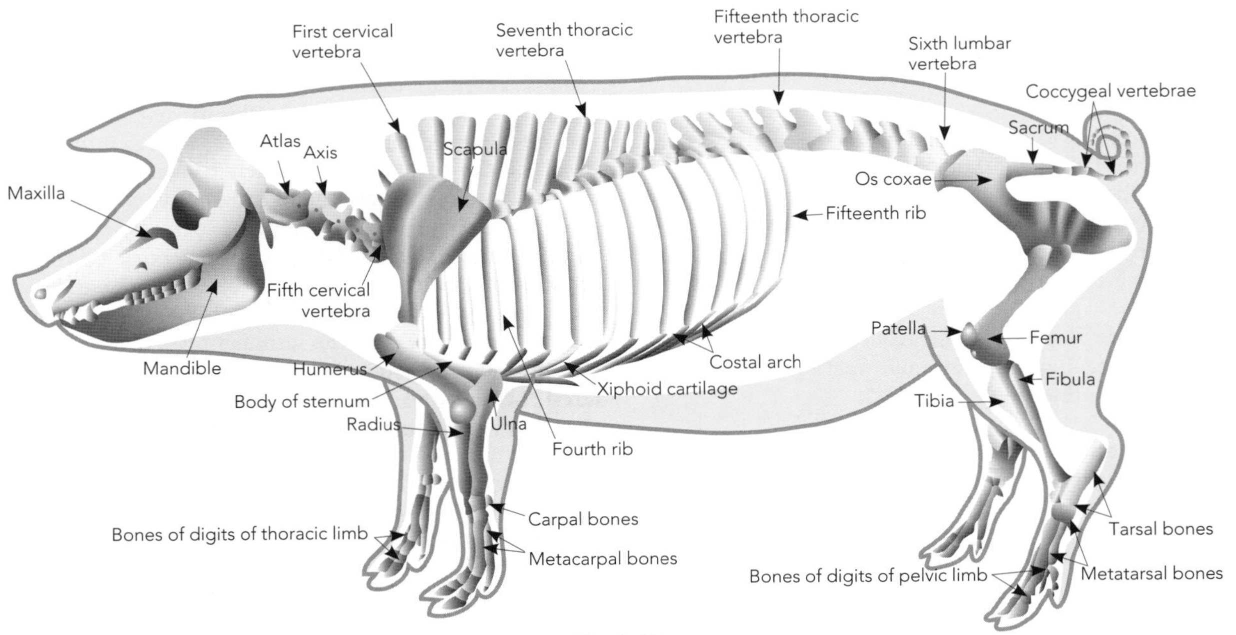

Fig 1-15

Fig 1-16. A joint consists of the ends of two bones held together by ligaments and muscles, surrounded by a strong membrane and covered with smooth cartilage which forms what are known as the articular surfaces. Cartilage is dense material that is shock-absorbing. The two articular surfaces are surrounded by a thin membrane called the joint capsule, the inner part of which is secretory and produces the joint fluid (synovial fluid). The muscles and ligaments surrounding the joint are attached to the periosteum, the membrane which covers bone. Beneath the periosteum is the layer of compact bone that provides the strength of the structure. The centre is composed of a spongy mass containing marrow, from which many of the cells circulating in the blood are produced. Near the ends of the bones are flattened areas of cartilage running at right angles to the bone called the epiphyseal plates, which by increasing their thickness cause bones to grow in length and width. The separation of bones at these plates is a common occurrence in leg weakness or osteochondrosis, particularly in young growing animals. Fast growth can increase the width of these growth plates, making them more unstable.

Bone is continually being broken down and rebuilt even in adults who have stopped growing. Thus they are able to repair fractures and respond to pressures. The main pressures are from muscle tone and exercise. Pigs that are able to exercise are likely to have stronger bones and joints than those that cannot. Thus sows kept in total individual confinement have softer, more brittle bones than sows kept in pens, yards, or outdoors.

In contrast, the articular cartilage when damaged and eroded cannot repair itself and is replaced by less effective fibrous tissue. This process can be progressive.

THE STRUCTURE OF BONES AND JOINTS

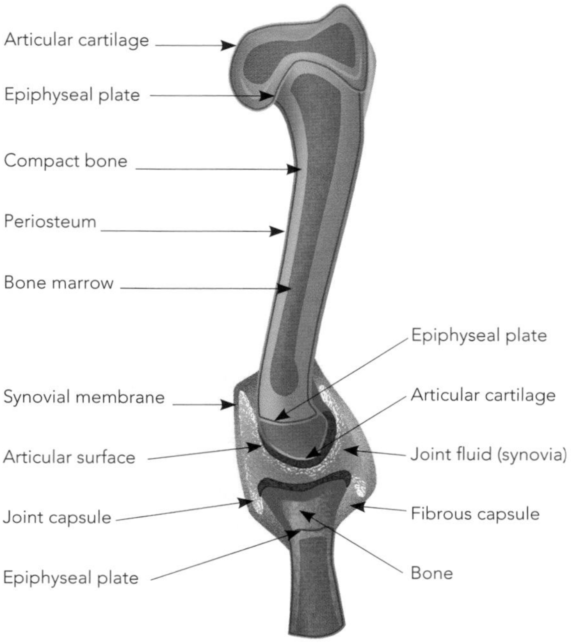

Fig 1-16

Effects of disease

- Incoordination.
- Lameness.
- Loss of social status.
- Poor feed intake.
- Poor gait.
- Poor growth.
- Unable to stand.

Some common diseases associated with the skeletal system

- Apophyseolysis.
- Arthritis.
- Bush foot.
- Epiphyseolysis.
- Fracture.
- Lameness.
- Osteochondrosis.
- Osteomalacia.
- Osteomyelitis.
- Periostitis.
- Rickets.
- Spondylitis.

Environmental considerations

- Draughts and chilling – causing piling and more injuries.
- Floor – poor construction.
- Gap between bottom gates.
- Slat width.
- Stocking density.
- Transport issues.
- Welfare – slippery floors.
- Wet floors.
- Width of passageway.

TERMINOLOGY

Adventitious bursa – A soft swelling containing fluid, resulting from a callus. Often found over the hock and elbow.

Apophyseolysis – Separation of the growth plate at the point of attachment of the main muscle mass on the back of the pelvis. It is commonly seen in the second parity female. The hind leg cannot be pulled backwards because the muscle attachment has been lost.

Arthritis – Inflammation of the joint. This can occur as a result of damage, but in the pig most cases are caused by infection resulting in increased synovial fluid, inflammation of the synovial membrane, sometimes erosion of the articular cartilage and sometimes the formation of pus. It is an extremely painful condition and makes the pig lame. Common infections causing this include *Mycoplasma hyosynoviae*, *Haemophilus parasuis*, streptococci, staphylococci, *Trueperella pyogenes* and *Erysipelothrix rhusiopathiae*.

Bush foot – Infection of the hoof and the bones in the foot. It arises from trauma and damage to the solar surface of the hoof. The claw is often swollen.

Bursa – A true bursa is a sac containing lubricating fluid but in pigs the term is often used to describe a fibrous lump beneath the skin covering bony prominences, caused by constant pressure.

Callus – An outgrowth of bone due to trauma to or irritation of the periosteum.

Chondrocytes – Cells found in cartilage. They form future bone.

Crepitus – The broken ends of bone rubbing together.

Epiphyseolysis – Separation of the epiphyseal or growth plate. It occurs as part of the leg weakness syndrome and fractures can occur, for example, in the ball and socket joint of the femur. In young growing animals separation of the plates in the vertebrae in the spine can result in spinal paralysis.

Foot rot – Infection involving the soft tissues between the two claws.

Laminitis – Inflammation of the soft sensitive tissues inside the hoof.

Leg weakness – A term used to describe conformation defects and abnormalities of gait in both fore and hind limbs. It is also used to describe osteochondrosis.

Osteomyelitis – Infection of the bone itself and the bone marrow in its spongy centre. It can occur after a septicaemia with organisms such as streptococci and erysipelas. It is often seen in the jaw bone after faulty teeth clipping.

Osteomalacia – This describes a softening of the bones and is caused by poor calcium and phosphorus deposition into the compact bone. This can be associated with the loss of these minerals during lactation or their unavailability in the diet.

Osteochondrosis (OCD) – This involves changes in the articular cartilage and the bone and it is very common. Most (if not all) modern pigs show such changes to bone structure at a microscopic level. Another term is leg weakness.

Periostitis – Inflammation of the periosteum. This is extremely painful and can arise through trauma or occasionally infection. The most common causes are mechanical damage to knees in sucking pigs, and swelling or leg calluses (particularly on the hind legs), seen on many animals that are reared on concrete floors.

Rickets – Soft bones due to a shortage of phosphorus or vitamin D.

Synovial fluid – Oily fluid in the joint.

Synovitis – Inflammation of tendon sheaths and joint capsules.

Tenosynovitis – Inflammation of tendons and tendon sheaths.

Urinary System

The kidneys are the organs in the body that filter out toxic and other waste materials from the bloodstream and maintain the body's fluid balance (Fig 1-17). Blood passes from the aorta into the kidney where it is cleaned by filtration and returned back into the bloodstream. The toxic products are then passed with fluid into the ureters which lead down to the bladder. Urine leaves the bladder via the urethra to the exterior.

Effects of disease

- Death.
- Failure to farrow.
- Vulval discharge.

DIAGRAMMATIC VIEW OF THE URINARY SYSTEM

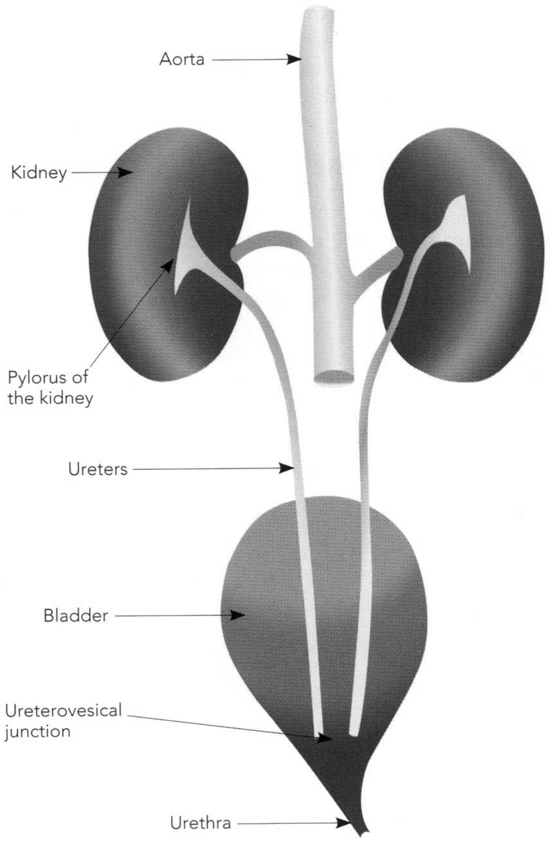

Fig 1-17

Some common diseases associated with the urinary system

- Crystalluria.
- Kidney worm.
- Nephritis.
- Pyelonephritis and cystitis.
- Urethritis.
- Postweaning ill-thrift syndrome/Periweaning failure to thrive syndrome (PFTS).

Environmental considerations

- Water intake.
- Water depth.
- Drinking position/design.
- Floor design – slippery floors.
- Water quality and availability.

TERMINOLOGY

Calculi – These may be seen as powder-like deposits on the vulva of the sow or as small stones in the urine. They are due to the crystallisation of mineral deposits and are not usually of any clinical significance. They are particularly striking in the kidneys of piglets which have died of TGE or greasy pig disease and are exaggerated in mercury poisoning.
Crystalluria – When a large amount of (usually white) crystal deposits are found in the urine. They may be seen attached to the vulval lips.
Cystitis – Inflammation, usually due to infection, of the lining of the bladder. The normal thickness of the bladder is approximately 10 mm but in severe cases it may be up to 50 mm. Haemorrhage often occurs and in such cases mortality in sows can be high.
Haematuria – Blood in urine. Always consider this as serious; a sign of severe cystitis/pyelonephritis.
Haemoglobinuria – Free haemoglobin in the urine.
Nephritis – Inflammation of the kidney. It can be associated with several different bacteria that are transmitted either via the bloodstream (septicaemia) or reflux from the bladder. Bacteria or their toxins can damage the delicate filtering mechanism of the kidney.
pH – The amount of acid in the urine. Normal urine has a pH of 6 to 8.
Proteinuria – Protein in the urine. Normal levels are 6 to 20 mg/100 ml. Levels are elevated in kidney disease.
Pyelonephritis – The ureters arise from the cup-shaped pylorus or collecting area in the kidneys. Infection of this area together with the kidney is called pyelonephritis. It is a common disease in the sow. Bacteria associated with this include *E. coli*, streptococci and *Actinobaculum suis*, the latter being the most common and important.
Pyuria – Pus in the urine.
Urethritis – Inflammation of the urethra.
Urolithiasis – Stones in the urinary system.

2 UNDERSTANDING HEALTH AND DISEASE

Chapter 2

2 UNDERSTANDING HEALTH AND DISEASE

Definition of Health

The term "health" means different things to different people. It is a state of physical and psychological well-being that allows the pig to exploit its genetic potential for maximising productivity, reproductive performance and lean meat production.

Good husbandry is the most important factor in preventing disease and maximising health and production.

The term "disease" means an unhealthy disorder of body and mind, sometimes with pain and unease that is likely to prevent the pig from exploiting its genetic potential resulting in lowered productivity.

The level of clinical disease is described by the term morbidity. Disease can be clinical (i.e. the affected pig shows clinical signs) or sub-clinical (the affected pig shows no obvious clinical signs). Sub-clinical disease can also have an adverse effect on productivity. It is important to distinguish between sub-clinical disease and sub-clinical infection. Every healthy herd, without exception, carries a multitude of potentially pathogenic organisms, mainly in the gut but also in the nose, throat, tonsils, skin and genitals, which are not causing disease either clinical or sub-clinical.

The absence of all "pathogenic" organisms is not possible. In fact it is not possible for the pig to survive without the presence of bacteria and other organisms within its intestines. It is an interesting concept that in fact 90 % of a pig's bodyweight may actually be bacterial! The concept of metagenomics looks at the total genetics of the pig – the pig and its resident microflora. This concept will become increasingly important in diagnostics and treatment regimes.

There is a delicate balance between these potential pathogens and the pig's immunity to them. Any physical or psychological disturbance of this immunity may render the pig susceptible.

Good husbandry, including good stockmanship aims to avoid such disturbances, provided the more virulent pathogens are absent; e.g. toxigenic *Pasteurella multocida*, progressive atrophic rhinitis (PAR) and transmissible gastroenteritis virus (TGE). Good husbandry means good housing, good nutrition and good management. Good stockmanship means care and attention to the pig's health and welfare.

This delicate balance between potential pathogens and the pig's immunity becomes even more precarious on a herd basis. By causing disease in small groups of pigs as a result of poor husbandry, the pathogenic organism multiplies up to a concentration that may overcome the more resistant pigs. The concentration again builds up and threatens to overwhelm the collective immunity of the herd (i.e. herd immunity) (Fig 2-1).

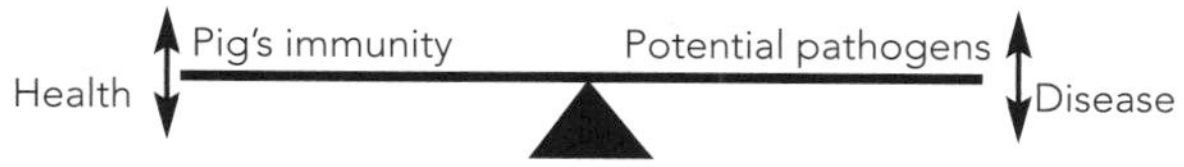

Fig 2-1

The Causes of Disease

It is likely that when considering the causes of disease you think first of infectious microorganisms/pathogens. You would be right to do so in that infectious disease plays a much bigger role in pig herds, particularly large pig herds, than in animals kept individually such as dogs and cats. Nevertheless there are other non-infectious causes of disease which may also be damaging to the herd and to productivity and profitability.

Most diseases have multiple causes.

Causes of disease are considered here under five infectious and five non-infectious main headings:

Infectious agents

- Viruses.
- Bacteria, including chlamydia and mycoplasma.
- Fungi.
- Parasites.
- Prions.

Non-infectious agents

- Trauma.
- Hereditary and congenital defects.
- Nutritional deficiencies and excesses.
- Toxic agents (poisons).
- Stress.

There are other causes, such as tumours, which in pig herds are much less important and will not be dealt with here.

When you think of disease it is often only in terms of a single cause. In some cases this may be right (e.g. a poisoning or a highly virulent virus infection such as foot-and-mouth disease). In most cases you would be wrong however, because clinical disease in pig herds usually results from the interplay between a number of predisposing, primary and contributory causes.

You should bear this in mind when thinking about how to suppress clinical signs and effects of diseases.

Infectious Agents

Viruses

What are viruses?

Pathogenic organisms that infect pigs can be listed in order of decreasing size and complexity (Fig 2-2). Viruses cannot be seen by the type of light microscope that is used for looking at bacteria. They can only be seen through an electron microscope.

Biggest and most complex	Protozoa
↓	Fungi
	Most bacteria
	Rickettsia
	Chlamydia
	Viruses
Smallest and simplest	Prions

Fig 2-2

At their simplest they consist only of nucleic acid (i.e. their genes) and proteins which are arranged around the nucleic acid in a geometrical design and which protect the genes (Fig 2-3). Some larger ones also have a loose outer coat (the envelope) which may contain lipids (fats) and carbohydrates which are derived from the host's cell.

Viruses contain their genetic code in the nucleic acid but they do not contain the full mechanisms for their own multiplication. They have no energy-generating systems and lack chemicals such as enzymes required for their own reproduction. They depend entirely on those of the host.

Viruses only multiply inside host cells. Viruses cannot multiply outside the host.

This is the first important point to be aware of. Outside the host viruses are inert. They have no metabolic activity. Inside the host's cell they behave like pirates. Their genes, using the host's nucleus, take over control of part or all of the cell's mechanisms. They programme these mechanisms to make many more viruses exactly like the ones that invaded. This activity usually damages or destroys the host cell.

How do they get into cells?

The proteins on their surface stick specifically to receptors on the cells' surfaces. The cells then engulf them as if they were taking in particles of food but the particles are destructive invaders. They quickly lose their outer envelopes and the proteins covering their genes. These released genes then take over from the cell's

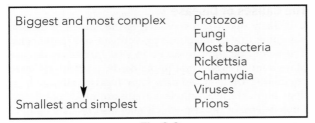

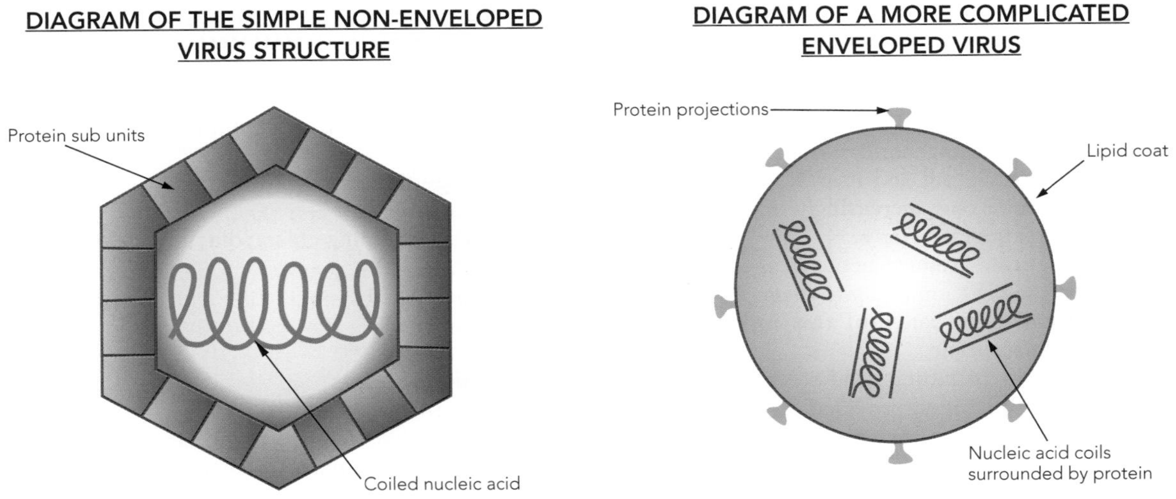

Fig 2-3

nucleus and direct the cell to make more viruses. RNA viruses replicate in the cytoplasm and DNA viruses replicate in the nucleus.

The classification of a virus

All animal, plant and bacterial genes are made of a nucleic acid called DNA (deoxyribonucleic acid) which is found in the central nucleus of the cells, and a further one, RNA (ribonucleic acid) found in the cytoplasm. Virus genes however only contain one or the other.

The first broad classification is therefore into DNA viruses and RNA viruses (Fig 2-4 and Fig 2-5). These are then classified into families based on their shape, size and structure.

Fig 2-6 illustrates the approximate survival times of viruses that cause pig diseases. Some are extremely fragile and survive only a matter of days whereas others can persist for months. The survival times given are only approximate because in any virus they vary widely depending on the material it is on (faeces, saliva, blood, water, dust etc.), the temperature, the humidity, sunlight, and relative acidity. In general, viruses survive long periods if frozen, fairly long periods in damp overcast cold weather, but only short periods in hot sunny dry weather. This is why viral diseases are more common in winter than summer. This information clearly is important in their control. Most but not all viruses are destroyed by strong acid or alkaline solutions and those that have a fat (lipid) coating are quickly inactivated by solvents/detergents.

FIG 2-4: DNA VIRUSES CAUSING/ ASSOCIATED WITH DISEASES IN PIGS

Virus Family – Genus/Species	Pig Diseases	Comments
Adenoviridae family		
– Adenovirus	Multifactorial respiratory problems.	Low grade, little known pathogen.
Asfarviridae family		
– Asfivirus	African swine fever.	Insect-borne virus.
Circoviridae family		
– Circovirus	Porcine circovirus associated diseases (PCVAD).	Two types, I and II.
– Torque teno virus	Non specifically identified.	Linked to circovirus and PCVAD.
Herpesviridae family		
– Pseudorabies virus (PRV)	Aujeszky's disease (AD).	Some countries/regions AD/PRV free.
– Cytomegalovirus	Inclusion body rhinitis.	
– Lymphotrophic virus (PLHV-1, PLHV-2, PLHV-3)	Malignant catarrhal fever.	PLHV-1 most common in pigs, also found in cattle & sheep.
Papovaviridae family		
– Papovirus	Genital papilloma.	Little known.
Parvoviridae family		
– Parvovirus	Porcine parvovirus (PPV).	Very common.
– Hokovirus	Unknown implications.	Newly emerging virus.
Poxviridae family		
– Suipoxvirus	Swine pox.	World-wide.

Some viruses can multiply sub-clinically in a pig for long periods of time (a carrier state) as in the case of Aujeszky's disease/pseudorabies virus (AD/PRV) and the cytomegalovirus that causes inclusion body rhinitis. Others can be carried sub-clinically for intermediate periods of 2 to 3 months such as PRRSV but others such as TGE are usually eliminated after a week or two. However, they may also persist in herds for long periods by successions of naive pigs postweaning becoming infected, thus maintaining the disease on the farm.

Viruses may be shed in saliva, urine, faeces, milk or in expired air from the lungs, or from vesicles on the skin. The transmission may be by direct pig-to-pig contact, or indirect contact on farm machinery or pig trucks, or by vectors such as the wind, birds, flies, or discarded human pig meat products. We probably know of about 5 % of the virus population so there are a lot more to find.

Viruses contain either DNA or RNA. Never both.

Clinical signs of disease

Viruses such as classical swine fever/hog cholera (CSF/HC) disseminate throughout the body and damage different organs thus producing a variety of clinical signs. Some such as TGE virus are more specific and disseminate sub-clinically initially to become concentrated in one or two target organs, which in this disease is the intestinal lining, causing diarrhoea. Some viruses can only multiply in one organ and swine influenza virus (SIV) for example is generally only found in the lungs and respiratory tract.

Common pig viruses and their associated symptoms are shown in Fig 2-7. The diseases marked by an asterisk (*) are the major ones that are notifiable in many countries and controlled by eradication policies. The diseases highlighted in **bold** are the most important at farm level.

Diagnosis of viral infections

It is more difficult and expensive to grow viruses than bacteria in the laboratory. They have to be grown in living cell cultures artificially in tubes or bottles or in the living cells of embryos in hens' eggs. Some cannot be grown by either of these methods. Previously, viral

FIG 2-5: RNA VIRUSES CAUSING/ASSOCIATED WITH DISEASES IN PIGS

Virus Family – Genus/Species	Pig Diseases	Comments
Arteriviridae family – Arterivirus	Porcine reproductive and respiratory syndrome (PRRSV).	Two strains: North American and European.
Astroviridae family – Porcine astrovirus	Associated mainly with gastroenteritis.	Poorly studied in pigs.
Bunyaviridae family – Akabane virus – Lumbo virus – Oya virus – Tahyna virus	Possible in appetence in adult pigs. Unknown implications. Unknown implications. Unknown implications.	Bunyaviruses have been identified in pigs in the last decade but minimal research has been conducted. Insect-borne.
Caliciviridae family – Swine norovirus – Swine sapovirus – Vesicular exanthema of swine virus (VESV)	Unknown implications. Unknown implications. Vesicular exanthema.	This group of viruses is not very well studied because they are difficult to culture.
Coronaviridae family (coronavirus) – Haemagglutinating encephalomyelitis virus – Porcine respiratory coronavirus – Transmissible gastroenteritis virus – Porcine epidemic diarrhoea virus	Vomiting wasting disease (HEV). Porcine respiratory coronavirus (PRCV). Transmissible gastroenteritis (TGE). Porcine epidemic diarrhoea (PED).	Viral disease of young pigs. PRCV is a mutant of TGE. Affects all ages of pig. Clinically, very similar to TGE.
Flaviviridae family – Flavivirus – Pestivirus – West Nile virus – Bungowannah virus	Japanese encephalitis. Classical swine fever. Unknown implications. Porcine myocarditis syndrome (PMC).	Mosquito-borne. Notifiable. Limited research. Identified on two Australian (NSW) farms.
Hepeviridae family – Hepatitis E virus	Unknown implications.	
Orthomyxoviridae family – Swine influenza virus	Swine influenza.	Several types based on H and N antigens.
Paramyxoviridae family – Blue eye paramyxovirus (Rubulavirus) – Menangle virus – Nipah virus	Blue eye disease Reproductive problems. Neurological and respiratory symptoms.	Significant in Central Mexico. Fruit bat-borne. Fruit bat-borne.
Picornaviridae family – Aphthovirus – Cardiovirus – Enterovirus – Teschovirus – Enterovirus – Sapelovirus – Seneca Valley Virus (SVV) – Kobuvirus	Foot-and-mouth disease (FMD). Encephalomyocarditis. Swine vesicular disease. Teschen/Talfan disease. SMEDI (stillbirth, mummification, embryonic death, and infertility). Acute diarrhoea, respiratory distress, and polioencephalomyelitis. Unknown implications. Unknown implications	Notifiable. The virus is transmitted by rodents. Similar signs to FMD. Notifiable. A fatal form of swine encephalomyelitis. Formerly enteroviruses. A result of two viruses. Formerly porcine enterovirus A. SVV isolates have been identified in pigs. Possibly associated with acute gastroenteritis. Possibly associated with acute gastroenteritis.
Reoviridae family – Rotavirus	Rotaviral enteritis.	Mainly type A.
Retroviridae family – Gammaretrovirus	Porcine endogenous retrovirus.	
Rhabdoviridae family – Lyssavirus – Vesicular stomatitis Indiana virus	Rabies. Vesicular stomatitis.	Insect-borne.
Togaviridae family – Getah virus – Sagiyama virus – Ross River virus	Associated with foetal death and abortion. Unknown implications . Unknown implications.	Rare, insect-borne. Rare, insect-borne. Rare, insect-borne.

FIG 2-6: APPROXIMATE SURVIVAL TIMES OF THE MAIN PIG PATHOGENS OUTSIDE THE PIG AND THEIR POTENTIAL AIRBORNE TRANSMISSION

Diseases Caused by Viruses	**Approx. Survival Time in Favourable Conditions**	**Limit of Airborne Spread**
African swine fever (ASF)	18 months	N
Aujeszky's disease/pseudorabies virus (AD/PRV)	14 days	4 km
Classical swine fever/Hog cholera (CSF/HC)	2 months	N
Foot-and-mouth disease (FMD)	8 weeks	300 km
Porcine parvovirus (PPV)	2 – 6 months	30 km
Porcine reproductive & respiratory syndrome virus (PRRSV)	4 days	4 km
Swine influenza virus (SIV)	A few days	T
Swine vesicular disease (SVD)	3 months	N
Transmissible gastroenteritis (TGE)	3 weeks	N
Diseases Caused by Bacteria/Mycoplasma – Common bacterium		
Actinobacillus pleuropneumonia (APP) – *Actinobacillus pleuropneumoniae*	2 weeks	T
Anthrax – *Bacillus anthracis*	Indefinitely	N
Arthritis – *Mycoplasma hyosynoviae*	2 days?	10 m
Brucellosis – *Brucella suis*	3 weeks	N
Cystitis-pyelonephritis – *Actinobaculum suis* + others	7 days	N
Diarrhoea/scour – *Escherichia coli*	6 months	N
Erysipelas – *Erysipelothrix rhusiopathiae*	Up to 8 weeks	N
Greasy pig disease – *Staphylococcus hyicus*	3 weeks	N
Mastitis – *Klebsiella*	4 weeks	N
Mycoplasma (enzootic) pneumonia – *Mycoplasma hyopneumoniae*	3 days	3 km
Rhinitis – Toxigenic *Pasteurella multocida*	7 days	T
Salmonellosis – *Salmonella*	6 months	10 m
Streptococcal meningitis – *Streptococcus suis*	5 days	10 m
Swine dysentery – *Brachyspira hyodysenteriae*	8 weeks	N
Tuberculosis – *Mycobacterium avium/M. intracellulare*	2 – 3 years	10 m

N = Not known to occur. T = Thought to occur but not proven.

FIG 2-7: MAIN VIRAL DISEASES OF THE PIG AND KEY CLINICAL SIGNS

Disease	Lameness	Diarrhoea	Respiratory	Nervous	Infertility	Misc.**
*** Aujeszky's disease/pseudorabies virus (AD/PRV)**			✓	✓	✓	
*** Classical swine fever (CSF), African swine fever (ASF)**	✓	✓	✓	✓	✓	✓
Congenital tremors (CT) (unknown type)				✓		
Cytomegalovirus (CMV)			✓			
Encephalomyocarditis (EMC)				✓	✓	✓
* Foot-and-mouth disease (FMD)	✓					✓
Porcine circovirus associated diseases (PCVAD) - porcine circovirus 2 (PCV2)		✓	✓			✓
Porcine epidemic diarrhoea (PED)		✓				
Porcine parvovirus (PPV)					✓	
Porcine respiratory coronavirus (PRCV)			✓			
Porcine reproductive and respiratory syndrome (PRRSV)			✓		✓	✓
Rotavirus		✓				
Swine influenza virus (SIV)			✓		✓	
Swine pox						✓
* Swine vesicular disease (SVD)	✓					✓
Teschovirus (formerly enterovirus)					✓	
Transmissible gastroenteritis (TGE)		✓				
Vomiting wasting disease				✓		

* Notifiable in most countries. **Bold** = Important at farm level. ** Miscellaneous – urinary, mastitis, skin, heart etc.

diseases were diagnosed by their clinical signs and post-mortem lesions and confirmed by serology. Serology requires two blood samples to be taken from each animal 1 to 2 weeks apart so that rising antibody levels can be demonstrated. A single sample is less helpful because you do not know whether the antibody levels are rising from a recent infection or are persisting from an old infection. Note it may take 7 to 10 days for any antibodies to be detected following infection.

Modern techniques have speeded up and simplified laboratory diagnosis because the virus does not need to be cultured.

Fluorescent-antibody techniques use antibodies labelled with fluorescent dyes. Direct fluorescent-antibody tests (DFA) are used to identify specific microorganisms (antigens). Indirect fluorescent-antibody tests (IFA) are used to identify the presence of antibodies against a specific antigen in serum. Test results are viewed via a fluorescence microscope or plate reader.

The enzyme-linked immunosorbent assay (ELISA) test is also commonly used in many diseases (e.g. foot-and-mouth disease) to provide a rapid accurate diagnosis. Techniques such as polymerase chain reactions (PCRs) have been developed to demonstrate the virus genetic material (i.e. the DNA or RNA) in samples; such tests for most pathogens are now available. Their advantage is that they can accurately detect tiny amounts of the DNA or RNA. Unfortunately they are expensive if done in small numbers.

Immunohistochemistry, the process of detecting the presence of a pathogen in cells of a tissue sample, can be particularly useful in diagnosing a viral infection.

In addition, the genetic sequence of a virus can now be obtained and allows for detailed epidemiology to be carried out, for example in the investigation of the source of PRRSV.

Interpretation of results

It is important to understand that no medical test is perfect. Sometimes test results can be positive even though the animal does not have the pathogen, a phenomenon called a false positive. Other times, the results can be negative when in fact they do have the pathogen – a false negative.

Very few tests are able to detect all positive cases, and sensitivity and specificity are used as measures of accuracy. The sensitivity of a test refers to how many cases of a pathogen a particular test can find. The specificity of a test refers to how accurately it diagnoses a particular pathogen without giving a false positive result. The higher the sensitivity and specificity of any given test the more accurate the result.

Treatment of a virus

Some new medicines are available for the treatment of a few viral infections in human beings but they are expensive and not available for use in pig diseases. Since viruses have no cell wall and no metabolism of their own antibiotics will not destroy them, although they may help by preventing secondary bacterial infection. Hyper-immune antiserums might be helpful in some virus infections if given by injection, but in general they are not commercially available. Tilmicosin has some effect on reducing PRRSV uptake in the alveolar macrophage and thus can be useful in an outbreak.

Bacteria

Grouped under this heading are chlamydia, mycoplasma and other bacteria generally.

Chlamydia

These and anaplasma used to be classified separately from bacteria but are now regarded as bacteria. Chlamydia are the first of the very small bacteria that live inside or on the surface of a host cell and they like viruses are obligatory parasites. They are relatively unimportant in the pig but can be associated with respiratory disease and heart sac infection. They are also associated with jaundice and poor growth although the bacteria can also be commensal parasites, in which case they cause little if any problems. Chlamydia can also cause abortion and conjunctivitis. In the pig they are mostly respiratory spread.

Mycoplasma

Mycoplasma are very tiny organisms less than half the size of other small bacteria (see Fig 2-8). They can only just be seen under a high power light microscope. They tend to live and multiply close to the surface of cells. They are respiratory spread and are mainly responsible

FIG 2-8: RELATIVE SIZES OF THE INFECTIOUS AGENTS

Infectious Agent	Means of Viewing	Approximate Size
Viruses	Electron microscope only	20 – 300 nm
Anaplasma	High power microscope	0.2 – 0.4 μm
Chlamydia	High power microscope	0.3 – 1 μm
Mycoplasma	High power microscope	0.5 μm
Bacteria	High power microscope	0. 5 – 30 μm
Fungi	Low power microscope	5 – 80 μm
Protozoa	Low power microscope	6 – 12 μm
Mange	Low power microscope	0. 5 mm
Lice	By eye/microscope	4 mm
Internal parasites	By eye/microscope	5 – 350 mm

1 metre = 100 cm = 1000 mm
10^2 or 1/100 metre = 1 centimetre (cm)
10^3 or 1/1000 metre = 1 millimetre (mm)
10^6 or 1/1,000,000 metre = 1 micrometre (μm or mcm)
10^9 or 1 billionth * metre = 1 nanometre (nm)
This gives you an idea of the actual size.
* Billion here 1,000,000,000 = 10^9

for mycoplasma (enzootic) pneumonia (*Mycoplasma hyopneumoniae*), and mycoplasmal arthritis associated with joint infections (*Mycoplasma hyosynoviae*).

Mycoplasma differ from some general bacteria by not having a cell wall. This is important, because penicillin type medicines work on the bacterial cell wall, therefore, the penicillin family (penicillin or amoxycillin for example) are totally ineffective against mycoplasma.

Mycoplasma can be grown on solid media; however, this is not easy and can only be done reliably in a few diagnostic laboratories.

Mycoplasma hyopneumoniae is found in the respiratory tract of pigs and generally survives for only a few hours outside the pig, but may survive a couple of days if protected in mucous. New herds can be established free of this particular organism. Then the main method of reintroduction of the disease is through the purchased carrier pig, by wind-borne spread from a nearby herd or a vehicle carrying pigs. Two other less important mycoplasma are *Mycoplasma hyorhinis* and *Mycoplasma flocculare*. The former can affect the smooth membranes that cover the joints causing lameness, arthritis and pericarditis (inflammation of the heart sac and may be associated with Glässer's disease). *Mycoplasma flocculare* causes small lesions that may be mistaken for mycoplasma (enzootic) pneumonia. It is often involved in pneumonia complexes. *M. hyopneumoniae* vaccine provides no protection to these other mycoplasmas.

Other bacteria

These can be readily seen under the microscope, particularly when they are stained. They are recognised by family group according to their ability to take stains, shape, size, biochemical characteristics, antigenic characteristics and recently by identification of their DNA using PCR.

The stain commonly used is called a Gram stain, and bacteria either stain positive (purple), or negative (red). Figs 2-9 and 2-10 show the different gram +ve and gram –ve bacteria and the diseases they cause.

Most pathogenic bacteria can be grown easily in nutrient liquids and on solid media containing nutrients set in agar gel. Most grow profusely within 24 to 48 hours, although some such as the tuberculosis bacillus can take up to 3 weeks or more. Some bacteria may be extremely difficult to culture – *Lawsonia intracellularis* for example. Bacteria tend to form little colonies consisting of billions of organisms and the shape and colour

FIG 2-9: GRAM NEGATIVE BACTERIA AND DISEASES

Causal Bacterium	Pig Diseases
Actinobacillus pleuropneumoniae	Severe necrotic haemorrhagic pneumonia in growing pigs.
Actinobacillus suis and equuli	Focal pneumonia in piglets.
Bordetella bronchiseptica	Mild reversible progressive atrophic rhinitis.
Brachyspira pilosicoli	Colitis.
Brachyspira hyodysenteriae	Swine dysentery.
Brucella suis	Abortion, arthritis and boar infertility.
Campylobacter coli, jejunum, hyointestinalis	Mild diarrhoea in piglets.
Escherichia coli (E. coli)	Piglet septicaemia and scour. Postweaning diarrhoea. Oedema disease. Mastitis and cystitis in sows.
Haemophilus parasuis	Glässer's disease.
Klebsiella species	Mastitis.
Lawsonia intracellularis	Ileitis (PIA, PHE, NE, RI).
Leptospira pomona	Infertility, stillbirths, weak piglets.
Leptospira bratislava/muenchen	Infertility.
Leptospira icterohaemorrhagiae	Haemorrhage, jaundice.
Pasteurella multocida (toxigenic)	Atrophic rhinitis (progressive).
Pasteurella multocida (non-toxigenic)	Pneumonia (secondary, opportunist).
Salmonella choleraesuis	Generalised disease with pneumonia, diarrhoea and fever.
Salmonella typhimurium, derby and others	Diarrhoea.
Yersinia species	Diarrhoea.

FIG 2-10: GRAM POSITIVE BACTERIA AND DISEASES

Causal Bacterium	Pig Diseases
Actinobaculum suis	Cystitis/nephritis
Bacillus anthracis	Anthrax
Brucella suis	Brucellosis
Chlamydia psittaci	Abortion
Clostridium difficile	Piglet diarrhoea
Clostridium novyii	Acute hepatitis Sudden death
Clostridium perfringens	Piglet dysentery
Clostridium tetani	Tetanus
Erysipelothrix rhusiopathiae	Erysipelas
Listeria monocytogenes	Abortions, encephalitis and septicaemia
Mycobacterium avium/intracellulare	Regressive tuberculosis
Mycoplasma flocculare	Mild pneumonia (small lesions)
Mycoplasma hyopneumoniae	Mycoplasma (enzootic) pneumonia
Mycoplasma hyorhinis	Secondary pneumonias
Mycoplasma hyosynoviae	Arthritis
Mycoplasma suis	Anaemia, infertility, poor growth
Staphylococcus hyicus	Greasy pig disease
Staphylococci – other	Abscesses and mastitis
Streptococcus suis type 1 & others	Arthritis and meningitis in piglets
Streptococcus suis type 2 & others	Meningitis in weaners, growers
Streptococcus suis type 7	Arthritis and heart lesions
Streptococcus other types	Various, Glässer's disease
Trueperella pyogenes	Abscesses and prevalent lesions

of these may be characteristic for a particular family or specific bacteria. A tiny smear from a colony spread on a glass slide and stained will give both the shape and the staining reaction of that particular bacteria.

This stain is of great help in carrying out primary observations. For example, the bacterium that causes meningitis in pigs, called *Streptococcus suis* type 2 is a small round organism in pairs or short chains that always stains positive or purple.

Each bacterium has a number of specific characteristics that are peculiar to itself. Some, for example *Bacillus anthracis*, form spores which can survive outside the pig for many years (see Fig 2-6). Organisms such as *Escherichia coli* (*E. coli*) and salmonella can remain viable outside the pig for up to 6 months. Like viruses the survival of bacteria is also dependent upon the material surrounding them (faeces, soil, pus, urine, blood etc.), temperature, moisture and exposure to ultra-violet light. Again like viruses, they can survive indefinitely when frozen and for long periods in cold damp dark weather but only for a short time in very sunny weather. Such knowledge is important in controlling the spread of the disease.

FIG 2-11: COMMON BACTERIAL DISEASES AND MAIN CLINICAL SIGNS

Diseases	Pathogen	Lameness	Diarrhoea	Respiratory	Nervous	Infertility	Misc.*
Actinobacillus pleuropneumonia	*Actinobacillus pleuropneumoniae*			✓			
Anthrax	*Bacillus anthracis*		✓				✓
Progressive atrophic rhinitis	Toxigenic *Pasteurella multocida*			✓			
Bordetellosis	*Bordetella bronchiseptica*			✓			
Brucellosis	*Brucella suis*	✓				✓	
Clostridial dysentery (piglets)	*Clostridium perfringens*		✓				✓
Cystitis/nephritis	*Actinobaculum suis*						✓
Mycoplsma suis	*Mycoplasma suis*					✓	✓
Erysipelas (growing pigs)	*Erysipelothrix rhusiopathiae*	✓				✓	✓
E. coli enteritis (piglets and weaners)	*Escherichia coli*		✓				
Mycoplasma (enzootic) pneumonia	*Mycoplasma hyopneumoniae*			✓			
Exudative epidermitis, greasy pig	*Staphylococcus hyicus*						✓
Glässer's disease (polyserositis)	*Haemophilus parasuis*	✓		✓	✓		
Ileitis (PIA, PHE, NE, RI)	*Lawsonia intracellularis*		✓				
Leptospirosis	Various species					✓	
Mycoplasma arthritis (gilts/boars)	*Mycoplasma hyosynoviae*	✓					
Oedema disease (*E. coli*)	*Escherichia coli*				✓		
Pasteurellosis	*Pasteurella multocida*			✓			
Salmonellosis	Various species	✓	✓	✓	✓	✓	✓
Spirochaetal diarrhoea	*Brachyspira pilosicoli*		✓				
Streptococcal infections	Various species	✓		✓	✓		
Sudden death in sows (clostridia)	*Clostridium novyii*						✓
Swine dysentery	*Brachyspira hyodysenteriae*		✓				
Tetanus	*Clostridium tetani*				✓		
Tuberculosis	*Mycobacterium bovis*						✓

* Miscellaneous - urinary, mastitis, skin, heart, sudden death etc.

Bacteria, like viruses, may attack specific parts of the anatomy or an individual system. For example, *Actinobacillus pleuropneumoniae* infects the pleura (smooth surface covering the lungs) and the lung tissue beneath. *E. coli* or salmonella invade the small intestine setting up an enteritis but *Salmonella choleraesuis*, the host adapted serotype of pig salmonellae, can affect the whole of the body including the lungs causing pneumonia and the intestines causing diarrhoea.

Most bacterial diseases are characterised by specific clinical signs and these are shown in Fig 2-11. The common routes of bacterial spread are by direct contact, close respiratory droplet contact, infected faeces or mechanical transfer on shovels, boots, vehicles etc. You will see in Fig 2-18 some of the distances recorded in the field when disease did not transfer between one farm and another. They can be surprisingly small. For depopulation and repopulation purposes however, those diseases with an airborne transmission of below 230 m should, in practice, be extended to at least 1000 m.

Freezing prolongs the survival of infectious agents and sunlight and drying kills them.

Fungi

See Chapter 13 for further information

Fungi (moulds and yeasts) are found in damp conditions such as in badly stored cereals. In the process of multiplication some species produce poisons (mycotoxins) which when eaten, are capable of causing a variety of clinical signs. The more important mycotoxic diseases are shown in Fig 2-12.

The following factors are important on the farm to prevent fungi multiplying and producing toxins:

- Do not store moist corn or cereals.
- Do not allow grain to ferment.
- Examine feed hoppers daily.
- Treat grain bins regularly.
- Check holding bins for leakages and bridged feed monthly.
- Do not allow feed to waste and ferment in feed troughs.
- Always examine your basic feed ingredients.
- Visually check the final feed prior to feeding.

Several different species of fungi (called dermatophytes) infect the skin of the pig and cause ringworm. Ringworm is uncommon, although it is a little more common in outdoor pigs than indoor pigs. It does no harm to pigs who seem unaware of it. It resolves spontaneously after a month or two and is unimportant. Occasionally fungi also cause abortion or mastitis in an individual pig but this is uncommon and has no overall importance.

Parasites

See Chapter 11 for further information

Parasites are organisms that either live in the body (internal parasites or endoparasites) or externally on or in the skin (ectoparasites). The smallest of the pathogenic parasites, coccidia, are found in the intestine. They invade and live in the lining of the small intestine. The major parasites of the pig are listed in Fig 2-13.

Parasites, unlike bacteria, have a life cycle which is the process of development from the egg through larval stages and finally to the adult. Some parasites require an intermediate host, for example the earth worm is the intermediate host in the life cycle of the lung worm. This type of cycle is called an indirect one. A knowledge of the life cycle is important in preventing parasitic diseases. The most effective and cheapest way of controlling parasites is to break the cycle either by good hygiene or by removing the intermediate host if there is one.

FIG 2-12: A GUIDE TO MYCOTOXIN LEVELS IN FEED: MILD TO SEVERE DISEASE

Fungus	Toxins	No Clinical Effect	Toxic Level	Clinical Signs
Aspergillus sp.	Aflatoxins	< 0.1 ppm	0.3 - 2 ppm	Poor growth Liver damage Jaundice Immunosuppression
Aspergillus sp. and *Penicillium* sp.	Ochratoxin & citrinin	<0.1 ppm	0.2 - 4 ppm	Reduced growth Thirst Kidney damage
Fusarium sp.	T2 DAS DON (Vomitoxin)	< 0.5 ppm	> 1 ppm	Mild pneumonia (small lesions)
	Zearalenone (F2 toxin)	< 0.05 ppm	1 - 30 ppm	Infertility Anoestrus Rectal prolapse Pseudopregnancy
			> 30 ppm	Early embryo mortality Delayed repeat matings
	Fumonisin	< 10 ppm	> 20 ppm	Reduced feed intake Respiratory symptoms Fluid in lungs Abortion
Ergot	Ergotoxin	< 0.05 %	0.1 - 1.0% ergot bodies by weight (sclerotia)	Reduced feed intake. Gangrene of the extremities. Agalactia due to mammary gland failure.

ppm – parts per million.
sp. – species - each of these fungi have several species, only some of which are toxic.
Note these toxic concentrations may be reduced when more than one toxin is present at the same time.

FIG 2-13: THE COMMON PARASITES OF THE PIG AND SITES OF INFECTION

Parasite	Scientific Name	Site	Clinical Signs
Coccidia	*Isospora suis*	Small intestine	Piglet scour.
Cryptosporidia	*Cryptosporidum*	Small intestine	Piglet scour.
Flies – house fly, black fly stable fly, blow fly horse fly screwworm fly		Skin	Skin lesions. Small papules.
Kidney worm	*Stephanurus dentatus*	Kidney Liver	Wasting. Blood in urine.
Large roundworm (Ascaris)	*Ascaris suum*	Small intestine Lungs	Liver damage. Reduced performance.
Lice	*Haematopinus suis*	Skin	Evident on the skin especially behind the ears but no lesions.
Lung worm	*Metastrongylus spp.*	Lungs	Coughing. Pneumonia.
Mange	*Sarcoptes scabiei var suis*	Skin	Irritation; skin rash. Thickened skin.
Muscle worm	*Trichinella spiralis*	Muscle	Very uncommon. Found at meat inspection.
Nodular worm	*Oesophagostomum spp.*	Large intestine	Reduced performance.
Red worm	*Hyostrongylus rubidus*	Stomach	Emaciation. Anaemia.
Thornyheaded worm	*Macracanthorhynchus hirudinaceus*	Small intestine	Nodules in small intestine.
Threadworm	*Strongyloides ransomi*	Small intestine	Diarrhoea.
Ticks		Skin	Evident but no lesions.
Toxoplasma	*Toxoplasma gondii*	Muscle	Abortion.
Whipworm	*Trichuris suis*	Large intestine	Diarrhoea. Dehydration.

Prions

The above list does not include prion diseases as these have not been reported naturally in the pig. Prion diseases include BSE in cattle and scrapie in sheep.

Non-Infectious Agents

Trauma

Trauma can be a major cause of disease on a pig farm and the common types of traumatic disease are shown in Fig 2-14. These are conditions that you will see constantly around pig farms. Most are preventable by good management and are dealt with in more detail in other Chapters under the specific disease.

Hereditary and Congenital Defects (Developmental Abnormalities)

Hereditary (genetic) and congenital diseases are quite common in pigs and cover a range of conditions. The term "hereditary" means that the condition was inherited by the piglet from the sow's or boar's genes. The term "congenital" means that it is present at birth but implies it is a development abnormality that occurred during the growth of the foetus while in the uterus, rather than a hereditary defect or abnormality. However some developmental abnormalities are not evident at birth, (for example, an inguinal hernia) and develop at a later stage. They are described as delayed developmental abnormalities. If a congenital defect occurs frequently and is related to a particular line or breed then it is likely to be hereditary.

Defects in the piglet can arise from nutritional causes, poisoning, infectious agents or spontaneously due to abnormal metabolism, as well as hereditary defects.

Most hereditary or congenital defects remain at a low incidence because breeding programmes cull affected animals. Sometimes however, a boar can be identified as being associated with a higher incidence of an abnormality than usual. A typical example would be umbilical hernia. (This condition can also be precipitated by abdominal pressure). Defects may have complex causes combining genetics and environmental and mycotoxin factors.

Embryo mortality varies considerably both between breeds and individuals and heredity plays an important but as yet ill-defined part. In most herds, records show that congenital malformations range from 0.5 to 2.5 % with an average of approximately 1.5 %. However, if all the defects were recorded then levels would approach

FIG 2-14: TRAUMATIC DISEASES

	Condition	Contact Source
Piglets	Bush foot	Trauma by sow's foot. Perforated metal slats.
	Face necrosis	Piglet teeth (fighting). Teeth clippers (poor technique). Poor sow milk supply.
	Fractured limbs/ribs	Sow trauma.
	Greasy pig disease	Housing defects. Faulty clippers or technique.
	Joint infections	Teeth damage (fighting).
	Knee necrosis	Floor pressure/de-tailing.
	Skin damage	Concrete – rough surfaces.
	Tail necrosis	Sow trauma
	Teat necrosis	Concrete – rough surfaces.
	Oedema of teats.	Any floor surface – Occurs within 48 hrs of birth.
Weaners	Greasy pig disease	Infection triggered off by fighting and skin trauma.
	Pig pox	
Grower Finisher	Bursitis	Floor trauma.
	Bush foot	Concrete – rough surfaces.
	Contact sores	Poor wet floor surfaces.
	Ear chewing	Other pigs.
	Flank suckling	Other pigs.
	Fractured limbs	Trauma by another pig. Housing defects.
	Greasy pig disease	Infection triggered off by fighting and skin trauma.
	Skin damage	Other pigs.
	Tail biting	Other pigs.
Breeding stock	Bursitis	Floor trauma. Poor slat or slats.
	Bush foot	Poor floor surfaces.
	Haematoma	Damage. Bruising.
	Leg weakness or Osteochondrosis	Sheer stresses on growth plates. Slippery floors.
	Long bone or pelvic fracture	Fighting. Trauma at service.
	Shoulder sores	Floor surface contact pressure.
	Skin abscess	Fighting.
	Vulval biting	Other sows.

3 %. Common developmental defects are described in Fig 2-15, the causes of which are often multifactorial in origin.

Another type of problem that occurs at birth is difficulty in farrowing associated with a small or abnormal development of the pelvis.

Nutritional Deficiencies and Excesses

See Chapter 14 for further information

Current knowledge on the nutritional requirements of pigs and the components of the different dietary ingredients has significantly reduced the problems associated with faulty nutrition. Deficiencies in the diet however do still occur from time to time and can be considered from four aspects: energy, protein, vitamins and minerals.

Whereas most problems arise due to deficiencies, diseases can also occur due to excesses. The clinical signs of both are shown in Fig 2-16 and Fig 2-17. A consistent feature of vitamin deficiencies is poor growth but this can also be associated with many other factors.

If there is a problem in your herd please refer you to the section in Chapter 9 on the differential diagnosis of poor growth in weaner and grower pigs under "Managing the Growing Pig….". Mineral deficiencies are not uncommon today, particularly where the demands of lactation in the rapidly growing offspring of modern genotypes are difficult to satisfy.

FIG 2-15: COMMON DEVELOPMENTAL DEFECTS

Condition – incidence	Possible Cause	Comments
Arthrogryposis – 0.2 %	Hereditary	Destroy piglets.
Artresia ani (no anus) – 0.4 %	Hereditary	Low heritability.
Bent legs	Unknown Hereditary Poisons Vitamin A	Low heritability. Possibly exposure mid-pregnancy to toxic agents. Auto recessive gene. Hemlock. Black cherry. Excess dietary levels or by injection.
Cleft palate – 0.2 %	Hereditary	Destroy piglets.
Congenital tremor	Swine fever virus Unidentified virus Sex linked in male Landrace Recessive gene in the Saddleback Trichlorvon poisoning Aujeszky's disease (pseudorabies virus) Organophosphorus poisoning	Type AI. Type AII. Type AIII. Type AIV. Type AV. Demonstration of virus to diagnose. Overdosing.
Dwarfism	Hereditary	Rare.
Epitheliogenesis imperfecta	Hereditary	Destroy if large.
Hermaphrodite	Hereditary	Low levels.
Inguinal hernia	Hereditary	Method unknown. Environmental influences.
Inverted teats – 20 %	Hereditary	Ensure good teat selection in boars and gilts.
Kinky tail – 1.5 %	Hereditary	Common.
Lymphosarcoma	Hereditary	Uncommon.
Malignant hyperthermia	Hereditary	Uncommon.
Meningocoele	Hereditary	Low incidence.
Naval bleeding	Hereditary	Associated with shavings.
Pietrain creeper syndrome	Hereditary	Uncommon.
Pityriasis rosea – 1 %	Hereditary in Landrace	Common.
Porcine Stress Syndrome (PSS)	Hereditary	Blood test and eliminate.
Thrombocytopenic purpura – 1 %	Antibody antigen reaction	Quite common.
Splay leg – 1.8 %	Hereditary Fusarium toxin Premature birth	Common in the Landrace. Mouldy feeds. Check prostaglandin use.
Umbilical hernia	Hereditary	Environmental influences.

FIG 2-16: MINERALS – CLINICAL SIGNS OF DEFICIENCIES AND EXCESSES		
Mineral	**Signs of Deficiency**	**Signs of Excess and Causative Level**
Calcium	Agalactia. Depressed milk yield. * Fractures. * Hypocalcaemia. Osteomalacia. * Osteoporosis. * Posterior paralysis in sows. * Rickets. Stillborn.	Changes in bone formation. If zinc is low (parakeratosis) more than 1 % in diet may cause problems. Reduced strength of bone.
Copper	Leg weakness. Loose faeces if suddenly withdrawn.	* Jaundice > 200 g/tonne in diet. Haemorrhage. * Black faeces. Death.
Iodine	Enlarged thyroid glands. Reproductive failure. Weak hairless pigs at birth.	Depressed growth rate > 800 mg/kg in diet.
Iron	* Anaemia. * Increased respiration. More prone to piglet diseases. Poor growth. Pale skin. Stillborn.	Death in piglets deficient in vitamin E. Muscle degeneration > 5000 mg/kg in diet.
Magnesium	Infertility (rare). Poor growth. Weak joints.	Loose faeces > 0.5 % in diet.
Manganese	Infertility (rare). Lameness. Poor growth. Weak piglets.	Inappetance > 2000 ppm in diet.
Phosphorus	Poor growth. * Rickets (see also calcium). Soft bones.	Changes in bone formation. Posterior paralysis in sows.
Potassium	Anorexia. Heart malfunction. Incoordination. Poor growth.	Loose faeces > 1.2 % in diet.
Salt (Sodium chloride)	Poor growth and feed efficiency. Unthriftiness.	* Common. Signs occur at any level if water is short. Death Fits. Incoordination. Thirst.
Selenium	* Mulberry heart disease. Muscle changes. Sudden death.	Diarrhoea. Feet deformity. Lameness. Respiratory distress. Sudden death > 5 g/tonne in diet.
Zinc	* Dry thick skin (parakeratosis). Poor appetite.	Reduced feed intake > 3000 g/tonne in diet. None up to 2500 g/tonne in diet.

* Likely to occur. Others uncommon or rare.

FIG 2-17: NUTRIENTS AND VITAMINS – CLINICAL SIGNS OF DEFICIENCIES AND EXCESSES		
Nutrient/Vitamin	**Signs of Deficiency**	**Signs of Excess**
Amino acids and protein*	A predisposition to disease. Poor growth. Lean tissue gain reduced. Poor growth. More prone to disease.	Digestive disturbances. Diarrhoea.
Biotin	Infertility. Anoestrus. * Lameness. * Poor hoof quality.	Unknown. Unlikely.
Choline	Poor litter size. Poor growth.	Unknown. Unlikely.
Cyanocobalamin (B_{12})	Poor growth. Infertility. Anaemia.	Unknown. Unlikely.
Energy *	* Infertility. Loss in weight. Poor fat deposition. Predisposition to: * Cystitis/pyelonephritis. * Postweaning enteritis. * Respiratory disease. * Villus atrophy and malabsorption. * Thin sow syndrome.	Deposition of excess fat.
Fat and fatty acids * (linoleic)	Dry skin in sows and piglets. Loss of weight in lactation. Poor growth.	Colitis. Digestive disturbances. Loose faeces.
Folic Acid	Anaemia. Poor litter size. Poor growth.	Unknown. Unlikely.
Nicotinamide (niacin)	Diarrhoea. Dermatitis. Poor growth. Paralysis.	Unknown. Unlikely.
Pantothenic acid (B_5)	Poor appetite and growth. Goose-stepping gait. Diarrhoea.	Unknown. Unlikely.
Pyridoxine (B_6)	Poor growth.	Unknown.
Riboflavin (B_2)	Infertility. Weak piglets.	Unknown. Unlikely.
Thiamine (B_1)	Poor appetite and growth. Sudden death.	Unknown. Unlikely.
Vitamin A	Rare but reports of: Infertility. Incoordination. Poor bone growth. Poor sight. Congenital defects, born blind.	* Epiphyseal plate changes. * Increased incidence of osteochondrosis. * Increased requirements for vitamin E. Joint pain. * Leg weakness. * Mulberry heart disease.

FIG 2-17: NUTRIENTS AND VITAMINS – CLINICAL SIGNS OF DEFICIENCIES AND EXCESSES CONT.

Nutrient/Vitamin	Signs of Deficiency	Signs of Excess
Vitamin D_3	Fractures. Lameness. Rickets. Rubbery bones or osteomalacia. Swollen joints.	Calcification of soft tissues.
Vitamin E *	Mulberry Heart Disease Agalactia. Discolouration of fat. * Gastric ulcers. * Liver, heart and muscle changes. * Predisposition to pathogens. * Reduced immune responses. Sudden death. Udder oedema.	Unknown. Unlikely.
Vitamin K	Enhances warfarin poisoning. Poor blood clotting.	Unknown. Unlikely.
Water	* All systems affected. Failure to thrive. Predisposition to disease.	Colic.

* Likely to occur. Others uncommon or rare.

Poisons – Toxic Agents

See Chapter 13 for further information

Poisoning by a variety of agents is still not uncommon in pigs today although less so than when herds were small and less intensive. Many substances if taken at excessive levels become toxic and cause disease.

Poisoning can occur in an individual pig or in a group or even affect a whole herd. In the latter two cases a number of animals will be affected at the same time, all showing similar clinical signs. A study of the history may indicate a common exposure to the poison by contact or ingestion.

If fed to excess many dietary components, including minerals and vitamins, can cause diseases. Many medicines are highly toxic if used above their therapeutic levels.

It is a common fault with stockpeople when treating animals to assume that twice the dose will act twice as well. This is a fallacy. Overdosing may well have the opposite effect.

Examples of common substances that may cause poisoning

- Antibacterial medicines – carbadox, furazolidone, monensin, sulphadimidine.
- Trace elements – iron, copper, zinc, iodine, selenium, arsenic, mercury, lead, fluorine.
- Coal tars.
- Gases – ammonia, carbon monoxide, hydrogen sulphide.
- Insecticides – organophosphorus, carbamates, lindane, dieldrin.
- Essential minerals – copper, iodine, iron, manganese, selenium, zinc.
- Rat poison – warfarin.
- Salt – if water is limited.
- Toxic plants.

Stress

Stress is a condition which occurs in all pigs when confronted with adverse management and environments. Better management of the environment has a beneficial effect on the health and the biological efficiency of the pig.

What does the pig do when stressed?

- It increases the leucocytes (white blood cells) in the blood.
- It increases output of hormones (cortisol) from the adrenal gland and this depresses immunity.
- It becomes more susceptible to disease.
- It eats and drinks less.
- Growth rates and feed efficiency deteriorate for a period.
- It requires an increase in environmental temperature.

Major factors that may cause stress

- Shortage of water supply.
- Shortage of trough space.
- Excessive stocking density.

- Low, high or variable temperatures.
- Draughts/chilling.
- Movement, mixing, fighting.
- Verbal or physical abuse.
- Poor light.
- Low levels of selenium or vitamin E may increase the susceptibility to stress.
- High levels of vitamin A.
- Inadequate or poor general nutrition.
- The act of farrowing or weaning.
- Transport.
- Changes in the environment e.g. changes in housing.
- Exposure to pathogens.
- Lack of sleep.

How Infectious Agents are Spread

To control infectious disease it is helpful to understand how pathogens are disseminated and gain access to the pig. Each organism has individual properties that determine how long it will survive outside the pig, how infective it is, and how easily it is transmitted.

The methods by which pathogens spread are listed below. Of these, the two greatest risks of contamination to your herd are neighbouring infected pig herds and the introduction of disease through the purchased pig (pig-to-pig contact).

- Locality – the location of your farm and any neighboring properties has an impact on disease.
- Airborne transmission in aerosol droplets.
- Direct contact between infected pigs, including newly purchased pigs, wild boars and infected pig meat and AI.
- Mechanical spread by vehicles and equipment, particularly pig transporters.
- Spread by people, boots and clothing.
- Spread by other animals – birds, rats, mice, insects, dogs, cats and wildlife (e.g. wild boars).
- Contaminated food, water and bedding.
- Environmental contamination on the pig farm.
- Medicines.

Locality – Neighbouring Herds

It is always difficult to quantify the risk from neighbouring herds but Fig 2-18 shows the results of a field study looking at the minimum distances between herds when pathogens did not spread and the length of time observed. These results demonstrate minimum distances over which specific infections appear not to have travelled but they should not be used as a basis for a decision in practice. The recommended distances given should be used as a minimum for decision making purposes. If high health herds are being established a distance of at least 3 km (2 miles) from other pigs is advised. However, other factors must also be taken into account.

Mycoplasma (enzootic) pneumonia, SIV and PRRSV are the three most difficult diseases to remain free from because they are wind-borne and are extremely common and widespread in the pig populations of most countries. The transfer of airborne mycoplasma is high within 800 metres (0.5 mile) of infected pigs but is reduced to almost negligible proportions provided there are no other sources of infected pigs within a 3 km radius (2 miles). The respective sizes of the two farms and their distance apart must also be considered. A 50 sow herd producing weaners is much less likely to spread infection than a 1000 sow herd of grower pigs because of the reduced numbers of infectious particles produced by the smaller weaner-only population. Another factor is the type of terrain in which herds are located; flat treeless countryside is worst, as trees break up aerosol plumes. Mapping programmes such as Google Earth may be extremely useful to evaluate the risk of pathogen spread.

FIG 2-18: MINIMAL DISTANCES (FROM FIELD DATA) BETWEEN TWO PIG FARMS WHERE TRANSMISSION OF DISEASE HAS NOT OCCURRED

Pathogen	Actual Distance (m)	Observed Time Scale	Recommended Minimum Distance (m)
Actinobacillus pleuropneumoniae (App)	500	5 years	500
Aujeszky's disease/ pseudorabies virus (AD/PRV)	500	4 months then herd infected	2000
Brachyspira hyodysenteriae (Swine dysentery (SD)	300	4 years	800
Mycoplasma hyopneumoniae mycoplasma (enzootic) pneumonia (EP)	150	10 years	3200
Porcine reproductive and respiratory syndrome (PRRSV)	800	3 years	2000
Toxigenic *Pasteurella multocida*, progressive atrophic rhinitis (PAR)	300	5 years	500
Transmissible gastroenteritis (TGE) vrus	400	4 months then herd infected	800
Sarcoptes scabiei var suis mange	100	5 years	500

Airborne Transmission

If the organism is exhaled from the pig in large droplets, as is common with respiratory bacterial infections, then the actual spread of the organism is limited to probably no more than 50 metres (165 ft) and field studies suggest it is often less than 5 metres (16 ft). Conversely, some viruses, which are extremely small, have been shown to be carried by wind for many kilometres under ideal conditions. For example, foot-and-mouth disease virus was shown to have travelled airborne over land 20 km (12 miles) and an incredible 300 km (190 miles) over water. Aujeszky's virus has been carried 9 km (6 miles) over land. Note these events are very rare.

Diseases/organisms spread through the air by aerosol droplets

Short distances – a few metres

- *Actinobacillus pleuropneumoniae* that causes severe haemorrhagic and necrotic pneumonia.
- Toxigenic *Pasteurella multocida* that causes progressive atrophic rhinitis.
- *Haemophilus parasuis* that causes Glässer's disease.
- PRRSV.
- *Mycoplasma hyosynoviae* that causes arthritis.
- Other *Pasteurella* that are involved in pneumonia.
- *Streptococcus suis* that causes meningitis and other conditions.

Intermediate distances – up to about 3 km (2 miles)

- *Mycoplasma hyopneumoniae* that causes mycoplasma (enzootic) pneumonia. This can carry 3 km (2 miles).
- Some influenza, probably unproven.
- Porcine respiratory coronavirus.

Relatively long distances – more than 9 km (6 miles)

- Aujeszky's disease/pseudorabies virus.
- Foot-and-mouth disease.
- Porcine parvovirus.

Not all viruses become wind-borne. The viruses of TGE, PED, PPV, CSF, ASF, EMC, or SVD are not known to spread on the wind. Theoretically one would expect that viruses causing respiratory infections would be most likely to be spread on wind.

Bacterial infections are less likely to be carried so far on the wind because of the larger size of the droplets. Furthermore it has been shown that the great majority of airborne bacteria in pig farms are dead (although their endotoxins can still cause problems). Nevertheless there is strong evidence that *Mycoplasma hyopneumoniae* which causes mycoplasma (enzootic) pneumonia sometimes travels at least 3 km (2 miles). Field experience has shown that if another pig farm with mycoplasma (enzootic) pneumonia is clearly visible from a newly repopulated pig farm which does not have mycoplasma (enzootic) pneumonia, then sooner or later the healthy pig farm will break down with the disease. There is little firm evidence about the ability or frequency of the other bacterial infections being wind-borne.

The frustration about wind-borne infection is that you have no defence against it except in the choice of the location in which you build your pig farm. Keeping pigs in totally enclosed buildings is no defence. If the ventilation involves fans the air inlets act like vacuum cleaners. Filters fine enough to filter microorganisms are available but impractical. Some filters are used to protect farms from PRRS viruses.

In hot dry summers the chance of aerosol spread between farms is lower, but this does not always hold true for the same regions in winter, particularly at night.

Pig-to-Pig Contact

Infectious agents enter and leave the pig as shown in Fig 2-19. The common methods are by mouth (ingestion) or by inhalation. In general inhalation requires smaller doses of the organism than ingestion to set up an infection and produce disease because the acidity of the stomach and the normal bacterial flora of the intestines inhibit the multiplication of the organisms. Even smaller doses are required if infected material is splashed into the eyes. Some organisms, notably leptospira, can penetrate the mucous membrane lining the mouth and lips. Infections may enter the body through skin abrasions (e.g. erysipelothrix) and some may multiply in and on the skin (e.g. staphylococci causing greasy pig disease). Infections may also enter at mating either via the semen and seminal fluids or mechanically from the boar's preputial sac, which is heavily infected with bacteria, or contamination. The penis may contaminate the vagina with contaminants from the female's dirty vulva and surrounding area.

HOW ORGANISMS ENTER AND LEAVE THE PIG

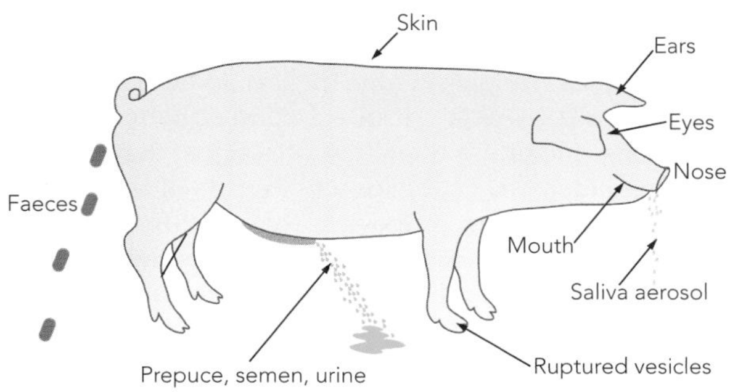

Fig 2-19

Within the pig different infections multiply in different organs, enteric pathogens in the gut, respiratory pathogens in the respiratory tract, leptospira and some viruses in the kidneys, etc., and some become generalised and multiply throughout the body. The route by which the pathogen is spread will depend on the organs it has multiplied in.

Many live in faeces, some in urine, vaginal discharges, semen, skin, saliva, nasal discharges and expired air.

Bad management means increased mortality, more treatment costs, welfare problems for the pig, poor growth rate, poor food conversion and LESS PROFIT.

The main method of spread of respiratory diseases is by airborne droplet infection. The transmission of pathogens by droplet spread in the air, is often restricted to as little as 10 metres (32 ft) but under the right climatic conditions can be many kilometres. Where pigs are in the same environment and particularly when in the same pen, or where the pens are separated by railings, organisms will be transferred by direct body or nose to nose contact. Whether clinical disease develops or not is dependent upon a number of factors including:

- The resistance of the pig to the pathogen.
- The survival of the organism outside the pig.
- The virulence of the organism.
- The numbers of the organisms to which the pig is exposed.

Spread of pathogen may be within or between herds and between countries.

The ways in which pathogens spread within farms are similar to those by which they spread between farms except that the relative importance of each is different. In spread within farms, rodents, flies, equipment, airborne dust and aerosols are most important.

There are also additional modes of spread within farms that are unlikely between farms. Often slurry or solid manure is moved or drained between pens which have a common defecating alley, and there is direct contact between pigs within pens and often between pens. Even if there is not direct contact between pens, infections, including intestinal infections, can move in the dust-laden air. The respiratory tracts of the pigs act as vacuum cleaners, removing the dust from the air, sticking it against the mucous of the respiratory tract and moving it to the throat by the mucociliary escalator so that it can be swallowed. This commences from birth. Thus newborn piglets take their first breath, inhale *E. coli* aerosol, which is caught in the mucus, thus enabling the bacteria to enter their intestines.

Spread of pathogens between different ages of pig within a farm

It was a commonly held belief, and still is in some places, that piglets pick up infectious disease from their dams. It is true for a number of potential pathogens such as *E. coli*, clostridia, fusiform bacteria, some streptococci and staphylococci but it is not true for many of the more serious enzootic pathogens. These are picked up after weaning from older pigs.

By the time a female pig has reached farrowing age, she has developed a powerful immunity to most of the serious pathogens enzootic in the herd and in most cases has thrown them off and is clinically free from them. She passes the full spectrum of her humoral immunity on to her piglets, mostly in her colostrum and to some extent also in her milk. This makes the piglets immune to the same pathogens. Before weaning, the piglets are either not encountering the enzootic pathogens because the sow is not shedding them, or if they are encountering them, by say, aerosol or from other parts of the farm on pig persons' clothes etc., they are in low numbers. Their maternally derived immunity can prevent infection. See Chapter 3: Immunity.

After weaning, they are exposed to an increasing array and concentration of pathogens shed by older weaned and growing pigs. This is at the same time as they are suffering the extreme stress of weaning, when their maternally derived immunity is wearing off and their own active immunity has not had time to fully develop. As they grow older the total weight of the group and stocking density increases and the need for maximum ventilation grows. This is often inadequate, and enzootic infections, particularly respiratory infections, begin to cause clinical disease.

Methods of disease control such as batch farrowing, medicated early weaning, partial depopulation, segregated weaning and multi-site production make use of these facts to break the cycle of infection and they are discussed in Chapter 3.

Wild boars

The wild boar is a vector of pig disease; classical swine fever/hog cholera (CSF/HC) has been found to be endemic in the wild boar population and contaminates domestic pig herds. The native pig species in Africa is the source that brings African swine fever (ASF) to domestic pig farms in Africa and to other countries. Feral pigs can be a threat. In the USA Aujeszky's disease is often spread from feral pigs to commercial pigs.

Pig food products

Human or animal food containing uncooked pork products must not be allowed onto the farm. FMD, ASF and CSF/HC can all be transmitted onto the farm via pork products. Note many pork products are only salted and are not cooked – ham for example. Viable PRRSV can be found in fresh uncooked pig meat.

Semen/AI

A number of viruses have been identified in boar semen thus it is important to ensure that bought-in semen is pathogen-free. This risk can be mitigated by ensuring semen is only purchased from an approved centre of known (or matching) pathogen status.

Vehicles and Equipment

There is no doubt that vehicles which carry pigs, particularly those that carry slaughter pigs to the abattoir, pose a serious risk, as do vehicles picking up dead pigs for disposal. The risk from feed trucks is less but should not be ignored.

It is obvious that a truck driver unloading pigs for slaughter immediately after diseased pigs have been unloaded from a previous truck is likely to contaminate his boots and clothing and his truck. Even if they wash the inside of the truck thoroughly and disinfect it, they cannot be sure that they have eliminated all the contaminating pathogens, including those in the driver's cab. When they call at the next farm to collect pigs and help to load them they may well contaminate the loading area and hence the herd. On many farms the water from the surface of the loading ramp drains directly back into the farm. Cross-contamination between herds is also likely when trucks are picking up slaughter pigs or weaners from several different herds to make up a full load for delivery. It was shown in the UK that pig trucks were major factors in the spread of both TGE virus and SVD virus so it is likely to be true for other pathogens, particularly highly infectious ones such as *Brachyspira hyodysenteriae* (swine dysentery).

Evidence for the spread of infectious disease by feed trucks is hard to find. In the USA it is thought that some farmers may carry grain to the local mill in farm vehicles that they have also used to carry pigs and that this is one source of contamination with TGE. It is the common practice of small mills to turn feed around quickly and therefore there is sufficient time for the virus to remain active. Storage would reduce this method of spread. The risk of a bulk feed truck transmitting faeces-borne pathogens on its wheels is very small but could occur over short distances. Diseases such as TGE, PED, swine dysentery, salmonellosis and SVD might, in theory, be spread in this way. To maintain a vehicle dip to the required strength of a specific disinfectant is a costly procedure. The risks versus costs of this must be equated against the reduction in the amount of contamination on the truck wheels as it travels along the roads together with the washing and the further diluting effect of a water dip. In general wheels are a negligible risk compared with other aspects of vehicular spread and wheel dips give a false sense of security. **Of greater risk however, is the feed pipe that attaches the truck to the bin.** This is carried from farm to farm and often becomes heavily contaminated with faeces. Each farm therefore, should have its own connecting pipe, particularly where the bins are not sited to the exterior of the farm. The siting of feed bins to the outside perimeter of the unit obviously reduces the risk.

The question is often asked, "How long should a vehicle be left empty after the transportation of other pigs before the transportation of pigs of a specific pathogen free (SPF) status?" This will depend on the pathogens that the conventional pigs are thought to have, how thoroughly the vehicle is cleaned and the particular pathogens that contaminated the vehicle.

The following cleaning procedure can be used as a guide:

- Remove all equipment and partitions from the vehicle.
- Completely remove all bedding and faeces.
- Soak the internal surfaces in water and detergent for at least 30 minutes at a temperature of 50 °C (122 °F).
- Pressure wash with hot water or use a steam cleaner.
- Pressure wash the exterior of the vehicle.
- Visually examine the efficiency of the above procedures. Repeat if there is contamination.
- Finally, spray both internal and external surfaces with a non-corrosive, rapidly acting, approved disinfectant at recommended levels.
- Document the cleaning procedure. This is important from a legal point of view.
- If the vehicle is to transport SPF pigs it must stand empty for 12 hours at least after disinfection provided the health status of the last occupying pigs was known. If not the period should be extended to at least 48 hours depending on the conditions of storage of the vehicle.
- Heat drying the vehicle can help reduce the risk of PRRSV spread.

Problems can arise during the winter where temperatures remain below freezing for long periods. Under such circumstances facilities must be made available to carry out efficient washing and disinfectant procedures and for holding the vehicle in an equitable environment not subjected to freezing.

A loading bay should be provided at the exterior of the farm and it should have the following features:

- Easily cleaned.
- Located well away from the unit perimeter.
- A contact bell.
- Clear instructions for drivers.

- Facilities for the truck driver to change boots and coveralls.
- A narrow passage for him to get around behind the pigs.
- Separate boots and coveralls for any farm personnel working outside the unit.
- A pig passageway leading to it from the pig buildings or a system of one-way doors.
- It should be elevated to mid-truck height unless the truck has hydraulic loading facilities.
- No bedding should be used on the ramp and water should be available for washing and disinfecting after loading.
- The loading ramp should drain away from the farm to a soak-away or sealed tank.
- A specified approved disinfectant should be used that has rapid activity.
- There should be a holding area near the vehicle, gated off from the farm, so that pigs can be moved into it without farm personnel entering.

People, Boots and Clothing

The role of people as a mechanical means of transmitting disease has evoked considerable debate over the years. Indeed, some of the severe restrictions such as 7 days pig freedom prior to entry onto the farm would perhaps suggest that the human is covered in pig pathogens. To try and clarify this a review was carried out on 122 pig farms covering a 15 year period, to determine the changes in health status that had taken place since their establishment (Figs 2-20 and 2-21). These changes were then related to the various entry precautions that were adopted on the different farms. It is interesting to note that in the 41 herds that required a period of pig freedom, there were 27 episodes of disease breakdown other than porcine respiratory coronavirus (PRCV) and only one herd remained free of the latter. However, in 20 of these herds the breakdown was due to *Mycoplasma hypopneumoniae* which we now know is only transmitted by carrier pigs or through the air. The provision of boots and protective clothing are the important criteria for disease control.

Recommendations for the control of people

- No visitors should visit the farm at any time unless wearing boots and coveralls provided by the farmer.
- Visitors may be required to take a shower on entry, including a hair wash, and to undergo a complete change of clothes. Showering for 3 minutes is advisable.
- On-farm footwear should be kept outside the showering area.
- A period of pig freedom (i.e. 1, 2 or 3 nights away from other pigs) may be required when visiting SPF farms.
- No farm clothing should be used off the farm.
- Foot dips should be used for entry to the farm and be properly maintained with a rapid acting disinfectant.
- No equipment having had contact with other pigs should be allowed on the unit unless cleaned, disinfected and/or fumigated followed by a gap of at least seven days.
- Staff should not own other pigs nor visit other pig farms.
- Staff should not live on other pig farms.
- Staff should not visit markets or slaughter houses.
- No pig products should be brought onto the farm.
- All human food should be eaten in a designated area separated from the pig pens.
- Human food should not be stored in the medicine storage fridge.
- Hands should always be washed before and after meals (to control zoonotic diseases) and after going to the toilet.

FIG 2-20: PATHOGEN/DISEASE APPEARANCE IN HERDS RELATED TO HUMAN/PIG CONTACT AND TIME OF ENTRY INTO THE HERD

	* 48 Hrs Pig Freedom		* 24 Hrs Pig Freedom or Same Day		* Boots and Coveralls Only		No Precautions		Totals & % Breakdown	
No. Farms	13		28		50		31		122	
Toxigenic *Pasteurella multocida*	0	(13)	2	(24)	1	(48)	1	(21)	4/106	3 %
Mycoplasma hyopneumoniae	9	(13)	11	(22)	1	(2)	0	(0)	21/37	56 %
Sarcoptes scabiei var suis	0	(13)	1	(24)	0	(10)	0	(3)	1/50	2 %
Porcine respiratory coronavirus (PRCV)	12	(12)	27	(28)	50	(50)	31	(31)	120/121	99 %
Streptococcal meningitis	1	(13)	3	(26)	2	(50)	1	(31)	7/120	6 %
Brachyspira hyodysenteriae	0	(13)	0	(24)	1	(50)	3	(31)	4/118	3 %

( ) = Number of herds susceptible. * On-farm boots and coveralls used.

The minimum recommendation for a high-health herd at a commercial level would be no pig contact on that day, a shower and the use of the unit's own protective boots and clothing. The necessity for a shower is often questioned. It exercises a discipline, fosters an awareness and good attitude and forces the visitor to make a complete change of clothes. It also eliminates chance pig infections from contaminated dust. The showers must be kept clean and warm or they will not be used. Personnel for the nucleus herd of a breeding pyramid should observe 24 to 48 hours of pig freedom. This is also often questioned. It is in fact difficult to argue a scientific case for it but there are several cases where veterinarians visiting successive herds have spread FMD and there is at least one clear case of a pig manager bringing back TGE from another farm. The minimum of an overnight gap, a shower and a change of clothes is probably adequate. It is strongly recommended that all farms should provide boots and coveralls for every visitor. Provided these precautions are carried out, field experience over many years shows that people are rarely implicated in the spread of pathogens.

The higher breakdown rates with mycoplasma (enzootic) pneumonia and PRCV reflect their wind-borne spread which none of these precautions would prevent. Streptococcal meningitis may be spread by flies, the introduction of carrier pigs and humans. Humans may introduce influenza viruses into the herd.

Other Aminals

Birds

The four porcine diseases that are definitely transmitted by birds are avian tuberculosis, TGE, erysipelas and salmonella, although it is likely that other infectious agents such as PRRSV may be carried on birds' feet or pass through their alimentary tracts into their droppings.

Birds visit pig farms mainly in winter time, late fall or early spring, to eat pig feed. Pathogenic organisms on their feet could conceivably contaminate the feed.

FIG 2-21: THE PERCENTAGE OF HERDS INFECTED WITH SPECIFIC PATHOGENS RELATED TO HUMAN/PIG CONTACT AND TIME OF ENTRY INTO THE HERD

	No Pig Contact on Day of Entry or Longer		No Time Limits	
Mycoplasma hyopneumoniae	20/35	57 %	1/2	50 %
Streptococcus suis	4/39	10 %	3/81	4 %
Brachyspira hyodysenteriae	.0/37	0 %	4/81	5 %
Toxigenic *Pasteurella multocida*	2/37	5 %	2/69	3 %
Sarcoptes scabiei var *suis*	1/37	3 %	0/13	0 %

More important, their droppings contaminate the feed, the floors, and sometimes stored bedding, such as straw or shavings. In temperate and warm climates, pig buildings are often open on one or more sides. Bird-proofing then involves extensive netting which is not often done in spite of the risk from birds. This is cost effective in the savings made in feed alone. In extreme climates pig buildings are usually totally enclosed so the risk from birds is much reduced. Even then, there is still the possibility of birds defaecating on stored or spilt food or stored bedding outside the buildings, the organisms in the faeces being carried into the pigs on people's feet as well as on the bedding or feed. Covering feeders reduces the risk of bird faeces entering the pig's feed.

In many countries the most dangerous birds are starlings. They tend to travel in large flocks, flying in a 30 km (20 mile) radius and landing on numerous different pig farms in a day. In a study carried out it was estimated that about 30 % of new outbreaks of TGE were due to starlings. When starlings ingest TGE virus, which is easily done when they are feeding on a pig farm during the acute stage of a TGE outbreak, they shed it in their droppings for up to 36 hours. Seagulls have also been implicated in the spread of TGE in the UK. Porcine reproductive and respiratory syndrome virus (PRRSV) has been shown to multiply in ducks.

Birds with avian tuberculosis shed vast numbers of *Mycobacterium avium* in their droppings, which, when ingested by pigs cause typical tubercle lesions in the lymph nodes of the neck and mesentery, resulting in condemnation of the head and offal at slaughter. Sometimes individual herds suffer long periods with high condemnation rates. Bird-proofing the pig farm and removing all the stored bedding in some cases may not have the desired effect. It must be remembered that *Mycobacterium avium* is not a uniform species of bacterium but covers a wide spectrum of variants some of which multiply readily saprophytically (outside the pig). Some strains multiply for example in peat used for bedding or in water tanks.

Birds are sometimes blamed for outbreaks of erysipelas in pigs. They can become infected by the causal organism, and may then shed it in their droppings. Most outbreaks however originate from the pigs themselves. The organism resides in the tonsils subclinically for long periods and becomes endemic in herds. For reasons that are not clear, probably related to stress, an individual carrier may develop clinical signs and then shed large numbers of virulent organism in the faeces.

Salmonellae can be carried and shed by birds in their droppings but often the serotypes are not those that would harm pigs. However, they may raise antibodies which can complicate slaughterhouse surveillance schemes. Highly infectious agents such as FMD could be carried for short periods mechanically on birds' feet but there is no firm evidence that this has played a significant role in spread. Seagulls may be involved in the carriage of materials contaminated by pig products, such as pig-meat wrappings from garbage dumps, into pig farms. (The wind may also blow such material into a pig farm if the garbage dump is nearby.)

Rats and mice

What pig disease might they carry? It has been shown that *Brachyspira hyodysenteriae*, the causal organism of swine dysentery, can infect mice and can be maintained in mouse colonies in pig farms. *B. hyodysenteriae*, isolated from mice on infected pig farms, has been shown to be pathogenic in pigs. Mice may provide one explanation why some pig farms break down with swine dysentery after they have been repopulated with clean stock or after attempts at eradication by blanket medication. Although rats have been infected experimentally with *B. hyodysenteriae*, they are not thought to play any role in its spread in the field.

The rat is the natural host of the virus of encephalomyocarditis (ECMV) which is implicated in outbreaks of so called SMEDI (stillbirths, mummification, embryonic deaths and infertility) in the USA. It is thought that outbreaks occur in localised areas when rats are allowed to build up in numbers. In some countries the virus causes myocarditis and sudden death in pigs and other animals (e.g. zoo animals). In Cuba, this form of the disease is regarded as a major problem in pigs and has also caused deaths in apes, porcupines and cattle.

Rats and mice also carry and shed *Salmonella typhimurium* and other salmonella serotypes which affect pigs, and dermatophytes, such as *Trichophyton mentagrophytes,* which cause ringworm in pigs. Rats seem to be resistant to infection with *Actinobacillus pleuropneumoniae* and *Streptococcus suis* type 2. Mice can be infected experimentally with large doses of *S. suis* type 2 but there is no evidence that they carry the organism under field conditions.

House mice remain resident in piggeries and do not normally travel between them. Field mice may travel short distances but are not likely to be significant in the spread of pig disease. Rats are much more mobile. Individuals will frequently cover 1 to 2 km (0.5 to 1 mile) in a night. However their movement between pig farms depends on a complex social relationship in the resident rat communities in farm buildings. Both rats and mice may be carried inadvertently between farms in vehicles such as feed trucks. Rats and mice and other rodents carry a variety of leptospira species which may infect pigs.

Flies

Flies are common on pig farms and have access to contaminated materials such as dead pigs, the secretions and excretions of diseased pigs, and faeces. They frequently travel up to 2 to 3 km (1 to 2 miles) between pig farms in breezy weather, particularly in summertime, and are attracted by smells that are slightly out of the line of the wind direction. When the common house fly (*Musca domestica*), was fed materials contaminated with *Streptococcus suis* type 2, the microorganism remained viable in the fly, probably in its crop, for up to 5 days, and the fly would then contaminate whatever it fed on. Before feeding, the house fly vomits its crop contents and enzymes onto the food which is then sucked up. Biting flies (*Stomoxys calcitrans*) have also been shown to carry other infections, including *B. hyodysenteriae*, the cause of swine dysentery, and PRRSV. It seems reasonable to assume that they can carry many more infections than those that have been studied and reported. However, we should be wary about concluding that because they can carry an organism they necessarily play a role in its spread. The dose of organism that they carry may be extremely small and in some cases, such as *B. hyodysenteriae*, may be below the infective dose required to establish the infection in a pig. Clinical observations suggest they play a role in disease in farrowing houses, such as greasy pig disease. Mosquitoes may play an important role in the spread of PRRSV.

External parasites

Mange mites, lice and ticks can be carriers of some infectious agents and act as mechanical vectors or indirect hosts in diseases such as ASF, Japanese encephalitis and *Mycoplasma suis*.

Domesticated/farm animals

What is the risk of keeping herds of pigs near other farm livestock, including poultry, or of keeping dogs and cats in such herds? In countries which do not have FMD, the risk in practice seems to be low. One can theorise about the contamination with *Salmonellae* from calves, toxigenic *Pasteurella multocida* from cattle and sheep, *Actinobacillus pleuropneumoniae* from cattle, sheep and deer, *Erysipelothrix rhusiopathiae* from poultry and *Streptococcus suis* (various serotypes) from cattle, sheep, goats and horses. In practice, provided they are physically separated from each other, there is little risk. However, outbreaks of disease caused by some of these organisms sometimes occur without obvious explanation. The source might be other animals. Aujeszky's disease/pseudorabies virus (AD/PRV) crosses species but the risk is mainly the other way around, i.e. from the pigs to the cattle and dogs.

Dogs can shed the virus of TGE for up to 14 days after eating contaminated material such as dead piglets.

This is clearly a risk when a nearby neighbour gets TGE but farm dogs do not usually wander great distances and there are so many other ways in which TGE can be spread between neighbouring herds that the importance of the dog as a common vector is probably small. Farm dogs have also been shown to carry *B. hyodysenteriae* but it is difficult to assess whether or not they play any role in the spread of swine dysentery. *Leptospira bratislava* can be shed in dogs' urine. If the dog is a good watch dog then its benefits in keeping intruders and other animals at bay may outweigh its risk of introducing TGE, swine dysentery or leptospirosis. Pig farmers in regions where CSF/HC, ASF and FMD occur should be a little more careful about dogs because of the danger of carrying bones into the farm.

Cats in pig farms are frowned upon as a potential risk of toxigenic *Pasteurella multocida* (causing PAR) but on the other hand in the many SPF herds in Denmark, cats are kept routinely to control mice and rats. If they stray from the premises they are killed or not allowed to return. Progressive atrophic rhinitis (PAR) is not a common cause of breakdown in the large Danish SPF programme. Toxoplasmosis is a serious disease risk from cats and this is likely to become an increasing zoonotic risk.

Other wildlife

Lawsonia intracellularis bacterium has been identified as the cause of ileitis (porcine enteropathies) and it has been transmitted experimentally to horses, hamsters, mice, ferrets and rabbits. It seems likely therefore that this organism could be spread by a variety of other species and it is not surprising that the disease syndrome occurs sometimes in the most secure, highest health herds.

A range of leptospira serovars are carried by wild animals but most are not normally pathogenic to pigs. Individual pigs may become ill with *Leptospira icterohaemorrhagiae* from rats. A main pig pathogenic serovar in North America and some other parts of the world is *L. pomona,* shed by hedgehogs and possibly mice and rats. Other serovars which affect pigs are present in wildlife in other countries however a serovar which is pathogenic in one country may not be pathogenic in other countries. The presence of *Brucella suis* in wild hares in Denmark and France has contaminated outdoor herds.

Tuberculosis may be transmitted between wildlife and outdoor pigs.

Contaminated Food, Water and Bedding

Water can become a major source of contamination in both outdoor and indoor pig-producing systems. Outdoors the fouling of wallows by faeces and urine from sows provides a potential medium for the survival of some organisms. Leptospira from rats and other wildlife may contaminate streams and be drunk by pigs. Feed may be contaminated by salmonella from vermin and birds or its ingredients may have been contaminated at source. Pathogens that can be spread by food and water include: FMD, ASF and CSF, *E. coli*, erysipelas, clostridia, salmonellosis and leptospirosis. Erysipelas can be transmitted via the water supply from fish farms infected with erysipelas.

If you use bedding, ensure the bedding has no contamination from other pigs.

Environmental Contamination on the Farm

Organisms such as clostridia, erysipelas, salmonella and *E. coli* may contaminate the environment, soil or concrete surfaces (for example in the farrowing houses) and if attempts are not made to reduce or eliminate them they may overwhelm incoming pigs.

Most respiratory infections die out of the environment fairly quickly but faecal pathogens tend to be more persistent.

When equipment moves between farms it can spread pathogens. If you purchase any secondhand equipment ensure that it is cleaned, disinfected and rested before placement on your farm.

Note needles and syringes can move pathogens between pigs and potentially between farms.

Pathogens/diseases spread by pig faeces

- *Actinobaculum suis*.
- Bacteroides.
- *Brachyspira hyodysenteriae* (causing SD).
- Campylobacters.
- Classical swine fever (CSF) virus.
- Clostridia.
- *Erysipelothrix rhusiopathiae* (causing erysipelas).
- *Escherichia coli*.
- Foot-and-mouth disease (FMD).
- *Fusobacterium necrophorum*.
- Klebsiella.
- *Lawsonia intracellularis* which causes ileitis.
- Parasitic diseases (internal ones).
- Porcine circovirus 2 (PCV2).
- Porcine epidemic diarrhoea (PED) viruses.
- Porcine parvovirus (PPV).
- Porcine reproductive respiratory syndrome virus (PRRSV).
- Rotavirus.
- Salmonellosis.
- Streptococci.
- Swine vesicular disease (SVD) virus.
- Teschoviruses (formerly enteroviruses).
- Transmissible gastroenteritis (TGE) virus.
- *Trueperella pyogenes*.
- Yersinia.

Skin trauma

There are many diseases on the farm that result from contaminated skin damage. These include tail biting, ear nibbling, greasy pig disease, trauma to the gums following teeth removal, necrosis of knees in piglets, joint infections, *Mycoplasma (Eperythrozoon) suis* (from needle contamination) and of course fighting. In addition, pathogens can be transmitted between pigs on the same farm and between farms, via needles and syringes. CSF and PRRSV are excellent examples.

Opportunist invaders

Opportunist invaders are those organisms that normally on their own would not cause disease but given the opportunity by additional factors will assume a pathogenic role. Typical examples include skin damage resulting in greasy pig disease, vulval discharges and endometritis.

Medicines

Live vaccines must be used with caution, especially those produced in cell culture – viruses for example. While pharmaceutical companies make every effort to ensure product safety, whilst remote, it is always possible that the cell culture contains unknown viruses which obviously cannot be tested for.

It is more likely however, that medicines will become contaminated with dirty needles. PRRSV and clostridia have been spread between batches of pigs this way. If these medicines are used on different farms the diseases can be spread. There have been a number of cases of CSF spread around Spain by contaminated medicines, needles and syringes.

Understanding Herd Pathogen Status

In an ideal world if the pig herd could be established and maintained completely free of all pathogenic organisms then from a disease viewpoint there would be few limitations to maximising production. Unfortunately we do not live in such environments and because of factors often out of the control of the farm, there will always be a variety of pathogenic or potentially pathogenic organisms present.

The purchased pig is the most important potential source of new infections and whilst the donor herd may be well monitored to determine its pathogen status, an infectious disease could be incubating at the time of purchase.

The situation is further complicated by the variety of terms that have been used to describe the perceived health status of a herd. They are imprecise and open to wide interpretation. There are three internationally recognised levels of herd health status:

1. **Germ Free (Axenic)** – This is a pig which is thought to be totally free from infection with microorganisms. Such pigs are produced by surgery (hysterotomy or hysterectomy) carried out on the pregnant sow near term. They are reared under completely sterile conditions, usually in small containers enclosed in plastic balloons with filtered air and sterilised food and water. They can usually only be reared to a maximum of about 6 weeks because they get too big and unmanageable to maintain. They are for research purposes only. In fact, it is now realised that a germ-free state is probably impossible to achieve. It is known that at least one retrovirus, which is inserted in the genetic DNA of every cell in the pig's body, moves to the offspring from the parents. There are probably other as yet unknown viruses that behave similarly. Several viruses, e.g. inclusion body rhinitis virus, PCV2, PRRSV and PPV may also pass from gilt/sow to foetus before birth.
2. **Gnotobiotic** – This term means "known life". It describes a pig which has been produced and reared initially as a germ-free pig but while continuing to be barrier-maintained is then deliberately infected with known microorganisms. Its microflora is thus clearly known and defined. Again they can only be reared to about six weeks when retrovirus and other unknown viruses may be present.
3. **Specific Pathogen Free (SPF)** – This term can be used to describe a pig or pigs, a herd or a pathogen control programme. It means that the herds are believed to be free from a short list of specified pathogens. The breeding companies draw up strict pathogen control regulations for their members to follow and regularly carry out a series of laboratory tests (mainly serological) to determine the presence or absence of specified pathogens. These usually include porcine reproductive and respiratory syndrome virus (PRRSV), Aujeszky's disease/pseudorabies virus (AD/PRV) toxigenic *Pasteurella multocida* (causning PAR), *Mycoplasma hyopneumoniae* (causing mycoplasma (enzootic) pneumonia), certain serotypes of *Actinobacillus pleuropneumoniae*, *Brachyspira hyodysenteriae* (causing swine dysentery), *Sarcoptes scabiei var suis* (causing mange) and *Haematopinus suis* (lice). The herd breakdown rate with *Mycoplasma hyopneumoniae* is usually high, PRRSV is moderate and there are also breakdowns with *Actinobacillus pleuropneumoniae* but at a lower rate. The other pathogens are generally more consistently kept out.

Other terms

In addition to these, you may also come across terms such as **Secondary SPF,** (herds set up with breeding stock from SPF farms), minimal disease (MD), high health status and defined health status. The use of high health status should be avoided as this is a non-specific term.

A specific pathogen free (SPF) health status declaration would generally be made by the company veterinarian following clinical examination of the herd along with the results of relevant pathological tests (Fig 2-22). Examinations would be carried out every 3 months.

The history of the herd supported by other tests could also allow certain other pathogens to be declared absent and added to the SPF status. It is emphasised that the assessment of SPF must always be carried out by a competent pig veterinarian so that the information and criteria upon which a declaration is made accords with current accepted practices.

Infectious pathogens/diseases that may be seen in a SPF status herd

Of course no herd will be free of all pathogens/diseases and the list below covers many of those that might be seen.

- Avian TB.
- Coccidiosis.
- Congenital tremor (all).
- Cystitis/nephritis.
- *E. coli* septicaemia/diarrhoea
- Erysipelas.
- Exudative epidermitis (greasy pig disease).
- Focal pneumonia (*Actinobacillus suis*).
- Glässer's disease.
- Ileitis.
- Internal parasites.
- *Lawsonia intracellularis.*
- *Leptospira bratislava.*
- Mastitis.
- Continued on next page.

FIG 2-22: HEALTH STATUS DECLARATION – SPECIFIC PATHOGEN FREE STATUS (*EXAMPLE*)

Name of Herd: Plantation Pigs Declaration Date: DD-MM-YY

Disease/Pathogen *	Declared Health Status Results of Clinical Examination **	Type of Test
Actinobacillus pleuropneumoniae (App)	Negative	Clinical, slaughterhouse, serological , ELISA, PCR.
Aujeszky's disease/pseudorabies virus (AD/PRV)	Negative	Serological, PCR.
Brachyspira hyodysenteriae (SD)	Negative	Clinical, PCR, culture.
Brucella suis (Brucellosis)	Negative	Serological.
Haematopinus suis	Negative	Clinical.
Mycoplasma hyopneumoniae	Negative	Serological, slaughterhouse, PCR.
Porcine epidemic diarrhoea (PED)	Negative	Clinical, PCR.
Pasteurella multocida (PAR)	Negative	Bacteriological culture, slaughterhouse.
PRRSV	Negative	Serological, PCR.
Salmonella choleraesuis	Negative	Clinical.
Sarcoptic mange	Negative	Slaughterhouse, serological.
Transmissible gastroenteritis (TGE)	Negative	Clinical, PCR.
Other government reportable diseases	Negative	Clinical and other tests.

Negative = Disease never diagnosed.
*** Specific serotypes** =Where there are different serotypes it may not be possible to test for all serotypes.
** Based on long term history supported by pathological tests as necessary.

- Mycoplasmal arthritis (*Mycoplasma hyosynoviae*).
- Oedema disease.
- Porcine dermatitis and nephropathy syndrome (PDNS) – sporadic form only.
- Piglet arthritis and meningitis (*Streptococcus suis* type 1).
- Piglet dysentery (*Clostridium perfringens* type C).
- Porcine parvovirus (PPV).
- Swine influenza virus (SIV).
- Teschovirus (formerly enterovirus).
- Tetanus.
- Vulval discharge.

Purchase breeding stock from an SPF herd or its equivalent.

Understanding Disinfection

Disinfectants are substances that kill both harmless and disease-producing organisms. They act either as bacterial poisons, to coagulate bacterial protein or as oxidising or reducing agents. Antiseptics normally prevent bacterial multiplication and are used for cleaning skin or wounds. Some disinfectants in a more dilute form may act as antiseptics.

Common antiseptics used include:

- Chlorine-based ones such as TCP.
- Quaternary ammonium compounds such as cetrimide.
- Chloroxylenol (Dettol).
- 1 % crystal violet.
- Iodine.
- Alcohol.
- 1 % salt solutions.
- Hydrogen peroxide.

Disinfectants have two prime functions. First to prevent infectious agents gaining access to the farm and second, equally important, to control those organisms already on the farm and that persist in large numbers in the environment. The process of disinfection can be considered in three stages.

1. The removal of the gross contamination within the building, that is the dried faeces, slurry and dust, by pressure washing preferably with hot water.
2. The use of detergents to assist in the final removal of the organic material. Only after completion of these two should a disinfectant be applied.
3. The use of the disinfectant.

Remember that the cheapest methods of disinfection are **the physical removal of contaminated material and its final removal by water**.

The effects of this are illustrated by a case of poor growth in finisher pigs. Two houses were involved, neither of which had been emptied or washed out for two and a half years. As an experiment, groups of pigs were split into two on entry into the houses and weighed in and out again at point of slaughter. One of the houses was divided into sections and used on an all-in/all-out basis. It was completely washed prior to the entry of each batch of pigs and the other house was used as a control with no changes. The growth curves of both groups were compared and by point of slaughter at 90 kg (200 lbs) there was approximately a 9 days difference, to the obvious advantage of the cleaned house. It is interesting in this case that only water and detergents were used without disinfectants. The all-in/all-out procedures will also have contributed to the performance.

This phenomenon of all-in/all-out production associated with cleaning has been well known for a long time and yet on many farms it is still not practised. The reasons for the improved efficiencies are now better understood. In this study the pigs were no longer coming into an environment in which there was endemic disease. This was maintained by aerosol droplet infection from other pigs in the building and enteric organisms on floors and walls. The cleaned building was empty for periods of four days during and after the cleaning process and all the respiratory droplet organisms had time to be ventilated or precipitated from the air space. Whenever a pig enters an environment with older pigs already in the house, the large numbers of organisms present challenge the immune system of the incoming pigs and this process uses large amounts of protein and energy. This has the effect of decreasing daily gain and food conversion efficiency whether serious specific diseases are present or not. This is a very important lesson to learn.

Pigs that are moved into houses that already have older pigs will always grow more slowly and be less efficient than those moved into an empty clean house.

Colitis (inflammation of the large bowel) in growing pigs is a common occurrence on pig farms. One of the major contributing factors is the continual use of pens, without washing and cleaning between batches.

Disciplines are almost invariably maintained in farrowing houses through all-in/all-out procedures and by washing and disinfection. The same principles should apply equally across all pig buildings on the farm and in particular the mating area and the finishing houses. On many farms the mating pens have been in continual use for 25 years and we often wonder why reproductive performance has dropped gradually over intervening periods from farrowing rates of 90 % to 80 %. Other areas that require constant cleaning are foot dips and hands after handling diseased pigs. When a litter of young pigs have been treated for scour, the hands and clothing become heavily contaminated by as much as 15 to 20 billion organisms; 0.5 m to 3 million will produce scour in the piglet.

Disinfectants

The ideal disinfectant should:

- Act quickly against a wide range of viruses, bacteria and fungi.
- Be safe to handle.
- Act in the presence of dust or organic matter.
- Have a long period of activity.
- Be a non-irritant, non-staining, non-toxic and non-corrosive.
- Be combined with a detergent or have such properties.
- Be capable of use as an aerosol.
- Be safe and effective when used in water systems.
- Be suitable for use through pressure washers.
- Be coloured.

There are 6 classes of chemicals used for disinfection:

1. Phenols.
2. Chlorine-based compounds.
3. Iodine-based compounds.
4. Quaternary ammonium compounds (QACs).
5. Aldehydes.
6. Peroxygen compounds.

1 Phenols

These are organic compounds that may or may not be combined with chlorine. They are usually effective in the presence of organic matter but do not normally have a high detergent action. They are not corrosive to metal but they can cause damage to plastic and rubber compounds. Their action is moderately slow. Tar acids may be combined with other organic acids such as acetic and sulphuric acids to increase efficiency by improving their effects against viruses.

Key facts about phenolic-based disinfectants:

- They are active in the presence of organic matter.
- Their activity persists for a long period of time.
- They are ideal for vehicle dips and concrete floors.
- They have no detergent activity.
- Their rate of activity is slow; two to twelve hours.
- They can be toxic and damage tissues.
- They are usually very effective against bacteria but not so good against viruses or spore-producing bacteria.
- They are usually quite cheap.
- Some phenols such as chlorxylenols contain chlorine which adds properties of quick action.
- They taint milk and processed meat.

2. Chlorine-based compounds

The chlorine-based compounds can be considered in two groups. Those without organic compounds such as the hypochlorites which depend on the liberation of chlorine for their disinfection action and those that contain organic substances. Chlorine disinfectants have a very quick action but are very quickly neutralised in the presence of dirt or organic matter.

Key facts about chlorine-based disinfectants:

- They can be very corrosive.
- They have a very quick action.
- They are inactivated by organic matter and hard water.
- They do not persist for very long periods of time.
- They have no detergent activity.
- They are very active against viruses and bacteria.
- They may cause tainting.
- They may be ozone unfriendly.
- They are very cheap.

3. Iodine-based compounds

This group of disinfectants includes substances called iodophors where the iodine is dissolved in a surface-active agent and then phosphoric acid is added. Iodine substances are very safe, have low toxicity and have almost no smell. However, when phosphoric acid is added to the iodophors the disinfectant becomes a little more irritant and corrosive.

Key facts about iodine based disinfectants (iodophors):

- They usually have a high detergent activity.
- **They are ideal for foot baths.**
- They are brown in colour when very active becoming straw coloured when losing their activity (they are used in foot baths for this reason).
- They are very quick in action.
- They are very effective against viruses and bacteria.
- They are moderately active in the presence of organic matter.
- They tend to be more expensive.

4. Quaternary ammonium compounds (QACs)

These compounds may be used for cleaning and sterilising water systems and equipment and they are very efficient, particularly if the organic matter has been removed. They are not usually suitable for the disinfection of premises on their own because of the large amounts of organic materials present that immediately neutralise them. QACs are not compatible with soaps and they should not be mixed with other detergents. Some are used as antiseptics. They are more active against gram positive organisms.

Key factors about QACs:

- They usually have little or no effect against fungi and bacterial spores.
- They are inactive in the presence of organic matter.
- They are inactivated by soaps and disinfectants.
- They have no activity against viruses.
- They are suitable for cleaning water systems and smooth surfaces.

5. Aldehydes

These substances such as formaldehyde, are very toxic but are good disinfectants in aerosol form. There are now alternative products available that are equally effective and much safer to use.

6. Peroxygen compounds

These are the new broad spectrum disinfectants that are highly active against most microorganisms. They are based on combinations of peroxyacetic acids or other derivatives, hydrogen peroxide, organic acids and anionic detergents. They are powerful oxygenating agents.

Precautions to be taken when using disinfectants

Always:

Follow the manufacturer's instructions carefully.
Wear gloves and eye protectors when handling the concentrate.
Wash concentrate off the skin immediately.
Ensure the dilution is correct for the purpose it is being used.
If there is contact with eyes, wash immediately with copious amounts of water and seek medical help.
Where foot baths are used, ensure that these are cleaned and replenished regularly.
Store in the original container, tightly enclosed.
Keep away from children.

What should you use on the farm?

This would obviously depend on availability but the following should be considered:

- Foot baths – use an iodine-based .
- General disinfection of houses – use a phenol or organic acid-based disinfectant.
- Water – use a QAC or chlorine-based disinfectant.
- Concrete surfaces – use a phenol or organic acid based disinfectant.
- Broken floor surfaces not easily cleaned – use an oil-based phenol type disinfectant.
- Virus infections – use iodophors or peroxygen complexes. Formalin fumigation is also effective.
- Bacterial problems – use iodophors or peroxygen complexes.
- Hands – use QACs or soaps.
- Loading ramps – use a government-approved disinfectant that is highly active against the major notifiable and transmissible diseases in your country.
- Aerosols – use formalin, chlorine, iodine or oxidising agents, preferably the latter.

Key points to consider when selecting a disinfectant (Fig 2-23)

- **Affectiveness against infections** – always use a disinfectant that has been independently proven and has been shown to be effective against a wide range of infections but particularly those that are present on your farm. In many countries there are lists of approved disinfectants. Ask your veterinarian or supplier.
- **Dilution rate** – always read carefully the instructions for use and in particular the amounts to be added to water for general disinfection purposes. Check also the amounts required for the highly infectious diseases such as TGE, Aujeszky's disease and PRRSV. From the cost of the concentrated disinfectant work out the cost of diluted chemical. A more expensive disinfectant with a dilution rate of 1:300 may be a better buy than a lower priced disinfectant with a dilution rate of 1:100.
- **Time to act** – there is always a minimum time before the disinfectant has killed microorganisms. Since most disinfectants are used at low temperatures always look at the killing time relative to this.
- **Effectiveness in the presence of organic matter** – this is important on the farm, because invariably under such conditions they are going to come into contact with large amounts of organic matter. Some disinfectants such as chlorine are very quickly neutralised in the presence of such materials. Foot baths contain

high levels of organic matter due to continual contamination.

- **Penetration** – it is very important that the disinfectant has the ability to penetrate organic matter (detergency). In most cases however it is more effective to apply a detergent cleaner prior to the use of the disinfectant.

Each group of disinfectants has their own special properties and an understanding of these will help you in your selection.

Detergents

These are cleansing agents which have good wetting powers and the property of penetrating surfaces. Detergents can either be acid, alkaline or neutral. The neutral ones tend to be those such as soaps and liquids that are mainly used to clean soiled materials. They are often combined with the disinfectant to provide a dual action.

Properties of a good detergent

- Is efficient in removing organic material.
- Acts well in pitted surfaces.
- Has a good degreasing action.
- Has good penetration.
- Is quick-acting.
- Works in the presence of the disinfectant.
- Leaves no residues.
- Is active with hard water.
- Is non-toxic.
- Does not make floors slippery.

A recommended routine for cleaning houses

1. Remove all muck and empty all slurry channels, tanks and gulleys.
2. Isolate the electricity supply.
3. Disconnect all moveable equipment, feeders, lamps etc. and open all inaccessible areas e.g. channels, fan boxes etc.
4. Brush down and sweep out the house.
5. Soak the complete building, roof to floor with a farm detergent or water, for 24 hours if possible.
6. Soak all moveable equipment and clean down.
7. Drain and flush out the water system, bowls, nipples, water tanks etc. and fill with a detergent steriliser. Leave for two hours, drain, and then refill with water.
8. Pressure wash the complete building using hot water or a steam cleaner.
9. Visually check the building.
10. Disinfect the complete house including all equipment and surrounds using a pressure washer or spray.
11. Follow this with fumigation using formalin gas where this is permitted, with suitable precautions. Alternatively use a suitable disinfectant or chlorine dioxide (see Chapter 15).
12. Place a disinfectant foot bath outside the house and use prior to entry.
13. Do not restock the house until dry (a minimum of 48 hours).
14. If you have to occupy the house before this then use a space heater to dry out the surfaces.

FIG 2-23: THE CHARACTERISTICS OF THE DIFFERENT DISINFECTANT CHEMICALS

Characteristic	Chlorine-Based	Peroxygen Compounds	Phenols (Unchlorinated)	Phenols (Chlorinated)	Iodophors	QAC-Compounds
Can be used in aerosols	A few	Yes	No	A few	Yes	Yes
Corrosive to metal/rubber	No	No	Yes	Yes	No	No
Detergent action	No	Yes	No	Some	Yes	Yes
Effectiveness in presence of organic matter	Moderate	Yes	Yes	Yes	Moderate	No
Good action against bacteria	Moderate	Yes	Yes	Yes	Yes	Moderate
Good action against viruses	Yes	Yes	Poor	Poor	Yes	No
Persistent residues	No	No	Yes	No	Poor	Yes
Speed of action	Quick	Quick	Moderate	Moderate	Quick	Moderate
Staining	Some	No	Yes	Yes	Some	No
Suitable for foot baths	No	Yes	Yes	No	Yes	No
Toxic or irritant	Yes	No	Yes	Yes	Some	No

You will see from the above that there are three chemicals required for the cleaning process: first a detergent, which helps remove dirt and soiled material; second, a detergent steriliser for cleaning and sterilising the water systems; and third, a disinfectant to complete the cleaning process.

In houses where the floors are worn, and in particular farrowing pens, it is good practice to brush the floor with lime wash (whitewash) after cleaning and disinfection. Carry out this procedure as follows:

- Wear goggles and gloves when handling this material since it is a strong irritant.
- Mix sufficient burnt builders' lime ($CaCO_3$) with water to produce a consistency of thin salad cream.
- Cover the farrowing house floor with the whitewash, using a soft household brush.
- Leave the surface for 48 hours to dry.
- Do not move the sow into the farrowing place whilst the lime wash is wet. If you have to move the sow in before this, do not brush the floor area where the sow's udder will make contact.
- Lime wash can also be used on all concrete floors and can be of value in pens where, for example, greasy pig disease has become a problem. It is a very cheap disinfectant and its value can be enhanced by adding 30 g of phenolic disinfectant to 4.5 litres of lime wash.

Guidelines for fumigation using potassium permanganate and formalin

Formaldehyde is a noxious and highly toxic gas, and exposure for only a short period can cause respiratory distress. Severe exposure can ultimately cause coma. (Its use is forbidden in some countries.)

Whenever fumigation is undertaken, there should be two people available; one in the house and one at the door to provide assistance if required.

- Wear protective goggles and a dampened face mask – these should be available to both persons.
- Use rubber gloves.
- Wet the house before fumigating and seal any openings.
- Always add formaldehyde to the potassium permanganate. Have the exact quantities to be added ready in separate containers and add slowly.
- **Use high-sided metal containers for mixing the compounds. Once the compounds are mixed vacate the house immediately.**
- Ensure no livestock will be exposed to fumes that might escape.
- If the person in the house gets into difficulty, the person at the door should turn on all fans, open all doors and immediately pull the person out.
- Distribute the metal containers evenly through the house.
- Add 1000 mls formalin (40 %) to 400 g potassium permanganate per 1000 m^3 of air space.
- Leave the building shut for 12 hours.
- Open up and ventilate for8 hours before use.
- Place a notice on the door warning people "Fumigation in process".
- Oxygenating disinfectants, sprayed as a fine mist, can be used as an alternative to fumigation with formalin. They are much safer and easier to handle.

Washing hands

Facilities for washing and disinfecting hands to prevent the spread of infection should become an important part of environmental control.

Disinfection of hands should always take place after handling livestock, particularly where there has been faeces or urine contact. This is not least for personal hygiene reasons. (See Chapter 17: Zoonoses safety data under Step 5.)

Hands should be washed following meals and after frozen or defrosted meat products have been handled because of the risk of transmission of diseases such as FMD, SVD and CSF.

Importantly, washing hands after handling diseased pigs can reduce the risk of spread of scour in the farrowing houses.

Summary

There is often a bewildering array of disinfectants available, each with attractive claims made as to their effectiveness. To make the best, cost-effective decision use the information covered in this section to help make your selection.

Water + detergent is the cheapest disinfectant you have on the farm.

The Costs of Disease

Economics and the SPF herd

The most important reason for the establishment of an SPF high health herd is to mitigate the severe effects of infectious disease. Most infectious diseases in the pig (excluding infections such as porcine parvovirus and leptospirosis) depress both the food conversion efficiency and daily liveweight gain together with increases in mortality.

The factors contributing to the cost of disease include:

- Pig mortality – this results in increased production costs.
- Pig morbidity – the number of culled animals unable to achieve full market value.
- Increased overhead costs.
- Increased feed costs.
- Loss of profit.
- Decreased feed efficiency.
- Increased stocking density.
- Slow growth – low daily liveweight gain.
- Less liveweight sold from the farm and a reduced throughput.
- An increased incidence of other diseases.
- Increased labour costs.
- Increased veterinary costs and medicines.

Fig 2-24 shows the mortality associated with specific diseases, based on field data, both in the early acute phase when the organism first enters a non-immune herd and after it has moved into an enzootic or chronic form.

Fig 2-25 shows the continuing effects on production once the acute episode has subsided.

Fig 2-26 shows the performance "costs" of outbreaks of disease based upon specific mortalities, loss of production and/or reproduction and feed efficiency. Clearly, actual costs depend upon the price of feed and market prices in your country but the figures place relative values on the costs of disease.

These "costs" relate to the acute phase of the disease and subsequent effects over 12 months and obviously such costs depend upon the severity of the outbreak and its period of continuation in that particular herd at that level of disease.

Maintaining Freedom from Disease: Biosecurity

You can establish a specific pathogen free status herd by building a new unit on a greenfield site or by depopulating your present farm, cleaning and disinfecting it thoroughly and repopulating it with pregnant gilts from a breeding company with the required pathogen status. Meanwhile you should consider carefully the location of the farm to be populated to decide what level of health status would be appropriate. Then you should decide on the source of the new stock together with what measures you should adopt to prevent your new herd becoming contaminated.

Freedom from the main infectious pathogens is a major contribution to the efficiency of production and to profitability. Thus the biosecurity measures that are taken to prevent new pathogens entering your herd are of paramount importance.

While the most important factor in biosecurity is the location, the way most diseases are spread is through contact with other pigs or their products. A secure location is one of very low pig density well away from any other pig herd, preferably in hilly or mountainous country and/or on the sea coast. You also need to control visiting people. If there are wild boar in the vicinity, a stout perimeter fence which goes into the ground is needed.

Of course, like most pig farmers, you may have no choice in your location because your farm is already up and running. With the help of your veterinarian, you should critically review the location your pig farm is in to assess what the risks are and what health status your herd can maintain. The worse your location, the more important it is for you to take strict biosecurity measures.

The checklist in Fig 2-27 will help in determining deficiencies. Fig 2-28 indicates the diseases that you should be able to maintain your herd free from and those diseases that are more difficult to keep out.

The minimum objective in establishing an SPF herd is to maintain the status quo for at least 2 years, reduce the days to slaughter (compared to the old herd) by 10 to 21 days and improve the food conversion ratio by between 0.1 and 0.4. Ideally, a farrowing rate of 85 % should be considered together with a grower mortality of less than 3.0 %. If you anticipate you cannot maintain this for 2 years, think again.

FIG 2-24: THE CONSEQUENCES AND SUBSEQUENT EFFECTS OF INTRODUCING A NEW DISEASE INTO THE HERD

Disease/Pathogen	Mortality Due to the Disease	
	During the Period of Acute Disease	During the Period of Chronic Disease
Actinobacillus pleuropneumonia (App)	3 – 30 %	2 – 4 %
Mycoplasma (enzootic) pneumonia	2 – 14 % sows	up to 3 % growers
PMWS/PCVAD	25 – 50 % growers	8 – 12 % growers
Porcine parvovirus infertility	0.5 – 4 pigs/litter	0.5 – 1 pigs/litter
Porcine respitatory disease complex (PRDC)	3 – 10 % weaners, growers	2 – 8 % weaners, growers
Progressive atrophic rhinitis (PAR)	1 – 5 % weaners	1%
PRRSV	5 – 30 % piglets	0 – 1 pigs/litter
Streptococcal meningitis	4 – 12 % weaners	1 – 5 %
Swine dysentery	1 – 4 % weaners, growers	1 – 1.5 %
TGE/PED	90 – 100 % piglets	1 – 4 %

FIG 2-25: THE EFFECTS OF DISEASE ON GROWTH AND FOOD CONVERSION EFFICIENCY (FCE)

Disease/Pathogen	Time of Acute Disease (weeks)		During the Period of Chronic Disease	
	Lost FCE	Lost Days to 100 kg	Lost FCE	Lost Days to 100 kg
Actinobacillus pleuropneumonia (App)	0.1 – 0.4	7 – 30	0.1 – 0.3	4 – 15
Mange	0.1	7 – 18	0.1	5 – 8
Mycoplasma (enzootic) pneumonia	0.1 – 0.4	10 – 21	0.05 – 0.1	3 – 21
Porcine respitatory disease complex (PRDC)	0.1 – 0.4	7 – 30	0.1 – 0.3	7 – 28
Progressive atrophic rhinitis (PAR)	0.1 – 0.2	4 – 15	0.1 – 0.2	4 – 15
Streptococcal meningitis (SM)	0.05	1 – 3	0.05	0
Swine dysentery (SD)	0.1 – 0.3	5 – 20	0.3	4 – 5
TGE/PED	0.1	4 – 10	0 – 0.15	0 – 3

FIG 2-26: PERFORMANCE "COSTS" OF DISEASE OVER A 12 MONTH PERIOD PER 100 SOWS AND THE PROGENY OF 2000 PIGS TO 100 KG

Condition or Disease	Mortality and Culls	Performance Loss	Recorded Disease Level	Target Level
Abortion	...	230 piglets.	11 %	1 %
Aujeszky's disease/pseudorabies virus	375 pigs	FCE, Dlwg.	High	Zero
Crushed by sow	76 pigs	Loss of margin over feed.	4 %	1 %
Farrowing rate	...	615 pigs.	70 %	89 %
Finishing losses	125 pigs	Loss of margin over feed + feed costs.	7 %	1.5 %
Foot-and-mouth disease	Slaughter policy	100 %	...	...
Infertility viruses	402 piglets	Loss of margin over feed. Reproduction.	High	Zero
Non-infectious infertility	410 pigs lost	Loss of margin over feed.	72 % farrowing rate	89 %
Non-pregnant sows	...	92 pigs.	6 %	2 %
Piglet mortality	...	177 pigs.	15 %	8%
Piglet scours	50 pigs	Loss of margin over feed.	2.5 %	0.5 %
Pneumonia complex	2 – 6 %	FCE, Dlwg.	Variable	Zero
Poor viability	163 pigs	Loss of margin over feed.	3.5 %	1 %
Postweaning losses	148 pigs	Loss of margin over feed + feed costs.	8 %	1.5 %
Progressive atrophic rhinitis	2 %	FCE, Dlwg.	High	Zero
Repeat matings	...	106 piglets.	17 %	6 – 8 %
Swine vesicular disease	Slaughter policy	100%	...	...
Swine fever	Slaughter policy	100%	...	...
Swine dysentery	1 %	FCE, Dlwg.	Low	Zero
Stillbirths	...	84 pigs.	8 %	5 %
Transmissible gastroenteritis	315 pigs	10 – 14 days growth.	High	Zero

FCE = Food conversion efficiency. Dlwg = Daily liveweight gain.

FIG 2-27: POSSIBLE SOURCES OF DISEASE ENTRY INTO YOUR FARM AND WHAT YOU SHOULD CONSIDER IN ORDER TO PREVENT IT – EXAMPLE FORM

Before the farm is populated with new pigs, check your farm for disease risk* :-

Is your farm in a pig dense area? .... ☐

What is the position of your pig unit relative to other infected pig herds around it? .... ☐
- How far away is the nearest infected herd?
- Do you have an uninterrupted view of it?
- Is the land around you flat or hilly, bare or wooded, on the coast or inland?

Your foundation stock – is it of appropriate health status for your location? .... ☐
- Can you rely on a future continuous supply in adequate numbers of replacement stock of the same health status?
- What precautions can you take against contamination?

When the farm is being populated and afterwards:-

Isolation .... ☐
- Do you have good isolation facilities for incoming stock?
- How long are they to be kept in isolation? Is it long enough?
- Do you check that there has been no disease outbreak in the herd of origin before you move them in?

Transport .... ☐
- Are the trucks that bring your pigs clean?
- Have they been to other farms?
- Have they been careful not to drive behind or park beside other pig trucks?
- Are the trucks that pick up your pigs for sale or slaughter empty when they arrive?
- Are they clean? Have they been disinfected?
- Does the driver wear clean boots and coveralls?
- Do you have a safe loading area?
- Does the water used to wash it run into or away from your pig buildings?
- Are the loading procedures safe? Can the driver contaminate your herd?

Visitors' book .... ☐
- Do you make visitors sign a visitors' book to affirm that they have not been near other pigs?

Mechanical transmission .... ☐
- Are clean boots and coveralls provided at the entrance to your unit for all visitors and your staff?
- Are there clean showers for visitors? Is there a proper changing area?
- Do you make anyone who may have been near other pigs recently, shower?
- Do you make anyone who may have been near other pigs recently have a gap of say, 24 hours before they visit your unit?
- Are toilets available for your staff and are they clean and hygienic?
- Do you ever need to use equipment on your pig farm that has been in contact with other pigs?
- If so, how do you disinfect it?

Bedding materials. .... ☐
- Are they from a known clean source?
- Are they contaminated by rats, mice or birds?

Feed source .... ☐
- Are you satisfied with its quality?

Feed trucks .... ☐
- Do they have to enter your pig compound?
- Do you have your own bulk feed pipe?

Water supply if not from the mains .... ☐
- Have you had it tested for bacteria?
- Are your water storage tanks clean and rat-proof?
- Should you chlorinate it?

Vectors .... ☐
- Flies – do you control them or are they a problem?
- Rats and mice – do you have a regular efficient rodent control programme?
- Birds – can they get to feed supplies and bedding stores? Should these be netted off?
- Do you have a perimeter fence (including building walls) that would deter stray animals, including wild boar and human curiosity seekers?

Human food .... ☐
- Do you allow it to be brought in to your pig buildings?
- Do you allow pork meat products to be eaten on your farm?
- Do you have a special area (canteen) where food must be eaten?

Dead animal disposal .... ☐
- Do you have safe procedures?
- Are they collected from your farm?
- What do you do with casualty animals?

You .... ☐
- If you are the unit owner or manager, do you visit friends' pig herds?

* Discuss these with your veterinarian and use the above as a check list.

FIG 2-28: PATHOGENS/DISEASES THAT YOU SHOULD BE ABLE TO KEEP OUT OF THE HERD		PATHOGENS/ DISEASES DIFFICULT TO KEEP OUT
Reliably	**Less Reliably**	**Difficult**
Actinobacillus pleuropneumonia (App)	*	*Actinobacillus suis*
Aujeszky's disease (AD/PRV)	**	Coccidiosis
Brucellosis	**	Congenital tremor (all)
Leptospirosis (pomona)		Erysipelas
Lice	*	*Escherichia coli (E. coli)*
Mange	**	Glässer's disease
Mycoplasma hyopneumoniae		Ileitis
OIE list diseases		Internal parasites
Porcine epidemic diarrhoea (PED)	**	*Mycoplasma hyosynoviae*
Porcine reproductive and respiratory syndrome (PRRSV)	**	Piglet dysentery (*C. perfringens*)
Progressive atrophic rhinitis (PAR)		Porcine circovirus 1 and 2
Swine dysentery		Porcine parvovirus (PPV)
Transmissible gastroenteritis (TGE)	**	Rotavirus diarrhoea
		Salmonellosis
		Streptococcal meningitis (SM)
		Streptococcus suis
		Swine influenza virus (SIV)

* In pig dense areas.
** When a major disease outbreak is occurring.

Selecting Your Source of Breeding Stock

Once you have determined your health status, it is then possible to organise how and where to obtain new genetic material.

The selection of the correct health status appropriate to your herd and location is vital before breeding stock are purchased. Your primary reason for purchase is to genetically upgrade your herd. Major requirements will be that they are available when you want them, in the numbers that are needed and at a price you can afford. But an overriding requirement is that they will not cause disease in your herd and lower your overall health status. At the outset therefore, consult with your veterinarian to determine at a veterinary level the information available about the proposed donor herd.

The investigations should include the disease history since its inception and those of any daughter herds that have been established from it; also the health status and disease history of other herds it supplies. All veterinary reports should be requested and examined together with the results of tests for specific diseases and the frequency of such tests. The breeding history on the farm should be checked together with any evidence of infectious reproductive disease. A detailed study of records of production parameters, and of growth and food conversion rates may be helpful. The biosecurity of the breeding pyramid should be checked along with details of the health programme. The biosecurity of the donor herd itself must be assessed including the methods by which pigs or genetic material are brought into the herd. Finally, a written veterinary statement should be obtained, indicating that on both clinical and pathological grounds those selected diseases that you wish to keep out of your herd have not been diagnosed in the donor herd.

Buying breeding pigs – the ground rules

Step 1: Select the source based on:
- Food conversion efficiency.
- Genetics (including fecundity).
- Number weaned per gilt and sow.
- Health.
- Market acceptability.
- Availability.
- Quality control.

Step 2: Determine with your veterinary advisor the pathogen status of your own herd.
Step 3: Request veterinary liaison with the suppliers' veterinarian and get clarification of the pathogen status of the donor herd.
Step 4: Assess the compatibility of pathogen status.
Step 5: Determine the isolation requirements for incoming stock.
Step 6: Decide on vaccination and acclimatisation procedures.

The donor herd

The suppliers may want to know the pathogen status you require and offer you a choice of sources.

Always purchase from a specific pathogen free herd or equivalent if available.

What are the methods and risks of pig movement

Since incoming pigs are probably the greatest potential source of infection to your herd, the methods by which they are introduced are vitally important; the same applies to other methods by which you improve the genetic potential of your herd

Five methods are available:

1 By introducing live pigs.
2 By segregated early weaning (SEW).
3 By hysterectomy and fostering.
4 By embryo transfer.
5 By artificial insemination (AI).

1. Introducing live pigs

a) Mature gilts and boars

Live pigs can be brought into your herd from a source herd of matching health status, or through SEW or hysterectomy and fostering if the source herd is of known but lower health status (depending on the pathogen to be eliminated).

If live pigs are brought into your herd with or without SEW it is advisable to hold them in isolation for a period before integrating them into your herd to check whether they develop disease and whether disease breaks out in the source herd. If the isolation premises are in a different site to your herd and not of the same biosecurity standards as your recipient herd, there could be a greater risk in holding them there rather than integrating them directly into your herd. The dangers of integrating them directly into your herd are obvious, namely, that if they are incubating an infectious disease sub-clinically then ultimately your herd will become infected (Fig. 2-29). Perfect separate quarantine facilities are rarely available to commercial herds, particularly smaller enterprises, but isolation that falls short of complete quarantine (e.g. on the same site) can be surprisingly effective. The incoming stock could be moved into a separate building on the same site, preferably over 50 metres distant, and this should be reasonably effective, provided separate boots and coveralls are used to tend the animals and provided the drainage from the building does not flow into your other pig buildings. If a separate building is not possible then a separate room sealed off from the main body of the herd is better than direct integration into the herd.

How long should the incubation period be? Here the importance of veterinarian liaison to match respective health status has already been highlighted.

If your herd is believed to be mycoplasma (enzootic) pneumonia (EP) free then it is advisable to place the incoming animals in isolation for a period of 8 weeks. At the same time sentinel pigs (i.e. pigs from your herd due for slaughter) should be moved in and blood-tested and/or slaughtered prior to the entry of the new pigs, and their lungs examined for EP freedom. If your herd is not free of EP, the length of isolation is debatable. Some veterinarians would advise 6 weeks but 4 is more practicable.

Should enteric or respiratory disease appear during the four-week period either in the pigs in isolation or in the source herd, the chances of preventing further damage by immediate slaughter would be reasonable.

b) Breeder-weaners

Instead of buying in mature replacement gilts and boars you could buy in so-called breeder-weaners, say, 30 kg (65 lbs) liveweight. This has the advantage of allowing them a long period of acclimatisation to your herd before you breed them. It also enables you to rear them yourself in the way you think best for future breeding gilts and allows you to carry out your own selection at slaughter weight. A disadvantage is that boars brought in this way cannot be performance tested. Also, if you sell your pigs at 25 to 30 kg (55 to 65 lbs) or at weaning, you probably do not have the facilities to rear such pigs.

The advantages of buying in breeding stock at a commercial level, compared to the selection of the home produced gilt are its low cost, the availability of gilts when they are required, the fact that the genetic potential is constantly improved, and the fact that if it is done carefully it presents few problems. Some farms however prefer to breed their own breeding females and thereby only introduce into the herd, a small proportion of grand-parent females and boars. This policy often fails because of the difficulty of rearing the future female replacements within a commercial operation, the poorer reproductive performance, and the fact that the gilts reared on the farm are often not available when required. This system is also a high cost one and often results in lower numbers of pigs reared. Extensive experience has shown that provided there is good health liaison and sensible practical procedures then the herd health status can be maintained with the purchase of breeding stock.

FIG 2-29: RISK LEVELS IN BUYING PIGS

Pigs Bought	Action	Risk Level
SPF pigs →	Moved via isolation into the herd with veterinary liaison and testing.	Very low. Risk confined to a possible prolonged incubatory state e.g. EP, PRRSV.
SPF pigs →	Moved direct into the herd.	Low risk.
Conventional health status pigs →	Moved in direct with veterinary liaison.	Moderate risk.
Unknown health status pigs →	Moved direct into the herd. No veterinary liaison.	High risk.

SPF - Specific pathogen free

2. Segregated early weaning (SEW)

This method of bringing in live pigs from another herd is through a modification of the medicated early weaning (MEW) technique, referred to as segregated early weaning (SEW). This is based on the principle outlined earlier under "How Infectious Agents are Spread". By the time females reach their first farrowing they have developed a strong immunity to the more serious enzootic pathogens in the herd and have eliminated most of them. The process relies on the transfer of maternal protection via colostrum. Therefore, any factor which reduced colostrum intake will reduce the effectiveness of this system. Thus if they are weaned immediately from the sow and moved to isolated premises at the appropriate age they will be free of the pathogens you wish to eliminate.

Thus if you wished to obtain future breeding stock from a particular herd but your veterinarian thought that the general health of that herd was below that of your own, you could obtain higher health status pigs free from the unwanted pathogen. If the pathogen you wished to avoid was *M. hyopneumoniae,* you could vaccinate the dams in the donor herd ahead of time to boost their immunity, put the sows and newborn piglets on an anti-mycoplasma medicine such as tylosin or tiamulin and wean the pigs at 10 days to the isolation facility on your farm. Isolation is necessary because if an unknown pathogen enters the donor herd it could go through the SEW system during the incubation period.

The SEW system is discussed in detail in Chapter 3.

3. Hysterectomy and fostering

This method of introduction of live pigs is through hysterectomy and fostering the piglets onto a newly farrowed sow in the recipient herd. This operation is carried out on day 113 of pregnancy when the sow is slaughtered. The uterus containing the piglets is either removed 50 m (165 ft) away to a pig-pathogen-free environment where the piglets are removed or it is passed through a disinfectant trap into a sealed room. The litter is then immediately taken into the recipient herd and suckled onto a newly farrowed sow. If done properly the mortality rate is less than 10 %.

The whole operation is synchronised using prostaglandins so that newly farrowed sows are available to act as foster mothers (see Chapter 15). Ideally the sow selected for the operation should be moved into isolation approximately 8 weeks prior to the due date and monitored for evidence of disease. At the same time it should be blood-sampled and tested. If carried out following written protocols, hysterectomy has produced many of the nucleus farms available today.

4. Embryo transfer

Embryo transfer has been used successfully in several countries for the introduction of new genes but it has not been widely adopted, probably because it requires two skilled teams, one to flush the fertilised eggs from the donor sow and one to insert them in the recipient sow. It has not been performed on anything like as big a scale as hysterectomy, therefore there is not the volume of field evidence to underline its safety, but in theory and on the limited evidence available, it is safe.

Its drawbacks are (1) it needs two skilled teams, (2) it requires immaculate synchronisation and timing, (3) the embryos cannot be kept viable for more than a few hours and (4) unless done expertly it results in a high failure rate and small litters. For practical purposes, SEW, hysterectomy and AI are much simpler.

5. Artificial insemination

Another method of introducing genes is by artificial insemination (Fig 2-30). It is known that viruses of swine fever, Aujeszky's disease, PRRSV, porcine parvovirus, PMWS and leptospira bacteria and *Brucella suis* could be introduced through AI mainly during the early stages of infection of the boar. If the boars first go through a true quarantine procedure and are screened for these infections, then housed in an isolated AI stud (i.e. one in a secure location), with high standards of biosecurity and hygiene during the production of semen, then field experience indicates that the risks are very small. The advent of frozen semen, which is possible at a company level, renders the use of AI much safer since the semen can be stored for a month or two, time enough to be sure that no new infection was incubating in the AI stud. The movement of genetics via frozen semen should be the only way genes are moved between nucleus farms. AI does however, have the disadvantage that only half the genes are introduced into the herd.

OPTIONS FOR THE TRANSFER OF GENES INTO A HERD

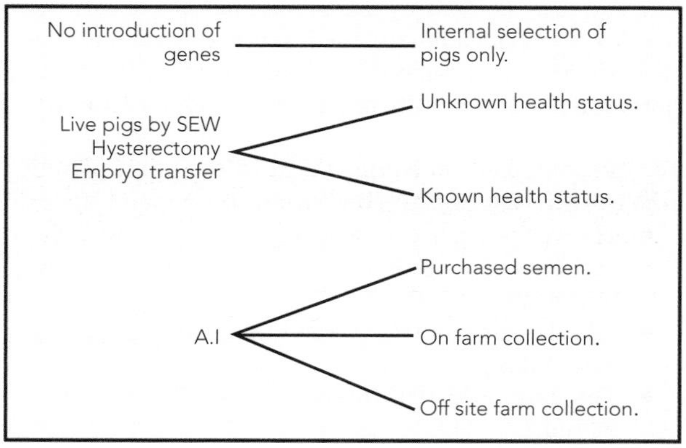

Fig 2-30

Depopulation and Repopulation

The major costs of establishing a high health herd by depopulation and repopulation are related to the loss in cash flow and ongoing overhead costs between removing the old herd and its sale of pigs, and the first pig sold at slaughter from the new one. In basic terms, the margin over feed per batch represents the approximate costs of depopulation (building/maintenance costs need to be budgeted for). The farm must be emptied for a minimum of 4 to 6 weeks (8 weeks if swine dysentery exists in the farm) and further time lost will depend upon the date of commencement of the mating programme for the new incoming herd. Depopulation can be carried out in one of two ways:

1. by complete removal of the herd at a fixed point in time, or
2. By gradual removal towards a predetermined date.

Repopulation of the herd can be carried out with maiden or pregnant gilts that are up to 2 weeks from farrowing.

The most cost-effective method is to remove the existing weaned and growing pigs to a totally separate, well-isolated site where their production can be completed. The donor herd should ideally have facilities to commence the new gilt mating programme and hold these to within two weeks of farrowing. Alternatively the pregnant gilts 21 days post-service could be moved to a holding site.

Fig 2-31 illustrates the actual programme used in the depopulation and repopulation of a 350 sow herd by method 2. The breeding farm producing 30 kg (66 lbs) weaners was a separate site 5 km (3 miles) away from the finishing farm. The new gilts were mated on the supply farm and then moved to a holding farm when they were 6 weeks in pig where they remained until 2 weeks prior to farrowing. The complete gilt herd of both maiden and pregnant animals was then moved into the depopulated farm over a period of 4 to 6 weeks.

The decision to use either method 1 or 2 is a complex one and it is dictated by such factors as the current economics of the pig industry, the levels and costs of disease, the availability of maiden or pregnant gilts, the type of herd and the availability of holding accommodation.

For the combined breeding and finishing farm, planning will be different and the decisions will depend upon:

- The age/weight of the youngest pigs to be sold and the point of depopulation e.g. 30 kg (66 lbs) liveweight.
- The date of the last farrowing to produce 30 kg (66 lbs) pigs.
- The date on which the last diseased sows are mated.
- The date when the new gilt mating programme commences. This will depend on whether the programme starts on the supply farm, a holding farm or the repopulated farm.

The following scenarios then exist:

Method B is by far the most cost effective but equally the most complex. Method A can be improved further by selling diseased sows up to 6 weeks in pig and weaners at an earlier age.

Depopulation and repopulation costs

These represent 14 weeks of loss on a 350 sow herd – probably the very best that can be achieved. This needs to be set against the improvements that should be achieved in the new herd of up to 0.4 in feed efficiency, 100 g increased daily gain and significant reductions in mortality and medicinal costs. The repopulation would also allow the most updated genetics to be used to further obvious advantage.

If the finishing accommodation on a combined breeding finishing unit can be used for multi-suckling then increasing the gilt mating programme and therefore the numbers of gilts farrowing is a very valuable procedure to consider. There are always a number of gilts which have to be culled after first weaning and this can also help reduce the herd to its proper size without the purchase of replacements in large batches. Farrowing a surplus number of gilts first time round costs more up front but gives a quicker return on cash flow.

Cleaning an existing pig farm is no mean undertaking and you should be aware of the hard work that is involved. Furthermore, considerable expense will be involved in refurbishing the farm since many maintenance requirements are not evident until the buildings are empty. It is also an opportunity for major alterations.

In carrying out the cleaning and disinfection procedures, efforts should be made to remove all faeces and manure from the farm. Slurry channels should be emptied and washed down and a visual and bacteriological check carried out once cleaning and disinfection have been completed.

It is vitally important that rats and mice are completely removed from the farm. Once all the houses have been cleaned, then each in rotation should be sealed and fumigated with formaldehyde gas, or sprayed with a suitable disinfectant. Dead pig pits should be covered with lime and finally buried.

Provided the farm is left empty for a period of 6 – 8 weeks, few of the important disease organisms will survive except for SVD, CSF, ASF and salmonella.

Over the years clean herds have been established successfully by depopulation and repopulation for the following diseases/pathogens:

- Actinobacillus pleuropneumonia.
- Aujeszky's disease/pseudorabies virus.
- Foot-and-mouth disease.
- Lice (*Haematopinus suis*).
- Mange (*Sarcoptes scabiei var suis*).
- Mycoplasma (enzootic) pneumonia (*Mycoplasma hyopneumoniae*).
- Porcine epidemic diarrhoea.
- Porcine multisystemic wasting syndrome.
- Progresive atrophic rhinitis (toxigenic *Pasteurella multocida*).
- Swine fevers.
- Swine dysentery (*Brachyspira hyodysenteriae*).
- Swine vesicular disease.
- Transmissible gastroenteritis.

Mycoplasma (enzootic) pneumonia and PRRSV have however proved to be the most difficult diseases to keep out. The purchased pig has been the main cause of reinfection.

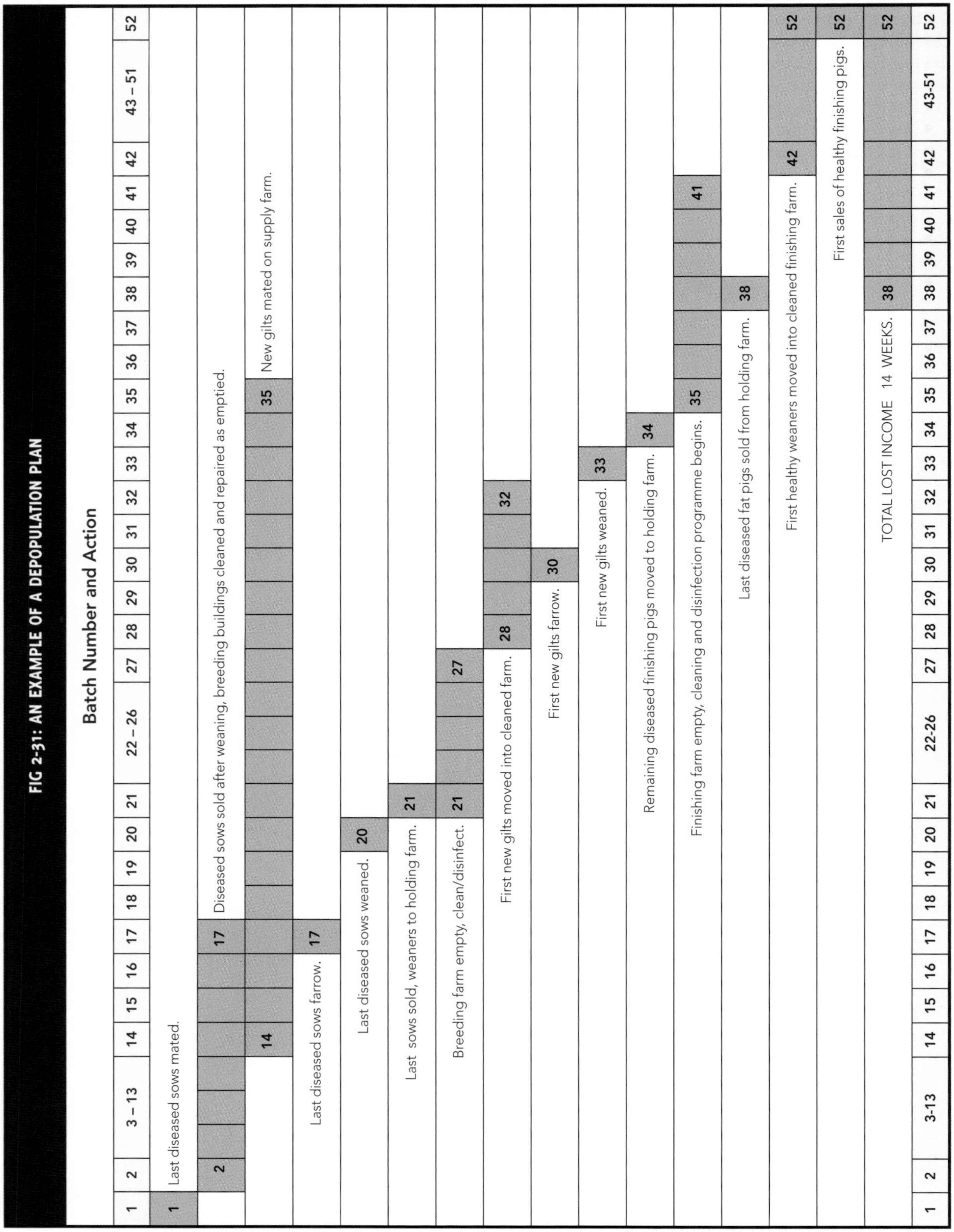
Batch Number and Action
1 2 3 – 13 14 15 16 17 18 19 20 21 22 – 26 27 28 29 30 31 32 33 34 35 36 37 38 39 40 41 42 43 – 51 52
Last diseased sows mated.
Diseased sows sold after weaning, breeding buildings cleaned and repaired as emptied.
New gilts mated on supply farm.
Last diseased sows farrow.
Last diseased sows weaned.
Last sows sold, weaners to holding farm.
Breeding farm empty, clean/disinfect.
First new gilts moved into cleaned farm.
First new gilts farrow.
First new gilts weaned.
Remaining diseased finishing pigs moved to holding farm.
Finishing farm empty, cleaning and disinfection programme begins.
Last diseased fat pigs sold from holding farm.
First healthy weaners moved into cleaned finishing farm.
First sales of healthy finishing pigs.
TOTAL LOST INCOME 14 WEEKS.
1 2 3-13 14 15 16 17 18 19 20 21 22-26 27 28 29 30 31 32 33 34 35 36 37 38 39 40 41 42 43-51 52
FIG 2-31: AN EXAMPLE OF A DEPOPULATION PLAN

3 MANAGING HEALTH AND DISEASE

3 MANAGING HEALTH AND DISEASE

Mycoplasma arthritisAn Introduction to Managing Health and Disease

There are many factors that affect the economic viability of pig production. For a business to be successful it must take account of all of them. Health is one that plays a significant role. The important factors that you might be able to control and improve are highlighted here. You should find it of value to consider each and assess them in relation to your own pig farm.

There are other factors, such as government regulations, which are completely out of your control and so are omitted from this checklist. Likewise, money available to the consumer is a national problem, rather than one at farm level. The important criteria on the farm, however, are the price of feed and the efficiency with which this is converted into liveweight gain and ultimately lean meat. The interaction between people, management, the environment and health dictates the efficiency of these conversions.

Additionally, there is a large social push towards improving the welfare of pigs on farm. Health and welfare are often intertwined and factors affecting welfare also need to be considered. These are covered in Chapter 16.

Factors that affect the economic viability of pig production:

- Application of new technology.
- Biosecurity.
- Continuing education/training.
- Control of pollution.
- Daily liveweight gain.
- Efficiency of feed conversion to lean meat.
- Efficiency of pen utilisation.
- Environmental restrictions.
- Equipment cost and quality.
- Genetic potential of the pigs.
- Health control and treatment of disease.
- Labour cost.
- Locality of farm and its neighbours.
- Management and motivation of people.
- Management decisions.
- Market outlets.
- Methods of feeding.
- Number of pigs sold per year.
- Nutrition and quality of the feed.
- Pig flow and batching.
- Planning.
- Price of feed.
- Reproductive performance.
- Size of the herd.
- Slaughter price.
- Stocking densities.
- Transport costs.
- Use of records.
- Wastage of feed around the farm.
- Weight and grade of pigs sold at slaughter.
- Welfare requirements.

Manipulate any of these and improve your profitability.

Management Components of Health Control

Chapter 2 discussed the various organisms that can produce disease. In many cases the causes are multifactorial, involving an interaction between the pig, the environment, the management and the organism. The components of this interaction are listed below. If the biological efficiency of your pigs is to be maximised it is necessary to give detailed attention to these eleven points.

1. Immunity of the pig. How the pig responds to infection.
2. Medicinal control of disease.
3. Eradicating pathogens.
4. Recognising and monitoring health at a clinical level.
5. Treatment of sick, compromised and casualty pigs.
6. The use of a veterinary consultant.
7. Staff training and education.
8. Understanding how to use records.
9. Planning for efficient production.
10. Management manipulation of the system and the environment.
11. Good nutrition.

These factors are placed into perspective in Fig 3-1, which also highlights how they should be adopted towards maximising pig health. Pig health is interpreted as maximum biological efficiency.

FACTORS THAT INFLUENCE "PIG HEALTH" AND PROFITABILITY

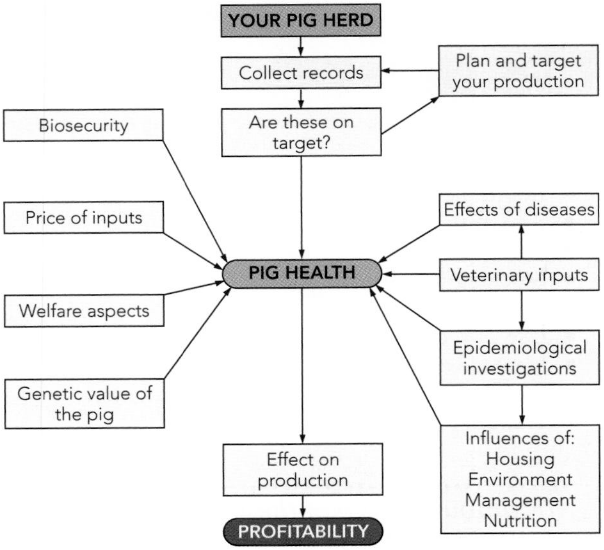

Fig 3-1

Maintaining an efficient, healthy pig farm starts with good planning. To achieve this it is necessary to have an efficient recording system. This can be a simple manual or a computer programme that identifies both production details and health levels and then measures results against predetermined levels of efficiency. It is also necessary to identify those diseases and pathogens that are present in the herd and take measures to prevent new pathogens from entering. The risks of pathogens being introduced by purchased pigs have been dealt with in Chapter 2, but the management and integration of the gilt and the boar when they first enter the herd also need to be addressed. Finally the services of an experienced veterinarian can have a major impact not only on the profitability of the farm but also on improving the *status quo*.

The infectious organisms that cause disease in pigs are viruses, bacteria, fungi and parasites. At a clinical level they can be divided into three groups:

1. **Primary non-indigenous pathogens** are not present in all herds but when they enter a susceptible herd they usually cause disease. They are responsible for major infectious diseases such as classical swine fever/hog cholera (CSF/HC), foot-and-mouth disease (FMD), transmissible gastroenteritis (TGE), Aujeszky's disease/pseudorabies virus (AD/PRV), mycoplasma (enzootic) pneumonia, progressive atrophic rhinitis (PAR), swine dysentery (SD) and mange. The effects of the pathogen also depend on the pathogenicity (virulence) of the strain (i.e. its capacity to produce clinical disease), the numbers of organisms that are presented to the pig, and the degree of immunity that exists in the herd to that specific infectious agent.
2. **Enzootic opportunist pathogens** are those that are always present in all herds and yet do not cause disease unless there are deficiencies in the environment. Such organisms include streptococci, responsible for joint infections or meningitis, coliform bacteria causing piglet diarrhoea, staphylococci that cause greasy pig disease, and klebsiella associated with mastitis, to mention but a few.
3. **Harmless commensals** are found on the skin and in the respiratory and alimentary tracts, vagina and prepuce. There are hundreds of species of bacteria in the large intestine alone, but some are fungi and protozoa.

Metagenomics describes the pig and its resident bacteria, viruses, fungi and parasites that live within and on the pig. Metagenomics considers the whole DNA package. This is becoming increasingly important in determining diagnostics and treatment options.

Group 1 – Primary pathogens produce disease in a susceptible herd. You are better off without them.

Immunity – How the Pig Responds to Infection

Infection is considered to have taken place when a virus, bacterium or parasite enters a pig and starts to multiply. If it is a potentially pathogenic organism it may change the normal structure and function of the pig. A study of these changes is called pathology and organisms causing disease are described as pathogens. Fig 3-2 shows the sequence of events that may take place when an infectious pathogen infects the pig.

Infection creates two scenarios. The first is when there is no clinical disease. The immune mechanism of the pig responds to challenge and the pathogen is either eliminated or remains within the body in a carrier state. The carrier pig may or may not shed the organism or may shed it intermittently.

The second scenario is that of clinical disease which can be "acute" (rapid onset and/or short-term), sub-acute (in-between) or chronic (persistent or otherwise long-lasting). The consequences of infection are varied. The pig's immune response may be sufficient to over-

HOW THE PIG RESPONDS TO INFECTION

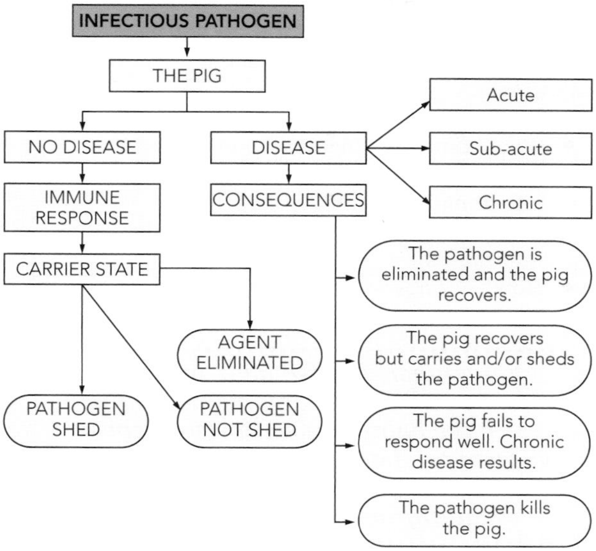

Fig 3-2

come the infection, and complete recovery and elimination of the pathogen follows. Alternatively the pig may recover but remain in a carrier state. Another alternative is that the pig's immunity may fail to adequately respond to the infection and chronic disease results. Under severe circumstances the infection may overwhelm the pig, resulting in death.

Whether the pathogen causes disease or not is dependent on its virulence (the capability of the organism to produce disease), how many organisms are present, what other concurrent infections are present, what protective mechanisms are available in the pig to prevent disease and what environmental or other factors are present that will lower the pig's immunity. For example, certain strains of *Actinobacillus pleuropneumoniae* are only mildly pathogenic and do not normally produce clinical disease. However, if they are in combination with PRRSV the presence of both could produce severe pneumonia. A non-pregnant gilt exposed to parvovirus infection only develops an immune response but no signs of clinical disease because the disease only develops in the foetus. Nevertheless the gilt would become serologically positive in laboratory tests. Understanding these events can be essential in developing successful health management programmes.

A successful and healthy pig farm is one in balance and harmony with the organisms that are present, both in the environment and the pig. It is the manipulation of this balance that is so important to health and ultimately to productivity and profitability.

Bad management allows opportunist pathogens to produce disease.

It follows that if the numbers of organisms in the environment can be maintained at a minimum then the threshold level of organisms necessary to produce clinical disease is unlikely to be reached. Examples here would be the excellent growth rates that can be achieved with all-in/all-out systems in nurseries or flat decks, when pigs are housed on weld mesh or slatted floors and they are divorced from their faeces and potential enteric organisms. Likewise in respiratory diseases, the more pigs there are in a common air space, and the smaller the cubic capacity of that air space relative to the number of pigs, then the greater will be the number of aerosol organisms and the more severe the disease.

To be a successful pig producer you should have some idea of how pigs resist infectious disease. Unfortunately, it is a complex subject and although here an attempt has been made to simplify it, it will probably still seem a little complicated.

The technical terms used in immunology (i.e. the study of immunity) are a major part of the problem. If you have not had a good grounding in biology they may be like a foreign language. If that is the case, or even if it is not, start by reading through the terminology below.

Disease is a numbers game: a minimum number of infectious organisms is required to cause disease. The number required depends on the virulence of the organism and also on the level of immunity in the pig.

TERMINOLOGY

Adjuvant – A substance added to an inactivated vaccine to make it more effective.

Antibodies – Complex large proteins (called gamma-globulins) which are produced by specialised cells in response to invading antigens. These stick specifically to the invading antigen, neutralising it or triggering off a destructive reaction.

Antigen – A foreign invading substance (i.e. a substance which is not normally part of the pig's body), usually consisting of protein or part of a protein, which stimulates the body to produce antibodies. Antigens exist on the surfaces of bacteria, viruses and parasites.

Antiserum – Serum with high antibody levels against a specific infection. It has usually been produced experimentally in laboratory animals by injecting the infection into them.

Blood sample – Whole blood sample taken hygienically with a syringe into a bottle or by a pin prick through the skin, absorbing the droplet of blood with blotting paper.
Colostrum – The means through which the piglet acquires antibodies and protective cells from its mother within 12 hours after birth.
Commensal bacteria – Bacteria that live permanently in or on the body without causing disease.
Epithelium – Cellular membrane (e.g. mucous membranes) containing epithelial and other cells.
Humoral immunity – Blood-borne immunity.
Hyperimmune antiserum – The same as antiserum above but emphasising its high titre.
Lymphocytes – Specialised defence cells in lymph nodes, other lymphatic tissue and the blood, which produce antibodies or take part in cellular immunity.
Micobiota – The group of normal bacteria living in or on a specific living surface. These can be protective.
Mucous membranes – Cellular membranes (e.g. those lining the gut) which secrete a sticky substance called mucus onto their surfaces.
Mucus – A clear sticky semi-liquid secreted by cells in mucous membranes.
Pathogenic infection – An infectious organism which has the potential to cause disease. This is in contrast to the many organisms that live normally in or on the body which never cause disease and are called commensals.
Phagocytes – Cells of the body whose special task is to engulf bacteria, viruses, or parasites in an attempt to destroy them. They are also called macrophages.
Phagocytosis – The process whereby the specialised cells of the body engulf bacteria, viruses or parasites in an attempt to destroy them.
Plasma sample – A whole blood sample taken hygienically with a syringe and mixed with an anti-clotting agent so that it remains liquid. The sample is spun fast in a centrifuge and the red and white blood cells sediment to a firm pellet at the bottom, leaving a clear liquid – the plasma.
Serology – Tests done in the laboratory to detect the level of specific antibodies in serum samples ("ology" means study of – so literally serology means "study of serum").
Serum sample – A whole blood sample taken hygienically with a syringe and allowed to clot. The serum is the clear straw-coloured liquid which can be drawn off with a pipette. It contains the antibodies.
Titre – The concentration of a specific antibody in a serum sample. It is expressed as the amount by which the serum has to be diluted before a serological test goes negative.
Virulence – The pathogenicity of an organism. Organisms with a high capability of causing clinical disease are called highly virulent.

The main components that make up the resistance of a pig to infection may also seem complicated but if you refer to Fig 3-3 as you read, it should also help you to understand the text better. You do not need to understand all the components, but should take note of the following:

- What antibodies are.
- What stimulates them to be produced.
- The importance of colostrum and milk in providing immunity.
- Why blood tests are done and how they are interpreted.
- How vaccines work.

Remember: expose your pigs to too many pathogenic organisms and you have compromised pigs – poor performers. Lower the level of pathogens in the environment and you have healthy pigs – high performers.

RESISTANCE OF A PIG TO INFECTION AND DISEASE

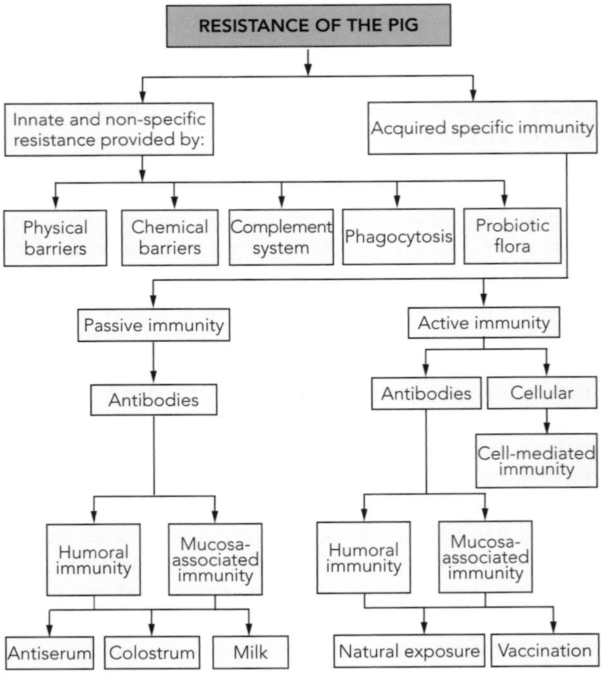

Fig 3-3

Resistance to infectious disease is key to good herd health (Fig 3-4).

MANAGING DISEASE: THE CHOICE IS YOURS

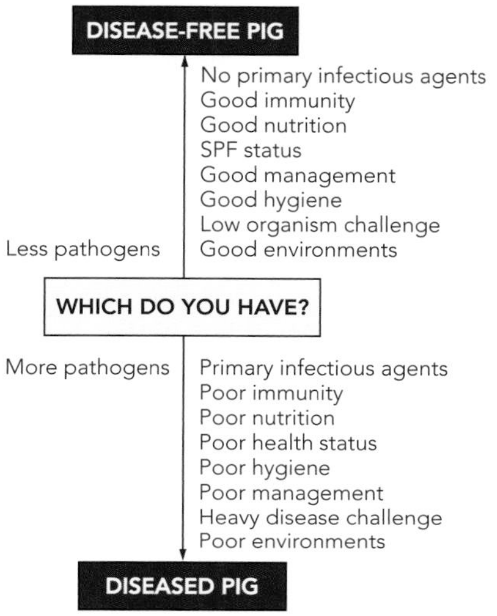

Fig 3-4

Innate (i.e. Inherited) and Non-Specific Resistance

Innate and non-specific resistance is that which all pigs are born with, or, in the case of some forms of non-specific resistance, that which develops as the pigs grow, regardless of what infections they are exposed to. It consists of:

Physical barriers – These are the external and internal body surfaces of the respiratory, alimentary and uro-genital tracts. The external surfaces are skin, horn and hair. The internal surfaces are: the membranes that line the nose, sinuses, throat and the tracheal and bronchial air tubes, the mouth, stomach and intestines and the vagina, uterus, bladder and urinary tubes. Other internal barriers include joint capsules and the blood-brain barrier.

Chemical barriers – The mucus on the surface of mucous membranes and some other body fluids such as saliva and tears contain antibacterial and anti-viral substances such as lysozyme and interferon. The acid in the stomach also inactivates bacteria and viruses to some degree.

Complement system – This is a series of many proteins (enzymes) in the blood that attack foreign cellular material in a sequential cascading manner, the first one or two acting and stimulating the next to act and so on. "Foreign" means non-self, i.e. material that is not part of the pig's normal body. Although complement acts non-specifically its action may be enhanced by some specific antibodies.

You may wonder how complement distinguishes between "self" and "non-self". It does so because the pig's own cells have a coating of special pig-protective protein that acts like Teflon in a non-stick saucepan. It specifically stops the pig's own complement from sticking to surfaces of its own cells to ensure they are not destroyed. If it cannot stick, it cannot destroy. Viruses, bacteria and parasites do not possess this special pig protein so they are not protected.

Phagocytosis – Certain defence cells, called phagocytes, can engulf foreign material such as an invading bacterium in an attempt to destroy it within the body's tissue or to carry it away (e.g. to the gut) in order to eliminate it from the body.

Phagocytes fall into two groups, polymorphs and monocytes.

- Polymorphs are part of the blood white cell population. They circulate in the blood but respond quickly when pathogenic infection occurs. They rapidly migrate out of the bloodstream to attack the infection. Some virulent bacteria are covered with a slippery capsule which inhibits the polymorphs from engulfing them. A specific antibody (called an opsonin) is then required to stick to the capsule, enabling the polymorph to engulf it. One type of polymorph (eosinophil) tends to attack parasites.
- Monocytes also start initially as part of the blood white cell population. They circulate in the blood to start with but then migrate into the tissues and onto inner body surfaces to become local tissue macrophages, or they wander through the tissues as wandering macrophages. These macrophages are usually capable of engulfing viruses and bacteria non-specifically, i.e. without the aid of specific antibodies.

Bacteria, viruses, parasites and toxins contain antigens. Antigens stimulate antibody production.

Macrophages engulf bacteria and viruses in order to destroy them or remove them from the body but they are not always successful. Some virulent bacteria and viruses can survive and multiply inside macrophages, often destroying the macrophages. For example, the PRRSV virus multiplies in the macrophages of the lungs and destroys them, thus compromising the lung immunity. Porcine circovirus 2 (PCV2) is frequently found in massive numbers in macrophages in affected tissues although its specific actions within the macrophage are unusually limited. Normally the macrophage would rapidly kill any organism within the cell.

Virulent strains of the bacterium *Streptococcus suis* type 2, and some other bacteria that cause meningitis, behave like the Greek legend of the Trojan horse. To get through the blood-brain barrier, which is normally resistant to penetration by bacteria, they hide in migrating monocytes. The monocytes then migrate through the blood-brain barrier to become brain macrophages where the bacteria break out to cause meningitis.

Normal microbiota flora – All surfaces of the body (the skin, mouth, stomach, intestines, vagina and prepuce) have a complex mixture of non-disease producing organisms, mostly bacteria, which have evolved over thousands of years to live in intimate relationship with each other. They are usually limited to one part of the host. Under normal circumstances these organisms provide a protective barrier preventing or reducing the attachment of other bacteria which may cause disease. Some of them are antagonistic to invading pathogenic bacteria and inhibit their growth. Examples in the gut are the inhibitory activities of bacteroides organisms against *Escherichia coli* (*E. coli*) or *Lactobacillus* against *Salmonella*. This 'indigenous flora' is not inherited, i.e. piglets are not born with it. However it is related to the pig's individual characteristics because it differs from the indigenous flora of other species. It is always the same mixture in any individual pig throughout life. As the pigs grow, this indigenous flora becomes increasingly complex.

Lactobacillus bacteria inhibit the multiplication of some acid-sensitive bacteria by secreting lactic acid. Lactic acid is available commercially to put in water or feed in order to reduce bacterial multiplication.

Various combinations of microbiota bacteria, (usually streptococci and/or lactobacilli) are also available commercially for feeding to newborn pigs or to restore the flora of pigs which has been lost through treatment with oral antibiotics. These are referred to as probiotics. Experimentally, probiotics have been shown to be effective but they are sometimes disappointing when applied on the farm. (See Chapter 14: Probiotics).

The study of all these organisms and their interaction with the pig is called megagenomics.

Acquired Specific Immunity

This is the immunity that the pig acquires as it goes through life. It may be obtained actively or passively.

Actively produced specific antibodies

When a foreign protein or part of a protein (called an antigen) enters the body, the body responds by producing antibodies in the lymphocytes of its lymphatic tissues. This is known as antigenic stimulation.

Antigens stimulate the production of antibodies which circulate in the blood and body fluids. They are secreted onto epithelial surfaces and specifically adhere to the antigen. The surface of the antibody fits snugly onto the complex corrugated surface of the antigen. The adherent antibodies may then have one of several effects. They may:

- neutralise the antigen (common with some viruses), or
- enable white blood cells to phagocytose it (such antibodies are called opsonins), or
- in conjunction with complement, disable and possibly destroy it.

  In some infections they may have little or no effect.

Active antibody production in pigs is mostly brought about by natural exposure of the lymphocytes to the surface proteins on viruses, bacteria and parasites or by exposure to toxins.

It can also be induced artificially by the inoculation with vaccines or toxoids (denatured toxins which have lost their potency but retain their ability to stimulate antibodies).

Or it can be stimulated by the deliberate feeding of substances such as faeces (i.e. so-called "feedback").

Delayed production

When a new pathogen invades a pig, the antibody response takes about 10 to 14 days to reach maximum levels. This delay is important in relation to natural infection and also to vaccination, since protection is not immediate. Pigs keep meeting new organisms throughout their lives, particularly when they are young. If a new infection is pathogenic it may be 5 to 7 days before specific antibodies against the pathogen start to appear, during which time the pig could succumb to disease.

Furthermore, if it occurs at a time when the pig is compromised, e.g. having just been moved and mixed in a different pen, or during a change of feed, the immune and antibody response may be partly suppressed.

When a pathogen which has infected the pig previously invades the pig again, the antibody response is much quicker because the lymphocytes which were primed by the first infection are still present. This is also true in a pig which has not been previously infected but has been vaccinated.

Antibodies are large proteins called immunoglobulins (Ig) of which there are three main types, IgG, IgM and IgA. They can be humoral (i.e. circulating in the blood and body fluids) and thus getting into all the body organs and tissues, or they can be local on the surface of mucous membranes such as those lining the respiratory and alimentary tract. These are called mucosa-associated antibodies.

Humoral antibodies

Over 80 % of the humoral immunoglobulin circulating in the blood of the mature pig is IgG, about 10 % is IgM and most of the rest is IgA.

If a pig is vaccinated by intramuscular, subcutaneous or intradermal injection against an infectious pathogen that invades the body such as erysipelas or CSF, the humoral antibodies produced are effective in blocking the invasion.

To produce maximum immunity in the intestines, the intestines themselves have to be antigenically stimulated.

Mucosa-associated antibodies

If a pig is vaccinated (e.g. into the muscle or under the skin) against an intestinal infection such as TGE virus or enterotoxigenic *E. coli* (ETEC) which multiply and do their damage in the gut, the humoral antibody response induced is unlikely to prevent the organism multiplying in the gut or to greatly reduce the diarrhoea that results.

To be fully effective in the gut, the immunising antigen must stimulate a local immunity, mainly by acting on the lymphatic tissues of the intestines ("Peyer's patches"). This results in the production of IgA just below the surface layer of the gut. The IgA passes through the cells of the mucous membrane and attaches to another molecule called the secretory component. It is then called secretory IgA and consists of two IgA antibody molecules joined together by the secretory component. This combination increases their potency, makes them resistant to digestion by gut enzymes, and more readily absorbed by mucus. Since the mucus coats the whole lining of the intestines and the respiratory tract, the secretory IgA acts as a shield against potentially pathogenic infections.

In contrast to humoral antibodies, therefore, mucosa-associated antibodies in the pig are principally IgA, in the form of secretory IgA.

Stress depresses the pig's immune response.

Cell-mediated immunity

There are two broad categories of immune reaction to infection:

1. Humoral immunity of which a major component is antibodies in the blood.
2. Cell-mediated immunity which need not involve antibodies.

Although they can occur independently, they usually both occur at the same time, with greatly enhanced benefit. Cell-mediated immunity is initiated by lymphocytes originating from the thymus gland (T lymphocytes), whereas humoral immunity involves lymphocytes that are derived originally from the bone marrow (B lymphocytes). Cell-mediated immunity also involves other types of cell such as macrophages and natural killer cells.

Whereas it is important for you as a pig farmer to have some understanding of humoral immunity because of its association with blood testing, maternal immunity (colostrum and milk) and vaccination, it is much less important for you to understand cell-mediated immunity, except to be aware that it exists.

Passively acquired immunity

So far we have been considering active immune reactions that result from stimulation of the pig's immune system by invading antigens, but immunity can be passive without the pig's immune system being stimulated. Passively acquired immunity, usually termed "passive immunity", is acquired naturally by the newborn piglet through the ingestion of colostrum and milk or artificially by the injection of antiserum or oral dosing of colostrum substitutes.

Colostrum

Unlike human babies and puppies, no antibodies are transferred through the placenta from the sow to her piglets before birth. Normally, piglets are born in a vulnerable state without any humoral or mucosa-associated antibodies and no acquired cell-mediated immunity. Fortunately, towards the end of gestation when the sow's mammary glands develop, the first secretion they produce, colostrum, is rich in antibodies representing the whole spectrum of the sow's own circulating antibodies, and cells from the mother's immune system. A first instinct of the newborn piglet is to find and suck a teat. Normally, a sow has voluntary control over milk let-down, but during farrowing this control is weak. The piglets nuzzle the teat and surrounding gland and then suck the teat. This results in a rapid let-down of colostrum. In the first 12 to 24 hours of life, the piglet's intestines are able to absorb whole antibodies and cells before the enzymes in the intestines digest them. Consequently, within a short period after a good first suck, the piglet's blood contains the full spectrum of its mother's antibodies, often at about the same level as that of the sow. In addition, immunity cells will pass from the mother to her own piglets – but not piglets which are fostered. These cells leave the stomach and migrate around the pig to establish themselves at vital points of defence in the piglet.

Four points must be emphasised.

1 Without maternal antibodies, the piglet is highly susceptible to infection. It is essential for the piglet's survival that it drinks colostrum soon after birth (ideally within 6 hours) before pathogenic microorganisms have had time to invade and multiply. It is also essential that it ingests enough colostrum to provide adequate protection until it has actively produced its own humoral antibodies.

Good hygiene in the farrowing house is essential for survival even if levels of colostral antibodies in the piglet's blood are high.

2 The ability of the piglet's intestine to absorb colostral antibodies and cells is short-lived, but is shortened still further when the piglet has drunk. Thus, if a piglet that has had no colostrum is to be cross-fostered onto another sow, or given substitute colostrum orally, it should be done in the first few hours of life and no other nutrients should be given in advance. The fostering sow must also still have colostrum available. Stressed piglets being chilled or having to sleep in a draught further reduce the ability of the pig to absorb and utilise colostrum adequately.

3 Being passively acquired, the amount of antibody in the bloodstream is finite and can be exhausted by exposure to excessive antigen. Put another way, there is a maximum amount of colostral antibody that a pig can absorb into its bloodstream. Overwhelming doses of bacteria will use it all up.

Maternal antibodies against different pathogens decline at different rates. Most have gone by 8 to 12 weeks of age.

4 The passively acquired colostral antibodies in the blood gradually waste away to about half the initial level by about 10 to 14 days (dependent on the pathogen), although they may persist at a reasonably protective level against most pathogenic antigens for 6 to 12 weeks (see Fig 3-5). The time taken to decline to ineffective levels varies depending upon the amount of colostral antibody taken in by the piglet and the type of infection or toxin against which the antibody acts. In some exceptional cases (e.g. against *Mycoplasma hyopneumoniae*, parvovirus and *Leptospira bratislava)* they may persist much longer, sometimes up to 4.5 months.

Milk

Mucosa-associated antibodies are present as IgA and secretory IgA in colostrum but at low levels relative to the other types of antibody (IgG and IgM). However, the normal milk which follows colostrum contains sufficient secretory IgA to get absorbed in the surface mucus and protect the piglet's intestines, provided the piglet sucks the sow every 1 to 2 hours. This is sometimes called lactogenic immunity.

High colostrum intake soon after birth is vital to the survival of the piglet.

Feedback

See Chapter 15 for further information

It is not surprising that mature sows provide better maternal protection to their piglets than first litter gilts. They are older and have had greater exposure to infections. The protection provided by gilts is frequently inadequate, for example, to cope with the challenge of virulent *E. coli*. Thus gilt litters tend to suffer from diarrhoea more often than sow litters. To boost the protection of their piglets it is good practice to expose gilts to farrowing or weaner house faeces at least 4 weeks prior to their anticipated farrowing. High dose levels are required however to be effective in the case of *E. coli*. Introduce these materials ideally within the acclimatisation period of the gilt and boar.

Diarrhoeic piglet faeces may be mopped up from the floor with paper towels or newspaper and put through a grinder along with the intestines of dead piglets. These should be untreated piglets in the case of *E. coli* diarrhoea because the antibiotics will neutralise the effect. The resultant emulsion is suspended in an equal quantity of water and a cupful poured on to the feed of sows and gilts which are in late pregnancy. This is done 2 or 3 times per week from about 6 weeks until 3 weeks before farrowing. Reproductive pathogens can be controlled using afterbirth or stillborn and mummified piglets fed to gilts and boars in their acclimatisation period. It is common practice to expose gilts to faeces from the floors of service pens 4 to 5 weeks before mating. Note, in some countries, including the EU, the feeding of animal protein (piglets, afterbirth etc.) to animals is banned.

Respiratory pathogens can be controlled using ropes. A rope is placed for a week in the nursery pens where the weaned pigs will play with the rope and cover it in saliva and tonsillar materials and organisms.

DECLINE OF MATERNAL ANTIBODIES AND RISE IN PIGLET ANTIBODIES

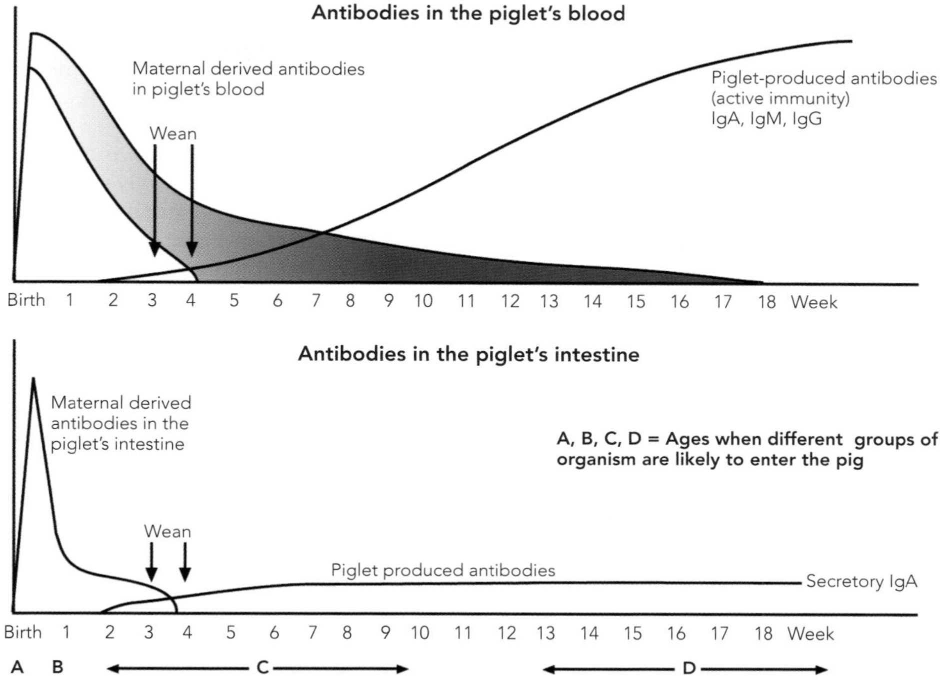

A = *E. coli, C. perfringens, S. hyicus*
B = Coccidia, *Streptococcus suis* types 1 & 7
C = Rotavirus, *streptococcus suis* types 2 & 14, App, AD/PRV, PRRSV, *M. hyopneumoniae, B. hyodysenteriae*, toxigenic *P. multicoccidia*, teschoviruses
D = Parvovirus (PPV), *M. hyosynoviae, Leptospira bratislava*

Fig 3-5

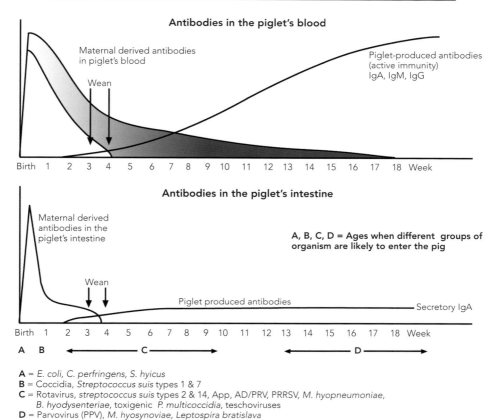

This rope is then placed in the gilt and boar pens during their acclimatisation period before breeding.

Feedback is usually an effective method of stopping an outbreak of neonatal scour in piglets, caused by viruses. There is of course a delay before the effect comes about because the pregnant sows and gilts need at least 10 days to respond fully. There is much less effect on scouring which occurs at 10 to 14 days of age compared to that which occurs within 5 days of birth.

It is not effective against piglet dysentery caused by *Clostridium perfringens* type C, nor does it protect against coccidiosis. Postweaning *E. coli* is generally not controlled by feedback.

Feedback must only be practised when the pigs are all on the same farm – the nursery together with the gilt breeding farm. Under no circumstances should materials be brought from different farms even in two-site production systems.

Feedback should be used in conjunction with vaccination programmes. But it has to be accepted that there are not vaccines for all pathogens and feedback therefore has an important role in protecting pigs from rotavirus, transmissible gastroenteritis (TGE), porcine epidemic diarrhoea (PED) and teschoviruses, for example. There are other clinical diseases in which the causal agent is not currently recognised – congenital tremor for example – and feedback is the only method of adequate control.

An effective alternative for *E. coli* diarrhoea is to grow the *E. coli* from the diarrhoeic piglets in milk and feed the milk to pregnant females. This method has been largely superseded by the introduction of commercial vaccines which are very efficient at stimulating immunity.

Antiserum

Antibodies can also be acquired passively by injecting antiserum. They will protect the pig for 7 to 10 days against the specific infection that the antiserum was prepared against. This used to be a common practice in pig medicine before antibacterial medicines and effective vaccines were widely available. This is rarely done now.

The Role of Passive Immunity in the Development of Actively Acquired Immunity

The newborn piglet is exposed to a vast array of antigens from the moment it is born. Its immune system is naive and immature but competent to respond and produce an active immunity. The maternally-derived immunity has to provide sufficient protection for long enough while the piglet gradually develops its own active immunity. This is illustrated in Fig 3-5. In the wild, a sow continues to suckle her offspring for several months. Weaning is gradual with plenty of time and opportunity for a wide range of antigenic stimulation. However, in pig production weaning is abrupt and at an unnaturally young age (e.g. 3 to 4 weeks). After weaning, the circulating humoral antibodies persist and continue to provide an effective protection against invasion of the pig's body. However at weaning, milk, the source of mucosa-associated antibodies is suddenly cut off. The antibodies present in the mucus decline within a day or so.

Serological Tests

Humoral antibodies that have been stimulated by infection can be used in blood tests in the diagnostic laboratory to diagnose what the infection is or to screen a herd for the presence or absence of an infection.

Numerous serological tests are available e.g. agglutination tests, conglutination tests, complement fixation tests, fluorescent antibody tests, the immunoperoxidase monolayer assay, ELISA and PCR. Different tests are useful for different infections. The laboratory and veterinarian have to decide which one is best in each case.

When carrying out such tests, the laboratory can either use commercially available antiserum to test against organisms that have been isolated, or they can use double serum samples (two samples taken over a period of time) from the sick pigs to test for specific antibodies using known antigens. Why double serum samples? Because if just one sample is taken and it is positive you do not know whether the antibodies are carried over from an old infection, which has long since gone, or whether they are associated with your current disease problem. The first sample is called "acute"; the second is called "convalescent" and is taken 7 to 14 days later. If the antibodies are due to the current infection, they will be rising from zero or very low to high. If they remain level or fall it is probably a past infection. When antibody levels rise it is called a "rising titre", titre being the term used to express the concentration of specific antibody in a given serum sample. This is measured as the amount by which the serum has to be diluted before the test becomes negative (i.e. the antibodies have been diluted to a non-detectable level). So if the titre rises from, say, 1:10 to 1:100, it means that the laboratory had to dilute the second serum sample by 10 times more, indicating infection. Most tests have a lower threshold titre below which the test is deemed negative. They may also have a narrow middle band which is deemed to be suspicious and a higher titre above which the test is deemed positive.

Vaccination

Vaccines contain antigens from viruses, bacteria, bacterial toxins, or parasites. They are given to pigs, usually by injection, to stimulate an immune response which will protect the pigs against later natural infection with the organism from which the vaccine was derived. Most stimulate both a humoral response and a cell-mediated response.

Vaccines can be live, containing living organisms which will multiply in the pig, or inactivated, containing only killed organisms which will not multiply in the pig.

In live vaccines the organism has usually been attenuated (i.e. its virulence has been reduced) so that although it will multiply in the pig; it will not normally cause any disease. Examples are the PRRSV vaccine (although some may cause mild reactions), Aujeszky's disease (pseudorabies) vaccines and CSF vaccines. Live attenuated vaccines have the advantage that because they multiply in the pig they give a bigger antigenic stimulus resulting in stronger, longer-lasting immunity. They have the disadvantage that they may die in wrong storage conditions (e.g. heat) or during dosing (e.g. by exposure to antibiotics, antiseptics or disinfectants) and are then useless. It is also important that they are stable and not able to return to full virulence even when exposed to natural viruses in the pig.

Inactivated (dead) vaccines may contain whole organisms, antigenic parts of organisms or antigens which have been synthesised chemically, for example PCV2 vaccine. An example of a commonly used whole organism vaccine is the erysipelas vaccine (in North America such vaccines are often called bacterins).

For example: erysipelas dead bacterial vaccine

1. This is made by growing the erysipelas bacteria in a liquid nutrient broth (several strains may be used).
2. The bacteria are then killed.
3. A liquid or adjuvant is added to the bacterial suspension.
4. This produces the vaccine.
5. A predetermined number of bacteria are injected into the pig. The first dose is usually 2 ml.
6. A second dose is required to complete the immune response, and is usually given 14 to 24 days after the first.
7. 7 days after the second dose, the pig is protected.

The immunity produced by inactivated vaccines can be enhanced by adding substances or adjuvants such as aluminium hydroxide or certain types of oil. You should take care, however, if you use vaccines with oily adjuvants because they can cause serious local reactions if you accidentally inject yourself, e.g. your hand.

Inactivated vaccines may also contain toxins which have been modified so that they still stimulate an immune response but are no longer toxic to the animal. Toxins which have been modified in this way are called toxoids. The classic vaccine of this type is the tetanus toxoid which is used commonly in horses but rarely in pigs. In pigs, some of the *E. coli* vaccines against piglet diarrhoea and the clostridial vaccines against piglet dysentery also contain toxoids.

Live vaccines may have the gene for the actual production of disease deleted (gene-deleted vaccines). The vaccine response can then be differentiated from the actual disease and thus carrier animals removed. An example would be Aujeszky's disease (pseudorabies) vaccine. This feature is very useful in eradication programmes.

Remember, however, vaccination is never 100 % effective.

FIG 3-6a: MAJOR VIRAL DISEASES THAT MAY BE CONTROLLED BY VACCINATION *

Aujeszky's disease/pseudorabies virus (AD/PRV)
Classical swine fever (CSF)
Foot-and-mouth disease (FMD)
Japanese encephalitis virus (JEV)
Porcine circovirus 2 (PCV2)
Porcine epidemic diarrhoea (PED)
Porcine parvovirus (PPV)
Porcine reproductive and respiratory syndrome virus (PRRSV)
Rotavirus
Swine influenza virus (SIV)
Transmissible gastroenteritis (TGE)

FIG 3-6b: BACTERIAL DISEASES THAT MAY BE CONTROLLED BY VACCINATION *

Actinobacillus pleuropneumonia (App)
Clostridial diseases
E. coli diarrhoea
Erysipelas
Glässer's disease
Greasy pig disease
Ileitis
Leptospirosis
Mycoplasma (enzootic) pneumonia
Pasteurellosis
Progressive atrophic rhinitis (PAR)
Salmonellosis
Streptococcal infections

* The availability of vaccines varies from country to country.

Autogenous vaccines

Autogenous vaccines are bacterial vaccines that are manufactured from the specific pathogenic bacteria isolated from the diseased pig. They are usually made under a licence for use only on that farm. You should consult with your veterinarian. They can be useful when serious disease outbreaks occur and standard commercial vaccines are not available.

Such vaccines could be made from most bacteria, including:-

- *Actinobacillus pleuropneumoniae* (App).
- *Escherichia coli* (*E. coli*).
- *Haemophilus parasuis* (Glässer's disease).
- *Pasteurella.*
- Salmonella.
- *Streptococcus suis*
- *Staphylococcus hyicus* (Greasy pig disease).

One drawback to vaccinating a herd is that you cannot then use blood tests to check whether the organism is present in the herd or not. All the pigs will test positive, which has obvious implications for an eradication programme based on blood tests, for example the eradication of swine fever or Aujeszky's disease (pseudorabies). To get over this, gene-deleted vaccines have been developed. A part of the organism's gene which codes for an antigen has been removed so that when the organism multiplies in the pig it does not stimulate antibodies against that antigen. Special blood tests can then distinguish between the array of disease antibodies and those stimulated by the vaccine. A new generation of such gene-manipulated vaccines, and possibly also synthetic polypeptide vaccines, can be anticipated.

Autogenous vaccines are those prepared with infectious pathogens from the herd which is to be vaccinated. The causal organisms have to be isolated, grown up, killed, and made into a safe vaccine form.

Vaccine usage

Fig 3-6a and b lists the pig diseases for which vaccines are available. This list is not exhaustive and some vaccines will be available in some countries and not in others. However they are used in most countries both to protect against disease and to assist in eradication programmes.

In the European Union, vaccination against CSF has been stopped in a programme aimed at stamping the disease out. Vaccination against FMD has also been stopped for a similar reason. AD/PRV is widespread everywhere in the EU except in the UK, Denmark and Ireland. With the exception of these countries, vaccination is widely practised. A blanket vaccination regime for all herds is being applied in some countries such as the Netherlands in an attempt to build up a national herd immunity resulting in the eradication of the virus.

North America is free from FMD, CSF and AD/PRV (commercially), so vaccination is not practised. Elsewhere in the world, the situation regarding these three diseases varies, so vaccination policies also vary.

The effectiveness of vaccines

This varies, because of the need to stimulate mucosal immunity locally. As mentioned earlier, vaccines given by injection against respiratory and intestinal disease are generally not as effective as those against systemic (or generalised) diseases. An exception to this is the vaccine for mycoplasma (enzootic) pneumonia (*M. hyopneumoniae*) because it stimulates cell-mediated immunity. If, however, the vacciens are fed or sprayed into the upper respiratory tract they may produce a stronger local immunity. The vaccine against piglet dysentery is a toxoid and if given routinely to sows in adequate doses is usually reasonably effective in providing passive protection via the colostrum.

Sometimes vaccines do not work particularly well on a farm and in such cases the following possibilities need to be considered:

- The vaccine was contaminated.
- The vaccine was not capable of producing the required immunity.
- The pig was already incubating the disease, or sick, when it was vaccinated.
- The vaccine had been incorrectly stored. High temperatures reduce the effectiveness. (Always keep vaccines in a refrigerator but do not freeze).
- The vaccine had been exposed to sunlight.
- The vaccine had gone out of date.
- The needle and syringe were dirty or faulty.
- Chemical sterilisation destroyed the vaccine.
- The animal's vaccination was inadvertently missed. This is particularly common with parvovirus vaccination in the gilt.
- Vaccine response was poor because there was maternal antibody present.
- The vaccine was deposited in fat and was not absorbed. Faulty injection techniques may have been used – needle length is a particular issue.
- With live bacterial vaccines antibodies were present in the pig.

The management of vaccines

- Check the expiry date.
- Store in a fridge running between 2 to 8 °C (35 to 46 °F).
- Monitor the temperature daily with a max/min thermometer. Freezing and heating destroys vaccines.
- Don't overstock the fridge.
- Don't store food in the fridge.
- Follow the instructions.
- Ideally use a fresh needle for each pig, but change at least every 5 pigs.
- Do not mix vaccines or medicines.
- Dispose of needles in a sharps box.
- Clean out syringes immediately after use.
- Only use vaccines licensed in your country.
- Clean bottle tops before and after use.

Immunosuppression

There are many factors that suppress both innate and acquired immunity levels, but in pigs infectious agents are the most common ones. These include:

- *Mycoplasma hyopneumoniae* – the cause of mycoplasma (enzootic) pneumonia.
- *Pasteurella multocida*.
- Aujeszky's disease/pseudorabies virus (AD/PRV).
- Swine influenza virus (SIV).
- African swine fever (ASF) virus.
- Porcine reproductive and respiratory syndrome virus (PRRSV).
- Porcine circovirus 2 (PCV2).

These agents may destroy the macrophages or lymphocytes, or delay or reduce the efficiency of the immune response, or damage the innate defences. Note that some vaccines may also cause problems with immunosuppression; CSF vaccines for example.

Weaning time is also a period of immunosuppression by withdrawal of the sow's milk that contains the protective mechanisms of IgA. When pigs are mixed, moved or stressed the plasma cortisol levels rise with an immunosuppressive effect, and the demand for vitamin E rises.

Maximising Health through Medicinal Control

See Chapter 4 for further information

Prevention

Specific details of this are discussed elsewhere under the relevant conditions/diseases. Clinical disease can be prevented or controlled effectively if treatment is applied during the incubation period – the period from the time of exposure to the organism to the onset of clinical disease. This is sometimes called prophylactic (i.e. preventative) or strategic medication. This medication must be given at the correct time which is determined by the pattern of the disease on the farm and by trial and error.

Eradicating Pathogens

Some pathogens can be eradicated by a combination of medication, vaccination and management or even by management procedures alone. Fig 3-7 illustrates

FIG 3-7: PATHOGENS (DISEASES) THAT CAN BE ERADICATED BY MEDICATION, VACCINATION AND/OR MANAGEMENT PROCEDURES

Actinobacillus pleuropneumoniae (App)
Aujeszky's disease virus (AD) – pseudorabies virus (PRV)
Brachyspira hyodysenteriae (SD)
Foot-and-mouth disease (FMD)
Haematopinus suis (lice)
Mycoplasma hyopneumoniae (mycoplasma (enzootic) pneumonia)
Porcine epidemic diarrhoea (PED)
Porcine reproductive and respiratory syndrome virus (PRRSV)
Postweaning multisystemic wasting syndrome (PMWS)
Salmonella choleraesuis (salmonellosis)
Sarcoptes scabiei var. suis (mange)
Toxigenic *Pasteurella multocida* (PAR)
Transmissible gastroenteritis virus (TGE)

pathogens that have been successfully removed from combined breeding feeding herds.

These proceedures are described under the specific conditions / diseases.

Recognising and Understanding Health on the Farm

Early recognition is the first priority for managing health on the farm. It is the responsibility of the stockperson using the senses of sight, sound, touch and smell to detect the abnormal animal and to differentiate it from the normal animals. Every day, a clinical examination of all pigs should be carried out. On a 100-sow farm, this could take up to half an hour per day, and on a large farm it becomes a major daily task, but can be split between department levels. How many managers, in organising their farms, allow such a time period for this function?

The Use of Sight

- Pigs generally like other pigs. Pay attention to any animal which seems to be separated from the main group.
- Inappetence is obvious where an animal is housed and fed as an individual. However in group-housed animals this is not easy to detect. The failure to eat, or a drop in feed intake in a pen of apparently normal pigs, must immediately arouse suspicions. The initial check should be for lack of water, which is usually the most important sudden cause of inappetence involving all pigs in a group. If the water supply is normal, look for other signs of disease.
- Listlessness, dull appearance or change of skin colour of the pig will be quickly detected by the good stockperson as early signs of illness. A very white pig may be an indication of a gastric (stomach) ulcer. A pig covered in faeces is likely to be hot and trying to cool down.
- Shivering and raised body hair is an important feature of disease and is one of the very early signs of streptococcal meningitis or joint infections in the sucking pig. Look for this sign next time you examine each individual in the litter. A pig laid on its belly and shivering with its hair on end compared to the rest of the group is either scoured or lame from a generalised septicaemia (bacteria in the bloodstream).

Examine all pigs carefully every day.

- Loss of bodyweight is a first indication of inappetence or dehydration due to diarrhoea or pneumonia.
- Discharges from the nose or eyes indicate an upper respiratory tract infection. Excess salivation from the mouth indicates an exotic disease such as vesicular disease. In sows, a discharge from the vulva could indicate vaginitis, cystitis, pyelonephritis or endometritis.
- Faecal changes can indicate a wide range of diseases but sloppy faeces can also be quite normal. Look for signs of mucus or blood indicative of swine dysentery, salmonella infections, gastric ulceration or proliferative haemorrhagic enteropathy. Constipation may be an important sign in the development of udder oedema and agalactia at farrowing.
- Vomiting can be a sign of diseases such as transmissible gastroenteritis, or in individual pigs it may indicate gastric ulceration. In the sucking pig, gastroenteritis associated with *E. coli* infections is often seen. Injections with long-acting penicillin may also cause pigs to vomit.
- Skin changes help in identifying diseases, typified by acute or chronic lesions of mange and lice. Erysipelas may not be evident by sight, but running the flat of the hand over the skin will indicate tell-tale lesions of raised areas. A blueing of the extremities could indicate acute viral infections, acute bacterial septicaemia or a toxic state, as seen in SIV, PRRSV infections or acute mastitis and metritis. Acute pneumonia or pneumonia associated with heart sac (pericardium) infection can give a similar picture.
- Scratch marks and skin lesions – fighting and increased aggression within a pen may be recorded by increased scratches and marks on the body. The position of the marks may also help diagnostically. During oestrus there should be scratches over the shoulder. Weaners' fighting scratches will be around the head, ears and neck. Finishing pigs will mark

each other's flanks and tail region.

- Respiration rates – if any of the above changes have been identified, cast your eye across the pen of pigs and compare the respiratory rates of both the normal and the suspect animals. Assess whether the breathing is a deep chest movement, due to consolidation of the lungs and a shortage of oxygen, or very shallow abdominal breathing indicative of pleurisy and pain. If the breathing rate is over 40 per minute the pig is likely to be over-hot.
- Dead pig – examine the posture and other features of the dead pig – scratch marks on the floor of a pig which died with meningitis, for example. Circumstances surrounding the death of a pig are an important observation, especially when backed up by post-mortem examination. The timing and place where pigs die in a herd relative to clinical observations can often help in identifying and understanding a problem.

Always examine the environment in a house at pig level.

The Use of Smell

The odour of a dead pig is one that we have all experienced from time to time. However, odours also occur with scour, bad feed or infected tissues. The smell of piglet scour on outdoor sows can help detect affected litters. The quality of the air through the sense of smell will highlight poor ventilation rates, high levels of gases, and high or low humidity. What is uncomfortable for us is likely to be the same for the pig.

The Use of Touch

It is essential to handle a compromised pig to detect changes in skin temperatures or any abnormal fluids or lumps on the skin. The limbs should always be palpated in cases of lameness for possible fractures or swellings in the joints. In the newly farrowed sow, always palpate the udder to detect any early changes of agalactia or mastitis.

Remember, a happy, healthy pig will usually approach and make contact with you.

The Use of Sound

Be extremely wary if there are no pig noises when you enter a building. A disaster could have occurred due to electrocution, suffocation, or high levels of toxic gases such as carbon monoxide or hydrogen sulphide. Remember toxic concentrations of hydrogen sulphide have no smell and kill quickly.

Pigs with meningitis may have a high-pitched squeal associated with laryngeal oedema. Acute SIV or App outbreaks may result in very quiet pigs when you enter their house and pen.

Changes in Behaviour

The pig is a social animal and in a healthy condition remains part of a group. In disease, however, it tends to rest on its own or often be rejected by the other pigs, even to the extent of being attacked. Altered lying patterns in a pen must always be regarded with suspicion. Conversely, where a number of pigs are ill or the environment is inadequate, huddling is common. The reluctance of pigs to rise or show an interest in the observer must always warrant a more detailed examination. Watch pigs using the water or feeders: watch how the pig moves, to ensure it is not in pain and lame.

Observation of the Group

Daily, regular time should be set aside for the examination of all pigs. Allow at least 5 to 10 seconds to observe each pen of pigs. The environment of the house must also be assessed by noting the following:

- Temperature.
- Humidity.
- Ventilation – hot or cold areas and draughts.
- Smell.
- Pig behaviour.
- Appetite.
- Reaction to humans.
- Ammonia levels as experienced through breathing and the effect on eyes.
- Abnormal changes in slurry and bedding.

Examination of the group

It is beyond the scope of this section to explain in detail all the aspects of clinical examination of an animal. However, there are major areas which merit discussion.

Key points to look for, indicating that there may be a problem within the group are:

- Animals that are separate from the group.
- Feed that is not eaten.
- Variation in the size/condition score of the animals within the group.
- Hairy pigs, particularly weaners and piglets, indicate illness.
- The consistency of the faeces.

- Coughing or sneezing.
- Signs of diarrhoea.
- Signs of lameness.
- Signs of respiratory distress.
- Any other abnormal clinical sign.
- Reduction in water use (needs to be monitored).
- Change in drinking pattern (needs to be monitored).

It is important to note that the ability to detect any abnormal behaviour depends upon the observer, stockperson or veterinarian being aware of what is normal. Examine the individual animals for:

- Skin changes.
- Presence of lumps.
- Configuration changes from normal.
- Change in expected growth.
- Vices in the animals.
- Prolapses.
- Changes in breathing.
- Changes in behaviour and locomotion.
- Discharges – ocular, nasal, aural, anal, vaginal, preputial.
- Refusal to eat.
- Vomiting.

Manage your farm by walking around it not by staying in the office.

Clinical Examination of the Herd

The daily clinical appraisal of the various sections of the herd should be carried out using a series of checklists held on a clipboard. The observations should combine all the senses described. Suggested checklists are outlined in Fig 3-8 a-e as examples. Use these as a basis for developing specific checklists for your farm.

FIG 3-8a: A CHECKLIST FOR THE FARROWING AREA: INDOOR PRODUCTION

	Comments
Batch farrowing target reached.	
Empty farrowing places.	
Sows' body condition at entry.	
Sow condition at weaning.	
Feed intakes.	
Water availability.	
Are any sows not in pig?	
Late or early farrowings.	
Farrowing problems.	
Mastitis or metritis.	
Quality of piglets at birth.	
Sow mortality, prolapses etc.	
Sow health problems.	
Sow feed intake at Day 18 of lactation.	
Litter sizes, still-births, mummified piglets.	
Fostering protocols.	
Litter weights.	
Weaning number reached.	
Weaning weights.	
Congenital conditions.	
Scour levels. Treatments.	
Room temps: Farrowing, Week 1, Weeks 2 – 4.	
Cooling system working correctly (if available).	
Quality of the environment, hygiene, dry floors,	
Creep temperatures.	
Health of piglets, evenness of growth.	
Iron injections completed correctly.	
Teeth clipping (if practised).	
Tail docking (if practised).	
Castration (if practised).	
Respiratory disease, rhinitis, pneumonia.	
Parasites. Fly control.	
Rodent control.	
Medication procedures: mange, worming, vaccination correct.	
Are other procedures being completed properly?	
All-in/all-out procedures correct.	
Use of footbaths.	
Building maintenance.	
Your additions.	

FIG 3-8b: A CHECKLIST FOR THE SERVICE/MATING AREA: INDOOR PRODUCTION

Item	Comments
Batch breeding targets being met.	
Condition of sows at weaning.	
Hair growth.	
Feed levels from weaning to service.	
Boar/sow ratio.	
Boar usage.	
Service procedure. Supervision.	
AI storage management: temperature and hygiene.	
AI hygiene.	
Fertility levels.	
Assess the litter size relative to service procedures.	
The environment at weaning: floors, drainage, temperature.	
Cooling systems in place and working correctly.	
Weaning to service intervals.	
Discharges, mastitis, diseases.	
Trauma, stress.	
Toys present.	
Lameness.	
Boar condition, health, service performance.	
Rodent control.	
Your additions.	

FIG 3-8c: A CHECKLIST FOR THE DRY/GESTATION AREA: INDOOR PRODUCTION

Item	Comments
Assess the environment – temperatures, humidity.	
Cooling systems in place and working correctly.	
Housing problems.	
Welfare problems.	
General sow contentment or restlessness.	
Toys present.	
Body condition.	
Feed levels, inappetence.	
Chronic mastitis.	
Vulval discharges.	
Sows not in pig.	
Sows repeating.	
Pregnancy diagnosis methods.	
Abortions.	
Oestrus abnormalities.	
Condition of the faeces.	
Sow feet conditions.	
Culls. Reasons for and causes.	
Mortality levels and causes.	
Prolapses.	
Vaccination programmes.	
Rodent control.	
Your additions.	

FIG 3-8d: A CHECKLIST FOR THE PIGLET WEANING AREA: INDOOR PRODUCTION

Item	Comments
Batch weaning number targets reached.	
Batch weaning weight target reached.	
All-in/all-out practised.	
Age and weight at weaning.	
Evenness of growth.	
Water availability.	
Feed: access, type and quality.	
Stocking density.	
Lying patterns of the pigs.	
Environmental temperature.	
Growth of the pig in the first 10 days postweaning.	
Respiratory diseases, rhinitis, pneumonia.	
Enteric diseases.	
Condition of faeces.	
Environment – ventilation, humidity.	
Insulation.	
Vermin control.	
Fly control.	
Rodent control.	
Skin conditions, mange, greasy pig disease.	
Lameness.	
Welfare aspects.	
Toys.	
Building maintenance.	
Your additions.	

FIG 3-8e: A CHECKLIST FOR THE GROWING AND FINISHING AREAS: INDOOR PRODUCTION

Item	Comments
Batch finishing targets reached (numbers, weights).	
All-in/all-out practised.	
Assess weight for age.	
Stocking densities.	
Evenness of growth.	
Nutrition and growth in different houses.	
Application of feed and types.	
Effect of movement of pigs.	
Type of flooring.	
Environment in different houses.	
Quality of environment: insulation, temperatures, humidity, draughts.	
Temperature fluctuations.	
Lying patterns of the pigs.	
Toys present.	
Feed conversion efficiency.	
Daily liveweight gain.	
Appearance of the pigs' skin – mange, hair growth.	
Appearance of faeces.	
Respiratory diseases.	
Enteric diseases.	
Other diseases, mortality and prolapses.	
Culls and compromised pig care.	
Parasites.	
Rodent control.	
Your additions.	

Assessing Health, Management and Disease in Outdoor Production

Successful outdoor production is dependent upon the interaction between breed, the variables of the climate, soil type and management. Routine clinical examinations and checks need to be made as detailed in Fig 3-9a-e.

If these lists are used daily, regular disciplines will be established in each area of the farm that will raise efficiency and awareness, and identify compromised/sick pigs early.

FIG 3-9a: DISEASES AND CONDITIONS THAT MAY BE EXPERIENCED: OUTDOOR PRODUCTION

Item	Comments
Clostridial infections, necrotic enteritis.	
Coccidiosis.	
Endometritis – vulval discharges.	
Internal parasites.	
Lameness:	
– Bush foot.	
– Erysipelas.	
– Mycoplasma arthritis.	
– OCD.	
– Physical damage.	
Leptospirosis.	
Lice.	
Mange.	
Mastitis.	
Parvovirus.	
PMWS – PCVAD.	
PRRSV.	
Summer infertility, abortions, embryo reabsorption.	
Sunburn/heat stroke.	
TGE/PED.	
Variable litter size.	

Refer to relevant chapters for information relating to specific diseases.

FIG 3-9b: A CHECKLIST FOR THE FARROWING AREA: OUTDOOR PRODUCTION

Item	Comments
Huts or arks	
Air flow, draughts, condensation.	
Bedding type, quality, amount, dryness.	
Door flaps.	
Environment.	
Fenders.	
Insulation.	
Level ground.	
Nose ring to prevent digging.	
Provision of solid floors.	
Siting against prevailing wind.	
Size for breed of sow.	
Soil type/nesting.	
Wallows/shade.	
Feeding/nutrition	
Ad lib/hoppers.	
Amount fed sow/year.	
Feed levels.	
Only dry feed.	
Ration composition.	
Wastage.	
Water availability.	
Sows	
Accuracy of recording.	
Behavioural problems.	
Body condition, body score.	
Born alive, dead, reared.	
Efficiency of breeding female.	
Gilt mothering qualities.	
Lactation ration used.	
Management efficiency.	
Mastitis, agalactia.	
Numbers born alive.	
Treatment required.	
Variability of body condition.	
Piglets	
Born dead.	
Effects of bedding on viability.	
Losses due to foxes, crows.	
Iron injections.	
Management at farrowing.	
Mortality %.	
Quality of piglets at weaning.	
Savaging.	
Scour.	
Stolen piglets.	
Teeth clipping, tail docking, castration.	
Treatments required.	

FIG 3-9c: A CHECKLIST FOR THE GESTATION/DRY PERIOD AND WEANING AREAS: OUTDOOR PRODUCTION	
Item	**Comments**
Huts or kennels	
Effects of weather.	
Environment.	
Ground conditions/bedding.	
Lying patterns.	
Siting – position, draughts.	
Soil type.	
Stocking density.	
Feeding/nutrition	
Feed intake – Service to 21 days. – Last 3 weeks of pregnancy.	
Feed used/sow/year.	
Type of ration – composition.	
Wastage.	
Water availability.	
Sows	
Barren returns.	
Body condition.	
Diseases evident.	
Fertility records.	
Grass types/mycotoxins.	
Lameness/mastitis.	
Management quality.	
Parity spread.	
Sunburn/heat stroke.	
Wallows/shades.	
Efficiency of identification tags etc.	

FIG 3-9c: A CHECKLIST FOR THE GESTATION/DRY PERIOD AND WEANING AREAS: OUTDOOR PRODUCTION CONT.	
Item	**Comments**
Boars, service paddocks	
Age.	
Boar usage.	
Boar's condition.	
Efficiency of boar type.	
Feeding/nutrition.	
Group sizes.	
Lameness – disease.	
Libido.	
Management and mating.	
Penis problems.	
Use of AI.	
AI management.	
AI storage.	
Wallows/shades.	
Gilts	
Acclimatisation.	
Feedback.	
Age, weight, oestrus.	
Anoestrus.	
Fertility.	
Flushing.	
Introduction to cobs.	
Lameness.	
Litter size.	
Management quality.	
Nutrition/back fat measurements.	
Vaccination.	
Weaners	
Bedding, dryness, ventilation, draughts.	
Environment in huts.	
Growth, daily liveweight gain.	
Health and disease at weaning.	
Management quality.	
Medication.	
Mortality.	
Nutrition and feeding.	
Respiratory, enteric diseases.	
Spacing of huts.	
Use one area per week.	
Water supply.	
Weaning weight body condition.	

FIG 3-9d: PREVENTATIVE MEDICATIONS: OUTDOOR PRODUCTION	
Item	Comments
Vaccinations to consider	
Progressive atrophic rhinitis.	
Clostridia.	
E. coli.	
Mycoplasma (enzootic) pneumonia.	
Erysipelas.	
Leptospira.	
Parvovirus.	
Other medications to consider	
In-feed medication.	
Water medication.	
Anthelmintics.	

FIG 3-9e: BIOSECURITY AND MANAGEMENT AUDIT: INDOOR AND OUTDOOR PRODUCTION	
Item	Comments
Bedding source, quality.	
Bird contamination.	
Boots, coveralls.	
Casualty pens.	
Casualty stock disposal.	
Control of substances hazardous to health.	
Dead stock disposal.	
Destruction of pigs.	
Feed storage.	
First aid box.	
Fly control.	
Foot dips.	
Health and safety.	
Loading ramp – disinfectant used.	
Medications:	
Cleanliness of equipment.	
Medication in feed bins.	
Records of use.	
Refrigerator.	
Storage of medicines.	
Use of syringes, needles.	
Withdrawal periods.	
Movement records.	
Perimeter fence.	
Pig freedom times.	
Rodent control.	
Shower facilities.	
Signing-in book.	
Siting of feed bins.	
Stocking rates.	
Transport and pig movement.	

The Management and Treatment of the Compromised/Sick Pig

Once a compromised/sick pig has been recognised, the following sequence of events is suggested:

- Identify the animal by spray. If the pig is over 60 kg (130 lbs), mark permanently by a tag so no mistakes are made with medicine withdrawal periods.
- Carefully examine the pig and its environment.
- What do you think is wrong with it? (If in doubt, seek veterinary advice).
- Take the rectal temperature.
- Is it necessary to treat the condition?
- What medicine has been recommended for treatment by the veterinarian?
- What nursing/welfare provisions are there?
- Should the pig be left in the pen?
- What method of medicine administration should be used?
- What dose level should be given and how often should the medicine be given?
- Determine the method of administration, the site of injection, syringe and needle type.
- Administer pain relief when required.
- Assess the response daily.
- Normal temperature is 38.6 °C to 39.5 °C (101.5 to 102.5 °F).
- Normal respiratory rate at 20 °C (70 °F) is 25 to 30 breaths per minute.

Having recognised the compromised/sick pig and the cause of the problem, a decision must be made as to whether to treat it in the pen or move it to a specialised "hospital pen". Treatment consists of 3 very important aspects: good nursing, good nutrition and necessary medicines. It is in the first of these that there is often a lack of awareness. Any compromised pig that cannot fend for itself should immediately be moved into a hospital pen.

Compromised pigs should only be left in the pen if they are still able to move around freely, have an uninhibited access to the drinker and are only inappetent for a maximum of 24 hours.

On every pig farm there should be 6 to 8 separate hospital pens per 100 sows, half for sick pigs and half for compromised pigs. At least 2 pens should be available for weaners, 2 for growers, 2 for feeder pigs and 2 for sows.

Your farm should have hospital pens.

Each pen should satisfy the following criteria:

- The floor should be solid and well-drained.
- It should be deep-bedded on straw, shavings or other suitable material.
- It should be well-lit so that examinations are easily carried out.
- There should be easy access to food and water, preferably by a water bowl.
- Where nipple drinkers are used, ensure height is correct for the age/size of pig.
- Provide ad lib feeders. Avoid bulk feeders.
- One person on the farm should be appointed responsible for all compromised pigs.
- There should be a maximum of 6 pigs per pen with a floor area of 1 m^2 (11 ft^2) per pig for pigs up to 100 kg (220 lbs) and 3 m^2 (32 ft^2) for bigger pigs.
- Adequate temperatures must be maintained in these pens and invariably this will involve either the provision of extra heating or the siting of the pens in a very warm building. To achieve this in weaners and young growing pigs, it is necessary to provide an insulated micro-environment within the building, consisting of an insulated floor, sides and roof with an infra-red bulb or alternative heat source controlled by a thermostat. Pigs will respond much more quickly if they are in a warm, well-bedded environment. On one regularly visited farm, all sick/compromised pigs, no matter how mild the sickness, were always moved into a series of 30 small hospital pens. The owner often related how many of these pigs reached slaughter weight days ahead of their healthy contemporaries.

Sick pigs should be moved to a hospital pen which should be warm and comfortable.

The compromised pig can be managed in 1 of 4 ways: (Fig 3-10 and Fig 3-11).

1. **Sell to a slaughter outlet**
   This assumes that the pig is destined for sale through normal outlets and would include a pig that has been in a hospital pen and has recovered, or one that is fit to travel, has no condition likely to render the carcass unfit for human consumption and has no medicine residues.
2. **Treat the pig**
   Treatment would be given on the assumption that the pig will respond and ultimately be fit for normal slaughter. Having made this decision, a careful

FIG 3-10: MANAGING THE COMPROMISED SOW	
Conditions	**Action**
Prolapse of the uterus.	Destroy.
Prolapse of the vagina.	Replace and retain by suture.
Prolapse of the rectum.	Replace, suture and casualty slaughter.
Rectal stricture.	Destroy as soon as noticed.
Open wounds.	Treat.
Cuts and wounds: Mild. Severe.	Treat. Sell when healed. Destroy.
Shoulder sores and ulcerated hocks.	Treat and move to a bedded area. Sell when healed.
Lameness: Off back legs. Acutely lame. Severely swollen infected joints. Not severe. Lame, no obvious cause but weight on all legs. No obvious wounds and no temperature.	 Destroy. Treat and assess. Treat and assess or destroy. Treat and assess. Casualty slaughter or treat. Treat.
Emaciated condition.	Destroy.
Dystocia (difficult farrowing).	Treat then review and retain only if sow expels pigs and recovers. N.B. If live pigs are present, consider on-farm hysterectomy. Or destroy. N.B. Never send a sow with retained piglets for slaughter; in almost all countries it will be condemned.

FIG 3-11: MANAGING COMPROMISED GROWING AND FINISHING PIGS	
Condition	**Action**
Lameness: Totally off the back legs. Acutely lame with swollen infected joints. Severe. Lame with no obvious cause, no open wounds and no temperature. Severely damaged claw. Recently broken legs. Severe sprains and dislocations.	 Destroy. Treat and assess. Destroy. Casualty slaughter or treat. Casualty slaughter or treat. Destroy or casualty slaughter on the farm. Treat and assess.
Injuries: Tail-bitten. Tail swollen, abscessed. Tail treated/recovered. Swelling without open wound. Severe traumatic injuries e.g. recent open wound. Ear-bitten, flank-bitten. Other recent wounds.	 Treat. Destroy. Sell. Treat. Destroy or casualty slaughter if fit to travel, or treat. Treat, move to hospital pen. It is essential that these cases are isolated immediately and during treatment.
Rectal prolapse.	Replace and suture, then sell for normal slaughter ASAP.
Severe rectal prolapse or rectal stricture.	Replace, then immediately slaughter or destroy.
Ruptures: Small. Large. Large with ulcerated skin.	 Sell for normal slaughter. Sell for normal slaughter at lowest possible weight. Casualty transport conditions should apply. Destroy – if unfit to travel, move to straw pen. Sell as normal pig when recovered or casualty slaughter if skin lesions still present at slaughter weight.
Runts and ailing pigs: Mild. Severe.	 Treat. Destroy.

review should be made of the progress on a day-by-day basis and if the pig is not responding, either the treatment should be changed, further advice sought, or it should be destroyed.

3 **Casualty slaughter**
The animal should be capable of being transported to the nearest available slaughter house without compromising its welfare. A veterinary slaughter certificate or owner declaration may be required, depending on the welfare rules and regulations. Such animals might be lame or with fresh rectal or vaginal prolapses, or may be slow-growing pigs. On-farm slaughter of the pig may be necessary for welfare reasons, for example a broken leg or acute severe lameness. This should only be carried out if the carcass is likely to be fit for human consumption.

4 **Destroy the pig (the 7 and 14 day rule)**
There should be facilities on the farm for humane destruction of all ages of pigs. (See Chapter 15: Slaughter). If the compromised pig has shown no significant response within 7 days, it should be destroyed. If the compromised pig responds slightly but has not recovered within 14 days it should be destroyed (the 7 and 14 day rule). However, note that as soon as you decide that the pig is unfit for human consumption, it should be destroyed – do not wait. In particular do not wait until the end of the month to keep mortality figures in check!

The Design of the Hospital Pen

The hospital pen should be the most comfortable, warm area on the farm, with easy access to feed and water, because the environmental requirements of the sick pig are exacting. For example the newly weaned pig affected with malabsorption will have lost most of its body fat and could require an effective temperature of 30 °C (95 °F). Hospital pens should cater for 3 groups of pigs: those in the immediate postweaned period, those in the growing and finishing period, and sows. In the weaning and the growing period there should be 2 types, one to handle the acutely ill pigs and the second to hold the recovered pigs.

This is similar to the straw-based weaner accommodation shown in Fig 3-49 later in the chapter, only smaller. It consists of an inner, well-heated chamber with strip curtains separated from a cooler outer section. In some designs the floor is heated as well and this provides an excellent environment. The complete pen is deep-bedded in at least 300 mm (12 in) of straw or other bedding so that the compromised/sick pig can select its required environment. This is vital for the recovery of those pigs who have lost body fat. The walls and roof of the accommodation should be insulated with 100 to 150 mm (4 to 6 inches) of foam or fibre glass depending on the temperature of the external environment. Provision should be made for a separate water tank leading to a water bowl so that medication can be applied as necessary. Feed should be readily accessible by open dishes in the case of weaners and well-sited hoppers in the case of growing pigs. The sow accommodation should provide a good grip for the feet on the floor, particularly for those sows that are lame (e.g. leg weakness). The stocking densities should range from 0.2 m^2 (2 ft^2) per pig for weaners through to 3 m^2 (32 ft^2) for sows. A kennelled area is very useful, for compromised pigs like to hide in the dark – where they feel safe.

Management features of the hospital pen

- For the acutely ill pigs it should contain no more than 5 or 6 pigs.
- Pigs should be examined at least twice daily and assessed.
- The pen should be well-lit and bedded.
- There should be no draughts and it should be warm.
- One person should be appointed responsible for these pens.
- Medication and electrolytes should be administered daily and recorded.
- There should be easy access to the pens for observation.
- Water should be available in a bowl at an accessible height for the smallest pig.
- Use the 7 and 14 day rule (described above).

Disposal of Dead Pigs

Dead pigs can be a source of continuing problems. They attract birds, rats and mice and are a breeding ground for flies. Vehicles collecting them pose a serious threat of pathogen introduction to your herd.

Give serious consideration to the disposal of dead pigs.

There are 5 options for the disposal of dead pigs:

1. A self-digestion pit dug into the ground and lined with concrete rings. This will cope with pigs up to 50 kg (110 lbs), but is only for use in ground with a low water table and in temperate climates.
2. Composting in a deep straw manure heap or using other materials. Pigs will decompose totally within 3 weeks provided they are placed into the centre of the manure heap and buried at a minimum depth of 1.2 m (4 ft). This is only of value in temperate climates. Make sure there is no access for foxes

and other animals. This method can be used for pigs up to 150 kg (330 lbs) weight.
3 Burial. This will depend on the water table and local restrictions.
4 Incineration on the farm.
5 Removal by a licensed person for incineration or disposal elsewhere (N.B. This is the only option allowed in some countries).

Poor pig disposal leads to the spread of disease.

Care needs to be taken for pathogen security at the collection site of dead animals. It should be out of the vicinity of the farm and at least 200 m (650 ft) away; 400 m (1300 ft) if live pigs are involved. There should be an entry to the collecting area on the farm side and an exit on the opposite side for collection by the disposing truck. An example of a reception area is shown in Fig 3-12. There are many variations of design. In some countries the design is stipulated by the authorities.

A RECEPTION AREA FOR THE COLLECTION OF DEAD PIGS

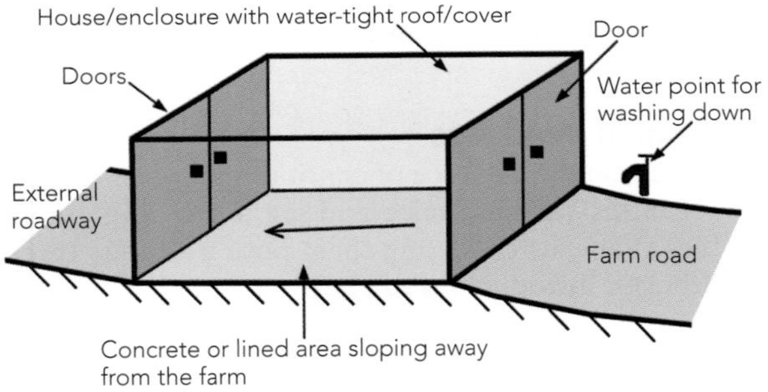

Fig 3-12

Vehicles collecting dead or sick/compromised pigs pose a disease risk to your herd.

The Consultant or Specialist Veterinarian

Traditionally, veterinary services involve attending compromised/sick animals, diagnosing disease and providing treatment. This fire brigade type of work is usually carried out by the general practitioner who often deals with various species in a multi-species mixed practice. As pig units have become larger and their techniques of production more sophisticated, the level of knowledge and expertise required from the veterinarian has increased substantially. As a result, specialised veterinary services have developed in areas of high demand. They differ in different countries in the way they have become organised, but the basic requirements are similar.

These services include independent pig specialists, who spend the majority of their time working amongst pigs on their client's farms within the framework of the general practice to which they belong. More specialised roles exist where vets spend all their time dealing with pig health, management and production. They not only attend herds within the practice but are also invited to advise on herds in other practices, often considerable distances away. A further role within the industry is the specialist veterinarian fully employed by large pig organisations or ancillary industries. This job perspective can often (but not always) become more of an administrative and decision-making one, rather than a "hands on" role on the pig farm.

Nevertheless, the actual resources required in different countries have some comparisons even if the organisations are different. The 3 most typical types of service are shown in Fig 3-13.

What services do you require from your veterinarian?

Veterinary services should be used to:

- Enhance welfare and wellbeing.
- Increase efficiency.
- Further educate and help understanding.
- Contribute to management.
- Control disease.
- Prevent disease.
- Reduce costs.

The fundamental needs of the modern pig unit are to maximise production and efficiency and thus the return on the investment. The achievement of maximum profitability, however, must be carried out within the economic constraints and accepted practices of animal welfare. Part of the service must include a routine visit to the farm every 2 to 3 months, or in the case of the large farm a shorter period, so that studies of the inter-relationships between management, the pig, the environment and organisms can be carried out, relative to the problems. The most successful farms are those which achieve high productivity and give attention to detail. They invariably have good teamwork, disease control and understand how to maximise biological efficiency. A major role of the veterinarian at the periodic visit must be to help with the understanding and awareness of the problems on the farm and identify the procedures to correct them. Fig 3-14 shows a format that could be used at the veterinary visit. It would start with a discussion and an examination of records followed by a clinical examination of the herd or vice versa. At each visit a special topic that relates to a problem area on the farm should be prepared for discussion and education. During the clinical examination of the herd the personnel responsible for each area of the farm should accompany the veterinarian to give their observations and discuss weaknesses and strengths. This also provides an opportunity for education and motivation.

FIG 3-13: VETERINARY PRACTITIONER SERVICES

Service Provided	General Practitioner	Specialist Working Most of Their Time With Pigs	Specialist Consultant Working Full Time With Pigs
Emergency services.	✓	≈	–
Diagnosis of disease.	✓	✓	✓
Treatment of disease.	✓	✓	≈
Disease control.	≈	✓	✓
Welfare advice.	✓	✓	✓
Production control.	–	✓	✓
Management control.	–	✓	✓
Epidemiology.	–	✓	✓
Diagnostic services.	–	≈	✓
Statutory work.	✓	≈	–
Computer technology.	–	≈	✓
Information technology.	–	≈	✓
Education.	–	≈	✓
Nutritional advice.	–	≈	✓
Environmental control.	–	✓	✓
Waste control.	–	≈	≈
Corporate decisions.	–	✓	✓
Profitability.	–	≈	✓
R & D.	–	≈	✓

✓ Yes – No ≈ Variable

FIG 3-14: A PROCEDURE FOR THE ROUTINE VETERINARY VISIT

- Pre-visit preparation.
- Assess previous recommendations.
- The clinical examination of the herd.
- The discussion and examination of records.
- An educational topic.
- A review of general problems.
- Actions to be taken.
- Preparation for the next visit.
- The report.

A written report of the observations and advice should be provided so that it can be used and acted upon pending the next visit. It is also a reminder of the discussions that took place and it can highlight agreed recommendations not acted upon.

The veterinarian can influence the relationships between people, management and pig and this communication, and the dissemination of knowledge associated with it, can be one of the main attractions for the purchase of veterinary services, particularly when the results are increased profitability.

Fig 3-15 shows some of these inter-relationships that occur, starting with the bank, the spouse or partner, the owner, the pig and finally profit. Within each group of people there must be constant interactions and dialogues with the ultimate aim of improving profitability.

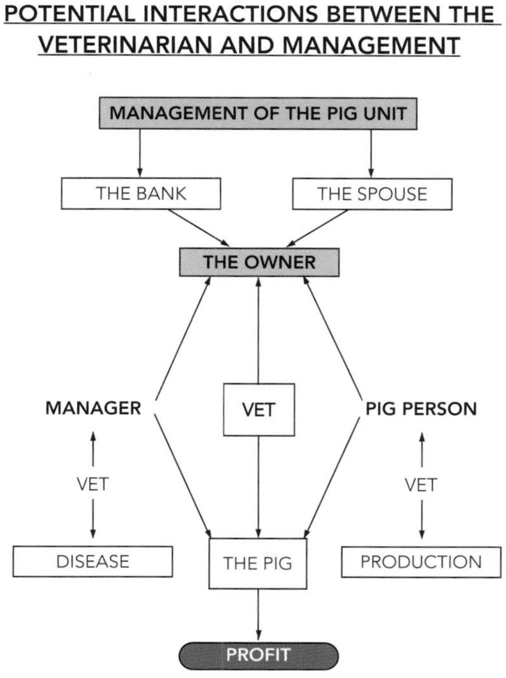

Fig 3-15

Fig 3-16 outlines detailed interactions of the specialist or consultant on the farm and some guidelines.

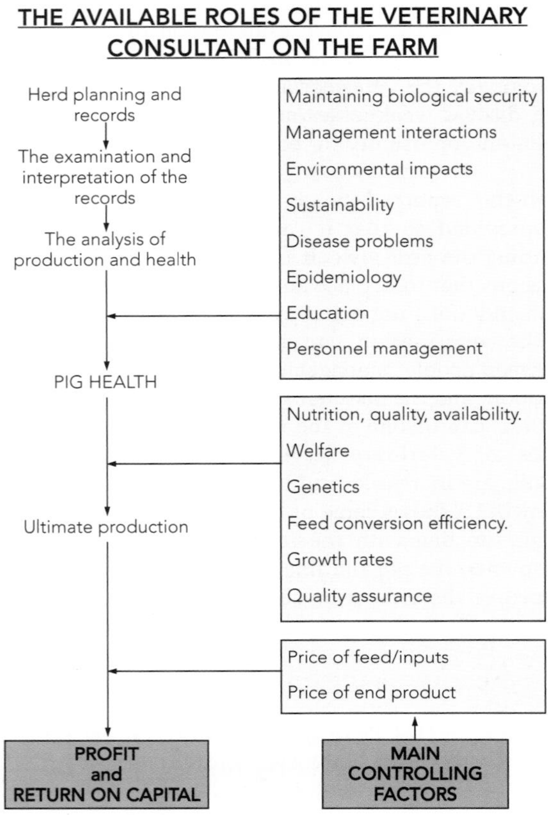

Fig 3-16

The frequency and time necessary for the veterinary visits relative to herd sizes is shown in Fig 3-17.

It is expected that the veterinarian/consultant will play an increasing role in quality assurance schemes, auditing the farm against required standards, the management of sustainability, reducing carbon footprint and other environmental impacts of the farm.

FIG 3-17: SUGGESTED VISIT FREQUENCY AND VETERINARY TIME ON THE FARM

No. of Sows	Visit Frequency (wks)	Hours per Annum
50 – 100	8 – 12	16
150 – 250	8 – 12	27
300 – 400	6 – 8	36
450 – 600	4 – 6	48
600 +	4 – 6	90

Staff Training and Education

Farms that have been highly successful for long periods of time invariably have a good management structure and rapport between pig people and pigs. In other words people management in all aspects is probably the most crucial part of successful pig farming. If there is a problem of production or disease, it usually arises from bad management decisions, or complete failure of the management (Fig 3-18).

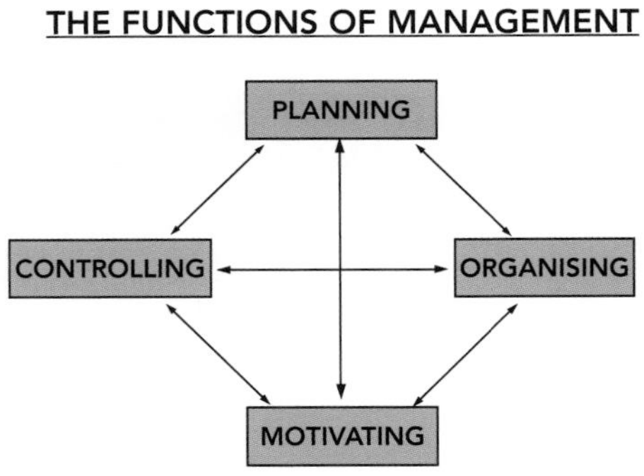

Fig 3-18

In order to achieve good management on the farm it is necessary to understand and satisfy the fundamental needs of the people at work. These needs, in a sequential order, can be listed as follows:

To satisfy basic biological needs

These in essence are the provision of an acceptable standard of living both at work and at home through an adequate wage and the availability of affordable medical services.

To be in a secure position

Satisfaction of the biological needs leads to the second requirement, that of security in the position of good employment. People at this level are constantly seeking reassurance about their job, are very conservative, do not take risks, and motivation can be very poor. It is in this group that education and training can have the greatest impact and allow development into the next level of needs.

To belong to the business

Through education, training and friendships, teamwork develops, where people ask questions and at the same time listen and are developing and becoming motivated in their job. At this point the managers or people in control play an important part by providing good relationships with the workers. From friendship and sense of belonging comes achievement. A sense of achievement on the pig farm involves recognition by senior management and in the process builds self-esteem. Furthermore it further stimulates motivation.

It costs nothing to say "well done".

Motivation is created on the pig farm by the feedback of information and the employee being recognised for their contributions. When people are given responsibility for a job, not only does it give them recognition, but it enables them to develop their own initiatives and thereby move into more challenging roles. From this a highly motivated team develops, with staff helping to solve problems.

To become confident

Through the development of a sense of achievement people will then develop confidence in their jobs and direct and help others. This is the point at which efficiency across the farm improves considerably.

To develop new skills

The final and ultimate part of the process is when a person's confidence reaches such a level that through further education and application, new skills can be developed. A typical example of this is when the farrowing house manager is promoted to under manager or the under manager is promoted to full manager. This must be the ultimate goal of the education process throughout the farm, though of course the pathways upwards will be limited by the capabilities of each individual and the size of the farm.

The objectives of training and education

These essentially can be grouped into 5 areas:

1. Education per se.
2. To provide a better understanding of the job.
3. To provide a better working relationship both between pig people and with their managers.
4. To provide a better environment for the pig and improve its welfare.
5. Finally, to increase efficiency of production and health, and the economic viability of the farm.

What is important to the employees in their work?

A number of surveys have been carried out in different occupations and the results, in order of preference, generally are very much the same. These include:

- An interesting job.
- Appreciation for the work that is done.
- Being involved in the job – this is important because it does provide people with considerable motivation, particularly if owners and managers are prepared to listen and take note.
- A good salary or wage for the job, which in itself provides a considerable amount of the basic security need.
- Confidence and trust – this further promotes motivation and in particular self-esteem.
- Incentives – these range from increased rewards financially, through bonus schemes to the sense of personal achievement and gratitude from the boss.
- A good environment in which to work.
- Good office facilities.

All the above are supported by education.

Are you providing your employees with the above? Why not ask them?

If as an employer or manager you provide education, then it is quite pertinent to ask the question, "How do I benefit?" Such benefits would include:

- More consistent standards of work.
- You have more confidence in the staff.
- The system of pig farming becomes more efficient.
- The system is much better maintained.
- There are fewer disasters or mistakes.
- There is a great deal more staff motivation and therefore work practices become more efficient.
- The staff become aware of their responsibilities when working with people and they are much more flexible.
- There is a much lower staff turnover.

The same question must then be asked, "How does the employee benefit from education?"

- There is more involvement in the job.
- There is better understanding of objectives and goals and greater achievement.
- There is a great deal more job satisfaction and more motivation.
- Confidence in abilities develops.
- There is a status of being skilled.
- Skills can be documented and thereby recognised.
- The prospects of achievement become apparent and through this, enhancement of the job position.
- There are fewer accidents at work.

FIG 3-19: TYPICAL STOCKPERSON DUTIES AND TRAINING LEVELS REQUIRED

Duties Required of Stockpeople		Level of Training: No. Indicate Level of Stockperson							
		UT	1	2	3	4	5	6	7
Biosecurity	– effluent removal and disposal.								
	– clean accommodation pens, fittings and equipment.								
	– clean and maintain protective footwear and clothing.								
	– apply control procedures.								
	– deceased stock disposal.								
	– clean and maintain buildings and equipment.								
	– maintain machinery and equipment.								
Building	– install new or replacement equipment and fittings.								
	– maintain sheds, fixtures and fittings, fences and surrounds.								
	– remove faulty or damaged equipment and fittings.								
Condition-score pigs									
	– be able to perform.								
Feed	– mix and mill feed.								
	– feed and water for all stock.								
Fire-fighting equipment trained									
Health	– recognise signs of bullying.								
	– administer medicines including injection.								
	– recognise signs of ill-health.								
	– care for sick or injured stock.								
	– recognise heat-distressed pigs.								
	– recognise lame animals and take appropriate action.								
	– maintain an adequate environment for the well-being of the stock.								
	– recognise ill-health and be able to take appropriate remedial action.								
	– recognise sows who are unsuitable for their accommodation.								
	– recognise adverse weather from a pig view.								
	– make medicine usage decisions.								
	– maintain herd health status at an acceptable level.								
	– post-mortem examinations of deceased stock.								
Husbandry	– routines for all stock.								
Identification	– systems can be administered.								
Move, draft and weight stock									
Order	– stores and equipment.								
Records	– maintain a recording system and interpret data.								
Water	– check drinkers.								
Perform other duties as required									
Specific Areas of the Farm Requiring Special Skills: Breeding Area									
Breeding	– artificial insemination.								
	– able to work with both boars.								
	– oestrus detection and mate breeding stock.								
	– pregnancy diagnosis.								
	– replacement breeding stock selection.								
Specific Areas of the Farm Requiring Special Skills: Farrowing									
Farrowing	– teeth clipping.								
	– tail docking.								
	– castration.								
	– assist sows and piglets at farrowing.								
Health	– recognise a piglet that has not eaten within 24 hours of birth.								

- Yes work unsupervised
- Work but needs supervision
- Direct supervision required
- Must not be carried out without specific training
- UT An untrained new employee

Training levels

It is important that staff are trained to the appropriate level for the tasks they are performing. Fig 3-19 lists a number of typical farm tasks and the appropriate level of training. Fig 3-19a defines the training level classifications used in Fig 3-19. These are obviously subject to the quality of on-farm training the employee receives.

FIG 3-19a: TRAINING LEVEL CLASSIFICATIONS

Untrained	One week experience
Level 1	First 3 months with on-farm training
Level 2	3 to 6 months with on-farm training
Level 3	6 months to 1 year on-farm training
Level 4	1 year on-farm and some off-farm training
Level 5	1 to 2 years on-farm and recognised off-farm training
Level 6	2 to 3 years on-farm and formal farm training
Level 7	3 years training and experience. Capable of all tasks.

Are you a successful employer or manager?

This is a very important question to ask yourself. It can be answered by the difficulties you have in obtaining employees or the time which they remain in your employment. It is interesting to look at the characteristics of the good manager. Check yourself against the list below and pick out the areas which you consider are weaknesses, then develop them into strengths.

Characteristics of a good owner/manager

- Manages people well with skill and understanding.
- Has as much interest in people as in their work.
- Demonstrates technical competence.
- Has good business and financial skills.
- Is able to motivate people and provide education.
- Is a good communicator and is aware of people's needs.
- Has good organisational competence.
- Is a clear decision-maker, having listened to the various relevant thoughts from people.
- Provides a good work experience.
- Gives employees an opportunity to contribute to debate.
- Always gives a perception of responsibility.
- Provides a challenge and encouragement.
- Rewards people through personal achievement, recognition, authority, status and pay.
- Provides a constant and enthusiastic environment within which the employees can work.
- Makes every effort to ensure that the employees are involved in all planning and decision-making and in particular has the quality to go and ask questions and listen.
- Involves people in their job. This is one of the highest priorities in most employees.
- Always creates an atmosphere of constant good relationships where employees are not frightened to communicate their ideas or indeed their feelings about their job.
- Provides a clear avenue for the expression of frustrations and any ongoing problems.
- Expects and receives excellent performance from the staff and conveys a belief that they are capable of carrying tasks out.
- Says, "Thank you – well done," often.
- It is interesting to note that the successful managers are those that have a sense of personal fulfilment. This also creates an excellent environment for employees.

Finally, there is a strong relationship between motivation and the belief that improved performance will lead to financial rewards. However, the methods by which the reward is determined needs careful thought and clarification. If you develop a bonus system, always assess it first on a wide range of theoretical scenarios before you commit yourself to it.

How should you improve education on the farm?

There are 4 or 5 clearly identified job specification for which programmes of education need to be developed. These areas can be categorised as follows:

1. **The trainee stockperson**
2. **The stockperson**
3. **The training manager or under manager**
4. **The manager**
5. **The owner**

At a farm level, training can be provided by the manager and/or the owner, the veterinarian and other people introduced to the farm for that purpose. The manager should play a pivotal role in this by constant daily instruction, by the assessment of various techniques, by staff meetings and by using their records. On-line training courses are also available and should be considered.

At a stockperson level, each farm should have a simple manual of the different daily tasks so that when instructions and training have been given and competency assessed, this can then be documented.

This type of training programme would be considered as basic and be undertaken over a 1 to 2 year period. It would be suitable for the new recruit or school-leaver.

The veterinarian would have a part to play in the basic training through promoting understanding and through the development of short seminars on the farm as part of the visit contribution.

Topics that should be Considered in the Training Programme

Basic training

- Technical pig production terms and their understanding.
- Aspects of safety on the pig farm.
- Biosecurity.
- Pig flow.
- Management of:
  - Boars.
  - Dry/gestating sows.
  - Farrowing houses.
  - Weaner production.
  - Feeder pig production.
  - The hospital pen.
  - Understanding reproduction.
  - Artificial insemination techniques.
  - Pregnancy testing.
  - Recording and the use of records.
  - Understanding genetics and breeding.
  - Recognising the healthy and diseased pig.
  - The disposal of dead stock.
  - The role of disinfectants in disease.
  - The management of medicines on the farm.
  - The administration of medicines on the farm.
  - Nutrition and the application of feed.
  - Slurry disposal.
  - The use of pressure washers, electricity.
  - Managing the gilt.
  - Welding and maintenance.
  - Understanding simple pig production economics.

Within each of these topics there will be further individual areas of education. For example in the farrowing house a list should be made of all the various tasks that are carried out, including:

- Preparing the house for occupancy.
- Recording farrowing details.
- Preparing the sow for farrowing.
- The signs of farrowing.
- How to assist at farrowing.
- Removing teeth and tails (if applicable).
- Injecting with iron.
- Injecting medicines.
- Managing the litter.
- Recognising disease.
- Recognising piglet diseases.
- Assessing the healthy and diseased udder.
- Castration (if applicable), tattooing.
- Feeding the sow.
- Controlling the environment.
- Moving the sow.
- Catching the litter in preparation for tasks.
- Handling pigs of all ages.
- Euthanasia of pigs of all ages.

Each of these would provide a short course of instruction followed by "doing", followed by assessment.

Use this book for training on your farm. It is written for that purpose.

Intermediate training

This area of training is aimed at the experienced stock-person or the person responsible for a section of the farm or a trainee manager. It should involve attending day courses that include personnel management and more advanced training on the basic topics. Veterinary seminars on the understanding of diseases should be an important part of this. This can be carried out either on the farm at the veterinary visit, if there are sufficient people to justify this, or alternatively by the development of seminars at the veterinary practice or at the agriculture schools. The following topics should form part of the veterinary training programme:

- Biosecurity.
- Pig flow/batching.
- Understanding infectious agents.
- Anatomy of the pig.
- The use and misuse of medicines.
- How diseases and pathogens are spread.
- The healthy and the diseased pig.
- The collection, understanding and use of records.
- The relevance of disease control to profitable pig farming.
- Understanding reproduction in the male.
- Understanding reproduction in the female.
- Non-infectious infertility.
- Infectious infertility.
- The process of farrowing.
- Approaching farm problems.
- Aspects of vaccination.
- Welfare of the pig.
- Notifiable diseases.
- Respiratory diseases.
- Problems of the dry/gestating sow.
- Nutrition, production and disease.
- Controlling parasites.
- Skin diseases.

Have a copy of this book accessible to all staff at all times. Do not lock it away.

Advanced training

This should be aimed at the under manager, the manager and the group farm manager. Instruction courses here would involve attending specific educational seminars.

At a veterinary level these would include the topics already mentioned for intermediates but they would now be dealt with in a more detailed and scientific way and include a greater understanding, particularly of epidemiology and the control of disease. Instruction techniques and the ability to teach people should be a major part of this training programme, together with people management, business management and communication skills. Training in the areas of business management and the use of computers form an essential part of pig production. The greatest complaint of people working on pig farms is the style of management and lack of appreciation. The education of the manager in this respect is important.

An Example of Management Failures and Disease

The following sequence of events highlights how bad management and decisions can result in disasters.

The farm was a 500-sow herd where the management made the decision to expand from 500 to 600 sows through the purchase of gilts. Unfortunately the health status of the purchased gilts was not checked out and they infected the herd with progressive atrophic rhinitis. To coincide with the herd expansion, a decision was also made to develop a building programme which was not completed in time. The number of breeding females increased and so did the stocking densities in the service area, resulting in poor hygiene and stress with increases in services and boar usage. The increased number of services resulted in high numbers of animals farrowing with no increase in farrowing accommodation. Finally, the resulting shortened lactation length, due to shortage of farrowing accommodation, caused ascending vaginal infection and endometritis post-service. This, together with the increased boar usage and poor hygiene, resulted in a major infertility problem. The increased throughput of sows through the farrowing houses resulted in major scour problems which, together with the poor hygiene, increased the severity of the rhinitis. The shortened lactation length together with the scour precipitated post-weaning problems, increased stocking density and further increased problems with rhinitis. The final insult was the fact that the shortened lactation length resulted in poor litter size. The end result therefore of the original management decisions was a herd with severe rhinitis, low litter size, increased pre-weaning mortality, heavy discharges and an infertility problem. The farm went out of business.

If you want a good farm,
have a good manager.

The Use of Records

The previous sections have illustrated some of the consequences of management failure, particularly in planning, that can lead not only to production problems but also to major disease breakdowns. The collection of records and their use are vital components necessary to develop management strategies.

Whilst there are many computer programmes and tools available for recording pig herds – and many of these are highly efficient – it is inevertheless mportant to understand the basic principles of using records for production and health control.

Recording Objectives

Before spending a considerable amount of money on a computer programme, and a lot of time entering data and producing a great deal of paper, it is important to answer a simple question. "What are we trying to achieve?" There are 5 reasons for collecting and producing information:

1. To improve overall efficiency.
2. To maximise profitability.
3. To produce end data that defines:
   - Production levels (kg of meat sold).
   - Reproduction levels.
   - Management achievements.
   - Economics – cost of production.
   - The use of feeds.
   - Growth performance.
   - The levels of disease.
   - The levels of medicinal treatment.
4. For epidemiological studies to understand problems.
5. Finally as aids for daily use by management.

Fig 3-20 shows the pathways for using data so that management control, production and health can be monitored and better understood.

There are 3 major areas crucial to management decisions, namely economics, production and feed. The total deadweight of pigs paid for by the slaughterhouse is fundamental to profitability. Every extra kilogram of deadweight sold increases the margin over feed with few extra overheads. Likewise, matching the type of pig to the best market is another important management decision. Note that some markets are still based on a liveweight system and this needs to be appreciated when comparing systems. In addition, be careful over the term deadweight – some areas do not include head weight or trotter weight, for example.

Production is monitored from individual records of the sow, sucking pig, weaner and grower-finisher and in each of these, disease, treatment, mortality and culling levels are recorded.

Feed usage, the third recording area, is the largest

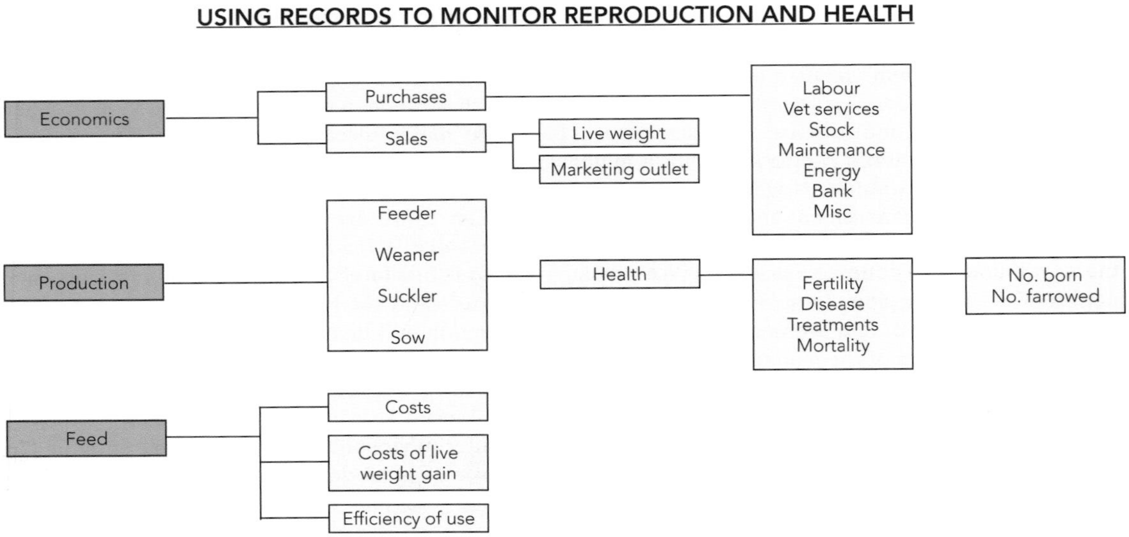

Fig 3-20

cost centre, and the monitoring of costs per tonne, costs of liveweight gain and efficiency of use are vital. It is here that the greatest use of records and computer technology can be made but it is also the area where there is often the least input.

Fig 3-21 shows a typical range, from farms around the world, of costs as a percentage of sales. It can be seen that the veterinary medicines costs were 2 to 3 % of the net sales. These might at first sight appear small but they do not take into account the costs of any disease, poor production and poor feed efficiency, which on some farms could be a further 8 to 10 %. Veterinary services in this respect can be highly cost-effective. Recorded information can be classified into 4 main categories for practical use:

1 **Action information** – This is used on a batch basis by the manager and staff. It should provide data relating to animals that are due for service, pregnancy testing, vaccination, and farrowing. Most computers forecast what is likely to happen in the weeks ahead.
2 **End data** – This is summary information produced as rolling averages of the previous batch, quarter or years. It indicates on a cumulative basis what has been happening during any defined period. This helps to monitor the efficiency of the production and the effects of disease and it is used against target figures to identify problem areas. End data does not, however, provide information for epidemiological analysis and problem solving. Months should be avoided as they are of variable time periods. The farm records should be coordinated around the batching programme used.
3 **Epidemiological information** – This is the detail that produces the end data and consists of a number of individual pieces of information. Unfortunately many computerised systems do not retain this, or if they do, its presentation is in such a vast amount of data that it can be extremely difficult for the pig farmer or the veterinarian to use and assimilate. A much more simple procedure is to identify the specific points of information that are required relative to the problem and analyse these manually. Using

FIG 3-21: GUIDELINES FOR BREEDING AND FEEDING HERDS

Assumptions: Pigs sold per sow per annum: 22 – 32
Sold deadweight: 70 – 90 kg (160 – 200 lbs)

	Approx. Percentage of Net Output
Feed costs (total)	65 – 70
Variable costs	
Veterinary/medicines.	2 – 3.5
Transport.	2
Electric/heat.	1 – 2
Water.	1
Bedding.	0.5
Miscellaneous*.	2 – 3
Fixed costs	
Labour.	8 – 14
Buildings, machinery, other.	2 – 10
Genetics.	4 – 6

* includes bank charges and other costs.

this method we can then utilise the observations and clinical examinations of the stockpeople, identify the animals which form the core group of the problem and then study them to identify any common features.

4 **Forecasting information** – This is used for planning and targeting and there are a number of useful computer programmes that respond to changing productions and disease information and allow more reliable decisions to be made. Whatever system of record keeping is used (or is proposed to be used) on the farm, consider the following questions:
- What are you trying to do?
- Is it possible to do it?
- How do you record it?
- Is the effort going to be worthwhile?
- Is the method of data collection and presentation simple and straightforward?
- Is the information to be collected accurate and reliable?
- How do you intend to use the information?

The consultant veterinarian has an important role to play in the analysis of recorded information and in its epidemiological use.

The following simple formats and examples have proved of value in both monitoring and investigating problems on the farm (see Fig 3-22 to Fig 3-32).

These records could be documented weekly and the format is valuable as a monitor for achieving efficient production. It allows remedial action to be taken.

Fig 3-22b shows the list of management targets for batch farrowing batches of 20 sows (based on the assumptions stated in Fig 3-22a). With weekly batching and an 85 % farrowing rate, a minimum of 444 sows are needed. This will rise if the farrowing rate falls (summer infertility). For instance, with a 75 % farrowing rate at least 462 sows will be required.

FIG 3-22a: ASSUMPTIONS

Batch farrowing number.	20 places
Farrowing rate.	75 %
Total born.	14.5 piglets
Weaned per place.	12 weaners
Weaner mortality rate.	3 %
Finisher mortality rate.	2 %
Replacement rate.	30 %/year
Batch time.	1 week
Weaning age.	4 weeks

FIG 3-22b: FARM MANAGEMENT TARGETS

Approximate No. breeding females 462 sows (75 % farrowing rate)

Batch number	1	2	3	4	5	6	7	8	9	10	11	12
No. Total matings (+gilts)	25	49	72									
Running target	27	54	81	108	135	162	189	216	243	270	297	324
Gilt matings	5	7	12									
Running target	5	10	15	20	25	30	35	40	45	50	55	60
Repeat matings	2	5	9									
Running target	5	10	15	20	25	30	35	40	45	50	55	60
No. farrowed	20	40	62									
Running target	20	40	60	80	100	120	140	160	180	200	220	240
No. total born	291	578	872									
Running target	290	580	870	1160	1450	1740	2030	2320	2610	2900	3190	3480
Pre-weaning mortality	24	50	74									
Running target	24	48	72	96	120	144	168	192	216	240	264	288
No. weaned	241	482	724									
Running target	240	480	720	960	1200	1440	1680	1920	2160	2400	2640	2880
Weaner mortality	6	15	23									
Running target	7	14	21	28	35	42	49	56	63	70	77	84
Grower/finisher mortality	6	9	15									
Running target	5	10	15	20	25	30	35	40	45	50	55	60
Finishers sold	228	458	680									
Running target	228	456	684	912	1140	1368	1596	1824	2052	2280	2508	2736
Sows sold	3	6	9									
Running target	3	6	9	12	15	18	21	24	27	30	33	36

FIG 3-23: RECORDING SOW PROBLEMS

Unit: Poplar Farm Total number of sows: 490 3-week batch 27-day wean

	Action	1	2	3	4	5	etc.	Target % *	Target No. per Batch
Matings Sow First time		71	62	50					64
Repeats		6	4	4				< 8	5
Gilt First time		10	16	12				< 5	15
Repeats		2	3	1				< 2	1
PD % Doubtful negative.		2	3	2				< 2	1
No. treated – hormones.		–	–	–				< 1	0
Anoestrus sows.	Culled	–	–	–				< 3	0.5
Anoestrus gilts.	Culled	1	–	1				< 5	0.5
Repeats/infertility.	Culled	2	1	1				< 2	< 2
Not in pig.	Culled	2	1	1				< 1	< 1
Abortions.	D/C	–	–	–				< 1	< 0.5
Discharge from vulva.	Treated	2	1	–				< 1	< 1
	D/C	1	–	–				< 1	< 1
Mastitis.	Treated	–	–	–				< 5	1
	D/C	–	–	–				< 2	2
No milk.	T/C	2	–	–					2 – 4
Lame.	Treated	3	4	2				< 20 p.a.	1
	D/C	1	–	1				< 12 p.a.	1
Miscellaneous problems.	Treated	–	–	1				2 p.m.	
	D/C	–	–	–				1 p.m.	
No. of farrowings.		70	64	67				73	73
No. of assisted farrowings.		–	2	1					< 3
Prolapse.	T/D/C	1	–	–				< 8 p.a.	
Fever.	Treated	1	–	–				1 p.m.	
	D/C	–	–	–				–	
Haemorrhage.	Treated	–	–	–				< 1 p.m.	
	D/C	–	–	–				< 3 p.a.	
Litters savaged.	T/C	1	–	–				< 2	
Low Nos. born/reared.	Culled	2	–	–				< 8	
Total normal sow sales.		4	8	12				15 p.m.	
Total sow deaths.								< 20 p.a.	< 2
Sows destroyed on the farm.	Died							< 4 p.a.	
Problem sows shipped or culled.	Culled							< 4 p.a.	

D/C = Died or Culled. **T/C** = Treated or Culled. **T/D/C** = Treated, Died or Culled. **p.m.** = Per month **p.a.** = Per annum.
* As a percentage of sows mated or farrowed in your defined period. **PD** = Pregnancy diagnosis.

FIG 3-24: RECORDING MATING

Batch/Week/Month.......*batch 1.*

Sow No.	Date Weaned	Date First Mated	First or Repeat Matings	No. Services	Boar Used	Lost Days	Date Due to Farrow	Comments
317	1 Jan 2013	6 Jan 2013	1	3	27/27/26	5	30 Apr 2013	–
27	1 Jan 2013	28 Jan 2013	1	2	3/3	27	30 Apr 2013	Mastitis
64	Gilt	2 Feb 2013	2	3	8/4/4	29	27 May 2013	Bled

FIG 3-25: RECORDING FAILURES TO FARROW

Batch: 3

Batch Mating Target: 100
TOTAL MATED: 104

CATEGORY: (See below)

Sow No.	Parity	Date First Mated	1st or Repeat Mating	AI Batch or Boar Used	Cause of Failure	Date of Failure	Days Interval	Result/History/ Stockperson/Comments
75	7	10 Dec 12	1	16	Dis	24 Feb 13	76	Discharge
161	5	12 Dec 12	2	4	–	2 Feb 13	52	Poor service
67	8	3 Jan13	1	60	D/C	27 Mar 13	83	Fighting
724	1	2 Mar13	1	24	Ab	4 May 13	63	Lame

Cause of failure: **Ab** = Abortion **D/C** = Death/Culled **Dis** = Disease **NIP** = Not In Pig **Rp** = Repeats

FIG 3-26: RECORDING AI AND/OR BOAR MATINGS AND SOWS THAT FARROW FROM THEM

AI Batch or Boar	Matings	Repeats	Non-Return Rate	Farrowed From the Matings	Farrowing Rate %	Average Total Born	Pigs per 100 Matings Farrowed
471	23	6	74 %	16	70 %	12.4	868
5	19	1	95 %	18	95 %	13.5	1282
7	12	0	100 %	11	92 %	14.6	1342
471	16	3	81 %	13	81 %	15.1	1223
10	24	1	96 %	22	92 %	14.5	1334

FIG 3-27: RECORDING SOW DEATHS

Sow No.	Date Served	AI Batch or Boar Used	Date of Death	State Pregnancy	Parity	Condition	Illness/ Treatment	PM Findings or Cause of Death	History
126	7 Sep 12	60	17 Sep 12	Unknown	6	Poor	Ill fever	Peritonitis	Bled at service

FIG 3-28: RECORDING VULVAL DISCHARGES

Identification of sow.	420	16		
Date of farrowing.	4 Jul 12	6 Aug 12		
Date weaned.	25 Jul 12	20 Aug 12		
Date mated.	30 Jul 12	29 Aug 12		
Parity.	7	2		
AI Batch or boar used.	20/20	14/147		
Date of discharge.	28 Aug 12	30 Sep 12		
No. days from mating.	24	32		
Description of discharge.	Mucus	White/pus		
Results of first pregnancy diagnosis.	-ve	+ve		
Results of second pregnancy diagnosis.	-ve	-ve		
Date of farrowing or otherwise.	Culled	Culled		

FIG 3-29: RECORDING LITTER SIZE DETAILS BY PARITY

Batch/Week:..

Sow No.	Litter No.	Date Farrowed	AI Batch or Boar	No. Alive	No. Dead	No. Mummified	No. Weaned	Days Suckled
604	7	24 Aug 12	23/23	14	3	1	13	19
906	1	29 Oct 12	29/471	3	3	3	9 (6 fostered)	–

FIG 3-30: RECORDING PIGLET DIARRHOEA FOR INVESTIGATION

Date Scour Noted	Sow No.	Parity	Date Farrowed	No. Pigs in Litter	No. Affected	No. Deaths	Comments: Vaccine Used, Treatment Given
23 Apr 13	52	2	22 Apr13	11	All	2	Not vaccinated. Neomycin.
27 Apr 13	604	1	26 Apr13	9	All	6	Not vaccinated. Trimethoprim.
28 Apr 13	606	1	26 Apr13	10	All	0	Enrofloxacin. Responded.

FIG 3-31: RECORDING OF WEANING/FINISHING PIG DISEASES

Batch	Action	1	2	3	4	5	6	etc.	Total	Target % *
Lameness.	Treated	12	2	1					15	< 1
	Died/Culled	1	–	–					1	< 0.5
Haemorrhage (pale pig).	Treated	–	–	1					1	< 1
	Died/Culled	1	–	–					1	< 0.5
Pneumonia.	Treated	24	36	52					112	1
	Died/Culled	1	4	6					11	< 0.5
Scour.	Treated	2	3	2					7	< 2
	Died/Culled	–	–	–					–	< 0.5
Prolapse.	Treated	16	14	14					44	< 1.5
	Died/Culled	–	–	–					–	< 1
Blown–up (rectal stricture).	Treated	4	3	1					8	< 1
	Died/Culled	2	1	1					4	All
Fever.	Treated	–	–	–					–	< 1
	Died/Culled	–	–	–					–	0
Stress (fighting).	Treated	2	–	–					2	1
	Died/Culled	–	–	1					1	0.5
Runt. Poor pigs.	Treated	8	7	9					24	< 1.5
	Died/Culled	1	1	1					3	< 1
Meningitis.	Treated	–	–	–					–	< 2
	Died/Culled	–	–	–					–	< 2
Miscellaneous. e.g. Middle ear.	Treated	–	1	–					1	< 2
	Died/Culled	–	1	–					1	< 1
Total treated									214	< 5
Total died									22	< 4
Total to cull pen										< 1.5
Pigs at risk.		2100	2010	2221						

* As a percentage of the population at risk in your defined period.

FIG 3-32: RECORDING WEANER OR FINISHER MORTALITY

Batch/Week:..

Date	Weight of Pig	Age of Pig	Name of House	Feed Being Used	Condition Good/Moderate/Poor	Cause	Comment/ Observation
3 Feb	60 kg	16 wks	G1	346	G	Blown up	Sudden
8 Feb	85 kg	20 wks	F3	1-64	G	?	Pale

Planning for Efficient Production and Health Control

One of the most difficult management tasks is to control the numbers of sows and gilts mated in any given batch (period of time). If for example the farm has 100 farrowing places in 5 rooms of 20 and practises 4 week (27 day) weaning, the farm pig flow programme requires 25 matings per week/batch. One week only 15 matings are carried out, followed by 35 in the next week. While the average for the 2 weeks remains 25, the chaos that will ensue will inevitably result in clinical disease because of the under-stocking of one batch and the overstocking of the next. The effects of these are shown in Fig 3-33.

Problems arise through:

- The overuse of boars in natural mating systems leading to infertility problems and variable litter sizes.
- Stocking densities increasing as more animals enter the housing system than it is designed for.
- A failure of the all-in/all-out system in the farrowing house resulting in early weaning and more fostering.
- An increased number of pigs being transferred into the weaner and grower accommodation at any one time with increased stocking densities, permanently populated houses and the movement of pigs from one house to another, thus disturbing the status quo.
- The end result of all these various changes is disease, typically diarrhoea due to chilled, understocked pigs or pneumonia due to hot, overstocked pigs. This eventually leads to: Poor growth depending on the severity of the mating control failure and pathogenic organisms present on the farm.

THE EFFECTS OF FAILING TO CONTROL THE BATCH MATING* PROGRAMME

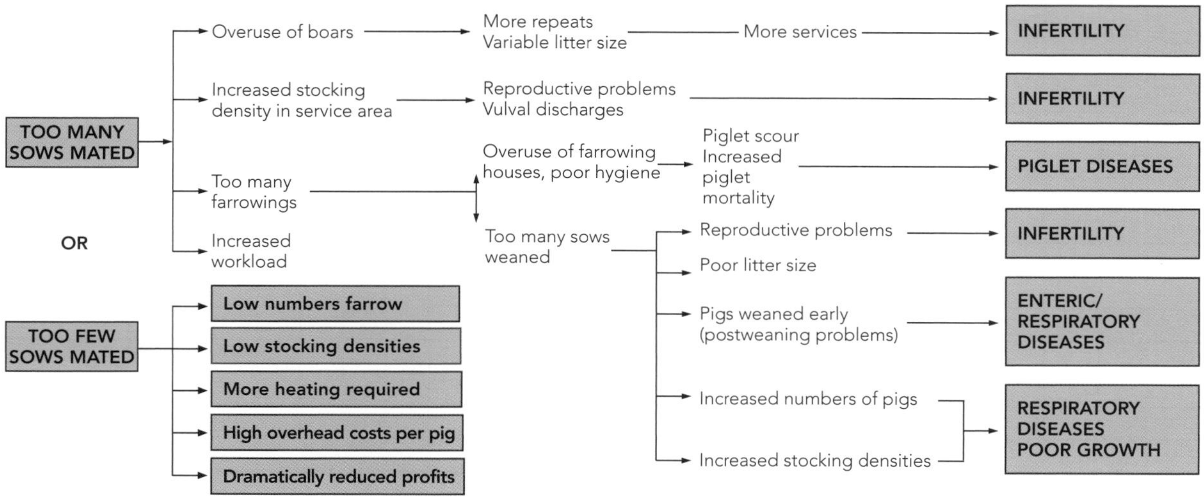

* The terms 'mating' and 'service' are often considered the same. Mating here means the complete process, i.e. it includes 1, 2 or more services.

Fig 3-33

Management Procedures for Maximising the Mating Programme

Managing the mating programme for maximum throughput without creating pig flow problems requires good forward planning to achieve consistency. The key is ensuring sufficient gilts are in the system.

Calculate the entry-to-exit time per farrowing place in days, e.g.:

28 days mean lactation length
3 days cleaning time
4 days entry to farrowing
Total = 35 days

Note: it is illegal to wean below 21 days in the EU. Other countries can wean at 3 weeks – average 20 days.

In theory therefore each farrowing place could be used 365 ÷ 35 = 10.42 times per annum. Calculate the actual number. This will indicate the efficiency of use. Maintaining regular batch farrowings each week, or each farrowing period, is dependent upon the following:

- Sows coming into oestrus regularly after weaning.
- A good conception and consistent farrowing rate.
- Knowledge of the per-batch farrowing rates.
- A planned input of gilts to replace culled sows.
- Anticipation of oestrus in the gilts.
- Planning at least 8 weeks ahead to determine the breeding females that will be weaned and due for mating.
- The effects of disease.

A number of management tools can be used:

1. Modern computer programmes will list out sows that fail to maintain a pregnancy on a batch-by-batch basis (Fig 3-34). This format can also be carried out manually from the weekly mating sheets. When sows reach their twelfth week of pregnancy (forecast week) those animals still pregnant or viable or those due to be culled because of age or disease can be identified. This will indicate those available for mating in 7 weeks time. The anticipated matings can be brought to target by planned gilt matings for that week.
2. Fig 3-35 shows an example of a farm operating on a batch basis where production is monitored against the cumulative targets. Such a method gives an opportunity to manipulate the system if there is a major loss and compensate for it. The fault on many farms lies in the use of averages. With batch production, the numbers must be whole animals and each batch must be the same. For example, Fig 3-35 implies that total matings per batch is 11.1 – whereas in reality it would be 12.
3. For small herds of up to 350 sows, a circular calendar can provide an excellent and simple visual method (Fig 3-36). Sows due for culling can be identified by a colour code. In this particular herd summer infertility is a regular phenomenon with increased matings carried out to compensate for sows that are anticipated to be culled.
4. Another option is to serve all sows weaned each batch and cull some sows after pregnancy checking to adjust the number of pregnancies required.

FIG 3-34: A FARM EXAMPLE

Batch	Target Matings*	Females Mated Actual	Sows Still in Pig (Gestation)																FR %		Farrowing House			
			Week No.																					
			1	2	3	4	5	6	7	8	9	10	11	12	13	14	15	16		17	1	2	3	4
1	12	12	12	12	12	11	11	11	11	11	11	11	11	11	11	11	11	11	92	11	10	10	10	10
2	12	12	12	12	12	12	11	11	11	11	11	11	11	11	11	11	11	11						
3	12	11	11	11	11	11	11	11	11	11	11	10	10	10	10	10								
4	12	12	12	12	12	12	12	11	11	11	11	11	11	11	11									
5	12	10	10	10	10	10	10	10	10	10	10	10	10	10										
6	12	11	11	11	11	11	11	11	11	11	11	10	10											
7	12	9	9	9	9	9	9	9	7	7	7	6												
8	12	11	11	11	11	11	11	11	11	11	11													
9	12	10	10	10	10	9	9	9	9	9														
10	12	10	10	10	10	9	9	9	9															
	etc.																							

* No. females mated (sows, gilts and returns). This may be higher in the summer and lower in the winter. **FR %** Farrowing rate.
The numbers in red indicate a sow or sows that will not farrow in that batch.
The numbers in **blue** highlight batches below farrowing target. More gilts will be needed in future.
Weeks 10/11: Check gilt requirements to maintain batch mating target.

FIG 3-35: AN ACTUAL EXAMPLE OF HERD TARGETS CUMULATIVE BY WEEK

Batch No.	1	2	3	4	5	6	7	8	9	10	11	12
Date	6 Apr	13 Apr	20 Apr	27 Apr	4 May	11 May	18 May	25 May	1 Jun	8 Jun	15 Jun	22 Jun
Total matings	14	30	43	56	64	75	83	91	103	111	127	134
Cumulative targets: (11.1)	**11**	**22**	**33**	**44**	**55**	**66**	**77**	**88**	**99**	**111**	**122**	**133**
Gilt matings	6	8	8	10	10	11	14	14	14	14	16	16
Cumulative targets: (1.9)	**2**	**3**	**5**	**7**	**9**	**11**	**13**	**15**	**17**	**19**	**20**	**22**
Repeats	0	2	4	6	6	8	8	8	8	8	10	10
Cumulative targets: (0.9)	**1**	**2**	**2**	**3**	**4**	**5**	**6**	**7**	**8**	**9**	**10**	**11**
No. Farrowed	9	22	25	36	47	56	65	72	84	95	105	116
Cumulative targets: (9.9)	**9.9**	**19**	**29**	**39**	**49**	**59**	**69**	**79**	**89**	**99**	**108**	**118**
Born Alive	100	242	280	413	541	640	743	820	945	1078	1195	1325
Cumulative targets: (118)	**118**	**236**	**354**	**472**	**590**	**708**	**826**	**944**	**1062**	**1180**	**1298**	**1416**
Weaned	110	218	328	415	522	588	689	748	895	999	1100	1224
Cumulative targets: (104)	**104**	**208**	**312**	**416**	**520**	**624**	**728**	**832**	**936**	**1040**	**1144**	**1248**

Target ( ) actual batch target; 220 sows; 2.35 litters/sow/year; Piglets born alive 11.9; Piglets weaned per litter 10.5; 24.7 pigs reared.

Management of the Environment

When faced with a problem it is always an interesting proposition to ask the question:

How might I make this problem worse?

Such an approach invariably highlights areas for corrective action. There are a number of common problems underlying loss of performance in the dry/gestating sows, lactating sows, sucking pigs and growing pigs. These call for management actions.

Environmental Factors Affecting Dry/Gestating Sows

- Catabolic state from weaning to 21 days post-service.
- Grouping sows at weaning. Fighting.
- Poor light intensity.
- Fluctuating lighting patterns.
- Fluctuating temperatures day and night.
- High air flow, draughts.
- House temperatures above the upper critical temperature.
- House temperatures below the lower critical temperature.
- Low mating house temperatures.
- Poor water availability.
- Small badly designed housing.
- Faulty floor surfaces.
- Wet poorly drained floors.
- Poor lighting patterns.

A CIRCULAR CALENDAR USED ON A FARM FOR MONITORING THE MATING PROGRAMME

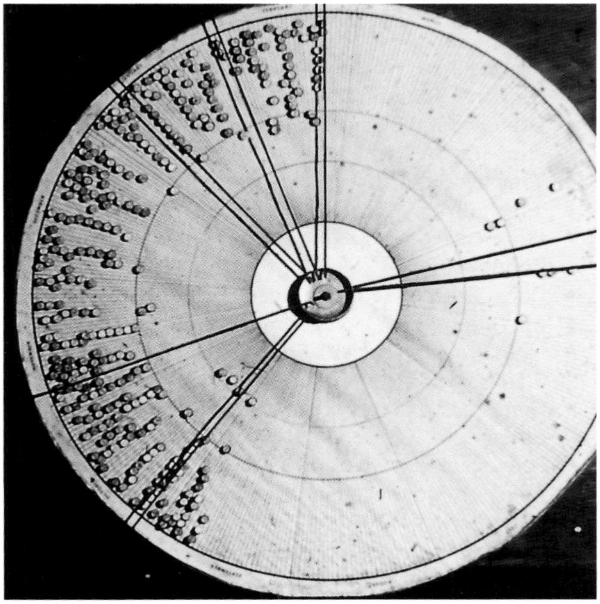

Sows start at the top, move to the left to farrow, and are then weaned. Weaned sows are on the right awaiting mating to start the cycle again.

Fig 3-36

Always ensure you have sufficient gilts in the system to meet mating requirements.

Environmental Factors Affecting Lactating Sows and Sucking Pigs

- Cross-fostering different age groups of piglets.
- Mechanical transfer of infections between litters and batches.
- High feed levels pre- and immediately post-farrowing.
- Fluctuating farrowing house temperatures.
- Failure to determine the best farrowing house temperature.
- Draughty creep areas, low creep temperatures.
- Draughts on the sow.
- Continuous throughput. No all-in/all-out system.
- Failure to wash, disinfect and dry between batches.
- Poor farrowing place design.
- Wet, poorly drained floors.
- Leaking feed troughs.
- Poor water supply.
- Poor lighting.
- Delayed faeces removal from behind the sow.
- Fly problems.

Environmental Factors Affecting Growing Pigs

- Incorrect house temperatures, particularly fluctuations.
- Pigs held below their lower critical temperature.
- High ventilation rates, air flow and draughts.
- Low or fluctuating humidity together with low temperatures.
- Poor insulation.
- Worn out environmental controllers and sensors.
- Floor types, poor drainage, wet floors, slats with draughts.
- Constant mixing and movement of pigs.
- Moving pigs too soon from one house to another.
- High stocking density.
- Small cubic air space.
- Continually populated houses with endemic disease.
- Feed changes, when pigs are moved from one house to another, together with lower levels of nutrition.
- Inadequate trough spaces or water availability.
- High slurry levels.
- Poor or inadequate nutrition.
- Continual exposure to faeces.
- High levels of toxic gases.

The first requirement of good management is to prevent the build-up of infection through the cleaning and disinfection of buildings and maintaining low stocking densities and good environments. Reducing stress from the effects of fluctuating temperatures, and the influences of humidity and ventilation on organisms in the air, is the second part. The third part is the provision of good nutrition and in particular, adequate levels of energy, protein and lysine.

If you have a problem, check through the relevant list.

Air Quality

The quality of the air in the pig building depends on a number of factors including the stocking density, the cubic capacity of the building, the lower critical and upper critical temperatures, concentrations of gases and levels of dust.

As the number of infectious organisms per cubic metre and the pigs' exposure to them increases, there is an increasing risk of clinical disease. Control of the environment therefore must constantly aim to reduce these levels.

Bacteria and viruses spread from pig to pig by direct contact, indirect contact (e.g. on walls or floors), on equipment and people, and by airborne dust and droplets. The latter mode of spread is obviously associated with air quality.

In dry, airborne dust most of the infectious organisms die quickly but their toxins (e.g. endotoxins) can still be harmful to the pig when inhaled. Aerosol droplets containing organisms dry out rapidly at low humidity and the organisms die. At the middle range of humidity the droplets do not dry and the organisms remain viable and infective. At very high humidities (> 90 %) droplets and dust pick up water, increase in size, and are precipitated out of the air.

The size of the dust particles or aerosol droplets has a bearing on the pigs' defences. When very small particles or droplets are inhaled they are sucked deep into the lungs, sometimes as far as the terminal air sacs (alveoli). Larger particles and droplets tend to be filtered out in the nose and throat or in the upper airways of the lungs.

Using systems that do not operate on an all-in/all-out basis, the incoming pigs meet with heavy doses of respiratory and enteric organisms which are being shed by the older pigs already there. The most harmful time for this to happen is in first stage and second stage weaner accommodation at a time when the antibodies from the sow's colostrum and milk are waning and before the pig has had time to fully develop its own. An overwhelming aerosol build-up of organisms also tends to develop in finishing rooms that contain heavier pigs.

The constant high challenge to the respiratory system results in major mobilisation of the pig's immune system which is costly in energy and lowers the pig's growth rate.

The situation may be worsened by suppression of the pigs' immune system caused by mixing, moving, stress, overcrowding, inadequate nutrition, and high levels of irritant slurry gases. Increased ventilation helps in reducing gases but does not have much effect on airborne dust and droplets. In fact it may make airborne dust levels worse. It also may create evaporative cooling over the pigs which will not help their resistance. Chilling tends to trigger disease.

Reducing stocking density and avoiding overcrowding have the biggest effect on improving air quality and reducing airborne organisms with a resultant boost to growth rate and efficiency of feed conversion to meat. This increased growth also serves to decrease the stocking density by faster throughput. Changing from dry feeding to liquid feeding reduces dust levels and respiratory disease.

Overcrowding leads to respiratory disease and poor growth rates which slow throughput, further increasing the overcrowding.

A checklist for maintaining air quality

- Assess the quality at the daily inspection.
- Check the humidity.
- Check condensation levels.
- Smell the air for levels of ammonia.
- Check the dust levels.
- Assess the lying patterns of the pigs. Are there draughts?
- Check temperature fluctuations.
- Check that fans and inlets are functioning.
- Strip and clean the inlets and outlets between batches of pigs.
- Test that controller systems are functioning correctly each week.
- Test the fail safe mechanism twice weekly.
- Assess levels of disease.
- Check the stocking density of the pigs relative to the cubic capacity of the house.
- Check that stocking densities are not above recommended levels.
- Check that feeders are covered.

Environmental Temperatures

It is important to maintain the pig within an equitable temperature range and this is called the thermo-neutral zone. It is dependent upon the type of floor, its insulation properties, the air speed and temperature and the insulation of the building.

This is particularly so if the pig is at a critical time in relation to disease challenge, or when under environmental stress, for example if it coincides with a move from solid concrete or straw-bedded floors to concrete slats. The temperature requirement for the pig might have been 20 °C (68 °F) before the move but could well be 25 °C (77 °F) for the first days in the new accommodation. Pigs that are within their comfort zone will lie on their sides barely touching their neighbours.

The point at which pigs must increase heat production to keep warm is called the lower critical temperature (LCT). Many factors affect this, including body weight, feed intake, age, insulation of the building and in particular the floor type. There is an upper critical temperature (UCT) and the range between the upper and lower ones is called the thermo-neutral zone (Fig 3-37).

Failure to keep the pig within its temperature comfort zone contributes to the development of disease.

Fig 3-38 gives guidelines to the temperature requirements at different phases in the production cycle.

Adverse temperatures cause detrimental effects at the following critical times:

- From birth to 48 hours.
- From 8 to 14 days of age.
- From weaning to 7 days postweaning.
- When pigs are moved to new accommodation.
- A change in the type of flooring e.g. solid to slats, bedded to non-bedded.
- A change of housing or nutrition.
- A change in stocking density.
- A move from dry to wet feeding.
- A move to poorly insulated houses.
- Movement into a wet house.
- Fluctuations in external temperatures due to ventilation and air speed.
- Faulty environmental controllers.
- Low energy diets or unpalatable feed.
- Restricted feeding.
- Failure to eat sufficient feed.
- Thin pigs.
- Disease.
- The sow during lactation.
- From weaning to 28 days post-mating.
- In the last 6 weeks of pregnancy.

AN EXAMPLE OF A THERMONEUTRAL ZONE

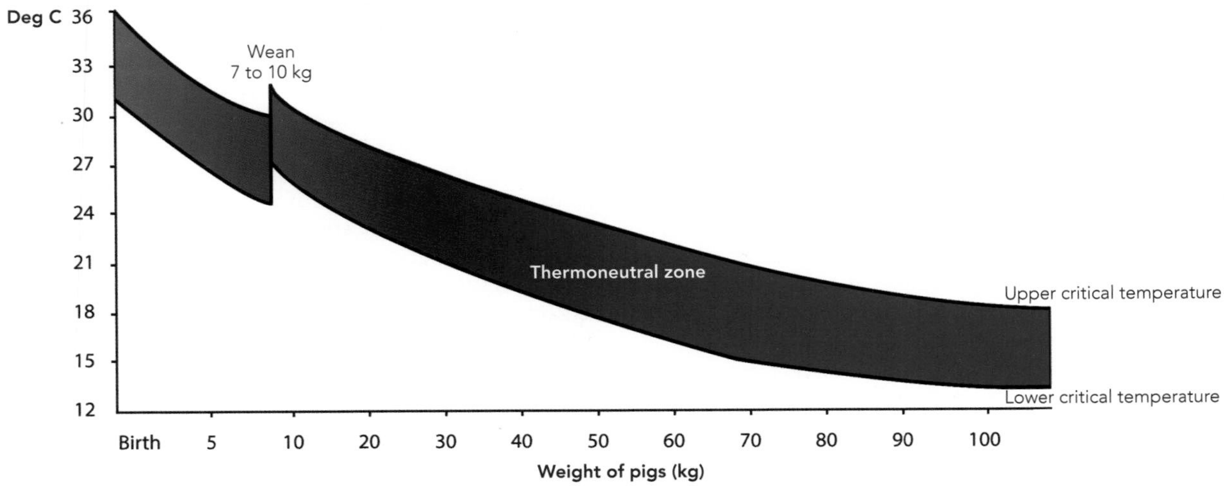

Fig 3-37

AN ILLUSTRATION OF COMFORT TEMPERATURES FOR LEAN GENOTYPE PIGS ON CONCRETE OR SLATTED FLOORS

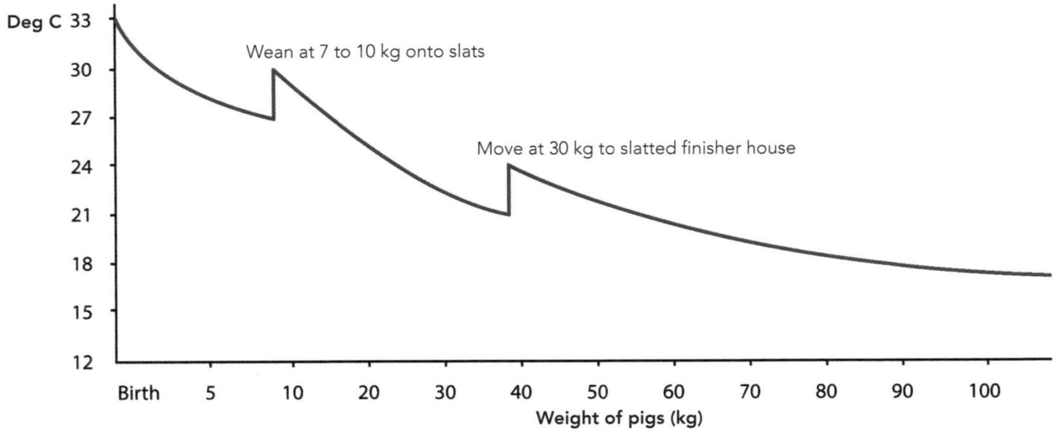

Fig 3-38

FIG 3-39: A GUIDE (ONLY) TO AIR TEMPERATURE ACCORDING TO FLOOR TYPE

		Floor Type							
Weight Pig		Straw		Concrete		Perforated Metal		Slatted	
kg	(lbs)	°C	(°F)	°C	(°F)	°C	(°F)	°C	(°F)
5	(11)	27-30	(81-86)	28-31	(82-88)	29-32	(84-90)	30-32	(86-90)
10	(22)	20-24	(68-75)	22-26	(72-79)	24-28	(75-82)	25-28	(77-82)
20	(44)	15-23	(59-73)	16-24	(61-75)	19-26	(66-79)	19-25	(66-77)
30	(66)	13-23	(55-73)	14-24	(57-75)	18-25	(64-77)	17-25	(63-77)
90	(198)	11-22	(52-72)	12-23	(54-73)	17-25	(63-77)	15-24	(59-75)
Lactating sow		16-18	(61-64)	16-20	(61-68)	16-20	(61-68)	16-20	(61-68)
Gestating sow		11-18	(52-64)	12-20	(54-68)	14-20	(57-68)	16-20	(61-68)
Boars		11-18	(52-64)	12-18	(54-64)	12-18	(54-64)	14-18	(57-64)

Because there are so many variables it is difficult to be categorical about specific temperatures for different weights of pig. Approximate guidelines for different floor types are shown in Fig 3-39 but the ultimate determinant is the pig itself, by its behaviour, lying patterns and performance. Outdoor, the sow's comfort zone is determined by the amount of straw available (to prevent chilling) and wallow management (to prevent overheating).

Adjust the temperature of the building by the lying habits of the pig. Then when the pig is comfortable note the temperature required.

For every degree below the LCT (lower critical temperature) a growing pig loses approximately 10 to 12 g of liveweight gain per day. A level of 1 °C (2 °F) below the LCT during the growing period could cost over 5.75 kg in extra feed per pig (see Fig 3-40).

Let the pig tell you if it has the correct environmental temperature.

Draughts at any age can result in the pig experiencing significantly lower temperatures than might be registered by an air temperature thermometer. This is the wind chill factor.

Draughts in pig houses are expensive.

It is essential that pigs are kept within their thermoneutral zone (Fig 3-41) where they are most comfortable, so that they do not need to cool themselves down, or to eat to keep themselves warm. Ensure there are min/max thermometers or temperature loggers in each house and check the readings regularly.

FIG 3-40: THE EFFECT OF TEMPERATURES ON GROWTH

Bacon Pig 18 – 100 kg			
	At Correct Temperature	°C (°F) Below LCT: 1 °C (2 °F)	°C (°F) Below LCT: 5 °C (9 °F)
Growth g/day	657 g	648 g	605 g
Days to slaughter	140	142	153
Extra feed	0	5.75 kg	31 kg

FIG 3-41: THERMO-NEUTRAL ZONES

Category of Pig	Temperature (°C)	Temperature (°F)
Sows	15-20	59-68
Suckling pigs in creeps	25-30	77-84
Weaned pigs (3-4 weeks)	27-32	81-90
Later weaned pigs (5 weeks +)	22-27	71-80
Finishing pigs (porkers)	15-21	59-70
Finishing pigs (baconers)	13-18	55-64

Stocking Densities

Stocking rate

There are two ways of looking at the stocking rate.

1. **From each pen alone:**
   Examine the size of the pen in square metres and remove any obstructions present within the pen, for example any feeder, and divide by the number of pigs in the pen.
2. **For the whole house:**
   Measure internal floor space of the whole house (length x width). Then subtract lost space – passageways, gating, feeders and any other floor obstructions.

This will allow a calculation of the number of pigs which could fit in the building and then the actual unobstructed floor space available. A pig flow model can then be constructed on this batch space available.

Figs 3-42a–d show the space requirements enforced by law within the EU, by a Government code in Australia and as industry standards (guidelines) in the USA and Canada.

In Canada, guideline stocking densities are determined with regard to the weight of the pig and the floor type. The guideline states that space requirements may need to be increased in hot weather by 10 to 15 %.

In Australia, stocking densities are calculated by a mathematical formula:

$$0.03 \times (\text{bodyweight})^{0.67}$$

Other countries are also looking at introducing formal stocking densities.

Disease levels, growth rates, feed efficiency and mortality are closely correlated to stocking densities but the threshold level will vary with the age and weight of the pig, its health status, the cubic capacity of the building and from farm to farm.

Factors to consider in assessing stocking densities

- Type of flooring and drainage.
- Age of pig and type of housing.
- Quality of insulation in the floor.
- Quality of insulation of the building.
- Type of feeding system.
- Availability of feed.
- Temperature requirements of the pig.
- Can the pigs produce sufficient heat?
- Are the pigs huddled?
- Air flow and draughts.
- Welfare requirements.
- Are mortality levels higher than the target?
- Is there disease?
- Is tail biting or vice a problem?
- Measure the daily liveweight gain and weight for age.
- Is there variation in growth within the pen?
- Levels of respiratory or enteric disease. A farm with respiratory disease should decrease stocking rates and may need to take these diseases into account when designing the pig flow model.

The stocking density should decrease with a change from:

Fully slatted → partially slatted → solid without bedding → solid with bedding.

For example: For a 100 kg pig on fully slatted floors, 0.65 m^2 is acceptable under EU requirements. However if the pig was housed on a solid floor, a space requirement of 0.97 m^2 is advised (Canadian guidelines).

Fully controlled ventilation → partially controled ventilation → artificially controlled natural ventilation → no ventilation control.

Large group → small groups.

All-in/all-out → continuous flow.

What are the best stocking densities for maximum efficiency of output?

The guidelines in Figs 3-42a–d could be used. But in order to measure true economic output in the most efficient terms it is necessary to measure kg of liveweight sold per unit of space per batch or year.

To determine this, identify the following:

- The total occupied floor area of the unit.
- The opening and closing period liveweights.
- The weight of the pigs entering the pens.
- The weight of pigs sold.

The weight sold per unit of space is then calculated as follows:

$$\frac{(\text{wt sold} + \text{closing period wt}) - (\text{wt entered} + \text{opening period wt})}{\text{Total pen space (square metres)}}$$

From Fig 3-43, it can be seen that whilst daily gain, feed intake and feed conversion were better at the lower stocking density level of 0.81 m^2. (8.8 ft^2) per pig the economic benefits in terms of meat sold were better at 0.49 m^2 (5.3 ft^2) as most meat was sold at this level.

Fig 3-44 shows predicted space for the best ADG and FCE.

To ensure that these stocking densities are not to be broken it will be essential to get your breeding programme properly coordinated, for example:
If you have 65 m^2 of finishing floor you can finish 100 pigs per batch in Europe. If you farrow 10 sows a batch there is little point breeding 20 for this batch as you will not have sufficient room to finish the pigs!

Health Control Procedures

The sow during her lifetime becomes strongly immune to most of the infectious diseases present in the herd in which she is kept. She passes on this immunity in the form of colostral antibodies and cells to her piglets in the first few hours of life and later also in the form of milk antibodies.

In addition because she has developed such a strong immunity the sow has thrown off many of the infections that commonly affect younger growing pigs, such as *Actinobacillus pleuropneumoniae*.

FIG 3-42a: EU STOCKING DENSITIES (LAW)

EU Legislation 91/630	
Average Weight of Pig (kg)	Minimum Space Requirement (m²)
≤ 10 kg	0.15
≤ 20 kg	0.20
≤ 30 kg	0.30
≤ 50 kg	0.40

1 kg = 2.2 lbs 1 m² = 10.76 ft²

FIG 3-42b: US STOCKING DENSITIES (GUIDELINES)

US Swine Care Manual			
Weight of Pig		Suggested Floor Type	
lbs	kg	ft²	m²
12 - 30	5.5 - 13.6	1.7 - 2.5	0.16 - 0.23
30 - 60	13.6 - 27	3 - 4	0.27 - 0.37
60 - 100	27 - 45.5	5	0.46
100 - 150	45.5 - 68	6	0.56
150 to market	68 to market	8	0.74

1lb = 0.454 kg 1 ft² = 0.0929 m²

FIG 3-42c: CANADIAN STOCKING DENSITIES (GUIDELINES)

	Minimum Space Required per Pig (m²)		
Pig Weight (kg)	Suggested Floor Type		
	Fully Slatted	Partially Slatted	Solid
10	0.16	0.18	0.21
20	0.26	0.29	0.33
50	0.48	0.53	0.61
75	0.62	0.7	0.8
90	0.7	0.78	0.91
100	0.76	0.85	0.97
110	0.81	0.9	1.03

FIG 3-42d: AUSTRALIAN STOCKING DENSITIES (GOVERNMENT CODE)

Weight (kg)	Space (m²)
10	0.14
20	0.22
30	0.29
40	0.36
50	0.41
60	0.47
70	0.52
80	0.57
90	0.61
100	0.66
110	0.70

FIG 3-43: STOCKING DENSITY AND DAILY GAIN

Av. Pen Space Per Pig		ADG		ADF		FCE	Weight Meat Sold
m²	ft²	g	lbs	kg	lbs		Tonnes
0.49	5.3	640	1.41	2.23	4.92	3.49	125
0.57	6.2	663	1.46	2.28	5.02	3.42	109
0.66	7.1	681	1.5	2.31	5.1	3.38	98
0.73	7.9	704	1.55	2.32	5.18	3.36	90
0.81	8.8	731	1.61	2.39	5.27	3.31	82

ADG = average daily gain.
ADF = average daily feed intake.
FCE = feed conversion efficiency.

FIG 3-44: WEIGHT, SPACE AND BEST PERFORMANCE

Weight of Pig		Space	
kg	lbs	m²	ft²
23 - 55	50 - 120	0.65	7
55 - 114	120 - 250	0.93	10

Note: Legislation requirements may well limit the ability to operate at maximum efficiency.

Whilst suckling the sow, piglets catch very few of the infections present in the herd.

Weaned pigs catch infectious diseases from older growing pigs whom usually have built-up immunity.

Two major factors are involved:

1. The sow is not shedding many infectious pathogens.
2. The piglet is strongly protected against infection.

In the majority of cases piglets start becoming infected after they are weaned, when the milk antibodies have stopped and the colostrum antibodies are wearing off. They become infected from the older pigs in the first and second stage weaner accommodation and further infected when they enter the grower accommodation. Thus in herds in which respiratory infections are endemic the pigs only start to show clinical signs when they are 7 to 10 weeks old.

This important principle is now used to control or eliminate disease.

TERMINOLOGY

As so often happens when new procedures are developed, a variety of different ill-defined terms are invented to describe them. So let's first list and then define the terms used for the procedures related to segregated weaning (see Fig 3-45).

Medicated early weaning (MEW)

(See also Chapter 4)

Pregnant sows are removed from the herd in small groups at about 110 days of pregnancy (gestation). They are washed and medicated against whatever infections are to be eliminated. The piglets are medicated from birth and weaned away to a separate clean site at 5 to 6 days of age. Farrowing is induced with prostaglandins at 113 to 114 days of age. The sows may also be vaccinated against 1 or more infections in the herd before being moved to the farrowing accommodation.

Aim: To produce high-health status breeding stock, free from the infectious pathogens present in the herd of origin.

Effectiveness: This has been used successfully by breeding organisations on a large scale. It is comparable to primary SPF repopulation.

It is not a practicable procedure for commercial operations. It will only be effective if the source herd does not have active disease.

Modified medicated early weaning (MMEW)

This is the same as MEW, except that the sows are not removed to isolated farrowing accommodation but are allowed to farrow in the herd of origin, the pigs being weaned away to a separate site at 5 to 10 days. The age of weaning depends on the infectious pathogens that are to be eliminated.

Aim: The same as MEW but less all-embracing. The range of infectious pathogens to be eliminated is not quite so comprehensive. MMEW can also be used to move pigs from a diseased herd to a healthy herd.

Effectiveness: Can be variable. Different pathogens require different weaning ages and some need medication and/or vaccination of the sows and/or piglets.

Different pathogens require different weaning ages and some need medication and/or vaccination of the sows and or piglets as illustrated in Figs 3-46 and 3-47.

The source herd must not have new pathogens. If it becomes infected with a new pathogen during the procedure, the pathogen is likely to break through the system. Because of this, if MMEW is used for moving pigs from a diseased herd to a healthy one, strict 4 to 8 week quarantine and testing has to be incorporated into the procedure.

Segregated early weaning (SEW)

This term covers Isowean, SW and MEW but may be done without the use of medications.

Aim and Effectiveness: The same as MMEW.

Segregated weaning (SW)

This is the same as SEW except that the piglets are weaned away from the farm at a more conventional weaning age, say, 21 to 28 days.

Aim: When used routinely its purpose is to produce healthy slaughter pigs.

Effectiveness: Obviously this is not as comprehensively effective as MEW, MMEW or SEW but it is more practicable for commercial production and is the basis of three-site production and where combined with all-in/all-out systems is highly effective in producing healthy high-performance weaners, growers and finishers.

Isowean

This term (short for isolated weaning) has become much broader in its meaning and is now used to cover MMEW, SEW, SW and even sometimes three-site and multi-site production.

Partial depopulation

This method applies the principles of SW but in the combined breeding/finishing herd. To start the system pigs are weaned for a period of 6 to 10 weeks into separate, naturally ventilated straw-based accommodation or outside farrowing arks. They should be far enough away from the endemically infected pigs to prevent droplet spread (> 15 m). The growing/finishing herd of endemically infected pigs is either then totally depopulated or each house is emptied sequentially. The segregated pigs are returned in to the housing system without contact with the infected pigs if these are still on the farm.

FIG 3-45: SEGREGATED WEANING PHRASIOLOGY

Original Terms	Alternative Terms
Medicated early weaning (MEW).	Classical MEW.
Modified medicated early weaning (MMEW). Isowean, segregated weaning (SW).	Segregated early weaning (SEW).
Partial depopulation.	Partial depop.
Two-site production.	Two-site segregated weaning.
Three-site production.	Farrowing, nursery, finishing.
Cooperative production.	Local farmers working together.
Multi-site production.	Multiple-farm segregated weaning.

FIG 3-46: SEGREGATED EARLY WEANING

The Oldest Age at Which Pigs can be Weaned to a Segregated Site and be Reliably Free From Contamination by Pathogenic Organisms Endemic in the Herd

Infection/Disease	Age (Days)	Medication/Vaccination * for added safety
Actinobacillus pleuropneumoniae (App).	28	Medication + vaccination.
Aujeszky's disease/pseudorabies virus (AD/PRV).	21	Vaccination.
Swine influenza virus (SIV).	16	None.
Mycoplasma hyopneumoniae/Mycoplasma (enzootic) pneumonia (EP).	10	Medication + vaccination.
Pasteurella multocida (toxigenic)/Progressive artophic rhinitis (PAR).	8	Medication + vaccination.
Porcine reproductive and respiratory syndrome (PRRS) virus.	16	None.
Salmonella choleraesuis.	16	Medication.

* Vaccination of the sow > 2 weeks before farrowing.
Medication of the sow and/or piglets with an appropriate medicine against the organism.
Medication and vaccination are not always necessary but increase the reliability.

FIG 3-47: ENTERIC DISEASES

The Oldest Age at Which Pigs can be Weaned to a Segregated Site and be Reliably Free From Contamination by Pathogenic Organisms Endemic in the Herd

Enteric Diseases	Age (Days)	Medication/Vaccination * for added safety
Coccidia.	–	Not possible.
E. coli.	–	Not possible.
Internal parasites.	14	Medication.
Porcine parvovirus.	28	Vaccination.
Transmissible gastroenteritis (TGE), Porcine epidemic diarrhoea (PED).	21	None.

* Vaccination of the sow > 2 weeks before farrowing.

This method can be applied to the breeding/finishing farm which cannot adopt a true SW system. If the farm has a problem with endemic respiratory disease it is first necessary to break the cycle. This is carried out by rearing the pigs into separate accommodation for a period of 6 to 10 weeks or so, whilst the remainder of the pigs are sold off the farm. (Fig 3-48). Such pigs are weaned either within the farm perimeter or its surrounds depending on facilities, into either straw-based kennels or outdoor arcs. They are not allowed droplet contact with endemically infected pigs in the existing houses and as far as possible are separated on a weekly basis. In practice the distance under natural ventilation need be as little as 15 m, and under fan-assisted ventilation, 35 m.

A Dutch straw barn can make ideal temporary accommodation for housing the weaners. Each batch of weaners should be separated by walls made of straw bales (or outdoor arcs 15 m apart). During the next 6 to 10 weeks as the endemically infected pigs are sold from the farm, each house is depopulated, washed and disinfected. Once the weaner-to-finisher accommodation has been emptied and cleaned (Fig 3-48) the segregated weaners re-enter the buildings.

Each building however must be split into sections, each to hold 1 batch worth of pigs (or part) so that an all-in/all-out management operation is established. This is important to its continuing success. The design and layout of temporary (or permanent) accommodation that can be used is shown in Fig 3-49 and Fig 3-50 and some farms with the availability of straw have continued to use this method very successfully.

Most combined breeding and finishing farms, even if established as specific pathogen free herds, ultimately become infected with 1 or more pathogens. These become endemic due to the continual use and management of pig houses (Fig 3-51) with worsening feed efficiency, reduced daily liveweight gain and poor profitability. Furthermore when a pig is moved into a house already occupied with other pigs it is exposed for the first time to a new range of organisms, both pathogenic and non-pathogenic. It is now recognised that this exposure, even though there may be no clinical disease, stimulates the immune system, with increased demands for energy and lysine and the result is poor growth. Segregated weaning systems break the cycle and production and health can be maintained.

Infections of the respiratory tract seem to have the biggest adverse effect on efficiency of lean meat growth. This is probably because a disproportionately large portion of the body's immune cells are in or associated with the respiratory tract. Every time the pig breathes in air it inhales bacteria and viruses along with dust and toxic gases. The effect of this on a respiratory tract that is already chronically infected and diseased is to repeatedly stimulate the immune system to fight off the insult. The energy required for this is very high. It has to be derived from the pig's food, to the deprivation of growth.

The respiratory tract is the main antigen sampler and immune stimulator of the pig's body.

To illustrate this, a trial experiment was carried out on a breeding-finishing farm where respiratory disease was endemic. 128 pigs were weaned at an average of

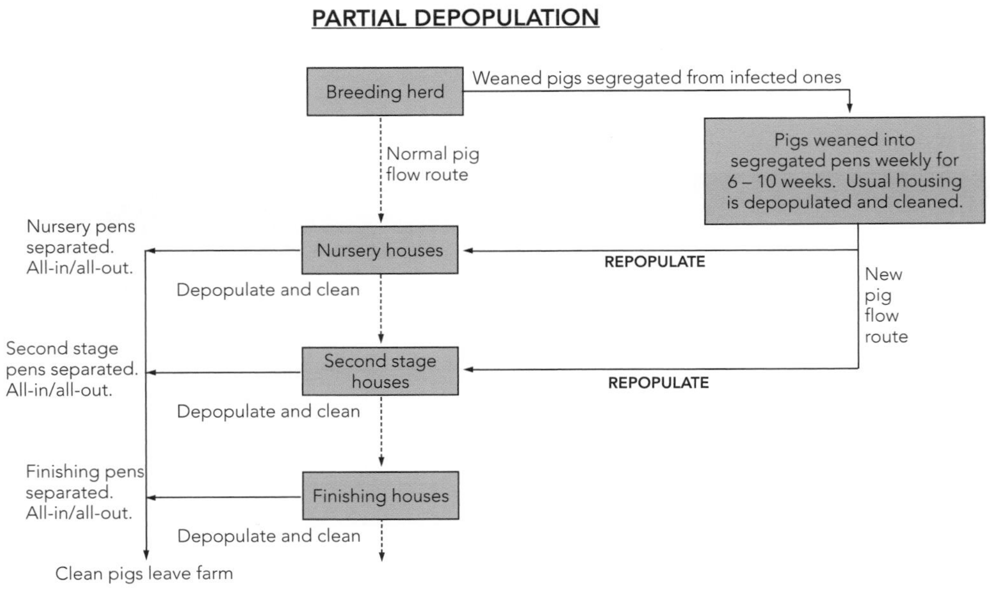

Fig 3-48

27 days of age and immediately moved off-site to a clean straw yard with no other pig contact. 143 pigs were left on the farm as controls (see Fig 3-52). The growth rate differences were spectacular, a 16 % improvement in feed conversion efficiency and a 74 % improvement in daily liveweight gain (DLWG) at the end of 49 days. Half the healthy pigs were then returned to the farm where FCE became worse by 21 % and DLWG by 40 % over the next 48 days.

It is significant to note that the off-site weaned pigs were still infected with mycoplasma (enzootic) pneumonia and PRRSV.

Partial depopulation can also be carried out by building, instead of total weaner/finisher depopulation, but the decision would depend upon the siting and distances between the buildings. Systems using partial depopulation in pig-dense areas would wean between 21 and 24 days of age. At these ages mycoplasma (enzootic) pneumonia and PRRSV will be maintained but other pathogens eliminated or controlled.

STRAW KENNELS FOR WEANERS – UP TO 75 X 7 KG PIGS

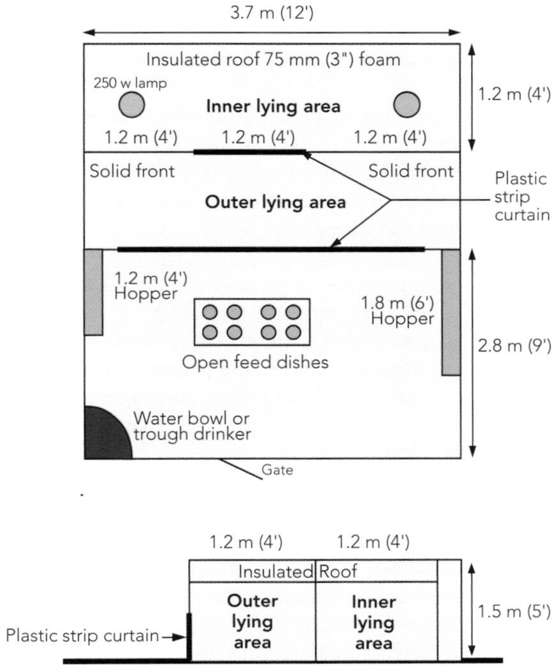

Fig 3-49

STRAW KENNELS FOR WEANERS SIDE VIEW

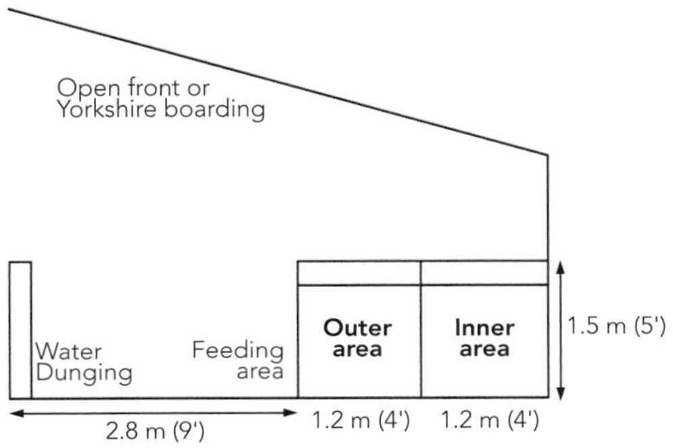

Fig 3-50

ENDEMIC RESPIRATORY DISEASES

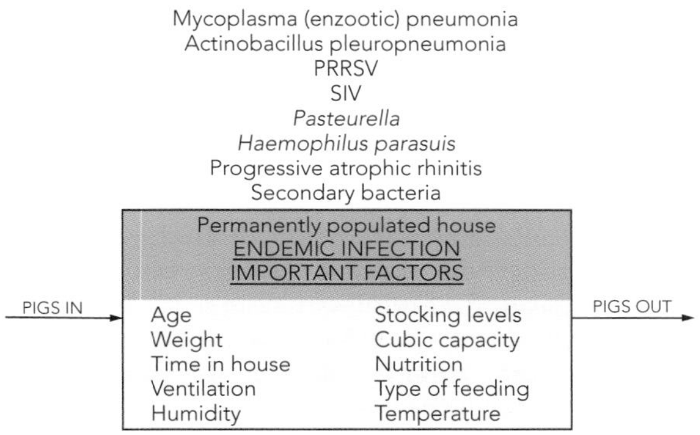

Fig 3-51

OFF-SITE WEANED PIGS 21 – 31 DAYS OLD

TESTED OVER 49 DAYS

	OFF SITE	CONTROLS
No. of pigs	128	143
Wt. start (kg)	9.29	8.34
Wt. finish (kg)	32.2	21.05
DLWG	438 g	252 g
FCE	1.72	2.05

½ CONTINUED OFF SITE → ½ RETURNED TO ORIGINAL FARM

TESTED OVER 49 DAYS

	½ CONTINUED OFF SITE	½ RETURNED TO ORIGINAL FARM
Wt. at finish	74 g	59.6 kg
DLWG	899 g	540 g
FCE	2.1	2.68

Fig 3-52

Factors to consider when adopting partial depopulation

- Discuss with your veterinarian the pathogen profile on the farm. It should be possible to eliminate a number of pathogens from the breeding herd; the following pathogens if they are causing severe clinical disease:
  - *Sarcoptes scabie*i var *suis* (mange) – by medication (see Chapter 11).
  - Toxigenic *Pasteurella multocida* (progressive atrophic rhinitis) – by sow vaccination after a 6 month period.
  - *Actinobacillus pleuropneumoniae* (severe pleuropneumonia) – by preventing droplet spread, and vaccinating and medicating the sow herd.
  - *Brachyspira hyodysenteriae* (swine dysentery) – by medication (see Chapter 9).
  - *Mycoplasma hyopneumoniae* (mycoplasma (enzootic) pneumonia (EP)) – will not be eliminated but it can be controlled effectively by vaccinating piglets at 1 and 3 weeks of age.
  - PRRSV – may remain but its effects will be minimal with endemic diseases removed and a segregated all-in/all-out system.
- The weaning age could vary between 0 and 28 days depending upon the level of pathogen control required.
- The partial depopulation programme is best carried out during warmer months of the year.
- All houses or sections of houses must be adapted for all-in/all-out use.
- All weaner and finishing pigs should be sold from the farm before the segregated weaners re-enter the building again. Pigs may enter the houses after they have been cleaned and empty for 14 days.
- A study of the pig flow should be made to see if by better use of houses, greater control of the system can be achieved.
- Allow a 2 week period between depopulation and repopulation of each house.
- If a *Brachyspira hyodysenteriae* (swine dysentery) eradication programme is being carried out, allow at least 4 weeks, depending on disinfection procedures (see Chapter 9).
- Consider batch farrowing to give better use of buildings.

Field experiences and the advantages of partial depopulation

- There is no loss in pig production and minimal loss in sale weight of pigs and cash flow.
- Respiratory and enteric diseases can be controlled effectively.
- Postweaning mortalities have been reduced from 12 to 4 % and maintained.
- Reduction of medication use by over 50 %.
- Continuous in-feed medication can be removed and only occasional strategic medication may be required.
- Increases in daily liveweight gains of up to 22 % have been achieved.
- Buildings can be altered and maintenance carried out.
- Significant improvements in health status can be maintained.
- Mange, progressive atrophic rhinitis, pleuropneumonia and swine dysentery may be successfully eradicated.

Successful systems have been maintained for at least 6 years.

Partial depopulation is an effective method of upgrading health status without total depopulation of the herd, which might not be advisable in a pig-dense area.

Two-site production

Traditionally, weaners have been reared to about 30 kg (65 lbs) on the same site as the dry sows and farrowing accommodation. However, the term "two-site" in the present context implies that the piglets are weaned at 3 weeks of age to a second separate site where they are housed to slaughter weight (Fig 3-53).

Aim: To produce healthy growers and finishers.

Here the breeding farm produces only piglets to weaning and moves these to a separate combined nursery and finishing farm.

The finishing farm would ideally have separate houses but more likely separate sections for each batch supply of weaned pigs. To reduce droplet spread infections, each house should be separated by at least 15 m.

In this system it would be advisable to vaccinate pigs against mycoplasma pneumonia and wean between 20 to 27 days of age depending upon the disease status to be achieved. In the European Union routine weaning under 28 days is not allowed unless for health reasons and only above 21 days if the weaners are moved into a clean building. It is not legal to wean below 21 days in the European Union.

Three-site production

The sows and boars are on the first site where the piglets are born and suckled. They are then weaned to a second site and at 25 to 30 kg (55 to 66 lbs) are moved to the third site for growing and finishing (Fig 3-54).

Aim: To break the cycle of infection.

Here as the name implies the system is separated into breeding, nursery and finishing sites. The latter site can be further adapted by housing separation as in two-site SEW to allow all-in/all-out on a batch basis.

Cooperative production

Groups of producers can combine to take advantage of three-site or multi-site production in many different ways including changing some farms to breeding-only, some to nursery production and others to finishing. Coperative production may also allow transfer of information regarding specific pathogens present on different farms, allowing for regional elimination of pathogens. This can be extremely beneficial in swine dysentery and PRRSV control.

TWO-SITE PRODUCTION

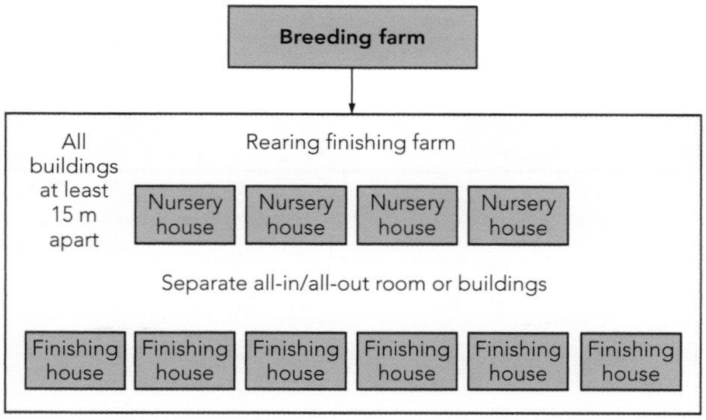

Fig 3-53

THREE-SITE PRODUCTION

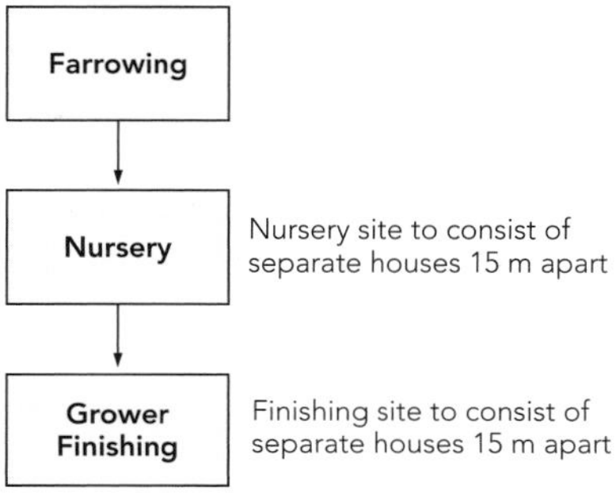

Fig 3-54

Multi-site production

The sows are all on one or more sites where they farrow. On the same day of the week the piglets are all weaned away to an all-in/all-out nursery where they are reared for 7 weeks. In each successive batch all the suckled piglets are weaned to different all-in/all-out nursery sites. After 7 weeks the first nursery is emptied, cleaned and disinfected and refilled with young weaners again. There are thus with weekly batch systems 8 separate nursery sites. Each successive batch when a nursery is emptied, the 30 kg (66 lbs) weaners are moved to separate all-in/all-out grower/finisher sites. There are 16 of these filled and emptied in rotation.

Aim: To break the cycle of infection and maximise growth and feed conversion.

Effectiveness: This is very effective on a commercial basis for large-scale commercial production. The all-in/all-out system mitigates against the occasional leak of single pathogens coming through, however they will only affect one nursery site and one finisher site. Producers have made the mistake of mixing up SEW and SW, weaning too young.

Using this system a number of breeding farms may cooperate and supply piglets all of similar ages together on a batch-by-batch basis to the separate weaning units (see Fig 3-55). For it to be successful, however, the age at weaning needs to be tightly controlled and pigs of older age must under no circumstances be fostered backwards. They should be removed from the system. Furthermore if there is a disease outbreak such as TGE on one of the breeding farms the result can be disastrous. The distances between breeding farm, nurseries, growing buildings and finishing houses should ideally be at least 3 km (2 miles) apart to take full advantage of SEW. The closer to each other they are, the greater the risk of windborne spread of diseases such as mycoplasma (enzootic) pneumonia, PRRSV, SIV and Aujeszky's disease. However, for droplet-spread diseases such as progressive atrophic rhinitis and actinobacillus pleuropneumonia,

MULTIPLE FARM MULTI-SITE PRODUCTION

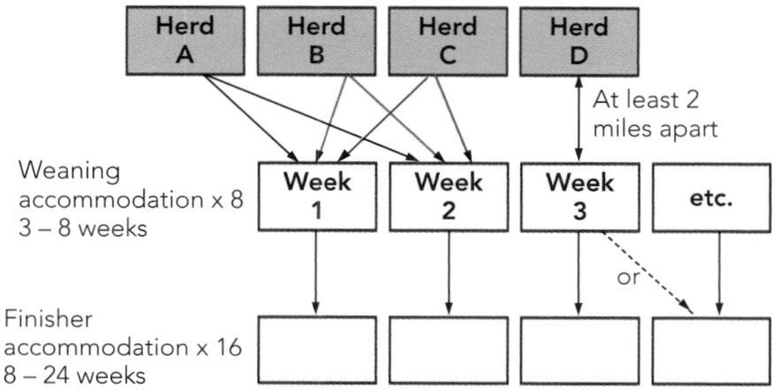

Fig 3-55

500 m is adequate, but clearly the risk is slightly increased.

This can be combined with batch farrowing so that each farm produces sufficient pigs to fill the nursery in one go. For example, 3 x 1000-sow units producing 333 pigs a week will fill a 1000-head nursery – but from 3 different sources. Whereas, the same 1000-sow units batch-farrowing every 3 weeks will produce 1000 weaners per batch.

Single sow herd multi-site production

This principle is illustrated in Fig 3-56. There are 8 separate nurseries holding pigs until 10 weeks of age. They are then moved to separate finisher accommodation each week, again all-in/all-out. For this to operate in practice, the sow herd has to have at least 1500 to 2000 sows.

It has the huge advantages of combining all-in/all-out production by site with those of segregated weaning. The development of outdoor sow herds provides an ideal base for multi-site production.

The SW principle can be adopted, with advantages to disease control, in many different ways, and that method best suited for the farming system must be identified.

It should be pointed out that SW is not a foolproof system and it can create considerable problems if there are a few piglets carrying disease amongst a naive population. Typical examples would be mycoplasma (enzootic) pneumonia, actinobacillus pleuropneumonia or TGE. Streptococcal meningitis is not controlled by SW and serious outbreaks have occurred.

SINGLE SOW HERD MULTI-SITE PRODUCTION

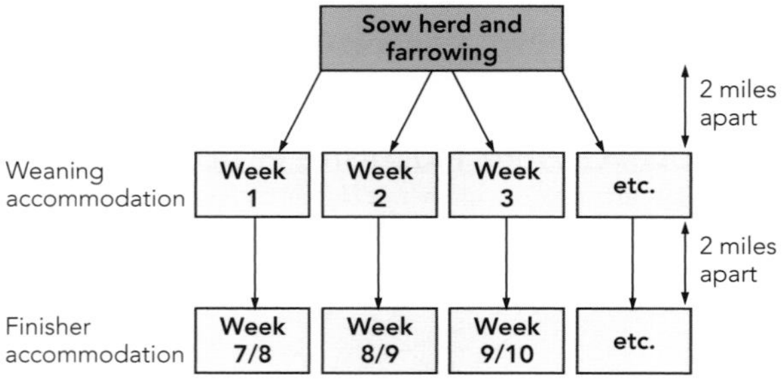

Fig 3-56

Parity segregation

Under this system the young breeding females are segregated from the remainder of the herd on the assumption that their immune systems are more susceptible to disease.

Gilts are bred and the first parity sows farrow on one farm with their offspring being raised on that farm. The second parity sows are transferred from this farm at around 6 weeks of gestation to sow-only farms where their offspring are finished separately.

The advantages are that the sow litters have a post-weaning mortality of 3 % compared with the parity 1 litters of 9 %, and the pigs reach slaughter weight 10 days earlier (Fig 3-57). The basis for this is that the sow litters have more resistance to pathogens due to a more developed immune system.

FIG 3-57: GILT AND SOW EXPECTATIONS

Expectation	Gilt	Sow
Litter weight kg	96	100
Finishing rate %	91	96
Days to 115 kg	162	154
Growth rate g/day	670	700
Feed conversion rate (FCR)	2.7	2.4

Nutrition and Feeding

See Chapter 14 for further information

Feed intake

(See the individual sections for each of the age groups)

Feed quality and ingredients

The palatability of the feed is vitally important and should be checked by tasting the feed. Where feed has been shown to be very unpalatable, weaner and growing pigs fail to eat substantial quantities of feed.

Feed consumption needs to be monitored, in particular in the farrowing area. A sow will look after her piglets if her feed is well-catered. Enhancing lactation feed intake reduces wean-to-service intervals and increases farrowing rates and subsequent litter sizes.

Nutrition and feeding provide an important tool for management manipulation in the control of disease. The quality of the feed, its methods of presentation and the amounts of feed provided are part of the health management process. More detailed discussion is given in Chapter 14, but it is relevant at this point to highlight the periods in the production cycle when feed levels and quality of nutrition should be carefully considered and monitored.

The gilt (see Chapters 7 and 8)

- Selection to the onset of puberty.
- Puberty to point of mating.
- Mating to 21 days post-mating.
- During the last month of pregnancy.
- Farrowing to 3 days post-mating.
- During lactation.

The sow (see Chapters 7 and 8)

- Weaning to service.
- Service to 2 days.
- 2 to 21 days post-mating.
- 7 days before farrowing.
- During lactation.

The piglet

- 7 to 21 days of age.
- During disease outbreaks.

The weaner (see Chapter 9)

- Weaning to Day 14.
- Dietary change unsuitable for the weight and age of the pig.
- Whenever a housing change is made.
- During disease outbreaks.

Grower and finisher pigs

- Any dietary changes.
- A change from dry to wet feeding.
- Changes in house temperature.
- Changes in the feeding system.
- After pigs are mixed.
- Environmental changes.
- When new cereals or corn are introduced.

Water

The ready availability of clean, fresh water is essential. Insufficient attention is given to this on many farms. It is useful to consider the role that water plays in the normal metabolic functions of the pig:

- It helps to maintain and control body temperature, through both the intake and during exhalation when the heat is dissipated from the pig. It is lost in 3 ways, either by respiration, in the urine or in the faeces.
- An imbalance between water intake and loss results in dehydration and increased concentration of urine. Clinical signs include very dry faeces, hollow eyes and a dehydrated skin.
- It is responsible for transporting food and waste products throughout the body. Waste products are eliminated via water through the kidneys.
- Hormones are transported around the body through the bloodstream.
- Water regulates the acid/alkali balance in the body through the controls exerted by the kidney.
- Water is used in protein synthesis. The digestive process will not function without it.

Any restriction of water therefore will affect the above vital functions. The farm should remotely monitor water use (via the internet) every 15 minutes. If this is not possible then monitor at least daily. Water usage can provide critical information on the health of the pigs, predicting feed intake and temperature requirements. An outbreak of SIV will result in a 10 % drop in water use even before the first sneeze! Water intake is a major driver of feed intake and thus daily liveweight gain.

The Piglet

Within 6 hours of birth, water should be made available in a shallow dish or a trough because fluid intake is so vital at an early age. An efficient dish used for both creep and water in the farrowing pen is shown in Fig 3-58. It is interesting to note how many piglets within 24 hours will drink small amounts of water when given the opportunity. Nipple drinkers are not a very attractive method of presenting the water to the piglet. Water consumption by piglets during lactation is also influenced by the farrowing house temperature, and at 28 °C (82 °F) in a warm creep area water requirements will increase dramatically. The provision of water to the piglet in the first week causes no harm and is more likely to be of benefit particularly if diarrhoea occurs on the farm as the pig will already be accustomed to water. For pigs from 1 to 3 weeks of age, clean water is best presented in an open type drinker rather than a nipple drinker. The water in dishes and drinkers must be clean and fresh.

A CREEP DISH FOR EITHER FEED OR WATER

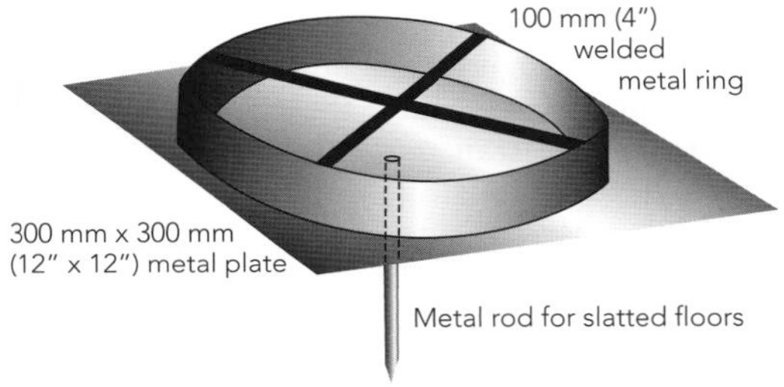

Fig 3-58

The Weaned Pig

The pig experiences dramatic changes at weaning by the sudden move from a liquid to a solid diet. The conditioned reflex, calling the pigs to suckle regularly, is also lost. Dehydration associated with poor water intake and marked villus atrophy is a common occurrence within the first 7 days of weaning. Ensure the flow rate is at least 0.7 litres per minute from nipple drinkers. It is advisable to offer water in small open drinkers or water bowls daily for the first 5 to 7 days postweaning. The loss of milk at weaning time and villus atrophy reduce the availability of liquid to the pig for the first 48 hours. Check the salt concentration in the weaner's water supply.

The Sow

The changes in water intake from pregnancy to lactation are considerable. Sows that have a lower water intake during lactation generally rear poorer litters and it is important therefore to encourage the sow to drink, the moment she enters the farrowing quarters. This is best carried out by giving 4.5 litres of water twice daily into the feed trough until 2 to 3 days post farrowing. The water flow through a nipple drinker for the lactating sow should approximate 1.5 to 2 litres per minute. Water intake in the dry/gestating sow varies from 9 to 18 litres per day and in lactation from 18 to 40+ litres.

Guidelines for water requirements, water flow rates, drinker heights and drinker to pig ratios are given in Fig 3-59 and Fig 3-60.

FIG 3-59: WATER REQUIREMENTS

Weight of Pig (kg)	Daily Requirements (litres)	Minimum Flow Rate Through Nipple Drinkers (litres/min)	Guideline Heights for Nipple Drinkers (mm)
Newly weaned	1.0 - 1.5	0.5	100 - 130
Up to 20 kgs	1.5 - 2.0	0.7 - 1.0	130 - 300
20 kgs – 40 kgs	2.0 - 5.0	1.0 - 1.5	300 - 460
Finishing pigs up to 100 kgs	5.0 - 10	1.2 - 1.5	460 - 610
Sows and gilts – pre-service and in-pig	9 - 18	2.0	610 - 760
Sows and gilts – in lactation	40 +	2.0	610 - 760
Boars	5.0 - 12	2.0	760 - 910

Nose/bite drinkers for lactating sows should be set at 760 to 910 mm

FIG 3-60: DRINKER TO PIG RATIOS

Type	Rationed Feeding	Ad Lib Feeding
Nipple	1 : 10	1 : 15
Bite	1 : 10	1 : 15
Bowl	1 : 20	1 : 30
Double sided trough (per 300 mm)	Up to 15 kg weight – 35 pigs per trough 15 kg to 35 kg weight – 30 pigs per trough Over 35 kg weight – 25 pigs per trough	

Water Supply Examination

The water supply needs to be examined from the entry point into the room through to the drinker. In some countries, the UK for example, water from the mains supply cannot be used directly because of the risk of backflow contamination and a non-return valve, often in the form of a header tank is used to prevent backflow. If present, examine the header tank hygiene and use. Is the size suitable for the number of pigs in the room? The header tank can be valuable for possible water medication routines. There should be a lid, acting as a light seal to reduce algae growth and to reduce aerial contamination of the water.

While walking around the room, examine as many drinkers as possible. In the farrowing house, all sow drinkers should be examined as a matter of course.

Examine the drinkers for:

1. Any leaking drinker – these are costly and a major cause of wasted water.
2. Cleanliness – a dirty-looking drinker is generally not working properly.
3. Location – can the pig gain access to the drinker properly?
4. Height and angle – is it appropriate for the pigs using the drinker? Classic mistakes occur when drinkers are not lowered in between batches of pigs.
5. Flow of water – this may need to be measured using a measuring jug and a stop watch.
6. Temperature of the water – measured by an infra-red gun on the collected water above.
7. The colour and taste of the water. Water high in iron is reddish and if high in sulphate can be foul-tasting.
8. Pressure of water from the drinkers.

Any drinker that fails to supply adequate water should be examined in more detail. This is likely to involve removing the drinker from its down-pipe bracket. When the drinker is removed, check that the flow of water is adequate from the down-pipe. If the flow is poor, examine the pipe in more detail. Partially blocked pipes from lime and other sediments, including biofilms, can reduce the internal bore.

Continue the examination by dismantling the drinker; in particular look for blocked and dirty filters, drinker settings and damaged seals.

Once the drinkers are fixed or replaced, reattach the drinker and ensure that the water supply is restored to the pigs.

Water Quality

The variables in water quality include organisms, the physical characteristics and the mineral content. Water can become contaminated with pathogenic and non-pathogenic bacteria and viruses. The presence of coliform bacteria (i.e. *E. coli* and related orgamisms) is an indication of faecal contamination and a potential source of disease (Fig 3-61).

The chemical quality of water can be assessed by determining the total dissolved solids, the pH (the alkalinity or the acidity), the iron content and the presence of nitrates or nitrites. Further testing would include levels of sulphates, magnesium, chloride, potassium, calcium, sodium and manganese. The total solids in water represent the amount of matter that is actually dissolved. If

FIG 3-61: GUIDELINES TO WATER QUALITY SUITABLE FOR PIGS

Nutrient	ppm (parts per million) Less than
Calcium	1000
Chloride	400
Copper	5
Fluoride	2 – 3
Hardness (calcium carbonate level)	< 60 soft water > 200 hard water
Iron	0.5
Lead	0.1
Magnesium	400
Manganese	0.1
Mercury	0.003
Nitrites	10
Nitrates	50
Phosphorus	7.8
Potassium	3
Sodium	150
Selenium	0.05
Solids dissolved	1000
Sulphate	1000
Zinc	40
Total viable bacterial counts (TVC) per ml 37 °C (99 °F) 22 °C (72 °F)	Low, but more important no fluctuation between samples. Target < 2×10^2 > 1×10^4 poor
Coliforms/100 ml	Zero

this level is less than 1000 ppm it is of no significance but once it reaches over 2000 ppm it becomes unfit for pigs. Generally if the total solid content is low it is usually of good quality and the water is safe to drink. The pH level of good water varies between 6.5 and 8.

Hardness of water is dependent on the levels of calcium and magnesium present but these have no effect on animal health at these levels. Hardness does however result in the accumulation of scale, causing pipes to gradually block, and the flow rate drops unnoticed. This is a common problem on farms that have metal pipes of at least 4 years' standing.

Iron can cause problems in water, with brown-coloured staining. Certain types of bacteria can grow and cause blockage of pipes.

High levels of nitrates and nitrites can interfere with the use of vitamin A by the pig and they may be responsible for high still-birth rates.

A summary of the effects of high mineral levels in water

Sodium and chloride

- If this is above 500 ppm then a brackish taste may develop.
- High levels of sodium chloride (salt) affect palatability and can adversely affect pig productivity and performance.
- Sodium sulphate is a laxative and mildly irritant.

Calcium and magnesium

- There are no effects on animal health unless there are high levels of the sulphates which result in the accumulation of scale (as $Mg(OH)_2$ and $CaCO_3$) and over a period of time the diameter of pipes is reduced with the poor flow rates.

Iron and copper

- High levels of copper have a catalytic effect on the oxidation of iron, and if the iron levels are high, precipitation of iron occurs when water is pumped, resulting in problems with the delivery system.
- Iron also supports the growth of certain types of bacteria, causing foul odours and blocked water systems.
- High levels of iron in the water may reduce lactation feed intake.

Sulphate

- High levels of sulphate in association with magnesium and sodium can cause diarrhoea.
- High sulphate may cause palatability problems.
- High sulphate may reduce lactation feed intake.

Manganese

- High levels promote oxidation leading to a reddish tinge in the water.

Nitrates/nitrites

- Nitrites can change the structure of the haemoglobin in blood, rendering it incapable of transporting oxygen. If levels are high the blood is a dark colour due to lowered levels of oxygen.
- Extremely high levels of nitrates/nitrites in water impair the utilisation of vitamin A in pigs and cause a reduction in performance – such levels however are very rarely found under practical conditions, but levels can be sufficiently high to increase stillbirths.

4 TREATING SICK AND COMPROMISED PIGS

4 TREATING SICK AND COMPROMISED PIGS

Understanding Medicines

You are advised to consult your veterinarian when assessing information given in this chapter.

The treatment of sick and compromised pigs in the pig unit is complex, with a wide range of medicines available for a variety of conditions. This chapter looks at some of the complexities and interactions involved in the use of medicines so that treatments can be carried out efficiently and without risk.

Specific treatments for diseases are discussed later in their relevant chapters.

Legal Requirements

Medicines must be used safely and correctly in food producing animals to ensure there are no residues. Most countries across the world have strict controls over both the methods of prescribing and the uses of medicines. In the EU for example medicines may be considered in 4 main categories, which are adapted in the UK as follows:

1. **AVM-GSL:** Authorised veterinary medicine – general sales list. This may be sold by anyone (formerly GSL).
2. **NFA-VPS:** Non-food animal medicine – veterinarian, pharmacist, Suitably Qualified Person. A medicine for companion animals which must be supplied by a veterinarian, pharmacist or Suitably Qualified Person (formerly PML companion animal products and a few P products).
3. **POM-VPS:** Prescription-only medicine – veterinarian, pharmacist, Suitably Qualified Person. A medicine for food producing animals (including horses), to be supplied only on veterinary prescription, which must be prescribed by a veterinarian, pharmacist or SQP (either orally or in writing) and which must be supplied by one of those groups of people in accordance with the prescription (formerly PML livestock products, MFSX products and a few P products).
4. **POM-V:** Prescription-only medicine – veterinarian. A medicine, to be supplied only on veterinary prescription, which must be prescribed (either orally or in writing) by a veterinarian to animals under their care following a clinical assessment, and which may be supplied by a veterinarian or pharmacist in accordance with the prescription (formerly POM products and a few P products).

When a medicine is supplied to the farm the following information should be available:

- A description of the medicine.
- The date of manufacture.
- The date of dispensing.
- The date of expiry.
- The client's name and address.
- The species to be treated.
- The date of withdrawal.
- The dose rate and instructions for use.
- The name and address of the supplier.
- The manufacturer's batch number.
- The name and address of the veterinarian prescribing.

A typical bottle label would appear as shown (Fig 4-1).

EXAMPLE: MEDICINE LABEL ON A BOTTLE

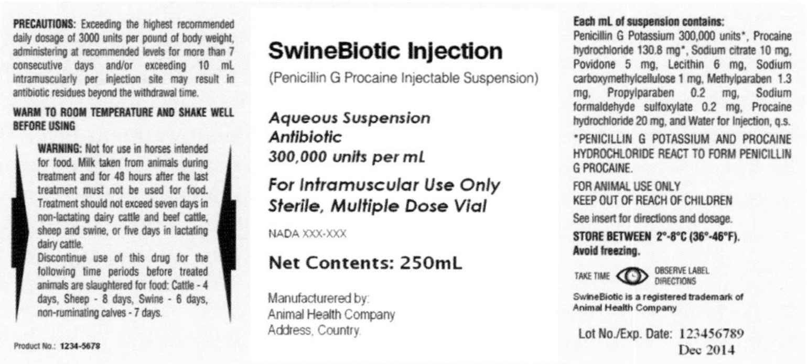

PRECAUTIONS: Exceeding the highest recommended daily dosage of 3000 units per pound of body weight, administering at recommended levels for more than 7 consecutive days and/or exceeding 10 mL intramuscularly per injection site may result in antibiotic residues beyond the withdrawal time.
WARM TO ROOM TEMPERATURE AND SHAKE WELL BEFORE USING
WARNING: Not for use in horses intended for food. Milk taken from animals during treatment and for 48 hours after the last treatment must not be used for food. Treatment should not exceed seven days in non-lactating dairy cattle and beef cattle, sheep and swine, or five days in lactating dairy cattle.
Discontinue use of this drug for the following time periods before treated animals are slaughtered for food: Cattle - 4 days, Sheep - 8 days, Swine - 6 days, non-ruminating calves - 7 days.
Product No.: 1234-5678
SwineBiotic Injection
(Penicillin G Procaine Injectable Suspension)
Aqueous Suspension
Antibiotic
300,000 units per mL
For Intramuscular Use Only
Sterile, Multiple Dose Vial
NADA XXX-XXX
Net Contents: 250mL
Manufacturered by:
Animal Health Company
Address, Country
Each mL of suspension contains:
Penicillin G Potassium 300,000 units*, Procaine hydrochloride 130.8 mg*, Sodium citrate 10 mg, Povidone 5 mg, Lecithin 6 mg, Sodium carboxymethylcellulose 1 mg, Methylparaben 1.3 mg, Propylparaben 0.2 mg, Sodium formaldehyde sulfoxylate 0.2 mg, Procaine hydrochloride 20 mg, and Water for Injection, q.s.
*PENICILLIN G POTASSIUM AND PROCAINE HYDROCHLORIDE REACT TO FORM PENICILLIN G PROCAINE.
FOR ANIMAL USE ONLY
KEEP OUT OF REACH OF CHILDREN
See insert for directions and dosage.
STORE BETWEEN 2°-8°C (36°-46°F).
Avoid freezing.
TAKE TIME
OBSERVE LABEL DIRECTIONS
SwineBiotic is a registered trademark of Animal Health Company
Lot No./Exp. Date: 123456789
Dec 2014

Fig 4-1

How Medicines are Prescribed

Most medicines have two names, one which describes the chemical which is the active principle, often referred to as the **generic name**, and the second, the manufacturer's own **trade name**. An example of a generic name is oxytetracycline hydrochloride (OTC) which is a broad spectrum antibiotic. Many animal health companies manufacture OTC and trade named OTC products and availability will vary from country to country.

You will be familiar with the trade names of medicines used on your farm but it is important to remember the generic names, because this will help you to understand how they function and how to identify them irrespective of a trade name.

Understanding Dosage Levels

All medicines have a recommended therapeutic range, expressed in milligrams per kilogram (mg/kg) of live bodyweight (kg). This range is used by the veterinarian to decide whether a higher or lower dose level is required. Dose levels may also change with varying bodyweights. Note in the US it may be recorded as mg/lb (milligrams per lb bodyweight).

For information:

1000 ηg (nanograms)	= 1mcg (microgram) also written as μg
1000 mcg	= 1 mg (milligram)
1000 mg	= 1g (gram)
1000 g	= 1 kg (kilogram)
1000 kg	= 1 T (tonne)
mg kg = g/tonne	= ppm (parts per million)
mcg/kg	= ppb (parts per billion)
1000 mcl (microlitres)	= 1 cc (cubic centimetre) or 1 ml (millilitre)
1000 ml	= 1 litre

For conversions to other measurements see appendix.

List the generic names of the medicines on your farm. You will then recognise their use in this chapter. The information will be on the bottle label.

In-Feed Medications

These are prescribed by grams (g) of active or generic substance per tonne (1000 kg) of feed. For example, 500 g per tonne (1000 kg) of oxytetracycline (OTC) means that there are 500 g of active medicine mixed in a tonne of feed. However, the manufacturer's product is normally available as a supplement – a mixture of the generic substance and usually a cereal base. OTC 10 % feed supplement means that OTC is present at a 10 % level. Thus 1 kg would contain 10 % or 100 g of oxytetracycline. To mix 500 g of active ingredient per tonne in this case would therefore require 5 kg of the feed supplement.

Examples of in-feed medication are shown in Fig 4-2. Feed can be analysed for the levels of antibiotic added. For technical and analytical reasons, do not expect 100 % recovery rate. It can often be as little as 40 to 60 %.

Injectable Medicines

Injectable medicines contain the active principle suspended or dissolved in a liquid. The label on the bottle indicates the actual amount of medicine usually as mg per ml. Each medicine has a recommended dose level expressed in mg/kg (or mg/lb in US) of liveweight and instructions as to its administration, frequency and any side effects or contraindications.

Example: OTC Q100 injectable solution contains 100 mg/ml of OTC. The therapeutic level is 10 mg per kg liveweight daily and thus the daily dose level becomes 1 ml per 10 kg of bodyweight.

In practice, instead of referring to mg/kg, it is normal for the veterinarian to prescribe on the basis of ml/kg of liveweight, but guidelines are also usually printed on the label.

Water Medication

Treating pigs via the drinking water involves the same principles, a daily intake of the active primary based on mg/kg of liveweight. In practice however, it is better to consider this by the active medicine required per day per tonne of liveweight of pig. For example, if the water soluble preparation consists of 100 % pure active medicine (in other words no carrier added to it), and the required concentration was 25 g/tonne of liveweight per day, then 25 g of powder would be dissolved in water for each 1 tonne of pig weight. If, however, the powder contains only 50 % of active principle (the other 50 % would be a carrier, usually a sugar or citric acid base), then the dose would be 50 g/tonne of liveweight or twice the amount.

The initial calculation, however, would be based upon mg/kg liveweight. As a guide, pigs drink 100 litres of water per tonne of liveweight per day (10 % of their bodyweight). But note drinkers also waste water – for example nipple drinkers often waste 20 % of the water. Thus the medicated amount needs to be increased 20 %. For example a treatment level of 10 mg/kg of active medicine and using a 50 % powder, 20 g/tonne would be required. In practice if the administration is by nipple drinkers the level is increased by up to 20 % to allow for wastage, thus 24 g of powder for every 1 tonne of pig is actually added to the water system.

FIG 4-2: A GUIDE TO IN-FEED ANTIBIOTICS AND THEIR USE

Medicine Active Principle (Guideline levels in g/tonne)	EP	App	CRD	PAR	SM	SD
Amoxycillin (200 – 500)		✓	✓	✓	✓	
Chlortetracycline (CTC) (300 – 800)	✓	✓	✓	✓	✓	
CTC (165) with sulphadimidine (164) & procaine penicillin (83)	✓	✓	✓	✓	✓	
Lincomycin (110 – 220)	✓		✓			✓
Lincomycin (44) Spectinomycin (44)	✓		✓			✓
Oxytetracycline (OTC) (400 – 800)	✓	✓	✓	✓	✓	
Phenoxymethyl-penicillin (200 – 300)		✓			✓	
Tiamulin (30 – 100)	✓		✓			✓
Tilmicosin (200 – 400)	✓	✓	✓	✓		
Trimethoprim/sulpha (250/750)		✓	✓	✓	✓	
Tylosin (40 – 100)	✓					
Valnemulin	✓		✓			✓

EP – Mycoplasma (enzootic) pneumonia, **App** – Actinobacillus pleuropneumonia, **CRD** – Chronic respiratory disease, **PAR** – Progressive atrophic rhinitis, **SM** – Streptococcal meningitis, **SD** – Swine dysentery.

Controlling and Storing Medicines

To achieve the maximum response to medicines and prevent any abuses, discipline should be maintained in their control, administration and storage. Consider all medicines to be dangerous. Many become potentially toxic if the recommended levels of treatment are exceeded or if they are given in the wrong way.

Some medicines have a very narrow range between treatment levels and poisoning. A good example is monensin, 100 g/tonne is the therapeutic level and 300 g is toxic.

Light and heat destroy medicines and freezing also has an adverse effect, particularly on vaccines. Tetracyclines may break down in light and produce toxic compounds. See Fig 4-3a and Fig 4-3b.

Checklist for medicines

- Provide a locked room or cupboard for all your medicines.
- Make sure that all medicines, syringes and needles are kept well away from children and people not on the staff. Note medicines should be kept away from other animals.
- Provide a refrigerator for vaccines and other medicines as required. Use a maximum minimum thermometer and record temperatures daily.
- Allow only certain designated people to have direct access to the main medicine store.
- Document all medicines in and out of the medicine store.
- Insist on empty bottles being returned before a second bottle is taken out. This prevents black market trade.

FIG 4-3a: MEDICINES THAT SHOULD BE STORED IN THE REFRIGERATOR 2 - 8 °C (36 - 46 °F)	FIG 4-3b: MEDICINES THAT SHOULD BE STORED IN A DARK, COOL PLACE* 18 - 25 °C (64 - 77 °F)
Iron injections.	Sedatives, vitamins and minerals.
All vaccines.	Stimulants.
Hormone injections, e.g. oxytocin.	Antibiotics.
Any bottles that have been opened and are in use.	In-feed and water soluble preparations.
Any other medicines where the label indicates this temperature requirement.	Disinfectants.

* Note – a dark cool place is not on a shelf at room temperature. In many parts of the world medicines would be too hot in the summer months. Pay particular attention to the temperature requirements of all medicine products.

FIG 4-4: A FORMAT FOR RECORDING TREATMENTS

Animal Treatments

Date Commenced	Animal	Identification	Condition/ Disease	Medicine Used/ Bottle Number	Dose per Day (ml) Day							Withdrawal Period (Days)	Date of Clearance	Administered by.
					1	2	3	4	5	6	7			
1 Jan 13	Sow	163	Mastitis	OTC	20	20	20					14	18 Jan 13	JC
2 Feb 13	60 kg	1246	Lame	Pen Strep	10	10	10					21	26 Feb 13	JC

- Agree with your veterinarian the minimum amounts that are required for a given period of time and follow the advice on usage. A month's requirement is typical.
- Make sure that all bottles are labelled for the correct use, that withdrawal periods are displayed and personnel are aware of them.
- Record the date the medicine bottle is first opened.
- Keep a daily record of all medicines used on the farm (Fig 4-4).
- Make sure you have safety data sheets to hand in case of accidents.
- Ask your veterinarian to check your storage and usage of medicines at each visit to ensure that the recommendations are being carried out.
- Check regularly that medicines are in date.
- Dispose of empty bottles, needles and syringes safely. Discuss with your veterinary surgeon.

Follow medicine storage instructions precisely.

Disposing of Medicines

This must be carried out with care to prevent environmental contamination and accidental human or animal contamination.

There are 3 ways of safe disposal:

1. Empty bottles

These should be placed into a plastic bag and disposed of within the local authority guidelines or rules. Return to the veterinary practice you obtained the medicines from.

2. Syringes

Needles must always be removed from the syringe, and the syringes placed in polythene bags, marked "Syringes only" and disposed of within the local authority guidelines or rules. Return to the veterinary practice you obtained the medicines from. It may be possible to incinerate them but check with your veterinary practice first.

3. Needles and needle holders

These should be placed into a Sharps box. Needles should be disposed of as required or ideally returned to the veterinary practice. A Sharps box is a very strong polythene box, with an automatically closing lid where the needles can be dropped through and retained safely (see Fig 4-5). When full these are sealed and disposed of by incineration. Needle guillotines are available to cut the tip off the needles so that they cannot be misused. Handle needles with care to avoid sharp injuries.

A SHARPS BOX

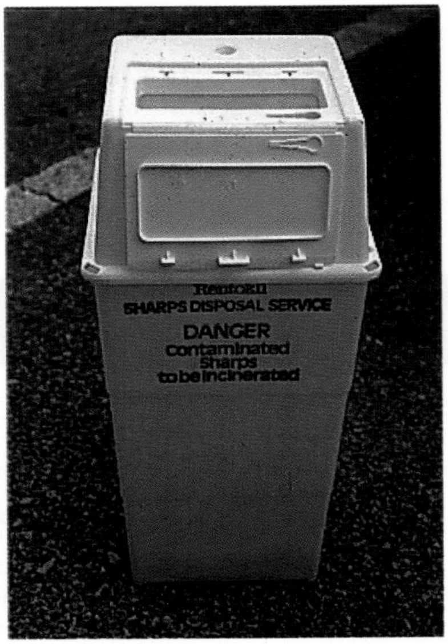

Fig 4-5

FIG 4-6: A DATA SHEET EXAMPLE

Pig-Medicine
Presentation Antibiotic Injection. A sterile white aqueous suspension containing Antbiotic 300,000 iu/ml.
Uses Pig-Medicine Injection is indicated in the treatment of systemic infections caused by or associated with organisms sensitive to antibiotic. When administered by intramuscular injection it will provide an effective therapeutic blood level for approximately 24 hours.
Dosage and administration Shake well before use.
By intramuscular injection only.
Routine daily dosage

Horses and cattle	10 – 20 ml
Calves, foals, sheep, pigs and goats	3 – 10 ml
Dogs	0.5 – 5 ml
Cats	0.5 – 1 ml

Contra indications, warnings etc.
Do not use in known cases of hypersensitivity to antibiotic. Do not inject intravenously.
Occasionally in suckling and fattening pigs administration of products containing antibiotic may cause a transient pyrexia, vomiting, shivering, listlessness and incoordination. Additionally in pregnant sows and gilts a vulval discharge which could be associated with abortion has been reported.
Withdrawal period: Milk taken from cows during treatment and for 48 hours (4 milkings) after treatment must not be used for human consumption. Meat and offal for human consumption 18 days.
Pharmaceutical precautions Store at room temperature not exceeding 25 °C. Protect from light.
Legal category POM-V
Package quantities Multi dose vials of 40 ml and 100 ml.
Further information Nil.
Pig-Medicine Manufacturer Ltd., Address
Marketing authorisation number Vm1234 /5678
GTIN (Global Trade Item No) Pig-Medicine for Injection
01234567891234

Types of Medicine and their Application

Medicines used in the pig industry can be grouped into 8 broad areas:

1. Antibiotics and antibacterial substances.
2. Minerals (for example iron), vitamins and electrolytes.
3. Sedatives and analgesics (painkillers).
4. Parasiticides to treat mange, lice and worms.
5. Vaccines and sera (and miscellaneous medicines).
6. Hormones.
7. Growth enhancers and probiotics.
8. Colostrum supplements.

For each individual medicament a manufacturer's data sheet similar to Fig 4-6 will be available, which gives guidelines as to its use, specific precautions and any contra indications.

Medicines can be administered to individual pigs or groups of pigs for treatment or to prevent disease.

Individual treatments are usually given by mouth in the case of piglets, by injection in older pigs and occasionally onto the skin or per vagina. In outbreaks of disease, group treatments are carried out by in-feed medication, injections or in the water. Treatment for mange or lice may involve the use of sprays. Where medicines are used to prevent disease, they can be used in a number of ways as illustrated in Fig 4-7. The most efficient and cost effective method of treatment is to administer medicines, either by injection or by mouth, to the individual pig. Sick/compromised animals do not eat much, and contrary to popular opinion, they do not drink much either. In outbreaks of disease therefore the treatment of sick pigs in the feed or water is not medically efficient.

Furthermore in a group of pigs affected with pneumonia for example, it is unlikely that more than 20 % of such animals have sufficient lesions to require treatment and 80 % of the group would therefore be treated unnecessarily. This cost must be added to that of the sick/compromised 20 %. However, group medication can be of value in preventing more disease developing. There can be practical problems with treating large numbers of individual pigs. Nevertheless, the response is much better. Medicines are administered in a variety of ways depending on the type of medicine and its availability. Some medicines are toxic by injection and may only be available by mouth, whereas others may be applied and absorbed through the skin. It takes a period of time for any medicine being absorbed into the system to reach concentrations sufficient to have a therapeutic effect

FIG 4.7: THE USE OF MEDICINES

Methods of Treatment	Methods of Prevention
The Individual Pig – By injection – Topical application – Oral – Per vagina – Per rectum **Groups of Pigs** – In-feed – By injection – In the water – By topical application	**Strategic medication** – In-feed – In the water – By injection **Continuous medication** – In-feed **Intermittent medication** – In-feed or in the water **Pulse medication** – In-feed or in the water **Medication and early weaning**

and then be excreted from the body. The frequency of treatment is determined using this knowledge.

The following methods of administration are used in the pig:-

By Injection – Intravenous, subcutaneous, intradermal, intramuscular and intravulval.

Topical – The medicine is applied to the surface of the body. An example would be the use of pour-on organophosphorus preparations or sprays for the control of mange.

Oral – Most injectable antibiotics are also available for oral administration.

Via the Uterus – Pessaries (small slow-melting tablets) can be placed into the uterus following interference at farrowing. Likewise, antibiotics can be deposited into the anterior vagina in cases of infection.

Via the Rectum – This is not a normal method for administration in the pig, although in cases of meningitis associated with salt poisoning and water deprivation, water can be dripped into the rectum to correct the imbalance. (See Chapter 15: Flutter valve.)

The method of administration will be indicated on the label of the bottle, and this should always be followed. For example, intravenous injections are used for anaesthesia; intradermal injections to test animals for tuberculosis; and subcutaneous injections for certain types of antibiotics or some vaccines. The most common route of injection however is intramuscular for antibiotics, iron injections, and oil based vaccines. Occasionally, injections might be given into the joint for arthritis or into the mammary glands for mastitis.

Always be aware of, and consult the label, for withdrawal times.

Antibacterial Medicines and their Uses

Antibacterial medicines are either produced from the fermentation of moulds (antibiotics) or they are synthesised chemically.

They act in 1 of 2 ways, by either killing bacteria, in which case they are called **bactericidal**, or by inhibiting bacterial multiplication, in which case they are called **bacteriostatic**.

Bactericidal antibiotics generally act quicker than bacteriostatic ones.

Bacteria often multiply after a primary virus infection and antibiotics are used to control these secondary infections.

Remember antibiotics will not kill viruses, only bacteria.

Antibiotics act in 1 of 3 ways:

1. They destroy the bacterial cell wall – e.g. penicillins, cephalosporins.
2. They interfere with the protein metabolism inside the cell – e.g. oxytetracycline, chlortetracycline, streptomycin.
3. They interfere with the protein synthesis of the cell nucleus – e.g. tetracyclines, macrolides.

Types of Antibacterial Medicine

Aminoglycosides

These antibiotics contain sugars and include:

- Apramycin.
- Framycetin.
- Gentamicin.
- Neomycin.
- Spectinomycin.
- Streptomycin.

They are very active against gram-negative bacteria such as *E. coli* and are used to treat piglet scours and to control bacteria in the digestive tract. They are bactericidal and are poorly absorbed from the intestinal tract. The use of streptomycin is banned in some countries.

Cephalosporins

Most of these medicines are poorly absorbed in the intestine and are thus given by injection. They include:

- Cephalexin.
- Ceftiofur.

Ceftiofur has a wide range of activity and is an excellent medicine for the treatment of respiratory disease.

Macrolides

This group includes:

- Erythromycin.
- Tiamulin.
- Tilmicosin.
- Tulathromycin.
- Tylosin.
- Tildipirosin.
- Lincomycin.
- Valnemulin.

They are mainly active against gram-positive bacteria and specifically act against mycoplasma such as *M. hyopneumoniae,* the cause of mycoplasma (enzootic) pneumonia. Tiamulin, tylosin and lincomycin are active against *Brachyspira hyodysenteriae,* the cause of swine dysentery. Tildipirosin is indicated for swine respiratory disease. Tiamulin, tilmicosin, tylosin, tildipirosin and lincomycin are generally bacteriostatic.

Penicillins

There are 3 types:

1. **Penicillin G. benzathine** – This is not used orally because it is destroyed in the stomach. It is used only by injection and is very active against gram-positive bacteria including staphylococci, streptococci, erysipelothrix and clostridia and has some activity against *Actinobacillus spp*, *Pasteurella haemophilus* and leptospira. It has a prolonged action. Penicillin G. procaine is slowly released giving a prolonged action.
2. **Acid resistant** – This is the oral form, it is absorbed from the digestive system and not destroyed by gastric juices.
3. **Semi-synthetic** – Ampicillin, amoxycillin and cloxacillin – all these have a wide range of activity against gram-positive and gram-negative bacteria.

All the penicillins are bactericidal. Penicillins have no effect against mycoplasmas as they have no cell wall.

Quinolones

These include enrofloxacin and are very active against gram-positive and negative organisms and thus of value in both respiratory and enteric disease.

Sulphonamides

There are approximately 30 different varieties available but the common ones used in pigs are sulphadimidine (also called sulphamezathine) and sulphadiazine. The former is recycled in the environment via faeces and can be responsible for tissue residue failures at slaughter particularly in the kidneys. Sulphonamides are often combined with synthetic substances called trimethoprim and baquiloprim. They are then termed potentiated sulphonamides and have a wider spectrum of activity. Sulphonamides are bacteriostatic but have a wide range of activity against both gram-positive and gram-negative organisms and they are also active against chlamydia, toxoplasma and coccidia.

Tetracyclines

These antibiotics are produced from *Streptomyces* fungi and are widely used in pig medicine.

- Chlortetracycline (CTC).
- Oxytetracycline (OTC).

Tetracyclines include oxytetracycline (OTC) and chlortetracycline (CTC). They have a wide range of activity against gram-positive and gram-negative bacteria. They are bacteriostatic at low levels but may become bactericidal at high doses and are used in respiratory diseases and secondary bacterial infections.

Florfenicol

Florfenicol is a broad-spectrum, primarily bacteriostatic, antibiotic with a range of activity similar to that of chloramphenicol, including many gram-negative and gram-positive organisms. It is indicated in the treatment of bacterial pneumonia and associated respiratory infections. Note that in many countries chloramphenicol is not allowed to be administered to pigs due to harmful residue concerns.

Other Antibacterial Medicines

Dimetridazole (Emtryl) – This may be used either in the feed or water. It acts mainly on anaerobic bacteria and it is used specifically in swine dysentery and colitis. This has been banned in many countries.

Nitrofurans – These are mainly active against gram-negative organisms found in the intestinal tract and are available for feed medication. This has been banned in many countries.

Antibacterial Sensitivity Tests

Some antibacterial medicines are more active against gram-positive than gram-negative bacteria and others are the reverse. Testing gives a guide to the choice of medicine to be used. Sensitivity tests carried out in laboratories give a better guide to a medicine's effectiveness against specific infections, but are not perfect. Some bacteria which are sensitive in laboratory tests are not so in diseased animals. They may multiplying in sites where the medicine cannot reach them, or the antibiotic is not reaching them in high enough concentrations.

In its simplest and commonest form the test is carried out by growing the bacteria in a growth medium and then suspending them in a saline solution. A thin film of this is spread over the surface of a culture plate and left to dry. Discs of cardboard impregnated with different antibiotics are placed on the surface and the plate is then incubated (Fig 4-8). The medicine diffuses out

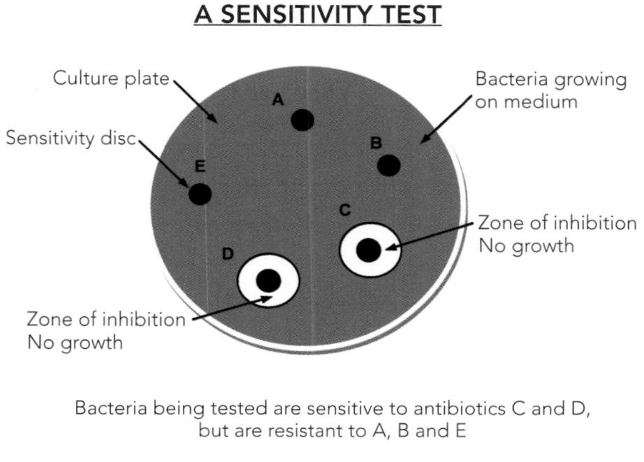

Fig 4-8

into the growth medium radially. If the organism is killed by the antibiotic there is a clear zone of no growth around the disc. If the organism is resistant to the medicine it grows right up to the disc. Such tests usually take 12 to 24 hrs. Some bacteria however take weeks to grow.

A guideline as to the use of antibiotics in specific diseases is shown in Fig 4-9 but you are advised to discuss these with your veterinarian.

Medicine use should be minimised while maintaining the welfare and health of the pigs. Various schemes are used to monitor medicine use on the farm. For example, Denmark uses a Yellow Card system, other countries measure "treatment days" or "treatment quantity" per pig.

Administering Medicines by Injection

An understanding of the basic anatomy of the skin and its underlying structures is helpful if medicines are to be injected efficiently. The outer layer of the skin consists of the epidermis or surface, beneath which is the thicker dermis consisting of living cells which multiply (Fig 4-10). The skin is attached to the underlying muscle by a combination of subcutaneous fat and fibrous connective tissue. An injection directly into the skin or the dermis (intradermal) requires a tiny needle less than 5 mm long. Fat itself has a poor blood supply and an injection into fat is poorly absorbed. Abscesses are also more likely to develop. Subcutaneous injections must only be given where there is a minimal amount of fat. This is difficult in the pig as the dermis is quite thin.

FIG 4-9: ANTIBACTERIAL MEDICINES WHICH MAY BE USED FOR SPECIFIC DISEASES

	Antibiotics																		
	Amoxicillin	Ampicillin	Ceftiofur	Cephalexin	Enrofloxacin	Florfenicol	Framycetin	Lincomycin	Penicillin/Streptomycin	Procaine Penicillin	Spectinomycin	Sulphonamide	Tetracycline	Tiamulin	Tildipirosin	Trimethoprim/Sulpha	Tylosin	Tulathromycin	Valnemulin
Actinobacillus pleuropneumonia	✓	✓	✓	✓	✓	✓			✓	✓		✓	✓		✓	✓		✓	
Chronic respiratory disease	✓	✓	✓		✓	✓			✓	✓		✓	✓			✓		✓	
Colitis					✓	✓	✓	✓			✓	✓	✓	✓		✓	✓		✓
Cystitis	✓	✓		✓		✓		✓	✓	✓			✓			✓			
E. coli diarrhoea	✓	✓	✓	✓	✓		✓				✓	✓				✓			
Erysipelas	✓	✓		✓					✓	✓							✓		
Generalised bacterial infections	✓	✓		✓	✓	✓			✓				✓			✓			
Glässer's disease															✓			✓	
Greasy pig disease	✓	✓		✓	✓	✓		✓	✓	✓			✓			✓			
Ileitis						✓			✓	✓			✓	✓		✓	✓		✓
Joint infections	✓	✓		✓		✓		✓	✓	✓			✓	✓		✓	✓	✓	
Leptospirosis	✓	✓				✓			✓				✓						
Mastitis metritis	✓	✓		✓	✓	✓	✓		✓	✓			✓			✓	✓		
Meningitis	✓		✓						✓	✓		✓				✓			
Mycoplasma arthritis					✓	✓		✓					✓	✓			✓	✓	
Mycoplasma (enzootic) pneumonia					✓	✓		✓				✓	✓	✓			✓	✓	✓
Pasteurellosis	✓		✓	✓	✓	✓			✓	✓		✓	✓	✓	✓	✓	✓	✓	
Progressive atrophic rhinitis	✓	✓	✓	✓	✓	✓			✓	✓		✓	✓		✓	✓	✓	✓	
PRRS (viral)														✓					
Salmonellosis	✓	✓	✓	✓	✓						✓					✓			
Swine dysentery								✓						✓			✓		✓

DIAGRAM OF THE SKIN

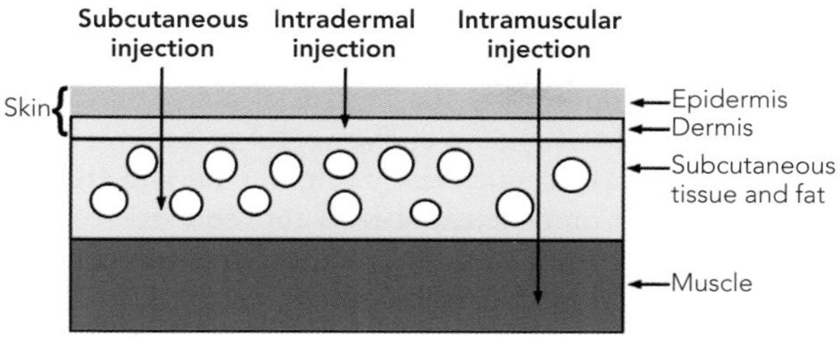

Needle use and size	
Local anaesthetic	25 x 0.6 mm (22 gauge 1")
Iron injection	16 x 0.8 mm (21 gauge ⅝")
Hormone injection	40 x 1.1 mm (20 gauge 1 ½")
Intramuscular	40 x 1.5 mm (16 gauge 1 ½")
Subcutaneous or intramuscular	25 x 1.5 mm (16 gauge 1")
Jugular bleeding	50 x 1.5 mm (16 gauge 2")
Vena cava bleeding	125 x 1.7 mm (14 gauge 5")

Fig 4-10

Using the Syringe and Needle

- On most pig farms 2 ml, 10 ml and 20 ml syringes are required.
- Always use disposable ones; they are sterile and easy to manage.
- Use syringes with a side rather than a centre nozzle because they break less easily.
- Use only 1 medicine in 1 syringe. Some medicines are incompatible when mixed.
- Use 1 syringe for 1 injecting session and then dispose of it.
- Always wipe the medicine bottle top clean with cotton wool and surgical spirit before use.
- Multi dose syringes must be kept in the refrigerator when not in use and cleaned and sterilised by boiling (10 minutes at 100 °C (212 °F)) between injecting sessions.
- Always keep part-used bottles in a refrigerator.
- Always use disposable needles that have a protective cap. This will keep the needle clean and prevent self-inoculation.
- Change the needle frequently and determine the frequency by the ease of penetration into the tissues.
- Do not inject a healthy pig after using the needle or syringe on a sick or compromised pig.
- Keep needles and syringes separate between batches of pigs.
- Do not move the needles and syringe from the isolation pens back on to the main farm.
- Always change the needle <u>immediately</u> if:
  - The end becomes burred.
  - You drop it on the floor.
  - It makes contact with the external environment.
- NEVER clean the needle with your fingers or wipe it with your clothing. Use fresh cotton wool and surgical spirit to clean the needle after 2 to 3 inoculations.
- Always use a separate needle for each individual animal when injecting breeding stock, to prevent spread of *Mycoplasma suis* by blood inoculation.
- The practical procedures for using a syringe and needle to administer iron are described in Chapter 15.
- See Fig 4-11 for the advantages and disadvantages of injections.

FIG 4-11: MEDICATION BY INJECTION
Advantages
It is the most effective method. It is most cost effective. The dose rate is accurately given. Treatment commences immediately after the injection is given. Medication is not dependent on water or feed intake. Sick/compromised pigs can be treated as soon as identified. The stockperson observes the pigs more efficiently. There is a better assessment of the response. Withdrawal periods can be accurately determined.
Disadvantages
Stress involved in handling the pigs. Extra labour costs. Practical difficulties for the staff. It is more expensive.

Needleless Injections

A range of needleless injections have being developed that deliver a tiny, high-pressure jet of medicine through the skin without the use of a hypodermic needle. Early models have tended to be cumbersome, but ongoing technological development will result in most medications being given by this method. The drive towards needleless injections is driven by the need to improve health and safety of stockpeople and to eliminate 'sharps' from meat products.

Self-Inoculation

If you inoculate yourself accidentally you should take the following steps:

- Report immediately to the person to whom you are responsible on the farm.
- Look at the label on the bottle. Does it give any emergency procedures?
- Read the leaflet or data sheets held on the farm, for example, in the UK, Control of Substances Hazardous to Health (COSHH) Safety Regulations. If not available ring your veterinarian, or your medical doctor.

- If you are using an oil based vaccine, (see the bottle label) go to the casualty department of a hospital immediately with the bottle. Such vaccines can cause blood vessels to go into spasm with potential loss of blood supply and consequent loss of tissue (e.g. a finger). The tissues usually require opening up and the injection flushing out.

Sites of Injection

Fig 4-12 shows the different sites of injection.

SITES OF INJECTION

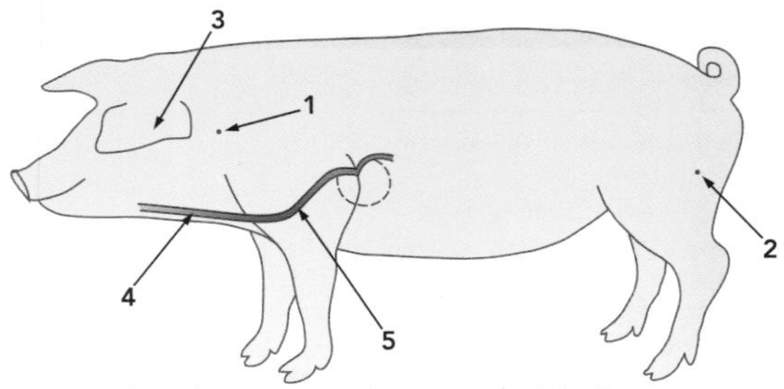

1=Site for subcutaneous or intramuscular injection.
2=Site for iron injection in piglets only.
3=Site of ear vein for intravenous injection.
4=Site of jugular vein for intravenous injection.
5=Site of anterior vena cava vein for intravenous injection.

Fig 4-12

Subcutaneous – The ideal site for the small pig is inside the thigh beneath the fold of the skin or, beneath the skin behind the shoulder. In the growing and mature animals, the best site is approximately 25 to 75 mm behind and on the level of the base of the ear, using a 25 mm needle at a 45° angle.

Intravenous – There are 3sites for injecting medicines directly into the blood stream: the ear veins, the jugular vein and the anterior vena cava or large vein that leaves the heart. The ear vein is the most common method particularly for anaesthetics and occasionally calcium injections. The skin over the outer part of the ear is cleaned with cotton wool and surgical spirit which also demarcates the veins. They are then raised by applying pressure to the base of the ear. The pig should be restrained by a wire noose or rope around the upper jaw and or by sedation. (For techniques see Chapter 15.)

Intramuscular – The common preferred site in weaners, growers, finishers and adults is up to 70 mm behind the base of the ear. Small piglets are often injected into the ham of the hind leg because there is not much muscle on the neck. This is not recommended in growers/finishers because of the possibility of abscesses.

Fig 4-13 lists some of the common antibiotics that are administered by injection.

The first column gives the mg/ml of active medicine and the second the mg/kg of liveweight to achieve therapeutic levels. How much bodyweight 1 ml will treat is then calculated and the number of ml required to treat the pig is then calculated. This varies with the concentration or mg/ml of the antibiotic.

FIG 4-13: INJECTABLE MEDICINES AND THEIR WITHDRAWAL PERIODS AVAILABLE FOR BACTERIAL INFECTIONS *

Antibiotic	Strength	Dose	Withdrawal Period ** (Days)
(Check availability in your country)	mg/ml	mg/kg	
Amoxycillin	150	10 – 20	18
Ampicillin	150	2 – 7	28
Ceftiofur	50	3	0.5
Cephalexin	180	7 – 10	2
Clavulanic acid plus amoxycillin	35/140	8.75	14
Enrofloxacin	100	2.5	10
Framycetin	150	5	49
Lincomycin	100	5 – 10	2
Oxytetracycline	100	10 – 20	15
Procaine pen plus dihydrostreptomycin	200/250	8mg Penicillin 10mg Streptomycin	21
Procaine penicillin	300	15 – 25	5
Spectinomycin	100	20	5
Streptomycin sulphate	150	10 – 25	21
Sulphadimidine	333	100	7
Tiamulin	200	10	10
Trimethoprim/sulpha	40/200	18 – 20	5
Tulathromycin	100	2.5	33
Tylosin	200	2 – 10	7

* Always refer to current data sheets. For guideline use only.
** Withdrawal periods: These may differ by country. Those provided are a guide only, always check the label.

Withdrawal period

This is the time between the last dose of medicine administered and the time when the level of residue in the muscle, liver, kidney, fat or skin is equal to or less than the maximum residue limit (MRL) allowed in carcasses. The MRL in the EU is the legally permitted maximum concentration allowed. Fig 4-13 gives guidelines as to the withdrawal periods for the different antibiotics but for the exact period it is necessary to refer to the current relevant data sheet.

Where a withdrawal period for pig meat is not specified a standard period of 28 days is recommended.

It is essential to follow the guidelines provided in the medicine data sheet regarding contraindications and warnings. Do not second guess. Discuss any concerns with your veterinary practice.

Administering Medicines Topically

Medicines can be applied to the nose, mouth, ears, eyes, skin and feet. Treatments to the skin are applied either by spray, liquid or immersion of the pig, and are used against skin parasites, such as mange or lice or greasy pig disease. Some medicines are poured onto the skin from which they may be absorbed and distributed throughout the body. Mange is treated this way using 20 % phosmet. (Topical administration of medicines is simple, cheap and causes the pig the least stress. As techniques develop it could become an increasingly used procedure.

Administering Medicines in Water

Group Treatment

The practical problems of giving medication in water can be considerable. On many farms, the water pipes are old, and if the antibiotic powder contains sugars or acids there is a tendency for mineral deposits to form and block the pipe, or for yeasts to multiply, producing a jelly which does likewise. It then becomes a major problem to clean them out. Water soluble antibiotics should be used in a pure form where possible or with a minimum carrier base. Water tanks can be quite small, and it is therefore necessary to introduce the antibiotic to the water 4 or 5 times a day. If water is taken directly from the mains supply, water medication is impossible unless a proportioner is used. Some water authorities prohibit the use of water directly from the mains supply. Up to 40 % of the antibiotic may be wasted through inefficient nipple drinkers. Small groups of pigs are best medicated using a large barrel (a wheelie bin is ideal) with a water bowl attached to the bottom. This is a very efficient method of administration but impractical on a large scale.

Fig 4-14 shows a range of antibiotics available for water use and dose levels per tonne of liveweight. For example amoxycillin powder 50 % contains 50 % of the active medicine amoxycillin, and 50 % of a carrier, either citric acid or a sugar such as dextrose. A dose level of 15 mg/kg liveweight of amoxycillin requires a dose of 30 g/tonne of liveweight per day of total powder but in practice this level would be raised to approximately 40 g/tonne to allow for the losses in the water through the nipple drinker.

Water medication can be added to the water used in gruel feeding for weaned pigs. In addition, it may be used in wet feed systems.

See Fig 4-15 for the advantages and disadvantages of medicating in water.

How to apply antibiotic powder to water in header tanks

1. Calculate the total kg of liveweight to be medicated in tonnes.
2. Calculate the g of powder required for the 24 hour period. Add 10 to 20 % extra if nipple drinkers are used.
3. **The water intake per tonne of liveweight per 24 hours will be approximately 100 litres.**
4. Calculate the total water used in 24 hours.
5. Divide the header tank capacity into the total water used which gives the times that the tank is emptied in 24 hours.
6. Divide the powder and add pro rata to the tank. Stir each time.

Example

Medication with amoxycillin:

- 50 % at 15 mg/kg. Water bowls used.
- 4 tonnes of liveweight at 30 g powder per tonne = 120 g.
- 400 litres of water consumed.
- If a 200 litre tank is available then 400 / 200 = the tank will empty twice in 24 hours.
- Place 60 g, or half the powder, in the tank early in the morning and the other half at the end of the day.

Individual Pig Treatment

This is most commonly used in sucking piglets for scours and other bacterial infections. Fig 4-16 lists medicines available.

When you decide to treat pigs orally (or by injection) ask yourself the following questions:

- Should the veterinarian be consulted?
- Have all individual affected pigs been identified?
- Is this condition one that has been reliably diagnosed before or is it new?
- Is it necessary to treat it?

FIG 4-14: DOSE LEVELS OF WATER SOLUBLE MEDICINES *

	Active Medicine	Dose mg/kg	g/tonne Liveweight	Withdrawal Period (Days)
Amoxycillin P	50 %	10 – 15	20 – 30	1
Apramycin P	30 %	7 – 12	23 – 39	14
Chlortetracycline P	100 %	10 – 20	10 – 20	10
Florfenicol L	23 mg/ml	10	550 ml/tonne	20
Lincomycin P	40 %	4.5 – 11	11 – 27	1
Neomycin P	70 %	11	16	14
Oxytetracycline P	80 %	10 – 30	12 – 37	5
Tiamulin P	45 %	8.8	20	1
Tilmicosin L	250 mg/ml	12 – 20	50 – 80 ml/tonne	14
Trimethoprim/sulpha P	2 – 10 %	24	240	3
Tylosin P	100 %	15	15	0

* For guidance only. Refer to current data sheet. **P**=Powder **L**=Liquid

FIG 4-15: MEDICATION IN WATER

Advantages

It is easy to administer.
It can be effective in a short period of time and can be given in the early stages of disease.
It can be introduced very quickly.
Large numbers of pigs can be treated at low cost.
It can be used strategically.

Disadvantages

Healthy pigs are treated.
Sick/compromised pigs often don't drink.
There is considerable wastage if used via nipple drinkers.
Water pipes tend to block up.
The design of the pipe system may not be suitable.

FIG 4-16: ORAL MEDICINES AVAILABLE TO TREAT PIGLET SCOUR *

	mg/ml	mg/kg wt
Amoxycillin	40	7 – 15
Apramycin	18	10 – 20
Enrofloxacin	5	1.5
Neomycin	50	35
Spectinomycin	50	50
Toltrazuril (coccidiosis)	25	6.25
Trimethoprim/sulpha	40/200	0.1

* Always refer to current data sheets. For guideline use only.

- Are medicines to treat this condition readily available?
- Are there any welfare or nursing implications?
- Should the affected pig(s) be moved to a hospital pen?
- What methods of administration should be used?
- What dose should be given? Is the correct information on this available?
- How often should the medicine be given and for how long?
- Are any adverse effects likely?

Then you should:
- Record when the treatment started and its progression.
- Assess the response on a day-by-day basis.
- If there is no response within 24 hours consult your veterinarian.

Administering Medicines In-Feed

The inclusion of antibiotics in-feed is the most common method of controlling and preventing diseases. See Fig 4-17.

In-feed medication can be an effective means of control when used over several weeks.

In-feed medication is wasteful in that it is inevitably given to healthy pigs that do not need it as well as to the diseased pigs that do need it. In-feed supplements can also be used as top dressings, that is, sprinkling small amounts over the feed to administer the antibiotic. Top dressing is a very suitable method for small groups of pigs and individually fed animals, such as sows and boars.

FIG 4-17: THERAPEUTIC MEDICINES THAT MAY BE AVAILABLE FOR IN-FEED USE *

Doses and some Disease Indications

Active Medicine	g/Tonne in-Feed	Common Uses	
Amoxycillin	200 – 600	Greasy pig disease. Streptococcal meningitis. Pasteurellosis.	Respiratory bacterial infections. Secondary bacterial infections.
Amprolium	125	Coccidiosis.	
Apramycin	100 – 150	*E. coli.* Postweaning diarrhoea.	Salmonellosis.
Arsanilic acid or sodium arsanilate	250 - 400	Swine dysentery. *Mycoplasma suis.*	
Bacitracin	275	Bacterial enteritis. Clostridial enteritis.	
Carbadox	55	Colitis.	Swine dysentery.
Chlortetracycline	300 – 800	Greasy pig disease.	Respiratory disease.
Chlortetracycline Penicillin Sulphadimidine	165 82 165	Enteric disease. Greasy pig disease.	Respiratory disease.
Dichlorvos	380 – 550	Internal parasite control.	
Fenbendazole	14 – 53	Internal parasites.	
Flubendazole	30	Roundworms.	
Furazolidone	200 – 400	Postweaning enteritis.	
Hygromycin	14	Control of ascaris, nodular worms and whipworms.	
Intagen	1.5 – 5 kg premix per tonne	Oral *E. coli* vaccine.	Prevent scour in piglets.
Ivermectin	2	Internal parasites, lice and mange.	
Levamisole	800	Internal parasites.	
Lincomycin	44 – 220	Greasy pig disease. Mycoplasma pneumonia and arthritis.	Respiratory disease.
Lincomycin Spectinomycin	44 44	*E. coli* infections. Mastitis. Mycoplasma pneumonia.	Salmonella infection. Swine dysentery.
Monensin	100	Colitis.	Swine dysentery.
Neomycin	163	Colitis. *E. coli* infections.	Enteritis. Salmonella infections.
Oxibendazole	40 – 100	Internal parasites.	
Oxytetracycline	200 – 800	Greasy pig disease. Respiratory disease.	Secondary bacterial infections.
Phenoxymethyl-penicillin	200 – 400	Actinobacillus pneumonia. Streptococcal meningitis.	Clostridial infections. Necrotic enteritis.
Pyrantel tartrate	100 – 880	Internal parasites.	
Salinomycin	30 – 60	Colitis.	Non-specific enteritis.
Sulphadimidine	100 – 300	Enteric disease. Greasy pig disease.	Respiratory disease.
Thiabendazole	50 – 100	Internal parasites.	

FIG 4-17: THERAPEUTIC MEDICINES THAT MAY BE AVAILABLE FOR IN-FEED USE * CONT.

Doses and some Disease Indications			
Active Medicine	**g/Tonne in-Feed**	**Common Uses**	
Thiophanate	168 – 500	Internal parasites.	
Tiamulin	30 – 100	Mycoplasma pneumonia & arthritis.	Swine dysentery.
Tilmicosin	200 – 400	Mycoplasma pneumonia. *Pasteurella*.	Actinobacillus pneumonia.
Trimethoprim/sulpha	Varies. See data sheets.	Progressive atrophic rhinitis. Colitis. Enteric disease.	Postweaning diarrhoea. Respiratory disease.
Tylosin	100	Mycoplasma pneumonia.	Swine dysentery.
Valnemulin	See data sheet.	Mycoplasma pneumonia.	Swine dysentery.
Zinc oxide	2500 g zinc	Postweaning enteritis.	

* For guidance only. Refer to current data sheet.

Factors to Consider when Using In-Feed Medication

- At the time of the disease outbreak, there may be no bin capacity available to hold the medicated feed and bagged food is sometimes required.
- If strategic medication is used, hold 1 bin for medicated feed only.
- If medicated feed is placed in a bin containing non-medicated feed, the time of the feed reaching the pigs and the withdrawal will be unknown.
- Bagged food is more expensive.
- There can be a delay in manufacturing and delivering the medicated feed.
- Sick/compromised pigs often do not eat or have reduced feed intake and therefore won't receive sufficient antibiotic.
- If the appetite is poor the medicine inclusion rate may need to be increased by up to 30 %, provided it is safe to do so.
- In-feed medicines may require a product licence for use in food producing animals and therefore the availability of medicines is narrow.
- Each particular medicine has its own withdrawal period and this may mean it is impracticable to use in pigs near the point of slaughter.
- Automatic feed lines make the application of in-feed medication to selective groups difficult.
- The bin containing the medicated feed should be marked with the date the medication was first added and the date when empty. Withdrawal times can then be calculated.

Strategic Medication

This method applies treatment at the anticipated start of the disease or during the incubation period. To carry this out, there are a number of essential components:

- The specific organism associated with the disease should be identified.
- The medicine sensitivity of the organism should be identified.
- The incubation period (the time between exposure to the organism and clinical symptoms) should be known.
- Last, and most important, the point at which the disease process starts and when it becomes clinically apparent should be determined. Medication can then commence prior to the latter.

Strategic medication (Fig 4-18) is usually carried out in the feed but it can also be applied in the drinking water or by injection using long-acting preparations. It can also be applied to eliminate disease from a group of pigs. A good example would be to prevent possible swine dysentery in purchased pigs on entering a finishing herd. In this case the feed could be medicated with either lincomycin (110 g/tonne) or tiamulin (100 g/tonne) for the first 14 days during isolation on the farm. If strategic medication is carried out routinely, medicated feed must be held continuously in a designated bin.

A typical example would be the use of high levels of oxytetracycline, (500 to 800 g to the tonne), in growing pigs to control severe outbreaks of mycoplasma (enzootic) pneumonia. The procedure would be to medicate all the pigs for 7 days, 1 week after entry into the houses or approximately 10 days prior to the commencement of clinical symptoms.

Thus each week a selected group of pigs would commence treatment to prevent the development of extensive lesions in the lungs and yet allow immunity to develop. An alternative strategy would be an injection of long-acting OTC given at a predetermined point. This method is of course more labour intensive. Strategic medication by injection is easier to apply in younger pigs. Fig 4-19 indicates some of the diseases where strategic medication can be used and the times of application.

The following on-farm case histories illustrate some of the uses for strategic medication:

Mycoplasma (enzootic) pneumonia (EP)

Farm A was a 200 sow herd producing pigs for sale at 110 kg. Its buildings were poorly insulated and due to the bad economic state of the industry at the time, no capital was available to improve the environment other than by managerial means. Severe bouts of coughing developed on this farm approximately 3 weeks after pigs moved from the first stage rearing accommodation into the finishing houses. Post-mortem and lung examinations showed extensive lesions of EP associated with secondary pasteurella infections, the latter shown to be sensitive to chlortetracycline (CTC). The first cases of clinical pneumonia became evident starting 10 days after the pigs moved into the finishing house. Commencing on Day 8 after entry, pigs were medicated with 600 g/tonne CTC for a period of 7 days. This had a dramatic effect on reducing the incidence of disease and in particular, the numbers of pigs requiring individual treatment. The variability in growth was reduced. Initially, this was a herd with young breeding stock but as they matured, the time of onset of disease changed to appear 5 weeks after moving in to the finishing houses. The strategic medication was therefore given 2 weeks later.

Progressive atrophic rhinitis (PAR)

Herd B was a 1200 sow unit showing clinical PAR at a visual level of 15 %. Toxigenic pasteurella and bordetella bacteria were isolated and were found to be sensitive to trimethoprim and amoxycillin. All piglets were injected with a long-acting preparation of the latter at 7 days. Pigs were further medicated with oxytetracycline in-feed for 14 days postweaning. For a period of 6 months following this, the incidence of PAR at a visual level dropped to 2 %. It then, however, started to rise again to a 7 % level. Apparently the stockperson had decided to inject the pigs at 10 to 14 days of age because this was more convenient. This allowed early establishment of pasteurella organisms in the nose and

FIG 4-18: STRATEGIC MEDICATION: SOME MEDICINES THAT COULD BE USED

	Respiratory Diseases	Enteric Diseases
Injections	Ceftiofur Florfenicol Lincomycin Long-acting amoxycillin Long-acting OTC Long-acting penicillin Tiamulin Tildipirosin Tulathromycin Tylosin	Amoxycillin Florfenicol Lincomycin Tiamulin Trimethoprim/sulpha
Water soluble antibiotics	Amoxycillin CTC OTC Tilmicosin	Apramycin Framycetin Lincomycin Neomycin Tiamulin Trimethoprim/sulpha
In-feed inclusions	Lincomycin OTC or CTC Tiamulin Tilmicosin Valnemulin	Apramycin Bacitracin Lincomycin Tiamulin

FIG 4-19: THE TIMING OF STRATEGIC MEDICATION USING EITHER LONG-ACTING INJECTIONS OR IN-FEED MEDICATION

Age (Weeks)	EP	App	PAR	SM	HPS
1 (Birth)			✓		
2			✓		✓
3		✓			
4			✓		
5 (Weaning)			✓		
6			✓		
7					
8		✓	✓	✓	✓
9 (Growing)		✓			
10					
11					
12					
13					
14	✓				
15 (Feeding)					
16+	✓				

✓ = Age when medication may be started.
EP = Mycoplasma (enzootic) pneumonia
App = Actinobacillus pleuropneumonia
PAR = Progressive atrophic rhinitis
SM = Meningitis caused by *Streptococcus suis* type 2 et al.
HPS = *Haemophilus parasuis* causing Glässer's disease

a considerable amount of damage. When the injections were moved back to the 7 day point the problem again returned to low levels. This example illustrates the importance of gathering the evidence to determine the critical point at which strategic medicine therapy should be given. At the time of this disease problem, vaccines to prevent PAR were not available.

Actinobacillus pleuropneumonia (App)

App can be a very difficult disease to control, especially in its severe form. Herd C was a 250 sow herd producing pigs for slaughter. At 9 weeks of age pigs were moved from the nurseries (flat decks) into a second stage rearing accommodation and within 7 days of entry, severe outbreaks of pleuropneumonia occurred. In view of the very short incubation period of this disease (12 to 48 hours) strategic medication can be difficult to apply. In this particular case, medication was applied immediately after the first clinical case became apparent. To obtain a very rapid response, chlortetracycline was placed in the drinking water header tanks for a period of 72 hours. The response on this farm was quite dramatic, but perhaps even more importantly, the trigger factors namely, a drop in energy intake and variable temperatures in the house, were then corrected. This reduced the strategic medication requirements which illustrates the importance of management.

Streptococcal meningitis (SM)

Farm D was a 350 sow herd previously free of clinical streptococcal meningitis. The introduction of disease caused severe problems in the nurseries (flat decks) where it regularly appeared approximately 16 days after entry into the house. All pigs were medicated through the drinking water with potassium penicillin V commencing on Day 10 through to Day 17. High numbers of pigs become carriers of the organism in their tonsils within 2 weeks of occupying nurseries and a variable number of such animals then develop the disease. The object of applying water medication at this stage was to reduce the level of infection. The disease reached its peak with some 15 % of each batch of pigs moved into the house requiring individual treatment. Strategic medication reduced this to less than 1 %.

Swine dysentery (SD)

Farm E was a finishing farm purchasing approximately 10,000 pigs per annum. It had previously been infected with swine dysentery and an eradication programme was successfully carried out. It was however, periodically committed to buying pigs from unknown sources, in response to market forces. In order, therefore, to protect the herd all incoming pigs were strategically medicated in groups of 200 when they entered the farm whilst in isolation premises. These premises were completely emptied and disinfected between batches. On arrival the pigs were medicated in the water with tiamulin for a period of 5 days, together with in-feed medication at 100 g/tonne for a further 7 days. These procedures prevented the appearance of disease.

These on-farm situations illustrate some of the methods by which strategic medication can be used to good effect to control disease and at the same time be cost effective. There are a variety of medicines available and Fig 4-18 lists some of these, together with in-feed dose rates. However, if you are thinking of adopting strategic medication discuss it with your veterinarian first so that they can advise you on the best type of medicine, the dose level and its timing. Fig 4-19 shows possible time applications for strategic medication.

Pulse Medication

Pulse medication is an alternative programme to continuous medication for the control of disease. The medication is given either in water or feed for short periods of only 48 to 72 hours.

Only medicines with short withdrawal periods can be used if pigs are near market weight. Treatment costs are reduced compared to continuous medication. Pulse medication does need careful control to ensure that withdrawal periods are observed and it is difficult to carry out where automated feeding systems are used. It is more suited to use in wet feeding systems.

Pulse medication has been used successfully in the control of pneumonia, using a combination of 300 g/tonne of CTC and 40 g/tonne of tiamulin. Such medication could be used for 2 days followed by 4 days off. Results on farms have shown an improvement in food conversion of 0.2, daily liveweight gain of 6.7 %, and considerable reductions in lung scores at slaughter.

Continuous Medication

Pigs are medicated continuously for periods of up to 12 weeks during the critical periods of exposure to disease. Dose levels are usually lower than those used for treatment. Fig 4-2 lists the in-feed levels that can be used for the various respiratory and other diseases.

Continuous medication can be expensive but in permanently populated houses with mixed ages of pigs it can work well with considerable growth enhancement which helps offset the cost.

Even with continual suppression of organisms, in severe endemic disease the system can break down. There may also be problems with withdrawal periods in finishing pigs. Continuous medication should therefore be regarded as a last resort. Always ask 3 questions if continuous medication is necessary:

- Why is there a problem?
- What has gone wrong with the management?
- CAN THIS BE CORRECTED?

Medicated Early Weaning (MEW)

See Chapter 3 for further information

This is a specialised technique for producing healthy as SPF breeding stock from a diseased herd. It is used mainly by breeding (seed stock) companies rather than commercial producers although systems such as three-site and multi-site production which have evolved from it are highly applicable to commercial production.

Basically, MEW breaks the cycle of infection by farrowing groups of sows in isolation, and weaning their piglets to clean premises at about 5 to 6 days of age. Medication of the sow and piglets, and sometimes vaccination of the sow are added safeguards depending upon what particular infections are to be eliminated.

This method produces excellent results in removing most bacterial infections. However, it is possible to produce a pig that is so devoid of pig organisms, that it cannot be acclimatised into conventional herds. Also if a new herd is established by this method it can be very susceptible to even low pathogenic organisms. The technique is used mainly to establish new SPF herds.

Procedures

- Sows are moved into isolated farrowing houses at least 5 days before farrowing, 800 m (2600 ft) from the nearest pig.
- An all-in/all-out system is used. Each batch of sows is washed prior to entry.
- Farrowing is induced with prostaglandin injections at Day 113.
- Sows are medicated from entry into the farrowing house until the piglets are weaned.
- The piglets are weaned at 5 days and reared in groups in isolated housing.
- The piglets are medicated from birth to 10 days of age.

A possible medication regime

Sows

- Injected with potentiated sulphonamide containing 40 mg trimethoprim, 200 mg sulphonamide/ml (TMS), on entry into the farrowing accommodation at the recommended treatment level.
- From entry to weaning the feed is medicated with TMS at therapeutic doses, and water soluble tiamulin is given twice daily in the water.

Piglets

- Injected daily with TMS until weaning.

Postweaned pigs

- Injected with TMS and dosed orally daily with tiamulin for 5 days.

Bacterial pathogens that can be eradicated by MEW:

- Actinobacillus pleuropneumonia.
- Toxigenic *Pasteurella multocida* – progressive atrophic rhinitis.
- *Mycoplasma hyopneumoniae* – mycoplasma (enzootic) pneumonia.
- *Sarcoptes scabiei var suis* – mange. (Treatment of the sow with ivermectin is required).
- *Brachyspira hyodysenteriae* – swine dysentery.

Anaesthetics, Sedatives, Analgesics

The indications for anaesthesia in the pig are limited but include castration, caesarean section, vasectomy, epididectomy and ovum transplants – operations that are carried out by a veterinarian. Most other surgical procedures can be carried out by using tranquillisers and local anaesthetics. Anaesthesia is carried out by intravenous injection, inhalation, spinal anaesthesia or local infiltration of tissues. The first 3 are only used by a veterinarian but local anaesthesia is frequently necessary to suture small skin wounds or replace rectal prolapses. Sedatives are frequently used by non veterinary pig people.

Anaesthetics

Medicines used for general anaesthesia

- Halothane or isoflurane – inhalation POM-V.
- Pentobarbitone – intravenous injection POM-V.
- Telazol + ketamine + xylazine – injection POM-V.

POM-V = Prescription-only medicine for veterinary use only.

Medicines used for local anaesthesia

- Procaine – topical POM-VPS.
- Lignocaine – topical/injection POM-VPS.
- Amethocaine – topical POM-VPS.

POM-VPS = Prescription-only medicine – veterinarian, pharmacist, Suitably Qualified Person. Check with your veterinarian for trade names in your country.

Sedatives

There are 3 medicines available for sedating pigs, acetylpromazine (ACP), azaperone and primidone.

ACP (10 mg/ml injection POM-V)
This medicine is used in animals to prevent travel sickness and occasionally in pigs as a general sedative at a dose level of 0.1 mg/kg liveweight. It is also useful for treating abdominal pain in cases of colic or to provide sedation together with local anaesthesia.

Azaperone (40 mg/ml injection POM)
This is a sedative and analgesic widely used in pigs,

and is very effective.
Indications for use:

- To prevent fighting.
- Sedation prior to anaesthesia.
- To examine pigs' feet.
- To prevent a gilt savaging her newborn piglets.
- To calm an excitable animal.
- Prior to mixing or transportation.
- To facilitate any manipulative procedure.

The dose level is 0.5 to 2 ml per 20 kg bodyweight. The effects of the medicine are dose dependent. When used at 2 ml per 20 kg the pig is completely sedated after 20 minutes and lies on its side. The lower level of 0.5 ml per 20 kg will prevent fighting when pigs are mixed.
Some guidelines to dose levels:

- Prevention of fighting in adult and growing pigs, 1 ml/50 kg.
- To prevent savaging, 2 ml per 20 kg.
- Sedation prior to anaesthesia, 2 ml per 20 kg.
- Sedation prior to manipulation, 1ml per 20 kg.

It is important not to disturb the pig for 15 minutes after injection. Distractions will reduce the effectiveness of the medicine.

Primidone

This is an excellent yet little used medicine for preventing the savaging of piglets particularly by gilts. One tablet per 12 kg bodyweight per 24 hours divided into 2 doses given am and late pm is advised. Treatment should commence at least 24 hours before farrowing and continue for at least 24 hours after farrowing. The tablets should be crushed onto the food.

Analgesics

Phenylbutazone (POM)

This medicine is very useful in treating painful conditions such as acute lameness and torn muscles, bush foot infections or acute mastitis. It can be given by injection, by powder or by mouth and its use will be advised by your veterinarian. In many countries it cannot be used in food producing animals.

Meloxicam (0.4 mg/kg injection POM)

An excellent pain relief injection, useful in all circumstances to relieve pain.

Salicyclic acid (asprin)

Commonly used for pain relief in feed and in water. Useful in mass medication and in outbreaks of PRRSV or SIV. May be useful in compromised pigs.

Parasiticides

See Chapter 11: Parasites for further information

Medicines to control parasite infections act variously on the adult worm, the egg or the larva.

Fig 4-20 shows the medicines available and the methods of application. Medicaments can be given by mouth in the water or feed, by injection or on the skin.

Vaccines

So far most of the chapter has dealt with the treatment of disease but therapeutics also includes the use of products to prevent disease and the most common of these are vaccines which stimulate the immune system. Vaccination involves exposing the pig to the protein components (called the antigen) of the infectious agent. Some vaccines contain living organisms that have been altered so that they cannot produce disease but still produce an immunity. Most contain killed or inactivated organisms.

The immune system responds by producing antibodies that destroy the infectious agents, usually in cooperation with specialised body cells or by neutralising the toxins that are responsible for the disease. This process of stimulating immunity is called vaccination.

Vaccination against gonadotropin releasing hormone (GnRH) which inhibits the production of testosterone or oestrogen is available. This can be used to chemically castrate males and may be useful to "spay" female pigs – such as pet pigs.

See Chapter 3 for further information on vaccination.

Remember, vaccination is never 100 % effective.

Hormones

These products are usually under the direct control of your veterinarian i.e. prescription-only medicines. As a general statement the use of hormones that act on the ovaries to stimulate oestrus should be avoided because the stage of the oestrus cycle cannot be accurately determined. However at specific times their use can be advantageous.

Hormones used in the pig include the following:

Prostaglandins

See Chapter 8: Controlled farrowings.

These are substances that following injection, cause the corpus luteum to regress. The corpus luteum is present in the ovaries during the middle period of the oestrus cycle and during pregnancy. Its removal may initiate either oestrus, abortion or farrowing depending on the

FIG 4-20: SOME MEDICINES AVAILABLE
*** TO TREAT PARASITES OF THE SKIN (ECTOPARASITES) AND INTERNAL PARASITES (ENDOPARASITES)**

		Parasite and Activity (See Fig 11-1 for alternative names)																
Active Medicine	**Presentation/Dose Levels ***	Eggs	Kidney worm	Large roundworm	Larvae	Lice	Lung worm	Mange mites	Muscle worm	Nodular worm	Pork bladder (tape) worm**	Stomach worm (red)	Stomach worm (thick /hair)	Thorny-headed worm	Threadworm	Ticks	Whipworm	**Withdrawal period (days) ***
Amitraz 12.4 %	Topical liquid concentrate. 40 ml to 10 l water.					✓		✓								✓		7
Amitraz 2 %	Pour on to skin.					✓		✓								✓		7
Doramectin	Injection. 1 ml/33 kg lw. (300 mcg doramectin/kg lw).		✓	✓	✓	✓	✓	✓		✓		✓	✓	✓	✓	✓	✓	28
Febantel	In-feed pellets.	✓		✓	✓		✓			✓		✓			✓		✓	35
Fenbendazole	Pellets for top dressing. In-feed for 1 day.	✓	✓	✓			✓		✓	✓		✓	✓	✓	✓		✓	5
Flubendazole 5 %	Powder. Top dress or in-feed for 10 days.	✓		✓	✓		✓		✓	✓		✓	✓	✓	✓		✓	7
Ivermectin 1 %	1 ml/33 kg. (300 mcg ivermectin/kg liveweight).		✓	✓	✓	✓	✓	✓	✓	✓		✓	✓	✓	✓	✓	✓	28
Ivermectin 0.6 %	Powder in-feed 330 g to 1 kg premix/tonne.		✓	✓	✓	✓	✓	✓	✓	✓		✓	✓	✓	✓	✓	✓	5
Levamisole 7.5 %	Injection.		✓	✓			✓			✓		✓	✓	✓	✓		✓	28
Oxibendazole 2 – 20 %	In-feed for 10 days or pellets for top dressing.	✓		✓					✓	✓		✓	✓	✓	✓		✓	14
Phosmet 20 %	Topical liquid pour onto skin 1 ml/10 kg liveweight.					✓		✓								✓		35
Thiophanate 22.5 %	Powder in-feed for 14 days.	✓		✓	✓					✓		✓	✓	✓	✓		✓	7

* See manufacturer's data sheets for further details.
** Difficult to treat.
Some bendazole compounds may have activity against muscle worm.
lw – liveweight

stage of the reproductive cycle.
Prostaglandins have 6 uses:

1. Given within 36 hours post-farrowing, to improve subsequent fertility and litter size.
2. To resolve endometritis or uterus infection.
3. To synchronise farrowing by injecting the sow from Day 113 of pregnancy. Farrowing usually commences within 24 hours.
4. To stimulate boar libido enough to encourage the boar to jump the dummy sow and be collected.
5. To cause an abortion. This can be used to assist batching.
6. To assist synchronisation of oestrus.

Prostaglandins are usually administered by intramuscular injection (Fig 4-21). Alternatively, the dose of prostaglandin at farrowing can be halved if injected intra-vulvally.

Prostaglandins are potentially hazardous to women and asthmatics and should never be handled by them. Fig 4-22 shows a format recommended for their control and use on the farm. This statement should be agreed to and signed by anyone who handles prostaglandins.

FIG 4-21: SOME AVAILABLE PROSTAGLANDINS

Active Principle	Route
Alfaprostol	Intramuscular
Tiaprost	Intramuscular
Dinoprost	Intramuscular
Cloprostenol	Intramuscular
Luprostiol	Intramuscular

Milk let-down products (Fig 4-23)

These are hormones produced by the anterior pituitary gland at the base of the brain. Their action is to release milk from the mammary gland and cause contractions of the uterus. They may be given to promote the farrowing process provided there are no mechanical obstructions. They are also useful in promoting milk flow when the udder is congested. Specific uses are discussed in Chapter 7.

FIG 4-23: SOME MILK LET-DOWN PRODUCTS AVAILABLE

Active Principle	Dose Intramuscular	Dose Intra-vulval
Oxytocin (10 iu/ml)	5 iu – 10 iu	2.5 iu – 0 5 iu.
Carbetocin (0.07 mg/ml)	1.5 – 3 ml	0.75 – 1.5 ml

Hormones Used to Control the Oestrus Cycle

These can be used to synchronise oestrus in groups of gilts or in sows after weaning.

Altrenogest (POM-V)

This is an oil based product containing the active principle altrenogest – a progesterone substance. Progesterone is produced by the ovary when the sow is in the middle of the oestrus cycle or pregnant. It suppresses oestrus in the non-pregnant female if it is given daily and when it is removed the sow or gilt will come into heat.

Its main use is to synchronise oestrus by medicating batches of gilts daily for 18 days, (5 ml per day per gilt is placed on the feed). At the end of this period following its withdrawal most gilts will be in heat within 5 to 7 days.

It can also be used in sows at weaning time to synchronise oestrus but this should not be necessary at a commercial level. Altrenogest should only be used ingilts that have shown oestrus. Some trials have shown an increase in litter size following its use.

Altrenogest can be administered individually to pigs via toasted bread or combined with apple juice and administered via a dosing gun.

Gonadotrophin (POM-V)

This product is based on a hormone (gonadotrophin) that stimulates the production and release of follicles from the ovary. It is used to stimulate gilts that have failed to show oestrus but in such cases it is not uncommon for only 50 % to respond, and come into heat.

Furthermore some gilts will be mated only to become pseudopregnant and not farrow. It is better to cull anoestrus gilts – they are telling you they are infertile.

Gilts treated with gonadotrophin should be served towards the end of the heat period.

Gonadotrophin based products can be used where there are anoestrus problems in first litter gilts. Litter size is often improved. Gilts are injected on the day of weaning and a normal oestrus follows. However if this is necessary, your management and/or nutrition in lactation is probably wrong and you should read Chapter 5.

This may assist gilts to cycle after administration of altrenogest.

If a gilt has not cycled by 240 days of age, cull it. It is infertile or subfertile. (Assuming it is not pregnant!)

FIG 4-22: A SUGGESTED FORMAT FOR THE SAFE USE OF PROSTAGLANDIN PRODUCTS

I .... understand and will carry out the following procedures for the safe use and administration of

Prostaglandin (*trade name* ....), for the induction of farrowing in sows.

I have read and understand the packing leaflet/data sheet.

The handling of this product is restricted to myself and .... who shall be bound by these rules.

At ALL TIMES the product will be stored in a nominated locked place.

No other person shall have access to the product or handle it.

It will only be administered to sows which are my property and on my farm/farms for the induction of farrowing as directed.

I will record the date of administration of each dose with records of the number of sows and dose volume given. These records will be available for inspection on request.

Supplies of the product will only be issued to me personally by a qualified veterinarian. I will sign for each consignment.

Empty containers will be placed in a sealed polythene bag, and returned to the issuing veterinary practice at which time a new supply may be issued.

I agree to receiving instruction as to the handling, storage, administration and recording of the use of this product.

A new sterile syringe and needle will be used each time the product is administered. After these injections have been given the syringes and needles will be rinsed out and either returned to the veterinary practice or safely destroyed on the farm as agreed.

Waterproof gloves will be worn by the operator handling the product. Accidental spillage will be washed off the skin immediately and in the event of accidental injection, medical advice will be promptly sought.

I understand that contact with prostaglandin products by women of child bearing age or by asthmatics is to be avoided.

These rules will be displayed where the product is to be stored.

I understand that failure to comply with these instructions at any time would result in withdrawal of stocks and no further issue of the product.

Signed .... Name and Address ....

Date ....

....

Site of storage of prostaglandin if different to above:

....

....

Countersignature of veterinarian Date

....

Vaccines against hormones

These vaccines contain a gonadotropin-releasing-factor (GnRF) which acts to stop gonadotrophin hormone production and thus immunologically castrates growing male pigs serving to reduce boar taint in the meat.

They can also be used to spay female pet pigs.

Growth/Health Enhancers

The aim of efficient pig production is to maximise growth as healthy, well looked after pigs, grow the quickest. Production can be enhanced by the inclusion of specific substances to the feed that stimulate growth. Remember however, that growth is also maximised by:

- The correct levels of vitamins in the diet.
- The correct levels of minerals, protein and energy.
- Metallic compounds such as copper.
- Some sedatives.
- Anabolic steroids.
- The effects of management, the environment, disease, genetics and housing.

How the farm is managed has the greatest influence on growth rates. Contributing factors include health status and the presence of specific diseases, pig flow and all-in/all-out production, the streaming of the pigs, general hygiene and the control of the environment. The effects of disease and, as a result, reduction in feed intake and immunosuppression, markedly affect growth.

In assessing whether to use specific products to promote growth the following should be considered:

- Ask to see all product trial data across a wide range of farm conditions.
- Ensure that the results are statistically significant.
- Seek veterinary advice as to the suitability of the product for your farm. For example, if you have a high health herd the effects may be minimal.
- Ask your feed supplier for his experiences and cost effectiveness of the products you are considering and obtain comments from other farmers using it.
- Beware of unsubstantiated claims.

Types of products that aim to promote growth

There are a number of different products that claim to promote growth in pigs (Fig 4-24).

Antimicrobial growth promoters (antibiotics)

The use of antibiotics for promoting growth is now widely banned, mainly due to concerns over antibiotic resistance. Where still allowed, approved products are becoming more limited and producers need to adapt their production methods to eliminate the use of these products in this manner.

Acids

These act either by lowering the pH of the intestinal contents and/or through an antibacterial effect.
The former include acids such as lactic, citric, fumaric and malic acids. Those having antibacterial effect include acetic propionic, and formic acids.

FIG 4-24: THE ADDITIVES TO ENHANCE GROWTH

Feed Additives	Efficiency
Copper sulphate	+++
Enzymes	+++
Fermentation	+
Immunoglobulins	++
Lactose	++
Mineral clays	?
Organic acids	+
Probiotics	+
Probiotics (fermentation)	++
Zinc oxide	++++

++++ Most efficient

Mineral clays

These are substances that have the ability to absorb certain toxins from the digestive system which may allow pathogenic bacteria to proliferate. Attapulgite would be a typical example. Their value has not been established.

Enzymes

These are many and varied and found in the digestive system of all mammals. Pigs have difficulty in breaking down the complex cell structures, particularly in wheat.

Enzymes assist in the breakdown of fibre and also enhance the efficiency of digestion of vegetable proteins, peas, beans and soya beans. Each enzyme is specific in its action.

The digestion of dietary fibre is enhanced by enzymes such as beta glucanase or xylanase which act upon fibres such as arabinoxylans and beta-glucans. Such fibres contain antinutritional factors that are also found in proteins. New enzymes have appeared on the market (proteases) to assist in the breakdown of such proteins and results have shown reductions in digestive disturbances particularly in weaned pigs. Phytase is an enzyme that assists the release of phosphorous from the feed.

Fermentation

Liquid diets provide the opportunity to create acid conditions by fermentation through the inclusion of certain microorganisms such as lactobacillus species prior to feeding. The establishment of the latter in the intestine, particularly in the weaned pig, helps to prevent the establishment of pathogenic bacteria. Fermentation also increases the digestibility of the diet with increased efficiency of use.

Metallic substances

The use of copper as a growth promoter is well documented. Zinc oxide in-feed has proved to be very efficient in preventing postweaning diarrhoea at levels of 2500 ppm/kg of elemental zinc, although its use is restricted in certain countries. The action of phytase in the weaner diet should be recognised.

Nutraceuticals

These are substances originating from different plants and whilst many claims are made for their effectiveness they are yet to be substantiated with confidence. Fatty acids and plants with high levels of vitamin E are claimed to increase the efficiency of the immune system. Claims are also made for garlic, ginseng, oregano, extract of cinnamon, aniseed, rosemary, peppermint and propolis extracted from honey.

Pre- and Probiotics

See Chapter 14 for further information

These are living cultures of bacteria such as lactobacillus, bacillus species and yeasts that are added to the diet. When given by mouth they help to maintain a balanced micro-environment in the gut and inhibit the multiplication of pathogens such as *E. coli*.

The exact way in which they promote growth is unclear and their efficiency appears to be most effective in the young growing pig.

In some cases they are claimed to improve feed efficiency, daily liveweight gain and health but evidence for this in most cases is not convincing.

Electrolytes

Approximately 55 % of a pig's bodyweight is made up of water. In young lean animals it is much higher (70 %) and in fat ones lower because fat contains little water.

The fluid balance in the body is regulated by the kidneys and in particular the concentrations of sodium (Na), chloride (Cl), potassium (K), hydrogen ions, bicarbonate, protein, calcium and magnesium in the tissues. These substances are called electrolytes.

A pig could lose 5 % of its bodyweight in fluid losses with little clinical effect but at 15 % it would die.

Such losses equate to 40 to 160 ml/kg of bodyweight. If the pig cannot maintain its fluid balance by intake relative to normal or abnormal losses dehydration will result (Fig 4-25).

Diarrhoea and vomiting are by far the most important problems in the young pig particularly during sucking and in the immediate postweaning period.

Certain types of *E. coli* and salmonella produce toxins which cause fluids to be excreted into the small intestine. This is called secretory diarrhoea.

Conversely viral infections such as TGE, PED and rotavirus destroy the villi in the intestine causing dehydration and a marked reduction in the absorptive capacity of the digestive system. This condition is called malabsorption with osmotic diarrhoea. It is also common in a milder form in pigs that have been weaned 2 days or more. It may then occur with or without the diarrhoea.

FIG 4-25: THE COMMON CAUSES OF DEHYDRATION IN THE PIG

Condition	Common Causes
Diarrhoea	**Viruses:** Epidemic diarrhoea Porcine epidemic diarrhoea (PED) Rotavirus Swine fever - classical and African Transmissible gastroenteritis (TGE) **Bacteria:** Campylobacter Clostridia Coccidia *E. coli* Salmonella *Brachyspira hyodysenteriae* causing swine dysentery
Vomiting	*E. coli* Haemagglutinating encephalitis virus PED TGE Rotavirus
Fever	Many viral and bacterial diseases Pleuropneumonia Pneumonia
Water shortage	Blocked water pipes Meningitis
Toxic conditions	Greasy pig disease Mastitis
Kidney failure	Cystitis Pyelonephritis
Haemorrhage	Ileitis Haemorrhagic enteritis Gastric ulceration Trauma

Rehydration by Mouth

This is the most practical method for use in sucking pigs and weaners. Glucose, water and electrolytes when combined with the amino acid glycine are well absorbed from the small intestine.

A typical electrolyte formulation would comprise:

- Glucose 67.5 %.
- Sodium chloride 14.3 %.
- Glycine 10.4 %.
- Potassium dihydrogen phosphate 6.8 %.
- Citric acid 0.8 %.
- Potassium citrate 0.2 %.

It is mixed with water at a rate of 30 g per litre for the first 24 hours and followed by 15 g per litre until the pig has recovered. The solution should be provided fresh, daily in easily accessible drinkers.

A number of commercial electrolyte solutions are available but all should contain glycine.

5 REPRODUCTION: NON-INFECTIOUS INFERTILITY

5 REPRODUCTION: NON-INFECTIOUS INFERTILITY

The Breeding Female

Problems associated with reproductive failure are often complex. In trying to resolve such problems it is essential to understand the important factors that maximise biological efficiency.

Key points to maximising reproductive performance:

- Use the breed of female that is suitable for your system.
- Ensure that the breeding female exhibits good hybrid vigour.
- Monitor the age of the sow and her continuing performance.
- Collect the required records and use them to understand the problem.
- Ensure a correct parity spread across the herd. Litters 3 to 5 are most productive.
- Use similar numbers of gilts per batch – avoid variation in parity structure of each batch.
- Use high levels of feed energy, protein and lysine in lactation and maximise feed intake.
- Provide an equitable environment with even temperatures. Cleanliness is essential.
- Compare the reproductive performance to the lactation length.
- Maintain good body condition throughout pregnancy but avoid sows becoming over-fat.
- Maintain sow health and immunity to disease.
- Assess the age of the gilt at first service, relative to life time performance in the herd.
- Assess the effectiveness of sow management from weaning to 28 days post-service.
- Make sure that the right type of stockperson is in charge of mating. The ability of the stockperson has a direct effect on reproductive performance.
- If you are mating naturally: Have a sufficient number of boars available so that they need only be used once every 24 hours at most.

Most of the major breeding (seedstock) companies have developed lines rather than standard breeds. The lines are generally based on breeds but have been developed differently. Thus one Line A may be based predominantly on the Large White breed but has been developed as a dam line by selecting for prolificacy (No. of piglets total born per litter and No. of litters born per year) and mothering ability (No. of piglets weaned per sow per farrowing place). Line D may also be based on the Large White breed but has been developed as a sire line by selecting primarily for growth rate, feed conversion efficiency and lean meat. This sire line development is often done at the expense of prolificacy and so if you are a commercial producer you should never select female back-crosses sired by such boars for breeding.

Efficient reproduction starts with females of high genetic potential for prolificacy and this includes good hybrid vigour (heterosis). The hybrid female is the progeny of the male of one breed or dam line (A), crossed with the female of another breed or dam line (B) to produce the commercial F1 gilt (AB). This animal is usually highly prolific in terms of pigs produced per sow per year. Some breeding companies cross the F1 with a third breed or dam line (C) to produce their commercial hybrid or 3 way cross. Such pigs are designated by a variety of numbers and names and Fig 5-1 illustrates the differ-

BREEDING AND SOW PRODUCTIVITY

The effect of breed on sow productivity

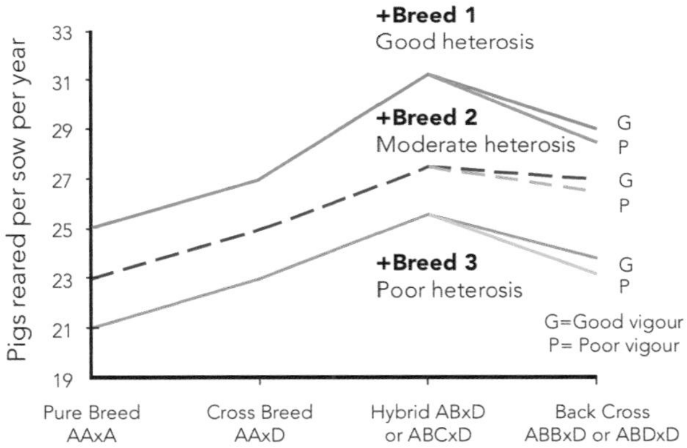

AAxA: Pure-bred female mated with a boar of the same breed on-line.
AAxD: Pure-bred female mated with boar of a different breed or sire line. This does not necessarily increase the numbers of piglets born but the improved vigour of the piglets means that more survive to weaning.
ABxD or ABCxD: Two way or three way crossbred females mated to a different breed or boar line, A, B and C. Should be prolific dam lines.
ABBxD: Back-cross female mated with a different breed or boar line. Both A and B are prolific breeds or dam lines.
ABDxD: Back-cross female mated with the same breed or boar line used in the back-cross. The back-cross has 50 % non-prolific boar line.

Fig 5-1

ences that may occur between crosses depending on the level of hybrid vigour. Pure breeding produces the worst scenario, hence the reason why at a commercial level no farmers should use pure bred females and boars of the same breed. In the past it has been common practice, commercially, to breed back from the F1 by using a boar of the same breed to produce a further breeding female that is $^{3}/_{4}$ pure breed (A), $^{1}/_{4}$ alternate breed (B). However, there is invariably loss of hybrid vigour and even in farms that are efficient, performance is usually less efficient than the F1.

Fig 5-2 lists the essential features of different breeding combinations and it is important when selecting the source of your gilts to remember that there can be a considerable difference in the prolificacy of the dam lines and in the expression of hybrid vigour between one breeding company and another.

The age of the sow also plays an important part in reproductive efficiency. In Fig 5-3 you will see that the best reproductive performance is in litters 3 to 6. Although by litter 7 the total number of pigs born may be greater there is often wide variation in the size of individual pigs at birth, an increasing number being too small or runts. There is also an increase in the numbers of stillborn with a reduction in live births. Furthermore, with age sows tend to be more clumsy and lazy, with higher levels of pigs laid on and an increased mortality, often 3 % or more above those of the efficient parities. Sows in litters 3 to 6 should rear at least 28 pigs per annum but by the 8th litter and above this often drops below 24. Levels of reproductive efficiency should also be compared to those being achieved by the incoming gilts. Culling decisions should be made in light of the performance of your first litter gilts, the maintenance of the mating programme and the availability of older sows as foster mothers.

The approximate parity distribution in the herd necessary to maximise production is shown in Fig 5-4.

Sow management from the day of weaning to 28 days post-service influences reproductive efficiency. If you have a problem, this is the first area that should be assessed.

FIG 5-2: CROSS BREEDING BENEFITS

- More pigs.
- Heavier and more even piglets at birth.
- Better mothering ability.
- Better production of milk.
- More pigs reared.
- Better fertility.
- More libido in the boar.
- Better conformation of teats and legs.
- Increased longevity.

FIG 5-3: AN EXAMPLE OF THE EFFECTS OF AGE ON REPRODUCTIVE EFFICIENCY*

	Litter Number			
	1	2	3 – 6	7+
Total pigs born	13	13.2	14.8	15
Alive	12.4	12.8	14	12.7
Dead %	5	5.5	6	15
Fostered (essential ones) %	3	4	8	10
Mortality				
Laid on %	3	3	3	5
Starvation %	1	1	1	2
Other	5	5	4	6
Total mortality %	9	9	8	13
No. weaned	11.1	11.6	12.7	11.5
Farrowing rate %	85	87	90	83

* Based on studies of field records.

PARITY DISTRIBUTION TO AIM FOR

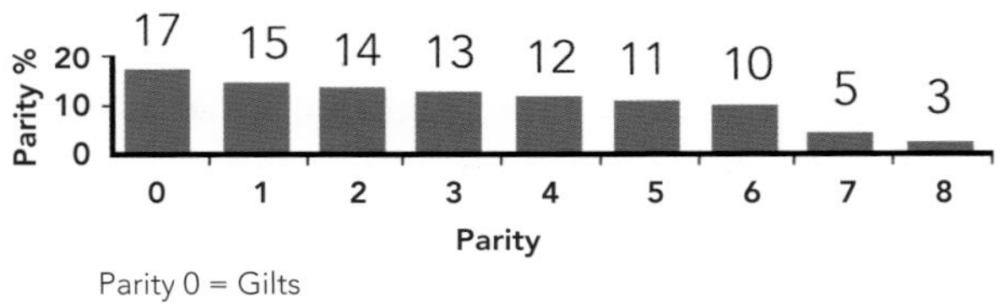

Fig 5-4

REPRODUCTION: THE CRITICAL PARAMETERS TO ACHIEVE THE MAXIMUM NUMBER OF PIGLETS BORN PER YEAR

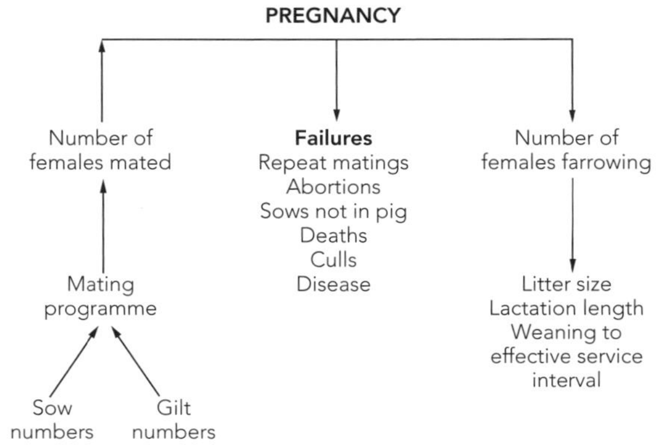

Fig 5-5

Consider the overall plan in Fig 5-5 which defines the critical parameters necessary to maximise reproductive efficiency.

The maintenance of gilt and sow numbers is the first critical factor. There must always be sufficient females available to maintain a mating programme.

A shortfall in numbers available is the most common failure on most pig farms.

From point of mating to point of farrowing there can be a variety of pregnancy failures which are dealt with later in this chapter. For every 100 sows served in a given period of time, fewer than 100 will farrow. This in percentage terms is described as the farrowing rate. Failures of animals to farrow are associated with repeat matings, abortions, sows not in pig, deaths, culls and/or disease.

$$\textbf{Farrowing Rate (\%)} = \frac{\textbf{No. of sows farrowed in a batch}}{\textbf{No. of sows served in a batch}} \times 100$$

How to approach a problem on the farm

Of all the problems on the pig farm, understanding and resolving a reproductive problem can be the most challenging. However, if a logical approach is taken, the causes of the reproductive failures can usually be worked out and corrective actions taken. The approach outlined in Fig 5-6 relates the performance of the herd to accepted levels of efficiency; this then defines the area of failure and finally the problem is identified. Further investigations are then necessary to understand what has precipitated the problem and these are carried out by examining the management procedures, records and performing appropriate pathological tests. An important objective here is to determine if the problem is of infectious or non-infectious origin or a combination of both. The end results are either poor numbers weaned, farrowing rate loss, a reduced litter size, anoestrus or any combination.

Remember: failure to maintain the optimum batch mating programme is the most common and most costly failure on pig farms today.

A PATHWAY TO UNDERSTANDING REPRODUCTIVE FAILURE IN THE BREEDING HERD

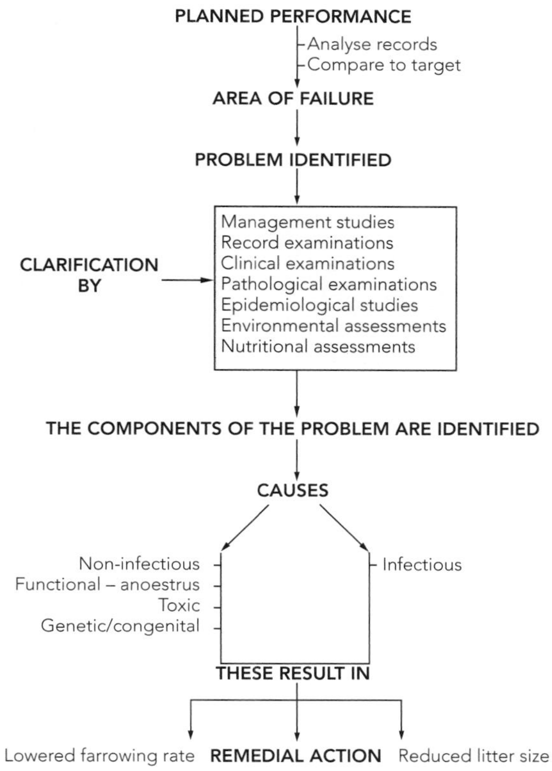

Fig 5-6

The Use and Interpretation of Records

The causes of infertility cannot be determined and corrective action taken without collecting reliable information and using it in a meaningful way. In order to understand reproductive failure, the records required for each breeding female and mating (AI or boar) include:

- The sow number.
- The parity of each sow (number of litters the sow has had).
- The dates of mating, farrowing and weaning.
- The number and type of services at each mating.
- Which AI (batch) or boar (number) was used.
- The service quality (rate each service as 1, 2 or 3 – good, moderate or poor).
- The lactation length.
- Mummified piglets and their size.
- The total numbers born (alive and dead).
- The weaning-to-service interval.

Records required for the breeding herd:

- Litter size variation* (scatter).
- Repeat matings – and their intervals in days.
- Abortions – and the age of foetuses.
- Females found not in pig.
- Discharges – the time when they occur post-service, observations and the outcome.
- Parity distribution.

FIG 5-7: REFERENCE DATA FOR REPRODUCTIVE EFFICIENCY IN A BREEDING HERD (27-day weaning)

A 100 Sow Module	Suggested Targets	Action Level
No. of gilts available for service at any time.	6	4
Age at first service (days).	240 +/- 10	< 220
No. of productive sows.	100	95
Average days from weaning to first mating.	6	7
Repeat matings.		
– Regular returns (18 – 24 days) %.	5	6
– Irregular returns (25 days +) %.	2	5
Non-productive days per sow.	12	14
Abortions %.	< 1	> 2
Sows not in pig %.	1	> 2
Sows culled pregnant %.	< 1	> 2
Deaths during pregnancy.	1	> 2
Farrowing rate %.	89	82
Vaginal discharge >5 days post-service.	1	> 2
Sows culled per year %.	36	42
Sow parity at culling.	6	8
Total born.	15	13
Pigs born alive.	14	12
Pigs born dead %.	5	7
Piglets mummified % < 100 mm.	< 0.5	1
Piglets mummified % > 100 mm.	1	1.5
Number of boars.	5	4
Mean age (months) boars.	21	24
Age at culling (years) boars.	3	> 3

FIG 5-8: FARROWING RATE LOSSES WITH NO INFERTILITY PROBLEMS

	Loss/100 Services	Lost Days
Normal repeats	5	105
Abnormal repeats	2	60
Abortions	< 1	40
Endometritis	< 1	21
Sows not in pig	1	90
Culls disease	1	30
Deaths	1	40
TOTAL LOSS	11%	386

Farrowing rate 89%.
Empty days from service to service = 10.
4 week weaning 2.38 litters per sow per annum.
3 week weaning 2.50 litters per sow per annum.

Litter size variation*

Calculate the litter size scatter for the herd. This is the percentage of litters where the total born are less than 8 (alive/dead/mummified).

$$\text{Litter scatter (\%)} = \frac{\text{No. of litters with 8 or fewer total born}}{\text{Total No. litters}} \times 100$$

Whilst the above information looks rather daunting, it is normally recorded in the mating book and on the sow and litter card on most farms. If a problem arises it is a simple procedure to identify the sows involved, collect the individual data and look for common features.

Fig 5-7 provides data against which your own farm performance can be related. These suggested targets and action levels for the various parameters are not absolute but rather broad guidelines. Alternatively records could be examined by batch farrowing place.

Monitor reproduction by farrowing rate loss. Record abnormalities in both the boar and the sow.

Understanding Farrowing Rates and Production Losses

In Fig 5-8 the components of farrowing rate loss are defined at the best levels of biological efficiency. A 7 % failure of repeat matings made up of 5 % at a normal interval of 18 to 24 days and a further 2 % outside this would be a good target to achieve. A continuing farrowing rate of 89 % for a minimum period of 12 months would be considered excellent and this would give approximately 10 non-productive days per sow from the first day of mating to the mating for the next pregnancy. Depending on the lactation length 2.3 to 2.5 litters per sow per year would result. It should be noted however that these figures do not take into account the period from the day of entry of the gilt into the herd to its first mating. Non-productive days in this respect often go unrecognised and can be significant.

Analysing the farrowing rate losses

Pregnancy losses from mating should be documented on a daily basis as they occur. Fig 5-9 shows a simple recording sheet that you could adopt. This type of approach is usually of more value than using computerised information, because the precise detail required is often not recorded in the programme.

This is particularly so in the column "Results/history /comments", where the stockperson's observations can

FIG 5-9: RECORDING FAILURES TO FARROW

Batch: 3

Batch Mating Target: 100
TOTAL MATED: 104

CATEGORY: (See below)

Sow No.	Parity	Date First Mated	1st or Repeat Mating	AI Batch or Boar used	Cause of Failure	Date of Failure	Days Interval	Result/history/ Stockperson/ comments
75	7	10 Dec 12	1	16	Dis	24 Feb 13	76	Discharge
161	5	12 Dec 12	2	4	–	2 Feb 13	52	Poor service
67	8	3 Jan13	1	60	D/C	27 Mar 13	83	Fighting
724	1	2 Mar13	1	24	Ab	4 May 13	63	Lame

Cause of failure: **Rp** = Repeats **Ab** = Abortion **NIP** = Not In Pig **Dis** = Disease **D/C** = Death/Culled

be very helpful in understanding the problem. For example, farrowing rate problems may not be associated with reproductive efficiency but be due to lameness and subsequent abortion or embryo absorption. Lame boars or lame sows can be associated with farrowing rate loss and if such information is not collected the core of the problem may be missed. Fig 5-10 shows the overall picture if you have a low farrowing rate.

MAJOR AREAS OF FARROWING RATE LOSS

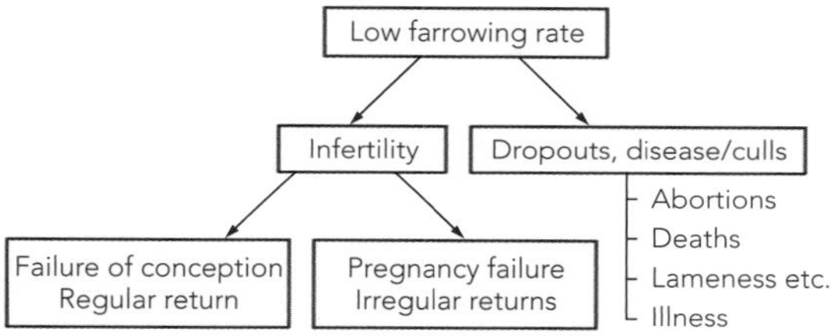

Fig 5-10

Embryo and Foetal Losses

To help you understand the processes of fertilisation and pregnancy Fig 5-11 and Fig 5-12 show the anatomy of the reproductive tract of the female. Note in Fig 5-12 that the opening of the urethra (tube from the bladder) lies on the floor of the vagina, approximately 70 mm inside. If a boar has very long back legs the penis enters the vagina in a downward movement and the tip of the penis sometimes enters the urethra. In such cases the sow shows pain and there is the potential for damage which can result in cystitis or infection of the bladder. In extreme cases this can even result in the death of the sow from rupture of the urethra. Two uterine arteries branch off the main aorta and supply blood to the uterus. In the pregnant state the changes in blood flow in these arteries are used to diagnose early pregnancy. See Chapter 15.

By studying records and carrying out clinical observations and pathological tests it is usually possible to determine precisely where reproductive failure has occurred. Such failures can be conveniently grouped into 6 categories related to stages in the reproductive cycle. In most cases failure can be narrowed down to 1 or 2 of these (Fig 5-13). However, to determine the causes of the loss it is first necessary to understand reproductive physiology, both in the sow and in the boar. The stages of development from fertilisation to farrowing and the consequences of each group failing are shown

ANATOMY OF THE FEMALE REPRODUCTIVE TRACT

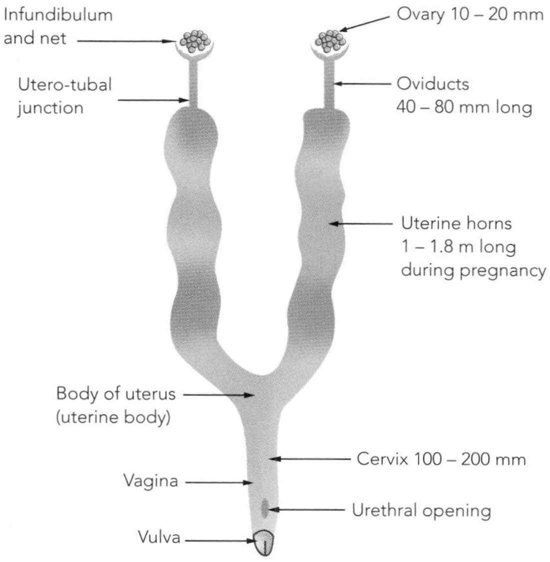

Fig 5-11

THE REPRODUCTIVE TRACT OF THE SOW

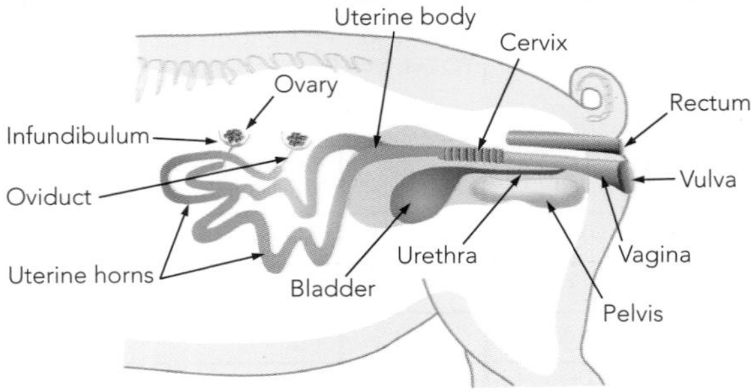

Fig 5-12

REPRODUCTIVE FAILURE: POTENTIAL AREAS OF LOSS

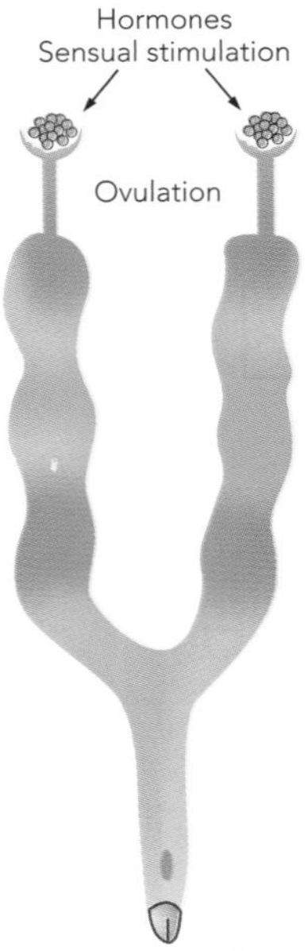

Areas of Loss	
Anoestrus	Group 1
Oestrus & Egg Production	Group 2
Fertilisation	Group 3
Implantation	Group 4
Maturation	Group 5
Farrowing	Group 6

Fig 5-13

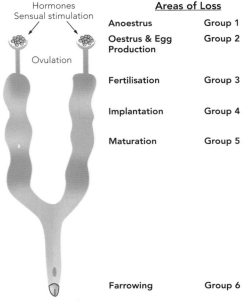

in Fig 5-14. You should spend time studying this to appreciate the procedures that are available for corrective action on the farm.

Group 1 Losses – Anoestrus

Anoestrus is common both in the maiden gilt and in the sow postweaning. Puberty is initiated in the gilt by complex hormonal mechanisms that are associated with growth and age. The hybrid gilt commences oestrus from around 160 days of age onwards but some breeds such as the Meishan come into heat much earlier.

Factors in the external environment stimulate the higher centres of the brain which activates the pituitary gland (Fig 5-15). This tiny gland at the base of the brain produces the hormones which preludes the oestrus cycle (the hormone surge). There are primarily 2 hormones involved: the follicle stimulating hormone (FSH), which as the name implies causes the ovaries to be stimulated and to produce follicles containing eggs; and the luteinising hormone (LH), which at the point of oestrus is excreted into the bloodstream to act on the follicles and cause the eggs to be released. The developing follicle at the time of oestrus is about the size of a small cherry. There may be up to 15 to 18 follicles in each ovary. After the egg has been released from the follicle and ovary the remaining tissues develop into a body called the corpus luteum. This produces the hormone progesterone and the levels in the blood rise as oestrus subsides. The hormone changes are depicted in Fig 5-16.

Oestrus is then initiated by the ovary which produces the other female hormone, oestrogen, and it is this that is responsible for the outward signs of heat (Fig 5-17). After ovulation, oestrogen levels drop as the progesterone rises. Females with high levels of progesterone in the first 3 days after mating are more likely to create an improved environment for the survival of the fertilised eggs. In the gilt, high levels of feed intake in the

EMBRYO AND FOETAL DEATH AND THE CONSEQUENCES

The Age and Position of Embryos	
Day 0	Ovulation
Day 0	Fertilisation
Day 2	Fertilised ova descend
Day 3	Embryo size 0.2 mm – 0.5 mm
Day 4 to 14	Embryos move between horns
Day 14 to 21	Implantation (pregnancy) 5 mm – 10 mm

The Failure	The Result
No ovulation	Anoestrus
No fertilisation	Normal 18 – 24 day return
Total embryo mortality prior to day 10	Normal 18 – 24 day return
Only 4 or less embryos present at day 17	Delayed return to oestrus at 25 – 35 days.
Implantation occurs but all embryos die before day 35 and are absorbed.	Sow not in pig, pseudo. pregnancy, returns around 63 days
1 – 2 embryos survive after day 35	Litter size 1 – 2 pigs born, delayed farrowing
All embryos die after day 35	Sow pregnant but does not farrow
An embryo dies after day 35	Mummified pig

Fig 5-14

first 2 to 3 days may lower progesterone levels and thereby result in an increase in embryo mortality. There is no effect in the sow.

OESTRUS STIMULATION

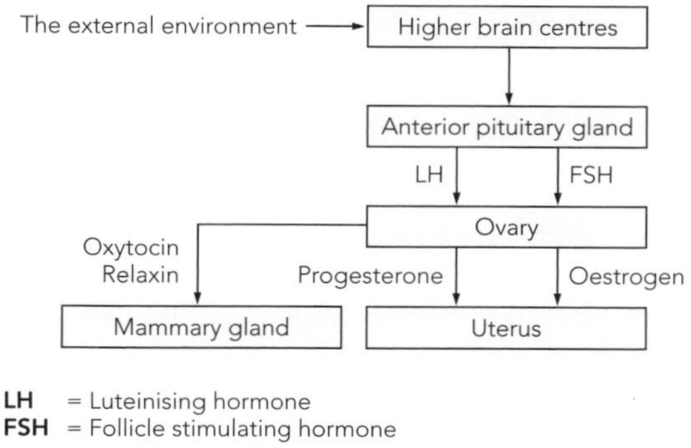

Fig 5-15

HORMONE CHANGES IN THE OESTRUS CYCLE

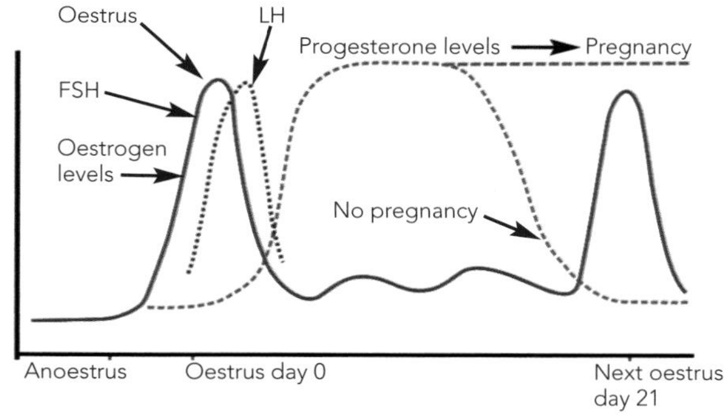

Fig 5-16

Key factors that stop the hormone surge and produce anoestrus in the gilt:

- Age – immaturity.
- Bad external environment.
- Bullying. Stress.
- Disease.
- High stocking density.
- Lameness. Pain.
- No contact with the boar and his pheromones – needs to be within 1 metre – saliva contact.
- Poor light.
- Poor nutrition.
- Low feed intake.
- Poor management.
- Sunburn. Skin damage.
- High temperatures – above 26 °C (79 °F).
- Low temperatures and chilling < 12 °C (55 °F).
- Draughts.
- Mycotoxins.
- Genotype.
- Flooring.
- Attitude of stockperson.

Approximately 9 kg (20 lbs) of pork is lost for each day of pregnancy failure. This increases with large litters. (12 pigs x 90 kg) / 115 days

Anoestrus problems in gilts or sows can be a major cause of economic loss, the significance of which is often unrecognised.

The most common problem is failure of the stockperson to recognise oestrus (Fig 5-17). Field experience shows that approximately one third of anoestrus gilts examined at slaughter have been cycling, one third are

FIG 5-17: DETECTING OESTRUS

Signs of Oestrus	Reasons for Failure to Detect Oestrus
Swelling of the vulva, redness and small amounts of mucus. Changes in vocal sounds. Standing immobile next to the boar. Standing still with ears pricked when back pressure is applied. Standing when mounted by other females. Seeking contact with a boar. Nosing the flank of other females or boars. Smelling the prepuce. Permitting the boars to mount and mate. Nipples become enlarged in gilts.	Lack of understanding and appreciation by the stockperson. Lack of stockperson experience. Stockperson overworked. Insufficient time for stockperson to observe animals. Lack of light. Dark environment. Poor lighting resulting in poor observation. Cold or excessively hot summer weather. Too many gilts in the group and poor observation. No boar stimulation. Apprehension from either pig persons or other animals. Sow in pain, particularly in joints and muscles. Lack of libido in the boar. Boar too heavy. Illness in the boar. Female not expected to be in heat.

pregnant and only one third are actually in anoestrus.

If there is an oestrus problem in your gilts then consider the following 5 areas and identify the potential failures in the system.

Key points to stimulating oestrus in the gilt:

1. Develop a successful gilt management system.
2. Create good health.
3. Acclimatisation of new gilts – create a low stress environment.
4. Provide good nutrition.
5. Satisfy the gilt's physiological needs.

1. Develop a successful gilt management system

It is important to develop a predictable method of management, an example of which is shown in Fig 5-18. Note this starts with a large group of gilts, 90 kg (200 lbs) weight, 180 days of age and given mature boar or V-boar (vasectomised or epididectomised) contact. In some herds however, particularly those with a very lean genotype, it may be better to feed a dry/gestating sow diet with a lower lysine level (0.7 %) to increase fat deposition. The rate of growth and nutrition also needs to be considered if leg weakness is a problem. The type of nutritional intake becomes an important variable. Does the gilt respond to the existing system of feeding and management? If she does not, then make a change. A P2 fat measurement of at least 18 mm should be achieved by point of mating. The gilt must not look thin (below condition score 3.5 – see Chapter 7). Heat is observed during the initial period in the holding pool and once gilts have been identified at their first oestrus they should be ear tagged and moved into the service area in small groups of no more than 6. The stocking density should provide for approximately 2.7 m² per animal.

Note: Control the boar exposure; do not just place the boar next to the gilts all the time.

The most common "causes" of anoestrus are either: failure to detect heat – or the gilt is already pregnant!

The stages of the oestrus cycle

DAY 0 – 1 Proestrus: This is the period just prior to standing for mating. The vulva reddens and the gilt becomes sexually active.

DAYS 1 – 3 Oestrus: The gilt stands firmly to the boar.

DAYS 3 – 6 Metoestrus: The corpus luteum develops and produces progesterone and the period of sexual activity disappears.

DAYS 6 – 18 Proestrus: The quiet period prior to the next oestrus.

DAYS 18 – 24 Oestrus again: But note the gilt is only receptive to standing for the boar for 10 to 15 minutes every hour.

Key points to a good gilt management system:

- Acclimatise the gilts to your farm.
- Ear tag all gilts and record them as they come into heat.
- Vaccinate gilts against parvovirus and erysipelas in the holding pool at around 90 kg (200 lbs) as you tag them.
- Move cull sows at weaning into this pool because as they come into heat urine containing oestrogen acts as a stimulus. An alternative technique is to spray urine from sows in heat onto the noses of anoestrus gilts.
- Preferably use an old vasectomised boar in the holding pool for intermittent periods of 10 days. If you use an entire boar – be sure he cannot mate gilts.
- Change the boar in the holding pool of gilts regularly every 7 to 10 days.
- Alternatively expose the gilts to the boar for 20 minutes daily under observation. Only use mature boars.
- A high stocking density of less than 1.4 m² per animal will have a marked depressant effect on oestrus.
- Provide a minimum of 14 to 16 hours of light per day of at least 250 lux (adequate to read a newspaper). Follow this by 8 to 10 hours of darkness (< 20 lux).
- Make sure pens are clean, dry, draught-free and not in heavy shadows.

A GILT MANAGEMENT SYSTEM TO PRODUCE OESTRUS

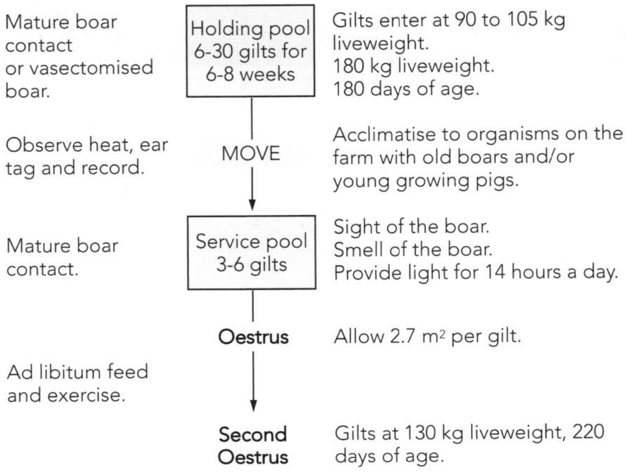

Fig 5-18

- Make sure there are no disease problems e.g. worms, mange, enteric or respiratory diseases.
- Determine the most responsive feeding programme and ad lib feed 2 to 3 weeks prior to expected matings.

2. Create good health

The health status and nutrition of the gilt is very important in deciding whether she will come into oestrus or not. Look at the group of gilts. Are they a good weight for age or is growth variable or poor? If this is the case, is there any clinical evidence of disease (mange for example) which may be responsible for anoestrus? Active respiratory diseases, or previous pneumonia, heart sac infection or pleurisy can inhibit puberty and the onset of oestrus. Water access problems can mimic pneumonia. Atrophic rhinitis or ammonia can destroy the sensitivity of the nose and therefore the response to pheromones (male hormones). If gilts are regularly moved into continually populated pens, heavy parasite burdens and coccidia can build-up that can interfere with the digestive process. Anoestrus problems have been associated with these. Poor flooring can make gilts reluctant to stand properly.

Remember, a happy comfortable gilt is a fertile one.

3. Acclimatisation of new gilts

Gilts that you introduce to your herd from elsewhere must be acclimatised over a period of up to 6 to 8 weeks prior to mating. This acclimatisation allows the gilts to adjust to your feed, housing and management system but above all to those pathogenic organisms enzootic in your herd to which they may have no immunity. Major problems arise when gilts from high health status herds (e.g. a herd free from mycoplasma (enzootic) pneumonia (EP), Actinobacillus pleuropneumonia (App), Porcine reproductive and respiratory syndrome virus (PRRSV), Aujeszky's disease/pseudorabies (AD/PRV), progressive atrophic rhinitis (PAR) are put straight into a herd in which these infections are present and active. The gilts become ill and fail to come in heat or fail to conceive. They may be made permanently infertile. In such cases vaccination of the gilts on arrival (against, for example, EP, AD/PRV, App and even in extreme cases *Haemophilus parasuis*), helps the acclimatisation and possibly allows it to be shortened. A period of medicated feed – at half to three quarters therapeutic level may also help, say, for the first 3 weeks. Even when the incoming gilts are from a herd thought to be of similar status to your own, a minimum of 3 weeks acclimatisation is necessary.

If possible (and it is not possible on many weaner producer farms) the incoming gilts should be segregated from the main herd for at least the first 3 weeks of acclimatisation. Some seedstock suppliers insist on this. It gives time to remove the gilts from your farm if disease breaks out in the source herd. The process of acclimatisation involves exposing the group of gilts to likely sources of virus or bacterial contamination. The exact procedures depend upon the pathogens present in your herd compared with the herd of origin.

The following should be considered:

- Identify the pathogens the gilt has been exposed to in the donor herd and those present in your herd.
- Decide whether to move the gilts into isolation or directly into the herd. If the latter, place them in a pen with separate drainage and air space for 4 to 6 weeks and use separate boots and protective clothing.

Feedback

Use feedback to expose new gilts to on-farm pathogens:

Enteric pathogens: Expose the gilts to weaner faeces (4 to 6 weeks of age). Place 3 kg (7 lbs) into the pen 3 times a week.

Reproductive pathogens: Expose gilts to placenta, stillborn and mummified piglets from the farrowing house. Note this is illegal in some countries.

Respiratory pathogens: Use rope which has been left hanging in the weaner pens for 3 days for the weaners to chew and play with; then place in gilt pens.

Fig 5-19 summarises the options for effective pathogen transfer by feedback.

FIG 5-19: PATHOGEN TRANSFER BY FEEDBACK

Pathogen	Weaner Faeces	Rope	Placenta etc.
App		YES	
Bordetella		YES	
Clostridia	YES		
E. coli	YES		
Haemophilus parasuis		YES	
Mycoplasma		YES	
Porcine parvovirus			YES
Pasteurella		YES	
Porcine Circovirus 2	YES	YES	YES
PRRSV		YES	
Rotavirus	YES		
Teschovirus	YES		YES
Transmissible gastroenteritis/ porcine epidemic diahorrea (TGE/PED)	YES		

Specific pathogens:

Aujeszky's disease – Vaccinate on arrival.

Congenital tremor – Expose the gilts to faeces from the farrowing house 3 times a week. Wipe the vulva of recently served sows with tissues to collect semen secretions and place these in the pen 3 times a week. In start-up units, obtain dead semen from the proposed AI centre and use this as part of a feedback programme.

Erysipelas – Vaccinate on arrival.

Escherichia coli – Expose gilts 3 times a week to weaner and/or farrowing house faeces. Vaccinate during pregnancy.

Leptospirosis – Vaccinate on arrival.

Mycoplasma (enzootic) pneumonia – If the incoming pigs come from a herd free from mycoplasma (enzootic) pneumonia vaccinate prior to arrival or well before if possible.

Parvovirus – Vaccinate the gilts 2 to 3 weeks prior to mating.

Porcine reproductive respiratory syndrome virus – In a sero-positive herd purchase PRRSV negative gilts then vaccinate the gilts whilst in isolation or on arrival or expose to an infected environment on the farm for 1 to 2 hours. A building containing pigs of 6 to 12 weeks of age is ideal. Expose the naive gilts to PRRSV at 6 weeks prior to mating. Ensuring adequate PRRSV exposure is extremely difficult. Tonsillar scrapes, autogenous vaccines or rope from the weaners should be considered. An alternative is to house a small number of pigs in the isolation premises. If your farm is PRRSV negative you must not purchase positive gilts nor should you use live PRRSV vaccines.

Porcine circovirus 2 – Place short pieces of strong rope in the weaner pens and allow them to chew and salivate onto the rope for 1 week. Move the rope into the gilt pens. If necessary vaccinate gilts during isolation and prior to farrowing to protect their piglets.

Progressive atrophic rhinitis – Vaccinate on arrival.

TGE/PED – If gilts are naive, vaccinate in isolation if your herd carries the virus or if no vaccine is available expose them to weaner faeces and any scour in the farrowing house soon after arrival.

Judge the result of acclimatisation by the response.

If a gilt has not come into oestrus by 260 days, consider culling her. She is telling you she is infertile.

Isolation

A period of isolation is well advised to ensure that incoming gilts are not incubating disease. The actual farm procedures will vary from one to another.

The increases in herd sizes, the adoption of segregated early weaning (SEW), 3 site production and the presence of PRRSV have all focused attention on the best way to introduce gilts into the herd. Recommendations have in some ways reached fever pitch with all kinds of complex systems, time periods, isolation premises, a myriad of recommendations, and pig farmers left in confusion. However the main objectives should be:

1. To prevent new pathogens from entering the recipient herd.
2. To establish procedures that result in a common immune status between the incoming pigs and the receiving herd without creating infection that disturbs the balance of infectious agents in the herd.

Ideally, do not serve on first oestrus unless you want a poor litter, although ensuring all the batch farrowing places are full takes precedence.

Preventing pathogen entry

- Assess specific pathogens present in the receiving herd. Veterinarians should assess the pathogen compatibilities of both groups of pigs at the onset.
- Assess their significance in relation to the health status of the incoming stock.
- Use Fig 5-20 to identify the common diseases/pathogens and the isolation periods that may be necessary.
- Isolation periods can be grouped into 4 time scales: none, 3 weeks, 8 weeks and 12 weeks, depending upon **known** incubation periods of different diseases/pathogens.
- Isolation accommodations should be run on an all-in/all-out policy. This might result in the need for 2 or 3 isolation facilities.
- All boots and coveralls used in the isolation facilities should be strictly separate from the on-farm clothing.
- If isolation is on-farm, incoming stock should be housed for a period of 21 to 56 days in their own air space and with separate drainage. This provides the opportunity (but a risk) for the pigs to be removed if the donor herd reports a new disease not compatible at the onset.
- Where off-farm (ideal) the facilities should be a minimum of 800 m (1/2 mile) from the herd with completely separate personnel.

FIG 5-20: RECOMMENDED ISOLATION PERIODS FOR INCOMING BREEDING STOCK

Disease/Pathogen	HEALTH STATUS		Suggested action
	Incoming pigs	Recipient herd	
Actinobacillus pleuropneumoniae (App)	-ve serology	-ve serology	56 days isolation then blood test sentinels.
	-ve disease history but +ve serology	-ve disease history but +ve serology	Discuss with your veterinarian.
	Disease history	+ve or -ve	Do not buy.
	-ve	-ve	21 days isolation then check donor herd.
Aujeszky's disease/pseudorabies virus (AD/PRV)	+ve	+ve or -ve	Do not buy.
	-ve	-ve	21 days isolation then blood test.
	-ve	-ve	56 days isolation then check donor herd. Blood & slaughter test sentinels.
Brachyspira hyodysenteriae Swine dysentery (SD)	+ve	+ve or -ve	Do not buy.
	-ve or no history of disease	-ve or no history of disease	21 days isolation then check donor herd.
Congenital tremor (CT)	-ve	-ve	21 days isolation then check the donor herd.
Leptospira pomona (LP)	+ve	+ve or -ve	Do not buy.
Mycoplasma hyopneumoniae Mycoplasma (enzootic) pneumonia (EP)	+ve	-ve	Do not buy.
	-ve	+ve	Vaccinate incoming pigs twice before arrival or do not buy.
	+ve	+ve	Buy.
	-ve	-ve	21 days isolation then blood test.
Porcine parvovirus (PPV)	Unknown	Unknown	Expose faeces and farrowing material from growing pigs to gilts for 3 weeks whilst in isolation.
Progressive atrophic rhinitis (PAR) Toxigenic *Pasteurella multocida*	-ve	-ve	21 days isolation then check the donor herd.
	+ve	+ve or -ve	Do not buy.
PRRSV	-ve	+ve	56 days isolation/acclimatisation.
	-ve	-ve	56 days complete isolation and blood test sentinels.
	+ve	-ve	Do not buy.
Salmonella choleraesuis (Salmonellosis)	+ve	+ve	Do not buy.
	-ve	-ve	21 days isolation then check donor herd.
Sarcoptes scabiei var *suis* Mange	-ve	+ve	Buy.
	+ve	+ve	Do not buy.
	+ve	-ve	Do not buy. Medicate with ivomectins if exposed.
Swine influenza (SIV)	Unknown	Unknown	21 days of isolation. Expose gilts to semen from the service area and to growing pigs.
Teschovirus	+ve or –ve or no disease history	+ve or –ve often unknown	21 days isolation. Could test serologically.
Transmissible gastroenteritis/porcine epidemic diarrhoea (TGE/PED)	+ve	-ve	Do not buy.
	-ve	+ve	21 days isolation. Expose pigs to faeces from young growing pigs.
	-ve	+ve	Vaccinate in isolation or on arrival.
Unknown viruses and bacteria			

Strategies for preventing pathogen entry

These need to be simple, practical and cost effective in meeting the objectives. At the outset there is no one method that can be adopted across all farms because of different organisms, herd sizes, facilities and breeding policies. The following will help to establish the best methods for your herd:

- Monitor the reproductive performance and health of the breeding stock as a guideline to success.
- Always buy replacement stock from the same source if possible and one with good biosecurity and health documentation.
- Determine the required isolation period.

No isolation – You are at the mercy of the donor herd. Diseases such as PRRSV and mycoplasma (enzootic) pneumonia have long incubation periods and they should at least be present in the recipient herd. Many farms adopt this strategy and introduce stock quite successfully. Sometimes it is necessary to medicate the feed with 300 to 500 g/tonne of chlortetracycline (CTC) or oxytetracycline (OTC) or 100 g/tonne tylosin over the first 3 to 4 weeks of introduction to suppress any development of disease.

21 days complete isolation – Allows for a check (albeit short) to see that no diseases are incubating in the donor herd. Pigs can be blood tested during this period. Not recommended if mycoplasma (enzootic) pneumonia or PRRSV are absent in the recipient herd as at least 56 days are required.

21 days on-farm isolation – It may be necessary to medicate in-feed during this period. Pigs can be blood tested but 21 days may be insufficient for some diseases.

56 days complete isolation – This is necessary for herds believed free of mycoplasma (enzootic) pneumonia and PRRSV. Sentinel pigs from the recipient herd can be slaughtered and tested as well.

85 days complete isolation – This is an extended period advocated by some veterinarians to allow time for PRRSV negative animals to become infected and ultimately become non-viral excretors. This is particularly so when live-vaccine is used in countries where it is available. Live vaccines are not available in some countries and therefore do not complicate procedures such as the introduction of new strains of virus.

Age of incoming animals

Traditionally this is around 180 days to allow 6 weeks of acclimatisation prior to mating although weaner gilts (30 kg/70 lbs weight) are becoming more popular. PRRSV has created major problems particularly if negative gilts are introduced directly into positive herds. Three-site production by breeding companies also reduces the availability of PRRSV positive animals. The age when gilts are purchased may drop to 28 days to allow for prolonged PRRSV exposure.

Experiences in the UK and in Europe confirm that the introduction of pigs either naturally exposed to mycoplasma (enzootic) pneumonia and PRRSV, or negative to both these diseases (mycoplasma (enzootic) pneumonia vaccinated) can be successfully integrated into similar herds, by adopting suitable isolation and acclimatisation protocols.

Suggested procedures

Discuss these with your veterinarian.

Procedure A – Specific pathogen free status pigs to another specific pathogen free status herd (indoor or outdoor)

Typical example being *Mycoplasma hyopneumoniae* and PRRSV negative in both donor and recipient herds.

- Weight of gilts 25 to 95 kg.
- Complete isolation required for 8 weeks with no farm contact.
- At the end of the first 7 to 10 days introduce sentinel weaner pigs 8 to 14 weeks of age either directly into the gilts or provide nose to nose contact. 1 sentinel per 4 gilts is advised.
- At the end of 8 weeks the sentinel pigs can be slaughtered and examined for disease.
- The group can be tested for mycoplasma (enzootic) pneumonia, PRRSV and other specified diseases.
- Check with the donor herd that pigs can move into the herd.
- Acclimatise the incoming gilts with feedback materials.
- When all is clear introduce the gilts to the service area.

Procedure B – Specific pathogen free status pigs to a lower status herd

Typical example is *Mycoplasma hyopneumoniae*, PRRSV negative pigs to an *Mycoplasma hyopneumoniae*, PRRSV positive herds

Indoor herds

- Vaccinate all incoming breeding stock at 6 weeks and 4 weeks and again on entry to the herd, against mycoplasma (enzootic) pneumonia and other diseases as required during isolation. If PRRSV vaccine is considered safe to use, the isolation period will need to be extended.
- Age of gilts 12 to 24 weeks.
- Complete isolation or on-farm isolation. Direct integration is not recommended.
- At the end of 7 to 10 days' isolation, introduce challenged weaner pigs (mycoplasma (enzootic)

pneumonia, PRRSV +ve) 8 to 14 weeks of age as for Procedure A. Introduce feedback materials.
- If PRRSV vaccine is not used medicate the feed with 300 to 500 g/tonne of CTC or OTC or 100 g/tonne tylosin for the first 5 weeks of isolation or as advised by your veterinarian. Medication may or may not be required.
- Blood sample 50 % (minimum of 5) of the gilts in Week 7.
- Check sentinels at slaughter if post-mortem examinations are required.
- Check with the donor herd that the pigs can move into the herd.

Outdoor herds
- Weight of gilts 25 to 90 kg.
- Because the exposure to respiratory organisms is low, medicated feed is not usually required.
- Vaccinate for mycoplasma (enzootic) pneumonia as for indoor herds.
- Mix challenged weaner pigs into the group whilst in isolation as for indoor herds. It may be necessary to change these pigs after 2 weeks. Introduce feedback materials.
- If PRRSV vaccine is used then a 12 week total isolation must be considered.
- Blood sample gilts for PRRSV in Week 7 in one batch to establish the success of the system.
- Move gilts to the service area.

Gilts should remain in their isolation pens for the full 8 weeks. In countries where PRRSV live vaccine is used gilts are totally isolated for 12 weeks to prevent excretion of the vaccine virus into the herd. In some breeding herds the use of vaccine has been associated with severe disease and its use in any case may not be advised for risks of incompatibility.

Checklists for the health of your gilts
Assess health and disease. Do you have:
- Variable growth?
- Coughing?
- Evidence of rhinitis?
- Mange?
- Pneumonia?
- Lameness and stiffness on movement?
- Failure of the gilt to stand to the boar – osteochondrosis or leg weakness?
- Poor feed intake or an incorrect diet?

4. Provide good nutrition
The methods of feeding the gilt and the composition of the diet will depend on the genotype. They will also depend on the environmental needs of the gilt relative to housing, temperature and insulation of the buildings. To determine the best system requires a degree of trial and error on the farm – the objective being to produce the second or third oestrus cycle within a predetermined time span so that the service programme can be accomplished satisfactorily.

Key points to success:
- Gilts should arrive on the farm at 85 kg (190 lbs) or less and be fed a gilt developer diet until 100 kg (220 lbs). This will also allow time for acclimatisation.
- It is necessary to increase backfat in the modern genotype to 18 to 20 mm at the P2 measurement. This is best carried out feeding a specified gilt rearing ration that may have low or high lysine ration depending on the principles being employed. Feed to appetite. Gilts are so important that a specific diet should be made for them. Consult with your breeding company for details of their specific recommendations.
- Approximately 2 to 3 weeks prior to moving into the service area for mating, ad lib feeding should take place. This is to maximise ovulation rate. Use a good lactator or grower diet containing 9.7 MJ NE/kg (14 MJ DE/kg) and 1 % lysine or a specialist gilt developer ration.
- Low protein, low energy, or poor quality diets in the period leading up to puberty will often produce a deep state of anoestrus that in some cases is permanent. Assess the response to the diet used.
- Do not leave gilts in a finisher house to point of service; many will never cycle and nutritional requirements may not be satisfied.

5. Satisfy the gilt's physiological needs
Oestrus should commence from approximately 165 to 200 days of age. Movement from one pen to another with controlled close boar contact provides the greatest stimulus. See: Develop a successful gilt management system earlier in this chapter.

Anoestrus in the sow
The act of weaning the sow or even removing just 3 to 4 piglets increases the pressure of milk in the mammary gland and this causes the udder or relevant gland to stop production. This mechanism causes the hormones that promote milk production to cease, thus allowing the development of the follicle-stimulating hormone and luteinising hormones, to bring the sow back into oestrus or heat. Providing the sow comes into heat and is served within 6 or 7 days of weaning then she is likely to have a high conception rate, good fertility and litter size. The length of the weaning-to-service interval is now recognised as being associated with reproductive effi-

ciency and some sows mated 7 and 14 days after weaning are sub-fertile with poor litter size and farrowing rates. A high proportion of single matings also occur at this time. If you have a reproductive problem associated with poor litter size and high numbers of repeat matings check the average weaning to first service interval across the herd (5.5 days is ideal) and analyse the results by parity or litter number. The farrowing rate loss in the sub-fertile period may be as much as 10 % and litter size may drop by up to a pig a litter. The first litter gilt is often a major problem in this respect and the delay in the weaning-to-service interval can compromise litter size by up to 3 pigs per litter. It is normal for the weaning-to-service interval in a gilt to be 6 days, but become concerned if this extends to 7 days postweaning. If the weaning-to-service interval is extended and associated with lowered fertility always check the newly weaned gilt, then follow through the checklist below to identify those factors that may be important on your farm (see also Fig 5-21).

FIG 5-21: A SUMMARY OF THE CAUSES OF ANOESTRUS

Cause	Relative Importance	
	Gilts	Sows
Large groups Badly matched groups High stocking densities	+++	++
Confinement housing	+++	+
No boar contact	+++	+
Genetic	+	–
Poor photo stimulation	+++	+
Catabolic state	+++	+++
High temperatures	+	+++
Poor water in farrowing	+++	+++
Overfat in gestation	+++	++

– not very important +++ very important

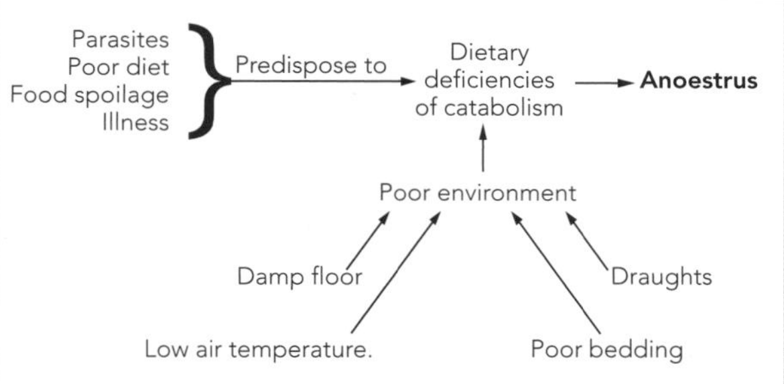

Anoestrus in the sow – a checklist:

- Identify from records whether this relates to a particular parity.
- Make sure during lactation that the sow does not lose bodyweight if possible and certainly no more than 15 kg (33 lbs). Weigh sows after farrowing and again at weaning to check this. Try to achieve weight gain. Monitor back fat – P2.
- The most common cause of anoestrus or delayed oestrus is loss of body condition, particularly in the first 2 to 3 weeks of lactation. Feed intake here drives the weaning-to-service interval and subsequent fertility and stimulates the primordial follicles to determine ovulation rate into the next litter.
- Feed the sow from 3 days post-farrowing to appetite, with a high energy diet, particularly if she is of a lean genotype.
- The diet should contain at least 9.7 MJ NE/kg (14 MJ DE/kg), 18 % protein and 1.1 % lysine. Make sure that overfeeding in lactation does not cause inappetence.
- Make sure that there is easy access to water, with a nipple flow of at least 2 litres per minute. Sows will drink 40 to 80 litres a day. Test the nipple drinkers and see how long it would take the sow to stand and drink. She could stand for 2 hours per day! Give the sow 4.5 litres twice daily into her trough at feeding time to encourage intake.
- Always remove uneaten feed from the trough because in a warm farrowing house fermentation takes place within 3 to 4 hours.
- If there is a second litter size problem, feed a weaner ration to the lactating gilts as half the daily total intake. Check gilt body condition at entry to the farrowing house. Many gilts are too fat.
- If the sow is correctly fed and managed during lactation, she should come into heat in the fertile period.
- Are sows comfortable in the farrowing house? Aim for around 18 °C (68 °F).
- Avoid any environmental factors that will cause the sow to lose weight, for example wet farrowing pens, high airflow and evaporative cooling, water shortage and inappetence associated with disease and spoiled feed.
- Avoid weaning/fostering more than 10 % of piglets from the sow during lactation because this may stimulate oestrus and poor ovulation with delayed oestrus again at weaning and sub-fertility. Feed the sow from weaning to point of service ad libitum with the lactator diet.
- Do not mix first litter females with older sows if group housing is practised.

- If group housing is practised, introduce a large quiet boar into the group on the day of weaning.
- Postweaning, maintain a temperature of 18 °C (64 °F), provide plenty of close nose contact with boars from Day 4. The boar should be no more than 1 metre from the sows and have plenty of saliva. Check there is no evidence of mastitis at weaning time and provide a dry environment with low air flow. Assess the comfort of the sow, by her looks and posture.
- For loose-housed sows allow at least 3.4 m^2 per sow at mixing and from weaning-to-service.
- If sows are exposed to temperatures above 24 °C (75 °F) feed intake becomes compromised. For continual temperatures above 26 °C (79 °F) it is necessary to adopt drip cooling or other evaporative cooling techniques.
- If you think a gilt or sow is on heat and she will not stand always try another boar but leave a gap of half an hour before doing so, because gilts in particular may only show a high, intense oestrus for 10 to 15 minute periods.
- If litter size is poor in the first litter gilt it may be worthwhile to skip or miss the first heat after weaning and serve on the next one, particularly if such animals are coming into heat in the sub-fertile period.
- Note and mark all sows that do not eat well in lactation.

Most reproductive failure is associated with compromised females.

Group 2 Losses – Ovulation and Egg Production

Group 2 losses are related to poor egg production or poor ovulation. If there is a failure here the changes will include a low litter size, litter size variation (an increase in litters with total born of less than 9), or a possible decrease in farrowing rate as a result of insufficient embryos to maintain the pregnancy. Use the farm records to identify the group loss by referring to the factors listed in Fig 5-22. These will then indicate which group or groups are involved.

Sows that come into heat during suckling are sub-fertile. Don't serve them.

Key factors to maximising ovulation rate:

- Ensure the gilt is fed ad lib for at least 3 weeks prior to service.
- Ensure the gilt has adequate easy access to a feeder 2 weeks prior to service.
- Keep breeding females that have the best hybrid vigour and that are from the most prolific dam lines. The pure bred female is less fertile than the cross bred. There are considerable variations in fertility between different ancestral lines and types of crossbreeds.
- Breed from females with records of high litter size and high live piglets at day 5 merit.
- Feed a good quality diet during lactation and in the case of the gilt for 3 weeks prior to mating.
- Do not serve breeding females which have any signs of disease.
- Check parasite levels. Heavy burdens can impair

FIG 5-22: THE STAGES IN THE REPRODUCTIVE CYCLE ASSOCIATED WITH INFERTILITY

Problem	Oestrus	Ovulation	Fertilisation	Implantation	Maturity	Farrowing
Litter size poor	YES	YES	YES	YES	YES	NO
Litter size variation * increased	YES	YES	YES	YES	NO	NO
Stillbirths increased	NO	NO	NO	NO	YES	YES
Mummies increased	NO	NO	NO	NO	YES	NO
Farrowing rate depressed	NO	YES	YES	YES	NO	NO
Abortions increased	NO	NO	NO	YES	YES	NO
W/S interval	YES	YES	YES	YES	NO	NO
Delayed returns increased	NO	YES Regular	YES Regular	YES Regular and Irregular	NO	NO

* % of litters with total numbers < 9. Normal levels <12 % in sows and 12 to 18 % in gilts.
W/S = Days from weaning to first service.

digestion and the uptake of nutrients and be responsible for poor body condition, catabolism and anoestrus.
- Ensure gilts are not over fed during first gestation and become too fat prior to farrowing.
- See the checklist for anoestrus in the sow above.
- Maximise feed intake in the first lactation.
- Avoid vaccinations and other procedures which may interfere with lactation feed intakes.
- Manage the sow so that she comes into oestrus during the early fertile period (Fig 5-23).

THE IMPORTANCE OF THE WEANING-TO-SERVICE INTERVAL

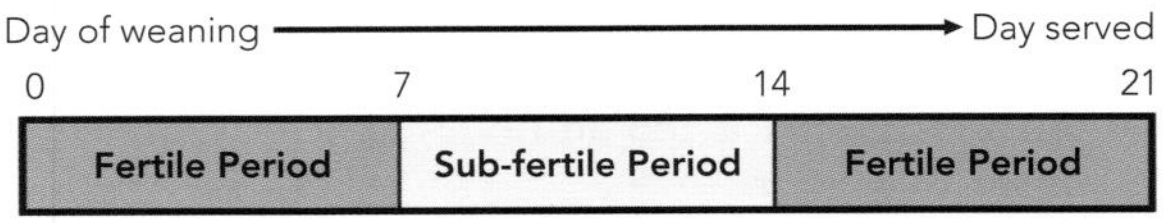

Fig 5-23

Group 3 Losses – Fertilisation

Group 3 losses are associated with failure or poor quality of fertilisation or conception. If total failure occurs then the sow will repeat on a normal cycle of 18 to 24 days. If however conception is poor, due for example to overworked boars, then either litter size or its variation will be affected, or a decrease in farrowing rate will be seen with increased regular returns, together with an increase in irregular returns.

Group 3 failures mean increased repeats at 18 to 24 days and/or poor litter size.

During service, contractions of the muscles in the uterus move the sperm rapidly up to the openings of the oviducts.

Sperm have to remain at the utero-tubal junction for about 8 hours and develop (capacitation) before they are fully able to fertilise the ova. Since the unfertilised ova can only remain viable in the oviducts for 8 to 10 hours it is important that mating is carried out at the correct time i.e. < 10 hours before ovulation (egg release). Ovulation takes place two-thirds through the heat period (70 %).

Standing heat is longer in sows with a shorter wean-to-service interval and shorter in sows with a longer wean-to-service interval.

Key factors in maximising fertilisation:
- A complete failure will obviously occur if the animal is not served. This is not as silly as it sounds. A stockperson may well think a sow has been served when in fact it has not.
- To ensure maximum meeting of sperm and egg, only carry out the first mating when the sow is standing "rock hard" or completely immobile.
- Do not serve sows in the early part of the oestrus period when the sow is not standing.
- Failure to place semen into the uterus results in poor fertilisation.
- To ensure maximum fertility and litter size, serve in the fertile period, that is, from 4 days postweaning up to Day 6.
- There is no advantage in farrowing rate or litter size in serving more than twice within a 24 hour period. Serving more than 3 times increases risk of 14 to 21 day post-service vulval discharges.
- Ensure that gilts are fed no more than 2 kg (4 lbs) of feed for the first 2 days post-service. High levels of feed may reduce progesterone levels and this may result in higher embryo mortality.
- On the other hand, feed sows well immediately after service to maximise placental size. When the sow is in oestrus, she is under the influence of high levels of oestrogens. Towards the end of the heat period these drop and progesterone levels start to rise. Sows with high levels of progesterone provide a good nutritious environment for the fertilised embryo and potentially good litters. Some sows with low progesterone levels will have increased levels of embryo mortality.
- Treat all thin sows at weaning as an emergency to get their body condition score back to a minimum of 2.5.
- The fertilised eggs remain in the oviducts for approximately 48 hours before they migrate down to the horns of the uterus. It is safe to change the sow's environment during this time.
- Any management change, mixing, or movement should therefore take place in this immediate post-service period only.

With natural service:
- Only use a boar once every 24 hours.
- Ensure that the boar's penis is entered into the vagina and locked into the cervix. Ensure that there is no leak back of semen.
- View the boar from behind and look above his testicles to ensure that he ejaculates, by evidence of the urethra and anal ring pulsating.

A return at 18 to 24 days indicates either Group 2 or Group 3 losses. If there is progressive loss in Group 3, then litter size will be low, there will be an increase in litter size variation and a decrease in farrowing rates. Fig 5-24 demonstrates the effects of the timing of insemination and ovulation and subsequent results. If insemination takes place in the early part of the oestrus period, even when the sow will stand to the boar, both conception rate and numbers born are compromised. Study Fig 5-14 again because it will help you to understand how records can pinpoint the specific time of the failure. With this information the management practices and contributing factors are identified on the farm and this allows corrective action to be taken. Note that even in the best oestrus, the sow and gilt will only be in standing heat for 10 to 15 minutes per hour.

In gilts, do not feed high levels for 48 hours post-service (> 2 kg), or you can expect some poor 1st parity litters.

THE TIMING OF MATING

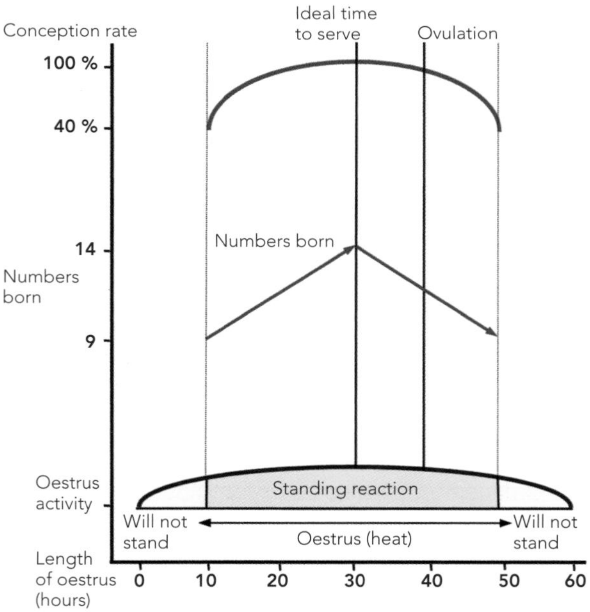

Time of ovulation:	38 - 42 hours after start of true heat
Life of unfertilised ova:	Min 10 hours/max 21 hrs
Time of service:	During the standing period
Arrival of sperm at oviduct:	Within 2 hours after service.

Fig 5-24

Fig 5-24 provides a good guide to the optimum time of mating but be aware that there is considerable variation between individual sows particularly in the time of ovulation and the time that they come on heat. Most sows come on heat in the early hours of the morning, around 4.00 am which means that the optimum time to mate them would be 30 hours later (10 am on the following day). Outdoor sows and hot sows will be in standing heat only in the cooler times of the day – often before dawn.

Feed gestating sows only once a day. Feeding in the late afternoon releases the morning for service and is good for urinary health.

Feed thin sows as an emergency.

Repeat matings on a normal 18 to 24 day cycle: Possible causes and corrective factors for Group 3 and some Group 4 losses

Returns at 18 to 24 days repeats indicate that pregnancy failure is occurring between service and approximately 10 days post-service.

Approximately 2 days after fertilisation, the embryos (which are dividing rapidly and becoming extremely complex) move down into the 2 horns of the uterus. Over the next 5 to 7 days they migrate from one horn to the other, depending on numbers, so that they are spaced evenly between the horns. On Day 7 the embryos hatch through the zona pellucida which acts like the shell of a chicken's egg. The embryos are now free-living and elongate rapidly. On Day 10 the free-living blastocysts release oestrogen sulphate into the uterine environment indicating to the sow, pregnancy. This is the first embryonic signal. This production of oestrogen sulphate stops the corpus luteum from regressing. Note that prior to this the corpus luteum is insensitive to prostaglandins. An assessment of management and mating procedures is therefore necessary, particularly to identify any form of stress on the sow.

If the embryos die by Day 10 there is no further development of the pregnancy and the sow returns on a normal cycle at 18 to 24 days – a normal return.

The target level of efficiency in this category would be 5 to 8 % of repeat matings. Fig 5-25 can be used to highlight areas of importance. The role and presence of the boar or AI is particularly important in Group 3 and 4 losses.

Basic rules for AI use:

- Use clean catheters.
- Ensure the AI is stored between 15 and 19 °C (59 to 66 °F).
- Only use fresh semen according to the instructions provided by the diluent. Most fresh

FIG: 5-25 FACTORS AFFECTING REPEAT MATINGS ON A NORMAL CYCLE (18 to 24 DAYS)

Management of Sows/Gilts

Factors resulting in increased return rate:
- Short lactation length of < 17 days for gilts and < 19 days for second parity sows.
- Fat sows/gilts in lactation (condition score 4+).
- Low feed intake in lactation.
- Loss of bodyweight in lactation.
- Weaning-to-service interval of 7 to 14 days.
- Excess stress at weaning, particularly group-housed sows.
- Poor environments at weaning, dampness, draughts.
- Low light intensity after weaning.
- Short light periods after weaning.
- Breed or type of sow e.g. Meishan cross breeds are highly fertile.
- Heavy parasite burden, catabolism and thin sows.
- Poor artificial insemination techniques.
- Poor checking of mating procedures.
- Old sows developing cystic ovaries.
- Failure to acclimatise gilts.
- Vulval discharges post-service.
- Bad management procedures for 28 days post-service.
- High stocking density in group-housed sows.
- Lack of boar contact post-service.

Actions to reduce return rate:
- Wean over 21 days.
- Consider culling all sows that return and pre-plan an increased intake of gilts.
- Maximise feed intake from Days 3 to 28 of lactation.
- Consider culling sows that come into heat at 8 days postweaning or later.
- Reduce stress at and after weaning.
- Improve postweaning and post-service environments.
- If sows are group housed provide > 2.7 m^2 floor space per sow.
- Maintain high light intensity postweaning of 14 hours per day.
- Check every service to ensure it is properly done.
- Check AI techniques.
- Feed a maximum of 2 kg (4 lbs) feed for 3 days post-service in gilts.
- Feed lactation diet ad lib from 3 days postweaning to service of at least 9.7 MJ NE/kg (14 MJ DE/kg), 1 % lysine.
- Feed to body condition thereafter.
- Check repeats by parity. Cull older sows that repeat.
- Cull repeat breeders.
- Prevent sows losing bodyweight in lactation to maintain the weaning-to-service interval at < 7 days.

Management of Semen

Factors resulting in increased return rate:
- Semen dead on arrival.
- Semen not stored properly.
- Semen too old.
- Semen handling too rough.
- Catheters dirty.
- Sow not in standing oestrus.
- Catheter not positioned properly.
- Semen forced into the sow.
- Semen abnormalities above 50 %.
- Boars collected more than 2 times a week.
- Boar has been sick in the last 6 weeks.
- Boar has been heat stressed.
- Management of the boar which would apply to the AI boar as well.

Management of Boars

Factors resulting in increased return rate:
- Mating the sow before she is fully in heat.
- Serving too early in heat.
- Using mature boars more than once a day every day.
- Multiple mating leading to overworked boars.
- Failure to identify and remove infertile boars.
- Using boars with low sperm counts and sperm abnormalities.
- Using young boars with poor fertility.
- Overusing young boars. No more than 3 times a week until 12 months old.
- Ill health, fever, lameness.
- Hot environment.
- Unsupervised services in which the boar does not serve properly.
- Dirty services resulting in vulval discharges.

FIG: 5-25: FACTORS AFFECTING REPEAT MATINGS ON A NORMAL CYCLE (18 to 24 DAYS) CONT.
Management of Boars (cont.)
Actions to reduce return rate: • Only commence mating after the sow has been fully in heat for 12 to 24 hours. • Check that every mating is properly carried out. • Ensure that every service is as hygienic as possible. • Mate the sow twice only, 24 hours apart. • * Do not use any boar twice in the same day. • * Check the fertility records of every boar. • Cull boars with poor fertility. • Introduce new young boars gradually to the mating routine. • Cross mate when using young boars for the first 4 services. • Match the age and weight of the boar to the female. • Examine every boar regularly for good health and physical fitness. • Maintain boar contact for at least the first 6 weeks of pregnancy.
Diseases/Conditions**
• Endometritis – uterus infection. • Infection of the oviduct (fallopian tubes). • Infection of the ovaries and surrounding tissues. • Generalised disease and sickness.
Specific Diseases
• Erysipelas. • Fever in the boar. This may affect sperm for a 6 week period. • Leptospirosis. • Porcine parvovirus. • Pneumonia. • PRRSV. • Swine influenza virus.

* Keep a boar calendar, ticking off every service against every boar.
** Diseases that can affect normal returns to heat are mentioned here but only for completeness.

AI semen needs to be used within 5 days of collection.

- Do not reuse AI bottles.

There are 3 golden rules relating to the use of a boar:

- Only use each boar once every 24 hours.
- Only serve a sow when she is standing absolutely "rock hard" or still.
- Ensure there is a good insemination, and note the quality of mating (1 = good, 2 = moderate, 3 = poor).

A normal return indicates embryo failure before Day 10 post-service. A delayed return indicates progressive embryo failure starting from 11 to 17 days post-service.

If you always cross mate (i.e. use 2 different boars on every sow) you will have difficulty identifying infertile boars. One option is to use pooled AI or a mixture of boar and AI.

Pregnancy failure from point of service to Day 10 is often due to ascending infection of the uterus (endometritis). In such cases a clear or opaque light discharge of the vulva will be seen 14 to 21 days post-service with such animals repeating. See Chapter 6: Endometritis and the Vulval Discharge Syndrome.

Examine the vulvas of sows post mating, between Days 14 and 21 for any tackiness or vulval discharge. Do such sows return? If so, read Chapter 6.

Group 4 Losses – Implantation

After Day 14 the embryos settle into the wall of the uterus and start the process of implantation.

If fewer than 4 viable embryos remain in the uterus (at least 2 per uterine horn) at or around 14 to 17 days of age (the time of implantation) they are insufficient in number to maintain a pregnancy (Fig 5-14). In such cases there is likely to be a delayed return of between 25 and 35 days (peak on Day 28). Between Days 14 and 17, the fertilised embryos become attached to the lining of the uterus with the development of the pla-

FIG 5-26: FACTORS AFFECTING REPEAT MATINGS ON AN ABNORMAL CYCLE (25 to 35 days)

Management of Sows/Gilts

Factors resulting in increased irregular return rate:
- As for those listed in Fig 5-21.
- Low ovulation rate.
- Poor conception with progressive embryo mortality.
- Implantation from Day 14 with progressive embryo failure.
- No boar pheromone contact during pregnancy.
- Management induced stress from Day 3 to 28 post-service (mixing and movement).
- Seasonal infertility. Poor light.
- Increasing/decreasing length of daylight.
- Effects of excessive sunlight.
- Embryo loss due to high levels of feed for 72 hours post-service in gilts.

Actions to reduce irregular return rate:
- As for those listed in Fig 5-21.
- Check feed intake in lactation.
- Provide boar contact for > 28 days post-service.
- Reduce stress from 2 to 25 days post-service.
- Provide a good intensity of light 16 hours a day.
- Reduce exposure to sunlight.
- Assess the effects of breed, particularly outdoors.
- Feed well from 2 to 28 days post-service.
- Assess records and history of affected sows.
- Assess the quality of nutrition.

Management of Semen

Factors resulting in increased irregular return rate:
- Semen dying on arrival.
- Semen not stored properly.
- Semen too old.
- Semen handling too rough.
- Catheters are dirty.
- Semen abnormalities above 50 %.
- Boars collected more than 2 times a week.
- Boar has been sick in the last 6 weeks.
- Boar has been heat stressed.
- Management of the boar, which would apply to the AI boar as well.

Management of Boars

Factors resulting in increased return rate:
- As for those listed in Fig 5-21.
- Most causes by now are due to maternal failure.

Actions to reduce return rate:
- As for those listed in Fig 5-21.
- Check boar records carefully.
- Check clinical records/observations of the boar.
- Make sure he has contact with pregnant sows from 3 to 30 days.

Diseases/Conditions*

- Endometritis – uterus infection.
- Vulval discharge.
- Infection of the oviduct (fallopian tubes).
- Infection of the ovaries and surrounding tissues.
- Generalised disease and sickness.
- Infection of the testes.

Specific Diseases

- Aujeszky's disease.
- Brucellosis.
- Teschovirus.
- Erysipelas.
- Fever in the boar. This may affect sperm viability for a 6 week period.
- Leptospirosis.
- Pneumonia – mycoplasma (enzootic) pneumonia for example.
- Porcine parvovirus.
- PRRSV.
- SIV.
- PCV2.

* Diseases that can affect delayed returns to heat are mentioned for completeness only.

centa. Oestrogen sulphate is then produced from the embryos which provide the second signal to the sow indicating that she is pregnant. The corpus luteum in the ovaries increases in size and continues producing the pregnancy hormone progesterone. This stops the next oestrus cycle and pregnancy continues.

Thus, if we have very small litters born of 3, 4 or 5 piglets and no mummified ones we know that there have been major problems occurring between approximately Days 14 and 30 post-service. Factors that need to be considered with sows found not to be pregnant with delayed repeats are highlighted in Fig 5-26. The normal level should be less than 3 % of matings.

Check the various points in Fig 5-26 for corrective action in your herd. Boar factors are usually of low priority for abnormal repeats because the failures are much more likely to be of a maternal nature. However, old semen or the overuse of boars resulting in poor conception and poor viability of embryos with progressive embryo loss should not be overlooked.

Any generalised disease of the sow could result in embryo mortality. Unless there are clinical signs in the herd, such factors will usually be of an individual nature. This highlights the importance of individual records and the need to collect data on the farrowing rate loss analysis sheet.

For specific information regarding the diseases refer to Chapter 6. For clarity between infectious and non-infectious causes also study Fig 5-27.

FIG 5-27: CLINICAL DIFFERENTIATION BETWEEN INFECTIOUS AND NON-INFECTIOUS CAUSES OF REPRODUCTIVE FAILURE

Type of Failure		Non-Infectious	Infectious
Anoestrus		+++	+
Repeats (regular)	– at 21 days and no discharge.	+++	+
	– at 14 to 21 days with discharge.	+	+++
Repeats (irregular)	– at 28 days.	++	++
Repeats (pseudopregnancy)	– at 63 days.	+++	+
Abortion	– sow in good health.	+++	+
	– with healthy foetuses.	++	++
	– with mummified or decomposing foetuses.	+	+++
Sow not in pig		+++	+
Mummified pigs	– small and variable in size.	+	+++
	– large.	++	++
Stillbirths	– increased within a normal litter.	+++	+
	– with increase in mummified pigs and reduction in live-born.	+	+++

+ Unlikely cause **+++** Likely cause

Group 5 Losses – Foetal Death and the Mummified Pig

From approximately 24 days of age through to 115 days, the foetus is maturing. All the organs have formed by Day 24 of gestation. After Day 35, bone starts to form in the skeleton and if the piglet dies, it is not completely absorbed and a mummified foetus remains. If the embryos die before Day 35, before bone is formed in the embryo, the sow can absorb all the piglets and returns after Day 50, normally around 63 days. This is termed a pseudopregnancy (some text books refer to this as long pseudopregnancy and a 25 to 35 irregular return as a short pseudopregnancy).

If the embryos are present after Day 35, then as long as 1 piglet remains alive, the pregnancy will continue to term. The embryos that die after this stage are passed out as mummified pigs.

The approximate age when a mummified pig has died can be determined by measuring the length from the crown of the head to the rump or tail base (Fig 5-28).

There is a simple equation which can provide an *approximation* to the age of the mummified piglets:

Approx. age in days =
(Crown of head to rump length in mm/3) + 21

There are 2 possible causes of mummified pigs. First, a piglet dies because there is a large litter and insufficient space in the uterus. Second, there is infectious disease, usually of a progressive nature, at any stage during the period of pregnancy (Fig 5-29). In the first case, a study of the records will show that the mummified pigs are occurring in large, normal litters. For example, a litter size of 14 alive, 1 dead, 1 mummified is of no significance provided the remainder of the litter is normal and healthy. If however, as in the second case, the litter

MUMMIFIED PIGS' SIZE AND AGE

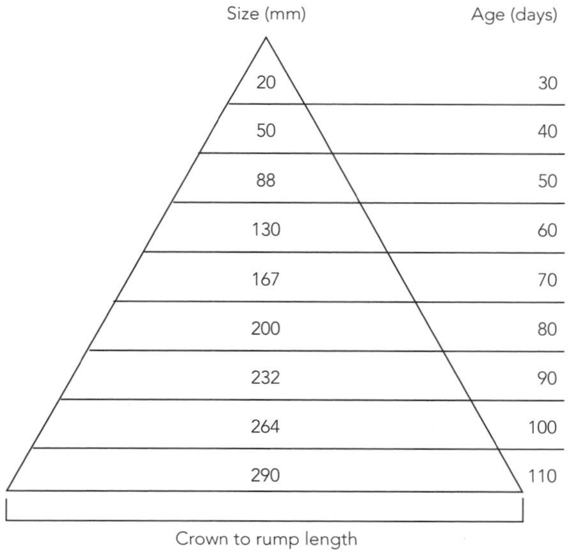

Fig 5-28

FIG 5-29: INFECTIONS THAT CAN PRODUCE MUMMIFIED PIGS
* Aujeszky's disease/Pseudorabies virus (AD/PRV)
Blue eye disease
Encephalomyocarditis virus (EMCV)
Erysipelas
Japanese encephalitis virus (JEV)
Other one-off infections
Porcine circovirus 2 (PCV2)
* Porcine parvovirus (PPV)
* PRRSV
Swine fever (African and classical)
Swine influenza virus (SIV)
Teschovirus

* Common causes in countries in which they occur.

size is 6 born alive, 2 dead and 4 mummified, this indicates disease during the pregnancy causing mortality. By measuring the size of the mummified pigs, we can determine at what stage in pregnancy the disease occurred. When all the mummified pigs are the same size, consider mycotoxins as a possible cause. Alternatively, if the mummified pigs are of a variable length, this is evidence of progressive disease over a period of time. Some viral infections affect the foetus from 35 days onwards and progressively spread during the period of pregnancy. Porcine parvovirus is a typical example.

Once the piglet inside the uterus reaches 70 days of age it becomes immunocompetent. This means that its immune system has started to develop and therefore can start to respond to any infection that challenges it and attempt to protect itself. For example, if pigs inside the uterus are infected with parvovirus beyond 70 days they respond by producing an immunity and thereby, no disease. However, they may be weak at birth. By sampling the blood of such piglets at birth, before they have suckled, we can tell from presence or absence of antibodies whether indeed this event had taken place.

An alternative to this would be PRRSV infection, which does not affect the foetus during the mid-period of pregnancy, but only after Day 60 as the virus cannot attack the foetus before this time. In this case, the virus actually kills the foetus and therefore there will be evidence of late mummified pigs, death having occurred at any time from Day 80 to the point of farrowing. Thus, by studying the numbers relative to total litter size and the time when death must have occurred, we can get a good indication as to possible disease or pathogen cause.

The significance of mummified pigs

- Large numbers of normal piglets born alive (> 10) and a few mummified pigs – unlikely to be due to disease.
- Small numbers of piglets alive (< 8) and large numbers of mummified pigs – more likely to be due to disease.

In Group 5 losses, total numbers born will be the same, but numbers alive will be reduced with corresponding increases in stillbirths and mummified pigs. Litter size variation will be increased, with farrowing rates decreased. Abortions, of course, can be a major cause of loss in this group and repeats will be irregular.

Group 6 Losses – Stillborn Pigs

Stillbirths are usually related to large litters, increasing age, slow farrowing, or farrowing difficulties. In herds with large litter sizes stillbirth rates are higher and the target level then ranges between 5 to 7 % of total pigs born. Increased stillbirths can also be associated with infections such as *Leptospira pomona* and PRRSV.

Stillbirths – target 3 to 5 %
Increases are associated with:

- Increasing age of sow.
- Fat sows.
- Large piglets – greater than 2 kg.
- Individual sows. Identify by litter size – more in large litters. Monitor their farrowing progress.
- Breed – more common in the pure bred sows.
- Lack of exercise – poor muscle tone at parturition, associated with individual confinement.
- Prolonged farrowings. Identify and investigate.
- Uterine inertia. Low calcium levels in the diet may be involved.
- High farrowing house temperatures.
- High carbon monoxide from old gas heaters.
- Farrowing place and floor designs which precipitate farrowing difficulties.
- Foetal anoxia (lack of oxygen) – resulting from uterine inertia. Check iron levels in sow.
- Early placental separation.
- Haematoma/bruising of the umbilical cord.
- Assisted farrowings.
- Diseases of the sow, fever, mastitis, etc.
- Porcine parvovirus infection (or just parvovirus?)
- PCV2.
- PRRSV.
- Teschoviruses.
- The use of certain boars.

The types of stillbirth:

- Pre-partum – the piglets died a few days before parturition – no lung inflation.
- Intra-partum – the piglets died during farrowing – no lung inflation. Usually maternal failure.
- Post-partum – the piglets show evidence of some lung inflation but fail to breathe properly.

Stillbirths may arise due to:

- Mechanical anoxia.
- Hypoglycaemia (low sugar).
- Hypothermia (low body temperature).
- Low viability.
- Maternal failure.

To reduce stillbirths:

- Identify the group cause by post-mortem examination.
- Do not let the age of the herd spread beyond the seventh litter.
- Identify problem sows from the previous histories and monitor their farrowings.
- Look at breed differences.
- Check sow's condition.
- Check farrowing house environment.
- Check farrowing pen design.
- Monitor farrowings.
- Interfere early in prolonged farrowings.
- Give good management at farrowing.
- Provide a heat source behind the sow at farrowing.
- Study herd records.
- Check haemoglobin levels in sows.
- Check parasite levels.
- Check for blood parasites.
- Check for diseases in the sow.
- Clean and service gas heaters.
- Check quality of sow feed, particularly minerals.
- Check water quality if using boreholes.
- Check calcium and iron levels in the sow.

Abortion

See Chapter 6 for infectious causes of abortion.

Embryo Loss and Abortion

Embryo loss occurs when there is death of embryos followed by absorption, or expulsion. Healthy embryos grow into foetuses after Day 35. Abortion means the premature expulsion of a dead or non-viable foetus. There is often alarm when an abortion is seen but it should be remembered that there can be loss of embryos at any time during early pregnancy, which often goes unseen.

Embryo loss or abortion can be considered in 3 main groups:

First, during the period from fertilisation to implantation; second, during the period of implantation at around 14 days post-service to 35 days; and finally, during the period of maturation. It can be seen therefore that losses can take place at any stage from approximately 14 days after mating, when implantation has taken place, through to 112 days of pregnancy.

The Maintenance of Pregnancy

Pregnancy is maintained due to hormonal changes initiated by embryonic signalling at Day 10 and between Days 14 and 17 of gestation. These changes allow the corpus luteum (the body from which the egg is released) in the ovary to develop and produce the pregnancy hormone progesterone. The presence of the corpus luteum is necessary to maintain the pregnancy throughout the whole of the gestation period. The loss or failure of the corpus luteum after Day 10, through any cause, initiates the farrowing process, hence an abortion, or if near to term a premature farrowing. A rise in body temperature (pyrexia) alone can result in prostaglandin release, causing regression of the corpus luteum after Day 10. Thus heat stress can result in abortion.

Abortions and their Cost

Natural biological failures of pregnancy due to a variety of causes occur across all species. In healthy normal sow herds abortions observed by the stockpeople are normally fewer than 1 per 100 pregnancies. The cost of an abortion can be calculated quite easily. If the abortion results in an empty farrowing place, losses include the fixed costs associated with the lost finisher pigs, the feed to maintain the sow to the next pregnancy plus lost profits on the finisher pigs which were not there to be sold.

Methods of Investigation

It is worthwhile monitoring the levels of abortion in your herd continually and comparing them to the normal levels. The following information should be recorded with each abortion:

- Sow number.
- Parity.
- AI source used.
- Boar used.
- Date of service.
- Date of abortion.
- Housing.
- Feed and amounts given.
- Clinical observations of the sow and any disease history.
- Condition of the aborted piglets – alive fresh, recently dead or mummified.

If you are using the farrowing rate loss analysis sheet illustrated earlier in this chapter (Fig 5-7) you will be doing most of this anyway.

It is important to study the herd history and environment. For example, is there a seasonal effect or an association with a particular area of the housing or management practice? You should also note the clinical state of the sow at the time of abortion. Does she show other clinical signs or is she apparently normal? You should examine the aborted foetuses. Are they fresh with no signs of any decomposition, or are they decomposing or mummified? Such observations, particularly if recorded over a period, may be of help to your veterinarian in leading to a possible diagnosis of the cause.

There are 3 parts to the investigations that must be carried out. First, collect information about the individual sows, then request post-mortem examinations and serological tests, and finally, assess the clinical evidence and feeding procedures in the herd. The object is to identify the area of failure and by management studies, examination of records, clinical examinations, and laboratory tests the cause may be identified.

Post-mortem and laboratory examinations

Fresh, aborted foetuses should be submitted to a competent diagnostic laboratory where examinations can be carried out for evidence of viral and bacterial infections, together with histological examinations and toxic studies. In many cases the end results of post-mortem and serological tests do not identify any particular infectious organism, which may seem disappointing. However, it is useful in telling us what is not present.

Clinical examinations

Of all the examinations carried out, clinical observations are perhaps the most important. Look at the environmental factors in Fig 5-30. By using the recorded information on individual cases and collating this to the problem group of sows, it then may become possible to differentiate clinically between an infectious and a non-infectious cause. Fig 5-27 indicates the likelihood of these for the different causes of reproductive failure.

Non-Infectious Causes of Abortion

There can be a number of non-infectious reasons for abortion. Fig 5-30 looks at the different factors.

Light

To maintain a viable pregnancy requires constant daylight length. Ideally this should be 12 to 16 hours per day. Light intensity experienced by the sow can be affected by a number of environmental inadequacies, for example, poor lighting in the first place, followed by fly faeces and dust on lamps gradually reducing the availability of light; high walls surrounding animals or automatic feeders in front of sows producing shadows. A simple tip here is to make sure that you can read a newspaper in the darkest parts of the building at sow eye level. If not, then problems may start. Painting the roofs and walls white to increase the reflection of light is one way of improving the environment and on a number of occasions abortions have ceased after such simple improvements.

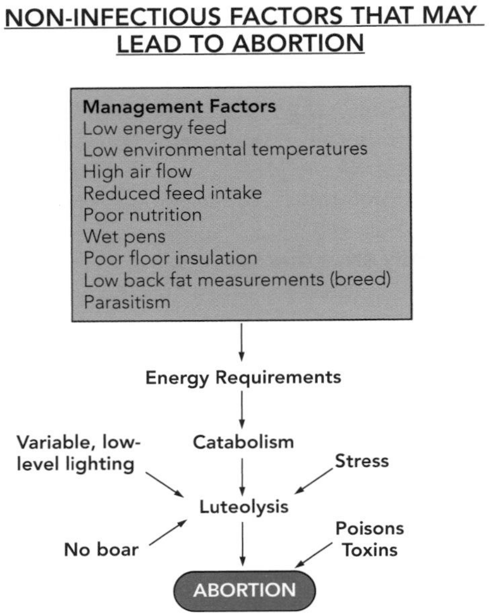

Fig 5-30

What you need to know about light

- Gilts exposed to 14 to 16 hours light (250+ lux):
  - Reach puberty earlier.
  - Are a lighter weight at puberty.
  - Show no difference in ovulation rate.
  - Are sexually more active, as are boars.
- **In lactation:**
  - 16 hours of light increases weaning weights.
  - There is no effect on numbers weaned.
  - Milk yield is increased.
  - 16 hours is recommended at a 150 lux level.
- The weaning-to-service interval is reduced if 16 hours of light are given after weaning with a continuous light intensity of 250+ lux.
- Sows are on heat longer when exposed to more light.
- Light has no effect on litter size but absorption of embryo and foetuses may occur in poor lighting during early pregnancy.
- During the dry period a minimum of 220 lux is

required for 14 hours a day.
- Fluorescent light is nearer to natural light than incandescent lighting.
- In a building 2.4 m (8 ft) high and 4.9 m (16 ft) wide a continuous row of fluorescent lights is necessary. As a guide 150 watts is required for every 1.5 m (5 ft).
- The light must be placed over the sows' heads.
- Make sure sows are not exposed to decreasing daylight lengths.
- Provide 8 to 10 hours of darkness (less than 20 lux) after a period of light .

Seasonal Infertility

Experiences have shown that 70 % of all abortions fall into this category. Because the sow historically only produced one litter per year, with farrowings during early spring, there is an in-built tendency for the animal not to maintain a pregnancy during the summer and autumn periods. This is well recognised with summer infertility and the autumn abortion syndrome, where environmental factors are likely to cause the corpus luteum to disappear.

Abortions, anoestrus and sows found not in pig commonly occur during the period of summer infertility when sunlight is intense and the weather is hot. This is particularly evident in outdoor sows where levels of pregnancy failure may reach 15 to 30 %. In such cases the abortions are so early that the foetuses are either not seen or there is progressive embryo mortality and a delayed return to oestrus. Look for slight mucous discharges from the vulva and if present refer to Chapter 6: Endometritis and the Vulval Discharge Syndrome.

The following factors are important, indoors or outdoors as applicable:

- Ultra-violet radiation may cause regression of the corpus luteum, particularly in white breeds. The outdoor breeding female should always be derived from at least 1 pigmented parent.
- Provide extensive shades so that the sows can protect themselves from the sun.
- Site the arks in the wind direction so that with open ends, cooling can take place.
- Provide extensive well-maintained wallows suitably sited so that sows do not have too far to walk to reach them.
- Always maintain boars within the sow groups for the first 6 weeks of pregnancy at least.
- Increase feed intake from Days 3 to 21 after mating.
- Increase the mating programme by 10 to 15 % over the anticipated period of infertility.
- Because boar semen can be affected, follow each natural mating 24 hours later by purchased AI.

The Boar

A further part of the equation involves the presence of the boar and his pheromones or male chemical hormones as some sows' pregnancies are very sensitive to pheromones.

Boar presence in the gestation/dry sow accommodation is recommended from the day of service through to the day of farrowing. The boar should be mixed in or have access to the group for at least the first 28 days of pregnancy. There is clear evidence that this will improve farrowing rates, particularly if they are associated with summer infertility. If sows are individually housed the boar should be allowed to move down the passages and make social contact daily.

A Catabolic State

If the metabolism of sows is allowed to progress to a negative energy or catabolic state, so that they have to use their body tissues to maintain the energy equilibrium, then some animals may abort. Clinical examinations will identify possible changes in the environment. For example, the removal of bedding, poor quality feeds, or a drop in feed intake. The latter may simply be associated with a change in stockpeople. Outbreaks of abortion may occur when there are changes from pellet feeding to meal feeding, or where feed is delivered by volume and not by weight. Wet, damp environments or high air movement, cause chilling and increase demands for energy. An important feature of environmental abortions is that the sow remains normal, often eating feed in the morning, and expelling the litter in the afternoon. Some people call these "farrowing abortions". The aborted foetuses are perfectly normal and the sow shows no signs of illness. The underlying initiating mechanism is regression of the corpus luteum.

Mycotoxins

See also Fungal Poisoning at the end of this chapter.

Abortion can also result from mycotoxins in a feed component (e.g. grain). These toxins are produced by fungi growing when the feed component was itself being grown or stored. Some of them cross the sow's placenta and cause abortions. Others affect the sow and result in abortion.

As well as mycotoxins in feed components, abortions can also be associated with mouldy feeds. To prevent this consider and act on the following:

- Check your feed bins, are they water tight?
- When were they last inspected internally?
- Do they contain bridged mouldy feed?
- Are the bins filled with warm feed? This precipitates mould growth.
- Do you regularly treat the bins to prevent mould?
- Are the bags of feed kept in a dry cool or wet warm place?

- If you practise home mixing and wet feeding, are the tanks and pipes mould free?
- Are you ever tempted to feed sows slightly mouldy food?

If you wet feed:

- Check mixing tank roofs to see whether feed splashed on to them has gone mouldy?
- Do you check the pipes?
- Do you check the source materials?
- Do you check the pH of the mix is correct?
- Do you let liquid components of the mix sit around in hot weather in storage tanks?

Empty and internally examine feed bins monthly. Examine feed daily. Treat bins regularly with a mould inhibitor.

Detecting Pregnancy

Carrying out pregnancy testing does not influence the outcome of an infertility problem and neither does it improve it, but it does help identify problems and allow corrective actions to be taken. Pregnancy detection will assist in identifying non-productive days at a much earlier stage in the pregnancy period, thus allowing management action. It will also give peace of mind.

Methods of Pregnancy Diagnosis (PD)

The list below highlights the different methods for diagnosing pregnancy:

1. Daily observation of oestrus.
2. Ultrasound scanning (real-time or audio).
3. Amplitude tests.
4. Vaginal biopsy.
5. Serum analysis.

A Checklist for Abortions

Use Fig 5-31 as a checklist for abortions.

FIG 5-31: A CHECKLIST FOR ABORTIONS

Abortion level. Is it less than 1.5 % sows served?	Review records and monitor situation. Probably within normal limits.
Abortion level. Is it more than 1.5 % of sows served?	– Take action.
Are sows ill?	– Probably disease.
Are sows otherwise normal?	– Probably non-infectious. Possibly mycotoxins.
	– Maternal failures.
Is the problem seasonal?	– Autumn abortion syndrome.
Do they occur in a particular part of the farm?	– Environmental.
Are the aborted pigs fresh or alive?	– Suggests the environment.
Are mummified pigs present?	– Suggests infection and/or mycotoxins.
Is the dry/gestation sow accommodation uncomfortable?	– Suggests the environment.
Are sow pens wet, draughty, poorly lit, close to a fan?	– Suggests the environment.
Does the ventilation system chill the sows?	– Suggests the environment.
Are there factors that place the sows in a negative energy state?	e.g.: High chill factors. Draughts. Low feed intake. A change in bedding. Availability of feed.
Are sows short of food?	– Check feed intakes by volume and weight.
Is the food mouldy?	– Check for mouldy feed.
Do the sows experience 14 hours of good light at eye level?	
Are the lights dirty, covered in fly dirt?	
Can you read a newspaper in the darkest corner?	
Do your sows have boar contact in pregnancy?	
Are any other diseases evident in the sows?	e.g.: Lameness. Cystitis. Kidney infections.
Are the abortions associated with stress?	

1. Daily observations for oestrus

After the eggs are fertilised in the oviduct (fallopian tubes) the embryos move around the 2 horns of the uterus to become equally spaced. Survival of these to Day 10 starts the pregnancy signal but a failure at this time results in a normal return of 18 to 24 days (average 21 days).

Implantation of the embryo commences during Day 14 and a minimum of 4 embryos are required for pregnancy to continue. If the pregnancy fails at this time the return to oestrus is delayed to 25 to 35 days (average 28 days) because pregnancy has already started.

If pregnancy is maintained to the completion of implantation and then fails totally with absorption, the sow becomes pseudopregnant for a varying period and then comes through not in pig after Day 50. In many cases there is a positive pregnancy test reading early on only to find the sow is negative later. The sequence of events and the results you can expect with pregnancy diagnosis are shown in Fig 5-32.

Take note that the loss of pregnancy between 15 to 35 days can give false positive test results.

FIG 5-32: INTERPRETING PREGNANCY TEST RESULTS

Time Post-Service	Observation	Outcome
2 – 17 days	Oestrus. Sow not pregnant.	Cystic ovaries, cull. Not in oestrus initially.
1 – 18 days	Not possible to detect pregnancy.	Sow may or may not be pregnant.
18 – 24 days	Sow comes into oestrus.	Sow not pregnant, embryos died at 1 to 10 days.
18 – 24 days	Sow not in oestrus. Methods 1, 2, 3 or 4 give -ve PD. Methods 4 or 5 give +ve PD.	Sow pregnant.
25 – 35 days	Sow in oestrus. Method 1 – ve PD. Methods 2 and 3 could give either +ve or -ve test results. Method 4 or 5 give +ve PD.	Sow now not pregnant, embryos died at 12 – 18 days but pregnancy started.
28 – 35 days	No oestrus. All methods give +ve PD.	Sow pregnant.
30 – 110 days	Oestrus. Methods 2, 3 and 4 could have given either +ve or -ve test results previously.	Embryos died at 15 to 35 days, a false pregnancy due to embryo loss.

2. Ultrasound scanning

There are different types of ultrasound scanners, some providing video results (like human scanners) and others that simply provide an audio tone.

Real-time scanning

Real-time scanners show video imagery of the womb and are extremely useful especially to control batch pig flow systems and should be used after Day 24 of expected pregnancy. Ensure you see the embryo as well as the uterine fluid. These machines are over 90 % accurate when the embryo is visualised.

Audio scanners

These scanners detect the sounds made in early pregnancy from the changes in blood flow that take place in the large arteries supplying blood to the uterus. Movement of the foetus and the placenta can also be detected together with the foetal heartbeat. Uterus infection, embryo absorption or early oestrus can give false positives and of course wrong interpretations of the sounds; also inexperience can give rise to wrong interpretations. Demonstration audio tapes are available with the equipment. At 28 days, these machines are around 65 % accurate.

3. Amplitude tests

These machines are only of value from 28 to 80 days of pregnancy. Beyond this they lose their sensitivity. Also false positives can often be detected if the bladder is full and scanning misses the uterus. These machines are around 50 % accurate.

4. Vaginal biopsy

This technique involves the removal of a small piece of the vaginal mucous membrane using a special instrument. The instrument is inserted into the vagina, 150 to 300 mm in, pressed into the membrane, and the end manipulated to cut off a small piece. The sample is placed in a small container with a special preservative and posted to a laboratory for histological examinations. It is time consuming, expensive and little used.

5. Serum analysis

This can be carried out after Day 22 by using a small stylette to puncture the ear vein. A thin capillary tube collects a spot of blood which is then tested for pregnancy hormones. It is time consuming, expensive and little used.

A practical format for on-farm pregnancy diagnosis

1. Do not move sows out of the service area until confirmed pregnant at Day 28 to 35.
2. From service to 18 days, maintain close boar contact. Provide daily boar pheromone contact throughout pregnancy. Do not place boars next to

sows for more than 1 hour a day. Constant boar contact will lead to acclimatisation and poor response to the boar's presence.

3 Observe sows for oestrus daily and particularly on 18 to 24 days post-service. No oestrus indicates a pregnancy has started. Maintain daily detection throughout pregnancy.
4 Check the vulva daily for tackiness from Days 14 to 21 and thereafter for discharges throughout pregnancy and possibly from Day 35 post-service.
5 Test the sow at 28 days post-service using an ultrasound machine.
6 A positive test – move the sow to the dry/gestation area or ideally leave her where she is.
7 A negative test – leave the sow in the service area.
8 A doubtful test – repeat 7 days later.
9 Pregnancy test the sow again at 35 days using an ultrasound machine. Follow as for 6.
10 Visually check the sow for abdominal enlargement, teat and mammary vein enlargement from Day 80 of pregnancy. If doubtful re-examine her with an ultrasound.

Records are important. They help us diagnose problems.

FIG 5-33: NUMBERS BORN

	Herd A	Herd B	Target
Born alive	9.8	10.3	14.1
Stillborn	0.3	0.8	0.7
Mummified	0.1	0.4	0.2
Total Born:	**10.2**	**11.5**	**15**

FIG 5-34: FARROWING RECORDS – HERD B

Litter No.	Gilt Litters			Parity 2		
	A	D	M	A	D	M
1	8	1	-	12	1	1
2	10	1	-	11	1	0
3	9	2	-	12	1	0
4	12	2	-	9	0	0
5	2	1	3	14	1	0
6	6	1	1	11	0	0
Total	47	8	4	69	4	1
Average	7.8	1.3	0.7	11.5	0.7	0.2

A = Born alive **D** = Stillborn **M** = Mummified

Low Litter Size

To study a problem of low numbers born alive look at the total born. Consider the following 2 scenarios of herds with problems (Fig 5-33).

Herd A had a problem of embryo loss before Day 35 of pregnancy. Herd B had a problem after Day 35, evident from the levels of mummified pigs. An analysis of records identified a management/service problem in A and an infectious problem in B.

Examination of the farrowing records for the last 6 gilt and second parity litters in Herd B gave the following results (Fig 5-34). The high numbers of stillbirths and mummies seen in the gilts is an indication of infectious disease – indicating probably parvovirus infection or PRRSV. The problem relates to only gilts in this herd as evident by the good performance in parity 2. The solution was to vaccinate gilts with parvovirus vaccine after demonstrating disease in the mummified pigs by the fluorescent antibody test. Of course, in the field you should look at more than 6 litters, as this is a simplified example.

Key points to consider with a litter size problem:

Sow effects

- Analyse at least 3 months of farrowing information.
- Analyse the results by born alive, dead, mummified; parity and boar used.
- Look at the distribution of good and poor litters. Is litter size generally depressed or are failures related to a few very poor litters?
- Is there anything common to those animals with poor litters?
- Determine if causes are infectious or non-infectious.
- Check the breeding of the sows – do they express good hybrid vigour?
- Assess the effects of lactation length. On average, 0.1 per pig is lost for each day lactation is reduced below 27 days, although this is also parity dependent.
- Assess wean-to-service intervals. Wean-to-service intervals of 7 to 14 days may be associated with a litter size drop of 2 to 3.
- If there is a gilt problem assess the management as described previously in this chapter.

- Assess potential losses in Groups 1 to 4.
- Assess key factors associated with poor fertilisation.
- Assess factors associated with normal repeats (Fig 5-25) and abnormal repeats (Fig 5-26).
- Check gilts are vaccinated against parvovirus.

Boar effects

- Young boars up to 9 months old produce smaller litters.
- Litter size differences exist between boars. Use records to identify.
- Check records. For each boar multiply the farrowing rate by the average total litter size. This gives a guide to numbers born per 100 services. A poor record would be 800, a good one more than 1200.
- Serve twice at 24 hour intervals.
- Always leave 24 hours between each completed AI or boar mating.
- Check the general health of boars. Lameness or stiffness may result in poor litter size.
- Look at the timing and frequency of services (Fig 5-24). The time of mating relative to ovulation is important.
- Litter size is maximised by the time the boar is 15 to 20 months of age. This is worth up to 1.9 pigs per annum extra compared to younger boars.
- Farrowing rate is also maximised in this age of boar.
- Have sufficient boars for each to mate 1 sow per week.
- Ensure fresh semen is used within 5 days of collection.
- Ensure semen handling is excellent and that semen is stored between 15 and 19 °C (59 and 66 °F).
- Ensure semen bottles are regularly turned.

Nutrition effects

- Feed the sow to appetite during lactation and to point of service with a high energy > 9.7 MJ NE/kg (14 MJ DE/kg) and > 1 % lysine.
- Feed gilts ad lib for 3 weeks prior to mating.
- Check the energy level of the diet relative to feed intake and to performance.
- Monitor body condition and fat depths.
- Avoid significant weight loss in the sow during lactation (> 10 kg). ½ condition score is 20 kg.

Management and environmental effects

- Assess the dry/gestation sow house environment for draughts, wet pens, no bedding, hygiene etc. particularly from weaning to 28 days post-service.

The Boar

Anatomy and Physiology

The male reproductive system (Fig 5-35) consists of 2 testicles, each of which is held almost vertically. The pig is unusual as the testes in the boar are actually upside down, with the head of the epididymis (the beginning) at the bottom and the tail of the epididymis (which moves into the vas deferens) at the top. The epididymis is the area within which all the mature sperm is stored and held until ejaculation. From each testicle, a tube, the vas deferens, carries the sperm into the abdomen via the inguinal canal. (If this hole is enlarged an inguinal rupture will be seen). From there it enters the neck of the bladder and continues in the groin down the penis to the exterior as the urethra. Thus from the neck of the bladder to the tip of the penis the urethra can carry either sperm or urine. There are 3 glands called: the seminal vesicles, the prostate and the bulbourethral glands. The seminal vesicles produce the bulk of the ejaculate (250 ml) and fructose to nourish the sperm. The prostate gland provides other nutrients and the bulbourethral gland the jelly that you often see at the end of mating.

During service the sperm in the epididymis are pulsated down the vas deferens to be joined by the seminal fluids. This is a continuous process during the period of mating. You can see it if you stand behind the boar and you should check it when you supervise services. This will appear as if the anus is winking.

The penis, which is long and rigid, has a sigmoid or S-shape in its top half and an anti-clockwise spiral at the end. It is 300 to 500 mm long.

The preputial sack is filled with very smelly fluids including pheromones and it also has a very high bacterial content. The bacteria are potential pathogens and emptying the sac by squeezing at service increases the risk of infection entering the uterus, particularly towards

THE REPRODUCTIVE TRACT OF THE BOAR

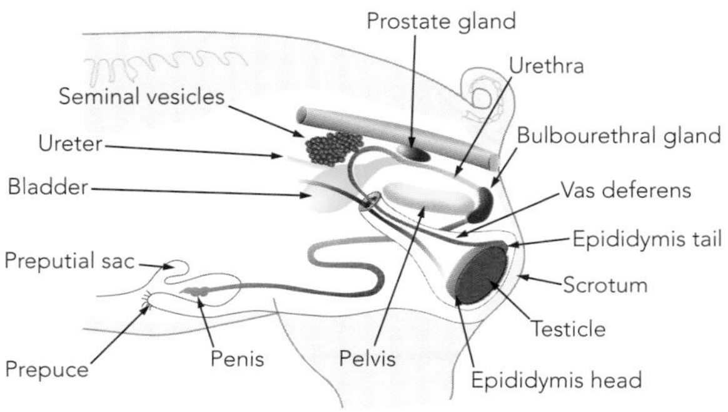

Fig 5-35

the end of the heat period.

Sperm is produced and matured under the influence of luteinising hormones (LH) and follicle stimulating hormones (FSH) and the whole process is controlled by the pituitary gland which is at the base of the brain.

Treatment for poor sperm quality is disappointing. While training AI boars to jump a dummy sow, an injection of natural prostaglandins may be used to enhance libido.

The effects of age

Boars reach puberty at around 5 months of age but the amounts and quality of semen are usually insufficient at this age to fertilise efficiently (Fig 5-36).

Frequency of mating or collection

AGE: 7 to 8 months, mate twice weekly, 48 hours apart. From 12 months onwards, mate 2 to 3 times weekly, 24 hours apart.

With AI boars collect once weekly. Keep a clear record of boar usage. If absolutely essential it is possible to collect twice in 10 days for short periods of time.

Facts about Semen

- Volume – 50 to 400 ml, mean 250 ml.
- Colour – creamy white.
- A good ejaculate is very milky.
- The ejaculate becomes watery if the boar is overused. If used am/pm/am the last ejaculate will be poor.
- Temperature of sperm – 37.5 °C (100 °F). The boar's core body temperature is 38.6 °C (101.5 °F).
- In hot weather > 30 °C (86 °F), sperm quality is affected.
- Sperm production – this takes up to 6 weeks. Thus high temperatures such as fevers will affect sperm production for this length of time and potentially affect fertility.
- Numbers of sperm per ml – up to 100 million.
- Sperm production time:
  - Up to 4 weeks of development in the testes and 2 weeks maturing in the epididymis.
  - 5 to 20 minutes for complete ejaculation.
  - Sperm remains viable in the sow for 30 to 48 hours.
- Ideal time for insemination is within 24 hours before ovulation.

The structure and shapes of normal and abnormal sperm are shown in Fig 5-37.

A sperm consists of a head and a tail. The head has a cap or acrosome which assists it to penetrate the egg. It appears normally as a crescent shape. The tail gives the sperm mobility to penetrate the egg and thus any abnormalities will decrease the possibility of penetration. Non-mobile sperm are incapable of fertilisation.

Never use a boar twice in 1 day.

As sperm matures, a small droplet – the cytoplasmic droplet – is produced just beneath the attachment of the tail to the head. However, once maturity has been reached it disappears and therefore if semen contains high numbers of sperm with droplets, fertility will be poor and it is probable that the boar is being over used. If a sample of semen is examined whilst warm under a microscope sperm appear to be swirling in waves. This is an indication of good viability.

CHANGES IN PREGNANCY RATE WITH INCREASING AGE OF BOAR

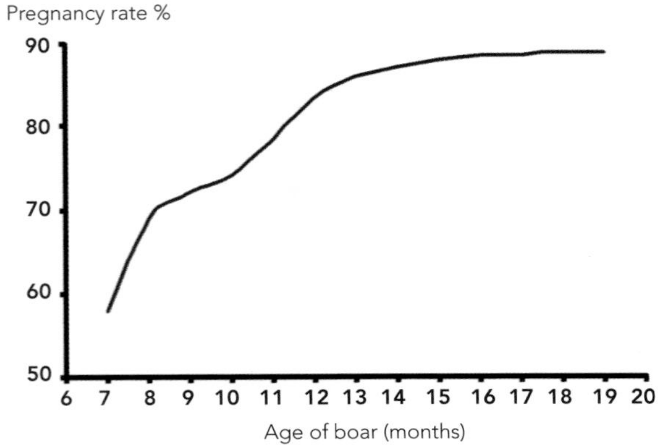

CHANGES IN LITTER SIZE WITH INCREASING AGE OF BOAR

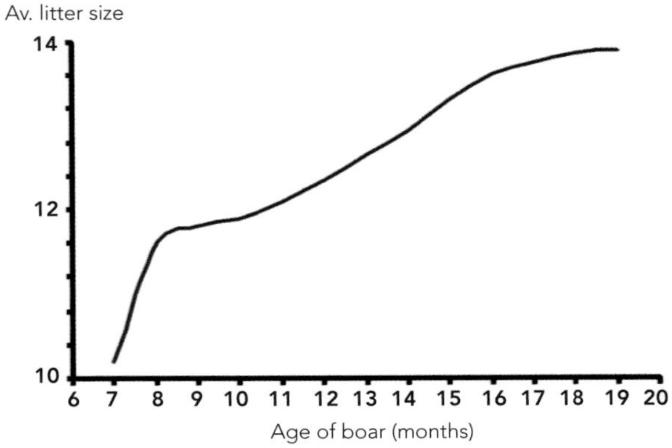

Fig 5-36

Bacteria and infectious agents

Semen in the epididymis normally does not contain any bacteria or viruses but during ejaculation contaminants from the preputial sac and elsewhere can cause high numbers to be present (Fig 5-38). Semen samples usually contain at least 100 bacteria per ml and often many more. It is common practice to add antibiotics to semen samples used for AI. These kill or incapacitate many of the bacteria but they have no effect on viruses.

At a practical level semen is not usually a cause of pathogen spread. Nevertheless the following viral pathogens may be present, particularly if the boar is ill or at the early acute stage of infection at the time of collection:

- Aujeszky's disease/pseudorabies virus.
- Cytomegalovirus.
- Foot-and-mouth disease virus (FMD).
- Genital papilloma virus.
- Parvovirus.
- PRRSV.

Many other viruses have been detected in semen but their significance is not known. Some important bacterial diseases can also be spread in semen during service, notably brucellosis and leptospirosis.

Fertilisation

It is important to appreciate that the natural mating process in the pig is a prolonged one, sometimes up to 15 minutes, during which time about 250 ml of fluid will be inseminated.

The penis spirals towards the vagina and into the cervix where it screws into the folds to become locked. This is vital because to establish an optimum pregnancy, sperm must be placed directly into the cervix.

Sperm is transported from the cervix to the top of the uterus by contractions of the muscles and normally it takes only minutes for them to arrive at the bottom of the oviducts (fallopian tubes), the utero-tubal junction. It is important for the sperm to transverse the uterus as quickly as possible for within the uterine horns the sow's defence systems are attempting to destroy the sperm and other bacteria and viruses.

Once the junction is filled with sperm, few more enter before ovulation. It is important to appreciate this because at the first mating the quality of the semen and timing probably decide the quality and success of fertilisation and subsequent embryo survival (Fig 5-39).

Libido

Failure of the boar to show sexual interest and activity is common in maiden boars. Sometimes up to 30 % of animals may be affected. Some pig farmers insist that any boar they purchase has been libido checked at least once before he arrives on the farm and some breeding

THE MORPHOLOGY OF SPERM

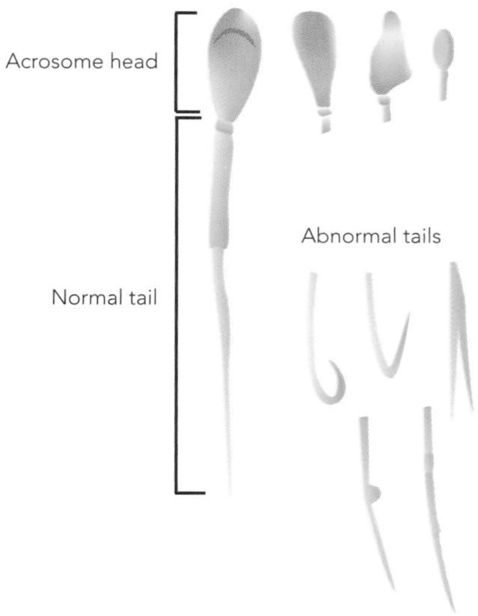

Fig 5-37

FIG 5-38: BACTERIA THAT COMMONLY CONTAMINATE SEMEN

Bacillus spp.	*Proteus*
Citrobacter	**Pseudomonas*
* *E. coli*	**Streptococcus*
* Haemolytic staphylococci	Various anaerobes
* *Klebsiella*	*Trueperella pyogenes*
Micrococci	

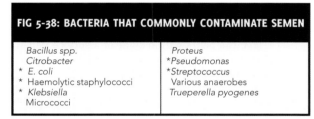

* Can be important.

THE INFLUENCE OF TIME OF MATING ON EMBRYO SURVIVAL

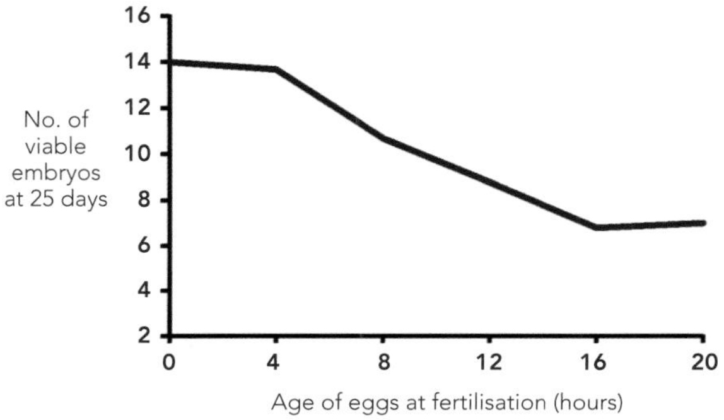

Fig 5-39

organisations carry this out. It is performed either by the boar mounting a gilt in heat or encouraging the use of a dummy. In spite of this, occasionally such tested boars will not work on the farm and careful coaxing and patience are required.

It takes 8 hours or more for sperm to mature (capacitation) inside the sow before they can fertilise.
With frozen semen the sperm becomes capacitated by the freezing process and thus this maturation is not required.

Procedures for libido testing

- Provide a well-bedded pen with a non-slip floor.
- Boars should be a minimum of 6 months old.
- Make good empathy with the boar.
- Be patient and show no aggression.
- Test boars in the morning when it is cool.
- Use a gilt firmly on heat. Always introduce the gilt to the boar pen.
- Ensure that the gilt in heat is not too timid.
- If no gilt is available use an AI stool and spray the stool with urine from a sow on heat.
- If there are poor responses:
  - Move the boars to different pens.
  - Have the boars within sight of a service pen.
  - Mix with a sow strongly on heat.
  - Leave a dummy stool in the pen.
  - Inject with prostaglandin.
- Ensure that the sow is not too dominant and does not bully the young boar.
- Cull boars if there is no response after 6 weeks.

Factors and actions to consider if a boar lacks libido

- Age of the boar. Has he reached puberty?
- Breeding. Are there any hereditary problems?
- Has the boar normal testicles? Measure their circumference.
- Is he housed in a poor social environment with little contact with other pigs?
- Boars reared in a total male environment can become infertile.
- Give plenty of female exposure.
- Is the boar apprehensive?
- Is he a timid or bullied boar? Has he been bullied by sows?
- Has their been poor stockmanship.
- Disease:
  - Look for any sign of respiratory disease, mange, sloppy faeces, pain or lameness.
  - Check for parasite burdens.
  - Lameness, specifically leg weakness or arthritis.
  - Inappetence.
  - A history of SIV, PMWS or PRRSV.
  - A previous history of pneumonia.
- Is empathy with the pig poor?
- Are lighting patterns poor? Ensure 14 hours of light per day.
- House the problem boar next to a mature good working boar.
- Allow saliva contact with working boars.
- Assess the response to an injection of prostaglandin or luteinising hormone but do not be too hopeful. Note synthetic prostaglandins do not work.
- Check for mouldy feeds (zearalenone toxin).
- Check for a damaged or defective penis. Sedate and examine.
- Provide a dry, well-bedded, well-drained pen.

Artificial Insemination (AI)

If you are starting AI for the first time, it is common practice to carry out 1 service, usually the first, with the boar followed up 24 hours later by AI. In other herds, all AI mating should be practised to maximise the genetic improvement availabilities.

The AI mating procedures should mimic the natural service techniques. It is always wise for the breeding team to be able to mate naturally to accommodate times when there are problems with semen delivery. A foot-and-mouth outbreak (for example) can cause considerable disruption to AI supplies. Be prepared to collect and process your own resident boars.

For herds with 300 sows or more it is cost effective to collect semen from boars on the farm daily and inseminate within 1 hour of collection (twice, 24 hours apart). The procedures and equipment necessary are simple and described in Chapter 15: Semen Collection.

Mating Procedures

Single Service (Supervised)

Large computer databases show that in most herds there is no difference in litter size or farrowing rate, whether the sow is single or multiple mated. If, however, records are analysed using conventional data, single services tend to show poorer litter size and farrowing rates. This is because a large proportion of single services take place in the sub-fertile period i.e. weaning-to-service interval 7 to 14 days, when litter size and pregnancy rates may be poor.

The advantages of single service

- Sows stand "rock hard". No continual trauma to the sow.
- There is a considerable reduction in management time.
- There is less capital cost in boars, or higher pointed boars can be purchased, thus improving performance of the progeny.
- In some herds it leads to an improvement in litter size and fertility.
- Litter size may improve in gilts and second parity.
- There is less trauma to the reproductive tract and less likelihood of ascending uterus infection.
- With natural service, boars are only used once in 24 hours.
- With natural service, semen quality is excellent.
- With natural service, the number of boars can be considerably reduced.
- With natural service, extra accommodation is available.

The disadvantages of single service

- In a few herds, the results may be inferior. Assess the effects on farrowing rate.
- The sow can be mated at the wrong time.
- With natural service, a boar may become infertile. However his failings will be detected earlier. This is particularly so if cross mating is used, when an infertile boar may not be detected. It is uncommon for boars to be infertile. Test the semen regularly by drawing it into a syringe at service using a stomach tube placed by the cervix.
- Sufficient boars must be held to allow for sudden illness or failure to work on farms practising only natural service.

Procedures for single service

- Make sure the sow is fed to appetite on the day of weaning e.g. give $^3/_4$ of the food 3 to 4 hours prior to movement from the farrowing quarters and the other $^1/_4$ after movement.
- Feed the lactation diet ad lib to point of service.
- **Only serve the sow when she is standing absolutely still** i.e. only mate her when you have to almost lift her into the pen.
- Ensure the service is of good quality i.e. a good lock and long insemination.
- Never single mate if the weaning-to-service interval is 7 to 14 days – i.e. the sub-fertile period. The time of standing oestrus in these sows is often 12 hours or less. Always mate these animals twice am/am.
- Double serve all repeat sows and especially gilts.
- Use AI plus a boar as a minimum. It is essential to get gilts pregnant.
- Ideally use pooled AI.
- Maintain excellent storage and handling of AI.

The common failure with single serving is to mate the sow too soon. If she is weaned on a Thursday the mating will usually take place on Tuesday morning (Fig 5-40). Sows on heat on the Monday will often have a long oestrus period and will still stand to the boar on the Wednesday. Thus if mating occurs on Monday, sperm will have to wait until ovulation some 48 to 72 hours later. Their viability could be much reduced. Many farms who wean on Thursday now do not bother to heat check on the Monday even if multiple serving.

To assess the efficiency of single service in your herd single serve 15 % of sows for a period of 6 weeks and compare the results to the rest of the sows. Fig 5-41 shows the results of single services in a 200 sow herd for a 2 year period compared to a 1 year period of multiple servings. There was no change in farrowing rates at 87 % but there was an increase in litter size.

FIG 5-40: SUGGESTED SINGLE SERVICE PATTERN FOR SOWS WEANED ON A THURSDAY

Day postweaning sow came into heat	+1	+2	+3	+4	+5	+6	+7	+8
	Friday	Saturday	Sunday	Monday	Tuesday	Wednesday	Thursday	Friday
S				H	S			
M				H	S			
T					S			
W						S		
T							S	
F								S

H= Do not serve – heat check only S= Serve

FIG 5-41: RESULTS OF SINGLE MATED SOWS AGAINST DOUBLE NATURAL, OR BOAR/A.1. AT 24 HOUR INTERVALS 200 SOW HERD					
Parity	**Alive**	**Dead**	**Mummified**	**Total**	
GILT 1	9.7	0.8	0.1	10.5	Double mated
	9.9	1.1	0.2	11.0	**Single mated**
2	10.5	0.9	0.1	11.4	Double mated
	11.1	0.9	0.1	12.1	**Single mated**
3	10.8	0.7	0.1	11.5	Double mated
	11.7	1.0	0.1	12.7	**Single mated**
4	11.5	0.9	0.2	12.2	Double mated
	11.6	1.2	0.1	13.1	**Single mated**
5	11.0	1.2	0.2	12.2	Double mated
	11.4	1.6	0.1	13.1	**Single mated**
6	9.8	1.0	0.2	10.8	Double mated
	11.8	1.8	0.1	13.6	**Single mated**
7	10.3	1.5	0.1	11.7	Double mated
	11.2	1.5	0.1	12.7	**Single mated**
8	10.4	1.6	0.1	12.0	Double mated
	11.2	2.2	0.2	13.2	**Single mated**
9	10.5	1.2	0.2	11.6	Double mated
	11.3	2.1	0.2	13.4	**Single mated**
10	10.6	2.3	0	12.9	Double mated
	10.5	1.7	0.2	12.2	**Single mated**
Total 488 litters	Av. 10.6	Av. 1.1	Av. 0.1	Av. 11.6	Double 12 months
Total 916 litters	Av. 11.2	Av. 1.4	Av. 0.1	Av. 12.6	**Single** 24 months

Test out single serving on a few sows first.

Double Services

Why should it be necessary to double serve the sow to establish a pregnancy? The simple answer is it is not, and single mating demonstrates this; but double services probably ensure a better timing of fertile sperm meeting a fertile egg and provide more latitude for variable service procedures (Fig 5-42). If double services give you excellent results do not make any changes. If, however, results are poor, it is worth assessing the effects of both changes in the number of services and their timing on litter size and conception rates. This may be particularly advantageous in ill-defined infertility problems. As an initial guide, serve am/am (Fig 5-43).

What should you do on the farm?

- Look at the results of your existing service procedure. Is it efficient? i.e. 14 plus total born and a pregnancy rate of 85 %. If so, don't make any changes.
- If results are poor then change the existing routine. Such a change could include:
  - Altering the number of times the sow is served (probably less rather than more).
  - Altering the time when the sow is first served (later).
  - Altering the frequency of the services e.g. from am/am to am/pm.
  - Mate the sow with pooled semen (semen from more than 1 boar).
  - Mate the sow with only 1 boar.
  - Mate the sow with 2 different boars.

If fertility is not being maximised then change your service routine. Remember however that if you have good fertility it does not matter if you serve 2, 3 or 4 times; **you do however risk continual trauma to the sow**.

There is no evidence that the 3rd or 4th mating contributes significantly to the breeding success.

FIG 5-42: SUGGESTED DOUBLE SERVICE PATTERN FOR SOWS WEANED ON A THURSDAY								
Day postweaning sow came into heat	**+1**	**+2**	**+3**	**+4**	**+5**	**+6**	**+7**	**+8**
	Friday	Saturday	Sunday	Monday	Tuesday	Wednesday	Thursday	Friday
S				S	S			
M				S	S			
T					S	S		
W						S	S	
T							S	
F								

Heat checking occurs at the same time as mating, but only 2 serves at 24 hour intervals either AI+AI, NS+AI or NS+NS.
S = Serves **AI** = Artificial insemination **NS** = Natural service.

FIG 5-43: DIFFERENT MATING OPTIONS

Regimes	Time of Mating			
	am	pm	am	pm
Single mating +	AI or B	–	–	–
Double mating	AI		AI	
	B	B	–	–
	B*	–	B	–
	B	–	AI	
	B*	–	AI	–

+ Only when the sow is standing completely immobile (rock hard).
B = Same boar or different boars. **AI** = Artificial insemination.

Key Points to a Successful Mating

- Ensure semen is stored correctly.
- Be prepared to mate the sows – do not introduce the boar and then have to collect the semen.
- Use clean catheters and clean techniques (Fig 5-44).
- Do not commence services too early.
- Make sure the sow is standing completely still and solid for the first mating.
- Observe that there is minimal leakage of semen from the vulva.
- Handle the sow and gilt quietly and patiently.
- Kindness and good empathy mean good fertility.
- Only serve once every 24 hours.
- Do not disturb the mated sows or gilts for 2 hours after mating.

FIG 5-44: SUGGESTED SERVICE PROTOCOLS USING A FOAM TIP CATHETER AND COLLAPSIBLE FLAT PACK OR TUBE

Preparation for service	When training tick off:	Tick
1	Remove the boar from sight and sound of the sow for at least one hour prior to breeding.	
2	Get prepared: semen pack, catheter and breeding belt.	
3	Parade a boar in front of the female outside the pen. If the signs of oestrus are demonstrated sit on her back and prove she will stand still.	
4	Only inseminate females which demonstrate the "standing still reflex".	
5	Place breeding belt onto sow's back.	
6	Wipe vulva clean with a dry tissue. If the vulva is clean do not clean with tissue.	
7	Place gloves on both hands.	
8	Select semen pack by checking it is from the required boar.	
9	Mix settled semen by gently rotating semen pack then place mixed semen in coverall top pocket.	
10	Take a new catheter. Do not use any catheter which is not clean. You do not need to lubricate a foam tipped catheter. If you do, avoid applying the lubricant to the hole.	
Insemination procedure		
1	Open the semen pack.	
2	Place the semen pack on the catheter.	
3	Part vulva lips and gently insert catheter at a 45° angle into the vagina, forwards and upwards to prevent the catheter entering bladder.	
4	Push the catheter straight forward until the ridges of the cervix are felt. Do not twist. Do not rotate a foam tipped catheter.	
5	Raise the semen pack to prime the catheter. Leave the catheter horizontal.	
6	Attach to breeding belt.	
7	Using your free hands, rub the female's flanks, and apply back pressure, during insemination to stimulate and maintain the standing still reflex.	
8	Allow the female to draw the semen.	
9	When the semen pack is empty slowly remove catheter by rotating clockwise.	
10	Record service and score quality of service.	
11	If this is the first insemination, re-inseminate 24 hours later if the female passes the standing still reflex.	

- Maintain the sow in dry warm housing for 28 days post-service.
- Do not mix sows from Days 4 to 28 post-service.
- Feed gilts 2 kg (4 lbs) per day maximum for 2 days post-service then to body condition to Day 28 (minimum of 2.8 kg per day).
- Feed sows to body condition score and a little more to enhance placental size.
- Feed a diet of at least 9.7 MJ NE/kg (14 MJ DE/kg) and 1% lysine from weaning to 28 days post-service.

Additional points regarding natural mating:

- Always introduce the sow to the boar.
- Always observe every service to completion in indoor systems. Examine boars clinically every week in outdoor systems and observe at least 1 service per sow as far as possible.
- Match the size of the boar to the size of sow.
- Serve in a pen that is dry, with no projections and a non-slip floor. The floor area should be at least 10 m^2. All sides should be equal in size.
- Make sure the supernumerary digits of the boar do not damage the sow's back. If so use a thin carpet or other protective material over the back of the sow.
- Assist entry of the penis into the vagina if necessary by cupping with a clean or gloved hand. Do not handle the prepuce; you will empty the preputial sac and cause heavy bacterial contamination.
- Ensure the penis is locked in and then observe between the testicles for the pulsation of the urethra to indicate that insemination is taking place.
- Always use a fresh unused boar for the first service.
- Never use a boar that is stiff or lame; you will risk a small litter.
- Avoid using a boar for at least 14 days if he has had a temperature of more than 40 °C (104 °F).
- Handle the boar quietly and patiently.
- Use 1 boar for 1 sow where possible and serve am and am.
- Only use each boar once every 24 hours.
- Give the boar a 48 hour rest between each complete sow mating if at all possible.
- Always house the boar in a clean dry pen.
- Mating will take 15 minutes.

Skip Services

Here the sow is not mated at the first oestrus post-weaning but at the second. This is a procedure worth considering if the second litter size is poor. Often the first parity weaning to first service interval is extended, which could be the cause of the smaller second litter. However a second litter size drop may occur even if the weaning-to-service interval is normal (less than 7 days).

There are no economic advantages in skipping services in sows after 2 litters.

Young sows are more prone to convert both fat and protein into energy and if this process of metabolism is a negative or catabolic one, (i.e. more energy is leaving the sow than is entering via feed) the hormone control of reproduction is affected. Skipped sows will be depositing protein and fat and are therefore anabolic at the time of mating with improved reproductive efficiency.

Is it worth skip mating?

- Probably, if there is a second litter size problem that cannot be solved by feed intake in lactation.
- If the increase in litter size is 2.1 pigs weaned then the 21 non-productive days lost and the consequences of this would be outweighed by an increase in 0.8 pigs per sow per annum.
- Practical problems such as housing and movement of the second litter female can sometimes create difficulties.
- Try the technique on a few sows first and see if it is worthwhile.

Studies have shown up to 2.6 extra pigs born (2.3 alive) in skip mated second litters and up to 2.6 extra pigs born (1.9 alive) in third litter females.

Fungal Poisoning – Mycotoxicosis

See also Chapter 13: Mycotoxins

Sometimes when moulds multiply on feeds such as wheat, barley, corn and cotton seed, mycotoxins are produced that can be poisonous (Fig 5-45). There are 3 factors necessary for their growth: an available carbohydrate source for energy; warm, moist conditions (10 to 25 °C (50 to 77 °F)); and oxygen. Other special conditions may be necessary for toxins to be produced.

An important toxin that affects reproduction is called zearalenone or F2 toxin, produced by the fungus *Fusarium graminearum.*

Zearalenone is an oestrogenic toxin and it is produced in high moisture environments in maize, well before harvest. Note distillers' grains (from ethanol production) may concentrate zearalenone and other mycotoxins in the feed at an average of 3 times the original concentration.

FIG 5-45: A GUIDE TO MYCOTOXIN LEVELS IN FEED: MILD TO SEVERE DISEASE

Fungus	Toxins	No Clinical Effect	Toxic Level	Clinical Signs
Aspergillus sp.	Aflatoxins	< 0.1 ppm	0.3 - 2 ppm	Poor growth Liver damage Jaundice Immunosuppression
Aspergillus sp. and *Penicillium* sp.	Ochratoxin & citrinin	<0.1 ppm	0.2 - 4 ppm	Reduced growth Thirst Kidney damage
Fusarium sp.	T2 DAS DON (Vomitoxin)	< 0.5 ppm	> 1 ppm	Reduced feed intake Immunosuppression Vomiting
	Zearalenone (F2 toxin)	< 0.05 ppm	1 - 30 ppm	Infertility Anoestrus Rectal prolapse Pseudopregnancy
			> 30 ppm	Early embryo mortality Delayed repeat matings
	Fumonisin	< 10 ppm	> 20 ppm	Reduced feed intake Respiratory symptoms Fluid in lungs Abortion
Ergot	Ergotoxin	< 0.05 %	0.1 - 1.0% ergot bodies by weight (sclerotia)	Reduced feed intake. Gangrene of the extremities. Agalactia due to mammary gland failure.

ppm – parts per million.
sp. – species - each of these fungi have several species, only some of which are toxic.
Note these toxic concentrations may be reduced when more than one toxin is present at the same time.

Clinical signs of zearalenone

The most striking clinical feature is the swollen red vulva of immature gilts. The other signs are dependent upon the levels present in the feed and the state of pregnancy.

The following may be used as guidelines to the symptoms that may be observed:

Boars – Semen may be affected with feed levels above 30 ppm, but not fertility. At higher levels poor libido, oedema of the prepuce and loss of hair may occur.

Gilts (pre-puberty (1 to 6 months of age)) – 1 to 5 ppm in-feed causes swelling and reddening of the vulva and enlargement of the teats and mammary glands. Rectal and vaginal prolapses also occur in the young growing stock.

Gilts (mature) – 1 to 3 ppm will cause variable lengths of the oestrus cycle due to retained corpora lutea and infertility.

Sows – Levels of 5 to 10 ppm can cause anoestrus, which may also be associated with pseudopregnancy due to the retention of corpus luteum. Note the F2 toxin mimics the 1st and 2nd embryonic signals produced at 10 and 14 to 17 days – thus the sow believes she is pregnant. F2 toxin will not, however, normally cause abortion. If sows are exposed during the period of implantation, litter size may be reduced. In lactation piglets may develop enlarged vulva and teats, but note that to a degree this is normal.

Effects on pregnancy – Low levels of 3 to 5 ppm do not appear to affect the mid-part of pregnancy, but in the latter stages piglet growth in-utero is depressed, with weak splay-legged piglets born. Some of these may have enlarged vulvas.

Effects on lactation – 3 to 5 ppm has no effect on lactation but the weaning-to-service interval may be extended.

Diagnosis

The clinical signs are distinctive. Rations that are suspected of contamination should be examined both for the presence of zearalenone and also other oestrogen-like substances.

Removal of the suspect feed will be followed by the regression of symptoms within 3 to 4weeks.

Treatment

- None is required, provided the toxin source is removed.
- Sows that are in deep anoestrus may respond to injections of prostaglandins.

Procedures for sampling feed

- Collect 8 x 1 kg (2 lb) quantities from separate areas of the feed.
- Mix all together.
- Take 4 x 1 kg (2 lb) separate samples from the bulked one into clean paper bags. The paper bags stop the feed from sweating.
- Number 1 to 4 and date.
- Seal and witness.
- Store at 4 °C (34 °F) or ideally in the freezer -20 °C (-4 °F) until transported to a laboratory.
- Retain 2 samples.
- Send 2 samples to the place of testing or use 2 separate laboratories.

Management control and prevention

Key factors leading to mycotoxicosis:

- The purchase of mouldy, damp or badly stored grains.
- The mixing of contaminated and uncontaminated grains.
- Holding cereals in moist, damp conditions.
- Allowing grains to heat.
- Prolonged usage of bins, feed bridging across the bin and development of moulds.
- Placing compounded feeds into bins whilst they are moist and warm.
- Poorly maintained bins that allow water to leak in.
- The bridging of feed in bins over long periods of time and their sudden descent.
- Prolonged use of automatic feeders and retention of mouldy feed.

If a problem of mycotoxicosis is suspected immediately stop using that particular feed and recycle it at a 1:10 dilution with other cereals into feeder pigs or destroy it. Sample all the feed ingredients and if these are required for future examinations, store in the correct conditions.

6 REPRODUCTION: INFECTIOUS INFERTILITY

Chapter 6

6 REPRODUCTION: INFECTIOUS INFERTILITY

Disease and Reproduction

The previous chapter has reviewed the various management and environmental factors that have an adverse effect on reproductive efficiency. The description of the diseases here concentrates on their effects on reproduction, but in some of them infertility is only part of a much broader picture. In such cases references to specific chapters are given. There is often an overlap between infectious and non-infectious infertility and in many cases the two are inter-related. It is important to determine if there is an infectious component to a problem in a herd because corrective measures may involve both treatment and management procedures. Fig 6-1 demonstrates a pathway that can be used to identify the causes of an infectious infertility problem. It asks the question of what diseases are present in the

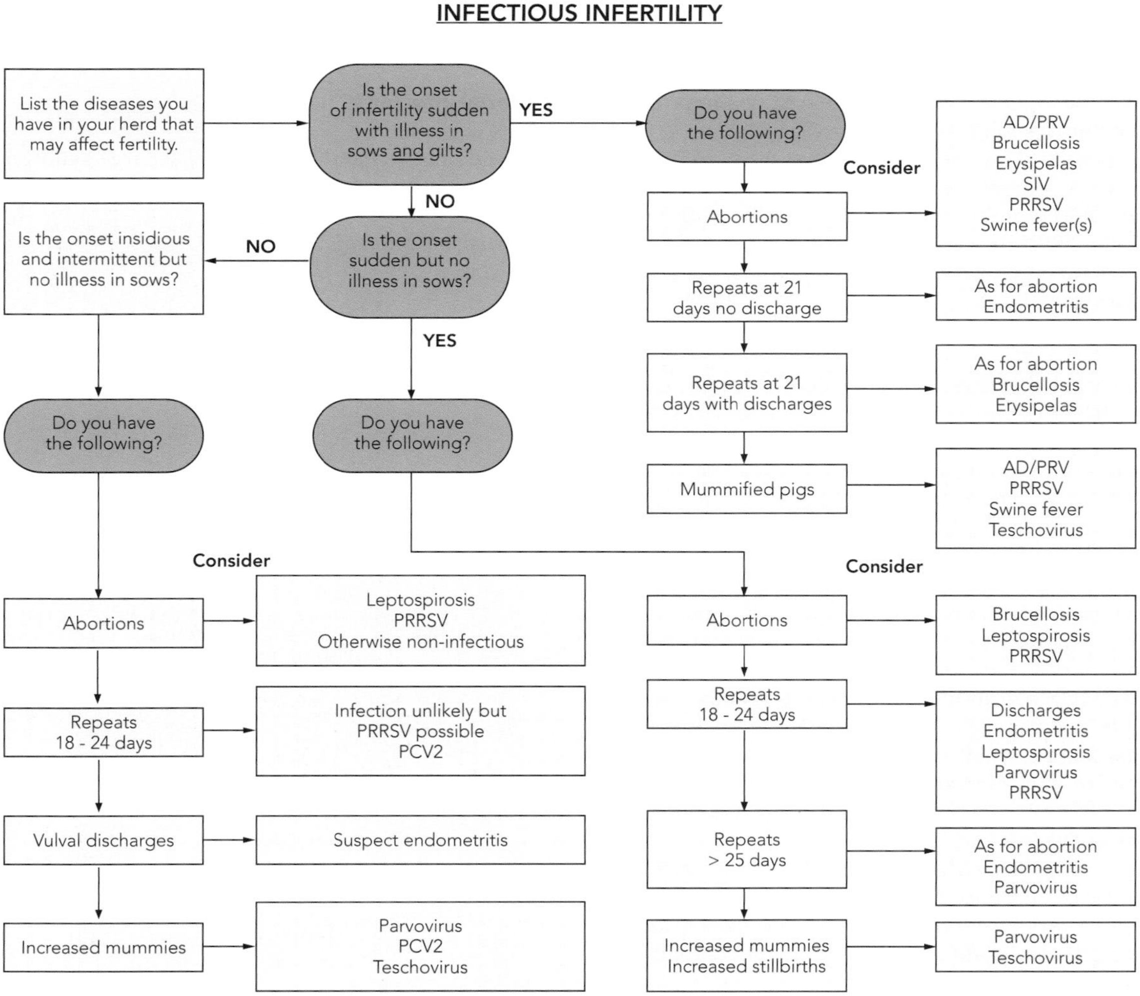

Fig 6-1

herd, because any one of these may have an occasional effect. Such diseases are listed in Fig 6-2 and Fig 6-3. For example, if a herd is infected long term with Aujeszky's disease virus, porcine parvovirus (PPV), porcine reproductive and respiratory syndrome virus (PRRSV) or swine influenza virus (SIV) then at varying intervals following the initial herd outbreak there may be reproductive failures, albeit in many cases at individual sow level. If a herd is free from these infections and then becomes infected with one of them, an acute episode of that disease will occur. This will be manifested in reproductive failure including increased numbers of abortions, repeats on a normal and abnormal cycle and in the case of those viruses that cross the placenta, there will be foetal death, mummified and stillborn piglets.

Diseases such as porcine parvovirus (PPV), PRRSV and leptospirosis may infect the sow without causing other clinical signs. Usually in such cases the disease picture is sporadic (unless it appears for the first time) and a detailed examination of records helps to clarify.

There are 9 main viruses that can cause reproductive disease, but only 5 of these are really important in those countries where they occur. The 5 are:

- Aujeszky's disease/pseudorabies virus (AD/PRV).
- Porcine parvovirus (PPV).
- Porcine reproductive and respiratory syndrome virus (PRRSV).
- Classical and African swine fever (CSF and ASF) virus.
- Swine influenza virus (SIV).

Bovine viral diarrhoea virus (BVDV), border disease virus (BDV), teschovirus and porcine circovirus 2 may cause sporadic reproductive problems, but in most countries disease would be uncommon. Encephalomyocarditis virus (EMCV) varies in its pathogenicity in different regions of the world. In Europe it is rarely implicated in porcine disease, in the Caribbean it causes heart problems but in North America it is associated with reproductive problems. Porcine circovirus 2 may play a variable but generally minor role in reproduction problems.

Most bacteria are opportunist invaders and usually only affect a few individual breeding females. A good example is erysipelas. The exceptions are leptospirosis and brucellosis which are herd problems.

Some of the clinical signs and pathological effects that viral and bacterial agents have on fertility are shown in Fig 6-2 and Fig 6-3.

Many pathogens when introduced into a naive herd can cause disturbance to reproductive performance. For example *Mycoplasma hyopneumoniae* introduction can temporarily devastate a herd's reproductive ability.

Diseases/Pathogens Affecting Reproduction

The following viral and bacterial diseases/pathogens can cause infertility and reproduction problems in the breeding herd. These are each covered in detail in this chapter.

Viral infertility

- Aujeszky's disease/pseudorabies virus (AD/PRV)
- Bovine viral diarrhoea virus (BVDV) and border disease virus (BDV)
- Encephalomyocarditis virus (EMCV)
- Porcine circovirus 2 (PCV2)
- Porcine cytomegalovirus (PCMV)
- Porcine parvovirus (PPV)
- Porcine reproductive and respiratory syndrome virus (PRRSV)
- Swine fever – classical/hog cholera (CSF/HC) and African (ASF)
- Swine influenza virus (SIV)
- Teschoviruses (formerly enteroviruses)

Bacterial infertility

- Brucellosis
- Endometritis and the vulval discharge syndrome
- Erysipelas
- Leptospirosis
- *Mycoplasma suis*

FIG 6-2: VIRUS INFECTIONS CAUSING INFERTILITY

	Embryo Death	Foetal Death	Abortion	Sows/Boars may be Ill	Virus may be in Semen *	Clinical Signs in the Breeding Herd
AD/PRV	+	+	++	+	+	+
BVDV and BDV #	+	+	+	—	?	—
EMCV	+	+	+	+	—	—
SIV	+	—	++	+	—	+
Parvovirus	+	+	—	—	+	—
PCV2	+	+	+ #	—	+	—
PRRSV	+	+	++	+	+	+
Swine fever	+	+	+	+	+	+ Severe illness
Teschovirus	+	+	—	—	+	—

# = Rare **+** = Yes **—** = No or uncommon **++** = Significant ***** In all diseases if the boar is viraemic **?** Unknown
BVDV = Bovine viral diarrhoea virus **BDV** = Border disease virus **EMCV** = Encephalomyocarditis virus **PCMV** = Porcine cytomegalovirus
PRRSV = Porcine reproductive and respiratory syndrome virus **PCV2** = Porcine circovirus 2

FIG 6-3: BACTERIAL AND FUNGAL DISEASES CAUSING INFERTILITY

	Embryo Death	Foetal Death	Abortion	Sows/Boars may be Ill	May be Spread in Semen	Clinical Signs in the Breeding Herd
Brucellosis	+	+	+	+	+	+
Endometritis #	+	—	+	+	Via preputial fluids	Discharges
Fungal diseases *	+	—	+	+	—	+
Mycoplasma suis	—	+ Stillbirths	—	+	—	+ Anaemia Jaundice Weak piglets
Leptospirosis	+	+ Stillbirths	+	—	+	Weak pigs at birth
Erysipelas	+	+	+	+	—	+
Any septicaemia or uraemia	+	+	+	+	+	+

= Non-specific bacterial infection of the uterus **+** = Yes **—** = No or uncommon ***** Including mycotoxicosis

Infertility: Viral Diseases

Aujeszky's Disease/Pseudorabies Virus (AD/PRV)

See Chapter 12 for detailed information on this disease

Aujeszky's disease (AD), also known as Pseudorabies virus (PRV) is an important disease of the pig and is caused by a herpes virus. The pathogen has been eliminated from a number of countries and regions so is not directly relevant to many pig farmers.

When introduced into a naive breeding herd the pathogen causes abortions and respiratory signs. Those females that are in the first third of pregnancy may absorb foetuses resulting in delayed returns to oestrus, but if infection occurs in the second, third or the last part of pregnancy, piglets are often stillborn and mummified or weak pigs are evident at farrowing. Once the acute phase of the disease has passed, a proportion of sows become carriers of the virus and shed it intermittently. Few clinical signs are then seen. Vaccination of the breeding stock is effective in preventing clinical disease. All aspects of this disease are covered in detail in Chapter 12.

Bovine Viral Diarrhoea Virus (BVDV) and Border Disease Virus (BDV)

These 2 viruses, which are in the same group of pestiviruses as classical swine fever virus (hog cholera virus) and which primarily infect cattle and sheep respectively, can get into pig breeding herds and cause reproductive problems.

Clinical signs

These include poor conception rates, a few abortions, foetal death, mummification, small litters and low birth weights. They rarely cause any other clinical signs in pigs.

Diagnosis

This disease is not a common cause of infertility in the sow and would be considered low on the list of priorities from a diagnostic point of view.

Problems can arise in regions where swine fever is endemic and routine screening of herds for swine fever is carried out, because pigs which have been infected with BVDV or BDV are positive to the common swine fever tests. Specialised tests have to be done to differentiate them, which take time. Meanwhile a standstill is enforced.

Treatment

There is no treatment and the infections are self-eliminating.

Management control and prevention

Prevent exposure to the pathogens. Transmission to pigs usually requires direct contact with cattle or sheep. Other possible methods by which infection may be introduced into the herd include exposure of pigs to cattle or sheep faeces, feeding of unpasteurised cows' milk, or via contaminated live-attenuated virus vaccines.

Encephalomyocarditis Virus (EMCV)

This is in the picornaviridae family. The main reservoir host is the rat although mice may also aid in its spread. It infects and causes disease in a wide range of vertebrate animals, but pigs appear to be the most susceptible of farm animal species. The virus is world-wide but differs in pathogenicity and virulence in different countries and regions. One strain, Type A, causes reproductive problems, a second strain, Type B, causes heart failure and other strains are mild or non-pathogenic. Types both A and B occur in Europe (e.g. Belgium) but in most countries of Europe, particularly those in the EU, it tends to be relatively mild or non-pathogenic and disease in pigs is rarely diagnosed.

In Australia the strains appear to be much more virulent for pigs than those in New Zealand. Virulent strains in Florida, the Caribbean and probably Central America damage the heart and cause death whereas those in the Midwest of the US tend to cause reproductive problems.

Clinical disease in pigs tends to occur when rat numbers increase to plague-causing levels. Pigs can be infected from rats or from rat-contaminated feed or water. It does not seem to spread very readily between pigs.

Clinical signs

In gilts and sows the first signs are often a few abortions near the end of pregnancy. Then, over a period of about 3 months, the numbers of mummified foetuses and stillbirths increase and pre-weaning mortality rises. The farrowing rate worsens. Affected females may go through a phase of fever and lack of appetite. In affected herds there are usually no clinical signs in weaned and growing pigs.

Diagnosis

To make a definitive diagnosis the virus has to be isolated and identified or rising antibodies demonstrated in blood samples taken 2 weeks apart.

Similar diseases

EMCV could be confused with AD (PRV), PPV and PRRSV although as you will see from Fig 6-2 there are distinguishing signs between these 4 diseases.

Treatment
- There are no methods of treatment.

Management control and prevention
- Check the source of incoming breeding stock for pathogenic strains.
- Reports of killed vaccines being effective have been documented.

Porcine Circovirus 2 (PCV2)
See Chapter 9 for further information
Porcine circovirus 2 is a very small resistant DNA virus which is extremely widespread in pig herds around the world. The virus has been implicated with abortion and reproductive failure in sows and gilts affecting the developing foetus resulting in heart muscle damage. The virus can be found in extremely large numbers in the heart muscles associated with myocarditis. The virus can kill the affected foetus and if sufficient numbers are affected a SMEDI condition will occur. SMEDI is stillborn, mummified, embryonic death and infertility. This will increase returns to oestrus and affect the number of piglets born alive. If sufficient numbers of foetuses die the sow may abort, but this is rare.

Confirmation of the role of PCV2 can be difficult as the virus is so common. Immunohistochemistry on the heart muscle may be useful.

Vaccination of gilts pre-breeding will help reduce the clinical signs if the farm has a sufficient problem. If a problem is identified in gilts, examine the gilt introduction programme. Normally an adequate gilt acclimatisation programme resolves the issue.

See Chapter 9 for further information on PCV diseases (PMWS and PDNS).

Porcine Cytomegalovirus (PCMV)
This is a herpes virus found in the tissues throughout the body including the nose of newborn piglets where it causes inflammation (rhinitis). PCMV is present throughout the world and exists in most if not all pig populations but most infections are sub-clinical and clinical disease is rare. Serology carried out in the UK, for example, indicates that over 90 % of herds have been exposed to infection.

The virus is excreted in discharges from the nose and eyes, urine and farrowing fluids. It is also transmitted via the boar through semen and crosses the placenta to infect piglets before birth.

The rhinitis produced by this virus is uncommon and mild and has no relationship to progressive atrophic rhinitis (PAR) caused by the toxin-producing bacteria *Pasteurella multocida*. In most herds therefore the infection is insignificant and apart from sometimes causing a mild sneeze has no major effect on the pig.

Clinical signs
Clinical signs are only seen if PCMV infects a sow for the first time when she is in late pregnancy. Signs include foetal deaths, mummified foetuses, stillbirths and weak piglets. The sow may run a slight fever and be off her food. Rhinitis in newborn piglets can be severe enough to cause haemorrhage from the nose. In herds in which PCMV is endemic there are no symptoms other than mild sneezing in sucking and weaned piglets.

Diagnosis
The presence of the virus can be confirmed by serological tests, fluorescent antibody tests and demonstration of inclusion bodies in tissue sections.

Similar diseases
The disease might be confused with progressive atrophic rhinitis or bordetella infection of the nose, however the effects are very short lived and there is no progressive atrophy or distortion of the nose.

PCMV rhinitis only occurs in newborn piglets and there is a tendency to assume that sneezing in piglets must be associated with atrophic rhinitis. Rhinitis means inflammation of the delicate tissues in the nose and is caused by dust, gases, bacteria or viruses, in fact any irritant. If *Pasteurella multocida* are present the inflammation persists with damage and progressive destruction of the tissues (atrophy). This is a serious disease. It can be differentiated from PCMV by swabbing the noses of sneezing piglets and testing for the presence or absence of the *Pasteurella multocida*. It is important to carry this out because if the tests are negative you have no worries (or expensive treatments).

Treatment
- None is required.
- If sneezing and poor growth occur postweaning, the creep can be medicated with antibiotics such as chlorotetracycline (CTC), oxytetracycline (OTC), trimethoprim/sulpha or tylosin for 14 days.

Management control and prevention
- Provide good environmental conditions in farrowing and weaner accommodation.
- Avoid fluctuating temperatures.
- Avoid dust.
- Maintain all-in/all-out management of farrowing and weaner houses.

Porcine Parvovirus (PPV)

This is a common and important cause of infectious infertility, particularly if vaccination of the herd is not practised. Porcine parvovirus is a small DNA non-enveloped virus that multiplies normally in the intestine of the pig without causing clinical signs. It is worldwide in its distribution. It can spread 30 km (18 miles) from a pig farm.

Whereas most viruses do not survive outside the host for any great period of time, PPV is unusual in that it can persist outside the pig for many months and it is resistant to most disinfectants. This perhaps explains why it is so widespread and so difficult to remove from the pig environment.

If you test for it in your pig herd it will almost certainly be present. It is therefore an infection you have to live with and manage. Most sows become infected by their second pregnancy and develop permanent immunity. As a result clinical parvovirus is most often seen in gilts.

To understand the role of PPV in reproduction it is important to realise that reproductive infection usually occurs without disease, but sometimes there is infection with reproductive disease. PPV is transmitted either by mouth or through the nose, passing into the intestine where it multiplies and is passed out in faeces. If a pig becomes infected for the first time when it is not pregnant there are no clinical signs.

It takes 10 to 14 days from first infection for PPV to reach the uterus. If the animal is pregnant and exposed, the virus crosses the placenta killing piglets selectively. If the embryo is infected at less than 35 days of age, before there has been an opportunity for bone development, death results followed by complete absorption. If infection is before Day 35 and the whole litter is killed the sow will return at 21, 28 or 63 days and the clinical sign will be infertility. If the whole litter is not killed, the litter size will be small.

If infection takes place in the foetus between 35 and 70 days of pregnancy the foetuses die and they become mummified. Do not assume that all mummified pigs are caused by PPV infection. This is often not the case.

From 70 days of age the immune system of the foetal piglet has started to develop and it can therefore respond and protect itself from the virus, although the piglets may be weakened and an increase in stillborn piglets may occur. Thus if pregnant animals are infected for the first time after approximately 70 days of pregnancy there will be little evidence of disease. This is quite different to PRRSV infection, which kills the foetus only after 60 days of age inside the uterus and therefore very late mummified pigs are seen in this disease. Once inside the uterus PPV spreads slowly from one foetus to another and as a result the sizes of mummified pigs will vary within the litter.

PPV is of minimal significance to the lactating sow other than for the fact that colostrum contains a very high antibody content and will protect the piglets from infection for up to 5 or 6 months of age.

Experimentally, PPV is important as a co-infection with PCV2 to create experimental cases of PMWS, but this doesn't occur in the field.

Clinical signs

Acute outbreaks of disease

Infection itself causes no clinical symptoms other than the presence of mummified and stillborn pigs at farrowing. In acute outbreaks of disease, the following occurs:

- Small litters associated with embryo loss before 35 days.
- Mummified pigs of varying size between 30 and 160 mm (1 to 6 inches).
- Increased numbers of stillbirths. These are associated with the delay in the farrowing mechanism which occurs because of the presence of the mummified piglet and the birth of weakened piglets.
- Abortions associated with PPV infection are very uncommon.
- There may be an increase in low birth weight piglets but neonatal deaths are not affected.
- The acute disease episode often lasts for up to 8 weeks, then wanes for 4 to 6 weeks, followed by smaller bouts of mummified pigs for a further 4 to 6 weeks.
- The virus can take up to 4 months to infect all sows in a susceptible, previously uninfected herd.

Sporadic disease in enzootic herds

This is seen in individual females which are infected for the first time. It is usually confined to gilts.

Records

Records can assist in the diagnosis of PPV disease and the differences between the normal herd and the diseased herd are shown in Fig 6-4.

In acute herd outbreaks of disease involving many animals, litter size is reduced with the percentage of litters totalling less than 9 piglets increasing from about 10 % up to 40 %. The numbers of mummified pigs, particularly associated with small litters, are elevated and sows not in pig may increase from 2 % to 6 %. Sows found not in pig are a result of either total embryo absorption before 35 days or complete foetal death followed by pseudopregnancy.

In some cases the sow reaches the point of farrowing with normal udder development, even to the extent of producing milk, but there are no live births. An injection of prostaglandin to bring about farrowing yields mummified pigs that have been present inside the uterus.

These animals would not otherwise farrow because a live foetus is necessary to initiate farrowing. The above picture is only seen at this level in a susceptible herd, that is, with 50 to 70 % of sero-negative breeding females. Such episodes are likely to occur in non-vaccinated herds every 3 to 4 years and arise because virus circulation ebbs and flows. During periods of low or no PPV activity a susceptible population gradually emerges.

Immunity

PPV infection results in high antibody levels in the serum which persist for long periods. You should appreciate that such levels do not necessarily mean that there is or has been a reproductive problem or a higher level of protection. For example, a titre of 1:2 will be equally as protective as a titre of 1:80,000 (Fig 6-5). Blood sampling all the sows in a herd on one occasion only indicates the percentage of animals that have been exposed to parvovirus at some previous period, which gives you an idea of the overall breeding herd immunity or susceptibility. Once an animal has been exposed to PPV it remains immune for the rest of its life.

FIG 6-4: PPV INFECTION
EXAMPLE OF REPRODUCTIVE DATA IN AN ACUTE OUTBREAK

	Normal Herd PPV Infected	Acute Disease
Total litter size	Normal	Reduced
Alive and dead	15	< 9.5
% of litters total born < 9	< 10 % sows < 18 % gilts	20 – 40 %
Stillbirths	4 – 7 %	7 – 12 %
Mummified pigs %	< 0.6 %	1 – 4 %
Sows not in pig	1.0 %	2 – 6 %
Delayed returns to oestrus	< 3 %	> 4 %
Weaning-to-oestrus interval	Normal	Normal
Other clinical signs	None Some disease in non-vaccinated gilts	None Disease in all parities

FIG 6-5: SERUM TITRES TO PPV

	Level/Significance	
Non-vaccinated female	Negative	Susceptible to infection and reproductive failure
Vaccinated female	1:2 to 1:160	Protected
Gilt with maternal antibody	1:4 to 1:320	Protected but will wane
Active immunity	> 1:640	Protected

From a practical standpoint, the breeding herd may be in 1 of 3 phases:

1. **Serologically negative.** In this situation all females are highly susceptible to infection and reproductive failure. This is an unusual situation but can occur occasionally in small herds, from which the virus can die out. In such herds if parvovirus is introduced there is a massive outbreak of reproductive disease: repeats, mummifications, not in pig and possibly a few abortions. Such herds should be vaccinated immediately.
2. **Endemic infection.** Here PPV is continually circulating and 50 to 90 % of animals are immune. However infection can take place in the early to mid-pregnancy period in any negative animals and therefore there is a variable amount of disease. This was the typical picture prior to the availability of vaccines in large non-vaccinated herds. Intermittent outbreaks of disease occur, particularly in gilts and second parity females. As viral activity increases, so does immunity across the herd. When there are large numbers of immune animals there is little infection and the herd immunity gradually drops as old immune sows are culled.
3. **Disease in replacement gilts.** This is common because at least 50 % of gilts at the point of mating may not have met PPV and therefore are susceptible. Up to a third of such animals may become infected in the first half of pregnancy, resulting in reproductive failure.

Key points to parvovirus infection

- The virus is widespread throughout all pig populations but it may disappear in small herds (< 100 sows).
- Infection is endemic (present all the time) in most pig units.
- Once a pig is exposed there is a lifelong immunity.
- Reproductive problems may appear every 3 to 4 years in a herd if vaccination is not carried out.
- Parvovirus infection in a susceptible female can cause death of the embryo with absorption, or death of the foetus with mummification.
- The major signs are therefore small litter sizes, mummified pigs of different sizes, and increases in pseudopregnancies and not in pigs.
- Abortion due to PPV is uncommon.
- Maternal immunity may persist up to 7 months of age but only in a few gilts. (This interferes with vaccine response.)
- Up to 50 % of gilts may be sero-negative at the point of mating.

Diagnosis

In the absence of any other signs of illness in the breeding females, PPV disease can be suspected by increases in mummified pigs and small litter sizes.

The important features are disease and death in the embryo and foetus from approximately 15 to 100 days of pregnancy. The mummified pigs can be examined by fluorescent antibody tests in the laboratory to confirm the infection. Serology will not help because many sows are positive and normal.

Key points for recognising PPV disease

- Small litters associated with variably sized mummified pigs occurring mainly in unvaccinated gilts or gilts vaccinated while still protected by maternal antibody and therefore vaccination was not fully effective.
- Increased percentage of repeats.
- No other signs of ill health in the breeding female or in individually affected animals.
- Some gilts or sows progress to the point of farrowing but produce no live pigs.
- A history of no vaccination programme in gilts.
- Examine all afterbirths from sows carefully to see whether there are small mummified pigs present which vary in size.
- Submit small mummified pigs – less than 150 mm (6 inches) – from small litters to a laboratory for fluorescent antibody tests. These will confirm whether the foetus has died from PPV infection or not.
- Check medicine storage. Freezing may destroy the vaccine. Ensure correct needle length used.

Similar diseases

An acute outbreak of PPV could be confused with AD/PRV, PRRSV, leptospirosis or certain forms of SIV. The differentiators with PPV are that there are no other clinical signs in adult breeding stock, newborn piglets are healthy and fully active, and there are few or no abortions.

Treatment

- There is no treatment.

Management control and prevention

- In an acute outbreak, immediately vaccinate the breeding herd to prevent infection in those animals that are still sero-negative. Discuss with your veterinarian. Remember it will take 10 days for the first dose of vaccine to take effect.
- If a sero-negative gilt is given a single dose of vaccine, the immune system is primed and a low antibody level is produced (1:64). Vaccination and stimulation of immunity by natural infection (which can be achieved by a controlled feedback programme) is sufficient to protect the litter from disease. It takes 10 to 14 days following infection for PPV to cross the placenta and infect the embryos or foetuses. If the infected breeding female has been vaccinated at some time in the past then when exposure to PPV takes place, there is rapid re-stimulation of the immune system (within 5 to 7 days). This is sufficient to prevent disease and to stimulate a permanent immunity.
- Fig 6-6 shows the levels of disease in 69 herds over a 10 year period, prior to vaccination and in 67 herds over an 8 year period following a single injection of parvovirus vaccine to gilts only.
- These results show that disease can be controlled by a single dose of vaccine even though serological tests demonstrated that PPV continued to circulate in all the herds. Discuss your vaccination policy with your veterinarian. Your situation may be different and he/she may advise 2 doses of vaccine followed by a once or twice yearly booster.

Eradication

Parvovirus cannot be eradicated from a herd.

FIG 6-6: PORCINE PARVOVIRUS

Results of Single PPV Vaccination	No. of Herds	Herds with Parvovirus Problems
Non-vaccinated herds over a 10 year period	69	58 (84 %)
Gilt-only vaccinated herds over an 8 year period	67	0

Once a pig is exposed to porcine parvovirus it gains a lifelong immunity. The diseases can be controlled through vaccination.

Porcine Reproductive and Respiratory Syndrome Virus (PRRSV)

See also Chapters 7, 8, 9 and 10 for additional information

Porcine reproductive and respiratory syndrome (PRRSV) is caused by an RNA virus from the genus Arterivirus and causes reproductive failure in breeding stock and respiratory tract illness in young and weaned pigs.

The disease was first reported in the USA in the mid to late 1980s and was called mystery swine disease or blue ear disease before being classified. Since then, outbreaks of PRRSV have been confirmed throughout the world with only a few countries such as Australia remaining free.

The PRRSV strains isolated from the European and North American outbreaks represent 2 distinct viral genotypes. Genotype 1 is predominantly seen in Europe and genotype 2 in North America. The genotype 2 virus has a high mutation rate and a variant of genotype 2 is often the cause of severe disease in Asia.

The PRRSV has a particular affinity for the macrophages found in the lung. Macrophages are part of the body's defences. They ingest and remove invading bacteria and viruses. Those present in the lung are called alveolar macrophages. In contrast to most other bacteria and viruses, macrophages do not destroy the PRRSV virus. Instead, the virus multiplies inside them producing more viruses and killing the macrophages. Up to 40 % of the macrophages are destroyed. This removes a major part of the body's defence mechanism and allows bacteria and other viruses to proliferate and do damage.

A common example of this is the noticeable increase in severity of mycoplasma (enzootic) pneumonia in grower/finisher units when they become infected with PRRSV. Another example is the alarming increase that can occur in clinical cases of meningitis in herds in which virulent *Streptococcus suis* type 2 is enzootic.

Once it has entered a herd, the PRRSV tends to remain present and active in the herd indefinitely although the clinical signs will rise and wane.

Viral persistence

The persistence of the virus within a herd is related to a number of factors:

- The regular introduction of naive gilts into the herd may allow the virus to persist.
- Infection is transmitted to recently weaned piglets (as their maternal antibody disappears) from older groups previously infected. This process is responsible for active enzootic respiratory disease continuing on many farms.
- It may take up to a year for all breeding stock, particularly in large herds, to become infected for the first time and although the virus appears to spread rapidly in a herd it may be some 4 to 5 months before at least 90 % of the sows become serologically positive to the virus. Furthermore, it is not uncommon for sow herds 1 to 2 years after infection to contain less than 20 % serologically positive animals. However, experience indicates this does not necessarily mean the serologically negative pigs have lost their immunity nor does it mean that they have stopped passing on immunity to their offspring.
- The American genotype 2 virus is very susceptible to mutation giving rise to multiple strains. Farms in the USA may be infected with multiple strains of the virus at the same time with immunity being poor between strains. This may be why the virulence of the disease appears higher in North America and Asia.

Methods of spread

The virus is spread by nasal secretions, saliva, faeces, urine and by airborne transmission. Field studies suggest PRRSV can be detected up to 3 km (2 miles) from a source of infection although transmission over 2 km (1 mile) is rare. A temporary carrier state exists in the pig that can last for 3 to 5 months and shedding of the virus can occur throughout this carrier state. In some individuals it is thought that it may last longer although the pigs may not be shedding virus. PRRSV infects all types of herd including high or ordinary health status and both indoor and outdoor units, irrespective of size.

The following are common methods of spread:

- Movement of carrier pigs.
- Airborne transmission up to 1 km (½ mile).
- Mechanical means via faeces, dust, droplets and contaminated equipment.
- Contaminated boots and clothing.
- Vehicles, especially in cold weather, where it can be transported in ice in the wheel hub.
- The mallard duck and probably other species of bird.
- Mosquitoes and biting flies.
- Artificial insemination but only if the boar is viraemic.

Artificial insemination (AI)

Artificial insemination can be a potential method of spread if semen is used when the virus is present in the blood (viraemia) and particularly during the first 3 to 4 week period following the breakdown of an AI stud. Boars may remain viraemic for up to 14 weeks after initial infection, or become chronic carriers. Outside this period, field evidence indicates the risk of spread in semen from previously infected groups of boars is very low.

Immunity

Field observations show that, after initial infection, the majority of breeding females become immune and do not succumb to further episodes from the same strain. However, if the virus continues to circulate in young growing pigs a few incoming negative gilts will become infected at some stage. If a new strain of PRRSV enters your herd, the current immunity may be insufficient to prevent clinical signs developing.

Adult pigs can shed the virus for up to 3 months and young (weaned) pigs for up to 5 months.

Clinical signs

See also Chapters 7, 8, and 9

The clinical picture can vary tremendously from one herd to another. As a guide, for every 3 herds that are exposed to PRRSV for the first time, 1 will show no recognisable disease, the second will show mild disease and the third will show moderate to severe disease. Field experience indicates that the higher the health status of the herd, the less severe the disease effects are.

The severity of an outbreak in a herd will depend to an extent upon other viruses and bacteria already present in the herd and their capacity to cause disease. Furthermore, when the pigs' immunity is compromised, excessive stocking density and the quality of the environment also become important. For example, PRRSV in a healthy pig may cause no detectable pneumonia, but if the pigs are already infected with respiratory pathogens and are housed poorly in an inadequate environment, severe disease may develop and persist.

Acute disease

When the virus first enters the breeding herd, disease is seen in dry/gestating sows, lactating sows and sucking piglets.

The initial phase of inappetence and fever will often take 3 to 6 weeks to move through the breeding herd. Cyanosis or blueing of the ears is a variable finding and less than 5 % of sows show it. It is transient and may last for only a few hours. Coughing occurs in some sows and a few individual cases of clinical pneumonia may occur. This acute phase lasts in the herd for up to 6 weeks, and is characterised by early farrowings, increases in stillbirths, weak pigs and an increase in the numbers of large mummified pigs that have died in the last 3 weeks of pregnancy.

In some herds, these problems may affect up to 30 % of the total pigs born. Piglet mortality peaks at 70 % in weeks 3 or 4 after the onset of symptoms and only returns to pre-infected levels after 8 to 12 weeks. The reproductive problems may persist for 4 to 8 months before returning to normal. In some herds performance may actually improve on the pre-PRRSV situation.

Occasionally, skin lesions are seen in PRRSV that are characterised by small discrete vesicles anywhere on the body but particularly around the nose and the shoulders at points of pressure (Fig 6-7). The vesicles rupture, become infected and dark coloured, and ultimately heal over a 3 week period.

Long term effects on reproductive efficiency

Longer term effects of PRRSV on reproductive efficiency are difficult to assess, particularly in herds of low health status. In some there are increases in repeat matings, vulval discharges and abortions, all of which may be blamed on PRRSV.

The effects of PRRSV on reproduction efficiency in herds in which the infection has become enzootic have been observed in the field for up to 12 months after disease has apparently settled. These are as follows:

- A 10 to 15 % reduction in farrowing rate (90 % of herds return to normality).
- Reduced numbers born alive.
- Increased stillbirths.
- Poor reproduction in gilts.
- Early farrowings.
- Increased levels of abortion (2 to 3 % but rises of up to 50 % have been reported).
- Inappetence in sows at farrowing.

Some herds report that PRRSV continues to be associated with reproductive failure and increases in repeats (5 to 10 %) on a normal and abnormal cycle for as long as 6 months after the acute episode has subsided. The disease can cause regular bouts of reproductive failure.

Clinical signs in boars

Clinical signs in boars can be mixed. There may be no clinical signs shown at all. Alternatively, signs may include inappetence, increased body temperature, lethargy, loss of libido, lowered fertility and poor litter sizes as a result of reduced sperm motility and acrosomal defects. The acrosome is a cap-like membrane that covers the head of a sperm. It contains enzymes involved in penetration of the ovum.

Diagnosis

If a herd has not previously been exposed to PRRSV then blood sampling a minimum of 12 adult animals (preferably those that have been off their food for at least 3 weeks) provides a reliable means of diagnosis. Serological tests available include the overlay or immunoperoxidase monolayer assay (IPMA) test, the fluorescent antibody test (FAT) and an enzyme-linked immunosorbent assay (ELISA) test. All these tests can give rise to false positives and false negatives on individual animals, but on a group basis, they are reliable in indicating whether the herd has been infected or not. Where disease persists, serum samples should be examined 2 weeks apart to demonstrate whether the virus is associated with a particular clinical problem. PCR tests on small blood samples taken in the early acute phase will demonstrate evidence of viral DNA. This is a good sensitive test and may be useful in examination of semen samples. RNA sequencing of the virus is now commercially available and is an excellent tool for identifying the similarity between isolates from different sites.

Check which type of PRRSV – European or American – is present in your herd as this could affect the approach you take to dealing with the outbreak. The American strain may not be present in all countries – the UK, for example, is free from this strain.

When disease first appears in a herd allow at least 3 weeks before carrying out IPMA or ELISA tests.

Similar diseases

When PRRSV first enters the herd, the clinical picture could be confused with Aujeszky's disease/pseudorabies virus (AD/PRV) but the absence of nervous symptoms in piglets and serological tests will differentiate between the 2 diseases.

Treatment

See also Chapters 7, 8 and 9

There is no treatment as yet available for animals against virus infections. With PRRSV, however, it is essential during the acute phase to prevent the multiplication of bacteria that normally would have been destroyed by the pig's immune system (macrophages). Antibiotic treatment including tiamulin should be given for 3 to 4 weeks to all sows and boars immediately the disease is diagnosed or suspected. Treatment should also include salicylic acid (aspirin). If necessary commence with water soluble antibiotics followed by in-feed medication. Prompt treatment usually reduces abortions, stillbirths, mummified pigs and early farrowings caused by secondary bacteria. Vaccinating sows with a PRRSV vaccine may help reduce shedding of the virus. Tiamulin may assist in a PRRSV outbreak as it may limit the ability of the PRRSV virus to enter the macrophage.

Management control and prevention

See also Chapters 7, 8 and 9

If your herd is in a region where it is unlikely to be infected directly from a neighbour it is important to determine by serology if your herd has or has not been exposed to PRRSV.

SKIN LESIONS OCCASIONALLY ASSOCIATED WITH PRRSV

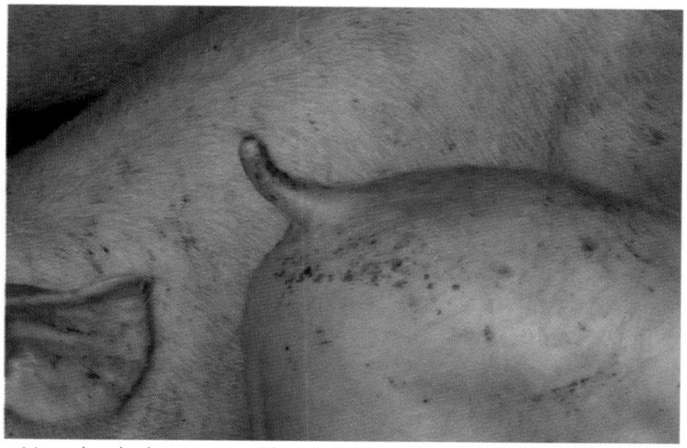

Note both the tiny and larger black lesions beneath the hair.

Fig 6-7

PRRSV-negative

If the herd is found to be negative, the following actions should be considered:

- Purchase breeding stock from herds believed free of PRRSV.
- Use semen from AI negative studs only. Ideally practise on-farm AI.
- Set up a quarantine system to hold pigs for a minimum of 8 weeks.
- As pigs arrive in quarantine, add 6 of your own known negative pigs and mix with the incoming stock.
- After 5 weeks of direct contact, blood sample the 6 sentinel pigs and 6 of the incoming stock.
- Check with the donor herd that it is still believed free of PRRSV.
- Make sure trucks do not come onto your farm with other pigs already on board.
- Remove all shoes before entering the farm.
- Provide boots and coveralls for all visitors.
- Do not borrow equipment from other pig farms.
- It is essential that the farm has excellent biosecurity programmes.

PRRSV-positive

In known infected herds, 1 of 2 breeding strategies can be adopted. If the virus is not circulating it may be advisable to buy in gilts and boars from negative herds and vaccinate them in isolation. Alternatively gilts and boars from known infected sources should be purchased and acclimatised for at least 6 weeks before mating. The method should be determined by the performance of the gilts. Both live and killed vaccines are available in some countries and control strategies will be determined by their known safety and efficacy. Seek advice from your veterinarian.

A control programme for the sows in the service area during acute disease

- Apply the same procedures as for dry/gestating sows.
- Do not cull any sows for the next 6 weeks (at least) to increase the mating programme.
- Accept that you will have a 10 to 15 % drop in the farrowing rate, and therefore plan to increase the mating programme by the use of more gilts and by retention of sows that would have been culled. Review batch numbers after pregnancy is confirmed and cull accordingly to maintain batch integrity.
- Buy in gilts if possible from a previously exposed herd; most, if not all, will be immune. If home bred gilts are kept, introduce these to the infection as early as possible.
- To do this, always move gilts into weaner or finisher accommodation for 3 to 4 days to expose them to the virus.
- Consider vaccinating incoming gilts whilst in isolation if available. Seek veterinary advice.
- Do not serve gilts until at least 6 weeks after exposure to infection or vaccination.
- Use one dose of AI to complement each natural mating for an 8 week period.
- Treat with water soluble aspirin.

Vaccination

Live vaccines are available for use in weaners and non-pregnant females (gilts) provided they are kept in isolation for 8 weeks. This has the advantage of stabilising herd immunity with minimal risk and the positive benefit of reducing postweaned viraemic pigs. It should be remembered that specific vaccines will not cover all types of PRRSV virus, thus it is important to ensure that the vaccine and virus match.

Feedback programmes can be developed to transfer the farm's virus strain to naive incoming gilts. A strong rope is placed in the nursery and left for 5 days. The weaned pigs then play and chew on the strong rope, covering it in saliva. Then place the strong rope into the gilt pen where it will be chewed by the gilts. The rope will then transfer the PRRSV to the naive gilts.

Alternative methods such as tonsillar scrapes and serum have been developed to act as autogenous vaccines.

But note all of these programmes may also transfer other pathogens, not just PRRSV.

When transferring feedback, the gilts must be going to the same farm site as the source material.

Elimination from the herd

The methods adopted to eliminate PRRSV depend on the status of the virus throughout the herd.

After a period of time following infection the virus may disappear from the sow and finisher population but remain endemic in the first and second stage rearing in pigs 3 to 12 weeks of age. By blood sampling sows, weaners and finishers, the status of the herd in this respect can be established. Where this is the case the following methods can be adopted to eliminate the virus from the herd.

Depopulation/repopulation

A farm can be depopulated, cleaned, disinfected and repopulated with PRRSV negative stock. Depopulation is very expensive, and before considering this investigate how the herd became infected in the first place and assess the chances of it being infected again. Repopulation should not be attempted in winter because as the temperature drops, the survival of the virus increases, e.g. 24 hours at 37 °C (99 °F), 6 days at 20 °C (68°F), 1 month at 4 °C (39 °F). The virus is stable when frozen for long periods of time. Control of PRRSV in weaners and finishers is discussed in Chapter 9.

Herd closure

This method is based upon the premise that over a period of 6 months or more (and provided no breeding animals are introduced, i.e. the herd is closed) the sero-prevalence of infection in the breeding herd usually drops to less than 15 %. Only sows and sucking pigs remain on the farm, weaned pigs being moved to an offsite facility. Once negative status is achieved only negative gilts and semen are introduced into the herd.

During the closure period it is important that farrowing houses are managed all-in/all-out with no back fostering of piglets. Do not allow piglets to remain on the farm after 28 days of age. At the end of the programme, the piglets will need to be weaned at 14 days of age for at least a month.

The following methods can also be used to eliminate PRRSV from the herd, but the risk of failure is high.

Partial depopulation

Where disease is only active in growing pigs up to 12 weeks of age they can all be removed from the farm together with the next 2 weeks' production of new weaners. The houses are cleaned and disinfected and left empty for 2 weeks. PRRSV disease can be eliminated by this method of partial depopulation.

Note other pigs on the farm may be non-clinical carriers of PRRSV, resulting in a failure of eradication.

Segregated early weaning (SEW)

PRRSV free pigs can be obtained in a similar way from a herd in which the virus has become enzootic, sow herd immunity is stable, and active virus infection is lower. In this process of segregated early weaning (SEW), the largest suckled piglets are weaned from the farrowing rooms at 5 to 10 days of age to be reared in isolated premises. The method, however, is not 100 % reliable and occasionally 1 or more piglets in a litter may come through already infected. To reduce this happening pigs should only be taken from sero-positive sows. To avoid all the pigs becoming infected, each group of piglets weaned should be kept in isolation until all the pigs have been tested. PCR tests are carried out on blood samples taken from the navels of the piglets at birth. It takes only a few days to get the results. To improve the system further, instead of farrowing the sows on the farm they can be farrowed in temporary isolated accommodation outside the farm.

Hysterectomy

Hysterectomies can be carried out on sows from stable herds. Fostering these piglets on to sows from a PRRSV negative herd may be an effective method of developing PRRSV negative stock. It must be remembered, though, that PRRSV can cross the placenta in recently infected gilts and sows. Therefore only pregnant sows that have been sero-positive for at least 4 months should be hysterectomised. Each newborn litter should be tested for infections using PCR. The foster sows and hysterectomy derived piglets should be kept in isolation until the results are known.

Swine Fever: Classical Swine Fever (CSF), Hog Cholera (HC) and African Swine Fever (ASF)

See Chapter 12 for detailed information on this disease

Classical swine fever is caused by a virus from the pestivirus family. The pig is the only natural host and in most countries, CSF is a notifiable disease. African swine fever (ASF) is the only member of a completely different virus family, Asfarviridae. However, both diseases have very similar signs and expert laboratory analysis is needed to differentiate. At a farm level what is seen is “swine fever”.

Both viruses are spread from infected or carrier pigs via discharges from the nose and mouth and in the urine and faeces and they are highly contagious. They get into herds by the introduction of a carrier pig or infected meat. ASF may also be spread through soft body ticks.

The viruses survive in frozen carcasses for long periods of time and can withstand curing and smoking processes.

Clinical signs

When first introduced into the breeding herd disease causes inappetence and high fevers in the sow. The virus can cross the placenta to invade the foetuses causing foetal death with mummification, abortions, malformations and increases in stillbirths.

An important characteristic of CSF is the birth of very weak pigs showing trembling, a form of congenital tremor. (In other forms of congenital tremor there are no clinical signs of illness in the breeding females). Convulsions may occur with death within a few hours and sows may lose the use of their legs. If the strain is of low virulence the only signs may be poor reproductive performance and birth of piglets with nervous signs. In the acute form CSF will have dramatic effects on reproduction. ASF and CSF are covered in detail in Chapter 12.

Swine Influenza Virus (SIV) – Swine Flu

See also Chapter 9

Swine influenza is a highly contagious respiratory disease caused by a number of closely related type A influenza viruses that are noted for their ability to change their antigenic structure and create new strains.

Each serotype is identified by surface proteins (antigens) referred to as “H” and “N”. The 3 common strains that affect the pig are described as H1N1, H1N2, and H3N2. There are also different strains within these serotypes with differing pathogenicity (capacity to produce disease).

SIV can be introduced by infected people, carrier pigs and probably on the wind although this has not been proved. Birds, particularly water fowl, are reservoirs of infection.

SIV is an important cause of infertility in the breeding herd which is covered here. SIV also has a significant impact on the weaner to finisher period which is covered in Chapter 9.

Clinical signs

The incubation period of the disease is very short – as little as 12 to 48 hours. When the virus first enters the herd 2 or 3 animals may be observed sick in this period, followed over the next few days by up to 40 to 50 % of animals going off their feed and looking very ill.

The effects on the reproductive system follow the sudden onset of a rapid spreading respiratory disease with coughing, pneumonia, fevers and inappetence. Acute respiratory distress persists over a period of 7 to 10 days (depending on the amount of contact between groups of sows).

There are 3 important periods when infection causes infertility.

- First, if sows are ill in the first 10 days post-service their developing embryos may not get established and an increase in 21 day returns results. If pregnancy has been established, normally 14 to 17 days after mating, and subsequently fails, returns at 28 days will be observed.
- Second, if infection occurs later in pregnancy, embryo mortality and absorption can occur. Where the embryo loss is total sows become pseudopregnant (with returns at 63 days) and not in pig. Otherwise, litter sizes are affected due to the reduction in viable embryos. Towards the end of the pregnancy period abortions or late mummified pigs at farrowing may be experienced.
- The third major effect is on the boar, where high body temperatures affect semen and depress fertility for a 4 to 6 week period.

SIV in large herds may become endemic with intermittent bouts of disease and infertility and different strains may also sequentially infect the herd. Immunity to influenza viruses is lifelong to that particular strain but there is little immunity between strains. However, immunity from vaccines is only of about 6 months duration. Therefore the immunity profile in the breeding herd varies considerably with time.

At the breeding herd level the following signs may be seen:

- A sudden, rapid onset of acute illness in sows.
- Coughing and pneumonia spreading rapidly.
- A return to clinical normality over 7 to 10 days.
- Delayed returns to heat postweaning.
- Increased repeats at 21 days.
- Increased repeats outside the normal cycle.
- Increased numbers of sows coming through not in pig.
- Increased numbers of abortions, particularly late term.
- Increased numbers of stillbirth rates and slow farrowings.
- Occasionally an increase in mummified pigs.

During the phases of high temperatures other diseases present in the herd may be triggered. A typical example would be an increase in abortions associated with leptospira infection.

If you are monitoring the water usage, swine influenza may reduce water intake by 10 % which may be helpful in predicting an outbreak.

Diagnosis
This can often be made reliably on clinical grounds because there are no other diseases that are so dramatic in their onset and clinical effects. Blood samples taken at the time of onset of disease from affected sows and repeated 2 to 3 weeks later show rising levels of antibody to the specific virus. SIV can be readily grown from nasal and throat swabs and identified in the laboratory. This is often the best approach to confirm the diagnosis. PCR tests on nasal swabs are possible.

Similar diseases
In acute disease the spread is so dramatic across all ages that little else can be confused with it. In endemic disease however differentiation from other viral infections can be difficult, but PRRSV, porcine respiratory coronavirus (PRCV), AD/PRV and erysipelas should be considered (see Fig 6-2 and Fig 6-3).

Treatment
There is no treatment specifically for SIV. However secondary bacterial infections may be involved and in such cases antibiotics can be used to control these.

- Individual breeding females or boars showing acute illness and raised temperatures, particularly with increased respiratory rate, should be treated with broad spectrum long-acting antibiotics for 3 days.
- Suitable medicines would include penicillin/streptomycin, long-acting OTC or synthetic penicillins such as amoxycillin. If the illness is severe then medicate the drinking water with either CTC or OTC at 25 g (100 % pure) per 1000 kg of liveweight per day, for 5 days.
- Vaccines, some of them multivalent, are available in some countries and may be given to stabilise the sow herd. Note that the immunity created by the vaccines only lasts 6 months. Immunity created by the natural disease is life-long – to that specific type of H and N strain.
- Medicate with soluble aspirin to reduce abortion risk and make the affected sows more comfortable.
- Medicate the water with soluble vitamins for 7 days.
- Medicate the feed as above for 2 weeks.

Management control and prevention
It is important to prevent any secondary bacterial infections and alongside treatment, good husbandry is critical.

- Keep sows within an environmental temperature of 20 to 23 °C (70 to 75 °F).
- Reduce all possible sources of stress, such as draughts.

- Maintain dry bedding and floor surfaces.
- Monitor boars carefully for evidence of illness.
- Identify boars that have been ill and cross-serve sows with another boar for the next 4 weeks.
- Serve sows by AI followed by a natural mating and consider a second dose of AI (am/pm/am).
- Purchase semen from a commercial boar stud.
- If periods of inappetence occur in boars, blood test them twice, 2 weeks apart, to establish a diagnosis.
- Because of the possible ways by which the virus may enter the herd it is extremely difficult to maintain populations free of infections. In some countries inactivated vaccines are available and appear to be protective. In herds in which the virus periodically circulates and causes disease, this route should be explored.
- If you believe your herd to be free from SIV (and this can be confirmed by serological tests) purchase breeding stock from herds that have a similar disease history and that are also serologically negative. The practicality of this, however, is not easy.

Teschoviruses (formerly Enteroviruses)

These are gut-borne viruses, host-specific to the pig. The teschoviruses are subdivided into serotypes of which at least 11 are known. Four of these, serotypes 1, 3, 6 and 8, have been implicated in reproductive problems in pigs. Serotype 1 is the Teschen/Talfan virus which can also cause paralysis (polio) in pigs. This might be a notifiable pathogen in some countries.

Usually, each pig herd has an array of different teschovirus serotypes which circulate in weaned and young growing pigs sub-clinically. The pigs are protected by circulatory antibodies derived from their sow's colostrum. By the time they reach breeding age they are solidly immune.

Reproductive problems only occur when a new serotype, to which the gilts are not immune, enters the herd and multiplies in the breeding females. The reproductive clinical signs include: stillborn piglets, mummified piglets, embryonic death, and infertility; commonly described under the name SMEDI. Similar clinical signs can also be seen in PPV and PCV2 infections.

It is interesting to note that since the introduction of parvovirus vaccine and the excellent results achieved, the effects from other SMEDI viruses would appear to be almost non-existent, suggesting that teschoviruses are not important as a cause of reproductive failure.

Clinical signs

Natural infection of teschoviruses takes place by mouth through the ingestion of infected faeces. They multiply in the small and large intestines, and in the absence of circulating antibodies escape from the intestine into the bloodstream to the uterus. They cross the placenta to produce the typical symptoms of embryo mortality, mummification and stillbirths.

In some cases infertility associated with absorption of embryos also occurs. If reproductive failure results there will be increases in embryo mortality, foetal deaths and mummified and stillborn piglets. Infection and disease only occur in non-immune sero-negative animals.

Diagnosis

This is carried out by serology and virus isolation.

Similar diseases

Teschovirus infections can be confused with PPV, PCV2 and PRRSV infection and occasionally with AD/PRV and leptospirosis.

Treatment

- There is no treatment but if a herd experiences problems with teschoviruses in incoming gilts, feedback management practices should ensure that gilts are exposed to infection at least 6 weeks before breeding, see Chapter 5 Acclimatisation of Gilts.

Management control and prevention

- Expose breeding females and boars to faeces from young growing pigs 8 to 14 weeks of age to immunise them (feedback). This is best carried out by introducing gilts at 90 kg (200 lbs) to faeces from pigs 15 to 50 kg (30 to 110 lbs) weight, twice weekly for 6 weeks prior to mating.
- The practise of feedback using faeces from the nursery to gilts in isolation will help to ensure the gilts are vaccinated before breeding.
- Vaccines could be made but are not indicated.

Placenta or material from farrowed sows for feedback helps to control the clinical signs of SMEDI viruses.

Infertility: Bacterial Diseases

Brucellosis

See Chapter 12 for further information

This disease, which is notifiable in some countries, is caused by the bacterium *Brucella suis*, which is one of the several different species of *Brucella*. It can be spread by venereal infection and the boar is a major source either by direct contact at mating or via artificial insemination. Brucellosis can be an important source of infection to the pig. The disease can be transmitted to people and is serious.

Clinical signs include infertility and abortions which may occur at any time. Diagnosis is carried out by serology and isolation of the organism. There is no reliable treatment and the disease is best eradicated either by depopulation, or testing and elimination of individuals.

This disease is covered in detail in Chapter 12.

Endometritis and the Vulval Discharge Syndrome

A discharge from the vulva 14 to 21 days post-service does not automatically mean there has been a pregnancy failure, but it will in most cases indicate infection of the vagina or cervix. Fig 6-8 shows the anatomy of the reproductive tract and the potential areas from which discharges could arise. These include the rectum, the vulva, the vagina, the cervix and the uterus. Discharges can also arise from infection of the kidneys (pyelonephritis) or the bladder (cystitis) with pus being passed in the urine.

It is important to record the time when discharges are first seen, their colour and composition and effects on the sow. Use the "Record of failures to farrow" sheet as shown in Chapter 5 (Fig 5-9) to record the observations.

THE ANATOMY OF THE REPRODUCTIVE TRACT

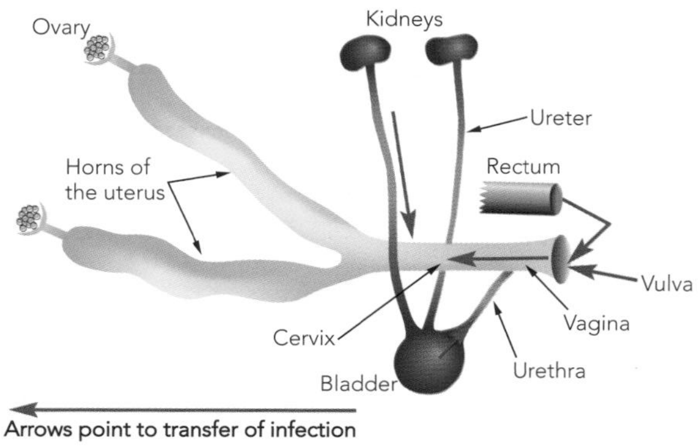

Fig 6-8

Clinical signs

Vulval discharges are common within 3 to 4 days of farrowing, when a thick viscous material may be excreted. If the sow is healthy, the udder is normal and there is no mastitis, ignore it. It is common practice to treat such sows, but it is not necessary under these circumstances. Always be mindful that a heavy smelling bloody discharge may be the result of a retained piglet or afterbirth.

The time in the reproductive cycle when the discharge is seen is important, particularly between 14 and 21 days post-service. The lips of the vulva of each sow should be parted daily and any tackiness or small discharge noted. The sow should be marked and if she repeats, a problem may be developing. Note also the periods when it is quite normal for the sow to show evidence of a slight discharge (Fig 6-9). Remember, discharging sows may be pregnant and always pregnancy test before culling. The types of discharge that might be seen are shown in Fig 6-10.

Fig 6-11 shows the reproductive performance of 42 sows in a herd that had a discharge problem. This shows that only 28 % of the sows farrowed with 69 % repeating. Most of the animals had 5 or more litters and parity was very unbalanced. Such low farrowing rates are not acceptable in a commercial herd and it is important that the old, repeating sows are culled as response to treatment is generally poor and not cost effective.

FIG 6-9: THE SIGNS OF VULVAL DISCHARGES IN HEALTHY SOWS

Time of Discharge	Significance
1 – 4 days post-farrowing	* Normal
> 5 days lactation	Abnormal
At mating	* Normal
Up to 2 days post-service	* Normal
14 – 21 days post-service	Abnormal
During pregnancy	Abnormal

* Unless heavy, increasing and continuous.

FIG 6-10: TYPES AND SOURCES OF DISCHARGE

Type of Discharge	Source
Thick, white, yellow pus	Vulva Vagina Cervix Uterus
Fluid, mucous, pus, blood & urine	Bladder Kidney Vulva Vagina
Chalky	Urine sediment from the kidneys & bladder
Pure blood	Internal ruptured blood vessel

Records

The changes that can take place during an outbreak of endometritis and the vulval discharge syndrome shown in Fig 6-12. The table shows how increases in repeat matings, sows not in pig and discharges, rise above normal levels. However it is important to remember that a proportion of sows will discharge and remain in pig. In these cases infection arises from either the vagina, the cervix or the bladder.

Doubtful or negative pregnancy tests at 28 and 35 days increase, but levels of abortions, litter size and sow health remain normal.

Observations

The bacteria causing reproductive tract infections are shown in Fig 6-13. The main organisms associated with endometritis and vulval discharges are opportunist invaders. In some herds no specific organism can be identified, although bacteriological tests may show one or more bacteria predominating either in the prepuce or vagina. A precise diagnosis can be difficult.

An initial diagnosis is made from observations of vulval discharges commencing 10 days post-service and associated with a loss of pregnancy. Leptospirosis caused by *Leptospira bratislava/muenchen*, particularly in gilts, can cause discharges post-service, particularly if there is early embryo loss (14 to 25 days) and absorption. Brucellosis also causes vaginal discharges following absorption.

Examination of records can help in differentiating between infectious and non-infectious causes of infertility, as shown in Fig 6-14.

Post-mortem examinations on sows with heavy discharges are not particularly helpful. Post-mortem examination of 47 discharging sows within 7 days of the event showed only 12 with an infection actually in the uterus (Fig 6-15). This is because by the time sows can be slaughtered and the uteruses examined they have come into oestrus which has the effect of resolving and removing the pus from the uterus. In spite of this, most sows still remain infertile.

Field studies have identified the following factors that do not appear to be associated with post-mating vulval discharges and infertility:

- Season.
- Breed of sow.
- Source of breeding female.
- Source of boar.
- Artificial insemination.
- Cystitis pyelonephritis.
- Discharges in lactation.
- Assistance at farrowing.
- Discharge at < 5 days post-mating.

FIG 6-11: HERD 20 PREGNANCY RATES WITH VULVAL DISCHARGES OBSERVED › 5 DAYS POST-MATING (42 SOWS UNTREATED)

No.	Farrowed	12 (28 %)
	Died	0
	Cystitis (culled)	1
	Repeated	29 (69 %)
Parity		**No. of Pregnancies**
	Gilts	2
	2 – 4 litter sows	4
	5 – 7 litter sows	18
	8+ litter sows	18

FIG 6-12: VULVAL DISCHARGES › 5 DAYS POST-SERVICE

	Problem Herd Ranges	Targets for Normal Herds
Repeats		
18 – 24 days	10 – 20 %	< 10 %
25 – 90 days	5 – 10 %	< 3 %
Sows not in pig with or without discharges	4 – 9 %	< 1 %
Pregnancy testing (doubtful or negative)		
% at 28 days	5 – 20 %	< 5 %
% at 35 days	5 – 20 %	< 5 %
Farrowing rate	60 – 85 %	> 87 %
Abortions	Normal level	< 1 %
Litter size	Unaffected	
Sow health	Unaffected	

FIG 6-13: REPRODUCTIVE TRACT INFECTIONS

Chlamydia
E. coli
Erysipelothrix
Klebsiella
Leptospira bratislava/muenchen
Mycoplasma suis
Pasteurella
Proteus
Pseudomonas
Staphylococci
Streptococci

FIG 6-14: CLINICAL DIFFERENTIATION BETWEEN INFECTIOUS AND NON-INFECTIOUS CAUSES OF REPRODUCTIVE FAILURE

Type of Failure	Clinical sign(s)	Non-Infectious	Infectious
Anoestrus		+++	+
Repeats	– at 18 – 24 days and no discharge	+++	+
	– at 18 – 24 days with discharge 14 – 21 days post-service	+	+++
Repeats	– 25 – 35 days	++	++
Abortion	– sow in good health	+++	+
	— with healthy foetuses	++	++
	– with mummified or decomposing piglets	-	+++
Sow not in pig		+++	++
Mummified pigs	– small and variable in size	+	+++
	– large	++	++
Stillbirths	– increased within a normal litter	+++	+
	– with mummified pigs	+	+++

+ Unlikely cause **+++** Likely cause

FIG 6-15: POST-MORTEM ANALYSIS OF VULVAL DISCHARGES

Results of Post-Mortem Examinations of 47 sows	
No gross lesions	21
Vaginitis	2
Vaginitis and endometritis	4
Endometritis	12
Endometritis and cystitis	6
Cystitis	2
Pregnant	0
Total	47

Vaginitis = inflammation of the vagina.
Endometritis = inflammation of the lining of the uterus.

Artificial insemination does not appear to be a major part of the complex because its use in problem herds does not necessarily solve the problem. Bladder/kidney infections do not cause pregnancy loss, unless the sow is ill, when abortion or death may occur. Generally, there is no relationship between discharges seen in lactation and those seen post-service with loss of pregnancy. Likewise, assistance at farrowing does not appear to influence a post-service problem.

During the period of oestrus and farrowing and whilst oestrogen levels are high it is difficult to infect the uterus. However, as oestrogen levels drop and the levels of progesterone rise the uterus becomes susceptible to infection.

It is likely that over a period of time in problem herds, certain opportunist bacteria gradually predominate in the preputial sac of the boar, the vagina and the environment.

If matings take place towards the end of the heat period with high levels of infection persisting at the cervix, the risk of ascending infection becomes much greater.

The major predisposing factors leading to the development of a vulval discharge problem include:

- Poor handling of the catheter prior to insemination.
- Herds with high numbers of old sows.
- A short lactation length (14 to 21 days).
- Multiple-matings.
- Matings towards the end of the oestrus period.
- Wet, dirty boar pens. Poor drainage. Continual use.
- Dirty, wet sow mating pens and continual use without cleaning.
- Inadequate space which causes the sow/gilt to adopt a dog sitting posture resulting in heavy contamination of the vulva.
- Heavy vulval contamination, for example in maiden gilts housed on slats where slurry spills over.
- Early embryo mortality.
- Re-mating discharging sows.
- Using old boars on young sows.
- Using young boars on older sows.
- Wet, dirty boar pens. Poor drainage. Continual use.
- Handling the prepuce at mating and squeezing the preputial sac.
- No supervision at mating. Cross-mating boars.
- Re-using AI equipment.

Management control is carried out by changing and improving the management factors considered important on the farm.

Diagnosis

Diagnosis is generally the result of studying the relevant records, as described above, and from regular observations of vulval discharges, particularly 14 to 21 days post-mating, to identify affected females.

It is possible to conduct bacteriological examinations of preputial and vaginal secretions, but this is an involved process and due to the fact that treatment is generally ineffective, on a commercial basis, affected animals are best culled.

Where bacteriological examination is chosen, all boars and 10 sows that have repeated should be tested. In the boars, swab the prepuce either while the boar is eating, confined in a stall or just before dismounting after mating. In problem sows the lips of the vulva should be cleaned with a tissue, parted and a swab inserted to its full length. The predominating organisms in the herd can then be determined and antibiotic sensitivity tests carried out. Discuss this with your veterinarian.

Treatment of the boar

Treatment of the boar has little effect on the pathogenesis of post-service vulval discharges, unless there is a specific causal agent, such as klebsiella, demonstrated from both the boar's prepuce and many of the sows' discharges. Where this is the case, you are advised to discuss aspects of treatment with your veterinarian. Treatment would be with suitable antibiotics administered either orally, by injection or by installing into the prepuce.

Treatment of the sow (assuming she is a particularly valuable or pet pig)

The most effective method to control 14 to 21 day post-service vulval discharges is to cull affected repeating sows. Discharges at other periods, not affecting pregnancy, should be monitored first and a decision made depending on her pregnancy status.

If treatment is preferred, the most effective method is to insert an antibiotic into the anterior vagina up to the cervix (but not actually into it). A 3 ml dose of most injectable antibiotics can be used, the selection being dependent upon the bacteriological examinations and sensitivities. A disposable AI catheter of small diameter or a cattle AI catheter can be used. Shorten it by approximately 150 mm (6 inches). Attach a syringe with an adapter. Fill the catheter and syringe completely with the antibiotic. The catheter should then be gently inserted as far as the cervix but not into it and the antibiotic can then be deposited easily. This can be carried out 6 to 24 hours after the last mating. It is important, however, when adopting this procedure, to always monitor the results for any adverse effects over the first 3 weeks in respect of both discharges and return rates.

An alternative method of medication is to top dress the feed of the sow from weaning to 21 days post-service. In herds with major problems it may be necessary to medicate all breeding females for a period of 10 days with in-feed medication using the appropriate antibiotic.

Management control and prevention

The most economical solution to this problem is through good management and husbandry. The following advice should be followed:

- Monitor the vulva for evidence of discharges.
- Do not serve towards the end of the heat period.
- Do not serve a sow more than twice.
- Do not re-serve discharging sows.
- Only mate the sow when she is totally receptive to the boar. If in doubt, wait.
- Do not be eager to serve too early.
- Consider single-day mating i.e. boar am, AI pm.
- Consider only one service.
- Multi-serve sows with the same boar.
- Avoid wet, dirty boar and mating pens. Clean and disinfect pens regularly.
- Avoid lactation periods < 21 days.
- Increase the number of gilts available for mating.
- Serve young boars to young females only.
- Prevent heavy faeces contamination of the vulva from weaning to 14 days post-service.

If repeat matings in the herd are high, check the vulvas of all sows 14 to 21 days post-service for tacky discharge. Do they repeat? If so, take action!

Erysipelas

See Chapter 7 for further information

Swine erysipelas is caused by a bacterium, *Erysipelothrix rhusiopathiae* that is found in most if not all pig farms throughout the world.

More than 20 % of animals may carry it in their tonsils. It is always present in either the pig or in the environment and is impossible to eliminate from a herd.

Erysipelothrix rhusiopathiae is also found in many other species, including sheep, poultry, fish and birds and can survive outside the pig for a few weeks and up to 6 months in light soils. Infected faeces is probably the main source of infection.

Non-vaccinated gilts and up to 4th parity sows are susceptible and may suffer fever, abortions or mummified piglets. Sows may show characteristic diamond shaped skin lesions. This disease is covered in detail in Chapter 7.

Leptospirosis

See Chapter 7 for further information

Leptospira are long, slender spiral-shaped bacteria, found in most mammalian host species. Over 160 serotypes are known, generally called serovars, with cross infections occurring between some species. Each serotype has 1 or more (usually only 2 or 3) reservoir hosts (Fig 6-16). A serotype can remain as a life-long infection in its reservoir host.

FIG 6-16: LEPTOSPIRA IN PIGS

Type of Leptospira that can Affect Pigs	Alternative Reservoir Host	Country	Signs
L. bratislava / L. muenchen	Hedgehog, vole	World-wide	Infertility, stillbirths
L. pomona	Skunk, rodents	Varies	Abortion
L. tarassovi	Wildlife	Europe, Asia	Abortion
L. icterohaemorrhagiae	Rat	World-wide	Haemorrhage and jaundice in young pigs

The pig is a reservoir host for *Leptospira pomona*, *L. tarassovi*, *L. bratislava* and *L. muenchen*, the last 2 being very closely related. It is not a reservoir host for *L. icterohaemorrhagiae* but can be infected from rat or dog urine and become severely ill with jaundice. It can also become infected by other serotypes from other animals' urine, for example *L. canicola* from dogs, *L. grippotyphosa* from horses and *L. hardjo* from cattle; but the infections are sub-clinical and do not result in disease. The pig is then an incidental host, i.e. does not perpetuate the infection and is only responsible for minimal spread.

The main serotypes that cause disease in pigs are:

- *Leptospira pomona* (see also Chapter 7).
- *Leptospira tarassovi.*
- *Leptospira bratislava/muenchen.*

Leptospira pomona

L. pomona causes important reproductive problems in female breeding pigs spreading slowly through the herd. It remains in the herd permanently unless steps are taken to eradicate it. It is not in the UK or Ireland and seems to have mainly disappeared from Western Europe but is widespread throughout the rest of the pig rearing world. Rodent forms of *L. pomona* may occasionally be identified and in Europe and America the skunk is an alternative reservoir host. This is covered in detail in Chapter 7.

Leptospira tarassovi

L. tarassovi causes a similar syndrome (i.e. a collection of signs and lesions) to *L. pomona* but tends to be milder and to spread more slowly. It is found in Eastern Europe and the Antipodes. It is thought that some wild animals are also reservoir hosts.

Leptospira bratislava/muenchen

The pig is also a reservoir host for certain subtypes of *L. bratislava* and *L. muenchen* which are widespread throughout the pigs of the world. They cause a different syndrome to *L. pomona* and *L. tarassovi* and affect mainly pregnant gilts and second litter females because they will not previously have encountered it.

Once these organisms are introduced into a herd the pigs become permanent carriers, with infection of the kidneys and intermittent excretion of the organism into the urine. *L. bratislava/muenchen* also permanently inhabit the oviduct (fallopian tubes) of sows and the reproductive organs of boars and are spread in semen.

Infection can enter the herd in 1 of 3 ways:

1. Introduction of infected gilts and boars.
2. Infection brought into the herd by other animals.
3. Exposure of the herd to indirect sources of contamination, e.g. contaminated water.

A herd can also be contaminated by *L. bratislava/muenchen* by AI if no antibiotics are used in the semen. Unless very stringent precautions are taken, most herds become exposed at some stage to *L. bratislava/muenchen*. *L. pomona* and *L. tarassovi* are more easily kept out.

Leptospirosis can be a very difficult disease to diagnose because pigs are often infected but there are no clinical signs to be seen. Thus if you carry out a serological test in your herd and the result is positive, for example to *L. bratislava*, this does not necessarily mean you have disease – only that the animal has been infected at some point in time and then responded by producing antibodies. If you carry out a serological test with positive results for *L. pomona*, this may be a cross reaction to other non-pig serotypes in the same group.

Conversely, if you have an infertility problem that clinically suggests leptospira as the cause, then test results would support a diagnosis of disease if there were rising titres of antibodies in the serum of the affected sows. However this may not be the case because pregnant females seroconvert early in the infection and by the time they abort or show symptoms the serum levels may be falling. Antibody concentrations are only maintained for a short period of time – months at best.

Methods of spread within the herd are as follows:

- Infection is by mouth, through the mucous membranes.
- Most leptospira are inhabitants of the kidney and found in urine. Pig-to-pig transmission via urine is common.
- Introduction and spread within a herd can occur by various forms of wildlife.
- Introduction and spread within a herd can occur through the introduction of carrier boars and gilts.
- Venereal infection is commonplace, particularly with *L. bratislava/muenchen*.
- Contaminated water, floor surfaces, pools and streams.
- Wallows used in outdoor production if there are pools of fresh urine around.

All leptospira require moisture, not only for indirect transmission, but also to survive. Desiccation kills them in 48 hours. When they invade the pig for the first time, there is rapid multiplication, and antibodies become evident 5 to 10 days later in the blood. Titres may rise as high as 1:1000, but they gradually decline to a low point or even become negative, although the animal may still carry the bacteria and excrete them.

Leptospira may become localised in the uterus during pregnancy, causing either abortions or increases in stillborn piglets. *L. bratislava* may persist in the oviduct (fallopian tubes) and uterus of non-pregnant sows, and in the genital tracts of boars. This may be an important medium for the maintenance of infection in the herd and be responsible for sows failing to conceive.

Clinical signs

In acute outbreaks inappetence and depression may be observed but chronic low grade disease is more common with abortions, stillbirths and an increase in poor, non-viable pigs. If abortions in a herd are at more than 1.5 % then investigations for leptospirosis should be considered. A reduction in farrowing rates and numbers of live pigs born per sow is also an associated factor, particularly with *L. bratislava* infection.

It is important to appreciate that many infections are sub-clinical but the organisms may persist in the kidneys and reproduction tracts to cause problems later.

Signs associated with acute *L. bratislava* disease:

- Repeating sows are common particularly in first and to some extent second pregnancy gilts.
- Infection often follows embryo loss and there may be copious vaginal discharges.
- Late term abortions.
- An increase in premature piglets.
- An increase in stillbirths.
- Mixed litters of live poor pigs and dead piglets at birth.
- An increase in mummified pigs.
- An increase in repeat breeding animals.
- Often there is a 2 year cycle of disease.
- Reproductive failure occurs in second litter females, rather than gilts, following their introduction to older carrier boars.
- Disease is less common in older animals.
- In long-standing carrier herds, disease can be difficult to recognise.

Diagnosis

This is carried out by assessing the antibody levels in a cross section of breeding females and the isolation of the organism from diseased tissues. The micro-agglutination/lysis test is carried out on serum, and recently affected animals will show titres of up to 1:1000 or more. At the onset of clinical signs a blood sample should be taken, and a further one taken 2 weeks later. If the second sample shows a rise in antibody levels of at least 2-fold, this would be indicative of leptospira involvement. Leptospira are difficult organisms to grow, and take a long period of time. *L. bratislava/muenchen* are especially difficult to grow and very few laboratories can culture them. There is, however, a method of detecting leptospira under the microscope using the fluorescent antibody test (FAT). Because of the difficulty of distinguishing between sub-clinical infection and infection with disease, an appraisal of the following will help. Do not rush into a diagnosis of leptospirosis; the majority of reproductive problems originate in poor breeding technique, not infectious causes. Consider the following as part of the diagnostic process:

- Records. Study the levels of abortions, repeats, stillbirths, weak piglets and the age of occurrence in sows and gilts.
- Study and evaluate the clinical picture.
- Eliminate other diseases.
- Eliminate non-infectious causes of infertility.
- Blood-sample suspicious animals and repeat 2 to 3 weeks later. Look for rising antibody titres e.g. 1st sample result 1:100, 2nd result 1:800. This would confirm active infection and indicate probable involvement.
- Blood-sample 10 females that have a history of infertility.
- Be aware that in chronic disease, the significance of titre levels is very difficult to assess.
- Test the aborted foetuses, urine or kidneys and oviduct (fallopian tubes) of slaughtered gilts by the FAT.
- Test thoracic fluid from stillborn pigs for the presence of antibody.

Similar diseases

The symptoms of leptospirosis can be mistaken for other causes of infertility, including:

- Chronic PRRSV.
- Endometritis.
- Non-infectious causes.
- Summer infertility.
- Management failures.

Treatment

- Medicate the feed with tetracyclines, either oxytetracycline or chlortetracycline, at levels of 800 g/tonne. Feed for a period of 3 weeks followed by a further course 6 weeks later, and repeat this for 4 treatment periods.
- An initial 3-week course of 800 g/tonne of tetracycline followed by a further 8-week course at 400 g/tonne.
- Strategic medication. Where there is a history of periodic infertility, in-feed medication can be targeted just prior to the expected time of disease.
- Inject sows at weaning time with streptomycin if available at 25 mg/kg. Boars should be treated with this medicine once every 6 weeks. Alternatively semi-synthetic penicillins could be used.

Management control and prevention

Control is achieved by vaccination and is reasonably effective. In many countries vaccines are available that cover 5 or 6 different serovars of leptospira. Where vaccines are not available, it is necessary to use antibiotic therapy. Also consider the following:

- Whilst it can be difficult to prevent *L. bratislava/ muenchen* from infecting the herd, nevertheless the more serious types such as *L. pomona* and *L. tarassovi* can be kept out of the herd by careful isolation of incoming stock, serological testing, veterinary liaison and a knowledge of the source herd.
- Check the serology of your herd. Do not buy in pathogens that you don't have.
- Check your sources of boars and gilts.
- Once leptospira are active and present in the herd, good hygiene, the constant removal of urine and good management become important methods of control. In outdoor herds, wallows could become sources of contamination especially if there are pools of fresh urine. The most effective method of control is to provide 2 wallows per paddock and use an electric fence which by movement will allow each to dry out and rest alternately.
- In indoor housing, poor concrete surfaces that allow the collection of urine and water are ideal sources for maintaining high levels of infection.
- If the sow is only exposed to low numbers of organisms, infection probably takes place with little disease.
- Make sure your diagnosis of disease is a correct one.
- Keep rodents under control, and particularly out of water supplies and storage tanks.
- Provide well drained concrete surfaces, particularly in defecating areas and boar pens.
- Remove slurry regularly.
- Identify problem parities and strategically medicate.
- Vaccinate if a vaccine is available.

Eradication

This can be done but it is unreliable and is probably contraindicated for *L. bratislava*.

Mycoplasma (Eperythrozoon) suis

See also Chapter 11 for further information

Mycoplasma (Eperythrozoon) suis is caused by a small mycoplasma bacterium. This blood parasite attaches itself to the red cells in the blood, sometimes damaging them and causing them to break apart. This causes a reduction in the red blood cell count, reducing haemoglobin (the substance which transports oxygen around the body) and thus diminishing the capacity of the blood to carry oxygen. When large numbers of red cells are damaged, haemolysis anaemia results and pigs look jaundiced.

A sow may carry *Mycoplasma suis* and yet remain quite healthy. However, if affected, it may become debilitated and pale (anaemia), with jaundice and bleeding into tissues. Infertility signs become apparent. This condition is covered in detail in Chapter 11.

Abortion and Disease

See Chapter 5 for information on non-infectious causes of abortion

Whilst abortion is an obvious part of infertility, the factors associated with abortion are complex, varied and often inter-related. This next section looks specifically at abortion.

Fig 6-17 shows the various factors that contribute to abortions, both infectious and non-infectious. In this chapter we focus on the infectious causes.

Infectious agents can bring about abortion in 3 ways:

1. They can invade and multiply in the placenta, causing inflammation (placentitis) and perhaps necrosis (tissue death), cutting off the nutrient and oxygen supply to the foetus.
2. They can invade the foetus and kill it.
3. They can multiply elsewhere in the body, causing fever and sometimes toxaemia (toxins in the blood) which may cause a generalised infection of the sow, making her feverish and ill resulting in abortion.

Infectious causes of abortion are listed in Fig 6-18. The major infectious bacterial diseases which spread through herds and cause abortion are PRRSV, brucellosis and leptospirosis. These pathogens spread to increasing numbers of sows in the herd. There is, however, a second group of pathogens which can be described as opportunist invaders which cause embryo mortality or abortion in individual sows and sporadically in small groups of sows. They do not spread through the herd like PRRSV, *Brucella* or leptospira. They are often mixed infections (i.e. several different species of pathogen are involved).

If, however, these opportunist bacterial infections occur in sufficient numbers of sows they can become a herd problem. They are often normal inhabitants of the vagina or the boar's prepuce and their identification following pathological examination needs careful interpretation. When high numbers of bacteria are deposited into the anterior vagina by the boar, particularly towards the end of the heat period, these bacteria can set up a vulval discharge. Such bacteria include klebsiella, streptococci, staphylococci and possibly leptospira. In such cases careful clinical examination of sows between 14 and 21 days post mating will sometimes reveal a tacky discharge on the vulva, which may not necessarily be very obvious. Such sows should be identified, and if they are returning out of cycle, it is likely that embryo loss is taking place.

Low grade or chronic infections such as cystitis (infection of the bladder) and nephritis (infection of the kidneys) occasionally result in abortion from ascending infection through the cervix.

FACTORS THAT MAY LEAD TO ABORTION IN THE SOW

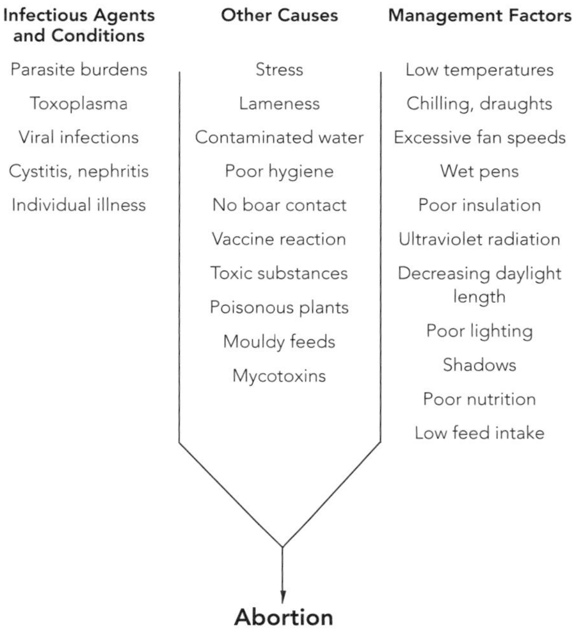

Fig 6-17

Lameness and pain, particularly coming from abscesses in the feet, or leg weakness (osteochondrosis) can also cause the corpus luteum to regress due to stress. Bullying and fighting are often forgotten as predisposing factors in individual sows. Clinical examinations and study of records are important tools for investigation. The stockperson's opinions and observations are often invaluable.

A checklist for non-infectious causes of abortion is shown in Fig 5-31 (Chapter 5). If you can satisfy all these criteria you will not have an environmentally induced problem. Fig 5-30 collects together all the precipitating management factors that may lead to abortion.

FIG 6-18: PATHOGENS/DISEASES THAT CAUSE ABORTION
Adenoviruses
African swine fever virus (ASF)
+ Actinobacillus pleuropneumonia (App)
Aujeszky's disease/pseudorabies virus (AD/PRV)
° Blue eye disease virus
Bovine virus diarrhoea (BVD)
* *Brucella suis*
Campylobacter
Chlamydia
* Classical swine fever (CSF)
E.coli
Encephalomyocarditis virus (EMCV)
* Erysipelas
Foot-and-mouth disease (FMD)
Klebsiella
* Leptospira
Mycoplasma suis
Pasteurella
Porcine parvovirus (PPV) (uncommon)
** Porcine reproductive and respiratory syndrome virus (PRRSV)
Pseudomonas
Staphylococci
Streptococci
Swine influenza virus (SIV)
Toxoplasma
Vesicular stomatitis

+ In severe outbreaks in naive herds.
° Mexico only.
* In countries in which they occur.
** Severe

7 MANAGING HEALTH IN THE GESTATION/DRY PERIOD

7 MANAGING HEALTH IN THE GESTATION/DRY PERIOD

General Clinical Signs of Disease

During the gestation/dry period diseases may appear as individual cases or in outbreaks involving a number of animals. Pathogens may attack more than 1 system of the body and the clinical signs then relate to the failure of that particular system and can include the following:

The digestive system

- Anaemia.
- Blood or mucus in faeces.
- Colic.
- Change in the faecal consistency – constipation, diarrhoea.
- Dehydration.
- Distension of the abdomen.
- Inappetence.
- Salivation.
- Variable body condition.
- Vomiting.

The locomotor system

- Abscesses.
- Fractured bones.
- Incoordination.
- Lameness.
- Paralysis.
- Swellings of joints, muscles and tendons.
- Trembling.

The nervous system

- Blindness.
- Fits and convulsions.
- Loss of balance, middle ear infections.
- Loss of leg function.
- Muscular tremors.
- Nystagmus (jerky eye movements), meningitis.
- Paraplegia/paralysis.

The respiratory system

- Coughing.
- Discharges from the eyes.
- Discharges from the nose.
- Heavy breathing – pneumonia.
- Sneezing.

The skin

- Abscesses.
- Cuts and bruises.
- Dermatitis (inflammation).
- Discolouration.
- Diamond markings. Erysipelas lesions.
- Excessive hair growth.
- Fly bites.
- Greasy skin.
- Haemorrhage.
- Jaundice.
- Lice.
- Mange. Scratching.
- Paleness – anaemia.
- Rodent bites.

The urogenital system

- Abnormal discharges from the vulva.
- Abnormal oestrus.
- Abnormal urine colour.
- Abortion.
- Blood, mucus, or pus in the urine.
- Crystals on the vulva.
- Fevers.
- Mucus indicative of an early abortion.
- Oestrus, anoestrus.
- Pregnancy failures.
- Wet, urine-stained vulva.
- Lameness and hunched back.

Management and Health

To achieve profitable reproductive performance, breeding males and females must be kept in good health. Each section presented below is part of the overall strategy required to achieve this. Compare the efficiency of your herd against the points raised.

Records

If productivity is to be increased it is necessary to identify those areas during the pre-mating and pregnancy period that are either inefficiently managed or adversely affected by disease. Disease and farrowing rate loss recording sheets are described in Chapter 5 and they are important aids to this identification process. Records will identify why reproduction failures have occurred and also provide information about culling and mortality. It is vital that records are accurate. Culling a sow in the 16th week of gestation because you have just found her not pregnant but recording this as a lameness cull is not helpful when investigating reasons for elevated "failure to farrow" statistics.

The purpose of the gestation area is to ensure that all farrowing places are filled batch after batch and the records should ensure they reflect this need. A 90 % farrowing rate is no good if there are not enough sows in pig to fill every farrowing place in the batch.

The records should include information on:

- Anoestrus.
- Culling policies relative to the slaughter price of the sow.
- Haemorrhage problems.
- Infertility including repeats, sows not in pig, discharges etc.
- Lameness.
- Mastitis, agalactia and udder oedema.
- Mortality and its specific causes.
- Poor conformation and its effects.
- Prolapse of the vagina or rectum.
- Savaging.
- Sows with poor litters.
- Specific diseases.
- The age profile of the herd.

Compare your records to the sow disposal targets in Fig 7-1.

FIG 7-1: TARGET FIGURES FOR SOW DISPOSAL

Reason	Percentage *
Abortion	1
Agalactia/Mastitis	1
Age only	8
Disease	4
Infertility	15
Lameness	5
Mortality	4
Miscellaneous	2
Poor production	5
Total	**45**

* Percentage of the sow herd per annum.

Reasons for Sow Disposal

Up to 50 % of all sows culled from a herd are usually associated with some form of reproductive failure or poor performance. Lameness and mortality also may be significant causes of failure. By collecting and studying such information the excessive areas of loss are identified. When farrowing rates are less than 85 % it should be possible to identify significant losses and take the corrective action. It has always been a point of debate how old the sow should be before she leaves the herd but the following should be taken into account when making a decision:

- The batch breeding target has been met.
- All batch farrowing places will be filled in the batch.
- Age and its effects on the individual sows' production.
- Individual record and history.
- Maintaining the mating programme.
- The availability of gilts.
- The performance of the sow compared to that of a gilt.
- Health of the sow – e.g. arthritis.
- Body condition.

With increasing age the performance of the sow will decrease, as shown in Fig 7-2. On most farms sows should be culled after they have had between 6 and 10 litters, the majority after the sixth pregnancy. But ensure that the batch farrowing place is (or will be) filled before making any culling decision.

FIG 7-2: PROBLEMS THAT MAY OCCUR AS THE SOW GETS OLDER

Agalactia.
Increases in smaller and larger sized litters.
Increased stillbirth levels.
Increased endometritis.
Increased levels of disease, particularly cystitis and pyelonephritis.
Increased mortality.
Increased number of blind teats.
Increases in prolapses.
Mastitis.
More fostering than necessary.
More lameness and arthritis.
More pigs laid on.
Poor milking.
Reduced farrowing rates.
Reduced fertility.
Variable size and weight of piglets at birth.

Reasons for Keeping Sows Beyond 6 Litters

Using today's sophisticated computer programmes it is easy to assess farrowing rate, litter size and pigs reared by pregnancy for each sow and the herd as a whole. This assists significantly in the decision to retain a sow for further breeding. Reasons include:

- To ensure that all batch farrowing places are filled.
- To maintain the mating programme. Losses here cannot be recovered.
- Insufficient gilts are available that batch for mating.
- Records show that in this herd, performance in parities 6, 7 or 8 is acceptable.
- Records show that the numbers born in older sows compare favourably with the mean of gilts and second litter females.
- Individual sow records show which sows have a good breeding and rearing history.
- Records show farrowing rates in 7th parity sows are above 86 %.

Where possible, mate all sows after weaning and only cull extra sows after the batch pregnancy rate has been exceeded. This would normally be after pregnancy checking at 28 days of gestation.

Reasons for Not Keeping Sows Beyond 6 Litters

- Variable litter size.
- Variable piglet size at weaning.
- Variable birth weights within the litter.
- A history of agalactia or poor milking.
- Chronic mastitis.
- Vulval discharges.
- Poor genetic potential.
- Other problem areas identified with increasing age.
- Poor fertility.

Despite these problems, it is better to farrow an older sow than have any empty farrowing places.

In indoor herds, levels of sows culled per annum should be between 38 and 45 % but in the outdoor herd this often reaches 50 %.

Key Points to Maintaining Longevity in the Breeding Female

- Use the correct female, that is, one expressing maximum hybrid vigour.
- Select or purchase females with sound strong legs, good teats and not too heavy hams.
- Avoid gilts showing any signs of leg weakness, e.g. standing on their toes, the back legs tucked under, the front legs bent or the pasterns dropped.
- Avoid allowing sows to develop ulcers on their lateral toes and long claws as these features are strongly associated with culling in early parities.
- Do not serve the gilt at less than 220 days of age. Gilts served too early will still be maturing into their second pregnancy. Equally do not serve too late (300 days) otherwise body size will increase.
- Breed at 130 to 140 kg (290 to 310 lbs) liveweight.
- Do not serve maiden gilts or first litter gilts with heavy boars. This can precipitate leg problems.
- Do not cull a sow if her first 2 litters have been poor. Review management procedures.
- Do not breed from a female that is too lean. Maintain at least 17 mm of fat at the P2 measurement by point of mating, particularly in the gilt.
- Do not allow excessive bodyweight to develop during the first pregnancy.
- Provide the pregnant gilt with exercise during the first half of the pregnancy if possible.
- Good nutrition is vital.
- Avoid mouldy feeds.
- Feed the sow to appetite from 3 days post farrowing, during lactation, and to point of mating.
- Identify sick or lame sows early and remove to a well-bedded hospital pen. Many will simply recover in a better environment.
- Breed all return gilts by AI and a boar to maximise their chances of getting pregnant.

Remember that management, feeding, housing design and a comfortable, well-lit environment are under your control and they will have a major effect on the viability, health and production of the sow.

Nutrition and Feeding

It is beyond the scope of this book to discuss nutrition in detail, but you should keep in mind that the quality of the feeds and the way in which they are fed are important in the management of disease. Aspects relating to nutrition in the lactating sow are dealt with in Chapter 8.

In the gestation/dry sow on the day of weaning, three quarters of the daily lactation feed should be given prior to actual removal of the sow from the farrowing house and the remainder given later on that day. The sow should be given the opportunity to eat at least half the amount of food on the 3 days postweaning as she ate during late lactation. This will ensure that she does not become catabolic (a negative energy state) with an extended weaning to mating interval and subsequent infertility. Such diets should contain net energy (NE) of at least 9.7 MJ NE/kg (14 MJ DE/kg), 16 % protein and 1.0 to 1.1 % lysine, particularly if a lean genotype is being used. It is advantageous to feed the sow to body condition using a gestation/dry sow ration over the next 21 days with 3 kg or more per day. Treat a thin sow (1 or 2 body score) as an emergency and feed accordingly. Consider restricting feed availability in gilts for 14 days post mating. She should already be in an excellent body condition at this point.

Many farms feed separate lactation and gestation/dry sow rations, the latter with net energy ranging from 9 to 9.5 MJ NE/kg (13 to 13.6 MJ DE/kg) and protein levels of 13 to 14 %. It is beneficial from breeding to 21 days post mating to optimise feed intake relative to the sows' demands, since this will satisfy the nutritional requirements for the development of the placenta. This is pertinent where farms have variable weights and quality of piglets at birth. The development of the placenta in the first 14 to 25 days post mating helps to determine the quality of the piglet at the end of the pregnancy. A small placenta will contribute towards a small pig. The availability and intake of feed in this respect is important, for example where sows are fed in groups. If there is a shortage of trough space, nutritional insufficiencies can occur in the under-privileged females. During the pregnancy period the sow should be fed to body condition but also to satisfy the environmental needs. Thus it is difficult to lay down specific levels of feed intake per day, they must be determined by the stockperson. The body condition of the sow can be assessed on a numerical rating of 1 to 5 and to do this the flat of the hand should be placed over the back bone just forward of the root of the tail and rolled laterally side to side. The condition of the sow can then be scored (see Fig 7-3).

As a guideline, sows approaching the point of weaning should score around 3. This should rise to 3½ in older sows by the time of the next farrowing. Sows scoring 2½ or less are moving into a problem area. If more than 5 % of sows at any 1 time score 2½ or less then feeding levels are wrong. If sows are allowed to farrow with a score of less than 3 there will be insufficient fat reserves to maintain lactation and they will use muscle as a source of energy. Such sows are then in danger of developing shoulder sores, being unable to maintain body condition with the demands of lactation. But fat sows will not eat in lactation so it is vital that the sow body score is well monitored.

In outdoor herds it is important that all sows should score over 3 before the onset of winter. It is difficult to improve body condition in cold weather. Feed levels should be higher during winter than in spring or summer.

Ideally the farrowing stockpeople should set and monitor the gestating/dry sow feed curve. These staff are in the best position to do this as they feed the lactating sow and have to cope with any problems which may occur because of poor feeding routines.

Housing

See Chapter 16 for further information

From weaning to mating the indoor sow can be housed in a number of ways, but this is dictated to some extent by the welfare regulations of your country. Where loose-housing systems are used on the day of weaning groups of sows are best mixed into a large yard with a minimum of 2.7 m² per sow together with a large old boar. The boar's presence will have 2 effects, 1 of stimulating oestrus and the other of reducing any fighting that is likely to occur. The grouped sows should be fed ad lib

ASSESSING BODY CONDITION IN SOWS

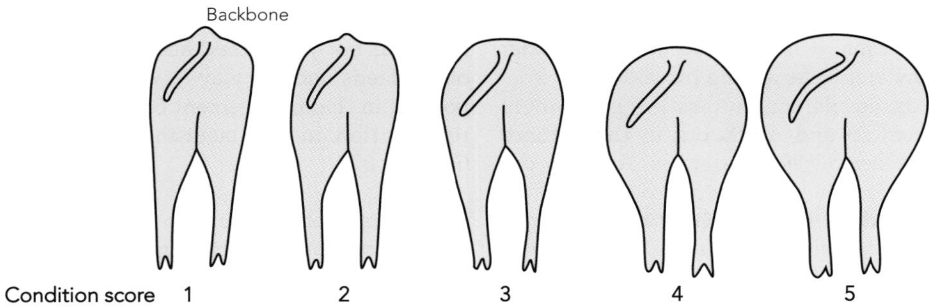

Score:
1. Emaciated sow, backbone very prominent.
2. Thin sow, backbone prominent.
3. Ideal condition during lactation and at weaning. Backbone just palpable.
4. Slightly overweight sow. Cannot find the backbone.
5. Over-fat sow, body rotund with flat back and sides.

Fig 7-3

from weaning to mating, from at least 2 well separated feeders. The boar should be removed from the group on Day 3 and where possible, sows should be confined to stalls over the next 48 hours during the period of oestrus. This is a welfare-friendly act because it reduces riding and stress as sows come into heat.

Where sows are confined they should remain in their stalls until 28 days post mating.

Keeping sows in stalls postweaning is not allowed in a number of countries and in these countries the sows must be returned to the loose-housed group after mating.

Loose-housed sows should be returned to the group as soon as mating is completed and a boar re-introduced until at least 28 days post mating. It is very important that the floor surfaces are dry and non-slip, the concrete not worn and there are no protrusions from sharp aggregate, especially in the mating pen. Slatted floors can be used for housing sows in the immediate postweaning period but should not be used in the mating or boar pen because they produce lameness and discomfort at mating. Remember the weight on the sows' hind legs and feet is multiplied 3-fold during natural mating and slats increase the risk of damage.

Ideally combine the mating and gestation area to minimise the movement and potential mixing of sows throughout gestation.

Hygiene

A clean environment is important from weaning to at least 10 days post mating. In the first 2 or 3 days postweaning the mammary gland is full of milk creating pressure which may predispose to mastitis. Contamination of the teat end with faeces and urine greatly increases this risk. The housing should be washed and dried between each batch of sows weaned. If this is not possible, it should be done at least once a month. If the sows are confined individually, the backs of the pens are best slatted to allow urine to drain away. Faeces should be removed daily. The rear gates should have a 100 mm gap above the floor so that faeces are squeezed out behind the sow and not allowed to build up and cause heavy bacterial contamination of the vulva. If solid floors are used they should have a fall of at least 1:20 in the last 450 mm behind the sow to allow for good drainage. Alternatively the pen should be well-bedded with straw daily. In loose-housing, the pens should be well drained and a deep bed of fresh clean straw used daily.

The vagina of the sow is susceptible to ascending infection in the immediate post farrowing period and for 2 to 4 days post-oestrus.

Temperature

The sow at weaning is removed from a comfortable warm farrowing house at around 20 °C (68 °F) into what is probably one of the coldest parts of the gestation/dry sow house, because there is such a low density of pigs. The environment should be controlled at an ambient temperature of 20 °C (68 °F) from weaning to 28 days post mating and, depending on the external temperature, this may involve the provision of heat or some form of evaporative cooling. This helps to maintain an anabolic (positive) energy status and provide a level of comfort necessary to achieve good reproductive performance.

Ventilation

Make sure that air flow in the mating area does not cause evaporative cooling to the animal during the critical period. A high air flow can lower the effective temperature for the sow by some 10 °C (50 °F) or more.

Water

See Chapter 14 for further information

A water flow rate of at least 2 litres per minute should be available. A commonly undetected fault is that drinkers often fall well short of this. If sows are provided with water in a trough it should have a depth of at least 3 cm (1 inch). A failure of water provision predisposes the sow to urinary tract problems.

Light

See Chapter 5 for further information

A minimum of 12 hours of continuous light followed by a period of darkness should be provided each day, both in the mating and gestation areas.

Stockmanship

Reproductive performance and litter size may be depressed by up to 6 % on farms where there is a lack of empathy between the stockperson and the pigs. This can be assessed by the way sows and boars approach people or withdraw. A calm atmosphere between the person and the pig, with a gentle pat here and there and a quiet voice, pays handsomely. The sow is a sociable animal, similar to the dog, a fact not always appreciated.

Stress

Stress is created whenever the sow has to respond to a noxious stimuli. Thus anything that causes fear or discomfort (a negative response) causes the adrenal gland to produce hormones called corticosteroids which depress the normal protective or immune mechanisms so that the animal becomes more susceptible to disease.

The Boar

The boar is the most important animal on the farm and good management is essential to maintain health and maximise normal reproductive function. When the young boar first enters the farm, allow him to make physical but not intimate contact with female pigs. He certainly must not be bullied by sows. Unless this physical contact takes place there is a risk of low sexual behaviour with poor matings, and poor shortened ejaculations. Such animals are often slow to mount and serve with reduced conception, farrowing rates and litter sizes. The same principles apply to the mature boar; he should have regular contact with females and regular mating activity. This should apply to boars used for heat detection and teasing. If you find a boar is becoming less interested at mating time, change the environment into a more stimulating one by moving him to another pen (assuming there are no problems of health). Lameness, stiffness, or difficulty in rising are often pointers to early arthritis, foot lesions or leg weakness and veterinary advice and treatment can often prevent 2 or 3 months of lowered fertility. When the young boar first arrives on the farm, always manage him carefully, without any aggression, and be extremely patient during the first 2 or 3 matings. The boar pen should be a minimum of 9.3 m^2 and for best results provide him with a solid, well-bedded floor of either shavings, peat or straw. He is then less likely to develop arthritis and callus formation and become stiff and lame.

The floor surface in the mating pen is important. It should be smooth, non-slip and well drained with no projections of aggregate. The pen should be pressure-washed and disinfected at least once a month. If the floor surface becomes slippery the judicious use of dry sand can be a short term solution. Do not use too much, however, because it can have abrasive effects on the prepuce and penis. (Sandy soil is a major cause of the penis bleeding in outdoor boars.)

Artificial Insemination (AI)

The storage and handling of the AI semen is a vital component to a successful mating programme.

Fresh semen should ideally be stored at 17 °C (63 °F), however between 15 and 19 °C (59 to 66°F) is acceptable. Do not trust the AI store thermometer. A max/min thermometer read at least weekly is useful. Ideally the AI store should be alarmed and incorporated into the farm's remote monitoring system. It is absolutely vital to maintain the fresh semen at the correct temperature.

Turn the semen regularly (twice a day) to ensure that the sperm and sperm diluent have plenty of contact.

Ensure that the AI store is clean and secure.

Identifying Problems in the Gestating/Dry Sow

If you have a problem in the gestation/dry period, refer to Fig 7-4 to identify the cause and then read the relevant section. Where a condition listed in Fig 7-4 is not covered in this chapter please refer to the index.

If you cannot identify the cause consult your veterinarian.

FIG 7-4: OBSERVATIONS AND CAUSES

Anoestrus (see Chapter 5)

- Age – immaturity.
- Bullying. *
- Disease.
- Genotype.
- High temperatures > 32 °C (90 °F). *
- High stocking density. *
- Lack of contact with the boar and his pheromones – needs to be within 1 metre – saliva contact. *
- Lameness. Pain.
- Low feed intake.
- Low temperature.
- Metritis.
- Parasites.
- Poor flooring. *
- Poor light. *
- Poor management. * Attitude of stockperson. Bad external environment.
- Poor nutrition.
- Stress. *
- Sunburn. Skin damage.
- Thin sows.

Coughing

- Actinobacillus pleuropneumonia (App).
- Aujeszky's disease (AD) – Pseudorabies virus (PRV).
- Dust. *
- Lung abscesses.
- Mycoplasma (enzootic) pneumonia. *
- Parasites.
  - Ascarids (roundworms). *
  - Lung worm.
- Pasteurellosis.
- Pneumonia.
- Porcine reproductive and respiratory syndrome virus (PRRSV).
- Swine fever (Classical and African).
- Swine influenza virus (SIV). *
- Toxic gases.

FIG 7-4: OBSERVATIONS AND CAUSES CONT.

Diarrhoea
- Colitis.
- Diet changes. *
- Draughts/Chilling. *
- Ileitis.
- Over eating.
- Parasites.
- Poor nutrition/biotin deficiency.
- Porcine epidemic diarrhoea (PED).
- Salmonellosis.
- Swine dysentery (SD).
- Transmissible gastroenteritis (TGE). *

Haemorrhage: Bloody Faeces
- Gastric ulcers. *
- Ileitis.
- Parasites.

Haemorrhage: Bloody Vulva
- Cystitis/pyelonephritis. *
- Dead foetuses.
- Ruptured blood vessel.
- Trauma. *

Lameness
- Arthritis – joint infections. *
- Biotin deficiency.
- Bursitis. *
- Bush foot/foot rot. *
- Erysipelas.
- Foot-and-mouth disease.
- Fractures. *
- Laminitis.
- Leg weakness or osteochondrosis (OCD). *
- Mycoplasma arthritis. *
- Osteomalacia.
- Poisons.
- Sand crack. *
- Spinal damage.
- Torn muscle. *
- Trauma. *
- Water deprivation (salt poisoning).

Mastitis
- Poor hygiene.
- Specific bacteria.
- Wet floors.

FIG 7-4: OBSERVATIONS AND CAUSES CONT.

Nervous signs
- Aujeszky's disease/pseudorabies virus (AD/PRV).
- Brain abscess.
- Meningitis.
- Middle ear infection. *
- Poisoning.
- Streptococcal infections.
- Swine fever (Classical and African).
- Talfan, Teschen diseases.
- Stroke.
- Water deprivation (salt poisoning).

Not eating – temperature normal
- Constipation. *
- Cystitis/pyelonephritis. *
- Faulty nutrition.
- Gastric ulcers. *
- Ileitis.
- Indigestion.
- Toxic conditions/poisoning.
- Unpalatable food.
- Water deprivation (salt poisoning). *

Not eating – temperature elevated (fever)
- Actinobacillus pleuropneumonia (App).
- Acute cystitis/pyelonephritis.
- Being in heat. *
- Dead foetuses.
- Erysipelas. *
- Glässer's disease.
- Mastitis.
- Porcine reproductive and respiratory syndrome virus (PRRSV). *
- Septicaemia/toxaemia. *
- Swine influenza virus (SIV).
- Uterus infection.

Mortality (pigs found dead)
- Abdominal catastrophe. *
- Actinobacillus pleuropneumonia (App).
- Acute cystitis/pyelonephritis. *
- Acute stress.
- Anthrax.
- Cancer.
- Clostridial diseases.
- Electrocution.
- Endocardiosis.
- Erysipelas.
- Gastric ulcers. *

FIG 7-4: OBSERVATIONS AND CAUSES CONT.

Mortality (pigs found dead) cont.

- Glässer's disease.
- Heat stress.
- Ileitis.
- Porcine dermatitis and nephropathy syndrome (PDNS).
- Poisons – gas poisons – H_2S, CO, CO_2.
- Prolapse of the bladder or uterus.
- Pyrexia.
- Retained piglets (metritis).
- Septicaemia.
- Stroke.
- Swine dysentery.
- Swine fevers.
- Torsion of the intestine.
- Water deprivation (salt poisoning).

Prolapse

- Bladder.
- Cervix.
- Rectum. *
- Uterus.
- Vagina. *

Skin diseases (See Chapter 10)

- Aujeszky's disease/pseudorabies virus (AD/PRV).
- Abscesses. *
- Anaemia.
- Dermatitis/biotin deficiency.
- Erysipelas. *
- Fighting. *
- Granuloma.
- Jaundice.
- Haematoma.
- Lice.
- Mange. *
- PDNS.
- Shoulder sores. *
- Sunburn.
- Swine fever (Classical and African).
- Swine pox.
- Swine vesicular disease.
- Vulva biting and other vices. *

Thin sows (See thin sow syndrome)

- Cystitis/pyelonephritis. *
- Draughts. *
- Low energy intake.
- Low feed intake.

FIG 7-4: OBSERVATIONS AND CAUSES CONT.

Thin sows (See thin sow syndrome) cont.

- Low environmental temperature.
- *Mycoplasma suis*.
- Old age.
- Parasites.
- Poor insulation of buildings.
- Poor nutrition.

Vomiting

- Fever.
- Gastric ulcers.
- Gastritis.
- Over eating.
- Parasites.
- Poisons.
- Poor nutrition.
- Porcine epidemic diarrhoea (PED).
- Transmissible gastroenteritis (TGE).
- Torsion of the stomach and intestine.
- Fungal toxins. *

* More likely to occur.

Diseases/Pathogens and Disorders Affecting the Gestation/Dry Period

Below is a list of the key diseases that are likely to be seen in the gestation/dry period. These are covered in detail in this chapter.

- Abdominal catastrophe.
- Abortion and seasonal infertility.
- Abscesses.
- Anthrax.
- Aujeszky's disease/pseudorabies virus (AD/PRV).
- Back muscle necrosis.
- Biotin deficiency.
- Botulism.
- Brucellosis.
- Bush foot/foot rot.
- Clostridial diseases.
- Cystitis and pyelonephritis.
- Dipped shoulder (humpy back).
- Erysipelas.
- Fractures.
- Gastric ulcers.
- Glässer's disease (*Haemophilus parasuis*).
- Haematoma.
- Ileitis.
- Jaw and snout deviation.
- Lameness.
- Laminitis.
- Leg weakness (osteochondrosis – OCD).
- Leptospirosis.
- Mange.
- Mastitis.
- Meningitis.
- Muscle tearing.
- Mycoplasma arthritis (*M. hyosynoviae* infection).
- Peritonitis.
- Pneumonia.
- Porcine epidemic diarrhoea (PED).
- Porcine parvovirus (PPV).
- Porcine reproductive and respiratory syndrome virus (PRRSV).
- Porcine stress syndrome (PSS).
- Progressive atrophic rhinitis (PAR).
- Prolapse of the rectum.
- Prolapse of the vagina and cervix.
- Salmonellosis.
- Shoulder sores.
- Sow mortality.
- Streptococcal infections.
- Swine dysentery (SD).
- Swine influenza or flu (SIV).
- Thin sow syndrome.
- Vulval biting (vice).
- Vulval discharge.
- Water deprivation (salt poisoning).

Abdominal Catastrophe

This is a common cause of sudden death in growing and adult pigs. Pigs are simply found dead and bloated. Typically the condition is characterised by a torsion/twist of 1 or more of the non-intestinal abdominal contents. A twist in the abdomen is most commonly associated with the intestinal tract at the mesentery root or associated with a mesenteric tear. This is known as torsion of the intestines or twisted gut and is covered in detail in Chapter 14 under that heading.

However, occasionally the spleen, a liver lobe or lung lobe can twist (this may or may not include the various elements of the intestines) causing an "abdominal catastrophe".

Abortion and Seasonal Infertility

See Chapter 5 for information on non-infectious causes of abortion and seasonal infertility. See Chapter 6 for infectious causes of abortion.

Abscesses

See Chapter 10 for further information

Abscesses are pockets of pus that contain dead cell material and large numbers of bacteria.

They commonly arise from fighting, particularly when sows are grouped at weaning. Initially there is a break in the skin which leaves an infected scar followed by swellings beneath. Abscesses can also arise as secondary infection to other conditions such as swine pox, PRRSV, pneumonia or tail biting and if they become widespread throughout the body, the result may be emaciation followed by death or condemnation of the carcass at slaughter.

Anthrax

See Chapter 12 for further information

This is an uncommon disease of pigs in most parts of the world including the EU where it is notifiable. Anthrax should be suspected if a sow is found dead and post-mortem examination shows copious blood-tinged mucus and large haemorrhagic lymph nodes under the skin of the neck and in the abdomen.

The source of the infection in sows is usually feed containing contaminated feedstuffs from a spore-contaminated area, although in sows kept outdoors in such regions the source may be contaminated soil or other dead animals. Care should be taken in handling diseased pigs or carcasses because the disease is communicable to people.

Aujeszky's Disease/Pseudorabies Virus (AD/PRV)

See Chapter 12 for information on this disease

Aujeszky's disease (AD), also known as Pseudorabies virus (PRV) is an important disease of the pig and is caused by a herpes virus. The pathogen has been eliminated from a number of countries and regions so is not directly relevant to many pig farmers.

When it is introduced into the breeding herd for the first time, abortions and respiratory signs can be seen. Those females that are in the first third of pregnancy may absorb foetuses resulting in delayed returns to oestrus, but if infection occurs in the second or third part of pregnancy, piglets are often stillborn and mummified or weak pigs are evident at farrowing. Once the acute phase of the disease has passed a proportion of sows become carriers of the virus and shed it intermittently. Few clinical signs are then seen. Vaccination of the breeding stock is effective in preventing clinical disease. All aspects of this disease are covered in detail in Chapter 12.

Back Muscle Necrosis

Back muscle necrosis is part of the porcine stress syndrome and in affected pigs degenerative changes take place in the back muscles along each side of the spine. It is usually seen in the young growing gilt although occasionally it occurs in the adult female.

Clinical signs

The onset is sudden with severe pain in the lumbar muscles with obvious swellings. The pig is reluctant to stand and often adopts a dog-sitting position. As the name implies there are muscle changes and death (necrosis) of muscle fibres with haemorrhages into the tissues themselves. The disease is relatively uncommon. It is usually initiated by sudden movement and is sometimes seen when maiden gilts are released from confinement to outdoor accommodation.

Diagnosis

There is a history of sudden lameness associated with movement and acute pain. The pig can be made to stand with difficulty but there is no evidence of fractures. Examine the lumbar muscles carefully; they will be swollen and painful on pressure. The temperature is usually normal but may be elevated.

Similar diseases

These include:

- Acute erysipelas.
- Fractures.
- Mycoplasma arthritis.
- Leg weakness (OCD).
- Spinal damage.

Treatment

- Inject with a suitable pain killer.
- Inject with corticosteroids provided the animal is not pregnant.
- If there is a temperature, give an injection of long-acting penicillin to cover the possibility of erysipelas.

Management control and prevention

- Breed from pigs that are free of the stress gene.

Biotin Deficiency

See Chapter 14 for further information

It is often thought that because biotin is present in most nutrient sources used for pigs and is synthesised in the gut, a deficiency is unlikely. However field experience indicates that biotin-deficient conditions can (and do) occur in breeding herds world-wide.

Widespread lameness will be a constant feature, particularly in sows. Detailed examinations should be carried out on at least 15 to 20 affected animals and the nature of the changes in the hooves documented. Examinations are best made when sows and gilts are at rest. The hooves will be soft over the walls and the soles will show slight evidence of haemorrhage. Dark transverse cracks will be seen on the hoof walls. Assess trauma from poor floor surfaces as a cause.

Pig farmers experiencing poor reproductive performance in their sow herds, associated with excessive loss of hair and severe foot lesions, should evaluate the biotin content of their sow diet and consider supplemental biotin.

Botulism

See Clostridial Diseases in this chapter

Botulism is rare in the domestic pig. It is caused by the toxins of *Clostridium botulinum,* found in food and decaying plants and animals.

Clinical signs

The pig becomes paralysed, with flaccidity of muscles, blindness, salivation and death.

Diagnosis

Laboratory examination for toxins.

Treatment

- There is no treatment.

Management control and prevention

- Avoid access to rotting carcasses.

Brucellosis

See Chapter 12 for further information

This disease, which is notifiable in some countries, is caused by a bacterium, *Brucella suis,* which causes abortion, infertility and lameness. Abortions occur at any time during pregnancy and when disease first enters the herd there will be a large percentage of sows and gilts repeating 30 to 50 days after mating. In chronically infected herds the clinical signs are ill defined and not easily recognised. Diagnosis is carried out by serology and isolation of the organism. There is no reliable treatment and the disease is best eradicated either by depopulation, or testing and elimination of individuals.

Bush Foot/Foot Rot

Bush foot results from infection of the claw which becomes swollen and extremely painful around the coronary band. It arises through penetration of the sole of the foot, cracks at the sole-hoof junction, or splitting of the hoof itself. It usually occurs in 1 foot only and is more commonly seen in the hind feet, especially the larger outer claws, which carry proportionately more weight. Infection sometimes penetrates the soft tissues between the claws, and this is referred to as foot rot.

As the infection progresses inside the hoof, the claw becomes enlarged and infection and inflammation of the joint (arthritis) often develops. The condition is important because of the effect on reproductive performance of the breeding female. Foot rot involves both superficial and deep infection of the soft tissues between the claws, often caused by fusiform bacteria.

Foot pain in the boar at mating causes poor ejaculation and a shorter mating time.

Lame sows are less likely to conceive; have poor litters and are more likely to abort.

Clinical signs

The pig is very lame with a painful swollen claw. Always try and examine the feet when the animal is lying down. In most cases a swelling will be visible around the coronary band which may form an abscess and burst to the surface.

Invariably only 1 claw is involved. With foot rot, the infection will be confined to the tissues between the claws.

Lame boars often have poor fertility and produce small litters.

Diagnosis

This is based on the clinical signs described above. Bush foot has to be differentiated from other forms of trauma and infection but the painful swollen claw is obvious.

Similar diseases

These include:

- Erysipelas.
- Glässer's disease.
- Leg weakness or osteochondrosis (OCD).
- Mycoplasma arthritis.
- Trauma.

Treatment

There is a poor blood supply to the infected tissues and therefore higher dose levels of antibiotics are required for longer periods of time.

- Antibiotics which can be used, depending on the advice of your veterinarian, include:
  - Lincomycin 11 mg/kg liveweight (gives a good response).
  - Oxytetracycline 25 mg/kg liveweight.
  - Amoxycillin 15 mg/kg liveweight.
- Inject daily for 5 to 7 days. If there is no improvement in 3 days, change the antibiotic. Complete recovery may take 3 to 4 weeks.
- Anti-inflammatory injections of cortisone may be given provided the sow is not pregnant.
- An anti-inflammatory medicine such as phenylbutazone or ketoprofen may be administered either by mouth or injection.
- If there is a herd problem, a foot bath containing either 1 % formalin (only use in the open air) or 5 % copper sulphate will help. Walk the sows through once each week on 2 or 3 occasions. However, if there are dry, cracked claws in the herd, this treatment might make them worse.

Management control and prevention

- Badly worn floor surfaces predispose.
- Sharp flint aggregates in concrete predispose.
- Pay particular attention to floors in boar pens, mating pens and loose sow housing.
- Check the quality of the floor surface around drinkers and feeders – particularly concrete slats.
- Use straw or shavings as bedding if practicable.
- Check the biotin levels in the diet.
- Wash and disinfect concrete surfaces regularly.

Clostridial Diseases

See Chapter 8 for other clostridial diseases that affect the piglet

Clostridia are large rod-shaped bacteria that also form spores and they persist in the environment for long periods. There are a number of different types. Disease in sows is associated with *C. novyi*, *C. chauvoei* and *C. septicum*. All these organisms produce toxins that may rapidly kill the host in a short period of time. The toxins are the main cause of disease, not the bacteria, but treatment must be given to prevent multiplication of the bacteria. The organism may enter the body through damage to the skin and underlying tissues and muscles. *C. novyi* spores also get carried from their normal habitat, the gut, to the liver, where they may lie latent and inactive for long periods. Clostridia are also responsible for rapid decomposition of the body after death. Whenever sow mortality in the herd is more than 4 %, death due to this disease should be considered.

Clinical signs

C. septicum, often accompanied by the other 2 types, causes malignant oedema, a form of gangrene, which is characterised by the appearance of painful and discoloured swellings. Fluid and gas are often present in the tissues. The most common disease in sows is associated with *C. novyi* which causes sudden death. This occurs when some unknown event cuts off the blood supply to an area of the liver, which provides a perfect medium for the *C. novyi* spores to vegetate and rapidly multiply producing toxins. These severely damage the liver and kill the sow. The course of the disease is extremely short and often the only sign is the finding of a good sow dead. A characteristic feature is the very rapid post-mortem changes, particularly in the liver, which is full of gas and turns a chocolate colour. Gas bubbles may also be present throughout the carcass.

Clostridial mastitis is rare.

Similar diseases

Anthrax must also be considered as a possibility and veterinary advice sought.

Diagnosis

It is most important that post-mortem examinations are carried out as soon after death as possible (within 2 hours) because of the difficulty of differentiating between the lesions caused by *C. novyi* infection in the live pig and post-mortem changes.

Diagnosis is made by examining impression smears made on to a glass slide prepared from a cut surface of the liver and a fluorescent antibody test is carried out to identify the species.

Treatment

This disease can be a major problem in outdoor pigs, and vaccination using either specific porcine vaccines or sheep vaccines containing the 3 clostridial organisms should be carried out. Two doses are given 3 to 6 weeks apart with a booster vaccination being given at each weaning time.

Vaccines are commercially available but it may be 6 to 8 weeks before sows are fully immune.

- Clostridia are very sensitive to penicillin. In-feed medication using 200 grams to the tonne of phenoxymethyl penicillin can be used for 3 to 4 weeks to control acute outbreaks whilst a vaccination programme is established.
- Long-acting injections of penicillin given in anticipation of disease may help in the short term.

Management control and prevention

- Vaccinate all breeding stock and if necessary repeat the vaccination every 6 months.

Cystitis and Pyelonephritis

This is an important cause of mortality in all ages of gestating/dry sows. Where disease is widespread, total sow mortalities can often exceed 12 % per annum. If a herd has an annual mortality of more than 5 %, some sows may be dying from unrecognised cystitis/ pyelonephritis. In such herds post-mortem examinations of all sows that have died without apparent cause should be carried out. Urine passes from the kidneys down the 2 tubes (ureters) to the bladder where they enter and continue along the surface for approximately 30 to 40 mm as straw-like structures called the ureteric valves. As the bladder fills up, pressure on them stops the urine being squeezed back towards the kidneys. In diseased sows, the ureteric valves are often shortened from their normal length to as little as 10 mm and if cystitis is present the bacteria can reflux back to the kidneys producing a very severe reaction. This causes the kidney function to cease (renal shunt) and death results in a matter of a few hours. Cystitis is inflammation of the bladder and pyelonephritis inflammation of the kidney and the 2 are often concurrent, particularly as sows get older. Occasionally it may be seen in gilts although this is uncommon unless there has been gross and prolonged faecal contamination of the vulva.

The bacteria associated with cystitis include *E. coli,* streptococci and in particular an organism called *Actinobaculum suis*. This latter organism is a common inhabitant of the preputial sac of the boar and occasionally the vagina of the sow. It has the ability to adhere specifically to the lining of the bladder and urinary tract and does not get easily flushed out with the flow of urine. It is thought that *A. suis* and sometimes *E. coli* may be deposited into the vagina at mating, and when certain conditions prevail (which are not clearly understood) the organism can ascend the urethra and gain access to the bladder. Reproductive failure is not associated with this disease unless the sow is ill and as a consequence either dies or aborts.

Clinical signs

Acute disease

The sow appears very ill and off her food with the mucous membranes of the eye injected and red. The area around the vulva is wet and soiled with evidence of blood and pus in the urine. Sows showing these clinical signs often die, or there will be a poor response to treatment with chronic cystitis developing. Disease can be so acute that death is the only sign. Post-mortem examinations will identify disease. It is more common in the first 21 days post mating because the urine of the sow becomes alkaline and both *A. suis* and *E. coli* will survive and multiply in alkaline urine.

Chronic disease

When nephritis caused by *A. suis* is present, the disease is usually rapidly fatal, but when cystitis occurs alone without progressing to nephritis the disease may be prolonged and not fatal. In these cases the appetite and the general condition of the sow can be normal, the only clinical signs being pus in the urine or a slight discharge clinging to the vulva. This should be distinguished from endometritis and vaginitis. (See Chapter 6: Endometritis and the Vulval Discharge Syndrome.)

Diagnosis

This is best carried out by post-mortem examination. In the live animal, diagnosis is based upon clinical signs and evidence of blood and pus in the urine. Urine can be tested for the presence of blood or protein, and the pH (acidity or alkalinity) by using paper strip tests. Urine can be collected in clean receptacles, especially if sows are made to stand 2 to 3 hours after feeding when they tend to urinate. Affected animals show evidence of blood and protein in the urine and a pH of 7 or more. (Normal urine is slightly acidic, that is, less than pH 7). Sows showing a pH of 8 or more have up to a 30 % chance of dying in their next pregnancy.

Similar diseases

Cystitis can be confused with a vulval discharge which comes from the vagina or the uterus and is usually of a salad cream consistency, whereas that from the urine contains pus and blood.

Treatment

- Antibiotic treatment is used to destroy the incriminating bacteria but it must be excreted in the urine.
- Lincomycin is effective at a dose level of 10 mg/kg. This medicine is active against *A. suis*.
- A more broad spectrum antibiotic may be required if coliforms or other bacteria are involved. In such cases either ampicillin or amoxycillin at 10 to 15 mg/kg should be given daily for 4 to 5 days.
- On a herd basis, treatment is best carried out using either CTC or OTC at levels of 600 g/tonne for a period of 14 days. It may be necessary to repeat this treatment every 4 to 6 weeks.
- An alternative method is to inject the sow at weaning or at mating with a long-acting single injection of penicillin or amoxycillin.
- Sows could also be medicated from weaning to 21 days post mating, the most susceptible period, by top-dressing with in-feed supplements. The dose used is based on the assumption that the sow will eat 2.5 kg of feed per day during this period and the amounts of top-dressing should be calculated on the basis of 600 g of active antibiotic to the tonne. In most cases, using a 10 % premix, this will be between 15 and 20 g of premix powder per day. In a herd with major problems, the prepuces of all the boars should be swabbed and forwarded to a laboratory to determine the isolation rate of *A. suis* bacteria and its antibiotic sensitivity. In a normal herd the isolation rate would be less than 30 % but in a diseased herd this can approach 100 %. In such cases attention to hygiene and management in the boar pens is required. Wet, poorly drained floors accumulate urine and this is an ideal environment for *A. suis* to multiply in. Avoid shavings or sawdust as bedding.
- Antibiotic mastitis tubes or liquid antibiotics can be instilled into the prepuce daily for 5 days to reduce the weight of infection.

Management control and prevention

The sow mortality rate from nephritis may become unacceptably high, particularly in herds in which the sows are in individual confinement throughout their breeding lives. The overriding reason for this is low water intake and infrequent urination. If the houses are (as they should be) comfortable, dry and free from draughts, if the ambient temperature is constant day and night, if the sows are fed a satisfying, balanced diet once a day and if they are otherwise undisturbed, most will tend to lie down for very long periods. They can develop a state of what is termed "passive withdrawal" or '"self-narcosis", and become too lazy to stand up to drink and urinate. The bladder is not flushed out regularly and fills with thick, turbid urine and salts. The female urethra (i.e. the tube leading from the bladder to the vagina) is short and negative pressures may result in faecal matter and other contaminated material being sucked up into the bladder, predisposing to cystitis. Contamination of the urethra and hence the bladder may also occur at mating. The nephritic organism, *A. suis*, is able to adhere to the lining of the urinary tract and to multiply there and it is not flushed out when the sow urinates.

Cystitis/pyelonephritis problems arise in a herd where there is a poor water supply or sows have a restricted water intake. If the disease is present as a herd problem check the points below to identify the problem areas:

- Ensure a good supply of clean, fresh water and always check the quality if there is any doubt.
- Feed sows in confinement twice daily to encourage animals to rise. Give the sows water at the same time.
- If sows are fed in a continuous trough always put a small amount of water in first before the feed to encourage intake.
- Wherever possible, but particularly in loose-housed sows, use water troughs rather than nipple drinkers.
- If once-a-day feeding is practised in confinement or in loose-housings, either convert to a twice daily feeding system or scatter small amounts of feed into the trough at watering time.
- Feed once a day in the evening – so the sows are up and urinating during the normal morning work activities.
- If there is a herd problem, look carefully behind each sow daily to identify any showing clinical signs of cystitis. Treat these and check the urine as described under diagnosis. Sows that are showing blood, protein or high pH in their urine are best culled.
- Check that drinkers are at the correct height and provide easy access. The ideal height for sows is 800 mm (31 inches). In loose-housing, provide 1 nipple drinker per 10 to 15 sows with the water flow rate of 2 litres per minute. Alternatively provide water troughs with a minimum depth of 30 mm (1 inch).
- Check the water supply daily.
- In herds in which the sows are kept in individual confinement and in which there is a high mortality from nephritis, the most important preventative factors are to induce the sows to drink and urinate more frequently than they might otherwise do. They may be induced to drink more by increasing the salt in the ration to 0.9 %. Walking a boar in front of the sow pens daily will encourage them to stand and urinate. These actions alone usually result in a gradual decline in nephritis deaths.
- Reduce or prevent contamination of the vulva with faeces, particularly from weaning to 21 days post mating. This occurs in pens when solid back boards drop down to the ground level. There should be a 100 mm (4 inch) gap between the bottom of the board and the floor to prevent faeces building up behind the sow.
- Sows that are too big for the pens often adopt a dog-sitting position with the vulva becoming heavily contaminated, allowing excessive bacterial multiplication.
- Badly drained boar and sow pens increase the risk of infection.
- The disease is more common in herds that have high numbers of old sows and boars.
- If group housing is used at weaning make sure the pens are well drained. Do not use sawdust for bedding, and wash and disinfect pens regularly.
- Do not handle the prepuce at mating. Squeezing the preputial sac increases the bacterial load transmitted to the vagina, (which may also result in increased returns to service).
- Always wear gloves at mating time to prevent the spread of the *A. suis* infection.
- Treat the boars' prepuces with antibiotic to reduce levels of *A. suis*.

Eradication

It is impossible to eradicate the organisms associated with this disease. They are present in every herd.

Dipped Shoulder (Humpy Back)

This is a not uncommon condition seen in both gilts and boars up to 6 months of age. Mild forms can sometimes be seen in individual mature boars in different herds. It has been described in pigs as young as 3 to 4 weeks of age where there may be a genetic predisposition.

As the animal approaches maturity its body changes shape and it starts to give the appearance of 2 different pigs joined together at the middle. The back above and behind the ribs becomes increasingly dipped and the back above the middle and rear abdomen becomes humped. The condition is abnormal and unsightly but, in the absence of other diseases, the pigs remain healthy and normal in every other way.

In some cases it is a developmental condition and probably environmental although what factors in the environment cause it are not known. The affected farms may have a significant proportion of their growing breeding stock with it but closely related breeding stock on other farms all develop normally. In herds where the condition is common there may be a boar conformation involvement. Record and observe progeny from suspect boars. It has also been associated with excessive riding of boars and back damage. Early onset of puberty has been implicated in boars.

In severe cases partial or complete paralysis may be seen. Such animals should be destroyed. Mortality is usually low. Administer pain relief to any suffering animal.

Erysipelas

Swine erysipelas is caused by a bacterium, *Erysipelothrix rhusiopathiae* that is found in most (if not all) pig farms with up to 50 % of intensively raised pigs considered to host the organism. It is always present in either the pig or in the environment because it is excreted via saliva, faeces or urine and is impossible to eliminate from a herd. *Erysipelothrix rhusiopathiae* is also found in many other species, including sheep, poultry, fish and birds and can survive outside the pig for a few weeks and for up to 6 months in light soils. Infected faeces is probably the main source of infection, particularly in growing and finishing pens. Disease is relatively uncommon in pigs under 8 to 12 weeks of age due to protection provided by maternal antibodies from the sow via the colostrum.

This disease is zoonotic (can be transmitted to man) and causes a skin problem called erysipeloid, but this is rare.

Strains of erysipelas vary in their capacity to produce disease, ranging from very mild to very severe. Adverse environmental changes, poor nutrition, fluctuating temperatures and movement and mixing can activate the disease. The incubation period is 24 to 48 hours. Erysipelas may become a problem in herds where PRRSV is endemic.

Once a pig has been infected it will build immunity and in many cases only mild or sub-clinical disease will be seen. The organism enters the body through the tonsils, naturally occurring breaks in the integrity of the small intestine, or through wounds associated with fighting. Pigs can become infected and show no signs of clinical disease.

Clinical signs

Peracute

Erysipelas can rapidly kill pigs, and in an outbreak the first clinical sign may be a dead pig.

Acute disease

The organism gains entry into the bloodstream and multiplies rapidly, causing a septicaemia. The onset is usually sudden and sometimes the disease will progress so rapidly that the first thing seen is a dead pig, although this is seen rarely in sucking pigs.

A consistent feature of the disease is a very high temperature, from 41 to 42°C (105 to 108°F). Some sows may appear very sick while others appear relatively normal. Stiffness and discomfort when walking and a reluctance to rise are common, indicating joint infection. Over the first 24 hours the organisms clump together and block the small blood vessels, particularly those beneath the skin over the back and sides of the body, causing thrombosis. These areas of skin then have inadequate blood supply. The restricted blood supply causes small raised areas called diamonds. These become red and finally black, due to dead tissue. Often these lumps can be palpated in the early stages before anything can be seen. In such cases pink to dark purple areas, often diamond shaped, develop on the skin (see Chapter 10). These can be palpated in the early stages, even before the colour changes start. These can be difficult to see on black-skinned pigs. Left untreated, these areas die/necrose and eventually slough off.

In non-vaccinated or inadequately vaccinated herds, infected sows with high temperatures may abort. In such outbreaks up to 20 % of animals in a group may be affected. Some of these animals farrow with high stillbirth rates and increased numbers of mummified pigs.

Erysipelas in the boar is a serious disease because the prolonged high temperature affects the development of sperm over its period of some 5 to 6 weeks. Thus fertility can be affected by small litter sizes and increased returns to mating at both normal and variable intervals.

Usually the disease is confined to 2 or 3 animals in any 1 outbreak, although in the non-vaccinated herd 5 to 10 % of animals could be affected at any one time.

Subacute disease

Here the animals are not as ill and the temperature is much lower, sometimes no more than 40 °C (104 °F), or even normal. In some cases the disease can be so mild as to be undetected. Some piglets may die in the uterus following subacute disease and become mummified. There may be skin lesions evident, and most animals will recover after 3 to 4 days.

Chronic disease

This may arise after acute or subacute disease, or without any other clinical signs. The organism settles in the joints, causing a chronic arthritis. There is a considerable amount of pain and loss of body condition, but more importantly, condemnations may occur at slaughter. Infection of the heart valves may result in growths (endocardiosis) and subsequent heart failure and death, for example during farrowing.

Sporadic disease is common in sows but if 1 sow in a group becomes infected the exposure is high from her urine and faeces and it is advisable to inject all contact animals with penicillin.

In the breeding herd erysipelas causes reproductive failure and signs include:

- Abortions during acute or sub-acute disease with ill sows and dead piglets.
- The death of piglets inside the uterus and mummification.
- Abortions with decomposing piglets.
- Absorption of embryos and delayed returns.
- Normal returns if infection occurs immediately post-service.
- Variable litter size.

If you have a sick sow running a high temperature, always run the flat of your hand down the sides of the body and over the ham. If you find raised lumps, think erysipelas.

Boars infected with erysipelas develop high temperatures and sperm can be affected for the complete development period of 5 to 6 weeks. Infertility is demonstrated by returns, sows not in pig and poor litter sizes.

Do not use a boar for at least 4 weeks after infection.

Diagnosis

This is determined by the clinical picture; inappetence, a very high temperature and the diamond shaped skin swellings which, if present, are diagnostic (Fig 7-5). If the diamond markings are not obvious to the eye they can be felt if the hand is run over the skin of the back or behind the back legs and over the flanks. This will assist in diagnosis. *Erysipelothrix rhusiopathiae* is easily grown in the laboratory: post-mortems and culture of the organism from the sudden deaths will confirm the diagnosis. Blood samples can be taken from the sow at the time of infection and again 2 weeks later and the antibody levels in the serum determined by the serum agglutinating inhibition test (HI). As a guide, titre levels of less than 1:60 would indicate subacute infection, low level exposure or a vaccine response. Titres of more than 1:320 would indicate recent exposure and a rising titre in tests (to say 1:640) 2 weeks apart would be indicative of disease.

Treatment

Erysipelothrix rhusiopathiae is resistant to tetracyclines but the following methods can be used to treat the disease.

- Swine erysipelas is very susceptible to penicillin, which is the medicine of choice.
- Affected sows should be treated with long-acting preparations unless disease is acute. Usually a single injection is adequate but in severe cases it is necessary to repeat this 2 to 3 days later.
- In acute cases, a quick-acting penicillin injected twice in the first 24 hours should bring about a rapid response. Continue daily injections for 3 to 4 days.
- Where a large number of sows are involved, water medication with amoxycillin or phenoxymethyl penicillin should be carried out. The dose level will depend upon the purity of the antibiotic powder used (see Chapter 4: Water Medication). Tylosin phosphate may also be administered via the water or feed to groups of at-risk animals.
- In prolonged outbreaks, in-feed medication using 200 to 300 g/tonne of phenoxymethyl penicillin for 2 weeks should control disease.
- In individual outbreaks, finishing pens should be washed and disinfected between batches. If wet feeding is implicated, the system must be cleaned out and disinfected.

Management control and prevention

- In the breeding herd, all the females and males should be vaccinated. Reasonably effective killed vaccines are available and 2 doses would normally be given 2 to 4 weeks apart. Ideally, breeding gilts should be vaccinated twice from 12 weeks of age onwards and given a third dose just prior to first mating. Re-vaccinate sows either 2 weeks prior to farrowing, or at weaning time, depending on the incidence and history of disease on the farm.
- Sows should be given a booster vaccination at each successive weaning. Killed vaccines are quite safe and have no adverse effects on the sow.
- In herds where there is a high challenge it may be necessary to re-vaccinate gilts and boars so that a third dose of vaccine is given 2 months after the second, often when the breeding animals arrive on the farm.
- Boars, likewise, should be vaccinated twice from 12 weeks of age and thereafter given a booster every 6 months. If a boar is ill with a temperature and shows skin lesions, treat immediately and do not use for mating for a minimum period of 4 weeks. Alternatively, cross

mate with boars that have no disease history or use AI.

- Birds can contaminate feed. Assess the levels of the exposure in your herd.
- In an outbreak, remember that water, faeces, nasal secretions, bedding and feed harbour the organisms.
- If disease breakdowns occur in spite of vaccination it is likely that the levels of challenge from the environment are high. Assess hygiene in breeding pens and move to an all-in/all-out method of housing.
- Wet feeding systems, particularly if milk by-products are used, can become major sources for multiplication of the organism.
- Check the water supply – fish may carry erysipelas and if the pigs are given river water to drink may contract the disease through the water supply.
- Disease is sometimes seen in vaccinated animals either because the challenge has been too great, vaccination has been missed or wrongly administered, or the strain of erysipelas is not covered by the vaccine. Efficient storage of the vaccine as per the manufacturer's recommendation is essential. If the vaccine becomes frozen it will be ineffective.

Vaccination – Prevention of this disease is easily achieved by vaccination. In continual outbreaks in growing pigs it may be necessary to vaccinate pigs after 10 weeks of age. If disease is occurring earlier than this, the age of vaccination needs to be reduced. Normally, however, pigs are not vaccinated before 8 weeks because colostrum antibodies reduce the vaccine response.

Fractures

Bone fractures are not uncommon in sows and gilts and are usually the end result of trauma and fighting, although spontaneous ones occur in bone disease such as osteomalacia, associated with calcium phosphorus and vitamins A and D, and osteochondrosis (OCD). Fractures are covered in Chapter 14. Also see Lameness in Chapters 7, 9 and 14.

Gastric Ulcers

See Chapter 14 for further information

Gastric ulceration in the sow can sometimes be difficult to diagnose.

Clinical signs vary according to the severity of the ulcer and whether it is bleeding or not. The feed intake can be variable, with occasional vomiting. If haemorrhage is occurring there will be dark-coloured faeces, the animal will have a tucked-up appearance, sometimes grinding its teeth (indicative of pain) and appear anaemic. This condition is covered in Chapter 14.

Glässer's Disease (*Haemophilus parasuis*)

See Chapter 9 for further information

This disease is caused by the bacterium *Haemophilus parasuis* (Hps), a small organism of which there are at least 15 different types. Glässer's disease normally affects the sucking and young growing pig, but can occasionally affect the adult pig. It is ubiquitous and found throughout the world but disease is rarely seen in the gestating/dry sow. However, if herds are set up using SPF or MEW techniques and are free from *H. parasuis* it can be devastating when the pigs first become contaminated, producing an anthrax-like disease with high mortality in sows.

Clinical signs include lameness in any of the legs with slight swellings over the joints and tendons. The organism has an affinity for the smooth shiny surfaces covering the joints, tendons and other tissues, including the pericardium (heart sac), lungs (pleura), intestines (peritoneum) and meninges, where it causes pericarditis, pleurisy, peritonitis and meningitis respectively, together called polyserositis. In the young gilt lameness and stiffness are the most common signs.

The definitive post-mortem findings of Glässer's disease is polyserositis, however it should be noted that polyserositis may also be associated with *Mycoplasma hyorhinis* and/or *Streptococcus suis*.

Haematoma

See Chapter 10 for further information

A haematoma is a pocket of blood that forms beneath the skin or in muscle tissue and is associated with a ruptured blood vessel. It usually arises from trauma, particularly over the shoulders, flanks or the hind quarters. Not uncommonly the ear of the sow may be damaged following fighting or head shaking and rubbing associated with mange. See Chapter 10 for details.

Ileitis

See Chapter 9 for further information

This disease is associated with changes in the intestines. Ileitis is seen in 4 forms, bloody gut or proliferative haemorrhagic enteropathy (PHE), porcine intestinal adenopathy (PIA), necrotic enteritis (NE) and regional ileitis (RI). All are uncommon in the mature female but outbreaks of PHE are not uncommon in maiden and pregnant gilts.

Jaw and Snout Deviation

This is a very common yet little recognised condition in the sow. When the jaw is at rest, a proportion of sows (often around 5 %) and particularly those housed in confinement, show a misalignment of the jaw to the left or right of centre. In extreme cases this can give the appearance of rhinitis. The condition is associated with a loss in height of the vertical part of the mandible or jaw bone and as a result the jaw swings over from one side to the other. The nose is always straight.

The vertical part of the jaw bone grows in height from the cartilage that forms part of the mandibular joint. During early growth, constant trauma from bar biting or the use of nipple drinkers interferes with the normal growth.

Occasional bending of the nose is seen where there has been infection of the bone as a result of faulty teeth-clipping in early life. Neither condition is of significance.

There is another rare condition associated particularly with sows derived from the Large White breed. It appears to result from prolonged feeding of very finely ground meal in narrow troughs which provide difficult access. The upper jaw and nose become shortened and flattened, in some cases to a grotesque degree, and the lower jaw protrudes forward several centimetres beyond the nose. There is no evidence of this condition in gilts but it gets worse with age. This is not atrophic rhinitis.

It should be remembered that teeth problems can be a significant cause of poor performance in older sows.

Lameness

See Chapters 9 and 14 for further information

There are numerous causes of lameness in the pig, but next to reproductive failure, lameness is the second most common cause of sows being culled. Most cases occur from weaning through to the point of farrowing. A lameness problem increases the culling rate, reproductive problems and the non-productive sow days, so reducing the litters and pigs weaned per sow per year. Often problems involve first parity gilts or second parity sows, just as they are reaching the most productive part of their life. Sows culled for severe lameness may have to be shot on the farm because on welfare grounds they should not be transported. Therefore they contribute significantly to the recorded sow mortality. In order to analyse a lameness problem on a breeding farm it is important to keep accurate records about each sow. These should include the following:

- Sow number.
- Parity.
- Breed and genetic line.
- Date of mating (AI or natural).
- Date of farrowing.
- Date of weaning.
- Date of lameness.
- Type of lameness.
- Housing area.

Alternatively you could use the farrowing rate loss sheet that is used in the dry/gestation period. An example is shown in Figs 5-9/3-25.

Lameness is generally triggered by tissue changes resulting from either infectious or non-infectious causes. Such tissue changes include:

Apophysitis (OCD) – Separation of the muscle mass from the growth plate on the pelvis.
Arthritis – Inflammation of 1 or more joints.
Damage to nervous tissue – Clinical signs vary (e.g. partial or complete paralysis of 1 or more limbs) depending on the site of the damage.
Epiphyseolysis (OCD) – Separation of the head of the femur.
Fractured bones – Common in the hip, hock and elbow joints.
Haematoma – Haemorrhage into the tissues.
Laminitis – Inflammation of the tissues connecting the hoof to the bone. It is not common.
Myositis – Inflammation of muscles.
Penetrated sole – Damage due to trauma.
Periostitis – Inflammation of the membrane (periosteum) which covers the bone.
Osteitis – Inflammation of bone.
Osteochondrosis – Growth plate and joint cartilage degeneration.
Osteomalacia – Softening of the bones due to calcium/phosphorus deficiency.
Osteomyelitis – Inflammation of all bone tissue including the spongy centre and bone marrow.
Osteoporosis – Weak bones which can be due to an imbalance of calcium and phosphorus in the diet.
Split horn – Poor hoof quality. Overgrown claws.
Torn ligaments or muscles – A common cause of lameness, particularly where muscles are attached to bones.

The causes of lameness in breeding animals can be separated into infectious and non-infectious. These are listed below.

Infectious causes

- Brucellosis.
- Bush foot/foot rot.
- Clostridial diseases.
- Erysipelas.
- Foot-and-mouth disease.
- Glässer's disease (*Haemophilus parasuis*).
- Mycoplasma arthritis.
- Salmonellosis.

- Swine vesicular disease .
- Streptococcal infections.

Non-infectious causes

- Fracture.s
- Laminitis.
- Leg weakness or osteochondrosis (OCD).
- Muscle tearing.
- Nutritional deficiencies.
- Porcine stress syndrome associated with the halothane gene.
- Toxic conditions.
- Trauma.

In the maiden gilt or during the first pregnancy, infectious lameness is usually due to erysipelas, Glässer's disease, mycoplasma infections and brucellosis in those countries where it is endemic. Clostridial diseases are rare in the gestating/dry sow but infections of the claws and hock areas due to trauma (bush foot/foot rot) are common causes. Foot-and-mouth disease and the vesicular diseases are discussed in Chapter 12. In such infections, a number of sows in both the gestating/dry sow area, the lactating area and indeed pigs across the unit will have varying degrees of lameness and blistering around the nose, mouth and feet. If lameness is a herd problem use Fig 7-6 to help identify the cause.

Lameness may also be caused by faulty nutrition and this is covered in more detail in Chapter 14.

Culling and lameness

Look for long main claws and long dew claws at the beginning of pregnancy and ulcers in the lower part of the feet at the end of pregnancy as these are strong indications of future culling.

Laminitis

This disease describes inflammation of the soft, highly vascular structures that connect the bone to the hoof. It is an uncommon but very painful condition causing animals to walk on their knees. The cause is unknown and affected pigs should be destroyed.

Leg Weakness (Osteochondrosis – OCD)

Leg weakness, as a result of trauma, is by far the most common cause of lameness in the gestating/dry sow from point of weaning to point of farrowing. Environmental trauma to the coronary band area and to the sole or wall of the foot results in penetration of the sensitive tissues, infection and lameness. These foot conditions are called bush foot and foot rot. Trauma however, more commonly arises indirectly from other causes within the environment that create shear forces on the muscles, tendons, bones and bone structures. Such changes associated with cartilaginous structures are referred to as leg weakness or osteochondrosis (OCD). The structure of the joint and bone is shown in Fig 7-7.

The term "leg weakness" is also used sometimes to describe poor leg conformation or describe a clinical condition associated with lameness and stiffness. It arises due to abnormal changes in the articular cartilage and the growth (epiphyseal) plates. These plates are responsible for the growth of bones, both in length and diameter. Whilst the exact mechanisms that cause these changes are not fully understood, they arise due to the pressure and shear stresses that are placed upon these rapidly growing tissues. This pressure reduces the oxygen supply, causing abnormal growth and consistency of the cartilage. Damage to the cartilage tends to be progressive and irreversible. The damaged cartilage

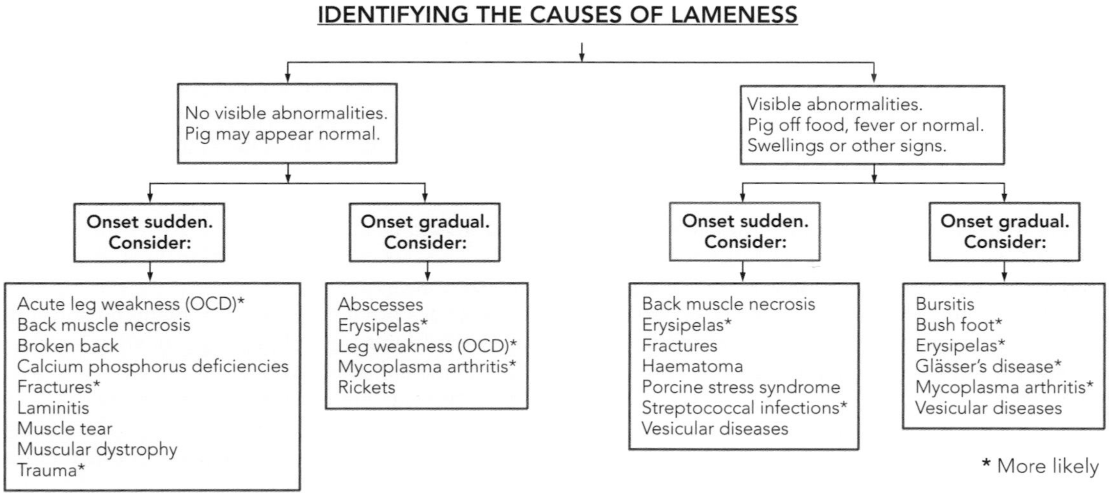

Fig 7-6

THE STRUCTURE AND DISEASES OF BONES AND JOINTS

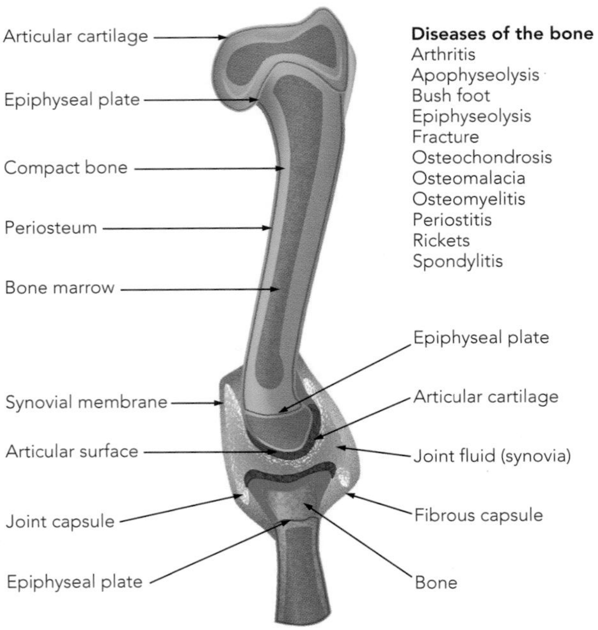

Fig 7-7

is replaced by fibrous tissue. This cartilage damage in turn produces shortening and bending of the bones near the joints and at the extremities of the long bones. Weak epiphyseal plates also have a tendency to fracture, and cartilage covering the joint surfaces splits and forms fissures. It is important to appreciate that such changes in the cartilage take place in most (if not all) modern genetic lines from as early as 2 months of age. In some cases many of these can only be detected under the microscope. It is interesting to note that such changes cannot be detected in the wild boar which takes up to 2 years to reach maturity. OCD is therefore a fact of life in commercial pig production but its severity and its effects depend largely on the environment. OCD results from the many years of selecting animals for rapid growth, large muscle mass, and efficient feed conversion and therefore much greater weight on the growth plates whilst they are still immature, together with the stresses of intensive methods of production. Conversion of cartilage to bone involves the deposition of calcium and phosphorus, and while the process of breaking down and reforming bone goes on throughout life, bone growth ceases when the sow is approximately 14 to 16 months of age.

It is not uncommon in breeding enterprises for 20 to 30 % of boars and gilts to be culled after completing the performance test, due to leg weakness and leg deformities.

The conformation of the pig is a predisposing factor in OCD. Fig 7-8 shows some of these traits.

Sows with good leg conformation show angulation of the bones at the hip, knee and hock joints. The bones below the hock slope slightly forwards and the feet are well-placed on the ground. Sows that are susceptible to leg weakness are straight-legged with little angulation of the bones between the joints, and the back tends to be arched. This alignment increases shear stresses on the growth plates.

Clinical signs

Acute disease

This is seen when there is a separation or fracture of the bones at the epiphyseal plate (epiphyseolysis) associated with sudden movement. The animal walks on 3 legs, the affected leg swinging freely. Crepitus or rubbing of the broken bones together can usually be felt. Sudden fractures can also occur in the knee and elbow joints, which are more common in the young growing pig. Fractures of the vertebrae in the spine occur particularly during lactation and immediately postweaning. In such cases the sow is in acute pain, often in a dog-sitting position with the hind legs well forward. Animals housed in farrowing places with slippery floors tend to slide the back legs forward and there is a risk of the hind muscles pulling away from their attachments to the pelvis (apophyseolysis). In such cases the sow will stand with assistance but cannot pull the hind leg backwards. When placed on to the ground it just slides forward. Such animals should be culled immediately.

Chronic disease

The onset is gradual. The pig shows abnormal leg conformation and gait with or without stiffness and pain. The temperature remains normal and joints will not be swollen unless there are fractures.

- Front legs:
  - These may be straight, with the pig walking with a long step on its toes.
  - The knees may be bent inwards or flexed which causes the pig to walk with short steps.
  - The pasterns may be dropped. This is common in old sows due to shortened bones and slack tendons.
  - The feet may be rotated or twisted.
- Hind legs:
  - These are straight with a swinging action from the hips as the pig moves. Avoid selecting such females for breeding.
  - The legs are tucked beneath the body.
  - The hocks turn inwards and are close together.
  - The pig walks with a goose-stepping action.
  - Again, in old sows the pasterns may be dropped (Fig 7-9).

GOOD CONFORMATION

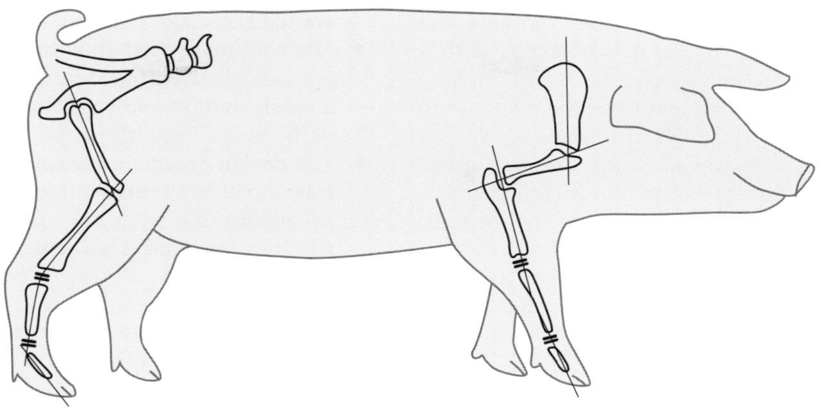

POOR CONFORMATION

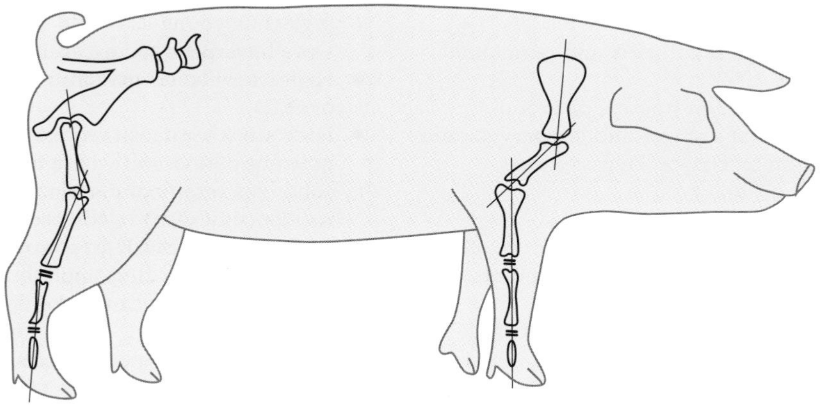

Fig 7-8

Abnormal gaits arise either from pain in the joints or abnormal movements in the hind legs from the hips, which give a swaying motion. The pain is associated with damage to the sensitive membranes around the joints resulting from either splitting or erosion of the cartilage in the joints or movement of the growth plates. Some pigs, however, may show severe clinical signs yet on post-mortem examinations the joints appear normal and vice versa. Joints may become inflamed (arthritis), particularly in the hip, knee and elbow. OCD may be seen within 3 months of gilts being introduced on to the farm, during their first pregnancy, in lactation or in the first 2 to 3 weeks postweaning.

Diagnosis

OCD is diagnosed on the clinical signs described. There are no serological or other tests and post-mortem examinations may be misleading because many pigs that are found to have joint lesions may not be lame.

LEG WEAKNESS OR OCD IN A SOW

Note the forward position of the legs and dropped pasterns.

Fig 7-9

Similar diseases

There are 2, mycoplasma arthritis (*Mycoplasma hyosynoviae* infection) and erysipelas. In both of these the disease is usually sudden in onset, sometimes with a raised body temperature and there is pain and swelling in the joints. A differentiating feature is the response to treatment. OCD does not respond to antibiotics whereas mycoplasma arthritis (*Mycoplasma hyosynoviae* infection) and erysipelas will respond within 24 to 36 hours, the former to lincomycin or tiamulin and the latter to penicillin.

Other diseases also result in lameness, such as osteomalacia, osteoporosis and rickets. See Lameness for more information.

Treatment

- There is no specific treatment for OCD; however, at an early stage the sow should be moved from its existing environment to a well-bedded pen where the foot can grip. If not, the lesions progress and ultimately arthritis and permanent lameness develop.
- Gilts that are confined in houses, or group-housed on floors that are wet and slippery should be moved as soon as clinical signs appear.

Management control and prevention

If OCD is causing a problem on your farm, check through the list below and identify those points that are important in your system.

- Determine the time and place when OCD first becomes evident.
- Look carefully at surfaces that cause the foot to slip and result in increased pressure on the growth plates.
- If the problem is in gilts, look carefully at the conformation of these animals and consider improving the selection procedure.
- Animals with a fine bone structure are more prone to leg weakness.
- Animals with heavy ham muscles and a disparity between anterior and posterior muscle masses have an increased horizontal pull on the growth plates. Such young gilts do not acclimatise well to confinement.
- If there is a problem in lactating gilts, look at the floor surfaces in the farrowing houses. Are these slippery and causing pressure on weakened bones? The use of dry sand in these pens once or twice a week may assist in the prevention of the condition until the floor surfaces can be altered.
- The modern hybrid gilt often suckles 10 or more pigs and produces a large amount of milk which is high in calcium and phosphorus. The bones are therefore more prone to calcium or phosphorus depletion and if the gilt is mixed at weaning time with older sows, damage and sudden fractures are likely to take place.
- The confinement of gilts during pregnancy can be a major contributing factor. Culling rates in such animals may rise towards 20 % if the floor surfaces are very slippery.
- The design of slats can contribute to OCD. Some slats slope to the edges from the centre and are so smooth that when the animals stand, the feet slip into the gaps. This causes constant twisting and pressure on the growth plates and a high incidence of OCD may result.
- Gilts in the first pregnancy should not be housed on slippery floor surfaces.
- High levels of vitamin A (> 20,000 iu/kg) particularly in the younger growing pigs can interfere with the normal growth of the epiphyseal plates. OCD lesions have been seen in piglets as young as 3 to 4 weeks of age where sows have been fed excessive levels of vitamin A.
- High lysine levels may also increase the incidence of OCD.
- High stocking densities, particularly in the growing period and where animals are housed on solid concrete floors or slats, will increase the incidence of OCD in the breeding stock. Such environments cause great stress on the legs and pens tend to be dirty and wet.
- Fat-sprayed diets can make floor surfaces slippery.
- Pigs treated with porcine growth hormone (this is banned in many countries) tend to develop severe lesions of OCD at an early age.
- New concrete surfaces or slats during the first few weeks of use may develop slippery surfaces after contact with faeces and urine. They should be pressure-washed with detergent on 2 or 3 occasions.
- Variations in nutrition within acceptable limits do not appear to influence OCD but conditions of osteomalacia and osteoporosis (soft and weak bones) occur where there is an imbalance of calcium and phosphorus in the diet, or excessive withdrawal of these minerals from the bones. Soft and brittle bones in such deficiencies often fracture across the mid-part of the long bones during lactation. Always check the calcium/phosphorus ratios of the diet and add additional minerals during lactation. Dicalcium bone phosphate could be used.
- Pain in the joints from OCD also gives rise to abnormal stresses on the muscles, particularly where they are attached to the bone. Tearing of these attachments, on the insides of the

shoulders and legs and from the muscle masses in the pelvis, is common, causing severe pain and periostitis.
- Smooth concrete surfaces that have been finished with a steel float can predispose to OCD.
- The addition of 0.5 % sodium bicarbonate/tonne to the ration has been shown to reduce the incidence.

Leptospirosis

See Chapter 6 for additional information

Leptospirosis can take several forms depending on the serotype involved. Over 160 serotypes are known, generally called serovars, with cross-infections occurring between some species. Each serotype has 1 or more (usually only 2 or 3) reservoir hosts. A serotype can remain as a life-long infection in its reservoir host.

Dry/gestating sows can be exposed to different strains of leptospira from the urine of rats, mice, other animals or pigs. This is an important infectious disease in the pregnant animal, causing abortions and pregnancy failures. Actual clinical illness is uncommon unless there are secondary infections and a prolonged carrier state exists.

The main serotypes that cause disease in pigs are:
- *Leptospira pomona*.
- *Leptospira tarassovi*.
- *Leptospira bratislava/muenchen* (see Chapter 6).

Leptospira hardjo is widespread in the cattle populations of the world and causes bovine abortion. It causes transient infections and antibody responses in pigs but no disease.

This section focuses mainly on *Leptospira pomona*. Chapter 6 covers *Leptospira bratislava/muenchen* in more detail.

Leptospira pomona Infection

This is enzootic in many pig-rearing parts of the world including North, Central and South America, Australia, New Zealand, South East Asia and Eastern and Central Europe. If you are involved in pig farming in any of these areas, your herd is at risk.

Porcine *L. pomona* is clinically not present in UK or Ireland pigs and is thought not to be present in other countries of Western Europe with the possible exception of Italy, although a rodent-hosted serovar may be isolated.

If you are involved in exporting pigs from countries of Western Europe, the pigs for export will probably have to be blood-tested for antibodies to *L. pomona*. You will probably be frustrated to find that a small number will be positive. This is because they have encountered the *L. mosdok* serovar, which is present in some species of wildlife, which sometimes cross-infects pigs causing little or no clinical disease. This is so closely related to *L. pomona* of pigs that it stimulates a positive immune response. It never develops into a herd problem. The pig can be thought of as an end host.

Key points:
- Once *L. pomona* establishes itself in a herd it is difficult to eliminate.
- It causes reproductive failure with consequent loss in production and income.
- It also infects people with a condition sometimes called "swine herds' disease", a severe flu-like condition, sometimes with meningitis.
- If you are involved in the supply of breeding stock from a farm that has just become infected you would be wise to stop sales.
- If you are involved in the export of breeding stock, the importing country will usually require you to either test serologically for *L. pomona* or to put all the pigs for export on a course of streptomycin to eliminate the carrier state. The problem with the former is that some other leptospira serovars that circulate in rodents and hedgehogs can cross-infect pigs, causing no disease but causing cross reactions with *L. pomona* and false positive results. The problem with the latter is that in some countries (e.g. the USA) streptomycin is not allowed to be used in pigs.

If you are working in a herd with L. pomona take care not to contaminate your skin, lips or eyes with pigs' urine. L. pomona causes a nasty disease in humans.

Clinical signs

The disease spreads slowly through the herd, causing infertility, abortions and the birth of weak, premature piglets. The sows which abort are not otherwise clinically ill. If pregnant cattle are in close contact with the pigs, some of them may also abort.

If any of the pig workers on the farm go down with flu-like symptoms and develop a bad headache, call the doctor immediately. Fortunately the disease in people responds to antibiotics if caught soon enough.

Diagnosis

This requires the help of a diagnostic laboratory. Send paired blood samples (i.e. sows sampled on the day they abort and 2 weeks later) and freshly aborted piglets. The piglets may yield a rapid diagnosis if certain tests are done. Serology on the paired samples inevitably results in at least a 2-week delay.

The laboratory is likely to carry out micro-agglutination/lysis tests to demonstrate rising antibody titres in the sows and the presence of antibodies in the piglets' serum. The latter is conclusively diagnostic. The laboratory may also try to demonstrate the leptospira in aborted piglets' tissues using fluorescent antibody tests (FATs) and sections stained with aniline dyes.

Treatment

- Medicate the feed with tetracyclines, either oxytetracycline or chlortetracycline at levels of 800 g/tonne. Feed for a period of 3 weeks followed by a further course 6 weeks later, and repeat this for 4 treatment periods.
- An initial 3-week course of tetracycline at 800 g/tonne of food, followed by a further 8-week course at 400 g/tonne of food.
- Strategic medication. Where there is a history of periodic infertility, in-feed medication can be targeted just prior to the expected time of disease.
- Inject sows at weaning time with streptomycin (if available) at 25 mg/kg. Boars should be treated with this medicine once every 6 weeks. Alternatively semi-synthetic penicillins could be used.

If sows are aborting and giving birth to weak litters, consider L. pomona. Call your veterinarian and tell your pig people to take great care of their hygiene.

Management control and prevention

Routine vaccination of breeding stock is practised commonly in countries in which the organism is enzootic, but not so much in fringe and free areas. The vaccine, like all bacterial vaccines (bacterins) does not provide a solid immunity but usually raises the resistance sufficiently to prevent clinical signs. Vaccines used are inactivated and contain an adjuvant. Other considerations include:

- Take care in the sources of your replacement stock to ensure that they are not carrying the organism.
- The organism is spread in urine from carrier animals. It can live for fairly long periods in water contaminated by infected urine. Do not provide water from questionable streams and keep your water supplies and tanks clean and free from contamination.
- Contamination from pig trucks are a minor risk as compared to other infections unless the truck has other pigs present.

L. tarassovi

This appears to be a pig-adapted strain that causes reproductive problems similar to (but milder than) *L. pomona*.

L. icterohaemorrhagiae

There is another condition in the human called Weil's disease and this is caused by *L. icterohaemorrhagiae*. This can enter a pig farm via rodents. Pigs that get *L. icterohaemorrhagiae* may die, with severe jaundice (yellow colouration to the skin). Control rodent populations, particularly if any are in the water supply.

Mange

See Chapter 11 for further information

Mange is a parasitic disease of the skin caused by 1 of 2 mites, either *Sarcoptes scabiei* var *suis* or *Demodex phylloides*, the latter of which is relatively unimportant in swine.

Sarcoptic mange (sometimes called scabies) is by far the most common and important in the pig because it is irritant and uncomfortable for the pig, causing it to rub and damage the skin, which becomes unsightly. Mange is an important condition as once infected a sow remains a carrier unless treatment is given to eliminate it. The constant rubbing to alleviate the itching can cause a lot of damage to housing.

Mastitis

See Chapter 8 for further information

Mastitis denotes inflammation and infection of the mammary glands. It is primarily a condition seen in the lactating sow. Nevertheless it can be a problem in the gestation/dry period with new cases occurring within 2 days of weaning. If left untreated it can become chronic, with thick fibrous scar tissue and large lumps which ultimately may ulcerate to the surface. Chronic mastitis may occur at weaning time when the udder is dried off, the gland having become infected in lactation. The disease can be caused by a number of different bacteria including streptococci, staphylococci, *E. coli*, klebsiella, pseudomonas and *Actinomyces spp*.

Clinical signs

Acute disease

These are seen within 2 to 4 days of weaning. The infected glands are enlarged, red and painful to pressure and the skin overlying is blanched or discoloured. The temperature is usually raised and the sow is off her food. In severe infections toxins are produced and sows may die within 24 hours.

Chronic disease
If there is a poor response to treatment or disease has not been recognised in lactation, the gland may be enlarged, often with fibrous tissue and small multiple abscesses that may ulcerate to the surface. Invariably at this stage there is permanent damage and if several mammary glands are affected the sow should be culled.

Diagnosis
This is determined by visual evidence of swellings and hard lumps in the mammary gland. The sow's udder should be examined routinely at 2 days after weaning and also while the sow stands at mating, to look and feel for chronic mastitis. Note that obtaining a sample can be difficult as each nipple is supplied by 2 or 3 separate mammae.

If the prevalence is more than 2 % of the gestation/dry sow population then an investigation should be carried out to assess the reasons why and to determine where the mastitis is starting. This could be in the immediate post farrowing period, during lactation or at weaning time. Preventive measures would then be carried out relative to that period.

Treatment
- This should follow the general guidelines as for mastitis in the lactating sow in Chapter 8. If the mastitis is being initiated from point of weaning, a long-acting antibiotic should be given on the day of weaning.

Management control and prevention
- Give long-acting injections of either penicillin, oxytetracycline or amoxycillin on the day of weaning.
- See also the procedures for the lactating sow.
- Wet, dirty pens that cause heavy contamination of the udder in the first 3 to 4 days after weaning predispose.
- Clean and disinfect weaning accommodation regularly.
- Many cases of "mastitis" are get misdiagnosed and are actually cases of udder oedema.

Meningitis
See also Streptococcal infections in this chapter. See Chapter 9 for further information
Meningitis is uncommon in the sow but it is sometimes secondary to middle ear infection or associated with water deprivation and salt poisoning. If an infectious disease enters a herd for the first time sporadic cases in sows may be seen.

Streptococcus suis type 7 may also cause meningitis and abscesses.

Clinical signs
The sow is inappetent and trembling with an unsteady gait. The temperature is raised, often as high as 42 °C (108 °F). As the meningitis develops in severity the eyes move sideways (nystagmus), fits develop, and the sow ultimately cannot stand.

Diagnosis
This is based upon the signs in an individual sow, or if there are a number of cases, a specific infectious disease. It may require a post-mortem examination, including histology of the brain and demonstration of the causal organism to confirm the diagnosis.

Similar diseases
These include:
- Acute kidney infection.
- Aujeszky's disease/pseudorabies virus (AD/PRV).
- Brain abscess.
- Glässer's disease (*Haemophilus parasuis)*.
- Heat stroke.
- Listeriosis.
- Middle ear infection.
- Poisons.
- Water deprivation.

Treatment
- This depends upon the cause. Always consider the common ones first which include middle ear infection, brain abscess, water deprivation and in some countries Aujeszky's disease/pseudorabies virus (AD/PRV). Refer to the treatment for these specific diseases.
- For bacterial infections use penicillin, penicillin/streptomycin or amoxycillin. Inject twice daily.
- Corticosteroids may also be required. Seek veterinarian advice.
- Move the affected animal to a clean, warm well-bedded pen.
- Provide easy access to water or dribble into the mouth 3 times daily by hose pipe.
- Ensure that the drip cooling system does not allow water to enter into the ear.

Muscle Tearing

This is a common condition in sows and first litter gilts whereby the muscle fibres are torn away from their attachment to the bone and periosteum. This occurs where muscles are attached to the inner surfaces of the elbow and knee joints, and at the points of attachment to the pelvis. Affected sows show considerable pain and often adopt a dog-sitting position. Inflammation of bone and periosteum result (periostitis). Torn muscles arise as a sequel to OCD, trauma and fighting, slippery surfaces and weak bones. Affected sows, as soon as they are identified, should be moved to a solid floor area that is deep-bedded, where the grip for the foot is firm. Soiled or grass areas are ideal. If this is done, most will recover. Administer pain relief where an animal is in pain.

Mycoplasma Arthritis (*Mycoplasma hyosynoviae* Infection)

Mycoplasma hyosynoviae is a ubiquitous bacterial pathogen and most, if not all, herds are infected with it. It is a respiratory-spread disease, the organism being found in the upper respiratory tract (nose and tonsils). It may be present in some herds and cause no clinical signs and yet in others cause severe disease. Infection with or without disease takes place in the young growing pig from approximately 8 to 30 weeks of age and particularly so in the gilt when first introduced onto a farm, or in the early stages of pregnancy. It is very uncommon in older sows because they develop a strong immunity resulting from repeated exposure to the organism. They pass this immunity to their offspring in the colostrum. This maternally derived immunity gradually disappears over a period of weeks. Infection then takes place as these pigs become exposed to older ones. *Mycoplasma hyosynoviae* infects joints and tendon sheaths rather than the respiratory system.

Clinical signs

Disease in the gilt or young boar is usually sudden in onset, the first signs being a reluctance to rise at feeding time. There is a considerable amount of pain and the affected pig will only stand for short periods of time. The temperature may be normal or slightly elevated. It is more common in the heavy-ham, straight-legged animal, and purchased gilts which have been reared in isolated grow-out units often become diseased 4 to 6 weeks after arrival.

Diagnosis

This is based on clinical signs and the response to tulathromycin, lincomycin or tiamulin therapy. Joint fluid can be aspirated and examined for antibodies and the organism isolated.

Serology is not much help because sub-clinical infection is common and so healthy animals often have antibody titres. Rising titres in blood samples taken 2 weeks apart together with typical symptoms strongly suggest disease.

In problem herds, post-mortem examination may be necessary to reach a definitive diagnosis.

Similar diseases

These include muscle damage, leg weakness or OCD, trauma, erysipelas, Glässer's disease and the major vesicular diseases.

Treatment

- *Mycoplasma hyosynoviae* is susceptible to tulathromycin, lincomycin or tiamulin injections. It is not responsive to the penicillin family of medicines.
- If the lameness is due to *Mycoplasma hyosynoviae* there should be a good response to antibiotics within 24 to 36 hours.
- Treatment is most effective if given early.
- Give in-feed medication strategically, commencing 7 days before the expected disease outbreak and continue for 14 days using either 220 g/tonne of lincomycin, 500 to 800 g/tonne of OTC, 100 g/tonne tiamulin or 200 g/tonne tilmicosin.
- An alternative strategy is to medicate the ration at half these levels and feed for 5 to 7 weeks.

Management control and prevention

Mycoplasma hyosynoviae can be a recurring problem in breeding gilts, particularly during the first 6 to 10 weeks after introduction to the farm. Consider the following:

- Identify the period of onset and apply strategic preventative medication.
- In-feed medicate susceptible groups over the critical period with either 500 to 800 g/tonne of OTC or CTC, 110 to 220 g/tonne of lincomycin or 100 g/tonne of tiamulin.
- Maintain pigs on ad lib feeding during the susceptible period.
- Assess the quality of housing – in particular low temperatures and draughts which act as trigger factors.
- Remember that this is a respiratory-spread disease and other factors need to be considered.
- Avoid mixing and fighting.
- Provide well-bedded pens.
- Avoid steps and rough edges which could damage joints.
- Avoid very smooth floors on which the growers slip and damage their joints.
- In outdoor herds, acclimatise gilts using cobs or

large nuts before they are introduced into the outdoor herd.
- Control mycoplasma (enzootic) pneumonia and other respiratory diseases if they are a coincidental problem.

Peritonitis

Peritonitis describes infection and inflammation of the peritoneum, the shiny membrane that covers all the internal surfaces in the abdomen. It can be caused by a ruptured gastric ulcer, a perforated bowel, penetration of the abdomen via mating, or a sequel to external trauma to the abdomen, a ruptured bowel or liver.

Diseases such as actinobacillus pleuropneumonia, migrating ascarid worms and miscellaneous generalised infections may also result in peritonitis.

Clinical signs

The onset may be sudden or gradual. Signs are associated with abdominal pain. The sow is reluctant to move, loses weight and has a tucked-up appearance. The mucous membranes are often pale. The most common time is 7 to 10 days post mating after damage by the boar at mating. A discharge from the vulva may then be apparent. The temperature may be normal or elevated and appetite normal or depressed.

Diagnosis

This is based on the clinical signs and history. A post-mortem examination may be required to confirm the diagnosis when females die or are destroyed.

Treatment

- This is by broad spectrum antibiotic treatments for 5 to 7 days. Inject with either: OTC, penicillin, streptomycin or amoxycillin.
- The response is usually poor.

Pneumonia

See Chapter 9: Respiratory Disease for further information

Pneumonia is normally uncommon in the mature herd but occasionally occurs in the gilt if immunity levels are low (Fig 7-10). However if swine influenza or PRRSV enters the herd for the first time or herd immunity wanes, periodic outbreaks involving a small number of sows may occur. When a new respiratory pathogen is introduced into the herd for the first time, for example a virulent strain of *Actinobacillus pleuropneumoniae*, severe pneumonia is likely to develop in all ages of animals.

FIG 7-10: PATHOGENIC ORGANISMS THAT CAUSE PNEUMONIA

- *Actinobacillus pleuropneumoniae*
- *Haemophilus parasuis* (Glässer's disease)
- *Mycoplasma hyopneumoniae* (Mycoplasma (enzootic) pneumonia)
- PRRSV
- Porcine respiratory coronavirus (usually very mild)
- Swine influenza virus

Clinical signs

These are seen at a herd level when new infections first enter. There is widespread coughing and up to 20 % or more of animals will be severely ill. The respiratory rate is elevated with some sows showing acute respiratory distress.

In herd breakdowns with mycoplasma (enzootic) pneumonia or actinobacillus pleuropneumonia mortality can be as high as 10 to 15 % if prompt treatment is not undertaken. If a clinical picture of widespread, sudden and progressive respiratory disease develops, then suspect a herd breakdown with 1 of the above organisms.

Diagnosis

This is based on the clinical signs of coughing, rapid breathing, a high temperature and post-mortem examinations. At an individual level sows may develop pneumonia due to infectious agents already in the herd. The introduction of swine influenza into a herd is usually dramatic, with large numbers of sows off their food over a period of 3 to 7 days. Widespread coughing and depression may be seen. In the case of a breakdown with mycoplasma (enzootic) pneumonia (in a herd that was previously free) the onset may be insidious with some inappetence but a gradual spreading cough over a period of 2 to 3 weeks. It may also appear to develop rapidly, affecting sows more severely. There is likely to be severe pneumonia and some mortality if the disease is not controlled. Laboratory tests involving serology and microbiology are necessary to identify the possible causes.

Treatment

- Usually pneumonia in the sow involves a mixed infection of viruses and secondary bacteria. Broad spectrum antibiotics such as tulathromycin, OTC, penicillin/streptomycin or amoxycillin are indicated.
- Inject individual cases daily for 3 to 4 days.
- For influenza with secondary bacteria:
  - Combine CTC or OTC in the water at the onset together with in-feed medication at a level of 600 g/tonne.
  - Antibiotics for at least 14 to 21 days.
- Mycoplasma (enzootic) pneumonia – If there is a herd breakdown, medicines specifically effective against mycoplasma are indicated:

- Lincomycin – In-feed, in-water, or by injection.
- Spectinomycin – Injection.
- Tiamulin – In-feed, in-water, or by injection.
- Tylosin – In-feed, in-water, or by injection.
- Chlortetracycline – In-feed or in-water.
- Oxytetracycline – In-feed, in-water, injection.

- It is important in the early stages of a breakdown to control the levels of infection, particularly in the numbers of organisms excreted into the air, until immunity has developed. This can be achieved by using 600 to 800 g/tonne of OTC or CTC in-feed for 2 weeks, reducing this to 200 to 300 g over the next 3 to 4 weeks.
- Actinobacillus pleuropneumonia – If there is a herd breakdown, early treatment of individuals is necessary together with preventative medication in-feed or in-water. Individuals should be injected with either: OTC, penicillin/streptomycin, tulathromycin, ceftiofur or sulphonamides.

Porcine Epidemic Diarrhoea (PED)

See Chapters 8 and 9 for further information

This is caused by a coronavirus which affects all ages of pigs. When it is first introduced into the breeding herd, the clinical picture in the gestation/dry sow area can vary considerably from very mild "cow pat" type faeces through to a watery diarrhoea. Sows normally recover over a period of about a week, usually without other effects. The virus is widespread throughout Europe and as a result it is uncommon to see disease in the breeding side of the herd due to the strong immunity that develops. The disease is much more important in the sucking pig. The condition is particularly problematic in Asia.

Porcine Parvovirus (PPV)

See Chapter 6 for further information

This virus is ubiquitous and present in almost all breeding herds. Once exposed, the sow becomes immune. The virus itself has no clinical effects on non-pregnant females when infection takes place. If, however, a sow or gilt is exposed for the first time in the first half of pregnancy the virus will cross the placenta and selectively kill off the developing foetuses in the uterus. Good control is achieved by vaccinating the gilt. This disease is covered in detail in Chapter 6.

Porcine Reproductive and Respiratory Syndrome Virus (PRRSV)

The information in this chapter relates specifically to PRRSV in the dry/gestating period. The main information on PRRSV, including methods of spread, immunity, diagnosis, similar diseases and elimination from a herd is covered in Chapter 6, with additional information in Chapters 8, 9, and 10. These chapters should be read to fully understand how this disease has an impact on the whole herd.

The effects of this disease in the sow are severe when infection is first introduced into the herd. The virus causes inappetence, abortions, late mummified piglets, high stillbirths and neonatal mortality.

Clinical signs

When the virus first enters the breeding herd, disease is seen in dry/gestating sows. The initial phase of inappetence and fever will often take 3 to 6 weeks to move through the breeding herd. Cyanosis or blueing of the ears is a variable finding and less than 5 % of sows show it. It is transient and may last for only a few hours. Coughing occurs in some sows, and a few individual cases of clinical pneumonia may occur.

Treatment

There is no treatment for viral infections. However, it is important to control secondary bacterial infections and antibiotic treatment should be given for 3 to 4 weeks to all sows immediately the disease is diagnosed or suspected. If necessary, commence with water-soluble antibiotics followed by in-feed medication. Tiamulin may assist in a PRRSV outbreak as it reduces the ability of the PRRSV virus to enter the macrophage.

A control programme for dry/gestating sows during acute disease

- Raise the sow house temperature to 21 °C (72 °F).
- Avoid night temperature drops.
- Avoid draughts.
- Medicate the sow's feed immediately with 500 g per tonne of tetracycline, either CTC or OTC.
- If sows are inappetent, medicate the water with OTC or CTC at the onset.
- Apply the medication to both gilts and boars.
- Maintain this for a 4-week period.
- Increase feed intake by at least 0.5 kg per day over the 4 weeks.
- Inject individual sows with long-acting OTC or penicillin during periods of inappetence or as advised by your veterinarian.
- Vaccinate with a PRRSV vaccine.
- Treat sows with a water-soluble aspirin to reduce rises in body temperatures and possible abortions.

Porcine Stress Syndrome (PSS)

See Chapter 9 for further information

This term covers a group of conditions associated with an autosomal recessive gene. It includes acute stress and sudden death (malignant hyperthermia), pale, soft exudative muscle (PSE), dark firm dry meat, and back muscle necrosis. Heavy muscle pigs are more likely to carry the gene. The gene can be identified by the pig's response to the anaesthetic gas halothane but recent developments have produced gene probes using blood samples that identify both the homozygous and heterozygous carriers.

Progressive Atrophic Rhinitis (PAR)

See Chapter 8 for further information

Rhinitis is inflammation of the tissues inside the nose and in its mild form it is very common. The term "atrophy" indicates that the tissues inside the nose, which become infected or damaged, shrink and become distorted. There are 2 forms of the disease, the more serious being PAR (progressive atrophic rhinitis). PAR is a serious condition in both sucking and growing pigs and is associated with the presence of toxin-producing strains of a bacterium *Pasteurella multocida*.

Clinically it is not important in the sow unless infection, as a growing gilt, has stunted her growth. However the sow might become a carrier, particularly if infected as an adult, and might spread it to other breeding animals and possibly to her litter. This condition is covered in Chapter 8 under Diseases of the Sucking Pig.

Prolapse of the Rectum

See Chapters 9 and 14 for further information

This is not uncommon in sows and occasionally outbreaks occur in herds. Information on rectal prolapse is provided in Chapter 9. Nutritional aspects of this condition are covered in Chapter 14.

Prolapse of the Vagina and Cervix

Prolapse of the vagina and cervix is more common prior to farrowing and may be seen in the last third of pregnancy including the pre-farrowing period in the farrowing house. It occurs normally in about 1 pregnancy in 200, usually in older sows from 5th parity onwards. It is a response to increased abdominal pressure together with a relaxation of the internal structures that support the cervix or the neck of the uterus. Older sows that are heavy in pig, have large litters and are in very good condition are also more likely candidates.

The following factors need to be considered as causal or contributing to the problem:

- It is much more common in older sows than young ones.
- When sows lie down there is an increased abdominal pressure which tends to force the cervix or vagina to the exterior.
- Fat sows are more prone, as are those carrying large litters.
- It is common in sows that are lying in confinement on a floor that slopes too steeply to the rear.
- High levels of feed intake, particularly food containing high starch materials, produce abnormal fermentation, excess gas formation and an increase in abdominal pressure.

Clinical signs

In the early stages, the protruding tissues appear between the lips of the vulva and return to their normal position when the sow stands. However with advancing pregnancy the prolapse may remain to the exterior and as soon as this occurs the animal should be removed from its existing environment and loose-housed. The tissues become swollen with time.

Diagnosis

The clinical signs are obvious but occasionally can be confused with vaginal polyps that may protrude from the vulva, and eversion of the bladder. Handling the tissues will differentiate the two.

Treatment

- Remove the sow to loose-housing.
- If the prolapse remains when the sow is standing, replace and pass a tape suture across the vulva.
- If the sow is at point of farrowing, the farrowing place floor should be raised to slope towards the feeding trough by using raised floor boards. When the sow then stands or lies down the weight of the piglets inside pulls the uterus forward to hold the vagina in. Under such circumstances the sow usually farrows normally.
- If the vagina remains prolapsed as farrowing approaches, the cervix will not open fully and both the sow and the litter are likely to be lost. In such cases a tape suture should be placed across the lips of the vulva to hold the prolapse in. As the sow reaches the point of farrowing it can be relaxed. This technique is described in Chapter 15.

Management control and prevention

- Consider the factors outlined above.
- Consider the factors predisposing to rectal prolapse.

Salmonellosis

See Chapter 9 for further information

Salmonellosis is mainly a problem in the growing pig, although it can affect sows. The organisms are found in the intestine and are excreted for long periods of time with little or no disease. *Salmonella* in the gut of the pig can contaminate carcasses during the slaughter process and their presence creates potential public health risks.

Infection is usually by mouth from contaminated faeces and the sow may continue to shed the organism for several months.

Disease is dose-dependent, that is, a minimum number of organisms are required before clinical signs occur. *Salmonella typhimurium* in the adult sow is mainly confined to the intestines, whereas *Salmonella choleraesuis* spreads throughout the whole system causing septicaemia, pneumonia, meningitis, arthritis and diarrhoea.

Treatment

This is rarely necessary in the sow.

Management control and prevention

- The most important control mechanism for preventing spread from sow to piglet is the maintenance of all-in/all-out procedures in farrowing houses.
- Recent work suggests that during sucking, maternal antibody prevents infection being established in the piglet and provided weaned piglets are not infected from older carrier pigs, good control can be achieved.

Shoulder Sores

See Chapter 10 for further information

Shoulder sores arise due to constant trauma over the bony prominences on the shoulder blade. Ultimately the skin breaks, there is an erosion and a large sore develops. It is generally associated with totally slatted or solid flooring accommodation and seen in individual sows that have a prominent spine to the shoulder blade. It is often present where the floors are slippery and the sow has difficulty in rising, thus constantly bruising her shoulder. This condition is covered in Chapter 10.

Sow Mortality

Sow mortality in some herds can be a cause of major economic loss. In approaching this problem, first categorise the causes and in particular differentiate between those animals destroyed on welfare or other grounds and those that actually die.

Target levels:

- 3 % to 6 % mortality in the herd. Up to 2 % of these may be destroyed on welfare or other grounds.
- The range seen in various herds: 3 % to 15 %.
- Levels of 7 % to 9 % are not uncommon and relate to specific contributory factors on that particular farm, which must be identified.

These include:

- Age of the herd.
- Culling policy.
- Welfare constraints (sows sold as culls and not recorded as deaths).
- Breed.
- Presence of the porcine stress gene (PSS).
- Body condition – sows in poor condition are more susceptible to disease.
- Availability of water and the incidence of cystitis/pyelonephritis.
- Susceptibility to leg weakness (OCD).
- Type of management system. Mixing of sows, fighting etc.
- Type of environment. Mortality levels are higher in poorly managed housing systems.
- Quality of the nutrition.
- Feeding system. This may predispose to ruptured or twisted intestines.

The causes of mortality

At least two thirds of all sow deaths occur in the gestation/dry period. Causes include:

- Abscesses.
- Cancer.
- Chronic disease.
- Clostridial infections.
- Cystitis/pyelonephritis.
- Dead piglets: uterus infection.
- Electrocution.
- Fighting.
- Fractures.
- Gastric ulcers.
- Heart failure.
- Internal abscesses.
- Internal haemorrhage.
- Lameness.
- Paraplegia.
- Peritonitis.
- Poor body condition.
- Prolapsed rectum.
- Prolapsed uterus.
- Prolapsed vagina.
- Strangulation associated with the environment.
- Torsion of the stomach or intestines.
- Vulval biting.

Follow the steps in Fig 7-11 to help identify a sow mortality problem.

FIG 7-11: PROCEDURES FOR IDENTIFYING AND CORRECTING A SOW MORTALITY PROBLEM

Step 1	Set up a recording system. (See Chapter 3).
Step 2	Analyse the results to identify the common causes.
Step 3	Refer to the specific disease once identified.
Step 4	Apply the recommended controls and prevention.

Sudden death

While alarming, sow mortality is generally sporadic with little pattern being evident. Ideally, less than 3 % of sows should die on the farm; an additional 3 % could be euthanised on farm for welfare reasons.

Such animals, which should not be sent to slaughter, include sows that cannot place all 4 feet on the ground and sows below a condition score of 2.

Sudden death in large numbers of sows – more than 3 sows die in a day – should always be investigated by the veterinarian. Consideration should be given to the following conditions:

Death with few obvious clinical signs

Anthrax – Generally does not kill the sow, but if they do die, may present with laryngeal swelling.

Erysipelas – The pig, which dies suddenly with erysipelas, they may present with no skin diamonds. Gross post-mortem changes may be absent. The most consistent post-mortem change seen is an enlarged spleen. Check the skin for diamonds which may be felt before they can be seen. Note in dark pigs the diamonds may be difficult to see.

Swine dysentery – When dysentery enters a naive herd, the adult stock may also become infected and present with similar signs to finishing pigs, including sudden death.

Swine fevers – Classical swine fever (hog cholera) and African swine fever. High mortality in all age groups. Generally, additional clinical signs (high temperatures, skin discolouration, abortions and increase in returns) are displayed in other stock.

Death with no obvious clinical signs

Electrocution – The sow may have died as a result of an electrical shock. Poorly maintained electrics are common on many farms.

Endocardiosis – Heart valve lesions of endocardiosis may be associated with streptococci or erysipelas and may result in heart failure.

Environmental injury – Poorly maintained environments may result in damage to the pigs. For example, poorly maintained floors – the sows can fall through and drown. Sows can strangulate themselves under gating if improperly fitted. Metal cladding can come away, forming very sharp points. Sows can be injured on drinkers in the mating arena.

Heat stress – If the ventilation system fails or in extreme weather, pigs will rapidly die from heat stress and suffocation.

Haemorrhagic bowel syndrome – The etiology is still unclear. Whey feeding may be associated with a bloat and sudden death. The condition appears as an allergic response, with haemorrhage into the intestines.

Internal haemorrhage – May occur for a number of reasons, for example gastric rupture associated with a gastric ulcer. This can be associated with trauma.

Intestinal disasters – torsions – Any area of the intestinal tract may undergo torsion – the stomach, spleen, liver, small and large bowel. Even the lung can come through the diaphragm.

Mating injuries – Boars and sows may be found dead shortly after mating associated with stress and heat stress. Occasionally the boar can tear the neck of the bladder and the sow bleeds to death.

Poisoning – Presents with sudden death in a group of sows. Investigate the dead sows very carefully. Hydrogen sulphide, when odourless, is lethal.

Retained piglets – Retained piglets may result in an acute metritis, toxaemia and septicaemia which can rapidly kill the sow. Even retained placenta at times will result in a fatal septicaemia. However, it is also not uncommon to find retained piglets as an incidental finding at post-mortem.

Salt poisoning – If the sow has no water for 3 to 4 days, she is likely to die of salt poisoning. If she has had repeated problems with her water supplies she may have damaged her kidneys and succumb rapidly to restriction in her water supply.

Death with few clinical signs (although post-mortem will generally reveal the cause)

Clostridial hepatopathy – Sow presents with no clinical signs. She swells rapidly and decomposes within hours of death. The organs often appear normal when the sow is initially post-mortemed but the organs decompose as the post-mortem progresses.

Cystitis and pyelonephritis – Cystitis will not kill the sow; however, if the infection spreads to the kidney, a fatal pyelonephritis may result. On several farms, this is a major cause of death. A review of the watering system is urgently required. The sow may appear lame before it dies.

Gastric ulceration – Gastric ulcers can result in anaemia severe enough to kill the sow. The ulcer may also erode into a large blood vessel in the submucosa resulting in haemorrhage into the stomach and death of the sow.

Prolapse – A variety of prolapses around farrowing may result in the death of the sow, even if surgical intervention is attempted. The prolapse is obvious on the outside of the pig.
Fighting – When mixed, strange pigs will fight, sometimes to the death. Young sows can be killed by fighting during oestrus. The pigs will have scratch marks over their bodies.
Glässer's disease – If young naive adults are introduced into a normal-health herd, an acute meningitis and polyserositis may occur with the resultant death of the gilt or boar. This is rare but very dramatic.
Ileitis – In young adults this may present as sudden death with the hemorrhagic form of ileitis. The removal of chronic medication from sows may initiate an outbreak of ileitis.

Other problems may include septic pleurisy, septicaemia and cancers. A post-mortem will reveal the cause of death.

Streptococcal Infections

See Chapter 9 for further information

Streptococcus can be implicated in a number of conditions. It may be responsible for the formation of abscesses. It may affect the uterus resulting in endometritis and even severe metritis. Occasional pyelonephritis may be recognised. Streptococci cause diseases in the sucking pig including meningitis, arthritis and septicaemias, and ascending metritis in the female.

The potential pathogens can be carried for prelonged periods both on the tonsils and in the respiratory tract. The skin and vagina are also other sources.

Swine Dysentery (SD)

See Chapter 9 for further information

Swine dysentery is caused by a small snake-like bacterium called *Brachyspira hyodysenteriae*. This organism causes a severe inflammation of the large intestine with bloody mucous diarrhoea (i.e. dysentery). It is a major disease in the young growing pig but the breeding female can become a carrier of the organism for a long period of time and therefore acts as a potential source of infection to the sucking pig. Clinical disease in the sow is unusual except where the organism is first introduced into the herd, when a mild sloppy or severe acute dysentery may be seen. In very severe cases the diarrhoea may also extend to the litter. The stress at farrowing can activate disease to produce sloppy "cow pat" faeces which expose piglets to infection.

Swine Influenza Virus (SIV) – Swine Flu

See Chapters 6 and 9 for further information

Swine influenza is a highly contagious respiratory disease caused by a number of closely related type A influenza viruses that are noted for their ability to change their antigenic structure and create new strains.

If a new strain is introduced into the breeding herd and there is no immunity, the respiratory disease that follows can be quite dramatic. Within 2 to 3 days, up to 40 to 50 % of animals may be off their feed and looking very ill. This can be quite alarming but the rapidity of the clinical signs is almost exclusively confined to this disease. The major risk to the pregnant sows is the high temperatures which cause abortions, embryo or foetal loss, or high stillbirth rates. Widespread coughing and pneumonia may also be seen. The course of the disease in individual sows is usually 4 to 7 days and within 14 days most animals in the herd have returned to normal. In large herds, SIV can become endemic, with disease appearing every 3 to 6 months. A vaccine is available in some countries. This disease is covered in detail in Chapters 6 and 9.

Thin Sow Syndrome

The thin sow syndrome occurs over a period of months, with gradual declining body condition, until 10 to 30 % of the animals have a condition score between 1 and 2. The syndrome arises due to inadequate nutrition or poor quality feeds failing to satisfy the bodily needs of the sow in that environment. During lactation, the sow is unable to maintain her body condition due to either an insufficient intake of energy, or increasing demands due to low temperatures or high milk output. The sow therefore uses her body fat to maintain the supply of energy and once this is used muscle protein is degraded. This process continues over successive lactations. It is exacerbated in sows kept outdoors in cold weather and by heavy worm burdens. In sows kept permanently outdoors, the stockperson should ensure that all the sows have a high body score before the start of cold weather.

Clinical signs

These will be evident by the appearance of a number of very thin sows. Each week during the clinical observations of the herd, an assessment of the overall body condition of sows in the gestation/dry sow area should be carried out. If there are more than 5 % of sows scoring body condition of less than 2 (see Fig 7-3) then the herd might be moving into problems. In such cases a more specific examination should be made of body condition at farrowing and at weaning time, the feed intake during lactation and pregnancy, the quality of the feed itself and evidence of any specific diseases.

Treatment

- This should be aimed at increasing the nutritional intake of the sow during the key periods of production and attending to any disease.
- Immediately raise feed intake across the herd by 1 to 2 kg a day for a period of 10 to 14 days.
- Sows that have become very thin should if possible be moved from their gestation/dry sow accommodation and housed in warm, deep straw pens in an environmental temperature of at least 20 °C (70 °F), and ad lib fed for 3 to 4 weeks using a lactating diet because the appetite in such animals is often depressed. It is essential to do this to reverse the catabolic or negative energy state and allow the sow to lay down body fat again. If the sow has become too lean (body score 1) the condition can become irreversible.

Management control and prevention

- Check the feed used per sow per year. This should be a minimum of 1.1 tonnes per annum for an indoor herd and for an outdoor herd 1.4 tonnes per annum.
- Check the quality and energy content of the ration in relation to the feed intake.
- Check the energy content of the ration during lactation. For the lean genotype this should be at least 14.2 MJ DE/kg of feed and 1.1 % lysine.
- Sows should be fed at least twice daily to appetite during lactation, commencing 3 days post farrowing.
- Assess the body condition of sows coming into farrow and at weaning time.
- Monitor temperatures in the gestation/dry sow accommodation. For indoor sows in houses with no bedding it should not drop below 17 °C (63 °F), preferably higher, and should not vary greatly day and night, otherwise it will be necessary to increase feed intake to compensate.
- Select at least 12 faeces samples from thin sows. Submit them to a laboratory for examination to eliminate coccidiosis, blood (from gastric ulcers) and parasites.
- Always carry out an immediate investigation if the body condition of sows is dropping because if it is not corrected there will be infertility, poor litter sizes, increased disease and mortality.
- Assess the comfort level of the sows last thing at night or first thing in the morning, particularly during severe spells of cold weather. Always increase feed intake when external temperatures drop significantly.
- Damp floors or draughts will increase the energy requirement of the gestating/dry sow. Once the body condition of the sow has dropped then it becomes more susceptible to a variety of diseases, including gastric ulceration, cystitis and pyelonephritis, PRRSV and other respiratory viruses.

Vulval Biting (Vice)

See also Chapter 9 for further information

Vice in the gestating/dry sow is confined to vulval biting, particularly in the last 3 to 4 weeks of pregnancy. This can be a major problem in loose-housed sows, and in badly managed systems there may be 80 % of all sows in a herd with the vulva completely bitten off. During the process of damage there can be severe haemorrhage with loss of life in a few animals.

Clinical signs

The vulva is a highly vascular tissue and trauma results in haemorrhage which further attracts sows. Extensive lacerations are common and evidence of blood on the skin and noses of the sows must highlight the possibility of the condition. Severely traumatised vulvas can heal with scar tissue and this can cause constrictions and difficulties at farrowing.

Diagnosis

This is obvious from the clinical evidence but an examination should be carried out to ensure the haemorrhage is not arising from the vagina, uterus or bladder.

Treatment

- Because sows continue to traumatise an already damaged vulva, it is most important that affected sows are removed from the group at the onset.
- In most cases once the sow is isolated the haemorrhage will stop and the tissues will shrink and heal.
- Occasionally it is necessary to stem the haemorrhage. To do this sedate or restrain the sow and apply pressure using a bandage as a tourniquet.
- Provide pain management where required.
- If haemorrhage continues, infiltrate local anaesthetic into the vulva and place 2 or more mattress sutures behind the bleeding points. (See Chapter 15).

Management control and prevention

- It usually occurs towards the end of pregnancy. The reasons for this are unknown but may be associated with the increased demand for food and perhaps the swollen vulva becomes attractive.
- Increase the feed intake and assess the response.
- Increase the salt levels to 0.9 % per tonne.

- It is more likely to occur if the stocking density is high. Allow a minimum of 2.7 m^2 per sow, particularly in the latter part of pregnancy.
- There is usually 1 offending sow in the group. If she can be identified, remove her.
- Where floor feeding is practised and the feed is placed in small areas, sows group together to feed. Any sow that is excluded quickly learns that a simple way to get in is to bite a vulva. To prevent vulval biting therefore it is important to spread the feed as widely as possible over the floor area, even to the extent that where automatic drop or dump feeders are used spread some feed manually as well over the feeding area.
- Vulval biting is also common when electronic feeder systems are used. It requires careful stockmanship and good pen design to prevent it occurring during the waiting periods outside the feeders and as sows leave them.
- There is a relationship between feed intake, the size of the feed pellet, the type of floor surface and the bedding used. A change from a small to a larger pellet or vice versa will often improve the situation because it allows a better feeding system on the floor surface. This needs to be carried out by trial and error.
- Vulval biting is much more common in pens that are long and narrow rather than those that are wide. There is less competition at feeding time in a wide pen.
- Where there are severe problems within a group, move them to a different type of pen with more floor area. Sometimes this will solve the problem, particularly in the last 4 weeks of pregnancy.
- Check the water availability and eliminate any aggression over drinking.

Other forms of vice are often prevalent in growing pigs including tail biting, flank biting, ear biting, ear sucking and prepuce sucking. Nutrition can again be associated and this is all covered in Chapter 9.

Vulval Discharge

In the gestation/dry sow, any evidence of discharge from the vulva must be viewed with suspicion as this could be due to a potential loss of pregnancy as a result of endometritis and the vulval discharge syndrome (see Chapter 6 for details) or could be evidence of cystitis or pyelonephritis, which are covered earlier in this chapter.

Water Deprivation (Salt Poisoning)

See Chapter 14 for further information

Water deprivation is unfortunately common in all ages of pigs and almost without exception is related to water shortage, either caused by inadequate supplies or complete loss. See Chapter 14 for information on this condition.

8 MANAGING HEALTH IN THE FARROWING AND SUCKING PERIOD

Chapter 8

8 MANAGING HEALTH IN THE FARROWING AND SUCKING PERIOD

How to Achieve High Numbers Weaned

Note: Indoor production unless otherwise stated

An efficient and well managed farrowing house should be able to achieve an average number of 12 pigs weaned per farrowing place. One objective of this chapter is to discuss the key factors that will achieve this target. Fig 8-1 identifies the major components of the management system that are relevant to high numbers weaned. A good recording system to identify the problem area or areas of failure is a prerequisite.

FIG 8-1: THE 7 KEY POINTS RELEVANT TO ACHIEVING HIGH NUMBERS WEANED

1.	Good planning.
2.	Using records to identify problems.
3.	Understanding and maximising the role of the sow.
4.	Farrowing house design.
5.	Good farrowing management, supervising farrowings.
6.	Good piglet viability.
7.	Good nutrition.

Good Planning

The planning and control of the mating programme, which ultimately dictates the farrowing programme, are important for good health because they determine the rate at which pigs move in and out of the farrowing rooms and indeed the whole of the farm. The quality and breeding of the sow/gilt and its health status also contribute significantly towards reducing piglet mortality, but the greatest influence arises from management techniques that are adopted around farrowing and during the first 72 hours after birth. A healthy viable piglet with a good birth weight must interact with a good farrowing house environment. Ensure each batch farrowing place is full. Maximising the number of piglets reared per breeding female per annum is a major contributing factor to ultimate economic viability of the unit. "Without pigs you cannot sell pig meat" is an obvious but little-appreciated statement. Fig 8-2 shows the effect on cash flow of increasing the pigs per sow per year from 18 to 25 and most of this economic output will be net profit because the overheads of the unit are not increased by any significant amount.

THE INFLUENCE OF INCREASING PRODUCTION

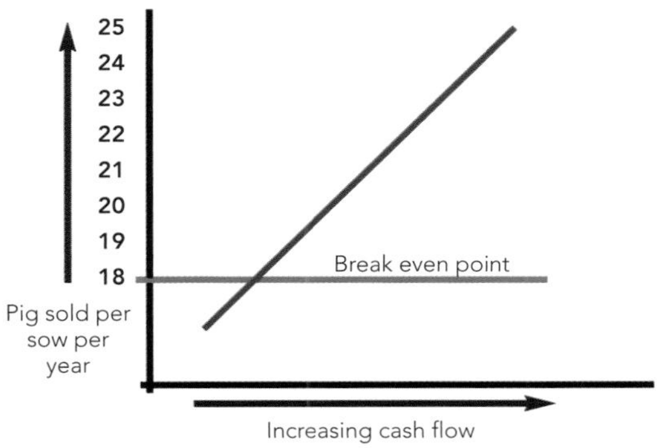

Fig 8-2

Pigs sold per sow per year multiplied by the weight of pig leaving the farm determines the kilograms of meat sold per annum and economic output. It is important to maximise this by achieving the maximum biological efficiency of the breeding female on the farm. However, over production initiated by too many matings in a given period of time can have a depressant effect on both health and the efficient use of feed. A study of Fig 8-3 highlights the progression of events that can take place. (See also Chapter 3: Management Procedures for Maximising the Mating Programme). Note the hidden effects that over use of boars and excessive numbers of matings may have on disease throughout the whole of the production system.

Using Records to Identify Problems

Records should provide 2 types of information: end data which identifies what is being achieved, and epidemiological data which gives the detailed components that have been responsible for this achievement. End data on its own is not of great value in understanding a problem. There must be the facility to collect, examine and learn from the components. For example, 3 % piglet mortality due to diarrhoea is of little value in helping to decide action without the knowledge of the sows involved, their parity and the age of the deaths. Within the management system of the farrowing house, detailed observations and time scales should be recorded on the sow and litter cards. Such information will help you to understand a problem if it arises. Records identify achievements that can be as-

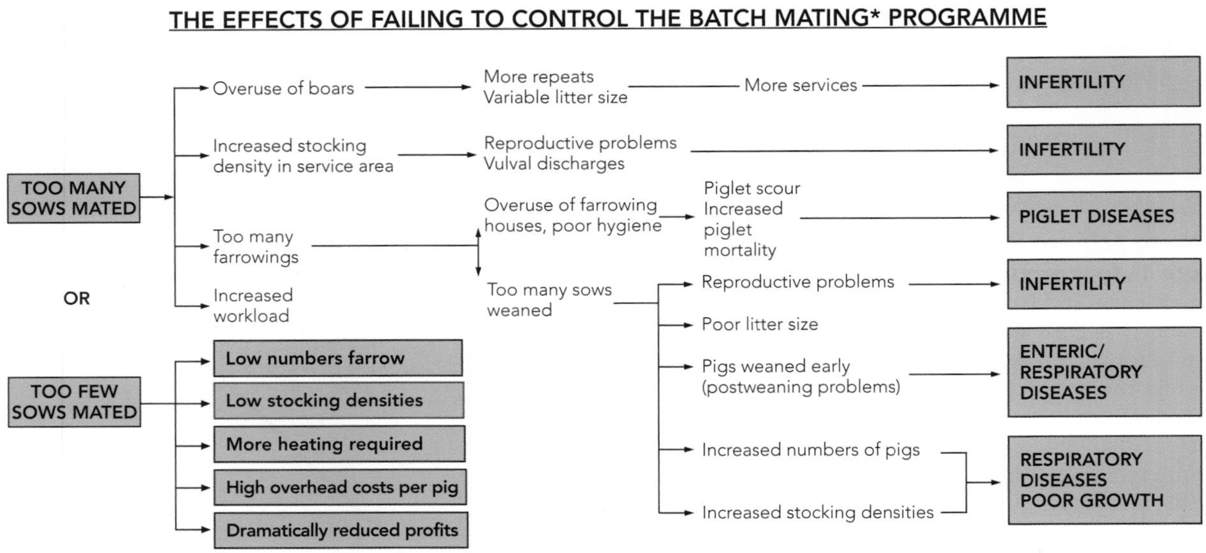

* The terms 'mating' and 'service' are often considered the same. Mating here means the complete process, i.e. it includes 1, 2 or more services.

Fig 8-3

sessed against target levels of efficiency.

Fig 8-4 shows targets that should be achieved using a modern breeding female in an indoor system.

If 12 pigs weaned is to be achieved it is necessary to both identify and document the reasons for the losses. A good management system produces lower mortality associated with poor viable piglets, starvation and in particular it controls losses from scour and miscellaneous conditions.

Information from indoor production (Fig 8-5) shows that piglets laid on and those of poor viability account for major losses followed by miscellaneous causes and scour. However, piglet mortality must also include deaths that occur both pre-farrowing and during the farrowing period.

Fig 8-6 shows a typical distribution of pre-weaning mortality and Fig 8-7 the typical time scale when this occurs. It is a sobering thought that 30 % of the pigs that die do so before or during birth and 44 % within the first 2 days of farrowing. This information highlights the importance of management inputs immediately before and for the first 48 hours after farrowing.

44 % of all the piglets that die do so within the first 2 days of farrowing.

The categorisation of piglet mortality as shown is based on observations, but accurate figures on born dead are difficult because some die in utero and others during farrowing. On some farms however, despite considerable

FIG 8-4: SUGGESTED TARGETS IN THE FARROWING HOUSE

Total litter size	14.5
Stillbirths/Mummified	< 10%
Born alive	13.2
Piglet mortality	10%
Pigs weaned	12

FIG 8-5: PIGLET MORTALITY. FIELD DATA

Causes of Piglet Deaths as a % of Total Deaths		% of Pigs Born Alive that Die	
		Actual	Target
Crushed/Laid on	38	4.1	< 3
Poor viability	18	1.9	1
Miscellaneous	15	1.6	1.5
Scour	10	1.1	0.5
Starvation	7	0.75	0.5
Not recorded	5	0.5	0.5
Deformed	5	0.5	0.5
Savaged	2	0.2	0.2

TOTAL PIGLET MORTALITY

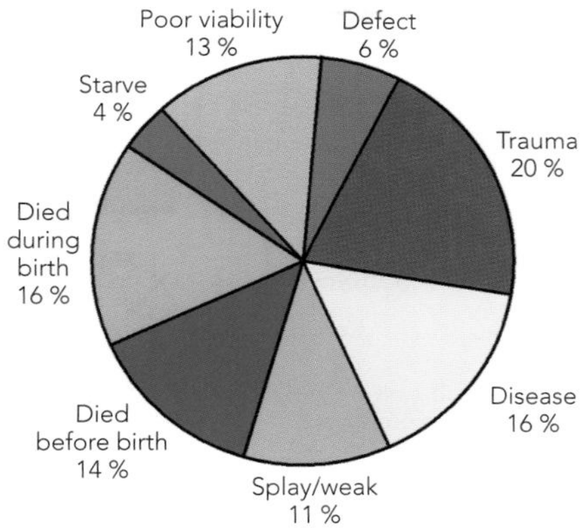

Fig 8-6

THE TIME PERIOD WHEN PIGLETS DIE

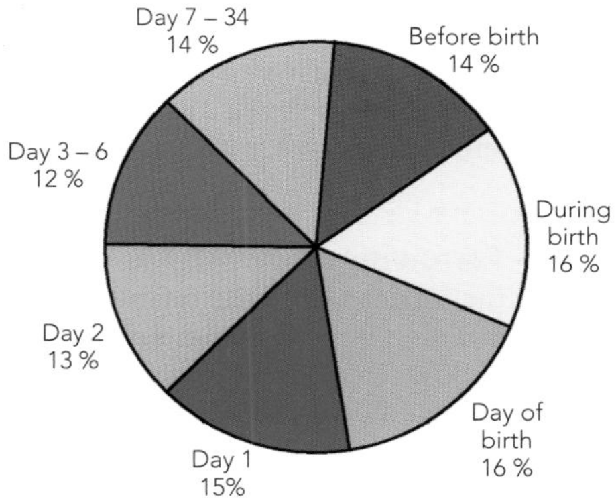

Fig 8-7

efforts, piglet mortality remains high and in these cases it is necessary to carry out a piglet mortality survey. This involves the post-mortem examinations of all the piglets that die over a period, for example 1 month, with each piglet being documented back to the sow, the litter size, and the fostering carried out. Clinical observations recorded on the sow and litter card together with the post-mortem findings then help in determining corrective action. It is also helpful to differentiate between a poor viability or a low birth weight pig and one that has the capability of surviving with assistance. If a sucking reflex can be felt by placing the little finger at the back of the tongue the piglet has a chance of survival.

Understanding and Maximising the Role of the Sow

Breed and Selection

The breed of sow has a vital role in the successful rearing of piglets and the selection of a good female is paramount. Breed from a hybrid or multiple breed female that through her mothering ability exhibits good hybrid vigour and one that is supported by historical data from the genetic base. This should show good milking ability, a good number of pigs reared, and low mortality. The larger the sow, the more difficult it becomes for her to rear pigs, maintenance costs increase and there is a relationship between girth size and piglet mortality. A large girth results in bad teat placements at farrowing and inadequate or delayed intake of colostrum in individual pigs. Increased body size may also produce sows too large for existing farrowing places.

Age

The age profile (parity) of the herd is important in relation to fertility, as discussed in Chapter 5 and the same comments are equally true of the performance of the sow in the farrowing house. Fig 8-8 shows typical results expected in gilts, and older parities in terms of litter size, mortality and pigs reared per sow per year. Stillbirth rates should be noted with increasing age of the sow and piglet mortality which will often exceed 12 % from parity 8 onwards. Also with increasing age, birth weights and litter sizes become more variable and there are greater losses associated with poor viable pigs.

FIG 8-8: AN EXAMPLE OF THE EFFECTS OF AGE ON REPRODUCTIVE EFFICIENCY *

	Litter Number			
	1	2	3 – 6	7+
Pigs Born	13	13.2	14.8	15
Alive	12.4	12.8	14	12.7
Dead %	5	5.5	6	15
Fostered (essential ones) %	3	4	8	10
Mortality				
Laid on %	3	3	3	5
Starvation %	1	1	1	2
Other	5	5	4	6
Total mortality %	9	9	8	13
No. weaned	11.1	11.6	12.7	11.5
Farrowing rate %	85	87	90	83

* Based on studies of field records.

However, older sows may have to be retained to maintain the mating programme and they can be useful in rearing surplus piglets. It should be noted that there are 4 distinct groups of breeding females in the herd: gilts, second parity, parities 3 to 8 and old sows, all with different performance levels. Each can become a specific problem associated with management and disease.

Parturition – Farrowing

To appreciate the intricacies of the farrowing process it is necessary to understand the anatomy of the pelvis and the reproductive tract at farrowing. As farrowing approaches the vulva becomes enlarged, together with the vagina, which leads to the cervix before opening into the uterus. The neck of the cervix opens into the 2 long horns of the uterus that contain the piglets. (Fig 8-9). The umbilical cord of the piglet terminates at the placenta which is attached to the surface of the uterus. Nutrients pass from the blood of the sow across the placenta and into the developing piglet. The placenta also extends around the piglet as a sac which contains fluids and waste materials, produced by the piglet during its growth. The placenta and the sac are referred to as the afterbirth.

How does farrowing start?

This is an intriguing mechanism activated by the piglet once it reaches its final stage of maturity, at approximately 115 days after mating. The sequence of events is depicted in Fig 8-10. The piglet activates its pituitary and adrenal glands to produce corticosteroids. These hormones are then carried via its bloodstream to the placenta. The placenta then produces prostaglandins which are circulated to the sow's ovaries, with the corpora lutea in the ovaries responsible for the maintenance of pregnancy. Prostaglandins cause them to regress, thus terminating the pregnancy and allowing the hormones that initiate farrowing to commence.

Length of pregnancy

The mean length in the sow is between 114 to 115 days with a range from 112 to 120. Gilts tend to have a shorter pregnancy. The variation within the range is influenced by the herd, environment, breed, litter size (it tends to be shorter in larger litters and longer in smaller litters), time of year and stockpeople. Note Day 1 of pregnancy is determined by the day of mating, not the actual time of ovulation.

The farrowing process

This can be considered in 3 stages: the pre-farrowing period, the farrowing process and the immediate post-farrowing period when the placenta (afterbirth) is expelled.

ANATOMY OF THE REPRODUCTIVE TRACT SHOWS ALSO THE BLADDER AND KIDNEYS

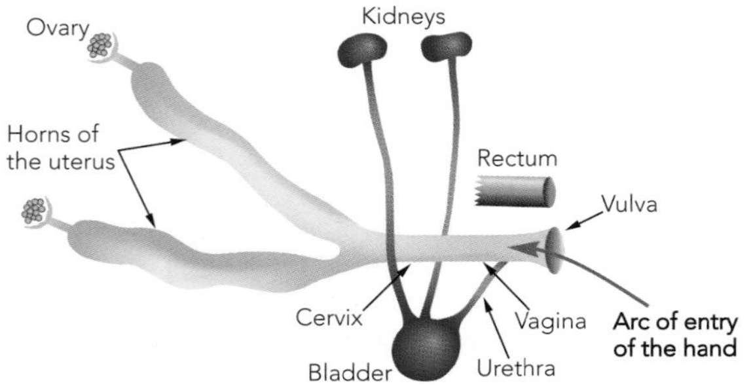

Fig 8-9

THE INITIATION OF FARROWING

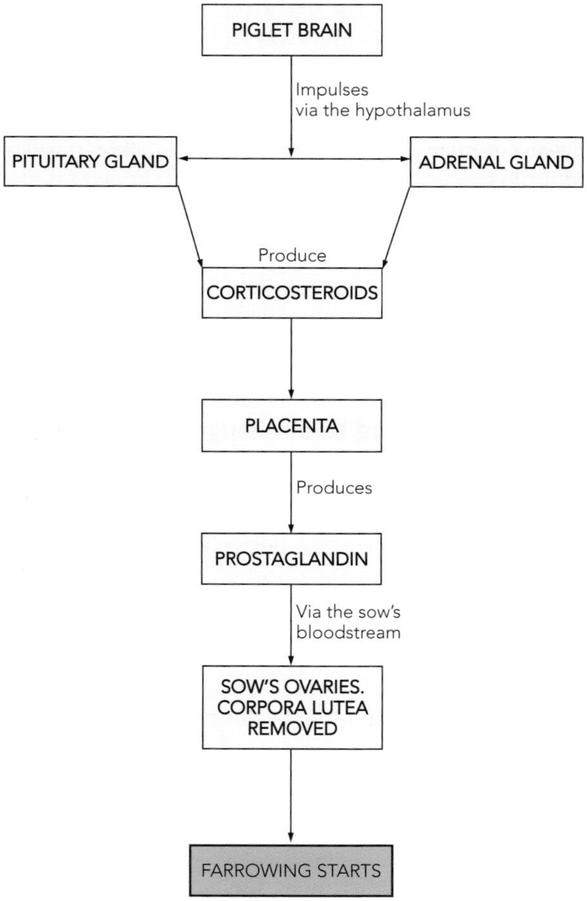

Fig 8-10

Stage 1 – The pre-farrowing period

The preparation for farrowing starts some 10 to 14 days prior to the actual date, with the development of the mammary glands and the swelling of the vulva. At the same time teat enlargement occurs and the veins supplying the udder stand out prominently. The impending signs of farrowing include a reduced appetite and restlessness, the sow standing up and lying down, and, if bedding is available, chewing and moving this around in her mouth. If she is loose-housed on straw she will make a bed. Within 12 hours of actual delivery of piglets, milk is secreted into the mammary glands and with a gentle hand and finger massage it can be expressed from the teats. This is one of the most reliable signs of impending parturition. A slight mucous discharge may be seen on the lips of the vulva. If a small round pellet of faeces is seen in the mucus and the sow is distressed, farrowing has started and it is highly likely the first piglet is presented backwards. This small pellet is the meconium or first faeces coming from the rectum of the piglet inside. **An internal examination is immediately required.** The final part of Stage 1 is the opening of the cervix to allow the pigs to be pushed out of the uterus, through the vagina and into the world.

Stage 2 – The farrowing process

This can range from 3 to 8 hours and piglets are usually delivered every 10 to 20 minutes but there is a wide variation. Consult the sow and litter card to see if there have been any previous problems at farrowing. For example, if a sow has had high stillbirth rates, monitor her more closely and take any necessary actions. There is often a gap between the first and second piglet of up to three quarters of an hour. The majority of pigs are born head first but there are more pigs presented backwards towards the end of the farrowing period. Immediately prior to the presentation of a pig the sow lays on her side, often shivering and lifting the upper back leg. This is an important point to take note of because it may indicate the presence of a stillborn pig. Twitching of the tail is seen just as a pig is about to be born. Her rectal temperature may be 0.5 to 1 °C (0.9 to 1.8 °F) higher than normal.

Stage 3 – Immediate post-farrowing period when the placenta is expelled

This usually takes place over a period of 1 to 4 hours and is an indication that the sow has finished farrowing although some placenta will sometimes be passed during the process of farrowing. Once the sow has completed the farrowing process there are certain signs that should be observed.

- She appears at peace, grunts and calls to the piglets.
- The shivering and movement of the top hind leg ceases. **If this is still occurring it is likely that a pig is still presented.**

After the placenta has been delivered there will be a slight but sometimes heavy discharge for the next 3 to 5 days. Provided the udder is normal and the sow is normal and eating well, ignore it; it is a natural post-farrowing process. Occasionally a pathogenic organism enters the uterus, causing inflammation (endometritis). This may cause illness, requiring treatment.

Problems at farrowing

Uterine inertia – This is where the uterus has just stopped contracting. Usually there will be 2 or 3 pigs waiting just beyond the cervix. If they are in an anterior presented position place the hand over the head with the first and second fingers around the nape of the neck (Fig 8-11). If the piglet is in a breech or backward position, raise both hind legs and clamp the hands around using the first and second fingers as leverage around the points of the hock (Fig 8-12).

ASSISTING THE PIGLET IN ANTERIOR PRESENTATION

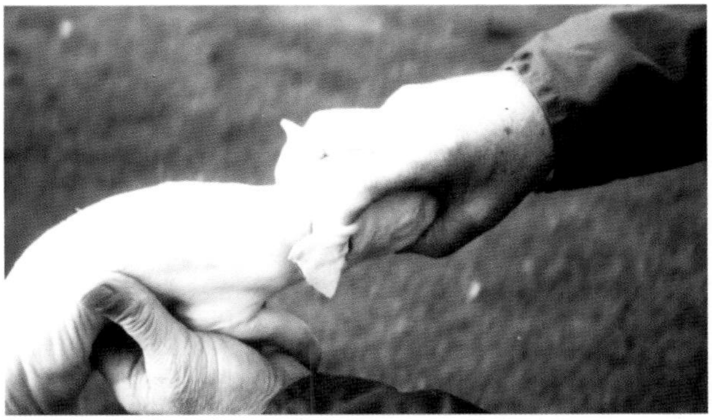

Fig 8-11

ASSISTING THE PIGLET IN A BREECH PRESENTATION

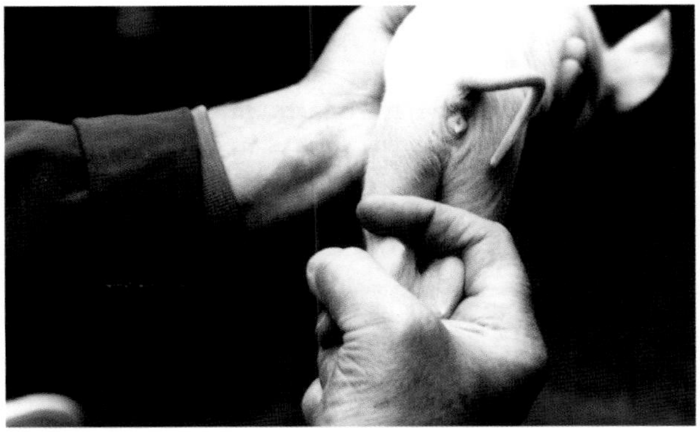

Fig 8-12

Difficult presentations – Occasionally (particularly in gilts) a large piglet is presented that is too big, but in most cases with gentle traction such a pig can be delivered. The best method is to use a piece of cord 2 metres long (clean disinfected nylon cord is satisfactory) and loop the centre of it around the end of the third finger. Using plenty of lubricant, pass the cord into the vagina to approximately 50 mm behind the head of the piglet. The cord is then placed behind the left and right ears and finally brought down beneath the jaw. Twisting it lightly under the piglet's chin may help to secure it. Traction can then be applied in a downward movement to bring the pig out. This is an excellent and simple technique and I would recommend that you familiarise yourself with it by cutting off the end of a wellington boot, place a dead piglet inside with its head presented to you and practise placing the cord around the neck (see Fig 8-13).

Rotation of the horns of the uterus – This sometimes occurs when very large litters are present. One horn crosses over the other. This distorts the cervix so that piglets cannot be pushed through and 2, 3 or 4 pigs form into a pouch below the cervix itself (many are presented backwards). When the hand is passed through the cervix (which has become elliptical) the pigs can be felt by reaching downwards and back towards yourself. In such cases it is necessary to take the arm full length into the sow (sow standing) and work hard to bring 3 or 4 piglets up. Once the piglets have been removed with the sow standing, use a closed hand on the side of the abdomen and swing it to try and realign the piglets and horns of the uterus. If the sow has not passed further piglets within half an hour re-examine.

Stimulating a piglet to breath – If a piglet is delivered and it fails to breath, take a small piece of straw and poke it up the nose. This will in many cases elicit a coughing reflex and remove mucus that has blocked the windpipe. Alternatively place the third finger across the mouth of the piglet with its tongue pulled forward. Place the rest of the hand around the head and hold the back legs. Swing the pig with a firm downward movement to propel any mucous from the back of the throat and the windpipe (Fig 8-14).

PLACING THE CORD BENEATH THE EARS OF THE PIGLET AND BENEATH THE JAW

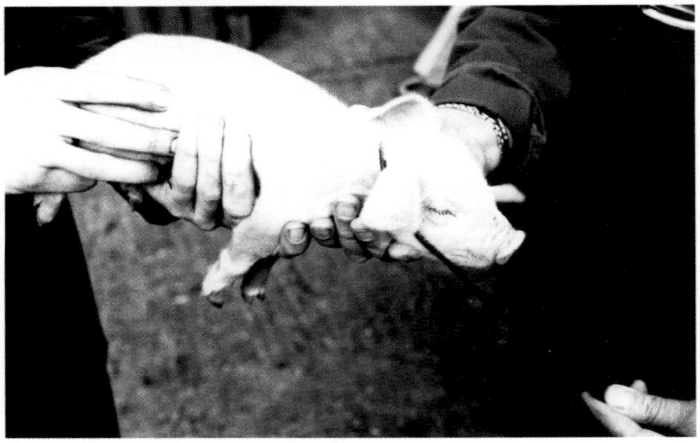

Fig 8-13

SWINGING THE PIGLET TO REMOVE MUCUS AT BIRTH

Fig 8-14

What to do when there are farrowing problems

Step 1. Recognise that the sow is in difficulty. This is shown either by lack of piglets being born, the sow panting heavily and being obviously in distress or blood and/or mucus at the vulva.

Failure to deliver the piglets can be due to the following:

- A large litter and inertia of the uterus.
- Very large piglets and a small pelvis.
- Two or more pigs presented in the birth canal at the same time.
- Illness of the sow, for example acute mastitis.
- Rotation of the uterus.
- Failure of the cervix to relax and open.
- Dead pigs inside the uterus.
- Mummified pigs.
- Failure of the uterus to contract (uterine inertia).
- Nervousness of the sow, excitement and distress.
- An over-fat sow.

Step 2. Investigate. Never carry out an internal examination without a container of clean warm water containing a mild antiseptic and use a soft soap or preferably a special obstetrical lubricant. Do not use detergents, they are irritant and **never** be tempted to try and force a dry arm into the vagina of the sow.

Step 3. Wash the hands and arm well and in particular ensure the finger nails are short. It is preferable to use a plastic arm sleeve because this reduces contamination from the hands.

Examine the sow as she is lying down on her side. It is easier to use your left hand if she is on her left hand side and your right hand if she is on her right side. Occasionally you may have to examine the sow in a standing position. Never farrow a standing sow with a bar in the way; if the sow suddenly lies down you could break your arm.

Hold the fingers of the hand together and introduce the arm into the vagina in an arc as shown in Fig 8-9. Progress to the cervix and beyond so that you can feel the entrance to each horn of the uterus. To do this your arm will have to enter up to the armpit.

Step 4. If after a manual examination you suspect some degree of uterine inertia (through fatigue or some other reason the uterus has stopped contracting strongly) or the sow appears to have given up trying, a small injection of oxytocin (0.5 ml) may be given. Normally it is not necessary because the pressure of the arm in the vagina stimulates further contractions. Well-grown piglets passing through the vagina have the same effect but small mummified piglets do not, hence a stillborn piglet may follow after a mummified piglet. Piglets suckling the sow's teats also stimulate uterine contractions so gentle massage of the udder and teats with your hand may be helpful.

Step 5. If an internal examination has been necessary and the farrowing process has been completed an injection of antibiotic should be given. An injection of long-acting penicillin (20 mg/kg) should be adequate to prevent any potential infection.

If there have been dead, possibly infected piglets present, 2 antibiotic pessaries should be deposited through the cervix at the end of the third stage.

Step 6. Always monitor the sow frequently over the next 24 hours to make sure that infection is not developing in the udder or uterus, that the placenta has been expelled and that the sow is suckling her litter normally.

Controlled Farrowings

If a sow is injected intramuscularly with the correct dose of prostaglandin from Day 113 days onwards farrowing will take place approximately 20 to 30 hours afterwards. Such medicines can be used to make the time of farrowing in an individual sow or a group of sows more predictable and allow better farrowing management.

- To carry this out on a group basis determine the mean gestation length in the herd.
- Inject the sows on either Day 113 or 114 depending on the herd gestation length, by intramuscular injection with the precise recommended dose. An intravulval injection is a cost effective alternative as only a half dose is required.
- Always check the mating and farrowing date of each sow.
- Assess the udder of each sow to ensure that she is close to farrowing in case mistakes have been made with the mating date. A good pointer here is that the back quarters of the udder tend to swell and fill up last.
- Inject the sows 24 hours before the required time of farrowing, which is usually early in the morning so that the sow commences farrowing predictably from mid-morning the following day. The normal distribution of farrowings and the results following injections are shown in Fig 8-15 and Fig 8-16.

NORMAL SPREAD OF GESTATION LENGTHS

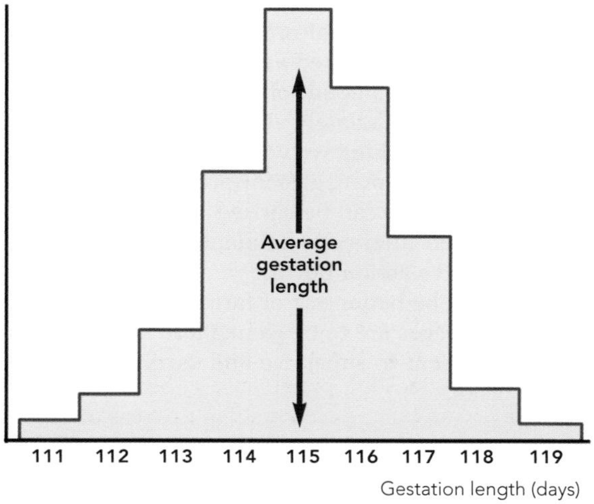

Fig 8-15

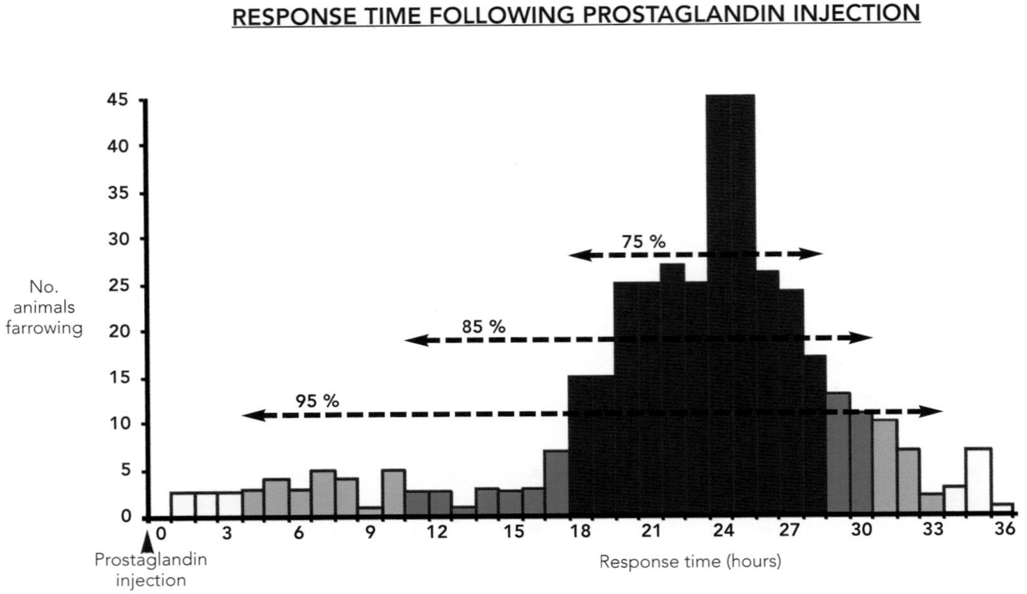

Fig 8-16

There are a number of advantages to synchronising farrowings:

- It is possible to prepare the farrowing pens at a more predictable time and increase throughput.
- The sow can be supervised before, during and immediately after farrowing.
- There can be a better use of labour particularly during the evening, nights and the weekends.
- The sow can be managed throughout the whole of the farrowing process and thus any difficulties, stillbirths, savaging or history of previous problems can be monitored.
- The sow can be treated as an individual.
- A considerable amount of care can be given to the piglets immediately after birth, particularly those not breathing well or of low viability.
- All the various management procedures to reduce piglet mortality can be carried out. Gilts which savage their litters can be identified and preventative action taken.
- There can be better use of farrowing houses.

If these advantages are to be gained it is important that people are present to supervise and carry out the management tasks.

There are disadvantages to using prostaglandins:

- Mistakes can occur and the sow can farrow early with weak pigs.
- In some herds there may be an increase in navel bleeding.
- Wrong timing of the injection may cause problems.
- If there is lack of quiet supervision mortality may be worse, because sows are farrowing during the daytime when there are a considerable number of distractions.
- Splay legs are more common with the smaller birth weight pigs.
- Lowers general birth weights which will add days to finish.
- Weaning on a Monday results in sows farrowing on a Tuesday and as this is towards the beginning of the working week there is less demand to induce sows which may increase birth weights and survivability.
- There is also a risk to the pig attendants of accidental self-injection which could result in abortion in pregnant women or temporary adverse side effects in non-pregnant women and in men. In some countries, prostaglandins can only be used under the supervision of a veterinarian and on veterinary prescription.
- Prostaglandins also pose a risk to asthmatics.

Prostaglandins should be stored in a locked cupboard under the manager's control and pregnant women should not be allowed to handle them. Procedures for the safe use of prostaglandins are described in Chapter 4: Hormones.

The decision whether to use prostaglandins or not must be an individual one based on veterinary advice and the results assessed on each farm. The economics are based on the fact that if 1 pig is saved per 10 litters farrowed (and this is easily attained by reducing still-births by 1 piglet) then the cost of the injections alone will be recovered. The important criterion however is that the extra time and effort should reduce piglet mortality by at least ½ a pig a litter.

Udder

A healthy functional well-formed udder is vital to piglet survival, not only to provide colostrum and milk but also to give all the piglets easy teat access.

Teat and udder conformation

There are 2 fundamental factors that decide whether a sow can rear 12 or more pigs: first, whether all piglets are able to get access to teats; and second, whether they can suck milk freely from them. These may seem obvious but how seriously are they considered when the gilt is being selected? It is not uncommon to see a gilt at farrowing with no functional teats at all, or a sow farrowing with say, 5 viable teats and the remainder non functional. But selection for good teats and udders is not as easy as it sounds, particularly if the number of females to choose from is small. Some teats that appear small and inverted at selection may develop and be fully functional at parturition and vice versa. Fortunately, in large batching herds the odd mistake can usually be mitigated by cross-fostering. Nevertheless, if you are a farmer who selects gilts from your own herd you should not forget that successful rearing of litters starts at gilt selection.

Teat conformation

A basic understanding of the anatomy of the teat is helpful if good functional ones are to be selected and their conformation can be classified from 1 to 5 (Fig 8-17). The perfect teat is elongated and pointed with 2 teat canals opening to the exterior. A Class 2 teat will not be so elongated but the teat end protrudes well down. Class 3 is the cut-off point for selection and this is where the teat sphincter (often appearing as a black dot) can still be seen when viewed at eye level. A class 4 teat is one where the teat sphincter is not visible, in other words the teat canal is shortened, resulting in an inverted teat. Such a teat should be considered non viable. A proportion of inverted teats will be drawn out by the piglet at suckling, but at least 50 % of them will remain blind. Why take the risk? A Class 5 teat is usually one where the teat has been rubbed off in the first 48 hours of birth (teat necrosis).

Teat numbers

The optimum or minimum number of functional teats on the breeding gilt is a debatable point.

Ideally the gilt should have a minimum of 16 functional teats. If, however, you are selecting gilts from your own herd, select 14 or more if possible. Remember the born-alive target is 13+ piglets born and each piglet needs a teat of its own.

Teat placement

The position of the teats on the udder is equally as important as teat conformation. It is no use having 14 perfect teats if their placement results in poor accessibility at birth. Teats should be equally spaced with no supernumerary ones and be in 2 parallel lines. When teats diverge they are poorly presented to the piglet at birth. Animals with large girths also exaggerate the teat placement (Fig 8-18).

Bad teat conformation is one of the major reasons why a breeding female will not rear 12 pigs. There may be a history in a herd of good litters born, yet by the time the pigs are 5 days of age, 2 or 3 begin to show signs of lack of milk, and they lose condition and have to be fostered. Two

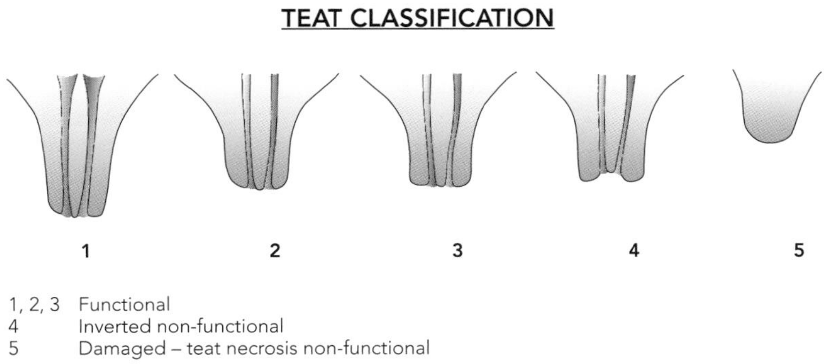

Fig 8-17

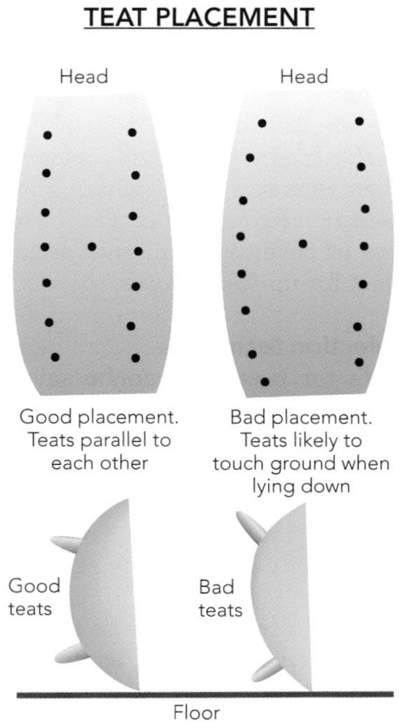

Fig 8-18

pigs can survive on 1 teat in the first 12 to 24 hours after farrowing but eventually the stronger pig takes over and the other is left with a onother teat that has now become accessible but may have started to dry off.

The placement of good teats on the boar that is used to produce breeding females should also be given due emphasis at selection.

Reputable breeding companies selling replacement breeding stock are fully aware of the importance of teat conformation, teat numbers and teat placement and make their examination an important part of the selection process. However, demand for gilts may be variable and selection criteria may vary. As a commercial producer buying gilts you should always check their underlines on arrival.

Teat necrosis

It has been recognised for a number of years that within 18 hours of birth some of the teat sphincters on those teats in front of the umbilical cord are traumatised by the floor surfaces. This causes the sphincter to become necrotic (die). Oestrogen produced by the sow in the farrowing process induces piglets to be born with swollen, oedematous teats and glands in both male and female piglets. The female piglet vulva is also swollen.

Such damage occurs on most floor surfaces but is obviously worse on rough floors and is almost complete within 24 hours of birth. The oestrogen only remains in the piglet for 24 to 26 hours, but by then the damage has been done.

Where gilts are to be selected from a litter, their teats should be protected from this trauma as soon as possible after birth. In some circumstances this can be helped by maintaining a deep bed of straw or shavings beneath the sow but in many cases this is not practical.

The alternative is to protect the teats by painting them with cow gum, (which is a rubber solution, often used for attaching photographs to paper), contact adhesive or covering them with adhesive zinc oxide elastoplast for up to 36 hours.

Selection technique

Gilts for breeding can be selected initially (and ear notched) at 5 days of age for conformation and 14 to 16 good pointed well placed teats. This will give an indication of the number of animals that are potentially available for selection at a determined future date. Up to 90 % (but allow for no more than 75 %) of these animals should be finally selected.

Potential breeding animals should be examined first in a confined space such as a weigh place to check the teats. If a 5 day selection has been carried out you will know that most gilts will have 14 to 16 teats. The ideal is to set a weigh place on a ramp so that the gilt's udder is 0.9 to 1.2 m from the ground. The observer can then carry out a detailed examination and at the same time assess the lateral displacement of teats. The final selection should be based upon a normal vulva, overall conformation and ease of movement to reduce the risk of leg weakness.

Recognising impending disorders and possible lactation failure

These must be determined at the onset and the following procedures will help:

- The udder of every sow at farrowing and 12 to 24 hours afterwards should be palpated. The palm of the hand is placed over each gland with the teat in the centre and pressure applied to a normal gland to the point at which the sow just responds. This standard is then used to detect any abnormal pain and changes in texture to the other glands.
- The presence of oedema or fluid in the vulva or in the surface tissues between the legs should be noted.
- A finger should be pressed hard into a gland to see whether a small impression is left behind. If so this is further evidence of the very early stages of oedema.
- The first detectable changes are usually seen 4 to 6 hours after farrowing but occasionally severe mastitis or infection of the gland would be evident before farrowing, in which case inappetence and failure to suckle are observed.
- The experienced stockperson or the veterinarian will recognise lactation failure by behavioural changes in the sow, lack of alertness and failure to lie over and suckle.
- Affected glands may be discoloured and swollen.
- The sow may be off her food with a fever and laid on her belly.

The first indication of lactation failure is shown in the piglets by raised hair and hollow flanks, and they actively seek food.

Disorders associated with the udder can be grouped into 4 conditions:

1. Udder oedema and failure of milk let down.
2. Mammary hypoplasia.
3. Agalactia.
4. Mastitis.

Udder Oedema and Failure of Milk Let-Down

See Chapter 14 for more information

This presents itself as a failure of milk let down associated with excess fluid in the mammary tissues and is a condition seen in both gilts and sows. It is characterised by a clinically normal animal with no fever or loss of appetite. Where the problem is in individual sows, treatment can be applied. If there is a herd level problem, feed and nutrition should be reviewed. Review pre farrowing feeding. In certain breeds over feeding protein to the pre farrowing sow predisposes this condition.

If the problem is suddenly seen in a number of sows check for mycotoxicosis. See Chapter 14 for details.

Agalactia – No Milk

Agalactia describes a shortage of milk supply in an otherwise healthy lactating animal. It is relatively uncommon as a prime condition but common as a sequel to extensive oedema of the udder and mastitis. It is also seen in older sows where hormone outputs are reduced. The major cause however is associated with inadequate water supplies.

Mammary Hypoplasia – Undeveloped Udder

This term defines failure of udder development and is relatively uncommon. It can occasionally be seen in gilts because the hormones that are responsible for the development of the udder have not been produced in sufficient quantity. Such animals should be culled. Occasionally herd problems are seen where poor nutrition, heavy worm burdens, chronic disease or mycotoxins may be implicated. The most likely mycotoxin to cause it is from ergot poisoning in pregnant gilts running in grass paddocks. Shortage of water is a common cause. Rarely, hypoplasia may be due to a genetic mutation.

Mastitis – Inflammation of the Mammary Glands

Mastitis is inflammation of 1 or more mammary glands caused by a variety of bacteria species or secondary to other diseases. It is a common condition that occurs sporadically in individual sows or sometimes as a herd outbreak associated with a specific infection. Disease starts at or around farrowing and becomes clinically evident up to 12 hours later. It can arise as a primary infection, that is, bacteria getting into 1 or more mammary glands for the first time around farrowing or it may be a flare-up of a sub-clinical latent infection, possibly present in 1 or more small abscesses, activated by the development of the gland and the flush of milk. It sometimes arises as a sequel to udder oedema possibly due to poor milk flow.

The route of entry of the bacteria is thought to be the teat orifice or injection into the gland by sharp piglets' teeth. Very occasionally mastitis may arise from a septicaemia. The bacteria considered to cause mastitis in the sow can be grouped into 3 broad categories: coliform bacteria, staphylococci and streptococci, and miscellaneous bacteria. Limited surveys suggest that coliform mastitis is the most common and usually most serious, staphylococcal and streptococcal mastitis fairly common and usually less serious and miscellaneous bacteria uncommon and varying in seriousness in the individual sow.

Coliform mastitis – Coliform bacteria are bacteria that are related to *E. coli*. The 2 commonest in sow mastitis appear to be *E. coli* itself and klebsiella species. These organisms can be responsible for severe acute necrotising mastitis. They release a toxin (endotoxin) which results in reduction in milk yield, a very ill sow and poor "doing" piglets. Marked discoloration of the skin over the udder and dark blueing of surrounding skin, ears and tail is a feature of the condition.

Major herd problems can develop because the normal habitat is the pigs' intestines and faeces, and bacteria may also be present, particularly klebsiella, in sows' urine. Consequently, they are everywhere in a piggery and can survive and build up in water bowls, pipes and header tanks. They can multiply rapidly in stagnant contaminated pools or films of liquid on the farrowing room floors and in wet bedding under the sow. Coliform mastitis may thus be regarded as environmental in origin.

Staphylococcal and streptococcal mastitis – These are usually less acute and less severe than coliform mastitis. They tend to occur sporadically in individual sows in 1 or 2 (or sometimes several) glands and usually do not make the sow ill. The exception is an acute severe staphylococcal infection, usually in a single gland, which becomes swollen, hard and discoloured. In the majority of cases, however, the sow remains normal with a hard gland which has reduced milk supply.

Unlike coliform bacteria, the source of these organisms is not usually the contaminated environment but the skin and possibly orifices of the sow herself. There is some evidence to suggest that as in the dairy cow and sheep some of these bacteria may persist sub-clinically in the udder and then flare up at or after farrowing.

Miscellaneous bacteria – These include organisms such as pseudomonas which can produce a serious mastitis and toxaemia and which are often resistant to antibiotic treatment. Fortunately such infections are rare.

Clinical signs

Acute disease

The sow is inappetent at farrowing, or before if mastitis is already developing; she is obviously ill and the mucous membranes of her eyes are brick red. There may be discoloration of the ears and the whole of the udder but particularly over the affected glands. In the early stages, palpation will identify the infected quarters but observation alone is often enough to detect swollen glands without carrying out an examination. The temperature ranges from 40 °C to 42 °C (104 to 107 °F).

Chronic disease

This follows acute episodes at farrowing or at weaning. The mammary tissue is infiltrated with abscesses and hard lumps that are usually not painful when palpated. They may ulcerate to the surface and thereby become a potential source of infection to other sows.

Diagnosis

The clinical signs are usually sufficient to diagnose mastitis. However if there is a herd problem with a number of sows affected, you should examine all animals clinically at farrowing and again at weaning, to determine the starting point of the mastitis. A sample of the secretions from the infected quarters should be submitted to a laboratory for examination. This is carried out by wiping the teat end with cotton wool soaked in surgical spirit, injecting the sow with 0.5 ml of oxytocin, and, once there is a good flow, squirt the milk on to a sterile swab. The swab should be immersed in a transport medium. Obtaining a representative milk sample can be very difficult as there are 2 or even 3 mammary glands supplying each teat. The non-infected glands may still be producing normal milk. It is very important that mastitis is diagnosed early and that prompt treatment is given.

Treatment

Treatment should consist of the following:

- Oxytocin to let milk down (5 iu).
- Antibiotics as prescribed by your veterinarian depending on the organism and its sensitivity.
- The following could be used: OTC, penicillin and streptomycin, trimethoprim/sulpha, semi-synthetic penicillins such as amoxycillin; framycetin, tylosin, enrofloxacin and ceftiofur.
- In very severe cases the sow should be injected twice daily.
- If the sow is toxic an injection of flunixin could be given.
- Corticosteroids may also be prescribed.
- In severe outbreaks the sow can be injected 12 hours prior to farrowing with an appropriate long-acting injection.
- If sawdust is used as bedding, stop using it because when soaked with urine it is an ideal medium for bacterial growth.
- Top dress sows' feed with antibiotic commencing 3 to 5 days pre-farrowing. Use OTC, trimethoprim/sulpha, amoxycillin or CTC depending on the antibiotic sensitivity.

Management control and prevention

There are 2 fundamental requirements for the development of mastitis. The first is the presence of the causal organism and the second is an ideal environment at the teat end for the organism to multiply and gain access to the mammary gland. Occasionally mastitis will arise from a blood-borne infection associated with the farrowing process.

The following factors predispose to mastitis and require remedial action:

- The continual use of farrowing houses.
- Poor farrowing pen hygiene, bad drainage, inadequate bedding, poor quality bedding.
- The use of sawdust or shavings for bedding that becomes soaked in water or urine.
- A warm temperature for the organisms to multiply.
- Worn, pitted farrowing house floors.
- Wet farrowing house floors.
- Contaminated drinking water.
- Adverse temperatures and ventilation in the farrowing houses that cause abnormal lying habits in the place.
- A build-up of faeces behind the sow. Faeces behind the sow should be removed every day.

Additional measures are as follows.

- If a klebsiella infection is the cause of a herd outbreak it may be necessary to clean out the watering system.
- If floor surfaces are poor these can be improved by brushing them with lime wash containing a phenolic disinfectant. This should be allowed to dry for 48 hours or so before the sow enters the farrowing house.
- The udder can be sprayed daily with an iodine based dairy teat dip, commencing 24 hours before expected farrowing. This spraying should continue once a day for the first 2 days post-farrowing. If a specific organism is identified and its antibiotic sensitivity is known, the sows' feed can be top-dressed from day of entry into the farrowing houses until 3 days post-farrowing with the appropriate in-feed antibiotic or injections of appropriate long-acting antibiotics at farrowing.
- Cull chronic infected sows.

If you have a mastitis problem on your farm and have read this section, now study Fig 8-19 and Fig 8-20 which summarise the various predisposing factors.

FACTORS CONTRIBUTING TO MASTITIS IN THE SOW

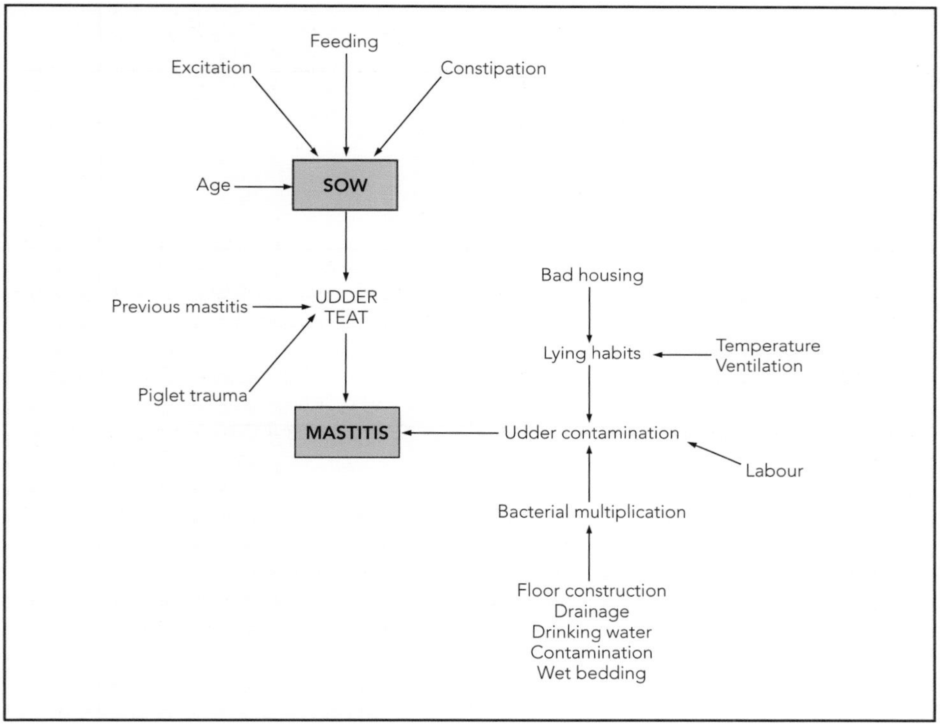

Fig 8-19

FIG 8-20: LACTATION FAILURE: SUMMARY

Udder Change	Causation Factors	Management / Action	Treatment
Udder oedema and failure of milk let down (congestion)	Constipation Excitement Fear High levels of feed Lack of exercise Lack of roughage Too much slope on the floor Udder oedema	Correct the cause Reduce feed level	Electrolytes for litter Oxytocin Penicillin Pain relief
Mammary hypoplasia	Age Breed Hormonal Individual animal Mycotoxins Water shortage	Correct the cause Cull affected animals	Foster litters
Agalactia	Age Excess body condition Poor place design Sequel to congestion Sequel to oedema Water shortage	Correct the cause Foster / cull Reduce feed level	Oxytocin Supplement milk
Mastitis, metritis and toxic agalactia	Contamination during assisted farrowings Contamination of udder Cystitis Diet change Farrowing High feed levels predispose Lack of roughage Metritis/vaginitis Nephritis Previous mastitis Prolonged farrowings Septicaemia Sequel to toxic agalactia Sequel to udder oedema and congestion Stress	Assess causal factors Carry out bacteriology of farrowing houses Check feed Check feed levels given See under mastitis	Antibiotics as for mastitis. Pain relief
Acute mastitis	Bad drainage Continuous antibiotic usage Faulty drinkers Piglet teeth damage Poor farrowing house hygiene Poor floor surfaces Presence of pathogenic organisms Shavings, sawdust as bedding Teat trauma Wet floors	Assess causal factors Check teat damage Check udder contamination Farrowing house hygiene/floor quality Keep individual records Lime wash pen floors (+ disinfectant) Look at all management procedures Water supplies	Antibiotics Pain relief Corticosteroids Diuretics Flunixin Framycetin Medicate pre-farrowing Neomycin Oxytetracyclines Oxytocin Penicillin Streptomycin Synthetic penicillin Tylosin

Management at Farrowing

Farrowing House Design

The fulcrum of efficient farm planning revolves around farrowing house design. This is covered in Chapter 3: Planning for Efficient Production and Health Control. The control of piglet diseases is dependent upon the design and the way it is managed. The design should enable an all-in/all-out management system to be operated with complete cleaning, disinfection and drying between each batch. This prevents the build-up of infection and reduces the continuity of pathogen exposure to viruses, bacteria and parasites.

- To achieve an efficient all-in/all-out system:
  - Each production space (farrowing room or rooms) should be of a size appropriate to the number of the sows in the herd and the number of farrowings planned for each batch.
  - There should be 1 production space per batch.
  - Each production space should have the same number of farrowing places.
- The floor and work surfaces should be made of non-porous, easily cleaned materials that dry quickly.
- The floors should be well constructed and drained so that no pools of liquid occur and they should be free from cracks and fissures that harbour infections.
- The buildings should be adequately insulated and the ventilation system should maintain even temperatures with no draughts around the sows or piglets. In temperate and colder climates the system should be mechanically operated.
- There should be passages in front of the farrowing pens as well as behind them for easy access to the piglets without climbing from 1 pen to another. This is particularly important when outbreaks of piglet diarrhoea occur.
- Biosecurity should be maintained between each production space. Stockpersons should wash hands between each production space; each should have its own equipment.

Examples of farrowing house layouts are given in Fig 8-21. Farrowing space and pen design are important in the management of the sow and litter. Over time a variety of pen and farrowing place designs have been developed, some good, some not so good, none perfect. Numerous pens have been designed that avoid total confinement of the sow or which confine the sow for the first few days of lactation and then allow her free movement (e.g. multi suckling systems) but they have all been found wanting and some have been disastrous. To date there is no system more productive, welfare friendly to both sow and piglets, and safe for the attendant, than full confinement of the sow for most or preferably all of lactation.

An example of a satisfactory farrowing pen layout is shown in Fig 8-22. Many other designs are also satisfactory.

The pen area is a minimum of 1.8 m wide by 2.4 m with the place offset to one side with a side creep to the larger side close to the front passage. Provided the management is good there seems to be little difference in piglet mortality whether the creep is in front of the sow place, to one side of the head of the sow (as in Fig 8-22) or further back, level with the udder. The position of the creep in Fig 8-22 gives the piglet contact to the sow's head, fostering a maternal bond. It encourages the piglets to remain at the head of the sow rather than at the udder where they may be laid on. In hot climates (including summer in the North of Europe) a heated front creep may prevent the sow cooling herself and drip cooling procedures become necessary.

That said, designs vary widely and serve several purposes:

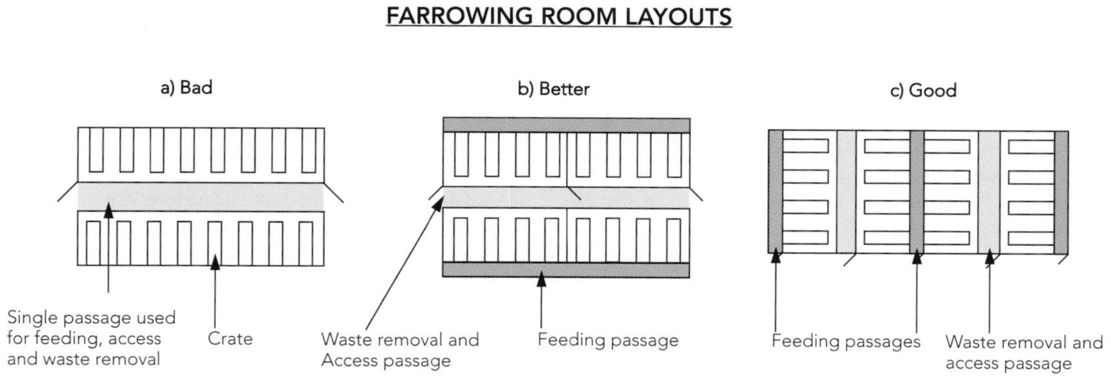

Fig 8-21

A FARROWING PEN FLOOR DESIGN

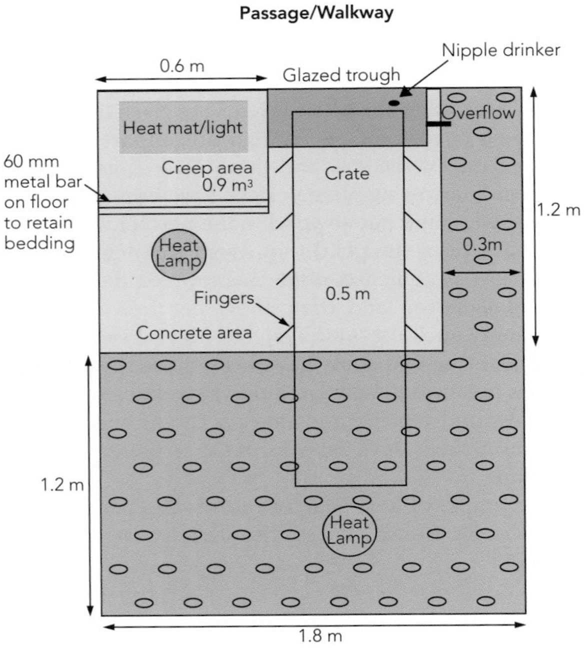

Fig 8-22

- Safety for the stockperson.
- Ease of management for such procedures as clipping teeth and tailing, examining the piglets and the sow's udder.
- Treating the piglets or the sow, feeding, changing creep feed and general hygiene.
- Provision of a simple safe creep close to the sow for the piglets.
- Reducing piglet mortality from crushing and disease.
- Economy of space in the farrowing room.

One guide to the efficiency of the design is the level of mortality achieved but other factors should also be assessed, such as the comfort and contentment of the sow, whether the places are big enough for the biggest sows and the availability of both rows of teats to the piglets. To facilitate this (and to reduce crushing against the bottom bar) "fingers" are incorporated (as in Fig 8-22) instead of a low bottom bar.

Various gadgets have been developed to reduce crushing when the sow lies down:

- Adjustable bottom rails for different sized sows.
- A hinged bottom bar that drops inwards when the sow stands up, making it difficult for her just to drop down when lying again.
- A fan that automatically turns on when the sow stands and blows cool air around her feet (blowaways) encouraging the piglets to return to the creep.

The floors of farrowing places are important to piglet survival and health. If solid they should be insulated and smooth enough not to cause abrasions on the piglets' legs but not too smooth and slippery to make it difficult for the newborn piglet to get to the udder and creep area.

Floors may be fully perforated, partly perforated (as in Fig 8-22) or only perforated at the back end of the sow. They may be raised to various heights above the passage level to help in manipulation by the stockperson, to deter the stockperson from climbing, and to raise the piglet away from draughts at floor level.

Whilst farrowing place design is important in assisting the sow to lie down gently, nevertheless its impact on pigs laid on is low if the management and the design of the house encourages the piglet not to lie in the danger dropping zone. The day by day management of the pen, the bedding and good drainage of the floor are vital components for success.

Good management in the farrowing house is the key to the successful rearing of healthy pigs, low pre-weaning mortality and maximum number weaned.

If your number weaned is less than 12 then consider in detail the following outlines.

Preparing the Farrowing House

- Sows should be moved into a dry, warm house about 3 days before the expected farrowing date.
- The house should have been completely emptied, cleaned, disinfected and, more importantly, dried.
- Check all drinkers, feeders and the ventilation system before any sow enters.
- As farrowing approaches, a second heat lamp should be placed opposite the sow's udder (if it is a side creep) to attract the piglet away from the sow.
- If the farrowing place floor is slatted and the sow is likely to farrow at night time, a lamp should be hung at the back of the sow.
- If a sow is farrowing on slats, the areas behind and to the side of her should be covered over with solid material during the actual period of farrowing and kept dry with shavings or use the rear passageway as a temporary farrowing nest.
- Make sure there are no draughts or high air flow across the house.
- Make sure all nipple drinkers are working correctly.
- Check the floors for any wear or tear or any loose

panels if slatted. (Failure to carry this out frequently results in the loss of a litter in the slurry).

Preparing the Sow

- Make sure that all health routines have been carried out, for example vaccinations, mange treatment and worming.
- Check the feeding of the sow.
- Regularly examine the udder as farrowing approaches.
- Examine the vulva of the sow twice a day for any abnormal discharges.
- Make sure the faeces is removed twice a day from behind the sow until 2 days post-farrowing, then once daily for 8 days.

Maternity Management – Supervising the Farrowings

The greater the management input from point of farrowing and for the next 2 days, the lower the piglet mortality. Thus, more pigs will be weaned. Remember most mortality occurs in the first 48 hours. Prostaglandin can be used to synchronise farrowings. Monitor the progress of the farrowings as previously described.

Acclimatising the Newborn Piglet to the Creep Area

This is a vital component of farrowing management and is the one procedure that dramatically reduces the number of pigs laid on. It is possible to teach the piglet within 4 to 6 hours of birth that the creep area is the most attractive and best place to lie. This is achieved by:

1. Removing the piglet as soon as it is born into a well-bedded comfortable creep area. (At least 50 mm of shavings is ideal). The piglet is fastened in the creep area for approximately 20 minutes, then introduced to the udder of the sow, so it is necessary to have a barrier to hold the pigs into the creep area for this period of time. Watch out for damaged navels and bleeding. If so use navel clips.
2. Once the piglet has had sufficient colostrum and has finished suckling it is placed back into the creep area for up to 1 hour.
3. Piglets during the period of farrowing when they are not suckling are gently moved and fastened in the creep area for up to 1 hour.
4. At any period of time thereafter if any piglets are seen lying next to the sow and not suckling they are immediately fastened in the creep for a short period.
5. As soon as the sow has finished farrowing, with the piglets fastened in the creep area, she is made to stand. She will often drink and then lie down, after which the litter is allowed out again. The following morning when the sow is fed, the piglets should be fastened into the creep for 40 minutes.
6. To carry out these acclimatisation procedures it is necessary to design the creeps with a simple and practical method of confining the piglets.
7. This system of management allows split suckling, whereby the larger pigs in the litter can be held in the creep area and the smaller ones given uninhibited access to the udder and then the procedure reversed.
8. A second creep lamp should be provided for 24 to 36 hours post-farrowing opposite the sow's udder.
9. Provide bedding beneath this lamp to attract piglets to it.

Provide a warm, comfortable well-bedded creep area. Both you and the piglet will be rewarded.

Maximising Colostrum Intake

- Maximising the intake of colostrum in the first 6 hours is vital to piglet survival.
- Once each pig has stopped suckling for the first time it should be marked and fastened into the creep area.
- Split suckling will increase the availability and intake of colostrum.
- As soon as piglets are born or have received adequate colostrum they may be split amongst other farrowing sows by number and even weights.
- Weak or poor viable piglets can be given colostrum by a syringe into the mouth. The colostrum is withdrawn from the sow as she is farrowing, into a small dish. (See Key points to managing the poor viable pig). A minimum of 10 ml of colostrum should be given to under-privileged small pigs. Piglets can drink 250 mls of colostrum in the first day of life.
- Assist weaker pigs to a teat.

Fostering Piglets

Post-mortem surveys carried out to investigate high levels of piglet mortality have shown that over 30 % of the piglets that die have no milk in their stomachs.

The baby piglet is usually born in a fairly precarious state with limited energy reserves and with no acquired immunity. It undergoes a marked drop in environmental temperature from 39 °C (102 °F), often down to as low as 18 °C (65 °F). It has no fat insulation, very little hair and poor thermo-regulating mechanisms. It is therefore very sensitive to temperature changes and is heavily dependant on a high environmental temperature to maintain its own body temperature. It has a disparity in size to the sow of approximately 1:200, which is rather an unbalanced situation to be presented with at birth.

It therefore has to satisfy 4 very important requirements:

1. The intake of antibodies from the colostrum, in particular IgG (immunoglobulin G) and IgA (immunoglobulin A). Without these it will die, having no protective mechanisms against the environmental organisms.
2. Transfer of cells from sow to her own piglets. These are important part of the piglets' future cellular immunity.
3. It must conserve heat to be able to utilise its scant energy resources to compete with litter mates and gain access to a teat.
4. It requires an immediate digestible source of energy (i.e. sows' milk).

Clinical abnormalities of the piglet at birth

- Low birth weight – immature.
- Hypoglycaemic – low blood sugar.
- Anoxic – short of oxygen.
- Defective – e.g. splay leg, cleft pallet.
- Anaemic.
- Diseased e.g. PRRSV, *E. coli*. (particularly toxins).
- Traumatised.

Fostering the piglet implies removing it from its own natural mother to another sow so that it is able to gain access to a teat, suckle, and thereby survive. There are a number of reasons why it is necessary to carry this out.

Reasons for fostering

Too many piglets – Some herds have the luxury of having too many pigs born alive and these surplus pigs if they are to survive must be given a new sow. The fostering of such pigs is a vital component of increasing output and when we see records of herds weaning 13 piglets per sow farrowed, invariably they are fostering to create new litter groups.

Variable birth weight – The mortality within any litter group is dependent in part upon the variation in birth weights; the greater the variations, then the higher the mortality in those piglets of low birth weight.

Fostering pigs at birth between sows is an important procedure in reducing piglet mortality.

Weak or poor viability piglets – As with small pigs, poor viable weak piglets are likely to die if left unattended. The grouping of 8 or 10 of these together onto a sow with good teat access is a part of efficient farrowing house management and increases the survivability of these piglets by some 80 %.

Mastitis or diseases in the sow – These may result in little milk being available. It may be necessary to foster the whole litter onto other good milking sows.

Savaging – This is common in intensive environments. Occasionally if a gilt or sow does not respond to sedation it is necessary to foster the whole litter.

Delayed weaning – Pigs that are held back from weaning because they are small or unthrifty are often moved to a foster sow. Be careful not to move sick pigs back to younger ones. The ideal is to have separate small farrowing rooms for such procedures. Foster sows should not be moved to the next room where sows are due to farrow (see "culled sows" below).

Starved piglets – These are often seen at 3 to 5 days of age because 1 or more quarters of the udder have stopped producing adequate milk.

Death of a sow at farrowing – Occasionally this occurs and it is necessary to foster the complete litter. This can also arise if an emergency hysterectomy is carried out on the sow due to farrowing difficulties.

The rules of fostering

If fostering is to be successful then there are certain ground rules which should be followed closely.

Timing – The ideal time to foster a pig is as soon as it is born or within 6 hours and this is the procedure often adopted to even up numbers across litters and birth weights when a number of sows are farrowing at the same time. Provided piglets are moved within this period onto another sow that is at a similar stage then there are no problems with incompatibility or intake of colostrum.

The second time period is when surplus pigs are being collected together to make a fresh litter. In this case they should not be moved from the sow until at least 6 to 8 hours after farrowing, when they have had a minimum of four, 40 minute periods of uninhibited access to the teat. This is to ensure maximum colostrum intake because the fostered pigs are going to be moved forward to a sow that will be suckling a litter of approximately 4 to 5 days of age and there will be no colostrum. The piglet should be totally dry when fostered.

This age factor is important if the establishment of a new litter is to be successful. Furthermore it is essential that only the biggest pigs are fostered forward to make up a fresh litter.

Procedures

1. Mix the sow's own litter and the foster one in the creep and hold there for approximately 30 minutes.
2. Move the sow's 5-day-old litter forward to a sow

suckling a litter of 10 days of age and repeat the mixing process. This acclimatisation is a valuable technique because it allows the piglets to intermingle and make the foster litter much more acceptable to the sow.

3 Repeat 2, and the 10-day-old litter is moved to a sow suckling at 15 days of age.
4 The 15-day-old litter is weaned early and there is no loss in non-productive days or increase in lactation length.
5 Fostering can also be carried out with poor piglets between 1 and 7 days of age. It is important to identify these pigs early because they tend to lose their suckling reflex and die. Again, the technique is similar. Find a sow that is suckling a litter 5 to 7 days of age, moving her litter forward and fostering on the poor pigs. For disease control purposes, isolated farrowing pens are best used. Give such piglets a 25 mg/kg injection of tulathromycin at the time of movement because they are compromised and susceptible to infections.

Selection of the sow and management – The success of fostering, particularly whole litters, depends on the number of days the foster sow has been suckling and keeping the age disparity between the foster litter and the sow's own litter to within 4 to 6 days. Always select a docile sow with a good teat profile, particularly if a litter of poor viable pigs are collected together. The gilt or second parity animals are best.

The sow's udder – Look carefully at the foster sow's udder and the quality of the piglets that are suckling. For example, if there are 10 good pigs suckling then that sow will receive 10 foster pigs. However, if there are only 8 good suckling and 2 poor pigs then do not expect 10 pigs to survive. Only foster 8.

Availability of water – Whenever a litter is fostered always make sure there is clean, fresh water available for the piglets. In some cases the sow may be reluctant and slow to accept the fostered litter and piglets quickly become dehydrated.

The movement of foster pigs – Wherever possible always foster within farrowing houses, or into a farrowing house with older pigs. It is bad policy to move piglets back into younger age groups due to the risk of spreading disease.

Culled sows – It is a useful technique to have a number of farrowing places set aside so that sows due for culling can be used for extra suckling. By removing them out of the mainstream of the farrowing houses they do not interfere with the important all-in/all-out procedures. These places are outside the normal farrowing rooms and are not to be considered part of the batch. They are part of the farm's hospital area. Assuming the pig flow has been met for the next batch, cull sows or gilts can make excellent foster mothers. The advantage of a gilt is that the extra lactation will normally increase her next and subsequent litter sizes.

If you have a litter size of a least 11.2 born alive and are only weaning 10 pigs per sow, I suggest you read this section again and look at the advantages that can be gained from fostering.

The Stillborn Pig

Stillbirths are usually recorded as such when they are found dead behind the sow. However this can be an erroneous assumption because there are 3 possible causes:

1 Death before farrowing.
2 Death during farrowing.
3 Death after farrowing.

If the pig dies before farrowing, then depending on how long before, it will show varying degrees of post-mortem or degenerative changes including discoloration of the skin and loss of fluids. If death occurs in the early stages of pregnancy, a fully-formed mummified pig will be seen. A cloudy eye indicates a pre-farrowing mortality which may indicate PRRSV as a cause.

A pig that dies during the process of farrowing or immediately afterwards will be fresh and normal. The 2 can be differentiated easily. The chest is opened and the lungs and the trachea examined to determine whether the pig had breathed, i.e. been born alive and then died. The lungs of the true stillborn pig are a dark plum colour, showing none of the pink areas associated with inflation and breathing. The stillborn piglet's lungs do not float in water. Pigs that attempt to breath during the process of farrowing will also show evidence of mucous obstructing the wind pipe. In a piglet which has breathed, the lungs will float.

Ask your veterinarian to show you the differences.

A good target level for stillbirths is 5 to 7 % of total pigs born. At this level there is no point in carrying out investigations because it is unlikely that external inputs can alter the situation. However once the level reaches beyond 9 % it is worthwhile carrying out an investigation by records and post-mortem examinations. The following factors need to be considered as causal or contributory to the problem:

- Stillbirths increase with the increasing age of the sow and beyond 5th parity may reach 20 %.
- Individual sows may be regular offenders and these can be identified by the sow litter card. The farrowing process should then be monitored. The sows and their cards should be clearly marked on entry to the farrowing house.
- Stillbirths occur in larger litters.
- They are more common in pure breeds.
- Sows that have prolonged farrowings (> 8 hours) will have a higher number of stillbirths.

- Farrowing house temperatures above 24 °C (75 °F) increase the risk of stillbirths due to the difficulties of the sow panting and resting during delivery.
- Sows with uterine inertia, particularly if it is associated with calcium or iron deficiency, produce high numbers of stillbirths. One sow can make the average look bad.
- High carbon monoxide levels in the air associated with faulty gas heaters can raise stillbirth rates significantly.
- Pigs found dead behind the sow can sometimes be related to specific farrowing places in certain rooms, associated with draughts behind the sow, with the pig dying shortly after birth due to hypothermia.
- An examination of records both by parity and total numbers born per individual litter will clarify whether the problem is one of individual sows or whether there is an infectious or common environmental component.
- Stillbirths are raised where there is a long gestation period and in such cases prostaglandin injections can be used. In some herds the use of prostaglandin has reduced stillbirths and yet in others it has increased.
- Lack of exercise may have an effect on the stillbirth rates.
- Diseases of the sow such as fever, mastitis, septicaemia, acute stress or haemorrhage can have an effect.

Diseases associated with the stillborn pig

- Anaemia.
- Aujeszky's disease/pseudorabies virus (AD/PRV).
- Teschovirus.
- Erysipelas.
- Leptospirosis.
- *Mycoplasma suis*.
- Mycotoxicosis.
- Porcine parvovirus (subsequent to the delivery of a mummified pig).
- Porcine circovirus (PCV2).
- Porcine reproductive and respiratory syndrome virus (PRRSV).
- Toxoplasmosis.

Where stillbirth levels are high it is necessary to eliminate disease as a possible cause and then identify the predisposing factors and their relevance. Most stillbirths in the absence of diseases or environmental faults are related to age, individual sows and large litters.

To reduce stillbirths

- Do not let the age of the herd spread beyond the 7th litter.
- Identify problem sows. Observe farrowing behaviour.
- Look at breed differences.
- Check farrowing house environments.
- Check farrowing pen designs.
- Monitor farrowings.
- Interfere early in prolonged farrowings.
- Give good management at farrowing.
- Provide a heat source behind the sow at farrowing.
- Study herd records.
- Check haemoglobin levels in sows.
- Check parasite levels.
- Check for blood parasites.
- Check for diseases in the sow.

The poor viability pig (often called low viable)

Poor viable pigs are usually classified as being small and less than 800 g (2 lbs) in weight, but they can also include those of good birth weight that are weak and lacking vitality. It is necessary to differentiate between the poor viable and the non-viable one. The latter is a pig on that farm with that management that has no possibility of survival. The rule of thumb is simple, when the body temperature has been brought up to normal and if the pig has no suckling reflex when the little finger is placed inside the mouth, it is unlikely to survive and therefore management time should not be wasted on it.

The size of the piglet is in part determined very early on in its life at around the time of implantation. While we do not understand all the mechanisms that are likely to produce a large or small placenta and thereby a large or small piglet, nevertheless, several contributing factors can be identified:

- Breed is important and in particular hybrid vigour. This is clearly seen in the difference between breeding from a pure-breed or pure line and a cross-bred female. There are different levels of hybrid vigour between different hybrid and breed combinations. The selection of a good breeding female should include the capacity of that animal to produce good even birth weights.
- Nutrition during the early part of pregnancy, particularly around implantation, may play a role. Unidentified growth factors contribute to the establishment of the placenta. Field experiences have shown that major problems of poor viable piglets (up to 40 %) tend to occur more in herds where milk by-products such as whey have been fed in the first 3 weeks post-mating. In such farms, when the ration was

changed to a cereal diet, the problems went away. The reasons for this are not known and one can theorise that dietary insufficiencies or unknown growth inhibiting substances might be present in some diets.

- Increase the daily ration during the last 3 to 4 weeks of pregnancy (to 3.5 kg a day) in order to increase the birth weight of all the pigs in the litter, particularly for outdoor sows in winter. This however, will not reduce the variation within the litter. **Note** if the increase in feeding continues after Day 110, it might induce udder oedema or mastitis.
- As the age of the sow increases, so do the numbers of poor viable pigs and there is a greater disparity in birth weights.

Diseases such as swine flu, PRRSV, swine fever and parvovirus (in fact any disease that can cross the placenta) can produce marked increases in poor viable pigs. If there is a herd problem, it is necessary to assess the overall clinical picture to identify any diseases that might be associated. Fig 8-23 shows the factors that contribute to poor viability.

Key points to managing the poor viable piglets

- Immediately place the piglet in a draught free environment at a temperature of at least 30 °C (86 °F) ideally in a well bedded box with an infra-red lamp above.
- Make sure that the lamp is not too far down - this may burn the skin.
- Poor viable pigs rapidly deplete their minimal energy resources if they are allowed to dry off in the normal farrowing house environment.
- Always make sure that the eyelids are prised open because some are born with eyelids stuck together.
- Provide the piglet with a rapid source of energy. Sows' colostrum is ideal, obtained at farrowing and given to the piglet by syringe.

If there is a poor viable problem consider lifting sows' feed intake from 3 to 21 days post mating.

- Do not use a stomach tube because it does not stimulate a suckling reflex and the sooner this is established the better. Do not syringe colostrum into the piglet until a suckling reflex is felt by the little finger placed in the mouth. Cow or goat colostrum collected soon after parturition and stored deep frozen can be used as an alternative source. It is thawed out in warm water (do not microwave) as and when required. Poor viable pigs should be given between 5 ml to 10 ml as soon as the body temperature has returned to normal and this should be repeated 4 to 6 hours later. Commercially produced artificial colostrums are available but they are expensive and no better than the natural products.
- A poor viable pig has a much smaller chance of

THE PIGLET WITH POOR VIABILITY

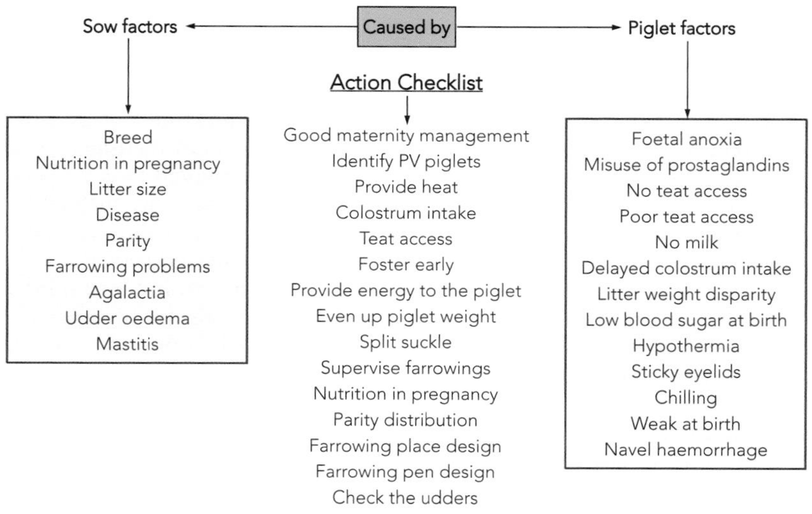

Fig 8-23

survival if it is left within the litter to compete with the bigger piglets. Where a number of sows are farrowing at the same time collect all the small pigs together to form a new litter so that they are given special attention and a much warmer, more comfortable environment. A newly farrowed sow with easy teat access should be selected to suckle these under privileged animals.

- Split suckling is useful if poor viable piglets have to be left on the sow. The litter is divided into 2 weight groups and the smaller weaker ones given uninhibited access to the udder on at least 2 separate occasions, as soon as they can be collected together after birth.

A creep temperature of 35 °C (95 °F) is required immediately at birth, provided the air flow is less then 0.15 m/sec. If the air flow is doubled, the temperature required by the piglet could rise by up to 6 °C (10 °F). A wet creep area and an uninsulated or unbedded floor could also increase the required temperature by 5 °C to 10 °C.

Nutrition

- Modern lactating sows are leaner than their contemporaries and they produce large amounts of milk. Some can produce 12+ litres (kg) of milk per day. This is comparable to many breeds of dairy cow on a bodyweight ratio and sows' milk is much richer than cows' milk. They also have a larger bodyweight relative to age and are more immature and still growing at the times of mating, farrowing, lactating and weaning as a gilt. These females therefore have higher maintenance requirements together with reduced feed intake in lactation. These changes make it particularly difficult for the gilt to consume sufficient energy to meet the demands of growth and maintenance, foetal growth and then milk production. As a result there may be a breakdown of body tissues (catabolism) to meet these requirements. Although mature sows are not growing, they often have larger litters and an increased demand for milk products. The following factors need to be considered when assessing gilt and sow nutrition, feed intake, production and disease:
- The breeding female should ideally not lose more than 10 kg (22 lbs) of weight during lactation.
- Losses above this will extend the weaning to mating intervals with fewer animals in heat within 10 days of weaning. Animals that have become catabolic may have poorer farrowing rates and litter sizes.
- Low feed intake during lactation can have a significant effect in depressing subsequent reproductive performance, primarily by extending the wean-to-service period.
- With a high feed intake body condition is maintained and milk production is increased.
- Growth rate in the piglet is maximised by converting feed into milk.
- The lean genotype female requires a high intake of lysine and the lactating ration should contain 1.1 to 1.2 % lysine with a protein level of 17 to 18 %. Energy levels should range from 9.4 to 9.8 to MJ NE/kg (13.3 to 14.8 MJ DE/kg).
- Remember the sow is an individual and the feed intake will vary from one animal to another.
- The first litter gilt is a particular problem because it has a restricted appetite and its energy and lysine requirements are greater than those of the sow. This can be satisfied by feeding an early grower diet (up to 1.3 % lysine, 9.8 MJ NE/kg (14.8 MJ DE/kg)) or giving it as half of the daily ration.
- An alternative is to top dress gilt lactation diets with 150 g of fishmeal per day.
- The palatability can be increased by adding 1.5 kg sugar per tonne of lactator feed. This may help to reduce the wean-to-service interval.
- Managing the feed intake is an art and sows from 3 days post-farrowing should be fed a lactation diet of the above specification to appetite but not to cause indigestion.
- Sows should be fed at least twice daily with sufficient amounts that are eaten within an hour and a half.
- Provide the largest feed in the evening so the sow has the cool of the night to eat.
- Many farms will feed 2 times a day until Day 14 then 3 times a day thereafter.
- Avoid providing any feed during the hot late morning and afternoon.
- Water flow should be a minimum of 2 litres per minute. Consider the type and accessibility of the drinker.
- There is considerable variation in feed intake between different genotypes during lactation. Manage your own herd to maximise feed intake but do not cause inappetence.
- Many sows will show a drop in their feed intake during the second and third week of lactation. This dip reduces milk production and hence

weaning weights. Make sure that it is not due to inadequate or slow water supply. Recent work suggests that controlling feed intake on a set rising scale improves weaning weights. This is a contentious area, however, and is dependent on the diet quality. You are advised to determine your own response in this respect.
- Maximising energy and lysine intake in the first 2 weeks of lactation stimulates the development of the primordial follicles in the ovary and ovulation rate in the next oestrus.
- Sows prefer to eat in the early morning and in the late evening.
- Sows should be eating 10 kg plus by Day 18 of lactation.

Factors that affect feed intake during lactation

- High environmental temperatures. Above temperatures of 24 °C (75 °F), feed intake may be reduced by up to 80 g per day for every increase of 1 degree. Cooling will be required above this temperature to maintain feed intake.
- Floor surfaces. Slatted floors are cooler than solid floors. Air flow, humidity and efficiency of insulation of the house can also affect the temperature of the environment.
- Some breeding females eat more than others.
- Sows eat more wet feed than dry.
- Heat lamps placed too near the sow increase the temperature and reduces feed intake.
- Low-nutrient-density, high-fibre diets will reduce the availability of nutrients to the sow, reducing milk production.
- Litter size.
- Lactation length.
- Fat depths at farrowing. If a sow has been fed too heavily for the 3 to 4 weeks pre-farrowing this will depress appetite during lactation.
- Overcondition sows (condition score 4+) will have depressed feed intake.
- Sow health – It is important to ensure that the preventative routines have been carried out, particularly worming, so that damage to the digestive tract does not impair the use of food.
- Palatability – A small pellet (5 mm) is more palatable than a large one, and this improves feed intake.
- Avoid vaccinations during lactation as this might interfere with feed intake.

Nutrition during lactation is ideal if:

- Sows maintain good body condition throughout.
- Average total litter size born is 14 or more.
- Weaning weights at 21 days average over 6.5 kg.

Do not make any changes if you are achieving this level of performance.

Identifying Problems in the Farrowing and Lactating Sow

If you have a problem in the gestation/dry period, refer to Fig 8-24 to identify the cause and then read the relevant section. Where a condition listed in Fig 8-24 is not covered within this chapter please refer to the index.

If you cannot identify the cause, consult your veterinarian.

FIG 8-24: OBSERVATIONS AND CAUSES

Blown up abdomen
- Constipation.
- Excess gas in large bowel.*
- Faulty nutrition.
- Too much food.*
- Torsion stomach or intestines.

Haemorrhage: Faeces/urine
- Acute cystitis.*
- Gastric ulcer.*
- Ileitis.
- Ruptured blood vessel.

Haemorrhage: Nose
- Ruptured blood vessel.
- Trauma.*

Haemorrhage: Vagina
- Dead piglets.
- Ruptured blood vessel.*

Haemorrhage: Vulva
- Haematoma.*
- Trauma.

Head on one side or nervous signs
- Brain abscess.
- Drip cooling misplaced – running into ear.
- Eclampsia.
- Haematoma of the ear.
- Meningitis.
- Middle ear infection.*
- Porcine stress syndrome (PSS).
- Water deprivation (salt poisoning).

FIG 8-24: OBSERVATIONS AND CAUSES CONT.

Inappetance over the farrowing period

a. Temperature normal

- Constipation.
- Cystitis/pyelonephritis.
- Gastric ulcers.
- PRRSV.
- Water shortage.*

b. Temperature elevated and/or the sow is toxic or ill

- Dead piglets inside the uterus.*
- Erysipelas.*
- Kidney infection.
- Mastitis.*
- Metritis (uterus infection).*
- PRRSV.
- Swine influenza.
- Torsion of the uterus.

Lameness, stiffness, paddling

- Acute stress.
- Arthritis.
- Bush foot/sandcrack.
- Erysipelas.
- Fractures.
- Glässer's disease in gilts.
- Leg weakness or OCD.*
- *M. hyosynoviae* in gilts.
- Torn muscles.*

No milk/ sow will not suckle

- Age.
- Discharges.
- Ergot.
- Excitation – especially gilts.
- Fever.*
- Mastitis.*
- Metritis (uterus infection).*
- PRRSV.
- Shortage of water.
- Sow ill or toxic.*
- Swine influenza.
- Trauma to teats.
- Udder oedema.

Mortality (sudden death)

- See gestation/dry sow.

* More likely to occur.

Diseases/Pathogens and Disorders Affecting the Farrowing and Lactating Sow

Below is a list of the key diseases that are likely to be seen in the farrowing/sucking period. These are covered in detail in this chapter.

- Abscesses.
- *Atresia ani* (no anus or no rectum).
- Aujeszky's disease/Pseudorabies virus (AD/PRV).
- Clostridial diseases.
- Cystitis/pyelonephritis.
- Diarrhoea or scour (enteric colibacillosis).
- Eclampsia.
- Electrocution.
- Erysipelas.
- Fat sow.
- Fever.
- Foot problems.
- Fractures.
- Gastric ulcers.
- Ileitis.
- Leg weakness – osteochondrosis (OCD).
- Leukaemia – acute myeloid leukaemia (AML).
- Lice.
- Mange.
- Metritis – inflammation of the uterus (womb).
- Mycoplasma (enzootic) pneumonia (EP) *Mycoplasma hyopneumoniae.*
- Osteomalacia (OM).
- Osteoporosis.
- Porcine parvovirus (PPV).
- Porcine reproductive and respiratory syndrome virus (PRRSV).
- Progressive atrophic rhinitis (PAR).
- Prolapse of the bladder.
- Prolapse of the rectum.
- Prolapse of the uterus (womb).
- Prolapse of the vagina and cervix.
- Savaging of piglets (cannibalism).
- Shoulder sores.
- Streptococcal infections.
- Torsion of the stomach and intestines (twisted gut).
- Vitamin E deficiency and iron toxicity.
- Vulva haematoma.
- Water deprivation (salt poisoning).
- Worms and internal parasites.

Abscesses

See Chapter 10 for further information

Atresia ani (No Anus or No Rectum)

This is uncommon, but in the gilt *Atresia ani* may be missed at selection. It is a result of the gilt developing a cloaca (combined vagina and anus) and may not be noticed until point of farrowing when only 1 orifice is identified as a result of the gilt having difficulty farrowing. An emergency caesarean is likely to be required.

Aujeszky's Disease/Pseudorabies Virus (AD/PRV)

See Chapter 12 for detailed information on this disease

Aujeszky's disease (AD), also known as Pseudorabies virus (PRV), is an important disease of the pig and is caused by a herpes virus. The pathogen has been eliminated from a number of countries and regions so is not directly relevant to many pig farmers.

Clinical signs are only seen in the lactating sow when the disease is introduced into the herd for the first time. In such cases sneezing and coughing are often the first signs with fever, inappetence, vomiting, abortion and nervous signs.

Clinical signs in piglets are usually only seen when the virus first enters the herd. The incubation period is short, only 2 to 3 days, after which sucking pigs become acutely ill, hairy, wander around aimlessly and listlessly and stop sucking. Nervous signs are common. Piglets go into fits and often paddle on their sides. Some piglets may adopt a dog-sitting position and develop vomiting and diarrhoea. During acute disease, mortality is very high, approaching 100 %. For a period of up to 3 months after the virus enters the herd, sows produce weak pigs at birth and there are high levels of mummification. Vaccination should take place as soon as the disease is diagnosed to boost immunity and mitigate the effects. All aspects of this disease are covered in detail in Chapter 12.

Clostridial Diseases

See Chapter 7 for further information

Clostridial infections in the sow during lactation are not common but occasionally do occur and the only sign is sudden death. If mortality is high, detailed postmortems must be carried out within 1 to 2 hours of death. The dead sow should be removed immediately from the farrowing house into a cooler environment, otherwise differentiation between disease and postmortem changes becomes impossible.

Cystitis/Pyelonephritis

See Chapter 7 for further information

Bladder and kidney infections are very common in the sow but disease is usually seen in the gestation/dry period. The stress of farrowing can occasionally activate disease and in such cases the response to treatment is usually poor. The disease is usually acute, the sow is very sick, toxic and continually passing blood stained urine, which often dribbles out from the vulva. Occasionally the more chronic form will be seen where the sow passes urine containing small amounts of pus, mucus and occasionally blood. Mortality can be high. See Chapter 7 for details.

Diarrhoea or Scour (Enteric Colibacillosis)

See Chapter 14 for further information

Diarrhoea may be seen in adults associated with TGE, PED, mycotoxins, sudden change of diet or water quality.

Eclampsia

This is an uncommon condition associated with low levels of calcium in the blood stream (hypoglycaemia) and it may occur at any stage during pregnancy but is most likely within 7 days either side of farrowing.

Clinical signs

These are sudden in onset with the sow becoming distressed and panting heavily. There is muscle trembling and shaking of the body. The sow is also reactive to external stimuli, both touch and sounds (hyperaesthetic).

Diagnosis

This is based on the sudden onset and the clinical signs presented but it can be confused with the porcine stress syndrome (PSS). The response to calcium injections if given early enough help to differentiate. Most animals with PSS die regardless of treatment.

Treatment

- This involves giving up to 100 ml of 40 % calcium boroglucinate by injection (this is the medicine used for treatment of the analogous condition in the dairy cow, which is very common). Ideally the injection should be given intravenously but this can be difficult. Alternatively 25 ml should be given by intramuscular injection at 4 separate sites in the neck. The muscles in the rump can also be used.
- Cool the sow by spraying with cold water.
- Administer pain relief where necessary.

Electrocution

Electrocution of sows and litters occurs sometimes in farrowing houses where electricity is used for heating. Farrowing places are often connected together throughout the house by various pieces of metal and because of this several animals may be killed – including piglets when they make contact with the sow.

Clinical signs

A large number of animals suddenly found dead in one house should immediately raise a suspicion of electrocution.

The skin will often be burned at the points where it has made contact with the metal, although piglets in contact with the sow may show no external signs. Blood and froth are commonly seen around the nostrils and mouth. Bones may fracture.

Diagnosis

Post-mortem examinations are necessary to differentiate electrocution from other causes of sudden death although the circumstances are almost diagnostic. Veterinary certification is usually required for insurance claims.

Prevention

- Trip out switches should be provided in the electricity circuits and the electricity lines and switches well maintained.
- A common cause, however, is damage by sows that escape from farrowing places. Make sure that gates into farrowing places are secure.

If you go into your farrowing house and find large numbers of dead animals, STOP and THINK: ELECTROCUTION or TOXIC GASES. BOTH CAN KILL YOU.

Erysipelas

See Chapter 7 for further information

Swine erysipelas is caused by a bacterium, *Erysipelothrix rhusiopathiae* that is found in most if not all pig farms. It is unusual to see clinical signs of erysipelas in the lactating sow but when it does occur there is a high fever and sometimes, but not always, skin lesions. In a susceptible herd the piglets may also be affected. If sows are ill, with very high temperatures, first check that there is no mastitis or uterus infection. In the absence of specific symptoms, treat with penicillin or amoxycillin and assess the results of this.

Previous infection with erysipelas can cause growths on the valves of the heart (valvular endocarditis) resulting in circulatory problems at farrowing or in some cases heart failure and sudden death. All breeding stock should be vaccinated against this disease and where it is recognised, early prompt treatment should be given to prevent such chronic lesions developing. This disease is covered in detail in Chapter 7.

Fat Sow

A fat sow should be considered a medical problem. A gilt or sow that farrows at body condition 4+ is likely to have increased farrowing difficulties and more importantly will eat poorly during her lactation. This affects the weaning weights and survivability of her piglets and disrupts her future reproductive performance.

The major cause of second litter size drop is gilts who fail to eat in their first lactation.

Fever

Fever means a high body temperature. It may occur with little or no other symptoms. The causes will in most cases be associated with bacterial or viral infections or, rarely, stress. Consider the following conditions in order of importance:

- Mastitis or metritis.
- Retention of a dead pig.
- Retention of afterbirth.
- A bacterial septicaemia (e.g. erysipelas).
- SIV or PRRSV.
- Secondary bacterial infections associated with SIV or PRRSV.
- Cystitis/pyelonephritis.
- Acute stress or eclampsia.
- Heat stroke.

Clinical signs

Usually the sow appears dull and sometimes shows a reddening of the skin. The respiratory rate may be raised. Clinical examinations will often indicate a cause - always look for the obvious first. Temperatures will range from 39 to 40 °C (103 to 109 °F).

Diagnosis

Examine the animal closely to see if any of the above conditions can be detected. If not and there are a number of animals involved, veterinary advice should be sought. Bear in mind that depending on where in the world your herd is located, fever may be the first clinical sign in such diseases as classical swine fever (hog cholera), African swine fever and Aujeszky's disease (pseudorabies).

Treatment

- In most cases, fevers in sows will be associated with bacterial infections and a broad spectrum antibiotic should always be used. Check the temperature at, and 24 hours after, treatment.
- Broad-acting antibiotics include oxytetracycline, trimethoprim/sulpha, amoxycillin and penicillin/streptomycin.

Foot Problems

Foot health is an important but often overlooked problem by the pig farming profession.

The ideal time for careful examination of a sow's feet is in the farrowing house. It is at this point that any foot trimming can be done to restore foot condition. Sows with long deformed feet are clumsier and more likely to crush piglets, resulting in a lower number of weaned piglets. This is simply addressed if given the right attention.

Fractures

Bone fractures are not uncommon in sows and gilts and can be the end result of trauma and fighting although spontaneous ones occur in bone disease such as osteomalacia, associated with calcium phosphorus and vitamins A and D, and osteochondrosis (OCD). Gilts may fracture their legs immediately postweaning due to lowered bone calcium levels as a result of excessive milk production during lactation.

Fractures are covered in Chapter 14. Also see Lameness in Chapters 7, 9 and 14.

Gastric Ulcers

See Chapter 14 for further information

Gastric ulceration in the lactating sow is probably common but difficult to diagnose. Many chronic ulcerated lesions are activated during lactation because of the high and continual intake of feed.

Clinical signs vary according to the severity of the ulcer and whether it is bleeding or not. The feed intake can be variable with occasional vomiting. If haemorrhage is occurring there will be dark-coloured faeces, the animal will have a tucked-up appearance, it will sometimes be grinding its teeth (indicative of pain), and it will appear anaemic. The sow may refuse to suckle and stay on her stomach. The stress of farrowing may aggravate a stomach ulcer. Getting sows to eat after farrowing is an important means to controlling the start of a gastric ulcer. This condition is covered in detail in Chapter 14.

Ileitis

See Chapter 9 for further information

In lay terms this condition is often described as bloody gut because there is acute haemorrhage into the lower part of the small intestine and, occasionally into the upper part of the large intestine (also called Proliferative haemorrhagic enteropathy PHE). Ileitis is common in maiden gilts after selection or when they are moved to new premises but it can occur in pregnant gilts and very occasionally in first litter lactating gilts. It would be rare to see disease in the sow. The disease is caused by a bacterium, *Lawsonia intracellularis*. The gilt with bloody gut may appear pale and weak with bloody or dark faeces or found dead. Post-mortem examination showing massive haemorrhage in the lower intestine is strongly suggestive of this disease. PHE and its related syndromes porcine intestinal adenopathy (PIA), necrotic enteritis (NE) and regional ileitis (RI) are grouped under the heading of ileitis.

Leg Weakness – Osteochondrosis (OCD)

See Chapter 7 for further information

This is more common in first and second litter females and it is also described under the term osteochondrosis (OCD). Leg weakness may result in separation of the head of the femur or tearing of the muscles from the pelvis to the leg bones. Fracture of the growth plates of the vertebrae may cause pressure on the spinal cord or on nerves leaving the spinal cord resulting in loss of leg function and acute pain. Major predisposing factors are sloping farrowing place floors or very slippery ones. When the young sow tries to stand up, the front legs are moved backwards and the back legs slip underneath. This creates enormous shear stresses on the young growth plates in the long bones, causing changes in the bone structure or even fractures. The lameness in many cases may only become evident at weaning time when sows are mixed and they fight and ride each other, or at mating due to the weight of a heavy boar. Thus if there are lameness problems in the postweaning period check and examine sows in the farrowing places carefully and in particular the relationships of the feet to the floor surfaces. The judicious use of fine dry sand on the floors daily can often significantly improve the situation in the short term until floor surfaces can be changed. Alternatively, because first and second parity sows are most susceptible, a small proportion of farrowing place floors can be altered and used specifically for these animals. See Chapter 7 for more details.

Leukaemia – Acute Myeloid Leukaemia (AML)

Myeloid leukaemia is a rare neoplasm in domestic animals with a few reported cases in pigs. It may be recognised in the farrowing area where a gilt is very thin and eats but fails to improve her body condition. There may or may not be signs of clinical illness. If not clinically sick, cull at weaning.

Lice

See Chapters 10 and 11 for further information

These are relatively uncommon in herds today, particularly if mange treatment is carried out, because this will also destroy the pig louse. They are blood-sucking parasites that are easily visible on the skin. The parasites can cause itching resulting in the sow not suckling properly. The constant itching can cause a lot of damage to housing.

Mange

See Chapter 11 for further information

Mange is a parasitic disease of the skin caused by 1 of 2 mites, either *Sarcoptes scabiei* var. *suis* or *Demodex phylloides,* the latter of which is relatively unimportant in swine.

Sarcoptic mange (sometimes called scabies) is by far the most common and important in the pig because it is irritant and uncomfortable for the pig, causing it to rub and damage the skin, which becomes unsightly. Mange may be seen in farrowing. It causes the sow to itch and become distracted and as a result she does not suckle properly. The constant rubbing to alleviate the itching can cause a lot of damage to housing.

Metritis – Inflammation of the Uterus (Womb)

Metritis in the immediate post-farrowing period is fairly common. During the process of farrowing a large amount of fluid, a varying number of piglets, and afterbirth have to be expelled from the uterus. At the end of this process the 2 horns of the uterus contract and squeeze the final contents out through the vagina. This process can continue for up to 3 to 4 days after farrowing and therefore it is not unusual or abnormal to see a slightly mucoid to white discharge from the vulva. However, discharges can also indicate the presence of an active infection requiring treatment. Metritis is more likely to occur where farrowings are prolonged or where there has been manual assistance. It can also be common in association with mastitis so that whenever discharges are evident carefully examine the udder (mastitis, metritis, agalactia syndrome).

Clinical signs

If the discharge is not heavy, it disappears after 3 to 4 days, the sow is eating well and there is no mastitis, ignore it. It is a normal biological process and no action is required. Alternatively if there are signs of mastitis, the sow's temperature is above 39 °C (102 °F) or the sow is off her food with bright red mucous membranes around the eyes, then treat her.

Diagnosis

This is based on a sow not eating and a fever. There will be evidence of a discharge from the vulva, usually a white or brown colour and sometimes associated with mastitis.

Treatment

- Give twice daily injections of antibiotics together with 0.5 iu of oxytocin each time.
- Treatment should be given for 2 to 3 days.
- If the sow has been assisted at farrowing then separate injections of long-acting penicillin and 5 iu of oxytocin (intra-vulval) is advised at the time to prevent infection.
- Antibiotics that can be used include OTC, penicillin/streptomycin, amoxycillin, ampicillin, framycetin, trimethoprim/sulpha.

Mycoplasma (Enzootic) Pneumonia (EP) *Mycoplasma hyopneumoniae*

See Chapter 9 for further information

Mycoplasma (enzootic) pneumonia is caused by the tiny pathogen *Mycoplasma hyopneumoniae*.

Clinical signs of mycoplasma (enzootic) pneumonia occur in the lactating sow only if the disease has been introduced into a fully susceptible herd for the first time. In such cases acute pneumonia may develop with severe respiratory embarrassment and sometimes high mortality. The breakdown of disease usually takes place over 6 to 8 weeks, with sows coming into the farrowing house continuing to be affected.

In negative herds, which are not part of a breeding company pyramid, vaccinate the adult herd and incoming gilts. This protects the herd from a major abortion outbreak, if the herd breaks. The vaccine is dead and does not affect the health declaration of the unit.

Osteomalacia (OM)

Osteomalacia (OM) and Osteoporosis (OP) are becoming more common in commercial production systems, particularly in the first litter gilt where the skeleton is still growing and there are heavy demands on calcium for milk production. Both diseases have very similar signs and x-ray analysis is required to differentiate.

Osteomalacia (OM) is a condition responsible for the downer sow syndrome – where sows cannot get up. Fractures of the long bones at the mid shaft and fractures of the lumber vertebrae are common, with the sow becoming paraplegic. The condition is due to inadequate levels of calcium, phosphorus and vitamin D in the ration. Sometimes sows cannot absorb sufficient micro-nutrients in spite of there being adequate levels in the diet. OM is also associated with immature skeletons, an imbalance of calcium and phosphorus and vitamin D and/or a failure of the sow to consume adequate feed and satisfy her nutritional requirements. Large amounts of calcium and phosphorus are excreted into milk from the bones resulting in weaker, less dense bone which predisposes to fractures. Bone mass is also lost due to lack of exercise during confinement in the farrowing place.

Clinical signs

The condition is common in first litter animals and up to 30 % of such animals may be affected. The history is one of sudden acute lameness, often with the animal completely off its legs. The lameness is usually precipitated when the sow is moved from the farrowing place, during mixing or when the boar mounts at mating. Other symptoms include a stiff gait, difficulty in rising, discomfort in the hind legs and a dog sitting position.

Diagnosis

This is based upon history, clinical signs and examinations.

Similar diseases

These include:

- Leg weakness or osteochondrosis (OCD).
- Spinal fractures.
- Torn muscles at their insertions into the bones.
- Mycoplasma arthritis – *Mycoplasma hyosynoviae* infection.
- Osteoporosis (OP).

Treatment

- In the early stages, move the sow to well-bedded loose-housing.
- If there are no fractures, inject with calcium and vitamin D3.
- If there is a problem in first litter females, inject them with vitamin D3 after farrowing and 7 days later.
- Supplement the diet with dicalcium bone phosphate (30 g/day of sterilised bone flour).
- If there are fractured bones affected, sows should be culled or destroyed.

Management control and prevention

Once OM has developed, treatment is of minimal effect, although injections of calcium, phosphorus and vitamin D may help. If your herd has a problem, consider the following:

- Feed a high dense diet in lactation 9.8 MJ NE/kg (14.8 MJ DE/kg) and 18 % protein.
- Check the levels of calcium and phosphorus (minimum 0.9 % and 0.75 %). In first litter animals it may be necessary to raise the levels to 1.2 % and 1 %. Check the levels of calcium and phosphorus in the diet. They should be 10 to 12 g/kg of calcium and 8 to 10 g/kg of phosphorus.
- Give up to 100,000 iu vitamin D3 by injection 10 days before farrowing. Inject pregnant animals with 50,000 iu vitamin D3 3 weeks prior to farrowing. Repeat again in the second week after farrowing.
- Keep gilt litters to 10 piglets or less.
- Top-dress the feed in lactation with calcium/phosphorus. Feed a good lactation diet during suckling and consider top dressing the diet daily with 20 g of dicalcium bone phosphate.
- Bone ratios of calcium/phosphorus in affected sows are often 3:1 (normal < 2:1). Check the ratio of calcium: phosphorus in bone ash. The normal ratio is approximately 2:1 or less. In problem sows this is often 3:1 or more
- Mate gilts younger – at 210 to 220 days. Only mate gilts from 220 days onwards and if the disease is a persistent problem in a particular genotype, change the source.
- Use a good lactation diet and feed through to 21 days post-mating.
- Provide non slip floors in farrowing places. Check that floor surfaces are not slippery.
- Wean first litter females singly and use a light weight boar.
- Provide exercise to the pregnant gilt. Increase exercise during pregnancy if possible.
- Check parasite levels to ensure no dietary insufficiencies arise.
- The problem is less common in outdoor herds.
- Outbreaks are often more apparent in new gilt herds. Selection of animals for good conformation is essential.
- Investigate the growth rates and nutrition and feeding in the gilt.
- Maximise feed intake to appetite during lactation.

Osteoporosis (OP)

Osteoporosis (OP) and Osteomalacia (OM) are becoming more common in commercial production systems, particularly in the first litter gilt where the skeleton is still growing and there are heavy demands on calcium for milk production. Both diseases have very similar signs and x-ray analysis is required to differentiate.

Bones affected with OP are quite normal in their structure but they become thinner, particularly in the dense parts and shafts of the long bones. As a result they become more prone to fracture. OP can arise due to a shortage of calcium in the diet and imbalance of calcium and phosphorus, poor or inadequate absorption from the diet, heavy losses during lactation and where there is a lack of exercise.

Osteomalacia (OM), covered under its own section in this chapter, is the adult form of rickets and is associated with a phosphorus or vitamin D3 deficiency. There is a failure of the mineral to be deposited in the bones, which become soft and either bend or fracture.

Clinical signs

These are most common in the first litter female and occasionally after the second litter.

The onset may be gradual with the pig having difficulty rising, and showing pain or sudden lameness associated with complete fracture of the long bones. Spinal fractures occur in some animals and they often remain in a dog-sitting position. Most pigs are affected in late lactation or shortly after weaning, associated with the onset of oestrus and the trauma that results from other animals, or the weight of the boar at service.

Diagnosis

This is based on clinical signs, a history in lactating and newly weaned sows and evidence of fractures of the long bones. If the herd has a problem it is necessary to examine the bones of an affected animal by x-ray to differentiate between OP and OM.

Similar diseases

These include:

- Leg weakness or osteochondrosis (OCD).
- Spinal fractures.
- Torn muscles at their insertions into the bones.
- Mycoplasma arthritis – *Mycoplasma hyosynoviae* infection.
- Osteomalacia (OM).

Treatment

- In the early stages, move the sow to well-bedded loose-housing.
- If there are no fractures, inject with calcium and vitamin D3.
- If there is a problem in first litter females inject with vitamin D3 after farrowing and 7 days later.
- Supplement the diet with dicalcium bone phosphate (30 g/day of sterilised bone flour).
- In cases of bone fracture the sow is best destroyed on humane grounds.

Management control and prevention

- Outbreaks are often more apparent in new gilt herds. Selection of animals for good conformation is essential.
- Investigate the growth rates and nutrition and feeding in the gilt.
- Increase exercise during pregnancy if possible.
- Check the levels of calcium and phosphorus in the diet. They should be 10 to 12 g/kg of calcium and 8 to 10 g/kg of phosphorus.
- Only mate gilts from 220 days onwards and if the disease is a persistent problem in a particular genotype, change the source.
- Check that floor surfaces are not slippery.
- The problem is less common in outdoor herds.
- Feed a good lactation diet during suckling and consider top dressing the diet daily with 20 g of dicalcium bone phosphate.
- Inject pregnant animals with 50,000 iu vitamin D3 3 weeks prior to farrowing. Repeat again in the second week after farrowing.
- Maximise feed intake to appetite during lactation.
- Check the ratio of calcium : phosphorus in bone ash. The normal ratio is approximately 2:1 or less. In problem sows this is often 3:1 or more.

Porcine Parvovirus (PPV)

See Chapter 6 for further information

This virus is ubiquitous and present in almost all breeding herds. Once exposed the sow becomes immune. The virus itself has no clinical effects on non-pregnant females when infection takes place. If however a sow or gilt is exposed for the first time in the first half of pregnancy the virus will cross the placenta and selectively kill off the developing foetuses in the uterus.

PPV is of minimal significance to the lactating sow other than the fact that colostrum contains very high antibody content and will protect the piglet from infection for up to 5 to 6 months of age

Good control is achieved by vaccinating the gilt. This disease is covered in detail in Chapter 6.

Porcine Reproductive and Respiratory Syndrome Virus (PRRSV)

The information in this chapter relates specifically to PRRSV in the farrowing and sucking period. The main information on PRRSV, including methods of spread, immunity, diagnosis, similar diseases and elimination from a herd are covered in Chapter 6, with additional information in Chapters 7, 9, and 10. These chapters should be read to fully understand how this disease impacts the whole herd.

Clinical signs in the farrowing and lactating sow

When first introduced into a herd PRRSV infection has a marked effect on all lactating sows over a 6 to 8 week period. They show varying degrees of inappetence and mild illness but the most striking sign is agalactia, not necessarily with oedema of the udder or mastitis, but just with a very poor milk flow. In part this is associated with inappetence and reluctance of the sick sow to drink. The temperature may be normal or elevated. Mastitis or urinary infections may also occur. Piglets will be born 2 to 3 days early and suffer from immaturity, anoxia and difficulty in suckling. Weakened piglets consume less colostrum and it is not unusual for severe outbreaks of piglet neonatal scour to accompany the PRRSV outbreak. Stress at farrowing may activate latent infections and this is a common experience 6 to 12 months after the initial outbreak of PRRSV.

Treatment

There is no treatment for viral infections. However it is important to control secondary bacterial infections and antibiotic treatment should be given for 3 to 4 weeks to all sows immediately the disease is diagnosed or suspected. If necessary, commence with water soluble antibiotics followed by in-feed medication.

Tiamulin may assist in a PRRSV outbreak as it reduces the ability of the PRRSV virus to enter the macrophage.

Management control and prevention

A control programme for the farrowing sow during acute disease:

- Inject sows 2 days before farrowing with antibiotics. Continue this in succeeding sows for a period of 6 weeks.
- Repeat 3 days later or until farrowed.
- Use long-acting preparations of antibiotics. Oxytetracycline or semi-synthetic penicillins are medicines of choice.
- Top dress the sows' food daily with in-feed antibiotics premixes. Use OTC, CTC or TMS. (Give 15 to 20 g of a 10 % premix per day).
- Continue this for 10 to 21 days post-farrowing.
- Raise the farrowing house temperature to 22 °C (75 °F).
- Deep-bed the pens with straw, shavings or paper if this is possible.
- Provide extra heat for piglets.
- Treat with water soluble aspirin.

Continue the above programme for 4 to 6 weeks or as advised by your veterinarian.

A control programme for the farrowing sow during chronic disease:

If it is suspected that a chronic reproductive problem may be associated with PRRSV, consider the following action plan to determine (Fig 8-25):

From this information the following possibilities exist:

- The herd is negative.
- There is no infection taking place in the sow herd and no evidence of infection in weaners. In other words the virus has died out and PRRSV is not a problem.
- The sow herd is serologically negative but there is evidence of persistent viral infection in younger growing pigs. Such animals will show evidence of rising titre levels. This continually exposes susceptible breeding stock to disease.
- Negative gilts become infected and disseminate the virus.

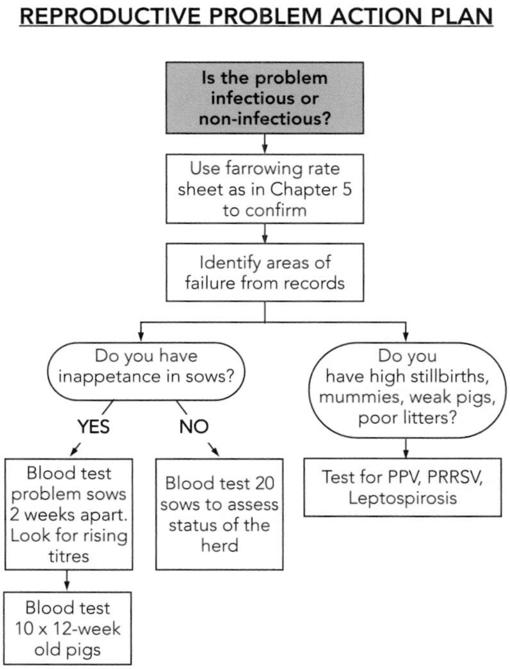

Fig 8-25

Because our knowledge of this disease is continually improving you are advised to discuss further controls with your veterinarian. Three procedures need to be considered:

1. Commercial vaccination.
2. Removal of the infected group of pigs, usually the first and second stage growers from the farm to eliminate the virus.
3. Expose gilts to maintain an immune breeding herd. This can be achieved via a well-planned acclimatisation programme including feedback and/or autogenous vaccination using tonsil scrapes.

Progressive Atrophic Rhinitis (PAR)

Rhinitis is inflammation of the tissues inside the nose and in its mild form it is very common. The term atrophy indicates that the tissues inside the nose, which become infected or damaged, shrink and become distorted. There are 2 forms of the disease, the more serious being PAR (progressive atrophic rhinitis) which is associated with the presence of toxin-producing strains of a bacterium *Pasteurella multocida*.

PAR is a serious condition in both sucking and growing pigs.

Clinically it is not important in the sow unless infection, as a growing gilt, has stunted her growth. However the sow might become a carrier, particularly if infected as an adult, and might spread it to other breeding animals and possibly to her litter. This condition is covered in detail later in this chapter.

Prolapse of the Bladder

This is an uncommon condition but rather confusing when it appears; the bladder turns inside out and protrudes from the lips of the vulva. It arises when there is a large urethral opening at the floor of the vagina and complete loss of muscle tone in the sphincter.

Clinical signs

The inside of the bladder appears as a large red mass about the size of an orange. It can be confused with an early prolapse of the uterus but examination will show that is like a small balloon.

Treatment

- The everted bladder can usually be returned to its former position using obstetrical fluid. The tissues are gently pushed back into the vagina and the bladder returned. This may be a task for your veterinarian.
- Give antibiotic cover by injections for 3 days.

Prolapse of the Rectum

See Chapters 9 and 14 for further information

This is not uncommon in sows and occasionally outbreaks occur in herds. Information on rectal prolapse is provided in Chapter 9. Nutritional aspects of this condition are covered in Chapter 14.

Prolapse of the Uterus (Womb)

This involves the complete eversion of both horns of the uterus which turn completely inside out. It usually takes place within 2 to 4 hours of the completion of farrowing but sometimes up to 24 hours afterwards. Prolonged straining causes a small part of the tube to be propelled outwards by uterine contractions.

Clinical signs

The prolapse occurs over a period of approximately 1 hour and commences with the appearance of the red congested lining of the uterus. This rapidly increases in size until the large everted mass is presented.

Diagnosis

This is obvious from the appearance.

Treatment

- This involves replacing the uterus inside the sow. It is often impossible or the sow dies from internal haemorrhage and a heart attack when toxins from the compromised uterus floods into the blood stream.
- The technique for carrying this out is discussed in Chapter 15.

In most cases on welfare grounds the sow should be destroyed. Uterine prolapses are uncommon but usually occur in old sows with large litters or where large piglets have been born. The supporting structures of the uterus become weak or the uterine wall becomes flaccid.

Prolapse of the Vagina and Cervix

See Chapter 7 for further information

Prolapse of the vagina and cervix is more common prior to farrowing and may be seen in the last third of pregnancy including the pre-farrowing period in the farrowing house. It occurs normally in about 1 pregnancy in 200, usually in older sows from 5th parity onwards.

Savaging of Piglets (Cannibalism)

This is a common condition in first litter gilts that may account for a 1 to 3 % increase in piglet mortality.

It occurs from time to time in many mammalian species after giving birth for the first time and is thought to be related, in part at least, to the major hormone changes that take place around parturition.

In the gilt a number of factors seem to predispose to it, including a harsh or alien environment, poor empathy between the gilts and the stockperson, nutritional deficiencies and the effect of being placed in total confinement for the first time. It may also be related to temperament and breeding. It seems to be more prevalent in some breeds than others. For example, it is more common in pure-bred Large White gilts than Landrace or Duroc; possibly their overlapping ears reduce visibility.

Major outbreaks have been experienced in new gilt herds where large numbers of the pregnant animals have been reared in extensive straw yards. In such environments (100 to 150 animals) no pecking order develops and one animal sees another as its enemy. In one particular herd a dramatic reduction occurred when the gilts were moved from the yards to sow houses for 3 weeks prior to farrowing. During this period the gilt learned to recognise another pig and become familiar with individual confinement. Sometimes a change of farrowing room attendants reduces the incidence.

Clinical signs

Offending gilts can often be identified by their nervousness and apprehension at the onset of farrowing. Such animals have a wild-eyed look. A careful watching brief therefore should be taken.

Treatment

- Try and identify potentially problematic gilts before farrowing.
- Constant supervision is required to identify offending animals at the onset of farrowing. To make this more feasible prostaglandin injection on the 113th day could be used on gilts, to give a more predictable time of farrowing.
- Watch all gilts carefully for the first 2 or 3 piglets and if there is any sign of savaging, inject with azaperone at a dose level of 1 ml per 12 kg weight. All the piglets should be confined to the creep area away from the sow for at least 20 minutes following injection, until she has settled down and rolled over on her side. The piglets should then be reintroduced. Most farrowings will continue normally thereafter.
- Providing the sow with a pint of beer can assist in calming her down.
- Having a radio on in the farrowing area has also helped to quiet young sows/gilts.
- Discuss with your veterinarian the possibility of treatment with primidone. This medicine is available in tablets containing 250 mg of primidone which is an anticonvulsant medicine but has the effect of reducing hysteria and nervousness. 3 to 4 tablets may be given twice daily, 24 to 48 hours prior to farrowing.

Management control and prevention

- Assess the gilts when they come into farrow.
- Try wherever possible to have a sow in a farrowing place next to a gilt.
- Give the gilt plenty of straw to eat pre-farrowing.
- If there is a major problem, hold the gilts in sow places for at least 7 days prior to entry into the farrowing places.
- Consider a different breeding female. Where there is a breed factor very often the savaging will progress into the second and sometimes even the third litter.
- Document any savaging episode on the sow card so that this can be noted into the next litters.
- Ensure that the farrowing houses are dimly lit, warm and comfortable and have no draughts.
- Play background music during the farrowing period, talk to all the gilts, stroke them and develop a good empathy starting 3 to 4 weeks before farrowing.
- With "the wild-eyed gilt", test her reaction to a 2-week-old pig placed in the pen before farrowing.
- Cut the top off a rubber boot and fasten this over the mouth of the gilt, using a retaining string fastened behind the ears.
- Try introducing a rabbit into the gilt farrowing pen for 48 hours prior to farrowing.

Shoulder Sores

See Chapter 10 for further information

These arise due to constant trauma over the bony prominences on the shoulder blade. Ultimately the skin breaks, there is an erosion, and a large sore develops. It is generally associated with totally slatted flooring and individual sows that have a prominent spine to the shoulder blade. It is first noticed in the farrowing house where floors are slippery and the sow has difficulty in rising, thus constantly bruising her shoulder. Affected sows should not be kept for future breeding. This condition is covered in Chapter 10.

Streptococcal Infections

See Chapter 9 for further information

Streptococcus may be responsible for the formation of abscesses. It may affect the uterus resulting in endometritis and even severe metritis. Occasional pyelonephritis may be recognised.

Torsion of the Stomach and Intestines (Twisted Gut)

See Chapter 14 for further information

Large amounts of feed eaten during lactation can result in sows suffering from this condition. Missing or changing the feeding routine may increase the risk. Early signs include a bloated abdomen but in most cases the sow is simply found dead. A post-mortem is necessary to confirm diagnosis.

Vitamin E Deficiency and Iron Toxicity

Iron toxicity occurs when the sow is deficient in vitamin E and piglets are born as a consequence with low levels. The routine iron dextran piglet injections become toxic and cause severe muscle reactions at the injection sites. Vitamin E deficiency in the sow occurs when fats in the diet become rancid or cereals or corn have fermented and spoiled and the vitamin E is destroyed. Piglets born from sows fed high levels of vitamin A may produce piglets with a low vitamin E status. This is a useful line of investigation where iron dextran problems persist in sucking pigs.

Clinical signs

Some 2 to 4 hours after injection piglets become acutely lame on the legs that have received the iron. The muscles are swollen and the piglets develop heavy breathing and look pale. Death occurs within 24 hours. At post-mortem the muscles are coagulated and appear like fish tissue due to necrosis of the muscle fibres.

Diagnosis

This is based upon the history of deaths within 24 hours of iron injections and swollen muscles at the site of the injections.

Treatment

- Inject sows due to farrow over the next 3 weeks with a vitamin E selenium preparation.
- Alternatively, use water soluble vitamin E.
- Inject all litters for a 3 week period with vitamin E or dose by mouth at least 2 days before iron injections are given.

Management control and prevention

- Clean out the cereals at the bottom of the storage bins.
- Assess cereals and corn for spoilage.
- Supplement the sow feed with 150 g/tonne vitamin E.

Vulva Haematoma

This is a condition where shortly after farrowing blood vessels inside the vulva rupture, due to stretching, pressure or trauma to the tissues. The vulva fills with blood. When this occurs the tissues become very fragile and if they are crushed, the vulva splits with severe haemorrhage. Vulval haematoma can also arise where a gilt has to be assisted at farrowing and damage occurs from a large arm.

Clinical signs

The vulva becomes swollen and very dark blue. If it ruptures it may bleed continuously. Blood clotting is poor and the animal becomes anaemic and ultimately bleeds to death. Always consider this as a serious condition that is life threatening and requires frequent monitoring over 24 hours.

Treatment

- The animal should be sedated and local anaesthetic injected around the tissues nearest the body of the sow just forward of the bleeding area. Three methods are then used for control:
  1. A piece of band or bandage is placed between the lips of the vulva and behind the bleeding tissues. It is then tightened to produce a tourniquet. This should be left for 24 hours.
  2. If this does not stop the bleeding then a series of mattress sutures should be passed through the vulva and tied to the exterior. See Chapter 15.
  3. If the haemorrhage still does not stop, the haematoma must be opened by a veterinarian who will use a pair of artery forceps to clamp the ruptured blood vessels and tie them off.
- Observe the effectiveness of the measures taken by placing a paper bag beneath the vulva so that any subsequent haemorrhage can be observed.
- Cover the tail gate of the farrowing place with a bag of straw or other suitable protective material to stop further crushing.

Water Deprivation (Salt Poisoning)

See Chapter 14 for further information

Water deprivation is unfortunately common in all ages of pigs and almost without exception is related to water shortage either caused by inadequate supplies or complete loss. See Chapter 14 for information.

Worms and Internal Parasites

See Chapter 11 for further information

Before farrowing it is not unusual for sows to be wormed and washed. This may release some worms in the faeces. As a general rule it is wise to check for intestinal parasites twice a year using faecal worm egg counts.

Diseases and Problems in the Sucking Pig

Whilst there are a large number of diseases that affect the sucking pig, only a few are important. Recognition of different conditions is not always easy, but Fig 8-26 highlights disease associated with clinical observations.

Identifying Problems in the Piglet

If you have a problem, refer to Fig 8-26 and then the index or relevant chapter. If you cannot identify the cause, consult your veterinarian.

The sucking pig is potentially exposed to all those disease-producing organisms that are present in the herd. The source of most of these infections is occasionally the sow, but more often the source is other diseased piglets or clinically healthy, usually older, piglets that become infected as the maternal antibody wanes. The levels of disease experienced in any one herd are dependent upon the management practices, the attention to detail and the various important predisposing factors that are listed under each specific disease. An overview of the diseases in the sucking pig is shown in Fig 8-27. Fig 8-28 highlights some of those diseases that may be controlled by vaccination of the sow to raise the level of antibodies in the colostrum. It is now recognised that provided the sow is not diseased and is passing antibodies via the colostrum, many of the diseases listed will not be transmitted to the piglet. However once maternal antibodies disappear, the piglet becomes susceptible.

FIG 8-26: OBSERVATIONS AND CAUSES

Diarrhoea 0 – 5 days of age
- Clostridial diseases.
- *E. coli* infections.*
- Low colostrum intake.*
- Mastitis.*
- PRRSV.*
- Rotavirus.
- TGE.*
- Udder oedema and poor milk supply.*

Diarrhoea 6 – 21 days of age
- Clostridia.
- Coccidiosis.*
- Cryptosporidia.*
- *E. coli* infections.*
- Low colostrum intake.*
- Low immunoglobulin A in milk.*
- Porcine epidemic diarrhoea (PED).*
- PRRSV.
- Rotavirus.

FIG 8-26: OBSERVATIONS AND CAUSES CONT.

Diarrhoea 6 – 21 days of age cont.
- Salmonellosis.
- Strongyloides.
- Transmissible gastroenteritis (TGE).

Laid on/trauma
- Inadequate temperatures.
- Poor farrowing place design.
- Poor environment.
- Poor farrowing house management.
- Poor farrowing house design.
- Splay leg.

Lameness (arthritis)
- *Actinobacillus suis.*
- Faulty teeth clipping and or tailing.*
- Glässer's disease.
- Poor environment, particularly floor surfaces.
- Specific infections.
- Splay leg.
- Staphylococci.
- Streptococci.*
- Trauma to knees, tail etc.
- Vitamin E deficiency and iron toxicity.

Mortality: Diarrhoea
- See Diarrhoea section.

Mortality: Generalised infection
- *Actinobacillus suis.*
- *E. coli* infections.*
- Erysipelas.
- Glässer's disease.*
- Meningitis.
- PRRSV.
- Swine Influenza.

Mortality: Poor viability
- Age or breed of the sow.
- Flu infection during pregnancy.
- Hypoglycaemia (low blood sugar).
- Low birth weight.
- Navel bleeding.
- Poor environment.*
- Poor nutrition/starvation (see below).
- PRRSV infection in the sow.
- Thrombocytopenia purpura (bleeding).

Mortality: Trauma
- Laid on.*

FIG 8-26: OBSERVATIONS AND CAUSES CONT.

Nervous symptoms cont.

- Actinobacillosis.
- Aujeszky's disease/pseudorabies (AD/PRV).
- Congenital tremor.*
- Glässer's disease.
- Hypoglycaemia.*
- Middle ear infection.
- PRRSV.
- Streptococcal meningitis.
- Swine fevers (hog cholera).
- Tetanus.
- Trauma.

Sneezing/coughing

- Ammonia.
- Actinobacillus pleuropneumonia.
- *Ascaris suum.*
- Aujeszky's disease/pseudorabies (AD/PRV).
- Bedding (dusty, mouldy).
- Bordetella infection.*
- Cytomegalovirus.
- Dust.*
- Mycoplasma (enzootic) pneumonia.
- Glässer's disease.*
- Progressive atrophic rhinitis.*
- PRRSV.*
- Rhinitis – non progressive.*
- Swine influenza.
- Virus infections of the nose.

Starvation/wasting

- *Actinobacillus suis.*
- Anaemia.
- Arthritis.
- Chronic diarrhoea.*
- Coccidiosis.
- Congenital tremor.
- Low birth weight.
- *Mycoplasma suis.*
- No milk available.*
- Pneumonia.
- Poor colostrum intake.*
- Poor teat access.*
- PRRSV.*

* More likely to occur

FIG 8-27: THE MAJOR INFECTIOUS DISEASES/PATHOGENS THAT MAY BE TRANSMITTED FROM SOW TO PIGLET

Viruses	
Aujeszky's disease/Pseudorabies	(V)
African swine fever	
Classical swine fever/Hog cholera	(V)
Foot-and-mouth disease	(V)
Porcine epidemic diarrhoea	
* Porcine parvovirus	(V)
PRRSV	(V)
* Swine influenza	(V)
Swine pox	
Swine vesicular disease	
Transmissible gastroenteritis	(V)
Bacteria and Mycoplasma	
* *Brachyspira hyodysenteriae*	
Brucellosis	
Haemophilus infections	(V)
Lawsonia intracellularis	
* *Mycoplasma hyopneumoniae*	(V)
Pasteurella multocida type D	(V)
Salmonella choleraesuis	(V)
Staphylococcus hyicus	
Streptococcal infections	(V)

(V) Vaccines available in some countries.
* Diseases unlikely to spread from the sow if the piglet is weaned at less than 14 days of age.

FIG 8-28: DISEASES AND CONDITIONS IN THE SUCKING PIG AND THEIR CLINICAL SIGNS

Disease or Condition	Born Dead	Low Viability	Enteritis (Scour)	Lameness	Respiratory Signs	Skin lesions	Nervous Signs
Actinobacillosis suis infection				✓	✓		
Arthritis				✓	✓		
Aujeszky's disease/pseudorabies	✓	✓			✓	✓	✓
Bordetellosis					✓		
Brucellosis	✓	✓					✓
Clostridia			✓			✓	✓
Coccidiosis			✓				
Colibacillosis			✓				✓
Congenital tremor							✓
Erysipelas	✓	✓		✓	✓	✓	✓
Exudative epidermitis						✓	
FMD disease				✓	✓	✓	
Glässer's disease (*Haemophilus parasuis*)				✓	✓		✓
Iron deficiency	✓	✓			✓	✓	
Leptospira	✓	✓					
Listeriosis					✓		✓
Mange						✓	
Mycoplasma infections				✓	✓		
Mycoplasma suis	✓	✓			✓		
Mycotoxicosis	✓			✓			
Pasteurellosis					✓		
Porcine epidemic diarrhoea			✓				
Porcine parvovirus	✓	✓					
PRRSV	✓	✓	✓		✓	✓	
Salmonellosis			✓	✓	✓	✓	✓
Streptococcal meningitis				✓	✓		✓
Swine dysentery			✓				
Swine influenza					✓		
Teschovirus infections	✓	✓					
Tetanus					✓		✓
TGE			✓				
Thrombocytopenic purpura						✓	
Toxoplasmosis	✓	✓					
Trueperella pyogenes				✓	✓		
Vitamin E deficiency	✓			✓	✓		

Diseases/Pathogens and Disorders Affecting the Sucking Pig

Below is a list of the key diseases that are likely to be seen in the sucking pig. These are covered in detail in this chapter.

- Abscesses.
- Actinobacillosis.
- Anaemia – iron deficiency.
- Arthritis – joint infections.
- *Atresia ani* – no anus or no rectum.
- Brucellosis.
- Bursitis.
- Campylobacter infections of piglets.
- Clostridial diseases.
- Coccidiosis (coccidia).
- Congenital tremor (CT) – shaking piglet.
- Cryptosporidiosis.
- Diarrhoea or scour (enteric colibacillosis).
- Epitheliogenesis imperfecta – (defective skin).
- Erysipelas.
- Facial necrosis.
- Glässer's disease (*Haemophilus parasuis* - Hps).
- Greasy pig disease – (exudative epidermitis).
- Hypoglycaemia – low blood sugar level.
- Inherited thick legs (hyperostosis).
- Leptospirosis.
- Mange.
- Middle ear infections.
- Mycoplasma (enzootic) pneumonia (EP). (*Mycoplasma hyopneumoniae)*.
- *Mycoplasma suis*.
- Navel bleeding/pale pig syndrome.
- Porcine epidemic diarrhoea (PED).
- Porcine reproductive and respiratory syndrome (PRRSV)
- Progressive atrophic rhinitis (PAR) and atrophic rhinitis (AR).
- Rotaviral enteritis (diarrhoea).
- Splay legs.
- Streptococcal meningitis.
- Swine dysentery (SD).
- Swine Influenza Virus (SIV) – swine flu.
- Teat necrosis.
- Tetanus.
- Thrombocytopenic purpura – bleeding.
- Transmissible gastroenteritis (TGE).
- Vomiting and wasting disease/Ontario encephalitis.
- Vitamin E deficiency and iron toxicity.

Abscesses

See Chapter 10 for further information

Abscesses are not uncommon in the piglet, generally being found over the joints (see Arthritis – Joint Infections Chapter 8), over the navel or on the jaw (from teeth roots as a result of poor teeth clipping technique). When the abscess has 'pointed' (become mature caseous liquid) it can be lanced (see Lancing an Abscess or Haematoma Chapter 15). The bacterial organisms that are involved in piglet abscessation are *Trueperella pyogenes* and *Streptococcus suis*. Both generally respond well to penicillin injections.

Actinobacillosis

This is caused by a tiny bacterium called *Actinobacillus suis*. This is present in most herds and lives in the tonsils of older pigs, particularly sows. It may enter the piglet via the respiratory system or via cuts and abrasions. It occasionally produces a septicaemia, that is, it invades and multiplies in the blood stream and settles out in various parts of the body, particularly the lungs and the joints. Here it produces multiple small abscesses. During the acute septicaemic phase of the disease, sudden death is often the only symptom. It can be precipitated by PRRSV. It is not a common disease. The organism may play a role within postweaning Glässer's disease.

Clinical signs

Sudden death in otherwise apparently healthy pigs is common, involving only 1 or 2 litters and only 1 or 2 pigs per litter. In less acute cases there may be discoloration of the skin, a very high fever, coughing and pneumonia. Occasionally skin lesions are seen that can resemble erysipelas. Some piglets may develop arthritis and lameness.

Diagnosis

If there is a history in the herd of sudden death then laboratory examinations are necessary to demonstrate the presence of the organism.

Similar diseases

Meningitis, acute *E. coli* infection, erysipelas, clostridial diseases and pigs that have been laid on can produce very similar symptoms.

Treatment

- The organism is sensitive to most antibiotics but in particular injections of ceftiofur, amoxycillin or ampicillin or procaine penicillin.
- Other antibiotics that can be used include OTC, lincomycin, ceftiofur, cephalexin and trimethoprim/sulpha.
- In persistent outbreaks, if the appearance of the

disease is predictable then preventive measures can be taken by giving long-acting preparations of the above medicines to all litters over a period of 3 to 4 weeks, after which the preventative medication then ceases and the situation is further assessed.
- In-feed medication with phenoxymethyl penicillin at 200 g/tonne to the sow for the first 3 weeks of farrowing has proved successful in problem cases.

Management control and prevention

- In persistent herd problems, the point of entry may be associated with teeth clipping or detailing, scrubbed knees from poor concrete, or respiratory spread from the sow. Assess the significance of these.
- Where there are rough concrete floors it may help to brush these over with lime wash containing 28 ml phenolic disinfectant to 4.5 litres of lime wash.

Anaemia – Iron Deficiency

See Chapter 14 for further information

Anaemia arises in the sucking piglet due to iron deficiency. This occurs because the sow's milk is deficient in iron and the piglet has minimal reserves. Iron forms an essential part of haemoglobin in the red blood cells and this is responsible for carrying oxygen. The lack of iron causes the piglet to become rapidly anaemic and susceptible to other diseases such as scours, unless iron is supplied either orally or by injection.

The piglet is born with limited supplies of iron and if it had been born in the wild would depend on supplementation to its diet from iron-bearing soils. Indoors, the pig has no access to iron other than to the sows' milk (which is deficient) until it starts to eat creep feed. It is therefore necessary to give extra iron either by mouth or by injection. The majority of piglets are born with a level of haemoglobin in the blood of around 9 g/100 ml to 13 g/100 ml and this rapidly drops down to 5 g/100 ml to 9 g/100 ml by 10 to 14 days of age. A shortage of iron results in lowered levels of haemoglobin in the red cells, (anaemia), a lowered capacity for the carriage of oxygen around the body and an increased susceptibility to disease.

Iron-deficient piglets appear pale from 7 days onwards, sometimes but not always with a slight check in growth. The colour of the skin may take on a slight yellow or jaundiced appearance. In severe cases breathing is rapid, particularly with exercise and there may be a predisposition to scour. Postweaning diarrhoea may be more common in piglets with iron anaemia. Inject piglets with 200 mg of iron dextran at 3 to 5 days of age. See Chapter 14 for further information on anaemia – iron deficiency.

Rarely piglets may be born deficient of vitamin E. In these cases the iron injection may cause irreversible liver failure and death in piglets a couple of days after the injection. See Vitamin E Deficiency later in this chapter.

Arthritis – Joint Infections

See also Lameness

Joint infections in the sucking piglet are very common. They are invariably infectious in origin, the sources of infection either being respiratory spread from the sow, or through the skin as a result of some form of trauma, which allows the organisms to enter the system. The common bacteria include *Actinobacillus suis, Haemophilus parasuis,* (Glässer's disease), *Trueperella pyogenes,* very occasionally mycoplasma infections, staphylococci, but most commonly *Streptococcus suis* serotypes. Most of these respond well to antibiotic therapy but treatment must be given early at the onset of disease.

Clinical signs

They are seen from 2 to 10 days of age. At the early onset the pig is laid on its belly, shivering slightly and with its hair stood on end. Stiffness or lameness involving 1 or more legs is evident. It is a common belief that the infection enters through the navel at birth, but this is unlikely. It is most likely to have gained entry via the tonsils or damaged/open teeth from teeth clipping. If the claws are involved, infection has probably arisen from damage by the sow, bad slats or poor floor surfaces. The hock and elbow joints are often visibly swollen. Whilst lameness is the most common symptom, if the organism gains access to the blood stream and a septicaemia results, death may occur before the arthritis develops.

Diagnosis

This is by clinical observation of lameness and the swollen joints. In well-managed herds the numbers of piglets requiring treatment should be less than 2 %. Where there are problems this could rise to as high as 10 to 15 %. *Streptococcus suis* type 14 can cause severe sudden outbreaks of arthritis with acute pain.

Treatment

- Treatment of the infected pig could include one of the following antibiotics: lincomycin, penicillin and streptomycin, oxytetracycline, amoxycillin, ampicillin, trimethoprim/sulpha, enrofloxacin, ceftiofur and framycetin. Inject daily for 5 days. Long-acting preparations can also be used and

these should be injected every other day. Antibiotic penetration of the joint is slow. The choice will depend on the organism, the antibiotic sensitivity and the best response obtained.
- Cortisone or other anti-inflammatory medicines and pain relief medications can be of value.

Management control and prevention

- Stop teeth-clipping piglets – There is little reason to teeth-clip, if the sow has sufficient milk.
- Check the mouths of the piglets to see that they are not infected following teeth-clipping.
- Check the teeth-clippers. Hold them to the light to make sure the edges are not damaged. If light is showing through, abandon them and use a new pair.
- Make sure the teeth-clippers are washed in warm soap and water between litters. They should be stored clean and dry between batches.
- Do not use the same instrument for removing both teeth and tails.
- Preferably remove tails either by gas burner, scalpel blade or sharp scissors to produce a clean cut to the surface. This will bleed a little but it will clot over with a minimum risk of infection. Tail dock before Day 3.
- Check that iron injections are carried out hygienically with a sharp needle. Only use 1 iron injection.
- Check for trauma to the piglet, particularly scrubbed knees, legs or tail.
- Review umbilicus hygiene and possibly dip in iodine shortly after birth.
- Practice split suckling and ensure all piglets get adequate colostrum.
- If there is a problem on the farm, submit samples to the laboratory to identify the organism.
- In severe cases preventative medication of the sow pre-farrowing may help, particularly if the organism is spread by the respiratory route.
- Where there are bad floor surfaces, brush these over with hydrated lime.
- Preventive medication – Administer a long-acting antibiotic injection 3 to 4 days prior to the expected onset of disease. Oxytetracycline, amoxycillin, ampicillin, ceftiofur or penicillin could be used. Most streptococci are sensitive to penicillin.
- Sometimes the skin of the sow is a source of infection. In such cases spray a skin antiseptic onto the udder 1 day before and 2 days after farrowing. Iodine dairy teat dips are ideal.

Atresia ani – No Anus or No Rectum

The piglet is born with no anus externally because it has developed a blind end to its rectum 5 to 10 mm from the exterior, inside the pelvis. The incidence in the mature herd is usually less than 0.5 % but it can be much higher in newly established gilt herds. The condition is heritable but of low penetrance. In a problem herd records will indicate whether there is an incriminating boar involved. The condition is self-limiting in that death invariably ensues. It is not worth attempting surgical repair because the artificial opening closes again. Make sure the female is mated to a different boar for the next litter.

Brucellosis

See Chapter 12 for further information

This disease is caused by the bacterium *Brucella suis*. Where a herd is infected with *Brucella suis* there is generally minimal effect on sucking pigs that are born viable and healthy at full term. However some weak piglets may develop infection of the spinal column (spondylitis) leading to posterior paralysis.

Bursitis

See Chapter 10 for further information

Bursitis is a common condition that arises from constant pressure and trauma to the skin overlying any bony prominence. It can commence in the farrowing house, particularly if there are bad floors, but it usually starts in the weaner accommodation on slatted floors. To address the problem, identify the point at which disease first appears and alter the floor surfaces or change the environment. This condition is covered in more detail in Chapter 10.

Campylobacter Infections of Piglets

Campylobacters are small curved rods which are present, sometimes in large numbers, in the small intestines and large intestines of most mammals, including pigs. There are several species in pigs. Whether or not they cause diarrhoea in naturally reared piglets is debatable because it seems likely that the antibodies in sows' milk would prevent this. However oral infections to newborn colostrum-deprived piglets may result in mild diarrhoea with mucus and sometimes blood in it. Spiral helicobacter species, related to campylobacters, may be found adhering to the stomach wall of pigs and in and around gastric ulcers. However, many other factors contribute to stomach ulcers in pigs and it seems unlikely that helicobacters are a primary cause.

Clinical signs
Campylobacters are associated with mild, sometimes creamy, diarrhoea in piglets lasting several days if untreated.

Diagnosis
This is difficult because campylobacters are so common in faeces along with other organisms which can cause diarrhoea, such as *E. coli*, rotaviruses, coccidia and cryptosporidia, and it is impossible to decide whether their isolation is significant in any particular outbreak.

Treatment
Fortunately, campylobacters are susceptible to oral antibacterial medicines such as tylosin, neomycin, tetracyclines and enrofloxacin, which are used for other diarrhoea-causing conditions.

Human diarrhoea
Acute campylobacter diarrhoea is common in people as a result of eating contaminated food, particularly poultry and milk, but although some of the campylobacters in people appear similar to those in pigs, no causal relationship has been found between pig meat products and campylobacter diarrhoea in people.

Management control and prevention
- Most creamy diarrhoeas in piglets are associated with coccidiosis and control procedures for campylobacters are similar to those for coccidiosis.
- Practice good hygiene and all-in/all-out.
- Management of farrowing houses is important.
- See also the section on control of *E. coli* infections under Diarrhoea or Scour.

Clostridial Diseases

See Chapter 7 for other clostridial diseases that affect the sow
Clostridia are large gram-positive spore-bearing bacteria that are present in the large intestines of all pigs. There are several species but 2 in particular affect the piglet. *Clostridium perfringens*, types A, B or C can under certain conditions produce severe diarrhoea with very high mortality. *Clostridium difficile* may be isolated in pre-weaning scour. Infection with *C.difficile* affects the large bowel causing inflammation or typhlitis (inflammation of the caecum, part of the large intestine).

Similar diseases
Anthrax must also be considered as a possibility and veterinary advice sought.

Clostridium perfringens
Cl. perfringens type C is by far the most important and if it gets into the small intestine and becomes established before colostrum is taken in, disease can result. Piglets are normally infected under 7 days of age and more typically within the first 24 to 72 hours of life.

Clinical signs
These are sudden in onset. Piglets rapidly develop rotten-smelling diarrhoea, which is often blood-coloured, 1 to 7 days after birth. Many piglets die. The lining of the small intestine sloughs off (necrosis) and this may also be observed in the scour. The disease caused by *Cl. perfringens* type A tends to be milder, less dramatic and more prolonged, but it can look similar to that caused by type C. Clostridial disease is common in outdoor herds.

Diagnosis
In typically acute cases the clinical signs and post-mortem lesions are diagnostic. If the abdomen of a dead piglet is cut open, the middle portion of the small intestine is often claret wine coloured and this can usually be seen without cutting into the intestinal wall. Bubbles of gas may also be seen in the wall of the intestine. In less striking cases, confirmation of the diagnosis must be carried out in a laboratory. It is necessary to submit preferably a live or very recently dead pig to the laboratory (within 3 to 4 hours) because the causal organisms multiply after death and cause rapid post-mortem changes. Toxin tests are carried out to determine the type.

Treatment
- In acute outbreaks lamb dysentery antiserum can be injected into the piglets at birth.
- Oral antibiotics and in particular amoxycillin should be given at birth and again at Day 2 or 3.
- The sows' ration can be medicated with 200 g/tonne of phenoxymethyl penicillin or the feed top-dressed daily with the premix, from 5 days pre-farrowing and during lactation.

Management control and prevention
- This is carried out by vaccinating the sow herd using either sheep vaccines made from *Clostridium perfringens* type C toxoids or pig vaccines containing these toxoids. At the onset of an outbreak, sows should be given 2 doses about 2 to 3 weeks apart, the last one at least 7 days before farrowing. The general principles of control of diarrhoea in the sucking piglet should also be considered as contributory factors. Improve hygiene, particularly outside. Consider using zinc bacitracin as an in-feed medication to the sows from 90 days of pregnancy until a week

post-farrowing.
- Autogenous vaccines can be produced against type A, for use in sows.

Clostridium difficile

Clostridium difficile is a normal inhabitant of the bowel, but in certain circumstances, particularly excessive use of antibiotics, the normal protective microbiota is disturbed and members of the normal flora become pathogenic.

Clinical signs

Diarrhoea in piglets 4 to 10 days of age which is unresponsive to most antibiotics, particularly penicillin-based (which is unusual for clostridia).

Diagnosis

Post-mortem reveals oedema in the loops of the large bowel – colon and caecum. Culture and examination of smears reveals large numbers of the organism without the presence of other pathogens.

Treatment

- Difficult. Review all current medications used in the pre-weaning programmes, particularly those which are penicillin based.

Management control and prevention

- Examine colostrum intake. Practise feedback of piglet scour and sows' faeces from the farrowing house to the gilts during their introduction and acclimatisation period. There are no effective vaccines.

Coccidiosis (Coccidia)

See Chapter 11 for further information

Coccidiosis is caused by small parasites called coccidia that live and multiply inside host cells, mainly in the intestinal tract.

There are 3 types, *Eimeria, Isospora* and *Cryptosporidia*. Disease is common and widespread in sucking piglets and is predominately caused by *Isospora suis*. Occasionally disease may also be seen in pigs up to 15 weeks of age and in young floor-fed boars and gilts that are housed in permanently populated pens.

Diarrhoea is the main clinical sign followed by secondary bacterial infections. Dehydration is common. This condition is covered in detail in Chapter 11.

Congenital Tremor (CT) – Shaking Piglets

This is a sporadic disease seen in newborn pigs.

Clinical signs

Signs are tremors and shaking of the muscles of the head and body. Usually there is more than 1 pig involved in a litter but the tremor is only seen when piglets are walking around and not when they are asleep. The condition decreases with age but if the tremors are too great for the piglets to find a teat and suckle then mortality maybe high. Mortality in an affected litter or in a herd outbreak could increase above the norm by 3 to 10 %.

The causes of the condition are varied and classified into 5 groups based on brain histology.

Group 1 – associated with a classical swine fever/hog cholera (CSF/HC).

Group 2 – most of the problems in the field are found in this group. Signs can include sporadic disease affecting the odd litter; an outbreak affecting a proportion of litters lasting 6 to 8 weeks, or and ongoing problem that only affects gilt litters. In these cases the disease acts like a virus, but the causal agent has not yet been recognised.

Groups 3 and 4 – associated with hereditary disorders seen in the Landrace (3) or Saddleback breeds (4).

Group 5 – Associated with organophosphorus poisoning. Tremors in piglets may also be seen with AD/PRV and Japanese encephalomyelitis virus.

It would be unusual to find a pig farm that some time in its history had not experienced 1 or more litters of trembling piglets. Outbreaks identified under group 2 appear widespread among most if not all pig populations, yet little disease is seen in most herds, presumably because an immunity is established in the sow herd. In gilt herds, however, there can be major outbreaks involving up to 80 % of all litters during the first parity. Note that once diagnosed, the problem may persist for 3 to 5 months, as all gilts that are exposed while pregnant are likely to produce trembling piglets. However, once exposed, the sow will not produce trembling in the second and subsequent litters – so there is no need to cull gilts based on them producing trembling piglets.

Diagnosis

This is based on clinical evidence although histological examinations in the laboratory can help to differentiate the groups.

Treatment

- There is no specific treatment for affected piglets but careful management will greatly reduce mortality.
- Ensure that piglets are given colostrum at birth and assisted to a teat.

Management control and prevention

This condition is a result of poor gilt introduction and acclimatisation. Attempts to immunise breeding stock should be carried out. The following may assist and the results should be documented for further studies where litters are continually affected:

- If there is a history of the disease on the farm, expose incoming maiden gilts to faeces from older animals and boars for 4 to 6 weeks prior to mating.
- At the time of mating, use tissue paper to wipe around the prepuce of the boar and the vulva of the mated sow. Expose the group of maiden gilts to the tissues. Do this 2 to 3 times weekly.
- Assess the results of using a vasectomised boar from one of your affected litters for a period of 6 weeks prior to full mating.
- Move all maiden gilts into the main mating area for a period of 7 days commencing at least 4 weeks before mating is due to start to expose them to any possible infectious agents.

Cryptosporidiosis

See Chapter 11 for further information

Cryptosporidia are parasites similar to coccidia that can also cause diarrhoea but at a slightly older age of 7 to 21 days. They can infect the human and this can be serious in immunosuppressed people. They are also found in other species such as rats and mice which can become a source of constant infection.

Infection may result in no clinical signs.

Diagnosis

Substances that are active against oocysts have an effect against cryptosporidia.

Diarrhoea or Scour (Enteric Colibacillosis)

See Chapter 14 for further information

Of all the diseases in the sucking piglet, diarrhoea is the most common and probably the most important. In some outbreaks it is responsible for high morbidity and mortality. In a well-run herd there should be less than 3 % of litters at any one time requiring treatment and piglet mortality from diarrhoea should be less than 0.5 %. In severe outbreaks levels of mortality can rise to 7 % or more and in individual untreated litters up to 10 % (in TGE it may reach 100 % overall).

There are several different causes of diarrhoea and these are shown in Fig 8-29. Four of the agents listed in Fig 8-29 are viruses: transmissible gastroenteritis (TGE), rotavirus, porcine epidemic diarrhoea (PED) virus and PRRS virus. The main bacterial causes are *Escherichia coli (E. coli)* and clostridia and the main parasite is coccidia.

Other agents such as adenovirus, astrovirus, Breda viruses, calcivirus, torovirus, picobirnavirus, norovirus, and chlamydia, have been identified in diarrhoeic faeces but their significance is mostly unknown. Adenovirus may be involved in pneumonia and is thought sometimes to worsen other pneumonias such as mycoplasma (enzootic) pneumonia. Breda viruses have been associated with loss of appetite, weakness, tremors and death in weaners, but they are extremely rare. Chlamydia causes conjunctivitis, coughing and arthritis (and in sows has been associated with infertility and stillbirths). None of these agents cause common diseases of any importance in pigs and are best ignored by the pig farmer.

This section deals principally with *E. coli* diarrhoea (colibacillosis). Diarrhoeas caused by clostridia, coccidia, TGE and PED are dealt with in more detail under their respective heading in this chapter.

At birth the intestinal tract is microbiologically sterile and it has little immunity to disease-producing organ-

FIG 8-29: THE MAIN CAUSES OF PIGLET DIARRHOEA

	Early Period Days		Late Period Days		Mortality Level
	0 – 3	3 – 7	7 – 14	15 – 27	
Agalactia	✓	✓	✓	✓	Moderate
Clostridial disease	✓	✓	✓		High
Coccidiosis		✓	✓	✓	Low
Colibacillosis (*E. coli*)	✓	✓	✓		Moderate
Porcine epidemic diarrhoea (PED)	✓	✓	✓	✓	Low
Porcine reproductive and respiratory syndrome virus (PRRSV)	✓	✓	✓	✓	Variable
Rotavirus	✓	✓	✓	✓	Low
Transmissible gastroenteritis (TGE)	✓	✓	✓	✓	High

isms. Organisms begin to colonise the tract quickly after birth, among them potentially pathogenic strains of *E. coli* and *Clostridium perfringens*. Immunity is initially provided by the high levels of antibodies in colostrum (IgG, IgM, IgA). After the colostral antibodies have been absorbed into the bloodstream, the immunity is maintained by the antibody (IgA) which is present in milk. IgA is absorbed into the mucous lining of the intestines. It is essential that the newborn piglet drinks sufficient colostrum soon after birth to prevent potentially pathogenic organisms multiplying against the intestinal wall and causing diarrhoea. It is also essential that the piglet continues to drink milk regularly after the colostrum has gone so that its intestines continue to be lined by protective antibodies.

The antibodies acquired passively from the colostrum and milk are finite and can be overwhelmed by large doses of bacteria present in the environment. The higher the number of organisms ingested, the greater the risk of disease. Environmental stress such as draughts causing chilling also plays a role because it lowers the piglet's resistance. There is thus a delicate balance between the antibody level on the one hand and the weight of infection and stress on the other.

Clinical signs

Scour in the piglet can occur at any age during sucking but there are often 2 peak periods, before 5 days and between 7 and 14 days.

Acute disease

The only sign may be a perfectly good pig found dead. Post-mortem examinations show severe acute enteritis, so sudden that there may be no evidence of scour externally. Clinically affected piglets huddle together shivering or lie in a corner. The skin around the rectum and tail will be wet. Look around the pen for evidence of a watery to salad cream consistency scour. In many cases there is a distinctive smell. As the diarrhoea progresses the piglet becomes dehydrated, with sunken eyes and a thick leathery skin. The scour often sticks to the skin of other piglets giving them an orange to white colour.

Prior to death piglets may be found on their sides paddling and frothing at the mouth.

Sub-acute disease

The symptoms are similar but the effects on the piglet are less dramatic, more prolonged and mortality tends to be lower. This type of scour is often seen between 7 to 14 days of age, manifested by a watery to thin salad cream consistency diarrhoea, often white to yellow in colour.

Diagnosis

The overall picture must be considered when making a diagnosis. Sudden outbreaks of scour involving large numbers of litters with acute diarrhoea and high mortality suggest TGE, PED or PRRSV. It always helps in differentiating these infections to know whether the herd had previously been exposed to any of these diseases or not. If exposure is for the first time the outbreak is likely to be explosive.

Rotavirus diarrhoea appears in waves in individual litters or groups of litters and normally in the second half of lactation. Mortality tends to be low. Coccidiosis has an incubation period of 6 days and is usually involved in diarrhoea complexes from 7 to 14 days of age. At less than 5 days of age the most common cause is *E. coli* with acute diarrhoea, particularly in gilts' litters because they pass on poorer levels of immunity. Clostridial infections also occur at this age.

Diagnosis is based on the clinical examinations, the response to treatment (viral diseases do not respond to treatment) and laboratory examination of the scour. Submit a rectal swab or a live pig to the laboratory for cultural examinations and antibiotic sensitivity tests.

Treatment

- Some antibiotics available are shown in Fig 8-30. Most of these are active against *E. coli* and clostridia but not the virus infections.

FIG 8-30: SOME ANTIBIOTICS AVAILABLE TO TREAT PIGLET DIARRHOEA

Medicine	Method of Dosing	
	Oral	Injection
Amoxycillin	✓	✓
Ampicillin	✓	✓
Apramycin	✓	
Ceftiofur		✓
* Chloramphenicol		✓
Enrofloxacin	✓	✓
Framycetin		✓
* Furazolidone	✓	
Neomycin	✓	
Spectinomycin	✓	
* Streptomycin	✓	✓
Sulphonamides	✓	✓
Trimethoprim/sulpha	✓	✓
Tylosin		✓

* Banned in some countries.

- In severe outbreaks of *E. coli* disease, the sows' feed can be top-dressed with the appropriate antibiotic daily, from entry into the farrowing house and for up to 14 days post-farrowing. This can be effective in reducing bacterial output in the sow's faeces.
- Observe litters for the presence of diarrhoea, both night and morning.
- Study the history of the disease on your farm. Is it sporadic, in 1 piglet in a litter, or in total litters?
- In light of the history, either treat the individual pig or at the first signs of disease, treat the whole litter.
- If a litter is badly scoured, dose night and morning for a minimum of 2 days.
- Assess the response to treatment. If there is no change within 12 hours then change to another medicine as advised by your veterinarian.
- Always treat piglets under 7 days old by mouth.
- For older pigs where the disease is less acute, injections are equally effective and easier to administer.
- Provide electrolytes in drinkers. These prevent dehydration and maintain body electrolyte balances.
- Cover the pen, the creep area and where the pigs defecate with straw, shredded paper, shavings or sawdust.
- Provide an additional lamp to provide an extra source of heat.
- Use binding agents such as chalk, kaolin or activated attapulgite to absorb toxins in the gut.
- Use pasteurized peat as a gut conditioner from 3 days of age, presented to the piglets in a simple pan. The peat needs to be pasteurised to reduce the risk of *Rhodococcus equi* transmission which may produce TB-like lesions in the neck lymph nodes which would still be present at slaughter.
- Collect the diarrhoea on newspaper and use as part of the feedback programme to your gilts.
- Minimise the number of gilts in the breeding pool.
- Ensure that the farm practises good batch production and all-in/all-out.

Management control and prevention

- Adopt procedures to prevent the spread of the scour:
  - Disinfect boots between pens.
  - Use a disposable plastic apron when dosing piglets to prevent heavy contamination of clothing.
  - Wash hands after handling a scoured litter.
  - Disinfect brushes and shovels between pens.
- Ensure that farrowing houses are only used on an all-in/all-out basis with a pressure-wash and disinfection between each batch.
- Farrowing pens must be dry before the house is repopulated. Remember that moisture, warmth, waste food and faeces are ideal for bacterial multiplication.
- Pen floors should be well maintained. Poor pen hygiene associated with bad drainage predisposes to scour.
- Look carefully at the part of the pen floor where there are piglet faeces. Is this poorly drained? Do large wet patches develop? If so cover them with extra bedding daily and remove. This is a most important aspect of control.
- Check nipple drinkers and feeding troughs for leakages.
- Ensure that faeces are removed daily from behind the sow from the day she enters the farrowing places until at least 7 days post-farrowing if the floors are slatted. Also remove faeces daily throughout lactation if they are solid concrete.
- Maintain creep environments that are always warm and comfortable. Fluctuating temperatures are a major trigger factor to scour, particularly from 7 to 14 days of age.
- Do not penny-pinch on your heating costs. Many cases of scour are precipitated by attempts to save on costs of energy.
- Check for high air flow and draughts. They predispose to scour.
- Consider vaccinating against *E. coli* (however, make sure first that this is the cause of the problem). *E. coli* vaccines only protect the piglet for the first 5 to 7 days of age.
- Ensure that vaccines are stored properly. Frozen vaccines are inactivated.
- Ensure feedback is carried out appropriately. In particular, gilts in isolation and acclimatisation need access to any diarrhoea from the farrowing house and faeces from the on-site nursery. Collect the diarrhoea on newspaper and use as part of the feedback programme to your gilts.
- Assess the environment of all the farrowing houses. Poor environments allow heavy bacterial multiplication and a much higher bacterial challenge is likely to break down the colostral immunity.
- Check the sow's health. Animals affected with enteric or respiratory disease, lameness or mastitis predispose the litter to scour.
- Avoid the use of milk replacers where possible. Their routine use, particularly if they are allowed to get stale or contaminated, may increase the incidence.

- Where farrowing house floors are very poor, pitted and difficult to clean, brush them over with lime wash containing a phenolic disinfectant. See Chapter 15.
- Ensure all rooms have their own equipment – brushes and scrapes. Change needles between rooms to stop cross-contamination.
- Scour is more common in large litters. Split suckling should be adopted.
- Minimise cross-fostering, which may interfere with colostrum management.
- Minimise the number of gilts in the breeding pool.
- Ensure that the farm practises good batch production and all-in/all-out.

Colostrum management

It is vital that the piglet receives the maximum amount of colostrum within the first 12 hours of birth. High levels of antibody are only absorbed during this period. Factors such as poor teat access, poor farrowing place design, and particularly the development of agalactia in the sow, associated with udder oedema, reduce intake.

In an outbreak of scour it is important to establish if udder oedema is present. It is more common in gilts and second parity than in older sows. If *E. coli* diarrhoea is a problem in younger aged females, this suggests that immunity levels are low and vaccination should be considered. Inject the sow twice, 2 to 4 weeks apart, the second injection at least 2 weeks before farrowing; but these times are variable depending upon the vaccine used. With good management it should not be necessary to vaccinate the sows, only the gilts. Review fostering management, particularly within the first 6 hours of birth.

Are piglets getting sufficient colostrum?

Eradication

It is not possible to eliminate organisms such as rotavirus, *E. coli* and coccidiosis from the herd and most (if not all) pigs will be infected with them. Herds can be maintained free of TGE, PED and PRRSV. All herds carry clostridia but other factors are required to cause disease.

A summary of the management factors associated with disease is shown in Fig 8-31.

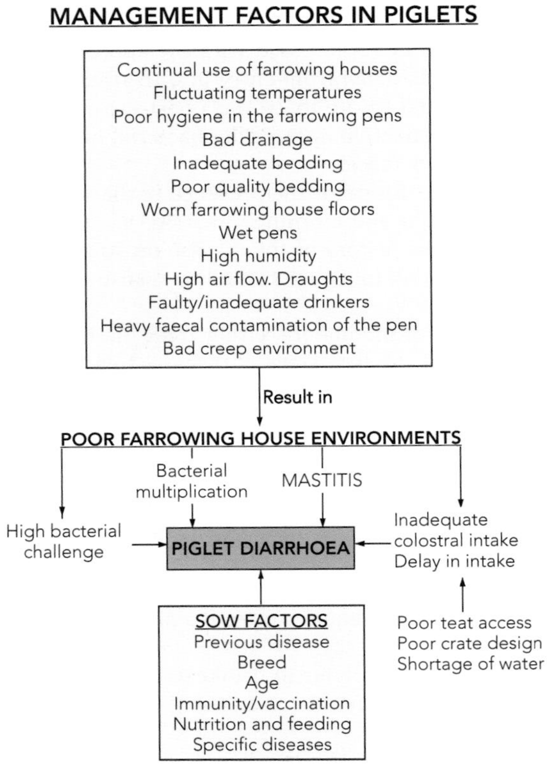

Fig 8-31

Epitheliogenesis imperfecta or Defective Skin

See Chapter 10 for further information

This condition is where a piglet is born devoid of areas of skin, which have failed to develop correctly. See Chapter 10 for details.

Erysipelas

See Chapter 7 for further information

Swine erysipelas is caused by a bacterium, *Erysipelothrix rhusiopathiae* that is found in most (if not all) pig farms. The disease is rarely seen in the piglet because it is protected from infection by colostral immunity. It occasionally occurs in gilt litters and the only sign may be a good piglet found dead from septicaemia. A laboratory diagnosis to demonstrate the organism is necessary to confirm disease. Treatment consists of injecting the whole litter with long-acting penicillin (sometimes piglets vomit when using procaine penicillin but this is not of any consequence). All breeding herds should be vaccinated for erysipelas. This disease is covered in detail in Chapter 7.

Facial Necrosis

This is a specific area of skin damage under the eyes on the face of the piglet associated with *Staphylococcus hyicus* (also associated with greasy pig disease). This condition however, rarely becomes progressive. *Staphylococcus hyicus* is a bacteria that can usually be isolated from the skin of a pig; its presence is not normally diagnostic.

However, if a sow is not producing sufficient milk, the piglets can fight over teat access causing scratches on the face. This enables the pathogen to enter the pig's system and cause facial lesions.

The presence of such lesions can be used as a diagnostic tool to indicate that there is insufficient milk available for the piglets. Under such circumstances the sow milking programme should be reviewed.

Many farmers teeth-clip/grind specifically to control this problem. However, when lactation is adequate, there should be no requirement for either teeth-clipping or grinding. A plentiful milk supply can be achieved through proper management of the sow before and during lactation. Studies have shown that including provision of enrichment and increasing space improves sow health, feed intake and milk production. In addition to minimising competition for teats and reducing injury, higher weaning weights can be achieved. When lactation is sufficient, teeth-clipping has a limited effect on mammary injuries in the sow and on piglet performance. Sows with large litter sizes or mastitis and young gilts will be at particular risk of low milk yields and piglet competition for milk, and management action needs to be taken accordingly.

Glässer's Disease (*Haemophilus parasuis*) - Hps

See Chapter 9 for further information

This disease is caused by the bacterium *Haemophilus parasuis* (Hps), a small organism, of which there are at least fifteen different types. Glässer's disease normally affects the sucking and young growing pig, but can occasionally affect the adult pig.

Outbreaks of disease are sometimes experienced in sucking pigs, particularly in gilt herds.

In the majority of herds in which the bacterium is endemic, sows produce a strong maternal immunity which normally persists in their offspring until 8 to 12 weeks of age and as a result, the effects of the pathogen in weaners are usually nil or very minimal. The pigs become sub-clinically infected when still protected and then stimulate their own immune response. If however maternal immunity is lacking or wears off before they become infected they may develop severe disease. It can however become a secondary organism where there are other major pathogens and in particular mycoplasma (enzootic) pneumonia.

Pigs with Glässer's disease become rapidly depressed, with an elevated temperature, stop eating and are reluctant to rise. *H. parasuis* also causes individual cases of arthritis and lameness with acute pain, fever and inappetence. It is respiratory-spread and a characteristic feature is a short cough of only 2 to 3 episodes. Sudden death in good sucking piglets is not uncommon in herds with a problem and in particular when immunity in gilt litters is low.

In chronic disease, sucking piglets are often pale and poor-growing and 10 to 15 % may be affected in a litter. Such pigs then continue into the growing period with poor growth. When long standing pericarditis is a feature, sudden deaths occur.

Greasy Pig Disease (Exudative Epidermitis)

See Chapter 10 for further information

This is associated with the normal skin bacterium *Staphylococcus hyicus,* which invades abraded skin causing infection. The disease is also called exudative epidermitis which describes the oozing of fluid from the inflamed skin. In the sucking piglet, disease is usually confined to individual animals, but it can be a major problem in new gilt herds and weaned pigs. This disease is covered in Chapter 10.

Hypoglycaemia – Low Blood Sugar Level

The newborn piglet can be born with low glycogen reserves in the liver and during the first few days of life it is unable to mobilise these to provide adequate levels of glucose in the blood. It is therefore dependent for energy on a regular intake of lactose from the sow's milk. If a piglet cannot obtain sufficient to maintain its energy output, the body temperature drops and it ultimately goes into a coma and dies.

Clinical signs

These progress from an animal laid on its belly, shivering and becoming very cold to eventually lying on its side, paddling, frothing at the mouth and becoming comatose. The eyes are sunken and the head bent backwards.

Diagnosis

Hypoglycaemia occurs in the first 12 to 24 hours of birth and the clinical picture is characterised by the symptoms. Examine the eyes to see that there is no evidence of lateral movements (nystagmus) which would indicate meningitis.

Treatment

- The condition must be recognised early if treatment is to be successful.
- Immediately remove the piglet to a warm, draught-free environment at a temperature of

30 °C (86 °F). A box well-bedded in shavings with an infra-red lamp above is ideal.
- Feed the piglet with warm sow's or alternatively cow's colostrum or 20 % dextrose solution by syringe or stomach tube every 20 minutes until it has returned to normal. Then introduce the piglet to a newly farrowed sow.

Management control and prevention

- Identify potential piglets at birth and treat as described under 'The Stillborn Pig/Key points to managing the poor viable piglets' previously in this chapter. See also Fig. 8-23: The Piglet with Poor Viability.

Inherited Thick Legs (Hyperostosis)

This is a very rare condition of newborn piglets which is thought to be inherited. Individual litters are born with bony thickening of the legs, most notably the front legs. The pigs cannot walk properly and usually fail to thrive. If such a condition arises, check that the boar has not produced other similar litters and do not mate the sow with the same boar again.

Leptospirosis

See Chapters 6 and 7 for further information

Sucking pigs can be exposed to different strains of leptospira from the urine of rats, mice, other animals or the carrier pig. Disease is uncommon in the sucking pig and would only involve individual animals.

When it does occur, piglets become ill and inappetent with jaundice and blood in the urine. Severely infected pigs die. Disease is caused by the *Leptospira icterohaemorrhagiae* serovar. This disease can be spread to humans if infected urine makes contact with broken skin or mucous membranes and causes Weil's disease. These are covered in detail in Chapters 6 and 7.

Mange

See Chapter 11 for further information

Mange is a parasitic disease of the skin caused by 1 of 2 mites, either *Sarcoptes scabiei* var. *suis* or *Demodex phylloides,* the latter of which is relatively unimportant in swine.

Sarcoptic mange (sometimes called scabies) is by far the most common in the pig.

The disease is rarely seen in the sucking piglet unless the sow is infected. However it is a common disease in the weaned and growing pig and adult animals.

Middle Ear Infections

See Chapter 9 for further information

This disease occurs occasionally in the sucking pig from 7 to 10 days of age. The middle part of the ear is responsible for balance, and infection causes the piglet to hold its head on the affected side and to lose its balance. Infection arises as a sequel to joint infections or septicaemia and the common organisms involved include: *Bordetella bronchiseptica, Haemophilus parasuis*, streptococci, and staphylococci. Early identification of the condition is essential to allow prompt treatment. This should be carried out using daily antibiotic injections of penicillin/streptomycin, OTC or amoxycillin. It is more common in the weaned pig.

Mycoplasma (Enzootic) Pneumonia (EP) *Mycoplasma hyopneumoniae*

See Chapter 9 for further information

Mycoplasma (enzootic) pneumonia is caused by the tiny pathogen *Mycoplasma hyopneumoniae*. The disease is not normally a problem in the sucking pig, because in herds in which it is endemic, immunity from the sow via the colostrum protects it. If, however, a herd has been free of the organism and becomes infected for the first time, severe disease may be seen in sucking pigs until the sows have had sufficient time to develop immunity.

Mycoplasma (Eperythrozoon) suis

See Chapter 11 for further information

Mycoplasma (Eperythrozoon) suis is caused by a small mycoplasma bacteria. This blood parasite attaches itself to the red cells in the blood, sometimes damaging them and causing them to break apart. This causes a reduction in the red blood cell count, reducing haemoglobin (the substance which transports oxygen around the body) thus diminishing the capacity of the blood to carry oxygen. When large numbers of red cells are damaged, hemolysis anaemia results and pigs look jaundiced.

A sow may carry *Mycoplasma suis* and yet remain quite healthy. However, it can cross the placenta and infect pigs in utero, causing weak piglets at birth. This condition is covered in detail in Chapter 11.

Navel Bleeding/Pale Pig Syndrome

At birth or within a few hours the piglet becomes extremely pale and in many cases dies. The condition arises in 1 of 3 ways:

1. Anoxia or shortage of oxygen inside the uterus during farrowing causes the piglet to pool its blood into the placenta. If it is born and the cord separated at this point, then it will be born very pale

and anaemic. This picture is seen when piglets are delivered by hysterectomy and they are removed from the uterus at a critical time before the piglet has time to recall its blood from the placenta. Affected piglets are more likely from old sows and in large litters.

2 Pigs are sometimes born with a haemorrhage or a haematoma in the cord itself. The cause of this is unknown but in some cases it is related to premature removal of the piglet from behind the sow at farrowing. The blood vessels in the cord bleed.

3 Continual bleeding from the navel during the first 3 to 4 hours after birth.

Clinical signs

Fresh blood on the floor of the pen arising from the end of the navel is diagnostic.

Treatment

- Early recognition of a bleeding navel is essential. The cord should be clamped approximately 13 mm (0.5 inch) from the skin using an umbilical clip. (See Chapter 15 umbilical cord applying a clamp). Those used for babies are ideal. Nylon or plastic ties used to bind together electrical wires are also good.
- As an alternate and in an acute emergency the navel can be tied in a knot.
- A ligature can be applied around the umbilicus but it shrinks and the bleeding often continues. The cord should be bent back on itself and re-tied in the shape of a "U".

Management control and prevention

- Navel bleeding is associated with the use of wood shavings as bedding. The reasons for this are unknown but wood preservatives or other substances may be responsible. Change the shavings to an alternate source or use straw for bedding.
- Warfarin poisoning can be responsible for haemorrhage.
- Vitamin C was thought to be involved and improvements by feeding sows with 1 g/day have been reported. However, experiences with this vitamin have been disappointing.
- Do not move pigs away from the sow immediately at farrowing. Allow the piglet to break the cord naturally. There is a particular part of the cord where separation takes place naturally without any haemorrhage. The navel cord is always slightly longer than the birth canal so that when the newborn piglet starts to rise and walk the cord is stretched, breaks and recoils to block the blood vessels. It should not be cut.
- Supplementing the diet with vitamin K can sometimes help.
- Mycotoxins from contaminated feed have been implicated.
- A riboflavin deficiency has been implicated.
- In some herds there appears to be an association with the use of prostaglandin to synchronise farrowings.
- Do not allow excessive trauma to the cord within 3 hours of birth. This may occur if too many piglets are fastened in the creep area.

Porcine Epidemic Diarrhoea (PED)

See Chapter 9 for additional information in the weaner/grower

Porcine epidemic diarrhoea is caused by a coronavirus somewhat similar to that which causes transmissible gastroenteritis (TGE). This virus is widespread in Europe and Asia particularly. In 2013 the pathogen was first recognised in the United States. The virus damages the villi in the gut, thus reducing the absorptive surface, with loss of fluid and dehydration. After introduction into a susceptible breeding herd, disease is followed by a strong immunity over 2 to 3 weeks. The colostral immunity then protects the piglets. The virus disappears spontaneously from small breeding herds but tends to be maintained in finishing farms due to the repeated introduction and subsequent infection of susceptible pigs.

Clinical signs

Acute disease

This occurs where the virus is introduced into a susceptible population for the first time. In such cases up to 100 % of sows may be affected, showing a mild to very watery diarrhoea. Two clinical pictures are recognised: PED type I only affects growing pigs whereas PED type II affects all ages including sucking pigs and mature sows. The incubation period is approximately 2 days and the disease episode lasts for 7 to 14 days. In sucking pigs the disease can be mild, although in Asia the disease can be severe with mortalities up to 40 %. The piglets may vomit.

Endemic disease

In large breeding herds, particularly if kept extensively, not all the females may become infected first time round and there may be the odd recurrence. This only occurs in piglets suckling from sows with no maternal antibodies and it is therefore sporadic.

Diagnosis

This can be suspected on the clinical signs but it cannot be differentiated from TGE. If acute diarrhoea is occurring in weaned and older animals on a growing/finishing farm with no symptoms in sucking piglets then this would suggest PED type I. PED type II would affect piglets. Virus particles from diarrhoea samples can be identified under the electron microscope but this would not differentiate PED from TGE. Blood tests can be carried out to look for rising antibody titres. An ELISA test is also available for examining diarrhoea samples or intestinal contents. When you examine the small intestine, check the pH of the contents. In cases of PED, the contents will be acidic.

Treatment

- Because this is a viral infection there is no specific treatment but often secondary bacteria complicate the picture and these can be treated by broad spectrum antibiotics such as neomycin, framycetin and apramycin or trimethoprim/sulpha.
- If the virus enters the herd for the first time it is important to ensure that all the adult animals become infected at an early stage to allow an early immunity to develop. This can be achieved by exposing sows to the diarrhoea 3 times, each 2 days apart, via the drinking water. Mix scour or contaminated material into a bucket of water and use this as the source.

Management control and prevention

Ensure that all gilts in isolation and acclimatisation have access to diarrhoea from the farrowing house and on-site nursery weaner faeces.

Porcine Reproductive and Respiratory Syndrome Virus (PRRSV)

The information in this chapter relates specifically to PRRSV in the farrowing and sucking period. The main information on PRRSV, including methods of spread, immunity, diagnosis, similar diseases and elimination from a herd are covered in Chapter 6, with additional information in Chapters 7, 9 and 10. These chapters should be read to fully understand how this disease impacts the whole herd.

PRRSV is normally only a problem in piglets when the virus infects the herd for the first time. It may continue infecting litters for up to 12 weeks until all sows have had sufficient time to develop an immunity and pass this protection on through the colostrum.

Clinical signs in the sucking pig

In the early stages of acute disease piglets are born 2 to 3 days early in a very weak condition and rapidly become hypoglycaemic because they are unable to get to the teat and suckle. Together with weak pigs there are high numbers of stillbirths and late mummified ones. Newborn piglets show sticky brown material over the eyelids and very occasionally small blisters on the skin. Scour, pneumonia and coughing are commonly observed but with increasing time the quality and survivability of the piglets improves. Starvation, splay leg, tremors, paddling and doming of heads may also be seen. Haemorrhage associated with trauma and tail docking due to reduce blood thrombocytes is also reported.

Treatment

This is aimed at preventing secondary infections, usually either respiratory or enteric until an immunity builds.

- Piglets should be injected with either long-acting OTC or amoxycillin on Days 3, 7 and 14 after farrowing.
- Electrolytes should be given to counteract dehydration.

Management control and prevention

Giving the piglets every chance to survive and recover is the priority. This is aided by the following actions:

- Raise the farrowing house temperature during the period of farrowing and whilst disease is active to 23 °C (75 °F).
- Provide extra bedding. Use shavings or other suitable materials to create the best environment for the piglet.
- Provide an extra heat lamp by the side of the sow.

Progressive Atrophic Rhinitis (PAR) – Atrophic Rhinitis (AR)

Rhinitis is inflammation of the tissues inside the nose and in its mild form it is very common. The term atrophy indicates that the tissues inside the nose, which become infected or damaged, shrink and become distorted. There are 2 forms of the disease:

Atrophic rhinitis

This form is mild and non-progressive where the infection or irritation occurs over a period of 2 to 3 weeks. The inflammation does not progress and structures in the nose called turbinate bones repair and return to normality. This is known as atrophic rhinitis.

All herds will show some degree of non-progressive atrophic rhinitis, the inflammation being of short duration. Organisms such as bordetella, haemophilus, non-toxigenic *Pasteurella*, other environmental organisms and dust or gases can produce non-progressive atrophic rhinitis in the nose.

Progressive atrophic rhinitis

The serious disease is progressive atrophic rhinitis (PAR) where toxin-producing strains of the bacterium *Pas-*

teurella multocida become involved causing continual and progressive inflammation and atrophy of the tissues and nose distortion. There are 2 types of *Pasteurella multocida*; A and D. PAR is a serious condition in both sucking and growing pigs.

For a herd to have PAR, toxigenic *Pasteurella multocida* must be present. They are carried in the nose and tonsils of the adult pig and there is always the risk therefore of buying them into the herd. This is the most common method of entry.

Spread of disease between herds is almost invariably by the carrier pig, the organism being found in the respiratory tract and the tonsils.

Spread within herds is by droplet infection between pigs or by direct pig-to-pig (nose-to-nose) contact. It can also be spread indirectly on equipment, clothes etc. When first infected, pigs can carry the infection for many months. Infection is usually picked up during the second half of the sucking period or after weaning and clinical disease may be evident from 3 weeks of age onwards. The toxin is absorbed into the system where it damages other tissues including the liver, kidneys and lungs, resulting in reduced daily gain and depressed feed efficiency.

Similar organisms may also be found in the cat, dog, rabbit, poultry, goat, sheep and turkey, but it is thought that these are host adapted strains and are unlikely to cause serious disease in the pig. The human may carry toxigenic *P. multocida* in the tonsils for a very short period of time, although the evidence for this is limited and there are no reports of transmission by people to pigs. Experience indicates that the main and probably only method of introduction into the herd is by carrier pigs, although occasionally unexplained outbreaks of disease may occur.

Clinical signs

In sucking pigs sneezing, snuffling and a nasal discharge are the first symptoms, but in acute outbreaks where there is little maternal antibody, the rhinitis may be so severe that there is haemorrhage from the nose. By 3 to 4 weeks of age and from weaning onwards, there is evidence of tear-staining and malformation of the nose associated with twisting and shortening.

Severely affected pigs may have problems eating. There is considerably reduced daily gain. In severe outbreaks pigs may not grow to market weight.

Diagnosis

This is based on clinical signs. However, do not assume if sneezing is occurring in young pigs that automatically it will be progressive atrophic rhinitis. The disease is easily identified by post-mortem examinations of the nose and culture of the organism from nasal swabs. (See Chapter 15: Swabbing the nose and tonsils). The snout is sectioned at slaughter at a level of the first premolar tooth and an assessment of the degree of atrophy of the turbinate bones made. The snouts are graded from 0 to 5, 0 being a perfect snout. Grade 1 would show a slight loss of symmetry of the nose, Grade 2 a slight loss of turbinate tissue and Grade 3 a moderate amount. It is only when Grades 4 and 5 are present, when there is severe progressive loss of tissue that PAR will be suspected (Fig 8-32).

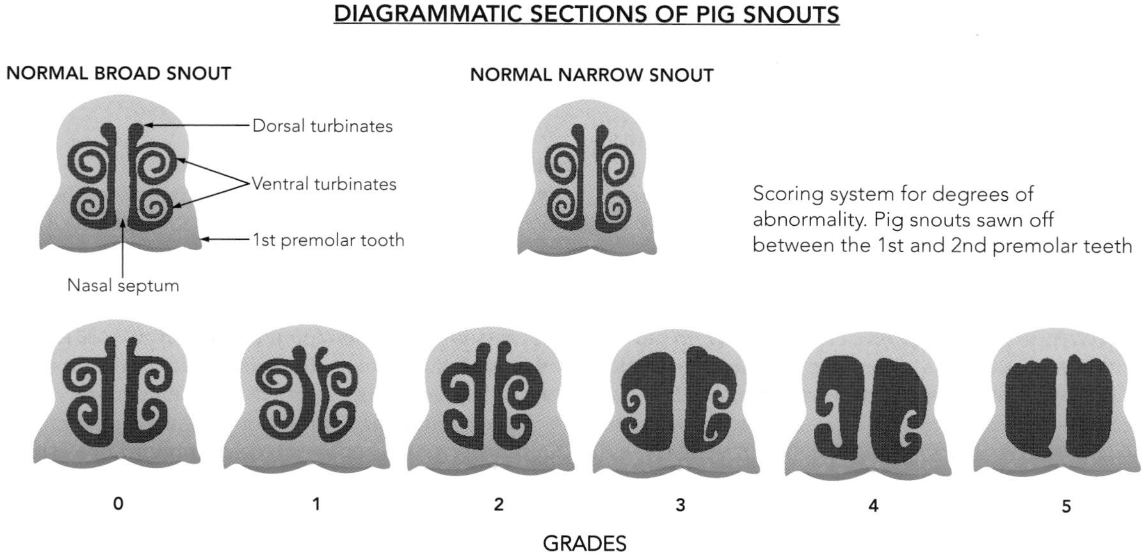

Fig 8-32

Similar diseases

Rhinitis may be caused by the following, but the distinction is less evident; less pigs are obviously affected and the turbinate bones will heal and regenerate.

- Air containing high bacterial counts.
- Aujeszky's disease/pseudorabies virus.
- *Bordetella bronchiseptica* infection.
- Chronic respiratory disease.
- Dust.
- Glässer's disease.
- High levels of ammonia > 25 ppm.
- Porcine cytomegalovirus infection (PCMV)
- PRRSV.

Treatment

Once toxigenic *Pasteurella multocida* have been identified, the complete breeding herd should be immediately vaccinated 6 weeks apart using a vaccine made from toxigenic *P. multocida*. It takes approximately 4 months for a total herd immunity to develop and it may be 9 months or more before the disease is brought completely under control. In the early stages of a herd breakdown the following could be recommended:

- The sow's feed should be top-dressed with either OTC or trimethoprim/sulpha (TMS) from point of entry into the farrowing house through to weaning (500 g/tonne).
- Medicate the creep rations with OTC or CTC at 500 to 800 g/tonne or use trimethoprim/sulpha combinations for 3 to 4 weeks postweaning.
- Inject all piglets with long-acting OTC, tulathromycin or amoxycillin on Days 3, 10 and 15 during sucking. Other antibiotics could be used depending on the bacterial sensitivity, such as penicillin, trimethoprim sulphas, tylosin and enrofloxacin.
- Inject pigs similarly at weaning time with long-acting antibiotic. This treatment programme should continue for a period of at least 2 months after all sows have been fully vaccinated.
- All weaned pigs should be medicated in-feed until the clinical outbreak has subsided.
- Sows should be given a booster dose of vaccine 2 to 3 weeks prior to each subsequent farrowing.

In acute outbreaks PAR is evident clinically in up to 25 % of pigs by the time they have reached 16 weeks of age. Four months after vaccination this should have dropped to around 10 % and after 6 months down to less than 1 %. Vaccination will usually prevent the establishment of infection up to approximately 8 to 12 weeks of age, until the pigs move into finishing houses. Here, unless the houses have been depopulated, the pigs will become infected but with few clinical signs. Infection, however, increases the predisposition to other respiratory diseases and depresses feed intake and performance.

Management control and prevention

Replacement breeding stock should be selected from herds that are known to be free of this disease. Such herds are monitored by regular examination of snouts at slaughter when the degree of damage to the turbinate bones is assessed, and also by nasal swabs taken from young growing pigs and breeding stock, cultured to demonstrate the absence of the organism.

- Disease is more common in young herds, particularly those with large numbers of gilts.
- Keep disease out by purchasing pigs only from known negative sources.
- Monitor presence of *P. multocida* toxin regularly through examination of nasal and tonsillar swabs.
- Monitor snout sections regularly in the slaughterhouse.
- If the herd is infected do not breed from home-bred gilts.
- Vaccinate sows.
- Maintain an older herd to produce good colostral immunity.
- Avoid continuously populated housing which may allow organisms to build up to a threshold level and initiate a disease outbreak.
- Adopt all-in/all-out procedures from weaning to slaughter.
- Avoid high stocking levels.
- Keep weaners to less than 120 per group.
- Poor ventilation and high humidity postweaning predispose herds to disease.
- Do not re-circulate air in flat decks.
- Avoid fluctuating temperatures in flat decks.
- Problems often flare up with new disease introduction, for example PRRSV or PMWS.
- Improve rodent control.

In the farrowing house

- Large, permanently-populated farrowing houses are ideal for the maintenance and spread of disease. Therefore split the houses into small modules of 6 to 10 places only.
- Operate an all-in/all-out system with pressure washing and disinfection between batches.
- Avoid more than 10 sows per farrowing room.
- If multi-suckling is practised then convert this to single suckling. The more pigs' noses make contact with each other, the greater is the spread of the pathogen.
- Make sure that divisions between farrowing pens are solid and at least 0.6 m high to reduce the risk of droplet infection.
- Outdoor rearing reduces the risk of PAR.

- Poor ventilation and low humidity together with dusty atmospheres and toxic gases predispose pigs to the disease.
- Poor colostrum management or udder problems such as agalactia result in piglets with poor immunity.

Always make sure that breeding stock is purchased from herds checked and believed free from toxigenic Pasteurella.

Eradication

It is possible by vaccination, segregated weaning or partial depopulation to eradicate the organisms from a breeding weaning herd. This is carried out by vaccinating the sows and after approximately 12 months of vaccination and no evidence of clinical disease, conducting a partial depopulation (see Chapter 3).

Piglets from vaccinated sows are snatch weaned at 8 hours of age and segregated from disease-carrying pigs until nursery accommodation has been depopulated and cleaned. The vaccinated "clean" pigs are not allowed contact with infected pigs and separate personnel are used between groups. PAR is only spread by close droplet contamination. An alternative and more successful method is to market all growing pigs over a 6 to 8 week period before bringing the "clean" pigs back into the system.

Problems however arise in the breeding finishing unit where the organism persists in the continually populated finishing houses and it is necessary to depopulate the weaning, growing and finishing houses. Total herd depopulation and repopulation is also successful.

Rotaviral Enteritis (Diarrhoea)

These viruses are widespread both in pig populations and most other mammals and there are a number of different types or groups. Group A is the common pig one, but B, C and E also occur.

Rotaviruses are ubiquitous and they are present in most if not all pig herds with virtually a 100 % seroconversion in adult stock. A further epidemiological feature is their persistence outside the pig, where they are resistant to environmental changes and many disinfectants. Maternal antibodies persist for 3 to 6 weeks, after which pigs become susceptible to infection, but exposure does not necessarily result in disease. It is estimated that only 10 to 15 % of diarrhoeas in pigs are initiated by a primary rotavirus infection.

The fact that the virus persists in the environment accounts for widespread infection and therefore a constant risk of disease.

Clinical signs

In a mature herd, disease appears after piglets are 7 to 10 days of age, with a watery profuse diarrhoea in younger animals. It is progressively less important with age. However if pathogenic strains of *E. coli* are present, severe disease can occur with heavy mortality. Villus atrophy is a consistent feature with dehydration and malabsorption, and diarrhoea usually persists for 3 to 4 days. Pigs look hollow in the abdomen, the eyes are sunken and the skin around the rectum is wet. Some pigs will vomit.

The role of rotaviruses in the post-weaned pig is probably less important although they are often identified when acute *E. coli* diarrhoea occurs in the first 7 to 10 days after weaning.

Diagnosis

Whenever there is a diarrhoea problem in pigs from 10 to 40 days of age, rotavirus infection (either as primary agents or secondary) must be considered. Laboratory examinations are required by electron microscopy and ELISA tests. Try the litmus test by soaking scour in litmus paper; *E. coli* infections turn blue, virus infections red.

Treatment

- There are no specific treatments for rotavirus infections.
- Provide antibiotic therapy either by injection, by mouth or in the drinking water, to control secondary infections such as *E. coli*.
- Apramycin, amoxycillin, neomycin, framycetin and enrofloxacin could be used.
- Provide dextrose/glycine electrolytes to counteract dehydration.
- Provide dry, warm and comfortable lying areas.

Management control and prevention

- Reduce the levels of virus in the environment by all-in/all-out procedures and effective disinfection. Leave the house empty for 2 to 4 days before pigs are moved in.
- Disinfect with peroxygen-based disinfectants or chlorine based ones.
- Reduce the spread of virus between affected and non-affected pigs. Use foot dips and clean clothing and wash hands after handling sick pigs.
- Apply control procedures outlined for coliform infections.
- If a persistent problem is diagnosed in sucking pigs, use feedback of piglet scour by collecting it in wet sawdust, or mix the faeces in water. Feed the contaminated material via the watering systems or into troughs 2 to 3 times weekly. Carry this out in Weeks 4 and 3 before farrowing. Pay particular attention to gilts and

1st parity sows.
- Ensure that the gilt in isolation and acclimatisation has access to the diarrhoea from the farrowing house and any on-site nursery faeces.
- Modified live vaccines are available in some countries.
- In the weaned pig, adopt all-in/all-out procedures with cleaning and disinfection in first stage flat decks. Pay particular attention to environmental stress and temperature fluctuations.

Splay Legs

This is a condition where the newborn piglet is unable to hold the front and/or (more commonly) back legs together. Up to 2 % of piglets can be affected. The mobility of the piglet is impaired, which makes teat access difficult. It is more common in the Landrace breed and males. Disease is caused by immaturity of the muscle fibres in the hind legs, over the pelvis and occasionally in the front legs which is not assisted by the piglet's failure to stand properly within 30 minutes of birth.

Clinical signs

The piglets are unable to stand with the hind legs deflected laterally and as a result they often adopt a dog sitting position. The condition is exaggerated when piglets stand on very smooth or wet slippery floors. Death usually ensues either due to starvation or crushing, because the pig cannot move away from the sow.

Diagnosis

This is based upon the clinical signs.

Treatment

- As soon as the affected pig is identified, use a 25 mm (1 inch) wide plaster strip and tape the hind legs together, leaving a gap of 50 to 80 mm (2 to 3 inches). The same procedure can be applied to the fore legs. The sticky tape should be passed around the legs just above the supernumerary digits. Never use string, as it will strangulate the legs if not removed (Fig 8-33).
- Hold the piglet up by both its hind legs (Fig 8-34) and vigorously massage the muscle masses over the pelvis and the front and rear of the hind legs. Repeat this 3 or 4 times during the first day .
- Assist the piglet to suckle 2 to 3 times daily.
- Dose the pig with 10 ml of sow colostrum or cow colostrum immediately after birth.
- Confine the strongest mobile pigs into the creep area for a period of 1 hour to allow splay leg pigs uninhibited access to the teats.

Management control and prevention

- If floor surfaces are smooth, these can be roughened by covering with lime wash, or increase the use of bedding, particularly sawdust, shavings or newspaper.
- It is essential that all piglets have a good footing during the first 30 minutes of life, otherwise splay legs will become a problem.
- It can also be beneficial to split suckle and ensure the piglets are dry before being given access to the teats.

TAPING THE LEGS OF A SPLAY LEG PIGLET

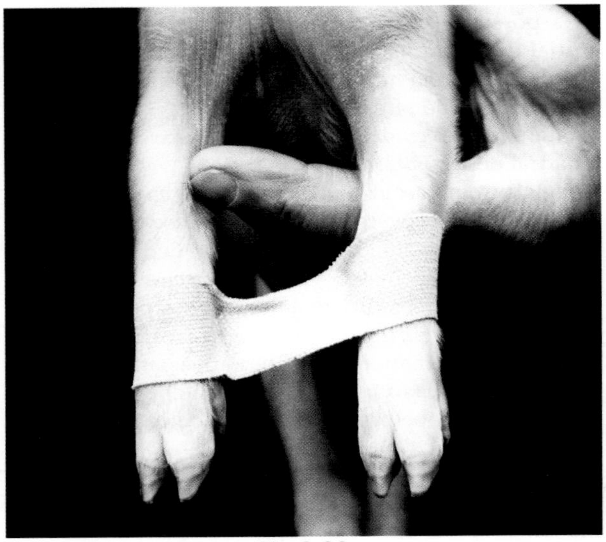

Fig 8-33

MASSAGING THE HIND LEGS OF A SPLAY LEG PIGLET

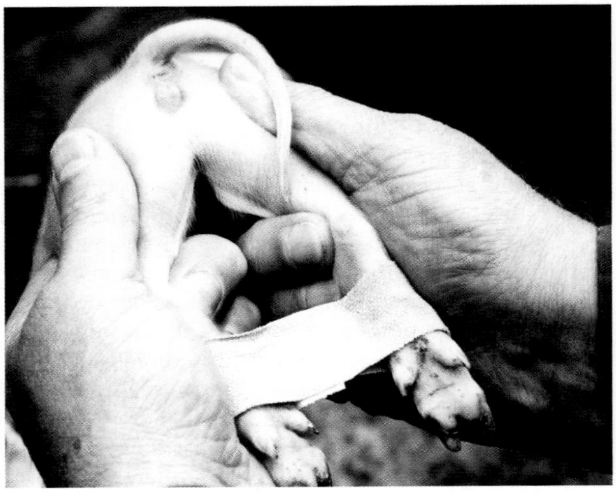

Fig 8-34

Streptococcal Meningitis

See also Chapter 9: Streptococcal infections for further information

Meningitis denotes inflammation of the meninges which are the membranes covering the brain. In the sucking piglet it is usually caused by *Streptococcus suis, Haemophilus parasuis,* or sometimes bacteria such as *E. coli* and other streptococci. *Streptococcus suis* is a zoonotic disease and thus can be transmitted to man. *S. suis* has many serotypes. In most countries *S. suis* type 1 is the main one in sucking piglets, but this may not be true in other countries. For example in Denmark it is type 7. *S. suis* also causes joint problems, particularly types 1 and 14. *S. suis* is carried for long periods in the tonsils and may be transmitted to the sucking piglet from the sow or from other piglets. The sow also provides a variable level of immunity in the colostrum. Streptococcal meningitis in sucking piglets is sporadic, occurring occasionally in individual piglets. *S. suis* type 1 occasionally causes meningitis but more commonly joint infections. Meningitis in piglets is more commonly caused by *Haemophilus parasuis* or Glässer's disease.

Clinical signs

These are rapid in onset, with the piglet lying on its belly and shivering. When presented with such symptoms, also consider scour or joint infections from a septicaemia.

Meningitis is characterised by a continual movement of the eyes from one side to the other (nystagmus) and this is an early diagnostic symptom together with shivering and shaking, paddling and convulsions. In acute cases the piglet may just be found dead. Streptococcal meningitis may be worse in sucking pigs when the organism has been introduced into the herd for the first time, or where it is secondary to infection with PRRSV.

With cases of meningitis in the farrowing area, review the feedback programme to ensure that the sow can protect her piglets through colostrum.

Swine Dysentery (SD)

See Chapter 9 for further information

This is a very important disease in the growing and finishing pig but not so in the piglet. It is unusual to see SD in sucking pigs except when the organism *Brachyspira hyodysenteriae* first enters the herd. In such cases there may be sloppy faeces in the sow, often containing some blood and mucus and a similar type of diarrhoea in piglets. In rare cases it may be much more dramatic, with very bloody faeces. In herds in which the organism is endemic the lactating sow may be a carrier and infect the older piglet, but there will be no disease in the piglet until maternal antibodies disappear.

Swine Influenza Virus (SIV) – Swine Flu

See Chapters 6 and 9 for further information

Swine influenza is a highly contagious respiratory disease caused by a number of closely related type A influenza viruses that are noted for their ability to change their antigenic structure and create new strains.

It would be unusual to see any signs of SIV in the sucking pig unless disease has entered the herd for the first time. Colostrum prevents infection during the sucking period.

SIV is manifested by coughing and short periods of illness and secondary pneumonia. Usually no treatment is required unless the pneumonia is severe. Good nursing and increasing farrowing house temperatures and bedding are important during the acute period. The disease becomes active in the young growing pig as immunity rises and falls. This disease is covered in detail in Chapters 6 and 9.

Teat Necrosis

Teat necrosis describes a condition where constant rubbing and pressure on the end of the teat causes the teat sphincter and delicate tissues to die (necrosis) and slough off. It is of no consequence in commercial herds which are buying in replacement gilts but it is a very important condition where breeding stock are being produced.

The condition occurs as part of the normal farrowing process, where the sow produces oestrogens. These cross the placenta and enlarge the teats (in both male and female piglets) and enlarge the female piglet's vulva. The oestrogens are metabolised (removed) within 24 to 36 hours of birth and the teats become small again. However, damage to the swollen teat is permanent.

Clinical signs

It first becomes evident 12 to 24 hours after birth. The teat end appears bright red, gradually becoming black. Trauma to the teats occurs on all floor surfaces but to a lesser extent on those that are well bedded with shavings or straw. The teats in front of the umbilicus are the ones at risk because these have the greatest contact with the floor during sucking.

The damage to the tissues can be severe, resulting in a blind or inverted teat.

Treatment

- There is no treatment for this condition and prevention is necessary.

Management control and prevention

- As soon as the female and male (on boar-producing farms) piglets are born and have dried off they should be held by the hind legs and the teats anterior to the umbilicus coated with a protective compound. Such compounds could include copydex, a white rubbery glue used for sticking carpets, cow gum (which is a rubberised solution often used for sticking photographs into albums) or a contact adhesive. These compounds gradually disappear over the next 3 to 4 days but protect the teats during the susceptible period.

Tetanus

Tetanus is caused by the bacterium *Clostridium tetani* which produces toxins that affect the central nervous system. The organism, which can form spores, lives in the large intestines and faeces of many mammals and in certain soils. It must enter through a dirty abrasion or a cut. In the sucking pig the most common source is castration. Tetanus spores are found in the soil and this disease can be a problem in outdoor pigs. The incubation period is from 1 to 10 weeks. It would be uncommon to see disease in the sucking piglet less than 2 weeks of age. The affected piglet is hypersensitive, shows stiffness of legs and muscles, an erect tail and muscular spasms of the ears and face. A multivalent clostridium vaccine containing tetanus toxoid is highly efficient in preventing disease and could be used in pregnant sows if a herd has a problem. Castration techniques should also be checked because unhygienic methods can lead to infection and tetanus 2 to 8 weeks later.

Thrombocytopenic Purpura – Bleeding

See Chapter 10 for further information

This is an uncommon condition seen only in young piglets from approximately 3 to 21 days of age. It arises when the sow's colostrum contains antibodies that destroy the piglet's blood platelets (thrombocytes). This condition is covered in Chapter 10.

Transmissible Gastroenteritis (TGE)

TGE is a very important and highly infectious disease in the piglet, caused by a coronavirus. It is similar in structure but quite distinct from the coronavirus PRCV that infects the respiratory system.

TGE virus enters the pig by mouth and multiplies in the villi (finger-like structures in the small intestine) and destroys them. This takes place in 24 to 48 hours and is followed by vomiting and a very severe acute diarrhoea with high mortality. When the virus enters the herd for the first time, mortality in piglets up to 14 days of age may be 100 %. This decreases in pigs over 3 weeks of age, but morbidity is high.

The virus multiplies in the intestine and is shed in large numbers in the faeces. Pig faeces therefore are the major source of transmission, either directly through the purchased carrier pig or indirectly through mechanical transmission. The virus is killed by sunlight within a few hours but will survive for long periods outside the pig in cold or freezing conditions. It is very susceptible to disinfectants, particularly iodine based ones, quaternary ammonia and peroxygen compounds.

Dogs and cats may shed the virus in their faeces for 2 to 3 weeks. Birds (and in particular starlings) may transmit the disease and management should ensure that feed is not exposed to attract these birds.

Read Chapters 2 and 3 on biological control of diseases entering the farm and the precautions necessary to prevent diseases spreading by faeces.

Clinical signs

Acute disease

The most striking feature of TGE when it is first introduced into the herd is the rapidity of spread. It affects all classes of pig on the farm, with evidence of vomiting and diarrhoea. Adult animals show varying degrees of inappetence and usually recover over a 5 to 7 day period. In the sucking piglet the disease is very severe and at less than 3 weeks of age there is very acute watery diarrhoea, with almost 100 % mortality within 2 to 3 days in piglets under 7 days of age due to severe dehydration and electrolyte imbalance. There is no response to antibiotic therapy. The most striking feature is the wet and dirty appearance of all the litter due to the profuse diarrhoea. Disease will persist in the farrowing houses over a period of 3 to 4 weeks until sows have developed sufficient immunity to protect the piglets.

Chronic or endemic disease

In herds of less than 300 sows the virus is usually self-eliminating provided there are good all-in/all-out procedures in farrowing houses and grower accommodation. In some herds, however, the virus will persist in the growing herd because piglets at weaning time, still under the influence of the maternal antibody, move into houses where the virus still persists. Once the antibody disappears the pigs become infected, allowing the virus to multiply. The pigs then shed the virus, contaminating the weaner rooms and infecting pigs being weaned after them. TGE can become endemic in herds in a mild form with high morbidity but low mortality.

Diagnosis

The clinical picture in acute disease is almost diagnostic. There are few other enteric diseases that spread so rapidly across all pigs. When you examine the small intestine, check the pH of the contents. In cases of TGE the contents will be acidic. The ultimate diagnosis of TGE must be made in the laboratory from the intestine of a fresh, dead pig using fluorescent antibody tests. Isolation of the virus is also carried out.

Similar diseases

The acute form, porcine epidemic diarrhoea (PED), could give a similar picture but is generally less acute and with less mortality in sucking pigs. That said, in Asia, PED can be explosive in a similar way to TGE in Europe and America. Rotavirus is similar, resulting in large outbreaks of scour, but the mortality is much lower. Where TGE has become chronic then differentiation from the other causes of diarrhoea must be carried out in a laboratory. If the herd has been infected previously with TGE and there are scour problems persisting it is necessary to determine whether the virus is still present or not.

Treatment

- There is no specific treatment for TGE.
- Antibiotic treatment by mouth in individual piglets may reduce secondary infections.
- Provide easy access to water containing electrolyte and an antibiotic such as neomycin. Make this available to the litters twice daily.
- Improve the nursing and environment of the litter by providing extra heat and deep bedding to reduce the weights of infection from the diarrhoea.

Management control and prevention

- As soon as disease is suspected, isolate those farrowing houses not infected by using separate personnel boots and coveralls. This is particularly important in piglets under 14 days of age. The longer the pathogen can be kept away, the more pigs will be reared and mortality reduced.
- If it is possible, move sows that are within 3 weeks of farrowing from the farm before they become infected so that they can farrow down in an isolated building or outside in arks and escape disease.
- It is essential to develop immunity in the gestation/dry sows as soon as possible.
- There are 2 methods: either squeeze the piglets' abdomens and collect the diarrhoea into a bowl, or use sawdust or shavings in the areas where the piglets are scouring. Paper towels can also be used to soak up piglet faeces. This material is then mixed with a bucket of water and fed to the pregnant sows (feedback) up to Day 90 of pregnancy. Do not feed sows beyond Day 90 as this can make the problem worse in these animals.
- A further method is to collect the small intestines from a number of pigs that have died and macerate them in a food blending machine. The liquid provides a rich source of virus and this can, if required, be preserved by deep freezing.
- The disease should be spread as soon as possible across the whole farm. The object is to get a good immunity developed in the shortest possible period of time. It will take approximately 3 to 4 weeks to achieve this, as this is the time it will take for colostrum levels to build-up to a protective concentration.
- Ensure that all gilts in isolation and acclimatisation receive feedback diarrhoea from the farrowing house and faeces from the on-site nursery.
- Once the infected period is over, ensure an all-in/all-out management system of the farrowing houses, weaner and finisher accommodation.
- Disinfection of pens between batches should be carried out using an iodine-based disinfectant or one highly active against viruses
- This cleaning process is an important one to ensure the virus does not linger on the farm and become endemic.
- If your herd has become infected with TGE, ask the questions why and how? Look at all your prevention procedures and biosecurity as discussed in Chapter 2. (Do this before you get TGE).
- Always provide boots and protective clothing for anyone entering your farm.
- Provide disinfectant foot dips at all entrances.
- Run each batch as a separate entity so each batch farrowing area has its own clearly marked, boots and equipment.
- Keep starlings and migrating birds away from the farm by not exposing them to feed.
- Do not borrow equipment from another pig farm.
- Site all bins to the exterior of the unit and always have your own feeder pipes to your own feed bins. This is a high risk source for the spread of enteric diseases.
- **Vaccination** – Live modified and killed vaccines are available in some countries. The results in the field are very variable. The objective is to maintain immunity in the colostrum. This can only be carried out by stimulating the gut of the sow to produce antibodies in the milk. Intramuscular vaccines give a very poor response.

Vomiting and Wasting Disease/Ontario Encephalitis

This is caused by a coronavirus called haemagglutinating encephalomyelitis virus (HEV). The virus is widespread in the pig populations of North America and is probably world-wide but is unimportant because clinical disease is rare. This is because most sows have been infected and are immune. They pass their immunity to their piglets in colostrum which protects them through the vulnerable period. Although the virus can infect susceptible pigs at any age it only causes clinical disease in newborn piglets.

Clinical signs
Although there is only 1 antigenic type of the virus there is a variation in virulence between strains, resulting in 2 different disease syndromes. Both start at around 4 days of age, are sudden in onset and affect whole litters. The piglets are huddled and hairy. They vomit bright green-yellow vomitus which is often mistaken for acute scour. In fact, they are constipated.

In the typical vomiting and wasting disease syndrome they lose their ability to suck or swallow, become very thirsty and stand with their heads over water but are unable to drink. They rapidly waste away, become severely emaciated and die.

In the typical encephalitis syndrome they froth and champ at the mouth, develop blueing of their extremities, their abdomens become bloated and they tremble. They have a stilted gait which rapidly progresses to partial paralysis of the legs. They lie down, go into convulsions, roll their eyes and die within 2 to 4 days of onset.

Diagnosis
The clinical picture in 4 day old piglets of vomiting and constipation is characteristic of the disease. If you open up their abdomens the appearance is typical; gas in the stomach and intestine but no food, only some brightly coloured liquid. There will be firm faeces and sometimes brightly coloured crystals in the kidneys. A blood test is available.

Treatment
- None is available.

Management control and prevention
- All the affected pigs will die so they are best destroyed. This is a one-off phenomenon which tends to occur in small herds in the litters of sows which have no immunity. The virus will circulate and immunise the herd. The disease will not occur again in these sows' litters or any others.

Vitamin E Deficiency and Iron Toxicity
Iron toxicity occurs when the sow is deficient in vitamin E and piglets are born with low levels as a consequence. The routine iron dextran piglet injections become toxic and cause severe muscle reactions at the injection sites. Vitamin E deficiency in the sow occurs when fats in the diet become rancid or cereals or corn have fermented and spoiled and the vitamin E is destroyed. Piglets born from sows fed high levels of vitamin A may produce piglets with a low vitamin E status. This is a useful line of investigation where iron dextran problems persist in sucking pigs.

Clinical signs
Two to 4 hours after injection most of the litter become acutely lame on the legs that have received the iron. The muscles are swollen and the piglets develop heavy breathing and look pale. Death occurs within 24 hours. At post-mortem the muscles are coagulated and appear like fish tissue due to necrosis of the muscle fibres.

Diagnosis
This is based upon the history of deaths within 24 hours of iron injections and swollen muscles at the site of the injections.

Treatment
- Inject sows due to farrow over the next 3 weeks with a vitamin E selenium preparation.
- Alternatively use water soluble vitamin E.
- Inject all litters for a 3 week period with vitamin E or dose by mouth at least 2 days before iron injections are given.

Management control and prevention
- Clean out the cereals at the bottom of the storage bins.
- Assess cereals and corn for spoilage.
- Supplement the sow feed with 150 g/tonne vitamin E.

Summary – 12 Key Points to Weaning 12+ Piglets

Piglet survival is increased by:

1. Maintaining a healthy sow.
2. Providing good hygiene at farrowing.
3. Providing extra heat lamps both at the side and behind the sow during farrowing.
4. Ensuring good observation at farrowing.
5. Maintaining a warm farrowing house with good insulated floors.
6. Eliminating all draughts at piglet level.
7. Ensuring the piglet dries off quickly after farrowing.
8. Ensuring immediate removal of the piglet to the creep area for acclimatisation.
9. Maximising early colostrum intake using split suckling.
10. Giving prompt assistance to weak pigs.
11. Fostering and even-up of weights at birth.
12. Maximising lactation feed intake to enhance the next litter size.

All the above procedures require a good stockperson. Remember that their qualities include a sound knowledge, recognition of the individual animals' needs, patience, awareness, ability to organise their work, immediate attention to detail and the most important, a sense of achievement. If you have a good stockperson, look after him or her!

9 MANAGING HEALTH IN THE WEANER, GROWER AND FINISHING PERIODS

9 MANAGING HEALTH IN THE WEANER, GROWER AND FINISHING PERIODS

Managing the Weaner for Health and Efficient Productivity

The production of healthy weaners is a complex interaction between pathogens, the environment and management. Management decisions and procedures however are the initiating factors that cause abnormal conditions in the immediate postweaning phase.

These factors include:

- The production and maintenance of a healthy sucking pig.
- An all-in/all-out system, pig flow and a good batch programme.
- An adequate weaning weight for the weaning system being utilised.
- An optimum age of pig for the weaning system being utilised.
- A correct diet for the age of the pig.
- Good feeding procedures.
- Good building design and maintenance.
- Good environmental control with the necessary temperature, ventilation and humidity for the age and weight of the pig.

Key points to producing a healthy weaned pig

- All-in/all-out pig flow.
- Providing good farrowing house management and hygiene (Chapter 8).
- Using a hybrid or cross bred female.
- Vaccinating the gilt and sow against *E. coli* diarrhoea.
- Bringing a healthy sow into the farrowing house.
- Satisfing the nutritional requirements of the sow both in pregnancy and lactation.
- Using a high energy and lysine diet in lactation to enable the sow to provide maximum nutrition for the litter.
- Achieving good birth weights.
- Ensuring each pig receives maximum colostrum at birth.
- Maintaining an even creep temperature.
- Providing fresh uncontaminated creep feed pre-weaning.
- Ensuring a draught-free piglet creep area.
- Maintaining a well-designed defecating area.

Age and weight of the pig at weaning

Successful weaning requires a combination of both minimum age and weight at the time of weaning to suit the weaning system. Weaning ages generally range from 14 to 28 days with most intensive farms having a mean of between 20 or 27 days. As weaning age is reduced it is important to appreciate the potential effects this might have, not only on the pig, but also on the sow.

The effects of reducing weaning age

- The younger the pig, the poorer its appetite at weaning.
- The younger the pig is weaned, the less efficiently it will adapt to and digest solid food.
- Poor feed intake results in lower daily liveweight gain.
- Highly specialised, more expensive, diets are required.
- More weaning accommodation is required.
- A more exacting weaning environment is required; more supplementary heat, more labour, more costly housing and higher creep costs.
- The piglet is more susceptible to enteric diseases.
- There is often an increase in postweaning mortality.
- A shorter lactation may reduce subsequent litter sizes and conception rates.
- The days from weaning to first mating interval may be increased, with more sows showing vulval discharges and found not pregnant.
- Pigs produced per sow per annum or pigs weaned per farrowing place can be increased but may reduce the farrowing rate and subsequent litter size.
- Pigs may be healthier if weaned away from the farm.

The question is "How do you determine the best age at which to wean on your farm?"

This is dependent on:

- The number of batch farrowing places.
- The milking capabilities of the sows, the breed and reproductive efficiency.
- Litter size.
- A good weight for age at weaning.
- The health of the piglets at weaning time.
- The quality and digestibility of the creep feed.

- Suitable weaning accommodation that will satisfy the pig's requirements.
- A weaning age that results – 2 weeks later – in a healthy weaner which has achieved good daily liveweight gain (Fig 9-1).
- A weaning age that does not depress efficient reproductive performance in the sow and in particular the first and second parity sow.
- The use of segregated weaning techniques to improve health status.

Feed intake after weaning

Optimum levels of feed intake postweaning maximise daily gain and feed conversion efficiency and reduce costs per kilogram of liveweight gain and energy requirements.

The heavier the pig at weaning the more efficient is the growth and feed conversion during the next 4 weeks and the quicker it reaches slaughter weight.

FIG 9-1: TARGET WEIGHTS FOR WEANERS

SUGGESTED TARGET WEIGHTS AT WEANING *	
Age (Average Days)	Target Weight kg
20	7.5
27	8.4
34	15

SUGGESTED WEIGHT TARGETS FOR 3 WEEK WEANING *			
Day	kg	ADG Previous Week (g)	FCE
21	6.5	–	–
28	7.3	140	1.25
35	9.5	357	1.33
42	13	420	1.38
49	17	571	1.4
56	23	630	1.41
70	30	660	1.8

SUGGESTED WEIGHT TARGETS FOR 4 WEEK WEANING *			
Day	kg	ADG Previous Week (g)	FCE
28	7.5	–	–
35	9.25	142	1.28
42	12.9	520	1.38
49	17	580	1.4
56	23	630	1.41
70	30	660	1.8

* Assuming no environmental challenges.
It is possible to wean 100 kg per batch farrowing place for 4 week weaning or 86 kg for 3 week weaning.

Feed intake postweaning is a critical control point in pig production. Pigs which eat even 250 g a day in the first week will finish 10 days quicker than pigs which do not eat. Gruel feeding is strongly encouraged during the first 4 days postweaning to teach the pig about eating and drinking in this critical period.

Changes in the intestine of the pig at weaning

Fig 9-2 shows the cross section of the small intestine of the weaned piglet to consist of many thousands of finger like projections called villi, which increase the absorptive capacity of the small intestine. During suckling they are continuously bathed by sows' milk which contains the immunoglobulin IgA. This becomes absorbed into the mucus, covering the villi surfaces and preventing *E. coli* and other organisms attaching to the fingers. If they are unable to attach they are unable to cause disease. The secretory IgA also helps to destroy bacteria. After weaning time, however, no more IgA is available, the levels rapidly decline and bacteria damage the villi, causing them to shrink. This atrophy reduces the absorptive capacity of the gut and the ability of the pig to use its food. The enzymes produced by the cells of the villi are likewise reduced. The changes result in malabsorption of food and poor digestion with or without the development of scour. The villi normally regenerate within 5 to 7 days after weaning from cells at their base called enterocytes, which multiply and migrate upwards causing the villi to return to their normal length. The rate of multiplication and regeneration is in part an environmental temperature and energy dependent phenomena. If the pig is weaned in an environment below its lower critical temperature (LCT), the rate of regeneration of the villi is reduced and in some cases ceases. (This results in the hairy pig that doesn't grow). Feed intake is a crucial part of the equation of maintaining villus health.

The greater the villus atrophy, the poorer the growth rate of the pig.

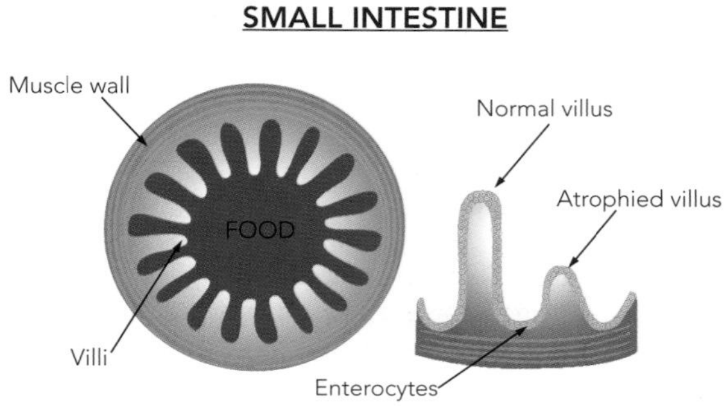

Fig 9-2

Before weaning, the piglet receives milk as a liquid feed at regular intervals. As a result the bacterial flora of the gut, although relatively simple compared with that of a mature pig, is stabilised.

At weaning, cessation of milk removes secretory IgA and there maybe a period of starvation, followed by irregular attempts to eat solid feed. This results in a change in the anatomy of the gut, with villus fusion and atrophy and a dynamic disruption of the microflora of the gut which may last for 7 to 10 days before re-stabilising. This bacterial disruption may also contribute to poor digestion and possibly scour, particularly when high levels of pathogenic strains of *Escherichia coli* (*E. coli*) are involved.

Before weaning the piglets lead an ordered life, being "called" with their litter mates to suckle and obtain small amounts of milk at regular intervals, sleeping between meals in a warm creep. All this suddenly changes at weaning; the pigs finding themselves in strange surroundings with strange piglets, and perhaps only solid feed. Psychological trauma is inevitable and is likely to affect some pigs more than others, resulting in impaired digestibility and lowered resistance to disease. The more this psychological stress can be minimised, the better. It is vital all pigs eat something on the day of weaning.

If poor growth is evident in the first 7 days post-weaning, the following options or variables need to be considered:

- Check that the weights of all pigs at weaning are to the target level.
- Check the ages of the pigs at weaning.
- Heavier but younger pigs will have a more immature digestive system.
- Group the pigs by weight or keep them in their litter groups.
- Use a highly digestible and palatable diet and mix and soak this for the first day or two with water.
- Use different diets to suit bodyweight and age.
- Use open dishes instead of troughs for feeding for the first 3 days at least. Have sufficient feed space so all the pigs in the pen are able to eat together. This equates to 10 cm per weaned pig.
- Feed small quantities of creep 4 to 5 times daily and remove uneaten stale feed.
- Talk to the weaners and act like their mum at these times.
- Provide easy access to fresh clean water.
- Use in-feed medication for the first 10 days postweaning.
- Check that the environmental temperature is constant and satisfies the pig's requirements, particularly in the first five days postweaning.
- Maintain a dry house without draughts in the sleeping area – max air speed < 0.2 m².
- Reduce any form of stress.
- If pigs are housed on slatted floors, provide solid comfort boards to lie on for the first few days.
- Remove the lowest 10 % of piglets from each pen after 7 to 10 days and place them together in one pen in the same room. Their diet can then be adjusted accordingly.

Key factors that dictate the degree of villus atrophy

- Age of the pig at weaning.
- Weight of the pig at weaning.
- The environmental temperature and its fluctuations.
- Feed intake and availability of feed.
- Digestibility of the feed.
- Quality of the proteins.
- Levels of milk proteins.
- Levels of bacterial and viral challenge.

Piglet nutrition

Only minimal amounts of solid food are eaten during the suckling period and very little before 10 days of age. The sow's milk provides all the nutrients the piglet requires. At weaning the piglet has its main source of food removed (milk) and needs to survive on a non-milk diet. Thus it is important that the piglet has got used to solid feed before weaning to assist this transition. There are a number of actions that can be taken to achieve this. See Chapter 14 for the key points to maximising feed intake, creep feeding options and the nutritional components of a good creep diet.

Water

Water requirements of the pig are covered in detail in Chapter 14. At weaning time the pig's diet changes abruptly from milk as its source of nutrients to water and solid feed and it should therefore be given encouragement to drink. Water is best provided in open cube drinkers, poultry drinkers or water bowls for the first 3 to 4 days. If nipple drinkers are the only source of water many pigs may take up to 24 hours before they drink adequate amounts and if the drinkers are not functioning correctly some pigs may never get enough water (Fig 9-3).

FIG 9-3: WATER REQUIREMENTS FOR THE WEANED PIG

Weeks Postweaning	Liveweight kg	Approximate Usage/Day Litres
Week 1	6 – 7	1
2	8 – 9	1.4
3	10 – 13	1.7
4	14 – 16	2

Housing

Good weaning accommodation should satisfy the following criteria:

- Be easy to clean and disinfect and be dry.
- Provide good observation.
- Require minimum handling of waste (faeces, slurry) and separate the weaner from its faeces.
- Have a simple feeding system that is adaptable for the first 3 to 4 days.
- Provide a draught-free, well-insulated environment that satisfies the needs of the weaners.
- Segregate one batch of pigs from another, ideally at least 10 m apart, with no direct air or faeces contact.
- Segregated weaning principles should be adopted (see Chapter 3).
- In nurseries and mechanically ventilated housing a fail-safe and alarm system should be in operation.
- Weaner kennels, deep-bedded in straw, provide an ideal postweaning environment in temperate climates.
- If slats are used these should preferably be made of plastic and self-cleaning.

Temperature requirements

Satisfying these is essential if a pig is to be weaned successfully (Fig 9-4). These requirements are dependent upon the weight of the pig, the feed intake, the quality of the feed, the floor type and the air flow. Whilst guidelines can be given for any particular age and group of pigs these should never be relied upon totally, but rather the temperature of the building adjusted to the observed requirements of the pig. A pig in its thermo-neutral zone (its comfort zone) lies on its side making little bodily contact with its contemporaries.

Temperature requirements of the pig are increased in wet, draughty pens, or where there is high air flow. Always measure the temperature at pig level using a maximum minimum thermometer or computer monitoring equipment, placed in a guarded position.

FIG 9-4: GUIDES TO LOWER CRITICAL TEMPERATURES IN THE SLEEPING AREA

Weight of Pig		Straw		Slats	
kg	(lbs)	°C	(°F)	°C	(°F)
5	(11)	27	(81)	30	(86)
6	(13)	25	(77)	29	(84)
7	(15)	22	(72)	28	(82)
8	(18)	21	(70)	26	(79)
9	(20)	20	(68)	25	(77)

Observe the pigs at the end of each day. If they are huddled and lying on their bellies, the environment is wrong. If they are laid apart on their sides, it is correct.

Stocking density

The stocking density for pigs weaned on fully slatted or partly slatted floors is approximately 0.1 m^2 per 10 kg liveweight. **Note** that there are legal requirements for space in some countries (see Chapter 3); for example if the nursery is intended to house pigs to 30 kg, the pigs will require 0.3 m^2 per pig at weaning. This should provide sufficient room for all pigs to lie down in the pen without body contact.

Do not mix different ages of pigs in one room. Avoid pens of 20 to 60. Either house in small groups or adopt big pen concepts. It is possible to wean up to 500 piglets into one pen if the accommodation is properly constructed so that all of the needs of the pigs are met.

Problems of over-stocking:

- There is a greater risk of disease developing, particularly greasy pig disease, postweaning diarrhoea, PRRSV, SIV and mycoplasma (enzootic) pneumonia.
- Growth rates are reduced.
- Vice increases, including ear and tail biting.
- Ventilation problems arise.

Ventilation

This is critical in the first 24 hours of weaning. Draughts must be avoided, otherwise the pig loses energy and becomes catabolic with a predisposition to the development of disease (Fig 9-5).

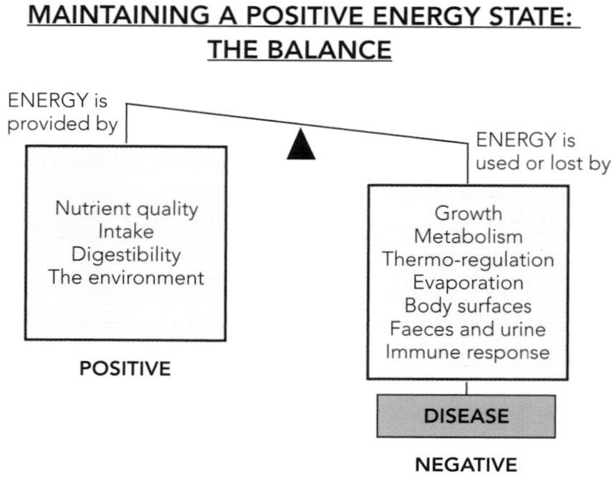

Fig 9-5

Fig 9-6 highlights the major diseases of the weaned pig and the major contributing factors. Take particular note of the importance of an adverse environment and bad management.

Managing the Growing Pig for Health and Efficient Productivity

To maintain efficient production in the growing pig it is necessary to understand the complex interactions that are involved. There are 5 important areas:

1. Management. The quality of this contributes to the health and biological efficiency of the pig.
2. Feed. This is the major cost component of the growing pig. The efficiency of use is vital and the nutritional value versus price is very important. How it is delivered and made available to the pig can increase feed intake and maximise feed efficiency.
3. The type of housing used and the quality of the environment.
4. The levels of disease and their economic effects. These are significantly demonstrated by the very marked improvements in daily gain and feed conversion achieved when pigs are segregated and early weaned, or produced under all-in/all-out batch management.
5. The genetic potential of the pig.

The pig must balance its energy requirements during the first 72 hours postweaning. If not, disease is likely to occur (Fig 9-5).

The differential diagnosis of poor growth

How do we identify problems of poor growth?

This is a most difficult area to clarify and understand. Up to 75 % of the total feed purchased is used from weaning to slaughter and it is surprising that so little emphasis is placed upon the efficiency of its use. The starting point is to identify those points along the pig's growth curve where targets are not met.

FIG 9-6: DISEASES OF THE PIG DURING THE FIRST 28 DAYS POSTWEANING

	Important Precipitating Factors				
Condition	**Stocking Density**	**Infection**	**Nutrition**	**Adverse Environment**	**Bad Management**
Actinobacillus pleuropneumonia (App)	✓	✓	✓	✓	✓
Aujeszky's disease/pseudorabies virus (AD/PRV) *		✓		✓	✓
Enteritis	✓	✓	✓	✓	✓
Glässer's disease	✓	✓		✓	✓
Greasy pig disease		✓	✓	✓	✓
Internal parasites - uncommon		✓		✓	✓
Ileitis		✓		✓	✓
Lice		✓		✓	✓
Malabsorption			✓	✓	✓
Mange		✓		✓	✓
Oedema disease		✓	✓	✓	✓
PMWS	✓	✓	✓	✓	✓
Progressive atrophic rhinitis *	✓	✓		✓	✓
Salmonellosis	✓	✓		✓	✓
Streptococcal meningitis *	✓	✓		✓	✓
Swine dysentery *		✓	✓	✓	✓
Transmissible gastroenteritis		✓			

* Prolonged carrier state.

First of all the factors that affect the growth and feed efficiency of the pig in an intensive production system must be considered, and these are listed in Fig 9-7a and b. Study these in relation to the factors that affect profitability in Fig 9-8. In many respects they have a great deal in common.

Note that planning and the use of records are considered 2 of the main factors because these are the starting points for the identification of poor growth rate, followed by clinical observations of the pigs on the farm.

The results of any tests or post-mortem examinations, either for monitoring purposes or disease investigations, must also be appraised. Fig 9-9 collects these together at farm level for practical use.

The pig

Profit comes from the margin between the price paid for the carcass and the costs that are incurred to produce it. The major cost is feed. The more feed required to produce a kg of meat then the less the profit. The genetic makeup of the pig is vital in this respect. To produce 1 kg of lean meat requires approximately 1.25 kg of feed, whereas to produce 1 kg of fat requires approximately 4 kg of feed. An animal that can convert more of its feed into lean meat is therefore much more profitable than one that converts it into fat. How often this obvious factor is neglected! Furthermore, excess fat at slaughter may be severely penalised. The rate of deposition of lean meat is dependent on the sex of the animal, its genetic back-

FIG 9-7a: FACTORS INFLUENCING GROWTH RATE

Disease	Diseases present Management of disease Therapy/prophylactics
Environment	Airborne dust Air speed Humidity Micro-organisms Noxious gases Temperature Toxins Ventilation
Feed and intake	Ad lib/restricted Availability Palatability Trough/floor feeding Wet feeding Water availability Growth enhancers
Housing	All-in/all-out management Floor type/bedding Floor space/stocking density/group size Insulation/ temperature Method of waste disposal Moving, mixing, stress
Management	Direction Education People quality
Nutrition	Amino acid/lysine levels Energy levels Protein levels and quality
Pig	Age/weight/sex Genetics

FIG 9-7b: FACTORS THAT MAY INFLUENCE FCE (MEAN 2.7)

(↑ Worse by up to. ↓ Better by up to)

Multiple sources of pigs (disease)	↑ 0.2
Increasing sale weight to 113 kg	↑ 0.2
> 500 pigs per air space	↑ 0.2
Cubic capacity per pig < 0.7 m^3 (25 ft^3)/100 kg	↑ 0.1
Continuous production	↑ 0.25
Feed waste	↑ 0.5
Mixing pigs	↑ 0.2
Temperature < 16 °C (60 °F)	↑ 0.01 per °C
Temperature > 30 °C (85 °F)	↑ 0.01 per °C
Add 1 % more fat to diet	↓ 0.05
Add a growth enhancer	↓ Varies
Reduce backfat by 10 %	↓ 0.05
Increase protein in diet by 10 %	↓ 0.1
Boars	↓ 0.1

ground, the type of feed used, the quantity fed and the disease and its effects on growth rate. A lean pig, however, is more susceptible to environmental change and disease.

There has been considerable emphasis on the selection of pigs with high lean tissue deposition that will continue through to the slaughter weight. The unimproved pig would maximise its lean tissue growth at around 40 kg (90 lbs). The modern pig breeds will maximise their lean tissue growth at the expense of fat at 60 to 90 kg (130 to 200 lbs), the boar being more efficient. Always use the best sires available i.e. those with rapid growth, good feed conversion efficiency, good killing out percentage or yield and high levels of lean tissue deposition. All these traits are highly heritable.

Records

The growing period is the most difficult section on the farm from which to gather useful information. It requires the extra burden of weighing, identifying pigs, recording feed usage, objective analysis of these factors in relation to the cost of feed (including medication), and carcass grading. However, such information is highly cost effective because it determines how the pig is growing during the different phases on your farm and identifies the inefficient and weak points. Monitoring is best carried out by tattooing a number in the ear for each week from birth, from which weight for age is easily determined. Alternatively, a minimum of 10 males and 10 females should be randomly selected every 1 to 4 weeks at point of weaning and each batch tagged with different coloured tags. Such selected groups of pigs can then be weighed and assessed against age at predetermined points. It is then possible to build up a growth curve and assess both the efficient and inefficient points. A typical example is shown in Fig 9-10, where there is increasing variability in the weight of pigs as age increases. There are 2 distinct dips in the growth curve, one at 80 to 90 days of age and another at 130 to 140. This graph identifies 2 problem areas that can then be investigated further. It should be noted that growth is not a straight line.

Computer programmes take a lot of hard work out of compiling information but don't always output the necessary information. The following are necessary to assess the efficiency of growth:

- Average liveweight gain per pig from weaning to point of sale.
- Daily gain related to age.
- Average numbers of days from weaning to point of sale.
- Amount of feed consumed per pig per day.
- Food conversion efficiency.
- The price of feed.
- Feed cost per kg of liveweight gain.

FIG 9-8: SOME FACTORS AFFECTING THE PROFITABILITY OF GROWING AND FINISHING PIGS

"Disease" – prevention/control/medication.
Education.
Environment – quality/control.
Feed conversion.
Genetic potential of the pig.
Grading or carcass quality.
Growth rate/throughput/pen utilisation.
Housing – type/design/stocking densities.
Management – decisions/work/detail.
Method of feeding.
Planning – use of records.
Price of feed.
Price of pig meat.
Quality of feed.

Efficient growth is dependent on many factors most of

DIFFERENTIAL DIAGNOSIS OF SUB-OPTIMAL GROWTH RATE

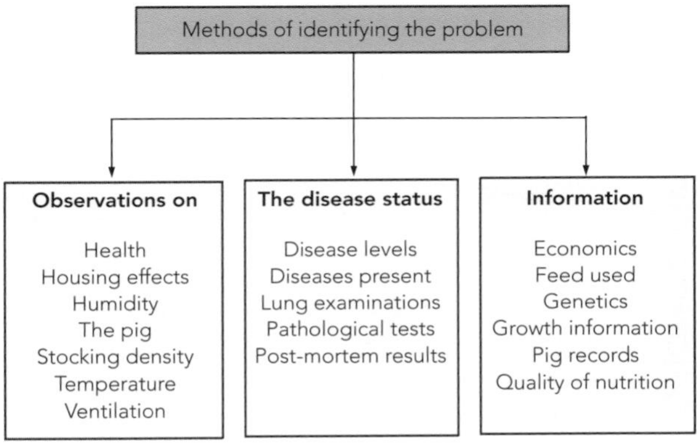

Fig 9-9

A GROWTH CURVE FROM A PROBLEM FARM

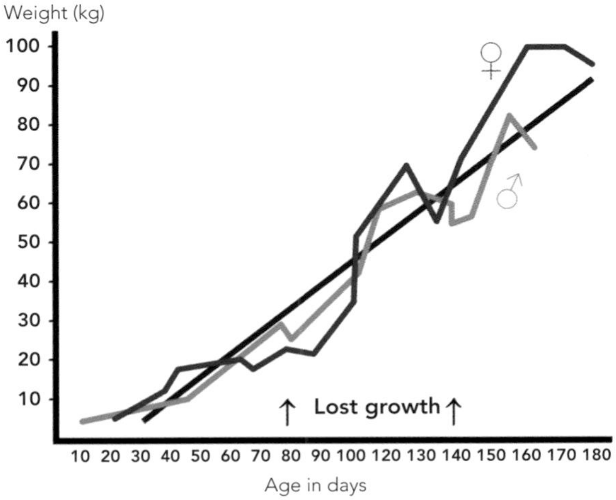

Fig 9-10

AN EFFICIENT GROWTH CURVE

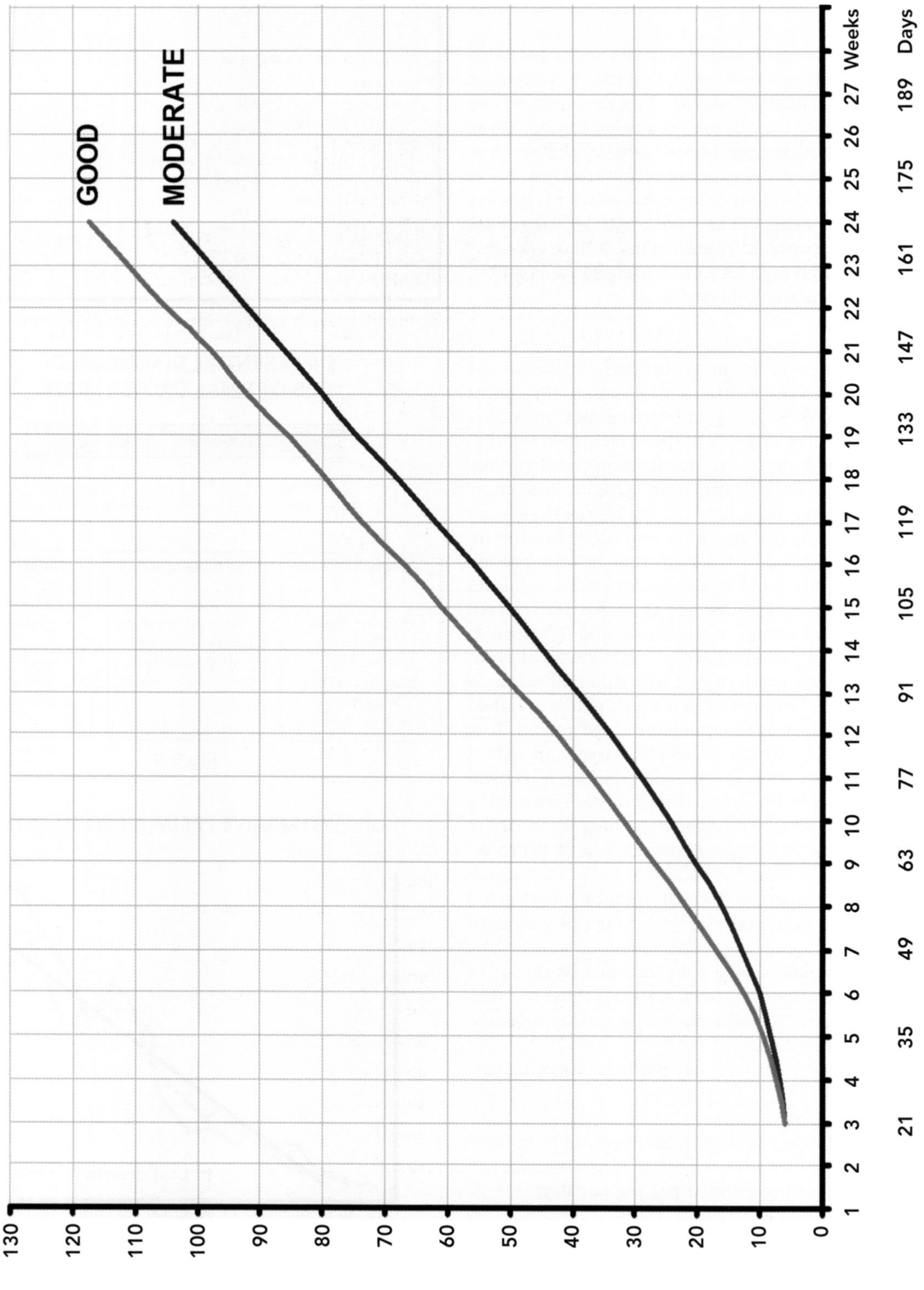

Fig 9-11

which have been listed in Fig 9-7. Fig 9-11 shows guideline growth curves that might be achieved under good conditions on the farm. Such figures can then be compared to targets set for the farm, as suggested in Fig 9-12 and the quick reference in Fig 9-13. The density of the diet fed and its effects on feed intake and feed efficiency may vary these figures slightly.

The feed intake in some genotypes will be higher with a slight worse feed efficiency but increased gain.

Computerised pig models can be used to map theoretical changes in feed, the environment, the genetics of the pig and the cost of feed. Results can sometimes be variable as there are often problems obtaining accurate input information from the farm. However as they provide valuable information to help identify inadequacies, look at "what if" scenarios and use them to help to make better decisions.

FIG 9-12: SUGGESTED REFERENCE DATA FOR EFFICIENT GROWTH (LEAN GENOTYPE, HIGH HEALTH STATUS)

Age	Weeks	Liveweight		DLWG (g/day)		DLWG	FCE		Protein %	Digestible energy (MJ DE/kg)	Net Energy (MJ NE/kg)	Feed Intake (kg/day)
		(kg)	(lbs)	From Birth	From 21 days	at age	At age	From 21 days				
At Birth	0	1.2	2.7									
21 days	3	5.75	12	216	N/A	230	0.9		23	15	11	0.22
28 days	4	7	15	242	178	202	1	1				
35 days		9.5	20	237	267							
42 days	6	12	26	257	297	379	1.4	1.25	22	14.5	11	0.55
49 days		16	35	302	366							
56 days	8	21	46	353	435	580	1.5	1.35			10	0.9
63 days		25	55	377	458				21	14.2		
70 days	10	31	68	425	515	694	1.8	1.45			9.9	1.25
77 days		35	77	438	522							
84 days	12	41	90	473	559	857	2.1	1.6			9.9	1.8
91 days		48	105	514	603							
98 days	14	54	119	538	626	933	2.25	1.9	19	14.1	9.8	2.1
105 days		61	134	569	657							
112 days	16	66	145	578	662	1106	2.35	2.1			9.7	2.6
119 days		73	160	603	686							
126 days	18	79	174	617	697	1120	2.5	2.25			9.6	2.9
133 days		85	187	630	707				18	14		
140 days	20	91	200	641	714	1132	2.65	2.3			9.6	3
147 days		98	217	663	736							
154 days	22	106	234	683	755	1000	2.8	2.45			9.6	3
161 days		114	251	701	772							
168 days	24	122	268	718	788	1100	3	2.5			9.6	3.2
175 days		129	285	733	802							
182 days	26	138	300	747	815	1100	3.2	2.6				

Note: If ractopamine is used at the end of the finishing phase, this will influence the feed intake, causing inefficiency.

FIG 9-13: PIG PERFORMANCE TARGETS

Age		Liveweight Target			
		Moderate		Good	
Week	Days	kg	lbs	kg	lbs
3	21	6	13	6	13
4	28	7	15	7.5	17
5	35	8.5	19	9.5	21
6	42	10.5	23	13	29
7	49	13.5	30	17	37
8	56	16.5	36	21.5	47
9	63	20.5	45	26	57
10	70	24.5	54	31	68
11	77	29	64	36	79
12	84	34	75	42	93
13	91	39	86	48	106
14	98	44.5	98	54	119
15	105	50	110	61	134
16	112	56	123	66	146
17	119	62	137	73	161
18	126	68	150	79	174
19	133	74.5	164	85	187
20	140	80	176	91	201
21	147	86	190	98	216
22	154	92	203	105	231
23	161	98	216	111	245
24	168	104	229	117	258
25	175	110	242	127	280
26	182	116	225	136	300

Nutrition

The way the pig utilises its feed is dependent upon its genotype, the level of available energy, the protein quality and quantity and the limiting effects of essential amino acids. Equally important are the temperature, ventilation and humidity of the environment in thehouse together with the availability of water.

It is important to provide adequate trough space to maximise growth and reduce any predisposition to disease (Fig 9-14).

Once the rates of growth in the herd have been established, assess whether target levels have been achieved and if not try to determine the reasons why. Use the checklists in Chapter 3 to provide additional information.

FIG 9-14: TROUGH LENGTHS, LIVEWEIGHT AND FEEDING METHODS

Weight of Pig (kg)	Trough Length Per Pig	
	Restricted Feeding (mm)	Ad Lib Feeding (mm)
5	100	33 (at weaning)
10	130	33
20	175	38
40	200	50
60	240	60
90	280	70
120	300	75

Single space feeder (350 mm wide), 1 per 10 pigs.

Observation

The pig

Observations by the experienced stockperson should identify healthy and diseased pigs and those with variable growth rates.

The following should be considered when carrying out the assessment:

- Were the pigs moved into the pen as an even batch, and if so is growth rate throughout the whole batch still even? This is important in relation to the presence of disease.
- Do the pigs exhibit any abnormal signs such as higher than usual levels of coughing, sneezing or skin irritation?
- Is there a history of disease or are there signs of disease?
- Do they appear settled in the pen or are there any signs of vice?
- Is the floor clean? Feed wastage can account for a loss of up to 0.4 on feed conversion efficiency on some farms, particularly those that are floor feeding. Pick up a handful of feed off the floor and smell it. Has it gone rancid?
- Check the feed hoppers. It is amazing how many of them have a hole in the bottom or are of such a design that the pig scoops food out down through the slats. Waste is one of the most important factors associated with "inefficient growth".
- Is the floor wet or are the pigs wet? This will have a marked effect on the way they use energy and feed.

Draught-free pigs grow faster on less feed.

The environment

- Next look at the pigs on a group basis rather than individuals. Are they huddled together in a corner, or, when at rest, lying on their sides not touching other pigs? If the latter is the case then you will know that the temperature is correct and that the airflow across the pigs is not chilling them.
- Do the skins of the animals appear shiny and pink with little hair or are they dull and dirty with excess hair growth? In the latter, the pigs may be below their lower critical temperature and therefore using feed to keep warm.
- Fig 9-15 indicates the different ranges of temperatures that may be required by the pig at differing weights on different types of floor surface. The thermo-neutral zone is the temperature range within which heat production is independent of air temperature. The limits of this range are described as the upper and lower critical temperatures (UCT, LCT). When pigs are housed below their lower critical temperature, a proportion of their feed is used to maintain body heat and they are more susceptible to disease. There is a marked increase in the lower critical temperature at point of weaning, due to low feed intake and the inability of the pig to satisfy its energy requirements.

An 80 kg (180 lbs) pig with restricted feed could have a lower critical temperature of 15 °C (59 °F) but if it was fed ad lib the LCT could drop to 11 °C (52 °F). At the other extreme, as pigs reach the upper critical temperature, which starts to take effect above 30 °C (86 °F), heat-relieving procedures come into play with fouling and wetting on the floor together with soiled food and loss of palatability resulting in reduced intake and therefore reduced growth rate. A checklist to determine the upper and lower critical temperatures is shown in Fig 9-16. If feed conversion efficiency is poor, check the factors in Fig 9-17.

A pig growing at 5 °C (41 °F) below its LCT could take 10 days longer to slaughter.

Disease

During the period of observation the veterinarian and experienced stockpeople can assess the levels and presence of different diseases and their influence on growth rate and economy of production.

Use the following checklist during the clinical examinations, as described in Chapter 3, to identify problem areas:

- Weight for age.
- Stocking densities.
- Evenness of growth.
- Effects of nutrition on growth in different environments.
- Time of feed changes and effects.
- Effects of pig movement.
- Environment changes.
- Quality of the environment, insulation, temperatures, humidity, draughts, temperature fluctuations.
- Wet pens, dirty floors.
- Spoiled feed.
- Undigested feed evident in faeces.
- Appearance of pigs' skins – mange.
- Appearance of the faeces, e.g. colitis – sloppy faeces.
- Respiratory disease.
- Enteric disease.
- Records of treatments/mortalities.

FIG 9-15: A GUIDE (ONLY) TO AIR TEMPERATURE ACCORDING TO FLOOR TYPE

Weight Pig		Floor Type							
		Straw		Concrete		Perforated Metal		Slatted	
kg	(lbs)	°C	(°F)	°C	(°F)	°C	(°F)	°C	(°F)
5	(11)	27-30	(81-86)	28-31	(82-88)	29-32	(84-90)	30-32	(86-90)
10	(22)	20-24	(68-75)	22-26	(72-79)	24-28	(75-82)	25-28	(77-82)
20	(44)	15-23	(59-73)	16-24	(61-75)	19-26	(66-79)	19-25	(66-77)
30	(66)	13-23	(55-73)	14-24	(57-75)	18-25	(64-77)	17-25	(63-77)
90	(198)	11-22	(52-72)	12-23	(54-73)	17-25	(63-77)	15-24	(59-75)
Lactating sow		16-18	(61-64)	16-20	(61-68)	16-20	(61-68)	16-20	(61-68)
Gestating sow		11-18	(52-64)	12-20	(54-68)	14-20	(57-68)	16-20	(61-68)
Boars		11-18	(52-64)	12-18	(54-64)	12-18	(54-64)	14-18	(57-64)

The sleeping area particularly requires the LCT temperature to be correct.

FIG 9-16: A CHECKLIST OF THE FACTORS THAT AFFECT CRITICAL TEMPERATURES

Lower Critical Temperature (LCT)	Upper Critical Temperature (UCT)
– Weight of the pig. – Feed intake. The greater the intake, the lower the LTC. **Weight of pig** / **Per day feed** / **LCT** 40 kg (88 lbs) / 1.5 kg / 16 °C 40 kg (88 lbs) / 2.0 kg / 13 °C – Energy content of the ration. – Fat depths. – Group size and stocking density. – Air speed – draughts. – Ventilation control. – Insulation of the building and floor. – Floor type, slats or solid. – Bedding. – Wet floors. – For every 1 °C below LCT there can be a loss in daily gain of 12 g. – Monitor temperature fluctuations.	– High levels of liveweight per m^2 of floor space. – High external temperature > 30 °C (86 °F). – High internal temperature > 24 °C (75 °F). – High energy intake. – Fat pigs. – Too high stocking density. – Poor ventilation. – High humidity.

FIG 9-17: CHECKLIST OF FACTORS AFFECTING FEED CONVERSION EFFICIENCY

Factors	Important Criteria	Methods of Improvement
Genetics	Select from a health-compatible source.	Replace with genetically improved stock.
Sex	Males / females / castrates.	Split Sexes. Do not castrate.
Age	Reduce days to slaughter. Feed efficiency: 1.1 : 1 at 4 weeks. 1.4 : 1 at 8 weeks. 3 : 1 at 24 weeks.	Feed Levels. Feed quality. Maximise growth rate. Good housing. Good health.
Feeding methods	Ad lib/ waste. Wet feed is better than dry. Pellets are better than meal. Trough feeding gives better FCE than floor feeding.	Prevent waste. Change dry to wet feed. Change meal to pellets or to wet feed.
Amount fed	Check waste / carcass quality and growth.	Better management.
Feed composition	Protein. Lysine. Energy. Growth enhancers. Quality.	Assess ration types according to genotype, environment and growth. Monitor ration quality.
The environment	Too hot or cold. Too draughty or high gas levels. Too dry or too wet.	Temperature. Ventilation. Humidity.
Levels of disease	Health control. Vaccination.	Veterinary and other advice. Management and housing control. Disease prevention.
Management efficiency	Attention to detail. Education of people. Maintenance of buildings. Purchasing good feeds.	Yourself.

- Prolapses.
- Number of pigs culled.
- Parasites.
- Examine hospital pens.

The changes in disease patterns in growing and feeding pigs in 63 intensive pig herds during a 40 year period are shown in Fig 9-18. These show the effects that changing production systems, disease and management practices may have over time and such changes may be relevant in your herd.

FIG 9-18: CHANGING DISEASE PATTERNS IN WEANED AND GROWING PIGS IN THE UK

Condition	1974	1984	1996	1999	2013
Actinobacillus pleuropneumonia (App)	–	–	++	+	–
Acute enteritis (diarrhoea)	+	++	++	–	+
Progressive atrophic rhinitis (PAR)	–	++		–	–
Oedema disease	+++	+	–	–	–
Mycoplasma (enzootic) pneumonia	–	–	+++	+	–
Exudative epidermitis (Greasy pig disease)	–	++	++	+	++
Malabsorption (villus atrophy)	–	++	++	+	+++
Mulberry heart disease	–	+	++	–	+
Other viruses (SIV)	–	–	++	+	++
Porcine respiratory coronavirus (PRCV)	–	–	–	–	–
Porcine reproductive and respiratory syndrome virus PRRSV	–	–	+++	+	+
PMWS (PCV2)	–	–	–	+	+++
Rectal prolapse	–	+	++	+	+
Rectal stricture	–	+	++	+	+
Streptococcal meningitis	–	++	++	+	++
Vice (abnormal behaviour)	-	++	+++	+	++

CHANGING DISEASE PATTERNS IN FINISHING PIGS IN THE UK

Condition	1974	1984	1996	1999	2013
Actinobacillus pleuropneumonia (App)	–	–	+++	+	+
Ascarid infections	–	+	+	–	–
Progressive atrophic rhinitis (PAR)	–	++	–	–	–
Aujeszky's disease/pseudorabies virus (AD/PRV)	–	++	–	–	–
Colitis	–	–	++	++	+
Mycoplasma (enzootic) pneumonia	+	+	+++	+	+
Lameness	–	–	++	++	+++
Mange	–	+++	+	+	+
Streptococcal meningitis	–	+	+	+	++
Mycoplasma arthritis	–	–	++	+	+
Ileitis	–	++	+	++	++
Porcine respiratory coronavirus (PRCV)	–	–	+	–	–
Porcine reproductive and respiratory syndrome virus (PRRSV)	–	–	+++	+	+
PMWS (PCV2)	–	–	–	–	++
Porcine dermatitis and nephropathy syndrome (PDNS)	–	–	+	–	+
Rectal prolapse	–	+	++	+	++
Rectal stricture	–	+	++	+	+
Salmonellosis	–	–	–	–	–
Swine dysentery (SD)	+	++	++	+	++
Transmissible gastroenteritis (TGE)	–	+	–	–	–

– Not significant to **+++** Most significant.

Identifying Problems in the Postweaning Period – 5 to 20 kg (10 to 45 lbs) Weight

If you have a problem in the postweaning period refer to Fig 9-19 and Fig 9-20 and then the index or relevant chapter. If you cannot identify the cause consult your veterinarian.

FIG 9-19: OBSERVATIONS AND CAUSES POSTWEANING PERIOD

Blown-up abdomen
- Constipation.
- No rectum (*Atresia ani*).
- Rectal stricture. *
- Recto vaginal fistula.
- Torsion intestine. *

Coughing
- Actinobacillus pleuropneumonia (App). *
- Ammonia. *
- Aujeszky's disease/pseudorabies virus (AD/PRV).
- Bordetellosis.
- Dust levels.*
- High endotoxin levels (causes constriction of the bronchi and breathing distress).
- Mycoplasma (enzootic) pneumonia. *
- Glässer's disease. *
- Parasites.
- Postweaning multisystemic wasting syndrome PMWS. *
- Porcine reproductive and respiratory syndrome virus (PRRSV). *
- Swine influenza virus (SIV).
- Water deprivation (salt poisoning)

Haemorrhage: Faeces
- Acute enteritis.
- Clostridia diseases.
- Gastric ulcers.
- Haematoma.
- Ileitis.
- Swine dysentery.
- Warfarin poisoning.

Haemorrhage: Nose
- Actinobacillus pleuropneumonia (App).
- Anthrax.
- Progressive atrophic rhinitis (PAR).
- Trauma.

Lameness
- Actinobacillus pleuropneumonia (App).
- Arthritis. *

FIG 9-19: OBSERVATIONS AND CAUSES CONT. POSTWEANING PERIOD

Lameness cont.
- Oedema disease – bowel oedema.
- Erysipelas. *
- Foot-and-mouth disease (FMD).
- Glässer's disease.
- Leg weakness/osteochondrosis (OCD). *
- Middle ear infection.
- Mycoplasma arthritis. *
- Streptococcal infections. *
- Trauma – muscles, joints, bones. *

Mortality: sudden death.
No signs and more than 1 %

Determine cause by post-mortem.
- Actinobacillus pleuropneumonia (App). *
- Acute enteritis.
- Anthrax.
- Bowel oedema – oedema disease.
- Clostridial disease.
- Glässer's disease.
- Mulberry heart disease (MHD). *
- Ileitis.
- Streptococcal meningitis (SM). *
- Torsion of the intestine. *
- Trauma.

Mortality: all causes.
More than 1.5 %, including after illness
- As for sudden death above.
- Chronic enteritis.
- Pericarditis. *
- Pleurisy.
- Postweaning multisystemic wasting syndrome (PMWS). *
- Pneumonia. *
- Ileitis. *
- Trauma.
- Vice (abnormal behaviour).
- Welfare causes.

Nervous signs
- Abscess spine.
- African swine fever (ASF).
- Aujeszky's disease/pseudorabies virus (AD/PRV).
- Classical swine fever/hog cholera.
- Oedema disease – bowel oedema.
- Glässer's disease. *
- Middle ear infection. *
- Poisoning.
- *Salmonella choleraesuis*.

**FIG 9-19: OBSERVATIONS AND CAUSES CONT.
POSTWEANING PERIOD**

Nervous signs cont.
- Streptococcal meningitis (SM). *
- Talfan, Teschen.
- Tetanus.
- Water deprivation (salt poisoning). *

Pale pigs
- Anaemia. *
- Actinobacillus pleuropneumonia (App).
- *Mycoplasma suis.*
- Gastric ulcers. *
- Haemorrhage. *
- Leptospirosis.
- Prolapse.
- Shortage of iron.

Pneumonia
- Actinobacillus pleuropneumonia (App). *
- Ascarids.
- Aujeszky's disease/pseudorabies virus (AD/PRV).
- Mycoplasma (enzootic) pneumonia (EP) *M. hyopneumoniae.* *
- Glässer's disease.
- Lung worm.
- Pasteurellosis.
- Postweaning multisystemic wasting syndrome (PMWS). *
- Porcine reproductive and respiratory syndrome (PRRSV). *
- *Salmonella choleraesuis.*
- Swine influenza virus (SIV).

Poor pigs, wasting, hairy
- Actinobacillus pleuropneumonia (App).
- Chronic enteritis. *
- Draughts.
- *Mycoplasma suis.*
- Glässer's disease.
- Inadequate temperature. *
- Poor nutrition. *
- Postweaning multisystemic wasting syndrome (PMWS). *
- Porcine reproductive and respiratory syndrome (PRRSV). *
- Salmonellosis.
- Swine influenza virus (SIV).
- Villus atrophy. *
- Water deprivation (salt poisoning).

**FIG 9-19: OBSERVATIONS AND CAUSES CONT.
POSTWEANING PERIOD**

Diarrhoea, scour or enteritis
- Campylobacter.
- Colitis. *
- Cryptosporidia.
- *E. coli* enteritis. *
- Postweaning multisystemic wasting syndrome (PMWS). *
- Poor environment. *
- Poor nutrition. *
- Ileitis. *
- Porcine epidemic diarrhoea (PED).
- Porcine reproductive and respiratory syndrome virus (PRRSV).
- Rotavirus.
- Salmonellosis.
- Spirochaetal diarrhoea.
- Swine dysentery (SD).
- Swine fever.
- Transmissible gastroenteritis (TGE).
- Villus atrophy. *

Skin diseases
- Erysipelas.
- Foot-and-mouth disease (FMD).
- Greasy pig disease. *
- Lice (visible).
- Mange (red spots). *
- Porcine dermatitis and nephropathy syndrome (PDNS).
- Pityriasis rosea (ringworm like).
- Porcine reproductive and respiratory syndrome virus (PRRSV) (small vesicles). *
- Purpura.
- Salmonellosis (blue colouration).
- Swine pox (round black lesions).
- Swine vesicular disease (SVD).
- Vice (abnormal behaviour).

Sneezing
- Ammonia. *
- Atrophic rhinitis non-progressive (AR). *
- Atrophic rhinitis progressive (PAR). *
- Bordetellosis.
- Dust. *
- Porcine cytomegalovirus (PCMV). *
- Porcine reproductive and respiratory syndrome (PRRSV). *

Vice; tail biting, ear biting, et al.
- Draughts.
- Fluctuating temperatures.

FIG 9-19: OBSERVATIONS AND CAUSES CONT.
POSTWEANING PERIOD

Vice; tail biting, ear biting, et al. cont.

- Greasy pig.
- High ammonia and carbon dioxide levels.
- High stocking densities.
- Poor environment.
- Uncomfortable pigs.

Vomiting

- *E. coli* gastritis.
- Gastric ulcers.
- Poisoning.
- Porcine epidemic diarrhoea (PED).
- Transmissible gastroenteritis (TGE).

* More likely.

Identifying Problems in the Growing Period – 20 to 110 kg (45 to 240 lbs) Weight

FIG 9-20: OBSERVATIONS AND CAUSES
GROWING PERIOD

Blown-up abdomen

- Chronic enteritis in the large intestine.
- Fermentation in large intestine.
- Peritonitis.
- Rectal prolapse.
- Rectal stricture. *
- Torsion intestine. *

Coughing

- Actinobacillus pleuropneumonia (App). *
- Aujeszky's disease/pseudorabies virus (AD/PRV).
- Ascarids. *
- Dust levels.
- Endotoxin levels – high.
- Mycoplasma (enzootic) pneumonia (EP). *
- High levels of ammonia.
- Lung worm.
- Pasteurellosis.
- Porcine reproductive and respiratory syndrome virus (PRRSV).
- Water deprivation (salt poisoning).

Haemorrhage: faeces

- Acute enteritis.
- Gastric ulcers.
- Ileitis. *

FIG 9-20: OBSERVATIONS AND CAUSES CONT.
GROWING PERIOD

Haemorrhage: faeces cont.

- Riding and mating issue.
- Salmonellosis.
- Swine dysentery (SD). *

Haemorrhage: nose

- Actinobacillus pleuropneumonia (App).
- Poisonings.
- Progressive atrophic rhinitis (PAR).

Lameness

- Arthritis. *
- Back muscle necrosis.
- Bush foot. Claw damage. *
- Classical swine fever/hog cholera (CSF/HC).
- Erysipelas.
- Foot-and-mouth disease (FMD).
- Glässer's disease.
- Leg weakness, osteochondrosis (OCD). *
- Middle ear infection.
- Muscle, bone, trauma. *
- Mycoplasma infections. *
- Poisoning.
- Poor nutrition.
- Riding injury.
- Swine vesicular disease (SVD).

Mortality: sudden death
No signs and more than 1 %

Determine cause by post-mortem.

- Actinobacillus pleuropneumonia (App). *
- Bloody gut (PHE).
- Clostridial diseases.
- Gas poisoning slurry.
- Glässer's disease.
- Hot temperatures > 30 °C (86 °F).
- Ileitis.
- Mulberry heart disease (MHD). *
- Pasteurellosis.
- Pericarditis.
- Poisons.
- Porcine stress syndrome (PSS).
- Riding injuries.
- Stress.
- Torsion intestine.
- Trauma.
- Whey bloat.

FIG 9-20: OBSERVATIONS AND CAUSES CONT. GROWING PERIOD

Mortality: all causes
More than 1.5 % and after illness

- Assess causes of sudden death.
- Ileitis.
- Parasites.
- Pneumonia.
- Postweaning multisystemic wasting syndrome (PMWS).
- Scour.
- Trauma.
- Vice (abnormal behaviour).

Nervous signs

- Aujeszky's disease/pseudorabies virus (AD/PRV).
- Glässer's disease. *
- High temperature.
- Middle ear infection. *
- Oedema disease – bowel oedema.
- Poisons.
- Stress.
- Streptococcal meningitis (SM). *
- Water deprivation (salt poisoning). *

Pale pigs

- Anaemia. *
- Actinobacillus pleuropneumonia (App). *
- Gastric ulcers. *
- Internal parasites.
- Ileitis.
- *Mycoplasma suis.*
- Warfarin.

Pneumonia

- Actinobacillus pleuropneumonia (App). *
- Aujeszky's disease/pseudorabies virus (AD/PRV).
- Mycoplasma (enzootic) pneumonia (EP) * *M. hyopneumoniae.*
- *Pasteurella.*
- Postweaning multisystemic wasting syndrome (PMWS). *
- Porcine reproductive and respiratory syndrome (PRRSV). *
- *Salmonella choleraesuis.*
- Swine influenza virus (SIV). *

FIG 9-20: OBSERVATIONS AND CAUSES CONT. GROWING PERIOD

Poor pigs, wasting, hairy

- Enteric disease – see diarrhoea. *
- Poor environment. *
- Poor nutrition.
- Postweaning multisystemic wasting syndrome (PMWS). *
- Respiratory disease * (see Pneumonia Sneezing).

Diarrhoea, scour or enteritis

- *E. coli.*
- Campylobacter.
- Classical swine fever/hog cholera (CSF/HC).
- Colitis.
- Ileitis.
- Porcine epidemic diarrhoea (PED).
- Postweaning multisystemic wasting syndrome (PMWS). *
- Salmonellosis.
- Swine dysentery (SD).
- Transmissible gastroenteritis (TGE).

Skin diseases

- As for weaners.

Vice (abnormal behaviour)

- As for weaners.

Vomiting

- Coughing.
- Fungal toxins.
- Gastric ulcers.
- Gastritis.
- Porcine epidemic diarrhoea (PED).
- Transmissible gastroenteritis (TGE).

* = more likely.

Diseases of the Weaned and Growing Pig

- Abscesses.
- Actinobacillus pleuropneumonia (App).
- Anthrax.
- Arthritis - joint infection.s
- Aujeszky's disease/pseudorabies virus (AD/PRV).
- Bordetellosis.
- Bursitis.
- Bush foot/foot rot.
- Classical swine fever/hog cholera (CSF/HC), African swine fever (ASF).
- Clostridial diseases.
- Coccidiosis.
- Colitis.
- Conjunctivitis.
- Diarrhoea – coliform infections and postweaning diarrhoea.
- Diarrhoea – enteric diseases.
- Erysipelas.
- Foot-and-mouth disease (FMD).
- Fractures.
- Gastric ulcers.
- Glässer's disease.
- Greasy pig disease (exudative epidermitis).
- Haematoma.
- Hepatitis E virus.
- Ileitis.
- Lameness
- Leg weakness – osteochondrosis (OCD).
- Leptospirosis.
- Lice.
- Mange.
- Middle ear infection.
- Mortality.
- Mulberry heart disease (vitamin E/selenium).
- Mycoplasma arthritis (*Mycoplasma hyosynoviae* infection).
- Mycoplasma (enzootic) pneumonia (EP) or *Mycoplasma hyopneumoniae* infection.
- *Mycoplasma suis*.
- Oedema disease (OD) – bowel oedema
- Parasites (see Chapter 11).
- Pasteurellosis.
- Pleurisy.
- Porcine dermatitis and nephropathy syndrome (PDNS).
- Porcine epidemic diarrhoea (PED).
- Porcine reproductive and respiratory syndrome virus (PRRSV).
- Porcine respiratory coronavirus (PRCV).
- Porcine stress syndrome (PSS).
- Postweaning ill-thrift syndrome (peri-weaning failure to thrive syndrome (PFTS)).
- Postweaning multisystemic wasting syndrome (PMWS).
- Postweaning sneezing.
- Progressive atrophic rhinitis (PAR).
- Prolapse of the rectum.
- Rectal stricture.
- Respiratory diseases and control strategies.
- Retroviruses.
- Riding.
- Rotaviral enteritis (diarrhoea).
- Ruptures or hernias.
- Salmonellosis.
- Spirochaetal diarrhoea.
- Streptococcal infections.
- Swine dysentery (SD).
- Swine influenza virus (SIV).
- Torsion of the stomach and intestines (twisted gut).
- Transmissible gastroenteritis (TGE).
- Tuberculosis.
- Vice (Abnormal behaviour).
- Water deprivation (Salt poisoning).
- Yersinia infection.

Abscesses

See Chapter 10 for further information

Abscesses are pockets of pus that contain dead cell material and large numbers of bacteria.

They commonly occur as a result of secondary infection following skin damage from trauma, fighting and tail biting. Tail bitten pigs should be removed immediately from the pen and given a long-acting antibiotic penicillin or OTC injection.

Affected pigs should not be sent for slaughter until the abscess has been lanced and drained (as described in Chapter 15) and antibiotic withdrawal periods have been satisfied.

Actinobacillus Pleuropneumonia (App)

There are 15 strains of the bacterium *Actinobacillus pleuropneumoniae*. Strains 1, 5, 9, 11 and 12 can be highly virulent whilst strains 3 and 6 are generally very mild. However, the clinical signs are heavily dependent on other stressors the pig is exposed too.

The pathogen is carried in the tonsils and respiratory tract and the incubation period is very short, from as little as 12 hours through to 3 days. Disease is dose-dependent i.e. the more bacteria the pig is exposed to the more severe will be the disease. The pathogen is transmitted by droplet infection and nose-to-nose contact between one pig and another.

The pathogen may survive in discharges, serum etc. for up to 5 days. It dies quickly if dried, but it may persist in water for 20 days or more. App can survive in the lungs and tonsils for long periods of at least 4 months. Contact with dead stock is therefore important from a biosecurity perspective. It is probably airborne for very short distances of around 5 to 10 metres.

Pigs may be infected with different serotypes simultaneously. PMWS, SIV, PRRSV and mycoplasma (enzootic) pneumonia can make the clinical disease worse. In a naïve herd, up to 30 % of animals may be affected. When App attacks the lungs, the toxins produced cause severe damage to the tissues which turn blue to black (necrosis) with extensive pleurisy. The chest cavity rapidly fills up with fluid.

Clinical signs

Acute disease

The organism may affect the pig from weaning through to slaughter but usually the age is from 8 to 16 weeks, once maternal antibody has disappeared. Sudden death is often the only sign, with blood and froth discharged from the nose. In the live pig a short cough may be heard with signs of severe breathing difficulties and blueing of the ears. Badly affected pigs are severely depressed. Body temperature is often high. Death is due to a combination of heart failure and the toxins produced by the organisms.

Sub-acute disease

This occurs at the same time as the acute disease with pneumonia characterised by abdominal breathing rather than chest breathing because the pleurisy is very painful. This abdominal breathing is used to clinically differentiate between actinobacillus pneumonia, where the coughing episodes are short (perhaps 1 to 3 coughs at a time), and the prolonged non-productive ones (7 to 10 coughs at a time) present with mycoplasma (enzootic) pneumonia.

Affected pigs may carry the organism for considerable lengths of time and are therefore a potential risk to younger pigs.

Chronic/sub-clinical

The majority of pigs are positive to the pathogen and show no clinical or pathological signs. The pigs grows as normal.

Diagnosis

This is based on clinical evidence, herd history and post-mortem examinations (including slaughter house checks and culture of the organism in the laboratory). The lesions in the lung are very characteristic with large red-blue areas in the upper diaphragmatic lobes with an overlying pleurisy. Serology can be used to identify different serotypes but in the absence of disease the interpretation can be difficult because of cross-reactions between serotypes.

Similar diseases

These include mycoplasma (enzootic) pneumonia, PRRSV, SIV, and *Salmonella choleraesuis* pneumonia.

Treatment

- In view of the acute course of the disease it is important to identify clinical cases very early and treat individuals by injection. Affected pigs stop eating or drinking so that water or feed medication is usually ineffective. App usually has a wide range of antibiotic sensitivity. On the first day, inject the pig twice, 8 hours apart. The following antibiotics are usually effective. Remember that injecting the pigs can be quite stressful so injections must be done quietly and with due consideration.
  - Amoxycillin.
  - Ampicillin.
  - Ceftiofur. This is a very rapid-acting medicine and gives a good response.
  - Enrofloxacin.
  - Tiamulin, OTC, LA. This can be used in more chronic cases. Repeat every 2 days.
  - Tulathromycin – this has the advantage of providing antibiotic cover for 9 days.
  - Penicillin.
  - Penicillin/streptomycin.
- It is important to determine when the onset of the disease is likely to occur, to assess adverse environmental factors and to apply strategic medication just prior to this time.
- In-feed medication during the period of risk could include:
  - Phenoxymethyl penicillin 200 to 400 g/tonne
  - Chlortetracycline 500 to 800 g/tonne
  - Trimethoprim/sulpha 300 to 400 g/tonne
  - Oxytetracycline 500 to 800 g/tonne
  - Tilmicosin 200 to 400 g/tonne for 7 to 15 days
- Water medication during the period of risk can be more effective in preventing disease. Treat for 4 to 7 days. Similar medicines to in-feed medication can be used.
- Preventative feed medication is not always effective, probably because of the rapid onset of disease and rapid loss of appetite. However tilmicosin in-feed at 200 g to 400 g/tonne has been shown to be very effective when used strategically.
- In an acute outbreak, examine the at-risk group 3 times daily to identify disease as early as possible. It may be necessary to inject or water-medicate the whole group. The decision to inject is a balance between effect and risk of more disease due to the stress of handling the pigs.

- During the acute outbreak, minimise the stress on the pigs as much as possible. Reduce the lighting and do not mix, tag or weigh the pigs.

Management control and prevention

This has 2 aspects:

1. Exclusion of virulent strains from the herd.
2. Prevention of clinical disease when virulent strains are present.

Exclusion from the herd

In breeding herds, the ideal situation is to have no highly virulent strains present but only mild or avirulent strains which then naturally immunise the herd. A naive herd, (i.e. one that has never been immunised by any natural infection) is a potential time bomb. However, in some countries this method of control i.e. a totally naive herd is being advocated. If a virulent strain gets in it will create havoc. Fairly effective vaccines are commercially available in most countries but they only immunise against homologous serotypes (i.e. the serotypes that are incorporated in the vaccine) and not against other serotypes. In contrast, natural infection tends to immunise against all serotypes.

In a breeding herd that is free from clinical disease (including absence of characteristic lesions in the lungs) it pays to try to keep virulent strains out. In pig disease areas where herds are close together and the level of infection is high, it may prove impossible to do so on a permanent basis. In more isolated herds it may be possible to maintain freedom from the disease for long periods (although even if extreme measures are adopted, breakdowns may occur, the sources of which are often unknown).

The following measures should be adopted:

- Prevent entry of virulent strains by checking that the herds which supply you with replacement breeding stock are screened on the basis of herd history, clinical inspections and absence of clinical signs, and regular lung examinations at slaughter.
- Check all visitors have not come direct from another diseased herd.
- Provide all visitors, including your veterinarian, with clean coveralls and boots and insist that they wear them.
- If you have a large, valuable herd, install a shower and make all visitors wash their hair, hands and beard if they have one.
- Build a loading bay in such a way that when trucks collect pigs the driver does not have to enter your building and you do not have to go on the truck.
- Avoid loading your pigs onto trucks which already have pigs on board from other farms. All vehicles should be empty and disinfected before arrival.
- Hold incoming breeding pigs in isolation, segregated from your herd, for a minimum of 3 and an optimum of 6 weeks and not only inspect them daily, but check that the source herd is still healthy before you bring them into your herd.
- Some people advocate testing the pigs in isolation serologically or micro-biologically (i.e. collecting nasal swabs and culturing for the bacterium) but these may be counterproductive because of false positive results and because the laboratory cannot always tell you whether the pigs have virulent or avirulent strains.
- In grower/finisher units which purchase 25 to 30 kg (55 to 66 lbs) pigs from weaner producers, it is difficult to maintain freedom from virulent strains of this organism unless you are purchasing pigs from a single known source or a limited number of known sources. The practice of all-in/all-out by building or preferably site may also help.
- The organisation of a multi-site system in which the 3-week-old piglets are weaned immediately from the breeding sow site into an all-in/all-out nursery before coming to the grower/finisher is also likely to result in freedom from this disease.
- Ensure that batching and all-in/all-out is correctly applied.
- Consider adopting partial depopulation techniques.

Prevention of clinical disease when virulent strains are present

In an infected breeding herd the most likely time to get clinical disease is in pigs over 15 kg (33 lbs). If it does occur, consider the following:

- Vaccinate for mycoplasma (enzootic) pneumonia and control PRRSV.
- Consider routine vaccination of sows and/or incoming gilts with App vaccine.
- Live attenuated genetically modified vaccines may be available which cause no disease but which immunise pigs against the resulting bacterial toxins.
- Intranasal vaccines are used in young pigs in Australia with good success.
- Operate all-in/all-out batching, at least by room, rather than continuous throughput production.
- Avoid stress and overcrowding.
- Avoid rapid temperature fluctuations.
- Avoid low humidity and low temperatures.
- Try fogging to decrease the numbers of organisms in the air.

- Increase the levels of vitamin E by 50 to 100 g/tonne.
- Maintain good ventilation and a warm air flow.
- Keep pigs warm, dry and draught free.
- Provide a plentiful supply of easily obtainable water. Temporary water deprivation will trigger disease.
- Consider strategic feed medication in advance of and during the likely time of onset of disease.
- Keep injectable antibiotics in a refrigerator ready for prompt treatment of sick pigs.
- Avoid diagnostics i.e. bleeding immediately before predicted breaks.

In infected grower/finisher units, consider all of the above but also do the following:

- Avoid introducing pigs from multiple sources.
- Do not mix pigs from herds with the disease and pigs from herds which are free from the disease.
- Practice all-in/all-out and not continuous flow.
- Consider prophylactic medication for a period after entry.
- Assess the results of vaccination.
- If pigs get sick immediately prior to slaughter, review handling practices.

Remember that when controlling the environment:

- Large airborne particles > 10 μm are retained in nasal passages.
- Particles of 0.5 to 3 μm penetrate deep into lung tissue (bacteria, App and mycoplasma).
- Low temperatures and high humidity produce large droplets that sediment quickly with less exposure.
- High temperatures and low humidity produce small droplets that sediment quickly with less exposure.
- Low temperatures and low humidity produce small droplets that stay airborne; a dangerous environment.

Anthrax

See Chapter 12 for further information

This disease is very uncommon in the growing pig unless contaminated food has been purchased. If there are sudden deaths with swollen discoloured necks or the passage of bloody faeces, anthrax must be suspected and veterinary advice sought. The disease is transmissible to the human.

Arthritis – Joint Infections

See also Lameness

Arthritis is common in the growing pig and if a problem exists it is necessary to identify the organisms or diseases responsible, by post-mortem and bacteriological examinations. The following need to be considered as possible causes (* common):

- Brucellosis (in countries where this exists).
- Glässer's disease (*Haemophilus parasuis*). *
- Erysipelas. *
- Mycoplasma arthritis (*Mycoplasma hyosynoviae* infection). *
- Leg weakness, osteochondrosis (OCD). *
- Streptococcal infection. *
- Trauma.
- Old age in pet pigs.

In the weaned and growing pig, erysipelas, *M. hyosynoviae* and OCD are the most common causes but in many cases the only clinical symptoms will be lameness. It is necessary therefore to consider arthritis under the general heading "lameness" and if you have a problem refer to this section to help identify the cause and then consider the specific diseases.

Aujeszky's Disease/Pseudorabies Virus (AD/PRV)

See Chapter 12 for further information

Aujeszky's disease (AD), also known as pseudorabies virus (PRV) is an important disease of the pig and is caused by a herpes virus. The pathogen has been eliminated from a number of countries and regions so is not directly relevant to many pig farmers.

In the growing period, affected pigs may show signs of fever, sneezing, coughing, pneumonia and high mortality with some nervous signs including incoordination and fits. Some strains of the virus cause severe respiratory disease and others severe rhinitis and complicate already existing respiratory problems. All aspects of this disease are covered in detail in Chapter 12.

Bordetellosis

Bordetella bronchiseptica is a bacterium found in most (if not all) pig populations. Some strains cause a mild and non-progressive rhinitis that heals spontaneously. The disease is clinically and economically of no consequence. However, if toxigenic *Pasteurella* are present in the herd then a combination of the 2 organisms can produce the severe disease progressive atrophic rhinitis (PAR).

Bordetella bronchiseptica can also be a secondary opportunist invader of the respiratory system.

Bursitis

See Chapter 10 for further information

Bursitis is a common condition that arises from constant pressure and trauma to the skin overlying any bony prominence. It can commence in the farrowing house, particularly if there are bad floors, but it usually starts in the weaner accommodation on slatted floors. Wire mesh, woven metal and metal bar floors can produce high levels of bursitis in weaner pigs in first and second stage housing. To address the problem, identify the point at which disease first appears and alter the floor surfaces or change the environment. This condition is covered in more detail in Chapter 10.

Bush Foot/Foot Rot

See Chapter 7 for further information

Bush foot results from infection of the claw, which becomes swollen and extremely painful around the coronary band. It arises through penetration of the sole of the foot, cracks at the sole-hoof junction, or splitting of the hoof itself. It usually occurs in one foot only and is more commonly seen in the hind feet, especially the outer claws, which are the larger ones carrying proportionately more weight. Infection sometimes penetrates the soft tissues between the claws and this is referred to as foot rot.

Classical Swine Fever/Hog Cholera (CSF/HC), African Swine Fever (ASF)

See Chapter 12 for further information on these 2 diseases

Clostridial Diseases

See Chapter 7 for further information

Clostridial infections are relatively uncommon in growers and finishers. They are usually manifested either by gas gangrene of musculature or sudden death, and the pig decomposes very quickly, showing a distended abdomen. If an outbreak of sudden death occurs in good pigs, post-mortem examinations should be carried out as soon as possible after death. Anthrax must also be considered as a possibility and veterinary advice sought.

Coccidiosis

See Chapter 11 for further information

Coccidiosis is caused by small parasites called coccidia that live and multiply inside host cells, mainly in the intestinal tract. There are 3 types, *Eimeria*, *Isospora* and Cryptosporidia. Disease, predominately caused by *Isospora suis,* is common and widespread in sucking piglets and occasionally in pigs up to 15 weeks of age. However, occasionally where finishing pigs are moved into continually populated pens to be retained as breeding animals, high levels of coccidia can persist and cause disease. This may be characterised by poor growth and sloppy faeces which may occasionally be tinged with blood. An examination of a faeces sample would help in diagnosis. Sows and older animals are generally host to *Eimeria* organisms, not *Isospora*.

Colitis

See Chapter 14 for further information

Colitis means inflammation of the large bowel and it is very common in some countries in growing pigs. It is characterised by sloppy "cow pat" type faeces, with no blood and little if any mucus but the condition may progress to severe diarrhoea. Affected pigs are usually 6 to 12 weeks of age and in any one group, 50% of the population may be affected. Detailed information on colitis is covered in Chapter 14.

Conjunctivitis

Conjunctivitis is probably caused by a combination of *Bordetella bronchiseptica* and *Chlamydophila psittaci* although the pathogenic role of these organisms is currently unclear. Exposure to irritants such as dust or ammonia may be implicated.

The condition affects older piglets, weaners and progresses into growing/finishing pigs and young adults.

It is likely that infection is spread by nose-to-nose contact. Organisms isolated from affected pigs are very common on all farms with or without the problem.

Clinical signs

Sneezing and runny eyes in piglets in the late farrowing house and nursery. There may be some rhinitis seen on post-mortem.

The conjunctiva of both eyes become injected and inflamed and third eyelid prolapses are seen in weaners, growers and finishing pigs. Often black tear-staining will be evident on the face. Once the eyelid becomes prolapsed, the condition seems to stabilise. The disease may only affect individual pigs; however on some farms all pigs in the pen are affected.

Treatment

- Review your progressive atrophic rhinitis vaccine programme; make sure it is fully effective.
- Test for toxigenic *Pasteurella multocida*; if present eliminate.
- Antimicrobials appear to have little effect.

Overuse of antimicrobials may actually encourage chlamydia as a problem.

Management control and prevention

- Review hygiene management of the farm buildings.
- Avoid chilling and draughts.
- Reduce ammonia concentrations in the air.
- Reduce dust levels in the grower and finishing house – Cover feeders, consider wet feeding.
- Minimise the clinical effects of PRRSV.

Diarrhoea – Coliform Infections and Postweaning Diarrhoea

The bacterium *Escherichia coli* (*E. coli*) is a common inhabitant of the intestine of the pig. There are 2 types, non-haemolytic and haemolytic, which describe whether or not the organism breaks down blood (haemolysis) on a culture plate. In some countries haemolytic types invariably cause disease due to the toxins that they produce but in others non haemolytic strains predominate. At weaning time the loss of sows' milk and IgA allow *E. coli* to attach to the villi of the small intestines, and the toxins cause acute enteritis and diarrhoea. Postweaning diarrhoea is a common cause of mortality and morbidity. To indicate the effect of this disease, for each day of diarrhoea, a pig loses about 5 days in growth.

Clinical signs

These are usually seen within 5 days of weaning. In severe cases a pig is found dead with sunken eyes and slight blueing of the extremities. Diarrhoea will not necessarily be seen but in less acute cases the first signs are often slight loss of condition, dehydration and watery diarrhoea. To identify the latter press the abdomen of a suspect pig and see whether diarrhoea is evident. Dehydration results in rapid loss of weight. The changes in the intestine can be so severe as to cause haemorrhage and blood, or black tarry faeces may be seen, but usually the pig dies. The diarrhoea varies in consistency from very watery to a paste with a wide range of colours from grey white, through to yellow and green. Colour is not of any significance. Fresh blood or mucus would normally be absent.

Diagnosis

This is based on the history of disease in the first week postweaning although diarrhoea can develop 10 to 14 days postweaning. Other causes e.g. rotavirus, can give similar symptoms and it is necessary to submit a live or recently dead untreated pig to the laboratory for bacteriological and virological tests to distinguish between them. Determine the antibiotic sensitivity to the *E. coli*. The *E. coli* can be differentiated by the adhesion factors – F4 (K88) for example.

Similar diseases

These include PED, rotavirus, TGE and *Salmonella* infections. A useful and simple test to differentiate between virus causes and *E. coli* diarrhoea involves the use of litmus paper to determine whether the scour is an alkaline or an acid consistency. Soak the paper in the scour; *E. coli* diarrhoea is alkaline (blue colour change) whereas viral infections are acid (red colour change).

Treatment

- It is important to know the history of the disease on the farm and antibiotic sensitivities to the bacteria present. Sick pigs should always be treated individually and group treatment applied to the pigs at risk, ideally by water medication (Fig 9-21).
- Add zinc oxide at a level of 2,500 ppm of zinc per tonne. Feed for 2 to 3 weeks. This is highly effective in controlling *E. coli* infection.
- If pigs become dehydrated, electrolytes should be provided in a separate drinker.

Management control and prevention

The principles of controlling this disease are common to the general management of the postweaned pig. These are discussed at the beginning of the chapter and you are advised to review these and adjust your control systems as indicated. If there is a problem on the farm, use the following checklist:

FIG 9-21: SUITABLE ANTIBIOTICS AND MEDICAMENTS FOR THE TREATMENT OF *E. COLI* SCOUR (POSTWEANING)

Injections	Water	In-Feed
Amoxycillin	Amoxycillin	Amoxycillin 300 g/tonne
Enrofloxacin	Apramycin	Apramycin 100 g/tonne
Framycetin		Combined CTC, penicillin sulphadimidine
Gentamycin	Neomycin	Furazolidone 400 g/tonne
Tiamulin	Sulphonamides	Lincomycin 44 g spectinomycin 44 g/tonne
Trimethoprim/sulpha (TMS)	Tiamulin	Neomycin 163 g/tonne
	TMS	Tiamulin 100 g/tonne
		TMS - variable levels
		Sulphonamides 200 to 400 g/tonne
		Zinc oxide 3.1 kg/tonne (Prevention only)

Pre-weaning

- Review the farrowing house systems.
- Assess health and body condition of the lactating sow.
- Are there respiratory or enteric problems during sucking? Adopt control measures.
- Are the weaning problems mainly in gilt litters? If so, consider *E. coli* vaccination.
- Are gilts and sows vaccinated against *E. coli*?
- Is an effective feedback programme in place?
- Consider aspects of farrowing house environment and hygiene as discussed in Chapter 7.
- Creep feeding: consider the type, frequency and age of introduction.
- Stop creep feeding before weaning and assess the effects.

On the day of weaning consider the significance of:

- Stress.
- Stocking density – group sizes.
- House temperatures and fluctuations.
- House hygiene.
- Water availability.
- Nutrition
  - Type: meal or pellets, wet or dry.
  - Feeding practices.
  - Quality of nutrition.

After weaning consider:

- Air flow – draught control.
- Temperature fluctuations.
- Ventilation, humidity.
- Creep feed management.
- Response to different creep diets.
- Disease.
- Age and weight at weaning.
- Floor surfaces – comfort boards.
- Rate and evenness of growth.
- Practise gruel feeding.

Diarrhoea – Enteric Diseases

See Chapters 8 and 14 for further information

A scour or diarrhoea problem in growing pigs is likely to be associated with one or more of the following diseases (common ones *).

- Anthrax (rare).
- Classical swine fever (in those countries where it is still endemic). See Chapter 12.
- African swine fever.
- Coliform infections and postweaning diarrhoea. *
- Colitis (non-specific). *
- Oedema disease (diarrhoea uncommon).
- Parasites.
- Porcine epidemic diarrhoea (PED). *
- Ileitis including PHE, PIA, NE and RI. *
- Rotavirus.
- Salmonellosis. *
- Spirochaetal diarrhoea.
- Swine dysentery. *
- Transmissible gastroenteritis (TGE).

Refer to the above specific diseases after a diagnosis has been made in a laboratory. Use the following flow diagram (Fig 9-22) to assist in interpreting the clinical picture. Refer to the relevant disease for further information.

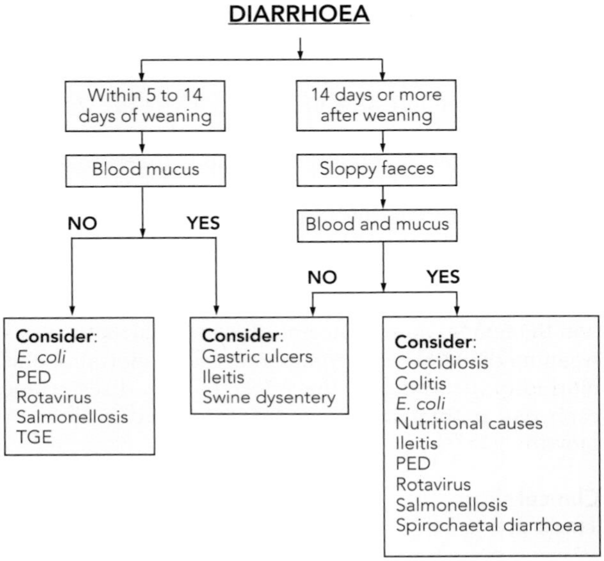

Fig 9-22

Erysipelas

See Chapter 7 for further information

Swine erysipelas is caused by a bacterium, *Erysipelothrix rhusiopathiae* that is found in most (if not all) pig farms. Whilst disease is relatively uncommon in pigs under 8 to 12 weeks of age due to protection provided by maternal antibodies, it can affect growing pigs. Infected faeces is probably the main source of infection, particularly in growing and finishing pens. Acute signs of disease are sudden and sometimes progress so rapidly that the first thing seen is a dead pig. Acutely ill pigs will be running a very high temperature of 41 to 42 °C (105 to 108 °F). Skin lesions may also be evident as large, raised diamond-shaped areas over the body that turn from red to black. They may be easier to feel than to see in the early stages.

Sub-acute disease is a more common picture with mild or few symptoms. Skin lesions are common and the pigs need not necessarily appear to be ill in spite of a temperature of up to 40 °C (104 °F). Over time the organism may settle in the joints, causing chronic arthritis and lameness. Joint problems can be responsible for condemnations at slaughter. Treatment is with penicillin or tylosin antibiotics.

This disease is covered in detail in Chapter 7.

Foot-and-Mouth Disease (FMD)

See Chapter 12 for further information

This disease should always be considered if sudden, widespread lameness appears. In all countries it is notifiable and must be reported to the authorities with urgency. As well as lameness, affected pigs salivate and blisters or vesicles are evident on the skin at the coronet at the top of the claws, and on the heels, nose and tongue.

Fractures

Bone fractures are not uncommon in growing pigs and are usually the end result of trauma and fighting, although spontaneous ones occur in bone disease such as osteomalacia, associated with calcium phosphorus and vitamins A and D, and osteochondrosis (OCD). Bone fractures may also occur in gilts when naturally mated with a large, heavy boar.

Fractures are covered in Chapter 14. Also see Lameness in Chapters 7, 9 and 14.

Gastric Ulcers

Gastric ulceration in the growing pig is common but can sometimes be difficult to diagnose.

Clinical signs vary according to the severity of the ulcer and whether it is bleeding or not. The feed intake can be variable with occasional vomiting. If haemorrhage is occurring, there will be dark coloured faeces. The animal will have a tucked-up appearance, sometimes grinding its teeth (indicative of pain) and it will appear anaemic. This condition is covered in Chapter 14.

Glässer's Disease (*Haemophilus parasuis* - Hps)

This disease is caused by the bacterium *Haemophilus parasuis* (Hps), a small organism of which there are at least fifteen different types. *H. parasuis* normally affects the sucking and young growing pig, but can occasionally affect the adult pig. It is ubiquitous and found throughout the world but disease is rarely seen in the gestating/dry sow. However, if herds are set up using SPF or MEW techniques and are free from *H. parasuis* it can be devastating when the pigs first become contaminated, producing an anthrax-like disease with high mortality in sows. Outbreaks of disease are sometimes experienced in sucking pigs, particularly in gilt herds.

In the majority of herds in which the bacterium is endemic, sows produce a strong maternal immunity which normally persists in their offspring until 8 to 12 weeks of age and as a result, the effects of the infection in weaners are usually nil or very minimal. The pigs become sub-clinically infected when still protected and then stimulate their own immune response. If, however, the maternal immunity is lacking or wears off before they become infected they may develop severe disease. *H. parasuis* can become a secondary organism where there are other major pathogens and in particular mycoplasma (enzootic) pneumonia.

Clinical signs

Acute disease

Pigs with Glässer's disease rapidly become depressed with an elevated temperature, stop eating and are reluctant to rise. *H. parasuis* attacks the smooth surfaces of the joints, coverings of the intestine, lungs, heart and brain. In young growing pigs meningitis or middle ear infections are common together with pneumonia, heart sac infection, peritonitis and pleurisy.

H. parasuis also causes individual cases of arthritis and lameness with acute pain, fever and inappetence. It is respiratory-spread and a characteristic feature is a short cough of only 2 to 3 episodes. Sudden death in good sucking piglets is not uncommon in herds with a problem and in particular when immunity in gilt litters is low.

If *H. parasuis* affects naive adults, it can kill very quickly, often with minimal gross pathology, although acute meningitis is often demonstrated on histology.

Chronic disease

Sucking piglets are often pale and poor growing and 10 to 15 % may be affected in a litter. Such pigs then continue into the growing period with poor growth. When long standing pericarditis is a feature, sudden deaths occur.

Diagnosis

This is confirmed by clinical observations, post-mortem examinations and isolation of the organism in the laboratory but it is not an easy one to grow, especially if the pigs have been on medication.

The definitive post-mortem findings of Glässer's is polyserositis. However it should be noted that polyserositis may also be associated with *Mycoplasma hyorhinis* and/or *Streptococcus suis*.

Post mortem findings may also appear similar to mycoplasma arthritis (*Mycoplasma hyosynoviae*) infection of joints and tendon sheaths.

Similar diseases

These would include:

- *Actinobacillus suis*.
- Anthrax.
- App.
- Mulberry heart disease.
- Mycoplasma arthritis (*Mycoplasma hyosynoviae)*
- Streptococcal meningitis.
- Streptococcal septicaemias.

Post-mortem and bacteriological examinations are required in order to differentiate.

Remember, *H. parasuis* is normally present and PCR is of limited value. Culture is generally difficult even in acute cases.

Treatment

- *H. parasuis* has a wide antibiotic sensitivity including synthetic penicillins, amoxycillin, ampicillin, OTC, sulphonamides, penicillin, ceftiofur and tulathromycin.
- Look for the very early signs of huddling and shivering and identify clinical cases.
- Treatment must be given early, particularly if cases of meningitis are occurring. It is important to differentiate this disease from streptococcal meningitis and this can only be done by isolating the respective organisms from the brain.
- Identify the onset of disease in sucking pigs and inject 3 to 4 days prior to this with long-acting penicillin to prevent disease.
- For the most effective treatment use injections of either penicillin/streptomycin, trimethoprim/sulpha, ceftiofur or synthetic penicillins.
- Medicate the water with amoxycillin or phenoxymethyl penicillin for 4 to 5 days over the period of risk.
- The response to a daily injection, given for at least 3 or 4 days, is usually good provided treatment is instituted early.
- When designing a treatment programme, one should consider the various organisms working synonymously.
- Because this disease can be very difficult to differentiate from the mycoplasma infections, a combination of treatments using penicillin and tiamulin injections used daily for 3 to 4 days will eliminate the various pathogens. Penicillin can be highly effective against *H. parasuis* and *Strep. suis* but has no effect on mycoplasmas as they have no cell walls. However, a new product, tulathromycin, will alone control the various organisms.

Management control and prevention

- Where the disease is a problem in sucking pigs, the sows' feed can be top-dressed daily 7 days before and 7 days after farrowing with phenoxymethyl penicillin.
- Alternatively sows can be injected with long-acting penicillin at point of farrowing.
- Vaccines are available. However, given the number of serotypes, the vaccine may fail. These vaccines are generally given to the weaner.
- Autogenous vaccines can be produced and given to the sow to stimulate immunity but the response is serotype-specific and in any one herd there may be a number of different serotypes. The vaccines need to be multivalent.
- The lactating and creep rations can be medicated with 200 to 300 g of phenoxymethyl penicillin.
- Apply the relevant general principles discussed for the control of respiratory disease (Chapter 9).

Greasy Pig Disease – (Exudative Epidermitis)

See Chapter 10 for further information

This is associated with the normal skin bacterium *Staphylococcus hyicus,* which invades abraded skin causing infection. The disease is also called exudative epidermitis, which describes the oozing of fluid from the inflamed skin. In the sucking piglet, disease is usually confined to individual animals, but it can be a major problem in new gilt herds and weaned pigs. This disease is covered in Chapter 10.

Haematoma

See Chapter 10 for further information

A haematoma is a pocket of blood that forms beneath the skin or in muscle tissue and is associated with a ruptured blood vessel. They are often seen in the growing period and the most common site is the ear, where large swellings may develop following fighting or trauma. See Chapter 10 for details.

Hepatitis E Virus

This caused some public concern in 1997 when it was isolated from pigs' livers which were suffering from another condition. Antibodies were found to be widespread in the pig population of the Mid-western United States. It was shown later that the pig hepatitis virus was distinct from the human one and there was no cross-species transmission. Hepatitis E has also been identified in pigs in Australia. This virus should not worry pig farmers.

Ileitis

This describes a group of conditions involving pathological changes in the small intestine, associated with the bacterial pathogen *Lawsonia intracellularis*. The disease is world-wide in its distribution with the infectious organism existing on most if not all farms. *Lawsonia intracellularis* lives inside the cells, lining the small and large intestines.

Ileitis is now recognised as having a serious effect on performance, including variations in weight gain, reduced feed efficiency, poorer grading and impaired gut function.

Disease occurs in 4 different forms;

1. Porcine intestinal adenopathy (PIA), which describes an abnormal proliferation of the cells that line the intestines.
2. Necrotic enteritis (NE), where the proliferated cells of the small intestine die and slough off (necrosis) with a gross thickening of the small intestine (hosepipe gut).
3. Regional ileitis (RI), or inflammation of the terminal part of the small intestine.
4. Proliferative haemorrhagic enteropathy (PHE). Here, there is massive bleeding into the small intestine, hence the common name bloody gut.

The exact mechanisms of spread are not known but the organism is found in other species including rabbits, birds and rodents. Infected faeces are the major vehicle for movement of the organism around the farm and those herds that have persistent problems are likely to have poor management of faeces, dirty pens and passages and heavily contaminated floor surfaces. Boars and adult pigs act as carriers with transfer of infection to piglets, thus maintaining the cycle of infection. The breeding herd can be a source of infection.

Studies suggest that the organism can survive outside the pig for 2 to 3 weeks. Infected pigs excrete organisms for 6 to 8 weeks and it can be found in the tonsils.

PIA and NE tend to occur in young growing pigs but sudden and severe outbreaks of PHE with high mortality can occur in pigs of 60 to 90 kg (132 – 198 lbs) weight, and maiden gilts. Gilts either already carry infection as they enter a farm and unknown factors trigger disease or they become infected on the farm for the first time. Ironically, PE is more common in high health herds but the reasons for this are unknown.

Clinical signs

These are dependent on the nature of the changes that take place inside the small intestine and in many cases disease is so mild that signs are not detected. With PIA the pig appears clinically normal and initially eats well but there is a chronic diarrhoea, gradual wasting and loss of condition, followed in some cases by a potbellied appearance. Necrotic enteritis gives a similar picture but acute disease is manifest by bloody gut or PHE and the pig may die suddenly or appear very pale and anaemic and pass black, bloody faeces. Secondary bacterial infections often increase the severity of the disease. PHE occurs frequently in young gilts, particularly within 4 to 6 weeks of arrival on the farm, at the point of service and up to the middle stage of pregnancy. Pigs with the chronic form of the ileitis recover over a period of 4 to 6 weeks. However there can be considerable losses in feed efficiency and daily gain of up to 0.3 and 80 g/day respectively. As a consequence there can be marked variations in sizes of pigs.

Diagnosis

This is carried out by the clinical picture, post-mortem examinations, histology of the gut wall and demonstrating the organism in faeces by an ELISA test. A serological test is also available. Test results can determine the levels of exposure in the herd. They are also an indicator of farm and pen hygiene. Examination of histological slides by IHC can be very useful, but it should be noted that almost all farms are positive.

Post-mortem examination showing massive haemorrhage in the lower intestine is strongly suggestive of this disease.

Treatment

The following antibiotics have been shown to be effective against the organism.

- Penicillin, tylosin, enrofloxacin and chlortetracycline. Intracellular antibiotics such as tylosin gain access to the organisms in the cells and can eliminate them.
- Tiamulin and tilmicosin (macrolides) also show activity.
- When gilts are introduced onto the farm, preventative in-feed medication using 300 to 500 g/tonne of tetracycline, or alternatively using 100 g/tonne tylosin over the first 4 to 6 week susceptible period, is effective means of control.
- Treat individual pigs with injections of tylosin or oxytetracycline and give 300 to 800 mg of iron dextran.
- Offer creep feed to affected pig for 1 to 2 weeks, fed as a gruel.
- In acute outbreaks, medicate the water for 2 to 3 days with OTC/CTC or tylosin. Follow this in-feed with 400 g/tonne OTC/CTC or tylosin 100 g/tonne for 2 to 3 weeks. If using tylosin as a preventative measure, medicate at 40 g/tonne continuously.

Strategic medication

This is applied at a selected time on the farm to pigs 6 to 8 weeks of age to prevent disease. Intracellular antibiotics are best used because the organism *Lawsonia intracellularis* lives inside the cells.

Vaccination

An alternative to antibiotic treatment is vaccination with a live attenuated isolate of *Lawsonia intracellularis* The vaccine is designed for active immunisation of weaned pigs from 3 weeks of age to reduce intestinal lesions caused by *Lawsonia intracellularis* infection and to reduce growth variability and weight gain losses associated with the disease.

The administration of the oral vaccine can be performed by drench or by drinking water.

Vaccination by oral application

Administer a single 2 ml dose orally to pigs (from 3 weeks of age), irrespective of bodyweight.

Vaccination via the drinking water

Water systems have to be cleaned and rinsed with untreated water to eliminate residues of antimicrobials, detergents or disinfectants. The final solution containing the vaccine should be consumed within 4 hours after preparation.

Management control and prevention

- Vaccination via a live vaccine through the water supply. Note the need to ensure the water supply is suitable for live vaccines. The vaccine is administered to weaned pigs about 25 to 30 kg (55 to 66 lbs) liveweight. Gilts and boars in isolation can also be vaccinated during their isolation and acclimatisation programme.
- Avoid over-stocking.
- Strategically medicate incoming gilts with tylosin or other antibiotics if there are problems, commencing 1 week before signs usually appear. Continue for 4 weeks as under treatment.
- Reduce environmental stress.
- Reduce the mixing of pigs.
- Make sure pigs have ample water.
- Wash out and disinfect gilt pens between batches. The organism is excreted via the faeces and continual use of pens increases the exposure rate allowing endemic disease to develop.
- Develop all-in/all-out procedures.
- Where problems continue in growing pigs, pens should be washed out and disinfected between batches and pigs strategically medicated.
- Keep pens as dry as possible.
- Carryover of infection between batches appears to be a significant part of the epidemiology.
- Remove any sick pigs to hospital pens and treat.
- Medicate hospital pens with tylosin or tetracycline in the water for 5 days.
- In severe continuous outbreaks it may be necessary to medicate all feed with 200 to 400 g per tonne CTC or OTC. If using tylosin, medicate for 3 weeks at 100 g/tonne followed by 40 g per tonne continuously.
- *Lawsonia intracellularis* is susceptible to quaternary ammonium compounds but not particularly so to phenolic disinfectants. Washing pens using a detergent is probably the best method.
- **Elimination of ileitis is unlikely.**

Lameness

See Chapter 14: Lameness for further information

There are numerous causes of lameness in the pig and it is prevalent in growing pigs with levels ranging from 1 to 5 %. It is generally triggered by tissue changes resulting from either infectious or non-infectious causes.

Such tissue changes include:

Apophysitis (OCD) – Separation of the muscle mass from the growth plate on the pelvis.

Arthritis – Inflammation of 1 or more joints.

Damage to nervous tissue – Clinical signs vary (e.g. partial or complete paralysis of 1 or more limbs) depending on the site of the damage.

Epiphyseolysis (OCD) – Separation of the head of the femur.

Fractured bones – Common in the hip, hock and elbow joints.

Haematoma – Haemorrhage into the tissues.

Laminitis – Inflammation of the tissues connecting the hoof to the bone. It is not common.

Myositis – Inflammation of muscles.

Penetrated sole – Damage due to trauma.

Periostitis – Inflammation of the membrane (periosteum) which covers the bone.

Osteitis – Inflammation of bone.

Osteochondrosis (leg weakness) – Growth plate and joint cartilage degeneration.

Osteomalacia – Softening of the bones due to calcium/phosphorus deficiency.

Osteomyelitis – Inflammation of all bone tissue including the spongy centre and bone marrow.

Osteoporosis – Weak bones due to an imbalance of calcium and phosphorous in the diet.

Split horn – Poor hoof quality. Overgrown claws.

Torn ligaments or muscles – A common cause of lameness, particularly where muscles are attached to bones.

In the growing pig, lameness may be caused by any of the following (common *):

Infectious causes

- Brucellosis.
- Bush foot/foot rot.
- Clostridial diseases.
- Erysipelas.
- Foot-and-mouth disease.
- Glässer's disease (*Haemophilus parasuis*).
- Mycoplasma arthritis.
- Salmonellosis.
- Swine vesicular disease.
- Streptococcal infections.

Non-infectious causes

- Fractures.
- Laminitis.
- Leg weakness or osteochondrosis.
- Muscle tearing.
- Nutritional deficiencies.
- Porcine stress syndrome associated with the halothane gene.
- Toxic conditions.
- Trauma.

Lameness can account for significant losses in growing pigs either because the pigs are unfit to travel on welfare grounds and are required to be destroyed, or they are part or totally condemned at slaughter. Early identification of lame animals and their removal to hospital pens for treatment is a vital part of the control and healing process. Stocking density and mixing are the 2 major factors that precipitate traumatic disease.

Infections can also account for considerable lameness losses, particularly from tail biting and septicaemias that arise during immunosuppressive diseases such as PRRSV, mycoplasma (enzootic) pneumonia and SIV.

Identifying the causes of lameness

If there is a lameness problem on the farm, it is necessary to identify the common problem and then refer to the relevant disease or diseases.

Consider the following (and also Fig 9-23) for identification purposes:

- If more than 2 % of pigs are recorded lame per month, further investigations are necessary.
- Keep records of the time lameness occurs, which house the pig is in and if possible the visual appearance of the lameness.
- If lameness involves the foot, look closely at floor surfaces.
- Look for marks or scarring on the skin that might indicate external damage due to fighting.
- Look for cuts or breaks in the skin related to sharp projections from the environment. The position of these on the body of the pig will indicate the height at which these are occurring.

IDENTIFYING THE CAUSES OF LAMENESS

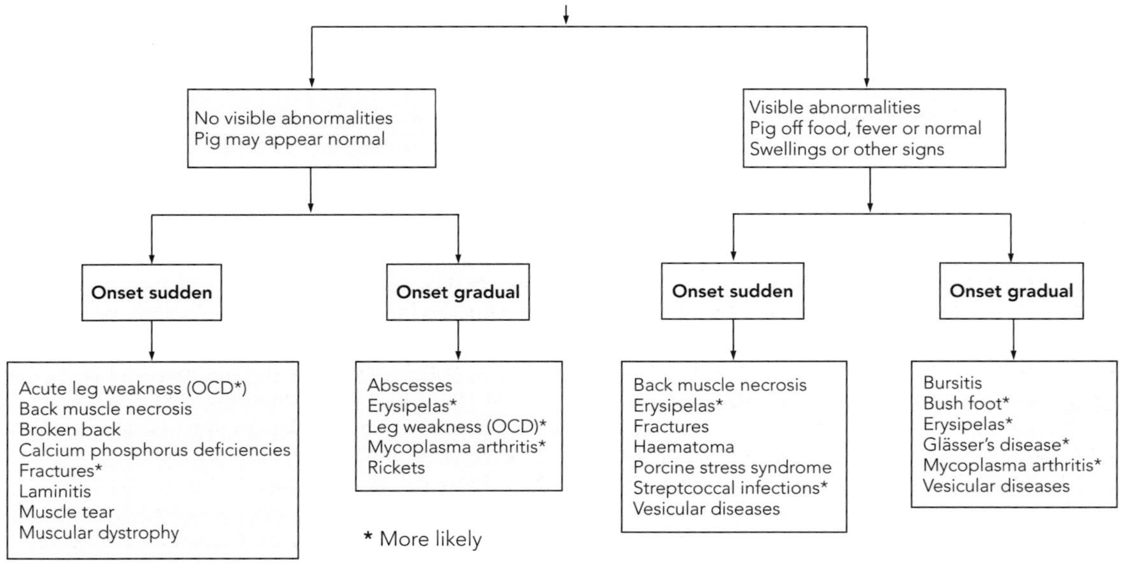

Fig 9-23

Typical examples are worn metal feeding troughs, worn metal pen divisions and bad slats.

- If there is a high incidence of leg sores associated with fractures, assess the conditions precipitating leg weakness.
- Consider specific diseases.

Lameness may also be caused by faulty nutrition, and this is covered in more detail in Chapter 14.

Leg Weakness – Osteochondrosis (OCD)

See Chapter 7 for further information

Leg deformities are common in the rapidly growing pig but are usually of no commercial consequence because they do not affect the daily gain or food conversion efficiency. However the conformation defects of leg weakness can restrict the sale of breeding boars and gilts. Separation of the head of the femur at the growth plate does occur as a problem in younger growing pigs on some farms. It is characterised by the sudden onset of acute lameness with the pig refusing to put the foot to the floor. Gentle examination of the leg will determine a fracture in the hip or knee joint. Similar fractures also occur in the elbow joint and at the attachment of the hind muscle mass to the pelvis. In the growing pig heavy stocking density, rapid weight gain and environmental factors that cause the foot to slip on the floor will predispose. See Chapter 7: Management control and prevention if you have problems.

Leptospirosis

See Chapters 6 and 7 for information

Growing pigs are often exposed to different strains of leptospira from the urine of rats, mice or other animals and they respond with positive titres in the blood. Disease, however, is uncommon. When it does occur it usually takes the form of acute jaundice, haemorrhage and rapid death caused by the *Leptospira icterohaemorrhagiae* serovar. This disease can be spread to humans if infected urine makes contact with broken skin or mucous membranes and causes Weil's disease. These are covered in detail in Chapters 6 and 7.

Lice

See Chapters 10 and 11 for further information

These are relatively uncommon in herds today, particularly if mange treatment is carried out, because this will also destroy the pig louse. They are blood-sucking parasites that are easily visible on the skin. They cause a certain amount of irritation and severe infestations can cause anaemia.

Mange

See Chapter 11 for further information

Mange is a parasitic disease of the skin caused by one of two mites, either *Sarcoptes scabiei* var *suis* or *Demodex phylloides,* the latter of which is relatively unimportant in swine.

Sarcoptic mange (sometimes called scabies) is by far the most common and important in the pig because it is irritant and uncomfortable for the pig, causing it to rub and damage the skin which becomes unsightly. Irritation and rubbing are constant findings, together with ear shaking. It is a major cause of loss of production and growth rate in growing pigs.

Middle Ear Infection

This is caused by a variety of bacteria including *Bordetella bronchiseptica*, *Haemophilus parasuis*, staphylococci and (mainly) streptococci that gain access to the middle part of the ear, which is responsible for balance. Infection likely arises from the tonsils at the back of the throat and travels down the eustachian tube to the middle part of the ear. The condition is sporadic but common and in some farms up to 5 % of weaner pigs may be affected. It must be recognised early and if treatment is prompt there is usually a good response. If treatment is delayed there is the risk that infection will spread from the middle ear into the inner ear and directly to the brain, setting up a meningitis or encephalitis (inflammation of the brain). Actinobacillus pleuropneumonia has also been identified in outbreaks of the disease and bacteriological examinations should always be carried out if abnormal numbers of pigs are involved.

Clinical signs

The pig stands with its head to one side, often shaking, with evidence of pain. As the disease progresses there is a gradual loss of co-ordination until ultimately the pig walks around in a circle, eventually falling over. Disease in the sow is often severe and such animals are best culled if the response to treatment is poor.

Treatment

- The response to treatment in the weaner is usually good when either penicillin/streptomycin or amoxycillin are used. In acute cases it is necessary to inject the pig twice daily for the first 2 days and then follow up with long-acting injections. Long-acting OTC can also be used.
- Cortisone injections are also of value, as advised by your veterinarian.
- Treatment must continue for 7 to 10 days and complete recovery may take up to 3 weeks.
- Administer pain relief if the pig is showing signs of being in pain.

Management control and prevention

- If there is a problem in your herd, identify the time of onset of the disease and study the environment and other diseases for predisposing factors. These could include mange, skin trauma, vice (abnormal behaviour), mixing and fighting, greasy pig disease, joint infections and PRRSV.
- PRRSV can initiate outbreaks in some herds.
- Consider preventative medication using amoxycillin long-acting injections given at the time just prior to disease onset.
- In sows, review the position of drip cooling, which may be dripping into the sows' ears.

Mortality

Excessive mortality in the weaning and finishing herd is a significant economic loss. It can be assessed realistically by assuming that the overheads on the unit, excluding feed, are going to remain constant and therefore the calculation of loss for each pig can be considered as follows:

(Cost of raising the piglet to the point of weaning)
+
(Cost of the feed to the point of death)
+
(Margin over feed that would have been made had the pig reached slaughter weight).

If the levels in the herd are above acceptable targets (Fig 9-24), then the reasons for the excess should be identified. For each pig that is found dead or destroyed, note the date, age and weight, the house in which it died, the believed cause of death (Fig-9-25), and any comments. Such a system can easily be recorded on cards by house (see Chapter 3). This is a simple method that is strongly recommended. The quality of nursing care given to sick pigs can significantly affect the target levels at the upper limit.

The movement to and reasons for sick pigs entering hospital pens should also be recorded.

FIG 9-24: TARGET MORTALITY FIGURES

Weaning to 3 weeks postweaning	1 %
3 weeks to 12 weeks postweaning	0.5 %
12 weeks to slaughter	1 to 1.5 %
Culls destroyed on welfare grounds	1 to 2 %
Total	**3.5 to 5 %**

FIG 9-25: RECORDED CAUSES OF DEATH
(i.e. those that may be readily diagnosed by an experienced pig person)

The Weaner	The Grower Pig
Acute enteritis	Abscess
Fighting	Bloody diarrhoea
Glässer's disease	Enteric problems
Meningitis	Erysipelas
Miscellaneous causes	Fighting
Oedema disease	Gastric ulcers
PMWS	Miscellaneous causes
Postweaning diarrhoea	Pale pig
Prolapse	PDNS
Respiratory disease	Rectal prolapse
Sudden death/stress	Rectal stricture
Vice (abnormal behaviour)	Respiratory disease
Welfare culls	Salt poisoning/water deprivation
Unknown	Sudden death/stress/torsion
	Vice (abnormal behaviour)
	Welfare culls
	Unknown

Are you making the correct decisions in the hospital pen?

Mulberry Heart Disease (Vitamin E/Selenium)

See Chapter 14 for further information

Problems associated with either the lack of availability of vitamin E and selenium, or absolute deficiencies in either, can cause major problems on some farms. These can arise due to high levels of polyunsaturated fats in diets, which are used as sources of energy. High levels of copper, vitamin A or mycotoxins in the feed can have a similar effect.

Clinical signs vary according to the system affected. Hepatosis dietetica (HD), muscular or nutritional dystrophy (MD) (also called a myopathy) and mulberry heart disease (MHD) (also called a myopathy) are usually associated with sudden deaths in rapid growing pigs without any prior clinical signs. Usually the best pigs in the pen, ranging from 15 to 30 kg (33 to 66 lbs) in weight, are affected. Diets being fed often contain high levels of fats and yet in many cases vitamin E levels appear within normal ranges. This condition is covered in detail in Chapter 14: Vitamin E/Selenium.

Mycoplasma Arthritis (*Mycoplasma hyosynoviae* Infection)

See Chapter 7 for further information

Mycoplasma hyosynoviae is a ubiquitous bacterial pathogen and most (if not all) herds are infected with it. It is a respiratory-spread disease, the organism being found in the upper respiratory tract nose and tonsils. It may be present in some herds and cause no clinical signs and yet in others cause severe disease. Infection with or without disease takes place in the young growing pig from approximately 8 to 30 weeks of age and particularly so in the gilt when first introduced onto a farm, or in the early stages of pregnancy.

Mycoplasma (Enzootic) Pneumonia (EP) (*Mycoplasma hyopneumoniae*)

Mycoplasma (enzootic) pneumonia is caused by the tiny pathogen *Mycoplasma hyopneumoniae*.

However, while *M. hyopneumoniae* is normally present, it is possible for the pig to have enzootic pneumonia (which is more of a description of the clinical and pathological findings) without the presence of *M. hyopneumoniae*. *M. hyopneumoniae* is widespread in pig populations and endemic in most herds throughout the world. It is transmitted either through the movement of the carrier pigs or by wind-borne infection for up to 3 km (2 miles) if the climatic conditions are right.

The organism dies out quickly outside the pig, particularly when dried. It can, however, be maintained in moist cool conditions for 2 to 3 days. It has a long incubation period of 2 to 8 weeks before clinical symptoms are seen. As an uncomplicated infection in well-housed and managed pigs, it is a relatively unimportant disease and has only mild effects on the pig.

However if there are other infections present, particularly *Actinobacillus pleuropneumoniae* (App), porcine circovirus 2 (PCV2), *Haemophilus parasuis* (Hps), *Pasteurella*, PMWS, PRRSV, streptococcus or SIV, the pneumonia can become more complex with serious effects on the pig. Fig 9-26 shows the basic structure of the lobes of the lungs. *M. hyopneumoniae* always attacks the lower, shaded areas of each of these lobes (anterior, cardiac, intermediate and anterior diaphragmatic) causing consolidation of the tissues.

The extent of this consolidation in each lobe is scored out of either 5 or 10 depending upon the lobe affected. Thus a severely affected pig with all lobes involved would score 55. Occasionally, particularly in disease breakdowns, the diaphragmatic lobes will be involved as well. If more than 15 % of lungs are affected it is highly probable that mycoplasma (enzootic) pneumonia is present in the population. Herds that do not carry *M. hyopneumoniae* rarely show consolidated lesions of more than 1 to 2 % and even then they are very small.

This scoring system can be used to assess the severity of disease and its effects on the pig (Fig 9-27).

PNEUMONIA
DIAGRAM OF THE LUNGS

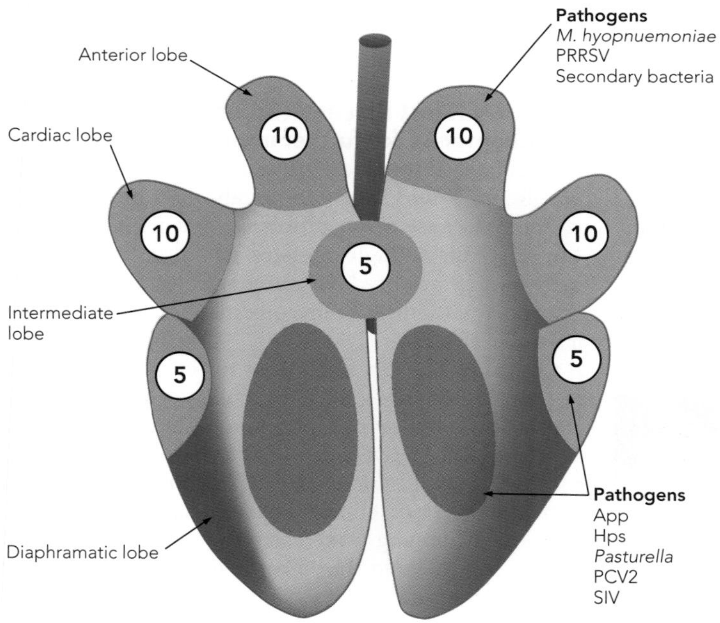

Areas of Infection and a Scoring System

Fig 9-26

FIG 9-27: A GUIDE TO THE EFFECTS OF MYCOPLASMA (ENZOOTIC) PNEUMONIA LESIONS ON DAILY GAIN

Total Score Max. 55	% Loss in Daily Gain
Zero	0
1 – 10	0
11 – 20	6
21 – 30	18
31 – 40	26
41 – 55	50

If mycoplasma (enzootic) pneumonia is not present in the growing population then the effects of the other respiratory pathogens are very greatly reduced. It is therefore considered a prime organism that opens up the lung to other infections.

Clinical signs

Acute disease

This is normally only seen when *Mycoplasma hyopneumoniae* is introduced into the herd for the first time. For a period of 6 to 8 weeks after entry there may be severe acute pneumonia, coughing, respiratory distress, fever and high mortality across all ages of stock. However, this picture is extremely variable and breakdowns are experienced when disease is mild or inapparent.

Chronic disease

This is the normal picture where the organism has been present in the herd for some considerable time. Maternal antibody is passed via colostrum to the piglets and it disappears from 7 to 12 weeks of age after which clinical signs start to appear. A prolonged non-productive cough, at least 7 to 8 coughs per episode, is a common sign around this time, with some pigs breathing heavily ("thumps") and showing signs of pneumonia. 30 to 70 % of pigs will have lung lesions at slaughter.

Diagnosis

In most cases this is based on the clinical picture and examination of the lungs of pigs at post-mortem or at slaughter, combined possibly with histology of the lesions.

However, these do not provide a specific diagnosis and in the herds supplying breeding stock or in special cases (e.g. litigation) it may be necessary to confirm the diagnosis by carrying out one or more of the following tests: immunoperoxidase, ELISA tests, serum tests for specific antibodies, fluorescent antibody test (FAT), polymerase chain reaction (PCR) tests and finally culture and identification of *Mycoplasma hyopneumoniae*.

A very high lung score may be indicative of a recently acutely diseased lung.

Similar diseases

Consolidation of the anterior lobes of the lungs at a low level can be caused by other respiratory pathogens including SIV, PRRSV, *Haemophilus parasuis*, PCV2, certain viruses and other mycoplasma. Laboratory tests are required to differentiate them. Furthermore, all or some of these may occur as mixed infections, together with *Mycoplasma hyopneumoniae*.

Treatment

In the herds in which the disease has become endemic, a decision to medicate feed should be based on the following considerations:

- Variable growth in pigs from 10 to 20 weeks of age.
- Ongoing pneumonia treatment of individuals at more than 2.5 % of the pig population.
- Lungs scoring more than 15.
- Active lesions – raised above the level of the lung surface and moist or wet.

Acute disease (herd breakdown)

Consider the following:

- Medicate pigs between weaning and 16 weeks of age for 4 to 8 weeks with 500 g/tonne of CTC or OTC and then reduce this to 200 to 300 g/tonne. Alternatively 200 g to 400 g/tonne tilmicosin for 15 days or tiamulin 100g/tonne for 14 days.
- Inject severely affected individual pigs with either long-acting OTC, tiamulin, lincomycin, valnemulin, tilmicosin or tulathromycin.
- If pigs become affected soon after weaning, inject with OTC LA at weaning time or 1 week prior to the onset of disease.

Penicillins have no effect against mycoplasma as it does not have a cell wall. However, it may be very effective against the secondary infections such as *Pasteurella*.

Chronic disease

- Identify the point at which disease is occurring and apply strategic medication, either in-feed, in-water or by injection using the medicines outlined above.

For strategic medication use tetracyclines 500 to 800 g/tonne, 220 g/tonne of lincomycin or 100 g/tonne of tiamulin and feed for 7 to 10 days; or 200 g to 400 g/tonne tilmicosin for up to 10 to 15 days commencing 1 to 3 weeks prior to the anticipated time of the disease starting.

Management control and prevention

The *M. hyopneumoniae*-free breeding herd?

- Most breeding organisations can supply breeding stock that is deemed to be *M. hyopneumoniae* free and so a new pig farm can be set up or an old one repopulated *M. hyopneumoniae* free. The question is whether it will remain so. One rule of thumb is that if you have an uninterrupted view of an infected *M. hyopneumoniae* herd, particularly if it is less than 3 km (2 miles) away, there is a definite chance that sooner or later your herd will become contaminated by wind-borne infection. If the herd you can see is a grower/finisher as distinct from a weaner producer than the risk is greater. The risk also increases the larger the infected herd. In pig-dense regions it is impossible to remain free. You are likely to remain *M. hyopneumoniae*-free indefinitely if your *M. hyopneumoniae*-free herd is in a region of low pig density. Likewise if the land is hilly or mountainous or on a sea coast.

To maintain a *M. hyopneumoniae*-free breeding herd:

- Keep it closed. Introduce genes only by AI, embryo transfer, or hysterectomy and fostering of pigs less than 10 days of age.
- Purchase only *M. hyopneumoniae*-free stock from a reputable seed stock supplier, if possible from the same source herd every time, or at least from the same breeding pyramid.
- Isolate incoming pigs for 8 weeks and check that the *M. hyopneumoniae*-free status of the donor herd has been maintained before moving them into your herd. (This is good practice whether your herd is *M. hyopneumoniae*-free or not).
- To be ultra-careful, mix sentinel pigs from your herd in with the new pigs in isolation, 1 to 2 weeks after their delivery. The sentinels can be slaughtered and their lungs examined before the pigs move in or they can be blood tested on entry to the isolation and again 5 weeks later.
- If you are not producing *M. hyopneumoniae*-free pigs, it is wise to vaccinate all incoming gilts and ensure that the adult herd is vaccinated (with a dead vaccine) because if the herd breaks, the pig flow can be devastated by the abortions and infertility that *M. hyopneumoniae* creates.

To maintain a *M. hyopneumoniae*-free grower/finisher herd

- Purchase only from an *M. hyopneumoniae*-free source but instead of an isolation period implement an all-in/all-out policy, by site if possible, or if not, by building.
- Again, the location is paramount.

The herd with endemic *M. hyopneumoniae*

- Purchase stock with or without *M. hyopneumoniae* but, depending on the health status of the herd, make sure the pigs are free from swine dysentery, mange and PRRSV. Ensure that the gilts in isolation receive vaccination against *M. hyopneumoniae*.
- Carry out isolation and monitoring procedures as above. Six weeks instead of 8 weeks is probably enough. If your herd has a low health status and is intensively housed it may be helpful to medicate the feed for the incoming stock if they are *M. hyopneumoniae* free.
- Use tulathromycin, lincomycin, tiamulin or tilmicosin for the last 2 weeks in isolation, or the first 4 weeks in your herd. This allows them to become immune without becoming ill. Discuss with your veterinarian.
- Keep a broad parity spread in your sows. Sows of second parity onwards are more immune than first litter gilts and pass a better immunity to their piglets.
- Check the faeces of the growing pigs for ascarids and if present keep them under control by routine worming and all-in/all-out housing procedures.
- If you are having clinical problems and poor growth later in growing pigs, vaccinate young piglets against mycoplasma (enzootic) pneumonia as per the manufacturer's instructions.
- Ensure that the water supply to the pigs is excellent. Poor water availability will exacerbate the clinical signs of mycoplasma (enzootic) pneumonia.

Increased disease is associated with the following (consider changes):

- Overcrowding and group sizes in any one environment of more than 200.
- Variable temperatures and poor insulation.
- Variable wind speeds and chilling.
- Low-temperature, low-humidity environments.
- Houses with poor hygiene and high levels of carbon dioxide and ammonia.
- High dust/bacteria concentrations in the air.
- Pig movement, stress and mixing.
- A shortage of trough space.
- Housing with a continuous throughput of pigs.
- Other concurrent diseases.
- Poor nutrition.
- Dietary changes at susceptible times.
- Slatted floors and liquid waste.
- Less than 3 m^3 air space/pig and 0.7 m^2 floor space/finishing pig.
- Houses that are too wide for good air flow control.
- Poor water availability.
- Gastric ulceration and anaemia.
- Lice and mange infestation.
- Presence of PRRSV.
- Presence of Aujeszky's disease.
- Presence of App.
- Presence of swine influenza.
- Vaccinate against PCV2
- Purchasing from different sources.

Therefore to keep *M. hyopneumoniae* and respiratory disease under control:

- Optimise stocking levels by pen and house. This will not only reduce clinical signs but the energy that had been required for immunity will be available for growth and result in faster throughput. Thus the number of pigs sold per year can remain the same as at higher stocking levels.
- Optimise the ventilation and improve the hygiene to reduce noxious gases.
- Improve the insulation if necessary and maintain constant temperature control.
- Reduce the dust and bacteria in the air by changing to wet feed or to less dusty dry feed. (Note that increasing ventilation does not decrease dust and may increase it).
- Check the nutrition and time of feed changes.
- Organise the grower/finisher stages so that moving and mixing is minimal.
- Enhance water supplies.
- Eradicate lice and mange infestations.
- Reduce gastric ulceration.
- Operate an all-in/all-out system wherever possible.
- **Vaccination**: Highly efficient dead *M. hyopneumoniae* vaccines can reduce lung lesions by up to 95 %. Some can be applied at 1 and 3 weeks of age and this provides excellent control during the growing and susceptible period. Maternal antibodies are a potential problem, but if sows are not vaccinated the vaccines are able to overcome any maternal antibodies with minimal effect on efficacy. Vaccine could also be applied to *M. hyopneumoniae*-negative herds if considered at risk.
  The use of *M. hyopneumoniae* vaccines in herds with complex respiratory problems has revolutionised respiratory disease control, particularly if used in conjunction with segregated weaning and partial depopulation control techniques – the latter on combined breeding/finishing operations.
  Vaccines may be combined with PCV2 to reduce the number of injections and time required to vaccinate the piglets.

If you have a problem on your farm, consider the following criteria in making the decision to vaccinate:

- The presence of *Mycoplasma hyopneumoniae*.
- A continual level of respiratory disease.
- Primary or secondary infections associated with PRRSV, SIV, AD/PRV and *Actinobacillus pleuropneumoniae*.
- Heavy bacterial challenge.
- The necessity for continual in-feed medication.
- Variable and poor growth associated with respiratory disease.
- Weaning to slaughter mortalities of more than 4 %.
- Finally, the cost of vaccination should be equal to or less than costs of the potential reduction in mortality and in-feed medication. Improvements in daily gain and feed efficiency become the bonus. In severely affected herds a cost benefit ratio of up to 5:1 has been achieved.
- Field experiences indicate that all herds should vaccinate growing pigs from 1 week of age onwards.

The *M. hyopneumoniae* vaccine is very specific and does not protect against other respiratory mycoplasmas or *M. hyosynoviae*.

Mycoplasma (Eperythrozoon) suis

See Chapter 11 for further information

Mycoplasma (Eperythrozoon) suis is caused by a small mycoplasma bacteria. This blood parasite attaches itself to the red cells in the blood, sometimes damaging them and causing them to break apart. This causes a reduction in the red blood cell count, reducing haemoglobin (the substance which transports oxygen around the body) and thus diminishing the capacity of the blood to carry oxygen. When large numbers of red cells are damaged, hemolysis anaemia results and pigs look jaundiced. In weaners, signs may include slow and/or variable growth and poor-doing pigs.

This condition is covered in detail in Chapter 11.

Oedema Disease (OD) – Bowel Oedema

This is also called bowel disease or gut oedema. It is caused by F18 *Escherichia coli* (*E. coli*) bacteria that produce a powerful toxin called verotoxin.

The specific *E. coli* are described as having a specific toxin – STx2e (Shiga-like). The type of *E. coli* which produces this toxin also has a specific attachment called fimbriae F18. The cell wall types which are associated with this fimbriae are O138, O139 and O141. Rarely F4 (K88) is also associated with this toxin.

These toxins damage the walls of small blood vessels including those in the brain and cause fluid or oedema to accumulate in the tissues of the stomach and the large bowel.

Blood pressure increases to 200 mm Hg (normally 110 to 140), which results in damage to the blood vessels in the brain and some of the characteristic signs of brain oedema, hence the neurological signs.

Disease is generally seen 1 to 4 weeks after weaning,

the peak being at 10 days.

The *E. coli* bacteria attach themselves to the finger-like villi in the anterior small intestine and produce the toxins. This mechanism is similar to that which occurs in postweaning diarrhoea associated with different strains of *E. coli*. During sucking the secretory IgA immunoglobulin component in milk prevents the bacteria from adhering. After weaning, when the IgA has disappeared, the pigs become susceptible to disease.

Clinical signs

Acute disease

Sometimes the only sign is a good pig found dead 1 to 4 weeks postweaning. A typical live affected pig will show a staggering gait, puffy eyelids giving a sleepy appearance, and an abnormal high pitched squeak. Pigs stop eating and in the later stages become partially paralysed and go off their legs, sometimes with nervous symptoms. Diarrhoea is not a consistent feature but breathing difficulties become evident. The damage to the brain is irreversible and most pigs die. Recovery in the few pigs that do not die takes up to 2 to 3 weeks. Certain breeds of pigs may be associated with disease, suggesting a genetic predisposition. The temperature is usually normal.

Diagnosis

This is made from the typical clinical signs, the sudden appearance of disease after weaning, post-mortem examinations showing oedema of the greater curvature of the stomach wall, coiled colon, and eyelids and isolation of the haemolytic *E. coli* serotypes from the duodenum (anterior small intestine).

Treatment

By the time the clinical signs are seen it is often too late and most pigs die. Treatment routines are aimed at preventing the organism establishing itself and also reducing the weight of infection. The general principles of controlling coliform infections and postweaning diarrhoea should be followed.

- Isolate the organism and determine the antibiotic sensitivity.
- Identify the stage (e.g. 10 days postweaning) when disease first appears and apply either in-feed or water medication 3 to 5 days before this.
- In-feed antibiotics of value include apramycin 100 g/tonne, framycetin 100 g/tonne, neomycin 163 g/tonne. Alternatively apramycin, neomycin or trimethoprim/sulpha can be used in the water.
- Individual treatments give a poor response but flunixin will help to reduce the effects of toxins and diuretics can be used to remove fluid.

It must be admitted however that the disease is most difficult to deal with.

Management control and prevention

- Certain types of breeding animals may be more susceptible and mortalities in such cases may rise to 25 % or more. Where this is the case, change the genetic make-up of the farm. There are resistant lines of pigs available.
- If vaccinating, use a non-toxigenic F18 *E. coli* vaccine. This is a live vaccine and given via the water supply. Treat the pigs at weaning, 1 week postweaning and at 4 weeks postweaning. Note the problems of using a live bacterial vaccine to the water supply regarding other antibiotics and chlorine in the system.
- Restrict feed intake postweaning.
- Assess the effects of different diets and feeding routines.
- Reduce the nutrient composition of the diet by increasing the fibre content by 10 to 15 %.
- Alter the environment at weaning time.
- Asses the effects of adding 3 % of milk powder to the diet.
- Add 2500 g/tonne zinc oxide to the diet.
- Consider acidification of the diet by adding 22 g/litre citric acid to the water supply.
- Increase the vitamin E to 250 iu/kg of feed to enhance the natural immunity.

Pasteurellosis

Pasteurella multocida bacteria are commonly involved in respiratory disease in pigs and they may be toxin-producing or non-toxin-producing strains. Either can cause pneumonia in their own right but the non-toxin ones are common secondary opportunist invaders associated with primary mycoplasma (enzootic) pneumonia or PRRSV or SIV infections. *Pasteurella multocida* type A causes pneumonia. PMWS has accentuated the role of *Pasteurella* in growing and finishing pigs.

Clinical signs

Acute disease

This is characterised by severe sudden pneumonia affecting all the lung tissue, high temperatures and high mortality. Pigs show rapid breathing and discoloured skin, particularly on the extremities of the ears, and the pigs are toxic.

Sub-acute disease

In this form the pneumonia is less severe but often complicated by pericarditis (heart sac inflammation) and pleurisy. Coughing and emaciation are also common clinical features. The condition usually affects pigs between 10 and 18 weeks of age.

Diagnosis
This is carried out by post-mortem examination and isolation of the organism from the lungs.

Treatment
- Because the organism is usually secondary to a more specific disease, antibiotic treatments should follow as for mycoplasma (enzootic) pneumonia.

Management control and prevention
- Carry out procedures as described under respiratory diseases and control strategies.
- Vaccines are available but are not very effective.
- *Mycoplasma hyopneumoniae* vaccination often prevents the *Pasteurella* invading the lungs.

Pleurisy
Pleurisy is a common result of pathogens/foreign material gaining access to the pleural cavity. Organisms are moved to the pleural cavity as part of the respiratory system's defence mechanisms.

It is not possible to differentiate the cause of the pleurisy by the gross examination of the carcass. The lesion is a healing scar tissue and generally sterile.

Pleurisy can be present for several months before resolution. Therefore, pleurisy seen in the finisher slaughterhouse may have occurred at any time in the last 5 months, i.e. the entire life of the pig.

Causes
Many infectious organisms, particularly bacteria, can cause pleurisy, including *Actinobacillus suis*, App, Hps, *Pasteurella* and Streptococci. *Ascaris suum* and *Metastrongylus apri* may rarely be associated with pleurisy. During infection, the pleura (the lining that surrounds the lungs and the chest cavity) become inflamed and rub together, potentially causing pain when breathing. Triggers can include vice (ear sucking, flank biting, tail biting et al.) and respiratory infections.

Clinical signs
Acute disease
With acute pleurisy the animal may appear reluctant to move and when examined closely, often painful abdominal breathing is noticed.

Diagnosis
Identification of pleurisy on the surface of the lung or attached to the carcass rib cage at slaughter.

Post-mortem findings are generally associated with a variety of pleurisy-causing pathogens, including:
- *Actinobacillus pleuropneumoniae.*
- *Actinobacillus suis.*
- *Haemophilus parasuis.*
- *Mycoplasma hyorhinis.*
- Septic pleuritis – pseudomonas – from a puncture wound.
- Streptococcus.
- Complicated enzootic pneumonia – *Pasteurella multocida.*
- *Trueperella pyogenes.*

Pleurisy can also be associated with other pathogens, for example:
- *Escherichia coli.*
- Swine influenza – localized areas.

Treatment
- There is no specific treatment for pleurisy. The lesion needs time to heal.

Management control and prevention
- Review causes of vice – fighting over feed, fighting over water.
- Review the flooring – rough floors – note new floors can be particularly problematic.
- Review postweaning housing and stress factors – especially air and water supplies.
- Control Glässer's disease.
- Control clinical actinobacillus pleuropneumonia. The majority of farms are positive to at least one serotype of App – but do not have any significant pleurisy issue.
- Control PRRSV and PCV2 on the farm.
- Avoid draughts.
- Poor pig flow.
- Remove any sharp objects within pig height.

Porcine Dermatitis and Nephropathy Syndrome (PDNS)
PDNS generally affects growing and finishing pigs, although it can occasionally afflict adult pigs.

There is an association with PCV2 and PMWS. However, rare cases of PDNS have been recognised for many years. With PMWS the incidence of the condition becomes greatly increased postweaning.

The cause is unknown. However, current thoughts suggest it is an immune-complex-mediated disease associated with abnormal stimulation of the immune system. Lesions in affected pigs present in the blood vessel walls throughout and if viewed microscopically suggest a hypersensitivity reaction to something in the bloodstream, possibly a bacterial toxin.

Clinical signs
PDNS occurs mainly in growers and finishers of 12 to 16 weeks of age, although it can occur sporadically in

other age groups. Within a herd, clinical signs may occur in a few pigs and the disease may then go undiagnosed. Alternatively, signs may occur in a bigger proportion of the herd and become economically damaging.

Growing pigs, generally above 40 kg (90 lb), will present with patches of purple discolouration of the rear quarters which may extend down the hind legs. The scrotal area is particularly affected. Pigs will often be in good body condition and may die a few days after showing signs. If not, they tend to start to recover after 7 to 10 days, although remain unthrifty.

Mortality in affected pigs may be around 15 %. However, mortality can rise much higher and in one study it reached 100 % in pigs aged 3 months and older. The majority of pigs that develop extensive skin blotching die. Pigs that recover are often permanently unthrifty. Affected pigs are reluctant to eat and also show anorexia, depression, incoordination and stiffness, a reluctance to move and a mild fever. A number may suffer with diarrhoea. Haemorrhages are seen in the skin, lymph glands, kidneys and spleen, hence the similarity to swine fever. Oedema or fluid may be seen on the limbs and around the eyelids.

Diagnosis

The clinical signs are strongly suggestive but not diagnostic; thus gross and microscopic post-mortem examinations are needed to make a firm diagnosis.

At gross post-mortem examination, lymph nodes, particularly those at the rear of the abdomen which are not usually examined, are reddened, enlarged and haemorrhagic with fluid. There is often fluid in the abdomen.

Consistent lesions are seen in the kidneys which are swollen, pale and mottled with many small haemorrhages showing through the surface. Gastric ulcers and haemorrhage in the small and large intestines will also be seen.

Tests can be conducted for high urea and creatinine levels in the blood, which indicate severe kidney damage. These tests may be negative if the kidneys are less severely affected. Microscopically, the lesions in the blood vessel walls are distinctive.

Since the cause is unknown, there are no specific diagnostic tests.

Similar diseases

Some outbreaks, clinically and at post-mortem examination resemble classical swine fever (CSF) (hog cholera), or African swine fever (ASF), which in most countries are legally notifiable and, if confirmed, herds are slaughtered out. If many pigs with PDNS are subjected to detailed post-mortem examination, virtually all the typical gross lesions of classical swine fever and African swine fever are likely to be found, not in a single pig but scattered through them.

Fortunately laboratory tests for CSF are rapid and accurate but tests for ASF may take several days to confirm a negative result.

Other diseases which might be confused with PDNS include erysipelas and *Actinobacillus suis*. Other kidney conditions may also be confused with PDNS.

Treatment

- Efforts to treat PDNS have proved unsuccessful. Antibacterial medication is usually ineffective. Immediate removal of affected pigs to a well-bedded warm hospital pen can improve the chances of survival.

Management control and prevention

- The vectors that are responsible for triggering porcine dermatitis and nephropathy syndrome are not known, therefore it is not possible to accurately quantify any measures that might prevent or control PDNS.

Porcine Epidemic Diarrhoea (PED)

See Chapter 8 for additional information in the piglet

PED is caused by a coronavirus infection of the small intestine. It cycles sometimes in weaned pigs in herds which have become immune because the protection of the maternal IgA disappears after weaning.

This condition can cause diarrhoea in adults. The clinical signs are very similar to transmissible gastroenteritis (TGE).

The disease is characterised by a sudden profuse watery diarrhoea that will last for 3 to 4 days and occurs when pigs are moved into environments where older pigs have succumbed to the disease, shed the virus and recovered. When the virus is first introduced on to the farm there is a rapid spread of diarrhoea across all breeding and growing pigs with almost 100 % morbidity within 5 to 10 days. The incubation period is 2 to 4 days. Type 1 virus causes diarrhoea in growing pigs and adults only; type 2 virus causes diarrhoea in piglets as well.

Clinical signs

There is an acute watery diarrhoea and no evidence of blood or mucus. In breeding to finishing farms, disease is usually sporadic, however PED is common where weaners are continually entering finishing-only operations. Groups of pigs become infected when they reach a certain age and as they enter a building where infection is endemic. Mortality is usually low but morbidity can be high, especially in Asia.

Diagnosis

This is based on the history, clinical symptoms and examinations of faeces samples for evidence of porcine epidemic diarrhoea virus by ELISA tests or electron microscopy.

Post-mortem examination of dead pigs and laboratory tests on the small intestine may be necessary to confirm the diagnosis. When you examine the small intestine, check the pH of the contents. In cases of PED, the contents will be acidic.

Similar diseases

TGE could give a similar picture, and live affected pigs are best submitted to a laboratory for differential tests. Rotavirus can cause a similar, but mild, disease.

Treatment

- The growing pig normally recovers without treatment unless there are secondary infections. In such cases antibiotics in the water or preventative medication in-feed maybe required.
- Use neomycin, apramycin, framycetin or trimethoprim/sulpha. Sometimes a good response is obtained with either lincomycin or tiamulin depending on the secondary bacteria present.
- Specific treatment is of no value since this is a virus infection.

Management control and prevention

- The disease may occasionally become endemic in finishing units as new weaners are introduced onto the farm. Under such circumstances it is necessary to break the cycle by stopping purchasing for 3 weeks or utilising partial depopulation methods (see Chapter 3).
- Batch, all-in/all-out procedures with disinfection will often break the cycle.
- The virus is easily killed by phenolic-, chlorine- or iodine-based disinfectants or peroxides.
- It is essential to develop immunity in the gestation/dry sows as soon as possible.
- There are 2 methods: either squeeze the piglet's abdomen and collect the diarrhoea into a bowl, or use sawdust or shavings in the areas where the piglets are scouring. Paper towels can also be used to soak up piglet faeces. This material is then mixed w[illegible]et of water and fed to the pregnant so[illegible]back) up to Day 90 of pregnancy[illegible]ot feed beyond [illegible] 90 as this can make [illegible]e problem worse [illegible]mals.
- A further method is to collect t[illegible]testines from a number of pigs that have died and macerate them in a food blending machine. The liquid provides a rich source of virus and this can, if required, be preserved by deep freezing.
- The disease should be spre[illegible]s soon as possible across the whole farm. The object is to get a good immunity developed in the shortest possible period of time. It will take approximately 3 to 4 weeks to achieve this, as this is the time it will take for colostrum levels to build up to a protective concentration.
- Ensure that all gilts in isolation/acclimatisation receive feedback diarrhoea from the farrowing house and faeces from the on-site nursery.
- Once the infected period is over, ensure an all-in/all-out management system of the farrowing houses, weaner and finisher accommodation.
- The cleaning process that all-in/all-out allows is important to ensure the virus does not linger on the farm and become endemic.

Porcine Reproductive and Respiratory Syndrome Virus (PRRSV)

The information in this chapter relates specifically to PRRSV in the weaner, grower and finisher periods. The main information on PRRSV, including methods of spread, immunity, diagnosis and elimination from a herd are covered in Chapter 6, with additional information in Chapters 7, 8, and 10. These chapters should be read to fully understand how this disease impacts the whole herd.

Clinical signs in the weaner, grower and finisher periods

Acute disease

When introduced first into a mycoplasma (enzootic) pneumonia and actinobacillus pleuropneumonia (App)-free growing herd there is usually a period of slight inappetence and mild coughing but in some herds there are no symptoms at all. If mycoplasma (enzootic) pneumonia and/or virulent App are present in the herd however, clinical signs may become severe with an acute extensive consolidating pneumonia and the gradual formation of multiple abscesses. Disease becomes evident within 1 to 3 weeks of weaning; pigs lose condition with pale skin, mild coughing, sneezing and increased respiratory rates. Mortality during this period may reach 12 to 15 %.

Chronic disease

Once the acute period of disease has passed through the breeding and finishing herd, PRRSV virus normally only becomes of significance in the early growing period, where severe endemic pneumonia can persist with periods of inappetence and wasting of pigs. Pigs become infected as maternal antibody disappears and then remain viraemic for 3 to 4 weeks continually excreting virus. Permanently populated houses maintain the virus at high levels, particularly in the first and second stage accommodation. Clinical disease is seen in pigs from 4 to 12 weeks of age and it is characterised by a fairly predictable time of onset, inappetence, malabsorption and wasting, coughing and pneumonia. In this postweaning period, mortality rates can rise to 12 % or more and persist in spite of antibiotic treatments. Secondary bacterial infections become evident in pigs at a later stage, from 12 to 16 weeks of age, from abscesses that develop in the lungs. These infections spread to other parts of the body, particularly joints with increased lameness.

Chronic respiratory disease

Chronic respiratory disease is caused by a combination of respiratory pathogens including PRRSV, SIV, PMWS, *Mycoplasma hyopneumoniae*, App, *Haemophilus parasuis*, *Pasteurella* and streptococci bacteria, particularly as they become additive. The diagnosis is then of multiple causes and individual pigs within the farm can have widely differing pathogen profiles.

The following technique has proved to be of considerable value in controlling chronic respiratory disease associated with PRRSV. The objective is control and not elimination. The essential components are:

- Removal of pigs at weaning (more than 21 days of age) on a weekly batch basis to totally separate housing for a period of 8 weeks.
- Vaccination of piglets at 1 and 3 weeks of age against *Mycoplasma hyopneumoniae*.
- Removal of all remaining weaners/growers/finishers.
- Adaptation of buildings to all-in/all-out and no droplet contact between batches.
- Return of the segregated pigs.
- Control/eradication of disease in the sow herd by vaccination/medication.

Treatment

- In the acute disease when PRRSV first enters the farm it is important to cover the period at risk, which is usually 6 to 8 weeks, with in-feed antibiotics or by individual injections and water medication.
- The broad spectrum antibiotics, tetracyclines, trimethoprim/sulpha, or synthetic penicillins are the medicines of choice. If mycoplasma (enzootic) pneumonia alone is involved, tiamulin or lincomycin may be used. If App is active, ceftiofur could be a medicine of choice for individual treatments.
- Tiamulin reduces the uptake of the PRRSV into the alveolar macrophage and may be useful in reducing the severity of an acute outbreak.
- In endemic disease, preventative medication over the period at risk, using 500 to 800 g/tonne of tetracycline or trimethoprim/sulpha at 400 g/tonne in-feed may be used but it would be advisable to identify the major bacteria involved and determine their antibiotic sensitivities.

Management control and prevention
See also Respiratory Diseases and Control Strategies in this chapter

- Consider using early weaning off-site to break the endemic cycle of disease.
- Refer to the principles of SEW and partial depopulation in Chapter 3.
- Section all the buildings so that they can be managed on an all-in/all-out basis and clean and disinfect between each batch.
  It is most important to adopt this principle at the onset of disease to prevent endemic infection becoming established.
- Review pig flow principles and avoid over- and under-stocking buildings.
- Consider the directions of pig movement around the farm. Change these to reduce droplet contamination from older pigs to younger pigs.
- Consider batch farrowing once every 2 to 4 weeks, thus giving an age break between groups and houses. This can be highly effective at reducing clinical signs.
- Consider depopulation of the first and second stage flat decks where virus is active. Before considering this step, however, first check by serology that the virus is no longer circulating in the breeding and finishing herds. Elimination of virus has been successfully carried out using this technique on a number of farms. The first and second stage houses are depopulated, washed with hot water and detergent, and left empty for 2 weeks. Pigs are weaned away from the farm whilst this is being carried out and then newly weaned pigs are introduced back to the nursery.
- Live vaccines are available that can be administered by intramuscular injection to the pig at weaning time. Note commercial vaccines will not protect against all serotypes of PRRSV.
- Dead vaccines are available but they generally offer little or no protection to naive animals. However, once a pig is infected these vaccines can boost the immune response and reduce the shedding time, which make them very useful in gilt introduction and acclimatisation programmes.
- Autogenous vaccines may be obtained via tonsillar scrapes and serum viruses. Autogenous vaccines have the advantage of using the strains present on the farm, rather than possibly introducing a new viral construct.

PRRSV-negative farms

The following points are vital in maintaining a PRRSV-negative status in weaning to finishing herds:

- It is essential that the farm has excellent biosecurity programmes.
- Do not use live vaccines in the herd.
- Ensure that off-farm boots are removed prior to visitors entering the farm.

Porcine Respiratory Coronavirus (PRCV)

PRCV first appeared in pigs in the 1980s in Europe. It is related to but distinct from transmissible gastroenteritis virus (TGE), which is another coronavirus.

PRCV is respiratory-spread and believed to travel long distances and because of this it is extremely difficult to maintain herds free from it and very few countries have not been exposed.

Clinically, it is almost non-pathogenic and field experiences have shown that herds exposed for the first time have few (if any) signs of disease.

It has been suggested however that it may have an effect on lung tissue when other respiratory pathogens are present in chronic respiratory disease complexes.

PRCV does however cross react with the serological test for TGE and it therefore can confuse the diagnosis. A differential test is available.

Porcine Stress Syndrome (PSS)

This term covers a group of conditions associated with an autosomal recessive gene. It includes acute stress and sudden death (malignant hyperthermia), pale soft exudative muscle (PSE), dark firm dry meat, and back muscle necrosis. Heavy muscle pigs are more likely to carry the gene. In suceptable pigs the gene is either homozygous-recessive or heterozygous-recessive.

The gene can be identified by the pig's response to the anaesthetic gas halothane but recent developments have produced gene probes using blood that identify both the homozygous and heterozygous carriers.

Clinical signs

When the homozygous state is present and following a period of muscle activity, there is a change in muscle metabolism from aerobic to anaerobic and biochemical abnormalities develop. The body tissues become acid with a marked rise in temperature to 42 °C (107 °F).

The onset is sudden with muscle tremors, twitching of the face and rapid respiration. The skin becomes red and blotched. Death usually occurs within 15 to 20 minutes. PSS is often precipitated by sudden movement. Back muscle necrosis is a more localised form. Whilst the gene produces a leaner carcass, growth rates are slower and the levels of sudden death increase.

The condition is associated with an imbalance of calcium (Ca) movement through the cell membrane.

Diagnosis

This is based on the sudden onset, symptoms, breed, susceptibility and the known presence or absence of the gene in the pig.

Examine the pig for the gene. Note that when taking samples for DNA analysis, always use a disposable system and a new needle.

In many cases the pig is just found dead and a post-mortem examination is necessary to eliminate other diseases. Rigor mortis (stiffening of the muscles after death) within 5 minutes is a striking feature.

Pigs dying during transportation to the slaughterhouse is not uncommon.

Similar diseases

These include the other causes of sudden death: twisted bowel, internal haemorrhage, mulberry heart disease and pyelonephritis. Hypocalcaemia in the lactating sow, although uncommon, can give identical symptoms to PSS.

Treatment

This is usually ineffective but the following are worth adopting:

- Spray the pig with cold water to control the temperature rises.
- Inject 50 to 100 ml of calcium gluconate (used in cows for milk fever) by intramuscular injections at 2 separate sites. Seek veterinary advice.
- Sedate the pig.
- Do not move or cause undue muscle activity.
- Give an injection of vitamin E 2 iu/kg.
- Dantrolene is the medicine of choice, but is very expensive and has a very limited shelf-life.

Management control and prevention

- Remove the gene from the population.
- Use a homozygous or heterozygous male on stress gene-free females if the gene is to be used to improve carcass quality.
- Maintain a gene-free herd.

Postweaning Ill-thrift Syndrome (Peri-Weaning Failure to Thrive Syndrome (PFTS))

This field condition seems to be the result of postweaning piglets failing to learn to eat and drink and then starving, sometimes to death. The condition occurs immediately postweaning and does not appear to be dependent on weaning age although is more common in piglets weaned before 17 days of age.

Clinical signs

Affected pigs fail to gain weight and become severely emaciated 10 to 15 days postweaning. The weaners are gaunt, dehydrated and lethargic, and often show signs of incoordination. The weaners may appear very hairy. Affected weaners often exhibit signs of vice – penile sucking, sham nursing. Unaffected weaners look normal.

Diagnosis

The clinical condition of the weaners is indicative. Postmortem is needed to diagnose and is indicative of the pig not eating. In the classic case, the stomach and small intestines will be empty of food. The stomach may be filled with fluid and possibly just straw (if housed on bedding), although if the weaner has recently figured out how to eat, signs of feed may be present. There is an absence of body fat and the superficial inguinal lymph nodes may be more prominent. The liver may be pale. Histological changes in the small intestine are typical of starvation and include severe villus atrophy and fusion.

Similar diseases

The sign are very similar to PMWS/PCVAD but this normally starts at an older age, around 15 kg in weight. Pigs with PMWS will also have learned to eat postweaning.

Treatment

- Gruel feeding.
- Ensure that gruel feeding does not continue beyond day 5 postweaning or a double weaning effect will occur.
- Feeding creep feed pre-weaning appears to have little impact on the progression of the condition as the pigs do not eat.

Management control and prevention

- Improve management of the postweaning period.
- Examine pig flow and weaning age. Increase weaning age if possible.

Postweaning Multisystemic Wasting Syndrome (PMWS)

Porcine circovirus diseases (PCVD)

PMWS is manifest as the name implies by wasting in pigs from 5 weeks of age to around 14 weeks.

The disease is associated in part with a porcine circovirus 2 (PCV2), so called because its DNA is in the form of a ring. It is an extremely small and hardy non-enveloped virus containing made up of a single-stranded, circular DNA. There are 2 serotypes of porcine circovirus: type 1 which causes no known disease: and type 2 which is associated with a number of diseases/syndromes in pigs including reproductive failure, postweaning multisystemic wasting syndrome (PMWS), porcine dermatitis and nephropathy syndrome (PDNS), and proliferative and necrotizing pneumonia (PNP). PCV2 can be found in the lesions of affected pigs and can be isolated in pure culture.

PCV2 is widely present through the global pig population, but only a small proportion of herds where the virus is present have a history of clinical disease. It seems that most infections are sub-clinical. It is believed that the main mode of transmission is via direct infected pig-to-pig contact.

Clinical signs

Weaners and growers

PMWS tends to be a slow and progressive disease with a resulting high fatality rate in affected pigs.

Signs start around 6 to 8 weeks of age; weaned pigs lose weight and gradually become emaciated. Their hair becomes rough, their skins become pale and sometimes jaundiced, and they are inappetent.

Postweaning mortality is likely to rise to 6 to 10 % but is sometimes much higher (30 % or more). In older pigs, mortality can rise to 10 %. Clinical cases may keep occurring in a herd over many months – 18 months or more with numbers affected varying from one group to another. Affected pigs do not respond to any treatment, but growth rates in unaffected pigs tend to be normal.

It is not unusual for cases of porcine dermatitis nephropathy syndrome (PDNS) to be seen in herds affected with PMWS.

PCV2 may also be associated with a granulomatous enteritis which may look very similar to ileitis.

Other signs include:

- Respiratory distress or laboured breathing caused by interstitial pneumonia.
- Discoloured ears.
- Incoordination.
- Occasionally, nervous signs may be seen.
- Diarrhoea may be a problem in 30+ % of cases.
- Good pigs are often found dead.

Sows and piglets

Mature animals, sows, boars and sucking piglets are not affected by PMWS and it is uncommon for newly weaned pigs to be affected before 6 weeks of age. PCV2 which crosses the placenta and affects the developing foetus can be associated with mummified, stillborn and weak piglets – it is part of the SMEDI complex of diseases. Foetal death may be associated with a myocarditis.

Causes/contributing factors

The following factors are believed to have an impact on the spread and severity of the disease:

- Exposure to infected faeces.
- Contaminated clothing, equipment, trucks etc.
- Birds and rodents.
- Mixing and stress.
- Continual production methods.
- High stocking densities.
- Spread through semen (AI).

Diagnosis

Since most herds have antibodies to PCV2, blood-testing a herd usually does not help.

The clinical signs are not specific (although the picture may be highly suggestive and include signs of wasting or ill-thrift) and to confirm a diagnosis it is often necessary to post-mortem several pigs.

Diagnosis is based upon the presence of PCV2 histological lesions in lung, tonsil, spleen, liver and kidney tissues. Immunohistochemistry is used to demonstrate PCV2 in tissues.

Post-mortem findings

- The carcass is emaciated and may be jaundiced.
- The spleen and many lymph nodes are usually much enlarged. However, the clinical picture with enlarged lymph glands is highly suspicious. The fat around lymph nodes may atrophy, making the lymph node more prominent.
- Kidneys may be swollen, with white spots visible from the surface.
- The gross post-mortem lesions are variable.
- The lungs may be rubbery and mottled with oedema. Microscopically these lesions are characteristic and diagnostic particularly if the circovirus is demonstrated in them. If affected pigs are suspended by their back legs, the inguinal lymph nodes appear enlarged, often the size of large grapes.
- Oedema or fluid may be seen in the chest and abdominal organs and tissues.

Similar diseases

Many conditions, such as starvation, malnutrition, lack of water, gastric ulcers, mycoplasma (enzootic) pneumonia, coliform enteritis, swine dysentery, PRRSV and other diseases can cause similar signs. These all have to be eliminated if a specific diagnosis of PMWS is to be made.

Treatment

There is no specific treatment for PMWS/PCVD. The diseases is managed through the use of vaccination and good husbandry.

Management control and prevention

The control of PMWS/PCVD is based on the use of PCV2 vaccines. There are 2 options, one includes vaccinating the sow, who passes immunity to the piglets. The other mode is to inject the piglets at around 2 to 3 weeks of age. In addition to significantly reducing mortality and wasting, vaccination often improves feed conversion, average daily gain and uniformity of growth.

Prior to the advent of vaccines, effective control measures focused on a number of key management procedures. These included limiting pig-to-pig contact, excellent hygiene, good nutrition and minimising stress, both physical and pathogenic. To this end, close attention was paid to preventing bacterial infections and controlling viral infections, especially porcine reproductive and respiratory syndrome (PRRSV).

Postweaning Sneezing

This condition is caused by multiple concurrent low-level bacterial and viral infections in the growing pig. Mixing of the 'normal' microflora present in the nasopharynx (nasal passages) between different litters of piglets at weaning can trigger the condition. The condition may be seen in environmentally controlled and straw buildings. The condition is spread via nasal contact. Mortality is minimal.

The pathogens involved may include:

Bacteria

- *Pasteurella multocida.*
- *Bordetella bronchiseptica.*
- *Haemophilus parasuis.*
- *Actinobacillus pleuropneumoniae.*
- *Actinobacillus suis.*
- Streptococci spp.
- Pseudomonas spp.
- Proteus and other environmentally originated bacteria.
- Various mycoplasmas.
- Chlamydia is involved in the conjunctivitis.

Viruses

- Inclusion body rhinitis.
- PRRSV.

Clinical signs

Signs typically seen are a group of weaned pigs that start with mild to severe sneezing, which may progress to middle ear disease in some of the animals. Conjunctivitis may be seen in several pigs and this can be severe in some. The symptoms will progressively reduce within 2 to 3 weeks. A reduction in feed conversion and growth rates is likely to result.

Diagnosis
This condition is identified by the clinical picture of sneezing pigs with conjunctivitis.

Similar diseases
This condition has similarities with PAR, although evidence of facial distortion will also be present.

SIV may also show similar signs, although SIV normally presents in older weaners.

Treatment
The condition is the result of mixed infection thus specific treatments, whilst supportive, are generally unrewarding. Postweaning sneezing is considered almost 'normal' on most farms.

Management control and prevention
- Practise all-in/all-out management.
- Maintain high standards of hygiene to minimise pathogen load.
- Review management of the farm building. Avoid chilling and drafts.
- Review your PRRSV stabilisation programme.
- Combine litters pre-weaning to raise immunity levels and enhance socialisation of the piglets pre-weaning.
- If sneezing is also in farrowing house, review atrophic rhinitis controls.

Progressive Atrophic Rhinitis (PAR)
See Chapter 8 for further information
Rhinitis is inflammation of the tissues inside the nose and in its mild form it is very common. The term atrophy indicates that the tissues inside the nose, which become infected or damaged, shrink and become distorted. There are 2 forms of the disease, the more serious being PAR (progressive atrophic rhinitis) which is associated with the presence of toxin-producing strains of a bacterium *Pasteurella multocida*.

PAR is a serious condition in both sucking and growing pigs and is covered in Chapter 8 under Diseases of the Sucking Pig.

Prolapse of the Rectum
See Chapter 14 for further information
This is a widespread condition occurring in good growing pigs from 8 to 20 weeks of age and is not uncommon in sows. The onset is sudden. The size of the prolapse varies from 10 to 80 mm and if small it will often revert to the rectum spontaneously. In most cases, however, the prolapse remains to the exterior and is often cannibalised by other pigs in the pen as evident by blood on the noses of the offending pigs and on the flanks of others. The fundamental cause is an increase in abdominal pressure which forces the rectum to the exterior.

Whilst the exact mechanisms are not fully understood, the following should be considered as contributory to the problem.
- A prolapse may occur following oestrus, associated with levels of oestrogenic hormones that are present at this time. Note Zearalenone mycotoxicosis may be implicated.
- It may be associated with constipation.
- Penetration of the rectum at mating is a common cause with prolapse occurring 24 to 48 hours later.
- Cases develop if sows are confined in houses where there is an excessive slope towards the back of the floor. Up to 8 % of sows have been affected where sows are confined to houses with sloping floors to the rear.
- Rectal prolapses are seen occasionally in sow pens where the retaining gate at the back consists of parallel bars. If these are of such a height that the sow can sit or rest with the tail over the back, pressure is placed on the anal sphincter. This causes a partial relaxation of the sphincter itself, poor circulation, swelling and ultimately the sow strains to prolapse.
- A small lying area with a step down to the defecating area causes increased abdominal pressure if the sows lay over it. This predisposes to prolapse.
- Prolapsed rectum may occur whenever there is an increase in abdominal pressure.
- Abnormal fermentation in the gut and the production of gas in the large bowel may predispose. In such cases the components of the feed and the method of feeding should be investigated.
- Mouldy feed or straw can be important causes of rectal prolapses due the presence of mycotoxins.
- Low-fibre diets can lead to constipation and rectal prolapse.
- High-wheat diets may create a higher prevalence. Enzymes may be added to the diet to alleviate the problem.
- When environmental temperatures drop, sows that are loose-housed group together to keep warm, thus increasing abdominal pressure.
- A water shortage may predispose.
- There is no evidence to suggest that genetic factors have a part to play in the disease.

Clinical signs
At the onset, the red coloured mucosa of the rectum protrudes from the anal sphincter and then may return

on its own. After a short period however it remains to the exterior and becomes swollen and filled with fluid. It is prone to damage and haemorrhage and where pigs are loose-housed, cannibalism often results, with evidence of blood on the skin. Pigs under 60 kg (132 lbs) in weight that develop a rectal prolapse and subsequently recover may progress to a rectal stricture.

Treatment

- Rectal prolapses must be recognised early and the pig removed from the pen.
- Replace the prolapse and retain it by a purse string or mattress suture. Return the pig to the pen. The technique for carrying this out is described in Chapter 15.
- Sometimes the prolapse is very swollen and it is necessary to gradually reduce its size by gentle pressure using hands covered in obstetrical lubricant. This can sometimes take up to 15 minutes.
- If the prolapse has been badly torn, still replace it, and consider moving the pig to a hospital pen and treating it with a long-acting antibiotic injection. In a proportion of pigs, the damaged tissues become scarred with constriction leading to rectal strictures. The incidence of this is reduced by replacing the prolapse and suturing.
- In some cases, the prolapse will be completely bitten off by other pigs. Here the pig should be left in the pen as most cases will progress to slaughter, although a few will develop with rectal strictures.

Management control and prevention

The following may be considered as causal or contributory when adopting control measures.

Disease

- Diarrhoea – excessive straining.
- Respiratory disease – excessive coughing, increasing abdominal pressure.
- Colitis – abnormal fermentation occurs in the large bowel with the production of excessive gas, increasing abdominal pressure.

The environment

- In cold weather the incidence of rectal prolapse increases. This is associated with low house temperatures and the tendency of pigs to huddle together, thus increasing abdominal pressure.
- Wet conditions and slippery floors, particularly those with no bedding, increase abdominal pressure.
- If stocking densities reach the level whereby pigs cannot lay out on their sides across the pen, the incidence may increase.
- Remove draughts in the pigs' sleeping area.

Nutrition

- Ad lib feeding – Feeding pigs to appetite results in continual heavy gut fill and indigestion. There is then a tendency for abnormal fermentation in the large bowel because undigested components of the feed arrive in greater amounts.
- High-density diets and in particular lysine levels increase growth rates and outbreaks may often subside either by a change to restricted feeding or using a lower energy/lysine diet.
- Water shortage – This can lead to constipation.
- Diets high in starch may predispose to prolapse – Try adding 2 to 4 % grass meal to the diet.
- The presence of mycotoxins in-feed – If there is a problem make sure that the bins have been well cleaned out. Examine the cereal sources.
- Change of diet – By studying the timing of the problem it is sometimes possible to identify rectal prolapses, not only with a change of diet but also a change of housing.
- There is an association with the concentration of wheat in the diet, particularly if the wheat is milled by a hammer mill. Consider changing to a rolling system.
- Field evidence does not identify breed as a causal factor.
- Increases in rectal prolapses have been reported in association with the use of tylosin but the evidence for this is unclear.
- Trauma.
- Tail docking – Docking tails too short can damage the nerve supply to the anal ring leading to a relaxation of the anal sphincter which can predispose.

Identify those factors on your farm from the above list and make changes where appropriate.

Consider tagging affected pigs and collecting the following information about each prolapse to see if common factors emerge:

- Age of pig.
- Comments and observations.
- Days in the house.
- Diet fed.
- House and pen.
- Number of pigs per pen
- Number of rectal strictures.
- Number of prolapses sutured.
- Outcome.
- State of prolapse.
- Tail biting.
- Weight of pig.

Nutritional aspects associated with rectal prolapse are covered in Chapter 14.

Rectal Stricture

This is a condition often considered to be a sequel to rectal prolapse. Approximately a finger's length inside the rectum the tissues gradually shrink, scar tissue develops and eventually the tube completely closes. Affected pigs in the early stage of the disease often show very loose diarrhoea that becomes projectile and a gradual increase in the size of the abdomen, with loss of condition.

The area where the stricture occurs is supplied by 2 tiny arteries that originate from the aorta. Some studies suggest that if these arteries are blocked or thrombosed by bacteria, a rectal stricture will result. Erysipelas, *Haemophilus parasuis*, streptococci and *Salmonella* have been implicated. If rectal strictures occur in large numbers at predictable times, consider infection as a cause and assess the effects of strategic medication by injection, water or in-feed. Possibly straining to defecate exposes the red internal mucosa of the rectum which is then bitten and a rectal stricture occurs as part of the scars during healing.

Treatment

- There is no treatment for this condition and as soon as affected pigs are recognised they should be destroyed on welfare grounds. Experience of trying to open up the stricture by palpation or surgery has been a total failure.

Management control and prevention

- Determine if there is a recurring time period when this first appears.
- Approximately 3 weeks prior to this look for trigger factors.
- If more than 2 % of growing pigs are affected apply in-feed medication at the predetermined time and assess the results. 500 g of OTC per tonne for 2 weeks may prevent the condition if an infection such as *Haemophilus parasuis* is the predisposing cause.
- Consider all the factors outlined for the control of rectal prolapses.
- Ear-tag rectal prolapses to see if they develop into strictures; often they do not.
- Replace all rectal prolapses immediately, suture and see whether this affects the incidence.

Respiratory Diseases and Control Strategies

Of all the diseases that affect growing and finishing pigs, chronic respiratory disease is the most economically important. It is extremely common and can be difficult to prevent and control. Growth rates and feed intake are depressed together with poor feed efficiency, and in some herds there is heavy mortality. The control of respiratory disease requires an understanding of the complexities and interaction between the organisms that are present, the pig and the management of the environment.

The prevalence of respiratory disease is affected by the following:

- The presence of respiratory pathogenic organisms.
- The virulence of the pathogens present.
- The level of the pathogens in the house environment.
- The immunity of the pig and the time of exposure to the organisms.
- The presence of secondary opportunistic bacteria.
- The interactions between management, environment, the diseases and the pig.

The presence of respiratory pathogenic organisms

The infectious agents that damage the respiratory system of the pig and the type of disease they cause are shown in Fig 9-28.

The severity of the disease will depend in part upon the number of infective agents present and the weights of infection that challenge the pig. PRCV is a relatively mild, almost non clinical infection.

PRRSV likewise on its own has little clinical effect on the respiratory system, but if it is combined with other pathogens, for example swine influenza, severe respiratory disease will be experienced.

Under good husbandry conditions mycoplasma (enzootic) pneumonia caused by *Mycoplasma hyopneumoniae* on its own is usually a relatively mild disease, but when complexed with PRRSV, SIV, PCV2 or *Pasteurella multocida* or combinations, serious chronic disease syndromes can develop.

Of the bacterial infections there are at least 15 serotypes (serovars) of *Actinobacillus pleuropneumoniae* but disease is dependent in part upon the virulence of the strains present, some causing no problems, others causing severe disease. Virulent serotypes vary from country to country. Individual strains within serotypes also vary. The chronic respiratory disease syndrome is summarised in Fig 9-29. The greater the variety of virulent organisms in the population of pigs, the more complex the picture. It is helpful when investigating a problem to determine which of these organisms are present.

FIG 9-28: INFECTIONS THAT CAUSE RESPIRATORY DISEASE

Diseases	Effects
Caused by viruses	
Aujeszky's disease/pseudorabies virus (AD/PRV)	Nervous symptoms, pneumonia, (reproductive failure in sows).
Porcine cytomegalovirus (PCMV) (Inclusion body rhinitis)	Non progressive rhinitis, sneezing in newborn piglets.
Porcine reproductive and respiratory syndrome virus (PRRSV)	Pneumonia, (reproductive failure in sows).
Porcine respiratory corona virus (PRCV)	Pneumonia (very mild), transient coughing.
Swine influenza virus (SIV)	Fever, generalised illness, pneumonia.
Classical swine fever (CSF)/hog cholera (HC)	Infertility, pneumonia.
Porcine circovirus associated diseases	
Postweaning multisystemic wasting syndrome (PMWS)	Wasting and immunity deficiency.
Caused by mycoplasma	
Mycoplasma (enzootic) pneumonia	Pneumonia.
Mycoplasma hyorhinis	Fever, heart sac infection, lameness, mild pneumonia, pleurisy, polyserositis.
Caused by bacteria	
Actinobacillus pleuropneumonia (App), (Necrotic pleuropneumonia), (pleuropneumonia)	Acute necrotic and haemorrhagic pneumonia, pleurisy, sudden death.
Actinobacillus suis	Abscesses, lung nodules, pneumonia, septicaemia.
Bordetellosis (bordetella rhinitis)	Non-progressive rhinitis, pneumonia.
Glässer's disease (In association with *Mycoplasma hyorhinis, Streptococcus)*	Fever, heart sac infection, lameness, mild pneumonia, pleurisy, polyserositis.
Pasteurellosis (*Pasteurella multocida)*	Pneumonia.
Progressive atrophic rhinitis (PAR) Atrophic rhinitis (AR)	Pneumonia, progressive atrophic rhinitis.
Streptococcal pneumonia *Streptococcus suis* and other streptococci	Arthritis, meningitis, pneumonia.
Salmonellosis *Salmonella choleraesuis* (*Salmonella typhimurium*, and other *Salmonella*)	Diarrhoea, pneumonia.
Caused by parasites	
Large roundworm (*Ascaris suum*)	Acute pneumonia, coughing.
Lung worm *(Metastrongylus)*	Mild pneumonia, coughing.

Immunity

See Chapter 3 for further information

The piglet at birth receives antibodies and immunity via the colostrum. Not only do these antibodies provide protection against enteric diseases but also against other infections endemic on the farm. These levels of passive or maternal immunity reach their maximum in the piglet a few hours after birth and they then gradually disappear over the next 3 to 14 weeks (Fig 9-30). The actual time of this disappearance differs from one disease to another; a vital piece of information that is used to manipulate the management of the system and control disease.

When investigating a respiratory disease problem it is essential to identify the point at which disease first appears, because this highlights where management changes could reduce the effects of disease.

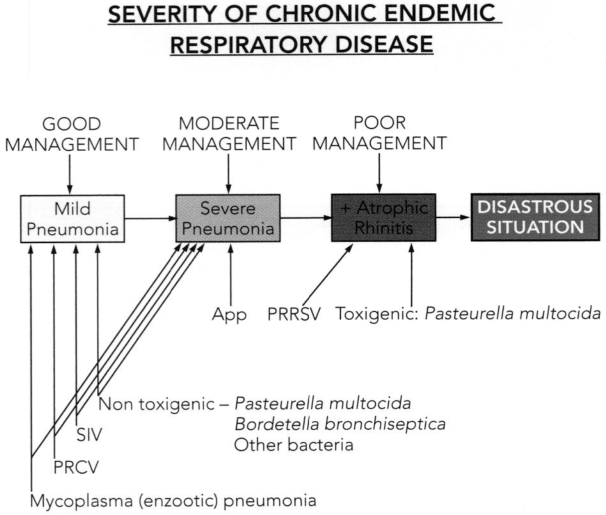

Fig 9-29

This biological fact can be used to control not only respiratory diseases but also many others, because if the piglet is removed from the sow at weaning to a totally clean environment the disease cycle may be broken. Fig 9-31 suggest the ages at which piglets would have to be weaned from the sow into isolated accommodation to prevent them becoming infected with different respiratory diseases. However in some herds these ages may be less because of the herd size and increased numbers of gilts which might be actively excreting organisms. Fig 9-32 also illustrates the similar ages for enteric disease for the sake of completeness. The ages given assume that the source herd contains only breeding females and sucking piglets.

Whilst the piglet is protected by maternal antibody it will not become affected with many of the infectious diseases endemic in the herd.

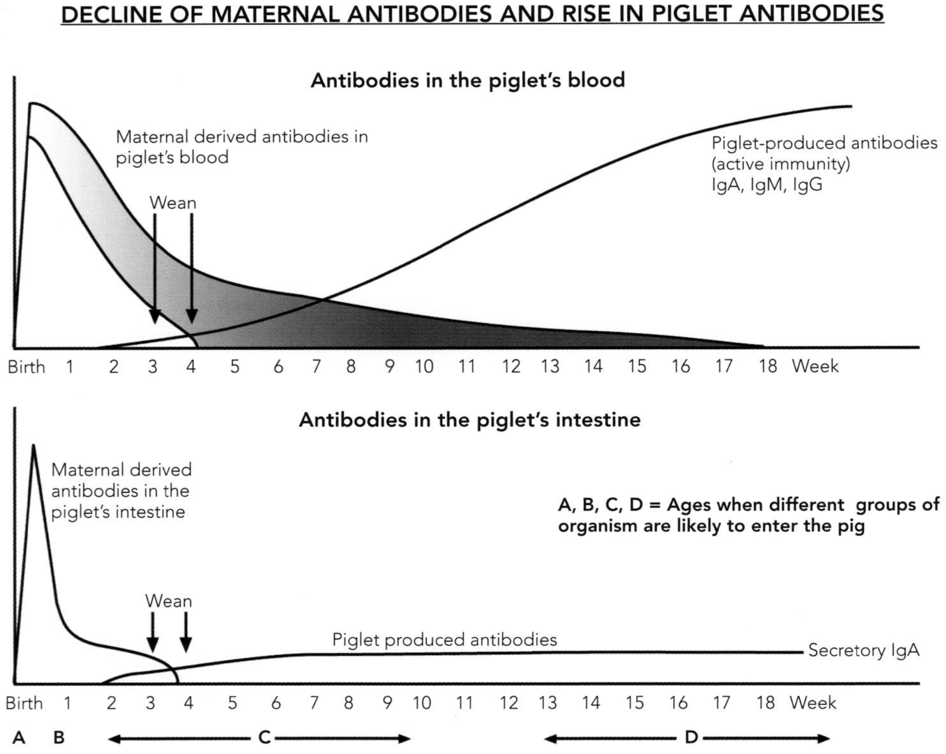

A = *E. coli*, *C. perfringens*, *S. hyicus*
B = Coccidia, *Streptococcus suis* types 1 & 7
C = Rotavirus, *streptococcus suis* types 2 & 14, App, AD/PRV, PRRSV, *M. hyopneumoniae*, *B. hyodysenteriae*, toxigenic *P. multicoccidia*, teschoviruses
D = Parvovirus (PPV), *M. hyosynoviae*, *Leptospira bratislava*

Fig 9-30

FIG 9-31: RESPIRATORY DISEASES

The Oldest Age at Which Pigs can be Weaned to a Segregated Site and be Reliably Free From Contamination by Pathogenic Organisms Endemic in the Herd

Infection / Disease	Age (Days)	Medication/Vaccination * for added safety
Actinobacillus pleuropneumoniae (App)	28	Medication + Vaccination
Aujeszky's disease/pseudorabies virus (AD/PRV)	21	Vaccination
Swine Influenza virus (SIV)	16	None
Mycoplasma hyopneumoniae	10	Medication + Vaccination
Pasteurella multocida (toxigenic) (PAR)	8	Medication + Vaccination
PRRSV	16	None
Salmonella choleraesuis	16	Medication

* Vaccination of the sow >2 weeks before farrowing.
Medication of the sow and/or piglets with an appropriate medicine against the organism.
Medication and vaccination are not always necessary but increase the reliability.

FIG 9-32: ENTERIC DISEASES

The Oldest Age at Which Pigs can be Weaned to a Segregated Site and be Free From Contamination by Pathogenic Organisms Endemic in the Herd

Infection / Disease	Age (Days)	Medication/Vaccination * of the sow for added safety
Coccidia	–	Not possible
E. coli	–	Not possible
Internal parasites	14	Medication
TGE	21	None

* Vaccination of the sow >2 weeks before farrowing.

Management control and prevention

Fig 9-33 shows the options that are available to control respiratory and other diseases. Obviously if the organism is not present on the farm there will be no disease. Disease freedom can be maintained at a national, country or farm level and it is therefore an important part of the control system.

Freedom from disease

- Whilst the piglet is protected by maternal antibody it will not be affected by many of the infectious diseases endemic in the herd.
- At a national level major diseases such as FMD, ASF, CSF, SVD, AD/PRV and brucellosis may be eradicated either by slaughter policies or testing and slaughter. Some countries declare freedom from some diseases. For example, Ireland is free from TGE and PED. The UK and parts of the EU from Aujeszky's disease and brucellosis.
- At a farm level, herds can be established and maintained free of App, AD/PRV, PAR,

FIG 9-33: RESPIRATORY PATHOGENS – METHODS OF CONTROL

Freedom from pathogens	On a national or regional basis. In integrated systems (e.g. breeding pyramid). At an individual herd level.
Eradication	Depopulate - repopulate with healthy stock (e.g. SPF pigs). Medicate and vaccinate. Medicate and early wean. Vaccinate/test and remove.
Control on the farm	Medication. Vaccination. Management – The environment – The pig – Nutrition Using partial depopulation.

mycoplasma (enzootic) pneumonia and PRRSV, but in dense pig areas breakdowns are common.

Eradication

- Many respiratory diseases can be successfully eradicated by total depopulation of the herd and its replacement by disease-free stock (see Chapters 2 and 3). In practice, the siting of the herd relative to sources of infected pigs needs careful consideration. PRCV and SIV will at some stage re-infect the herd. Mycoplasma (enzootic) pneumonia and PRRSV will be transmitted if infected pigs are within 2 to 3 km (1 to 2 miles) and in such cases it may be advisable to restock with exposed animals and vaccinate them.
- PRRSV can be controlled by selective depopulation of first and second stage nurseries provided the disease is stabilised in the breeding and finisher herds.
- Aujeszky's disease/pseudorabies can be successfully eradicated through the use of gene-deleted vaccines, and testing for field virus carrier pigs and slaughtering them.
- By using the techniques of partial depopulation, a new disease-free population can be established on another site from an existing diseased herd.
- Progressive atrophic rhinitis can be eliminated by vaccination and partial depopulation.

Control on the farm

- Medication and vaccination – Most if not all farms have at least one or more respiratory organisms present that require continual control procedures. The use of medication strategies for respiratory disease has been dealt with in Chapter 4 and you are referred to this for further information. The use of vaccines can be limited by cost and the practical problems of administration to every pig, and often 2 doses are required. However vaccine programmes to control AD/PRV, mycoplasma (enzootic) pneumonia and PAR have given excellent results in the field. The use of PRRSV and SIV, App and *Haemophilus parasuis* vaccines are more problematical. Some vaccines are effective through the sow alone to control disease and the overall strategies that can be adopted are shown in Fig 9-34.
- Management – This involves the manipulation of the environment, the pig and nutrition, giving attention to detail and identifying the important areas and making adjustments.
- Identify the time of commencement of the disease. This is absolutely crucial to understanding the problem on the farm. If the onset of disease is identified and the incubation period calculated, then the time the pig first meets disease and the environmental factors contributing can be determined.
- Keep the pig above its lower critical temperature. Creating an environment where the pig becomes catabolic (using more energy than it is consuming) is the ideal method to trigger respiratory disease. The LCT is commonly breached by faulty ventilation systems, a change in feed from one house to another or a change in stocking density. Draughts, shortage of water, wet floors, dirty pens and low stocking densities will have a similar effect. Look at the weight of the pig entering the house. Is it heavy enough to cope with the environment, the type of housing and floor surfaces?
- Assess the lying patterns of the pigs early in the morning and last thing at night.
- Avoid rapid changes in the environment in the house. This is particularly important where high-speed ventilation systems may cause chilling. The pig will experience similar changes whenever it is moved from one house to another, particularly if this is associated with a change in floor type, from a solid to a slatted floor. Contrary to popular opinion a high air flow will have little effect on the transfer of organisms between the nose of one pig and another.
- A floor space < 0.6 m^2 per finished pig is associated with more pneumonia.

FIG 9-34: POTENTIAL USES FOR VACCINES

Disease	Sow	Piglet	Growing Pig
Actinobacillus pleuropneumonia (App)		✓	✓
Progressive atrophic rhinitis (PAR)	✓		
Aujeszky's disease/pseudorabies virus (AD/PRV)	✓		✓
Mycoplasma (enzootic) pneumonia		✓	✓
Glässer's disease (Hps)	✓	✓	
Porcine cytomegalovirus (PCMV)	not necessary		
Postweaning multisystemic wasting syndrome (PMWS) (PCV2)	✓	✓	
Porcine respiratory corona virus (PRCV)	not necessary		
Porcine reproductive and respiratory syndrome virus (PRRSV)	✓	✓	✓
Salmonella choleraesuis	✓		✓
Swine influenza virus (SIV)	✓		✓

- Pneumonia levels are less in environments that provide more than 3.6 m^3 of air space per finished pig.
- Bacterial levels above 100,000 m^3 will predispose to pneumonia.
- Keep faeces contamination in pens at a low level because gut organisms produce endotoxins that can damage lung tissue and predispose to pneumonia.
- Avoid low temperature and low humidity environments, because the particles in the air are very small and remain suspended. Such particles are inhaled to the very bottom of the respiratory tract where the organism can attack the lung tissues.
- Maintain either a high temperature/low humidity, or low temperature/high humidity environment. Here the particles of water vapour are large and they attract the organisms from the air, preventing them penetrating the deep parts of the lung.
- As dust levels increase in the house, so does the severity of both pneumonia and rhinitis. Dust levels of more than 4 mg/m^3 start to have a harmful effect on the lungs.
- There is less pneumonia in extraction-type ventilation systems than in ones that create positive pressure.
- Natural ventilation, within large cubic capacity air spaces, is ideal, because the organisms rise away from the nose of the pig by natural convection.
- Adapt all-in/all-out housing systems that accommodate each batch production. This reduces the development of endemic disease.
- After houses have been washed out, use a space heater to bring the temperature back to normal. This is essential for pigs from 7 to 14 weeks of age. It may be necessary in certain climates to provide heat for 7 to 10 days when pigs move from a solid to a slatted floor and when there is a change from dry to wet feeding.
- Keep ammonia levels at less than 25 ppm, carbon monoxide at less than 30 ppm, hydrogen sulphide at less than 5 ppm and carbon dioxide at less than 2000 ppm.
- Fogging – this has been used in an effort to control respiratory disease within a populated house but the results are not convincing. Fogging with water periodically will increase the humidity but it also has the disadvantage of cooling the pig by evaporation. It requires fogging at least every 5 to 10 minutes to maintain sufficient water vapour in the house.
- Make sure health control procedures are in operation that eliminate the risk of introducing new pathogens into your herd.
- Avoid purchasing pigs from multiple sources, particularly finisher pigs.
- The larger the herd, the greater the problem.
- The more pig movement there is around the farm, the more likelihood there will be of pneumonia.
- As stocking density increases there will be a greater incidence of pneumonia and this is certainly true once more than 200 pigs occupy a common air space.
- Avoid continual throughput and the mixing of different ages of pigs. The best way to maintain pneumonia on the farm is to mix pigs into a house that contains older ones that have already incubated disease and are passing out large amounts of infectious agents.
- Study the complexity of organisms on the farm in relation to the time when pneumonia occurs. This can be a sequential phenomenon, that is, 3 or 4 diseases affecting the pig over a range of 4 to 12 weeks of age.
- Does the housing satisfy the environmental requirements of the pig at the different ages and weights?
- Ensure the water supply is excellent. Many cases of respiratory disease are associated with poor water availability.

Nutrition

- Identify the time when disease first becomes evident.
- Note when nutritional changes take place.
- When pigs move from one house to another there is often a drop in the nutrient density of the diet. This also coincides with a drop in intake for the first 2 to 7 days.
- This reduced feed intake and change can result in catabolism and the pig dropping below the LCT.
- Maintain high energy diets at critical times. Do not change the feed for at least 5 days when pigs move from one house to another.
- Provide adequate water at all times. Look at the type of nipple drinkers and their availability and accessibility to the pig, particularly when it moves from one house to another. Monitor the flow and use of water.
- Look at the methods of feeding. Are there any features of design or feeder placement that inhibit access to the feed?
- Is there sufficient hopper space?
- Make sure there are adequate levels of vitamin E in the diet. Add an extra 50 to 100 iu to the tonne if there is a disease problem.

Segregated weaning and partial depopulation

- When pigs are moved from one building to another already containing pigs they are exposed to a range of viruses and bacteria, many of which they may not have encountered before. Some of these may produce clinical disease and others may result in sub-clinical infection. The end result is a large uptake of protein and energy diverted to satisfying the demands of the immune system. This significantly depresses daily gain and food conversion efficiency. Segregated weaning to all-in/all-out buildings on a batch-by-batch basis eliminates or reduces such a challenge. The improved performance is likely to far exceed that of weaning into a continuously operated nursery containing different ages of pigs, but in addition it is of great benefit in the control of respiratory disease. These procedures are discussed in detail in Chapter 3.

The costs of respiratory disease are proportional to the numbers of infectious agents present on the farm, the management and environmental conditions that allow endemic disease to develop, and high levels of exposure to susceptible pigs.

If there is a respiratory disease problem on the farm, check all the key points and identify the weak ones.

Retroviruses

These are thought to occur within the genome of every cell in every pig's body. The virus genomes have become part of the genome sequence of the pig and are passed on through the cells of the ovum and foetus. They are of no consequence to the pig or the pig farmer and are only of concern if the pig has been reared as a source of organs or tissues for human patients. The concern is that a retrovirus may pass from the transplanted organ or tissue into human cells and cause disease. The overriding fear is that this virus might then spread to other people causing an epidemic.

Riding

As the finishing pig matures, they may become reproductively active.

With entire males being present, this can lead to an explosion of fighting and aggression associated with the developing sexual drive. In the event of increased sexual activity, pigs may become traumatised and even killed. Food conversion and growth rates can be significantly affected. The boars may be more interested in fighting than eating. The trauma may lead to carcass damage and condemnation. In future breeding stock the penis can become damaged and permanently traumatised.

Separation of males and females may assist with reducing some of this aggression, but groups of males will still fight among themselves.

The use of chemical castration may be considered as part of the control programme.

Rotaviral Enteritis (Diarrhoea)

See Chapter 8 for further information

These viruses are widespread both in pig populations and most other mammals and there are a number of different types or groups. Group A is probably the common pig one, but B, C and E also occur. Rotaviruses are ubiquitous and they are present in most if not all pig herds with virtually a 100 % sero-conversion in adult stock. Clinical signs can appear in piglets 7 to 10 days of age through to postweaning.

Ruptures or Hernias

Of many congenital abnormalities, ruptures at the umbilicus or the inguinal canal are most common. They are considered to be developmental defects yet have a very low heritability. Umbilical hernias can sometimes be traced back to a particular boar in which case he should be culled. Environmental factors can increase the incidence of umbilical hernias and if there is a problem (more than 2 % of pigs) consider the following:

- Are prostaglandins used to synchronise farrowings? If so check that piglets are not being pulled away from the sow at farrowing and the cord stretched abnormally.
- Is navel bleeding occurring on the farm? Are navel clips being used to prevent bleeding? If so, make sure they are not placed close up to the skin, otherwise the tissues will be damaged and weakened.
- Identify the precise time when the ruptures appear. Do these coincide with a change of housing?
- In veranda-type housing (where the pigs pass through a small hole to the faecal area), sudden severe abdominal pressure, as a result of pigs getting stuck in the hole, may cause ruptures.
- Are stocking densities high, thus increasing abdominal pressure?
- In cold weather, do the pigs huddle, thereby increasing abdominal pressure?
- Check records to see if the boar and the sow are related.

- If the rupture is large and the pig is on a concrete floor or slats it should be moved to a soft-bedded area so that the overlying skin does not become sore and ulcerated.
- Examine navels at births and 2 days later to see if there are any abnormalities.
- Pigs with ruptures which are larger than 300 mm (12 inches) in circumference and/or ulcerated, or which are in contact with the ground should be culled.

Inguinal ruptures are not as important a problem unless they become very large. Where castration is the farm policy, a minor surgical operation needs to be performed. This is described in Chapter 15.

Salmonellosis

Salmonella bacteria are widespread in human and animal populations. Some of them cause food poisoning in man or disease (salmonellosis) in animals. There are many serotypes of *Salmonella* but the ones that are most likely to cause clinical disease in pigs are *Salmonella choleraesuis*, and *Salmonella typhimurium* and to a lesser extent *Salmonella derby*. Other "exotic" *Salmonella* serotypes may infect pigs and be shed in the faeces for limited periods but they usually remain sub-clinical. *S. choleraesuis* is the specific host-adapted pig serovar and can cause major generalised disease. *S. typhimurium* and *S. derby* are more likely to cause a milder disease, the main sign of which is usually diarrhoea. Pigs may become long-term sub-clinical carriers of *S. choleraesuis,* the organism surviving in the mesenteric lymph nodes draining the intestine. Many such carriers do not shed the bacteria in faeces unless they are stressed. Some, however, may become sub-clinical carriers of *S. typhimurium, S. derby* and other serotypes. They may be intermittent or continuous faecal shedders but the carrier state is usually relatively short - weeks or a few months - and it is self-limiting.

If a pig is infected with a large dose of *S. choleraesuis,* it is likely to develop severe generalised clinical disease starting with a septicaemia (blood infection) followed by severe pneumonia and enteritis. Subsequently the organism may settle out in a variety of tissues including the central nervous system, resulting in meningitis, and the joints, resulting in arthritis. *S. typhimurium* and *S. derby* may also cause septicaemia and become generalised and involved in pneumonia, but in most pigs these are transient and enteritis is the only persistent manifestation.

Salmonellosis can occur at any age but is most common in growing pigs over 8 weeks of age. Severe *S. choleraesuis* infection occurs typically at around 12 to 14 weeks.

Many species of *Salmonella* may be found in pigs without causing any clinical signs. However, these may be considered significant by meat processors.

Clinical signs

Acute septicaemia and pneumonia result in fever, inappetence, respiratory distress and depression. The skin of the extremities (i.e. tail, ears, nose and feet) becomes blue. On the ears, in *S. choleraesuis* infection, there is often a clear line of demarcation between the blue and normal skin. Foul-smelling diarrhoea, which may be blood-stained, is a common feature. Jaundice (yellowing) may result from liver damage and lameness from arthritis. Meningitis results in nervous signs. If untreated, mortality may be high. *S. choleraesuis* should be considered in pigs demonstrating both severe diarrhoea (often golden) and pneumonia (coughing) at the same time.

Diagnosis

It is necessary to submit to the laboratory either fresh faecal samples from untreated pigs, or where available, a dead or live untreated pig. This is essential to demonstrate the presence of the organism and differentiate it from other causes.

The post-mortem lesions are strongly suggestive of *S. choleraesuis*, particularly the generalised pneumonia, the appearance of the lining of the small and large intestine, the congested spleen and multiple small haemorrhages.

Similar diseases

Severe salmonellosis caused by *S. choleraesuis* can occur alone but it also commonly occurs at the same time as classical swine fever (hog cholera) in those countries in which this disease still occurs. In such countries it is important to ensure by serology and laboratory tests that swine fever is not the primary cause (NB. swine fever usually also affects sows and sucking piglets and also causes mummified litters and abortions).

Severe PRRSV in herds with endemic mycoplasma (enzootic) pneumonia may give the appearance of salmonellosis. However PRRSV also causes abortions and stillbirths, and precipitates scouring in piglets.

Treatment

- The response to the treatment of *Salmonella* infections is often poor. It is necessary to determine the serotype and the antibiotic sensitivity, and treat individual pigs at a very early stage.
- Preventative medication or strategic medication both in-feed or in-water are important and the medicines used are dependent on the bacterial sensitivity.

- Live vaccines for *S. choleraesuis* are used in more countries and may be effective even in the face of an outbreak.

Management control and prevention

The severity of clinical salmonellosis is dose-dependent. This is true to a greater or lesser extent of all infectious diseases but is particularly so of salmonellosis. The overall aim, therefore, is to get the concentration of the organism in the environment down to below the disease-producing threshold. The second aim is to reduce the spread of infection.

- Improve hygiene by frequent and thorough removal of waste.
- Optimise stocking density. Overcrowding predisposes.
- Reduce stress by reducing moving and mixing.
- Develop an all-in/all-out system with cleaning and disinfection between batches.
- Avoid the movement of pig faeces between one batch and another.
- Vaccines against *S. choleraesuis* are available in some countries and can be used to help in elimination programmes if this type is present. The principle is to vaccinate the sow to produce high maternal antibody and then carry out segregated early weaning or partial depopulation programmes.
- Disinfect boots between houses.
- Control vermin and flies.
- Prevent contamination of feed by birds, rats and mice.
- Monitor raw feed ingredients and final product.
- Cleanliness and hygiene are important.
- Organic acids added to feed also help.
- In some countries, formalin is added instead.

Many countries (for example Denmark, Sweden and Eire) have national control policies and other countries are considering developing them. The Swedes monitor herds mainly by culturing faeces. They try to eliminate *Salmonella* from all herds. An ELISA test to check blood or meat juice for antibodies to the commonest *Salmonella* serotypes in pigs is routinely used. On the basis of the results they calculate a score for each herd. Herds with high scores are penalised and made to adopt control measures until their scores come down to an acceptable level.

Note: *Salmonella* in a final pork product may have very little to do with the on-farm situation. *Salmonella* organisms may migrate from the lairage floor to the hind gut and lymph nodes within 20 minutes of arriving at the slaughterhouse.

Zoonoses

It is important to remember that *Salmonella* organisms are transmissible to people and are one of the commonest causes of food poisoning. This has 2 implications for you as a pig farmer.

First, you should ensure that everybody working with the pigs adopts a high standard of personal hygiene so that they do not become infected. Second, it is important that pig farmers and pig meat products have a good public image for safety and do not get linked with outbreaks of human disease. It is therefore imperative that you get *Salmonella* under strict control by the particular use of batching all-in/all-out procedures.

Spirochaetal Diarrhoea

This is a disease associated with spirochetes distinct from those that cause swine dysentery (*Brachyspira hyodysenteriae*). It occurs mainly in young pigs, appearing very similar to colitis, and the possibility of *Brachyspira pilosicoli* and *Lawsonia intracellularis* (ileitis) playing a major part cannot be discounted.

Spirochetes are common inhabitants of the large intestine and caecum, and disease is often associated with changes in the diet such as the inadvertent removal of copper or the withdrawal of antibiotics from the diet.

Clinical signs

A mild to moderate diarrhoea develops 2 to 6 weeks postweaning and persists for a few days, with dehydration and loss in growth. Most cases resolve in 7 to 10 days but in some pigs chronic diarrhoea results. The disease can be difficult to differentiate from the bacterial infections, particularly colitis.

Diagnosis

This is difficult because specific organisms cannot usually be identified. If there is an on-going problem on the farm, live diseased pigs showing typical signs should be submitted for post-mortem and bacteriological examinations to eliminate swine dysentery and demonstrate large populations of spirochetes associated with an enteritis.

Treatment

- In-feed medication. Lincomycin, tiamulin, monensin, valnemulin and tylosin all have good activity against spirochetes and can be used for both prevention and treatment.
- In acute outbreaks, lincomycin, tiamulin or tylosin could be used in the water.
- Inject individual pigs with either lincomycin, tiamulin or tylosin and assess the response.

Management control and prevention

- Adopt all-in/all-out procedures.
- Ensure that floor surfaces are kept clean and dry.
- Provide clean sources of water.
- Avoid draughts and chilling.

Streptococcal Infections

Streptococci are common organisms in all animals, including people. They are broadly but not entirely species-specific. The main species is *Streptococcus suis*, which is widespread in pig populations. It is associated with a variety of conditions, including meningitis, septicaemia (infection of the blood), polyserisitis (inflammation of the lining of the abdominal, chest, heart, joints and brain surfaces), arthritis, endocarditis (infection of the heart valves) and pneumonia. It has also been isolated from cases of rhinitis and abortion. The pattern and relative importance of the different syndromes vary in different countries.

The situation is not as simple as it may first appear. *S. suis* is sub-divided into at least thirty-four serotypes. They vary in their pathogenicity and the diseases they cause, both between and within types. Some types appear to be non-pathogenic and have been isolated mainly from healthy pigs; some are mainly associated with lung lesions. Type 2 occasionally also causes septicaemia and meningitis in people, mainly in people handling pigs or pig meat products. Fortunately, human cases are rare in the West but a little more common in parts of S.E Asia. There is at least 1 report of an isolation of type 14 from a human brain. People who recover may be profoundly deaf.

Different types predominate in different countries, with type 2 among the commonest in most countries and the type most often associated with disease (one exception is Denmark, where type 7 is the most common). In the UK, type 2 is the main cause of serious acute meningitis, along with polyserisitis and arthritis in weaned and growing pigs, and it is rarely associated with pneumonia. Type 1 occurs fairly commonly in most countries and causes sporadic arthritis and occasionally meningitis in sucking piglets, usually around 1 to 2 weeks of age but sometimes up to 6 weeks. It is a relatively unimportant condition. Type 12 causes massively swollen hock joints in 10-day-old pigs. *S. suis* type 14 is associated with acute severe outbreaks of arthritis in both sucking and weaned pigs.

The syndrome that is important and worrying to the pig farmer is persistent endemic meningitis caused by type 2.

S. suis 2 is spread from one pig to another by direct nose-to-nose contact. It can also spread within a herd by indirect contact and in confined space by aerosol infection. Clinically healthy pigs can carry the organism in their tonsils for many months and a carrier state exists in some sows. Once a serotype has entered a herd, no techniques are yet available to remove it and it becomes established in the tonsil as part of the normal flora. The most likely source of entry into the farm is the purchase of carrier boars or gilts. There is little documented evidence to suggest that the human will transmit disease from pig farm to pig farm. *S. suis* is quickly killed by disinfectants in common use on farms, including phenolic disinfectants and chlorine- and iodine-based ones; detergents will also kill the organism in thirty minutes. Outside the pig, in very cold and freezing conditions it may survive for 15 weeks or more but at normal room temperatures it dies within 1 to 2 weeks. It survives long periods in rotting carcasses. The sow passes on antibody through colostrum to the sucking pig and the disease is therefore uncommon in this group of animals unless it is introduced into the herd for the first time. It is much more common in the immediate postweaning period, often starting 2 to 3 weeks after weaning and continuing through to approximately 16 weeks of age. In nurseries, almost 100 % of pigs become carriers within 3 weeks.

S. suis type 2 is carried from the base of the tonsil to the brain, joints and serosal surfaces inside migrating white blood cells called monocytes (analogous to the "Trojan horse"). Pathogenic strains can survive and multiply in these cells whereas non-pathogenic strains cannot. There are also strains of low pathogenicity which may be activated by PRRSV infection. PRRSV may also raise the incidence of meningitis caused by pathogenic strains when it first enters a herd. Although PRRSV alone does not affect the brain, it has been shown experimentally that many more pigs are affected with meningitis when they are infected with both *S. suis* type 2 and PRRSV than when they are infected with *S. suis* alone. Species of streptococci other than *S. suis* may sometimes cause disease in pigs. For example, *Streptococcus equisimilis* causes sporadic cases of septicaemia and arthritis in sucking pigs, infection of the heart valves in growing pigs and ascending infection of the uterus in sows. *Streptococcus porcinus* causes throat abscesses and septicaemia and is sometimes isolated from pneumonia. Note that humans can carry *Streptococcus suis* 2 without any clinical signs. This makes it impossible to keep off the pig farm.

Clinical signs

Acute disease

Weaners may just be found dead. In very early stages of meningitis the pig is laid on its belly, with it's hair standing on end, and shivering. Within 2 to 3 hours there are lateral jerky movements of the eye (nystagmus). The animal then lies on its side, paddling and frothing at the mouth. The organism invades the bloodstream and is carried around the body where it may cause arthritis and pneumonia.

Diagnosis

A history of the presence of recurring meningitis in weaned pigs is highly suggestive and is confirmed by the isolation of the organism from the brain and its specific identification, which not all diagnostic laboratories are capable of.

Because of the existence of strains that are non-pathogenic or only mildly pathogenic, the isolation of *S. suis* type 2 from the tonsils of a pig is difficult to interpret. Isolation from the brain of a pig showing signs of meningitis is more conclusive.

Similar diseases

Similar nervous signs may occur in Aujeszky's disease, Glässer's disease, bowel oedema or water deprivation (salt poisoning).

Treatment

This must be carried out as soon as disease is recognised.

- Remove the affected pig from the group to a hospital pen. Meningitis is extremely painful and the recovery rate is increased substantially through good nursing.
- Provide warmth and bedding and trickle water into the pig's mouth from a hosepipe every 4 to 6 hours. Alternatively, water can be given by inserting a narrow hosepipe gently into the pig's rectum or using a flutter valve (see Chapter 15).
- Give intra-muscular injections of penicillin 2 to 3 times daily for the first 24 hours.
- Use a quick-acting penicillin for the first 24 hours followed by long-acting penicillin. Alternatively, use ceftiofur. Time is of the essence with this disease. Trimethoprim/sulpha would be an alternative medicine to use.

Strategic medication

This is a method to adopt on farms where disease levels remain high. The following options are available:

- Identify the onset of disease and apply strategic medication 2 to 3 days before that time.
- Strategic medication can be applied in the drinking water using phenoxymethyl penicillin, tetracyclines, synthetic penicillin (particularly amoxycillin), or trimethoprim/sulpha.
- In-feed medicate continuously from day of weaning through to 6 weeks postweaning. Phenoxymethyl penicillin at 300 g/tonne is the medicine of choice. TMS could also be used as an alternative.
- Inject all pigs with long-acting penicillin at weaning time.
- Assess the response to injecting pigs with long-acting penicillin 5 days before weaning.
- Review the environment of the weaned pig; particularly avoid any level of poisoning by raised gas concentrations.
- Ensure excellent pig flow.
- Remove all draughts and chilling of the sleeping areas.

Management control and prevention

- If your herd is free from this disease, try to keep it free.
- Check out your sources of replacement stock before purchase.
- Do not purchase pigs from herds with clinical cases of meningitis.

If you have the disease endemic in your herd, the incidence increases with:

- High stocking density in nurseries.
- Continuous production systems which perpetuate infection.
- PRRSV infections may activate *S. suis* already in the herd.
- Mixing of pigs postweaning.
- A small cubic capacity air space per pig. Provide at least 0.8 m^3 per pig at weaning.
- Poor ventilation and high humidity.
- High dust levels.
- Stress.
- Damp pens.
- High carbon monoxide (CO) concentrations associated with gas heaters.
- High slurry levels under perforated metal floors and the damaging effects of gases to the respiratory system.
- Weighing pigs.
- Tattooing, ear notching and extra stress at weaning.
- Changes in nutritional status at critical times.
- Low vitamin E in the diet. Assess the response to adding 50 to 100 iu/kg to a concentration of 200 to 250 iu/kg for weaned pigs to 30 kg liveweight.

Swine Dysentery (SD)

Swine dysentery (SD), which is generally caused by a small snake-like bacteria called *Brachyspira hyodysenteriae* can be one of the most expensive diseases of the growing pig. There are other similar bacteria that can also cause clinical signs. These include *B. hampsonii*, *B. pilosicoli*, associated with colitis and *B. innocens* which is non-pathogenic but may still be associated with dysentery. Disease is common in pigs from 12 to 75 kg (26 to 165 lbs) weight and occasionally in sows and piglets. It is spread entirely by the ingestion of infected faeces or from carrier pigs that shed the organism in faeces for long periods. It may enter the farm through the introduction of carrier pigs or mechanically from infected faeces via equipment, boots or birds. The disease has a marked depressant effect on feed conversion and daily gain with symptoms appearing intermittently often associated with stress, such as the movement of pigs which increases the output of organisms. The organism survives in slurry for up to 8 weeks. Continual exposure to infected faeces is a major factor in the maintenance of disease in a herd.

The costs of disease are associated with mortality (low), morbidity (high), inefficient production and continual in-feed medication. Disease often appears in cycles and previously affected pigs will still transmit infection to susceptible ones for at least 10 to 12 weeks. SD can be spread by flies, mice, birds (starlings), cats and dogs; all these species shedding the organism for up to 21 days. Mice will carry the organism for 180 days. SD will survive outside the pig for up to 8 weeks in moist conditions, particularly in cold slurry but it dies out in 2 to 3 days when allowed to dry.

Clinical signs

The incubation period is normally 2 to 14 days but can be much longer. The pig may develop a sub-clinical carrier state first and then break down with clinical disease when put under stress or when there is a change of feed.

The early symptom is a sloppy diarrhoea, which stains the skin of the perineum under the anus, and a rapid hollowing of the flanks. Initially the diarrhoea is light brown and contains jelly-like mucus. As the disease progresses, blood may appear in increasing amounts, turning the faeces black and tarry. The pig's appetite is decreased and it rapidly loses condition, becomes dehydrated and takes on a gaunt appearance. Spread through the herd is slow, building up in numbers as the dose rate of the causal agent builds up in the environment. Sudden death sometimes occurs, mainly in heavy finishers.

Post-mortem examinations show the lesions are confined to the large bowel and sometimes the greater curvature of the stomach.

Diagnosis

This is based on the history, the clinical picture, post-mortem examinations, gram-stained faecal or colonic smears, fluorescent antibody tests on faecal smears and the isolation and identification of *B. hyodysenteriae*. Specific identification of *B. hyodysenteriae* to distinguish it from other similar colonic spirochetes requires specialised procedures which may not be available in every laboratory.

Similar diseases

These include non-specific colitis, bloody gut (PHE) and acute *Salmonella* infection, particularly by *S. choleraesuis*. Heavy infections of the whipworm *Trichuris suis* can also simulate swine dysentery.

Treatment

- The following medicines have an activity against *B. hyodysenteriae*:
  - Carbadox.*
  - Chlortetracycline.
  - Lincomycin.
  - Metronidazole.*
  - Monensin.*
  - Ronidazole.*
  - Salinomycin.
  - Tiamulin.* (some strains are now resistant)
  - Tylosin.*
  - Valnemulin.*

  Some of these products may not be available in your country. * Very active.

  NOTE Some strains of *B. hyodysenteriae* have become resistant to some of these medicines.
- With the first signs of disease, medicate the drinking water with either metronidazole, lincomycin or tiamulin for at least 7 days.
- Inject badly affected individual pigs daily for 4 days with either lincomycin, tiamulin or tylosin.
- In-feed medication may be used to suppress clinical disease for treatment. The following medication could be used: lincomycin – 110 g/tonne, tiamulin – 100 g/tonne or valnemulin 75 g/tonne could be used.
- It may be possible to combine tiamulin and oxytetracycline in the control programme.

Management control and prevention

- If the herd is not infected with swine dysentery, carry out all the biosecurity measures discussed in Chapter 2 to prevent its entry. Transmission usually requires a moderate dose of infected faeces so it can usually be kept out successfully even in pig-dense areas. Rodent control is a priority.
- In infected herds, control is aimed at preventing the movement of the organism between groups of pigs. Management procedures should therefore be directed towards all-in/all-out systems with cleansing and disinfection between each batch. The organism is sensitive to most disinfectants, and in particular phenol-based ones.

The following specific management procedures should be adopted:

- Develop a batch all-in/all-out housing system with disinfection. Slurry channels should be separate.
- Where solid faecal passages are used it may be necessary to depopulate the whole house, clean and disinfect and bring in new pigs at weaning time that have not been contaminated.
- Control flies; they can transmit the organism from one group of pigs to another.
- Carry out strategic medication, for example postweaning using either lincomycin (110 g/tonne) or tiamulin (100 g/tonne) for a period of 3 weeks. This should produce a dysentery-free pig, and provided this pig moves into further accommodation which has not been exposed to the organism, it becomes possible to break the cycle. Segregate the herd into believed clean and infected areas and prevent transmission of faeces between them as much as possible.
- Use liberal disinfectant foot baths dispersed around the farm, particularly when personnel move from one batch of pigs to another. Change into room-specific boots.
- Reduce the movement and handling of pigs to as little as possible particularly at weighing time when transfer of faeces is likely to take place and stress is likely to cause clinical flare-ups.
- Do not overcrowd pigs and endeavour to keep a dry environment; the organism will die out quickly on drying. Poor sanitation and wet pens enhance the disease.
- One of the greatest risks is the introduction of the disease into a finishing system. This is likely when pigs have been brought in from unknown sources. Isolate incoming pigs for a period of 3 weeks and strategically medicate with lincomycin at 220 g/tonne or tiamulin at 100 g/tonne for a period of 2 weeks. Water medication could also be used. This should eliminate the organism if it is present or at least lower it to a manageable level.
- A common source of infection is from infected pigs that are present on the vehicle which calls at your farm to collect your pigs. Infected faeces are carried back into the buildings on boots.

Eradication

Swine dysentery is a disease that no producer can afford to live with. There are therefore 2 options in the medium to long term: either to depopulate and repopulate the herd or attempt to eradicate the disease without depopulation. Eradication has been carried out on a number of farms successfully, although depending on farm circumstances, the success rate may only be as high as 80 %. In the breeding-finishing herd, eradication commences with the treatment of the sow herd to eliminate the organism and produce and wean a dysentery-free pig. Such pigs must not be allowed to come into contact with any other infected pigs or their faeces on the farm. There are a number of programmes that can be adopted from this principle (Fig 9-35).

There are however some rules that need to be noted:

1. Do not start an eradication programme until clinical disease is reduced to an absolute minimum by management control and continuous preventative in-feed medication; 60 g/tonne tiamulin, 55 to 100 g/tonne lincomycin or 100 g/tonne monensin could be used.
2. Maintain the preventative medication throughout the whole period of the eradication and for 2 months afterwards.
3. The numbers of growing pigs on the farm should be reduced to a minimum, thereby reducing the

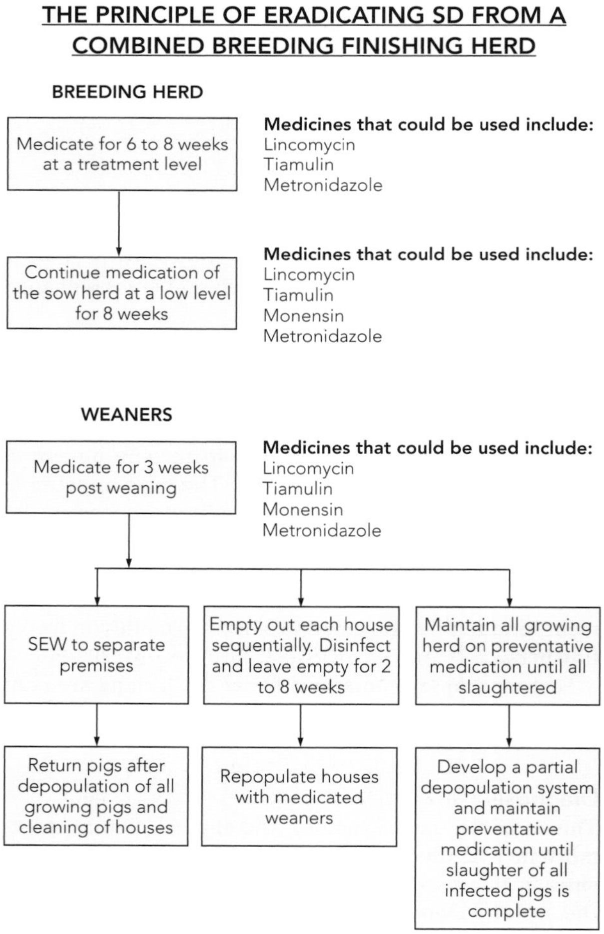

Fig 9-35

susceptible population. Ideally this should start at weaner weight, with all the grower pigs being sold off the farm. Practise a partial depopulation.

4 An alternative would be to carry out segregated weaning for a period of weeks depending upon the availability of accommodation. Whilst this is going on the pigs from weaning through to slaughter can be marketed at their suitable weights.
5 One or more weaner houses should be totally emptied, washed, disinfected and left empty for 4 weeks to give a break between clean and dirty pigs. This gap is maintained throughout the programme. Strict isolation procedures must be adopted between cleaned and infected houses.
6 As each house is emptied it should be completely cleaned down, all evidence of faeces removed and faeces channels totally emptied. Slurry channels become the most dangerous area because the organism can survive for up to 8 weeks. However in practice disinfection can be carried out with individual buildings left emptied for a period of 4 weeks only, but this is a less reliable method.
7 Always attempt eradication during the warm summer months since the organism dies out in a matter of hours under dry conditions, whereas it can survive for weeks in cold, wet conditions.
8 Prior to the commencement of the eradication programme, the sow herd should be medicated for at least 6 weeks to remove any possible carrier state.
9 As the elimination procedures continue, mice, rats and flies should be dealt with accordingly. The organism can persist for long periods of time in mice.

Medicines that have been used successfully for the eradication of swine dysentery:

Sows – **Period of medication: up to 8 weeks.**
Tiamulin 100 g/tonne of feed.
Lincomycin 110 g/tonne of feed.
Monensin 100 g/tonne of feed.
Metronidazole 200 g/tonne of feed.

Weaners – **Period of in-feed medication: 3 weeks.**
Lincomycin 110 to 220 g/tonne of feed.
Tiamulin 100 g/tonne of feed.
Metronidazole 200 g/tonne feed.

Growers – **Period of medication: until the weaner supply is negative.**
Lincomycin 55 to 110 g/tonne of feed.
Tiamulin 60 to 100 g/tonne of feed.
Monensin 100 g/tonne of feed.
Metronidazole 200 g/tonne of feed.

One note of warning: eradication is not always successful but it can be more cost effective than total herd depopulation and repopulation. Discuss this with your veterinarian.

Swine Influenza Virus (SIV) – Swine Flu

See also Chapter 6

Swine influenza is a highly contagious respiratory disease caused by a number of closely related type A influenza viruses that are noted for their ability to change their antigenic structure and create new strains.

Each serotype is identified by surface proteins (antigens) referred to as "H" and "N". The 3 common strains that affect the pig are described as H1 N1, H1 N2, and H3 N2. There are also different strains within these serotypes with differing pathogenicity (capacity to produce disease). The virus is highly prone to mutation. It is thus possible for the pigs to be infected by one virus and develop disease and then some 2 to 3 months later be infected by a different serotype and develop disease yet again.

In large herds of over 300 sows the virus may circulate in young growing pigs, disease becoming more active in the winter time, associated with reduced ventilation rates. In the growing and finishing pig, disease may not be seen until maternal antibody has disappeared, sometime between 7 and 12 weeks of age. SIV can interchange between man, pig and birds and the carrier state can exist in the pig for some 2 to 4 weeks. SIV is also spread by birds, particularly ducks. It is also thought that the viruses may spread up to 3 km (2 miles) on wind but this has never been demonstrated. Thus it is virtually impossible to guarantee or maintain a population of pigs that is free of this disease.

Clinical signs

Acute disease

The incubation period is short; as little as 12 to 48 hours. The onset can be extremely rapid and dramatic. In growing pigs the classical picture is a house full of pigs that are normal one day and the following morning most of them are prostrate and breathing heavily. Severe coughing and laboured breathing will be observed. You may think most of the pigs are going to die but rest assured most of them survive, and provided the herd does not have a history of ongoing pneumonia, the pigs will recover on their own, but it is always difficult to predict the outcome. Severely affected individuals or groups of pigs are therefore best given antibiotic cover.

Many pigs die with what appear to be unrelated flu problems – pericarditis, peritonitis and pleurisy. The influenza is selectively killing the weakened pigs within the group. These pigs may be saved by the use of antibiotics. If you are monitoring the water usage, swine influenza may reduce water intake by 10 % which may be helpful to predict an outbreak.

Endemic disease

This is where the virus continually circulates through the herd, infecting individual pigs within groups. SIV causes severe pneumonia on its own but when it is combined with other infections a chronic respiratory disease syndrome can develop.

Chronic respiratory disease

Chronic respiratory disease is caused by combinations of respiratory pathogens including PRRSV, SIV, PMWS, *Mycoplasma hyopneumoniae*, App, *Haemophilus parasuis*, *Pasteurella* and streptococci bacteria, particularly as they become additive. The diagnosis is then of multiple cause, and individual pigs within the farm can have widely differing pathogen profiles.

The following technique has proved of considerable value in controlling chronic respiratory disease associated with PRRSV. The objective is control and not elimination. The essential components are:

- Removal of pigs at weaning (more than 21 days of age) on a weekly batch basis to totally separate housing for a period of 8 weeks.
- Vaccination of piglets at 1 and 3 weeks of age against *Mycoplasma hyopneumoniae*.
- Removal of all remaining weaners/growers/finishers.
- Adaptation of buildings to all-in/all-out, and no droplet contact between batches.
- Return of the segregated pigs.
- Control/eradication of disease in the sow herd by vaccination/medication.

Diagnosis

In acute disease the rapidity of development and spread, together with typical clinical signs, are diagnostic. No other disease will affect so many pigs so quickly.

In the chronic respiratory disease syndrome it is necessary to carry out serological tests and virus isolation to determine the presence and serotype of the virus.

Treatment

There is no treatment specifically for flu viruses. However secondary bacterial infections may be involved and in such cases antibiotics can be used to control these.

- In-feed medication in acute disease is a waste of time because pigs do not eat. Although labour-intensive, it is far more efficient to treat individual pigs that have secondary pneumonia with long-acting antibiotics, such as oxytetracycline or amoxycillin.
- Vaccines, some of them multivalent, are available in some countries and may be administered to the finishing herd after 6 weeks of age (once maternal antibodies have fallen). Note that the immunity created by the vaccines only lasts 6 months. Immunity created by the natural disease is life-long to that specific type of H and N strain.

Management control and prevention

It is important to control serious secondary bacterial infections and alongside treatment, good husbandry is critical.

- Use the management procedures outlined for the other respiratory diseases in this chapter.
- Reduce the weights of exposure to other organisms.
- Vaccination is used in some countries, with mixed results.
- Avoid buying pigs from sources where SIV is active.

Torsion of the Stomach and Intestines (Twisted Gut)

See Chapter 14 for further information

This is one of the most common causes of sudden death in the growing pig and it is often one of the best pigs in the group that dies. There are usually no other clinical signs. A post-mortem is necessary to confirm diagnosis.

Transmissible Gastroenteritis (TGE)

See Chapters 8 and 12 for further information

This is a highly infectious disease which in the weaning and the growing pig is clinically indistinguishable from porcine epidemic diarrhoea. If the virus is introduced into the finishing herd for the first time, there is a rapid spreading illness with vomiting and a watery diarrhoea affecting almost 100 % of animals. Disease disappears spontaneously over a 3 to 5 week period. Mortality is usually low but morbidity can be very high. The virus usually then dies out of the population unless there are large numbers of pigs on-site. This disease is covered in detail in Chapter 8.

Tuberculosis

Tuberculosis is a disease affecting human beings, mammals and birds. The causal organism *Mycobacterium tuberculosis* is sub-classified into types based on the species of host usually affected: the human type generally referred to as *M. tuberculosis* affects people and primates, the bovine type *M. bovis*, affects cattle, badgers and other wild herbivores and sometimes people; and the avian type, *M. avian/M. intracellulare* complex, which affects mainly birds. Pigs are susceptible to all 3 but in practice are rarely infected by the first two. Most TB in pigs is caused by the avian/intracellulare complex which causes small nodules in the lymph nodes of the neck and those that drain the small intestine. In the great majority of cases the lesions are non-progressive; they do not spread through the body, do not make the pigs ill and the organisms are not shed.

The disease does not therefore spread between pigs and is rarely diagnosed in living pigs. Similarly, the *M. avium/intracellulare* complex causes non-progressive infection in normal healthy people. The main concern is that the *M. avium/intracellulare* could cause more serious disease in immunosuppressed people and people with AIDS, and the lesions in the pig's carcass at slaughter cannot be distinguished from human and bovine TB which would cause progressive disease in otherwise normal people. Therefore, in most countries if lesions are found in the neck at slaughter the whole head is condemned and if they are found in the mesenteric lymph nodes which drain the intestines, the offals are condemned. If they are more widespread, the whole carcass may be condemned or require cooking. If small lesions are missed by the meat inspector, cooking will normally destroy the organism.

The sources of infection to the pig include:

- Outdoor pigs – grazing land that has been treated with poultry manure even up to 1 year previously, or that which has been grazed by infected cattle or badgers infected with *M. bovis*.
- Avian TB, as the name implies, is found in wild birds and in particular starlings. The organism is shed in large numbers via droppings and therefore food or grain contaminated by birds becomes a potent source.
- Sawdust/shavings can be a major source of infection.
- Peat often contains *M. intracellulare* and is capable of causing lesions at slaughter. Peat is used both for bedding and gut stimulation in the young piglets. It should only be used if it has been pasteurised.
- Water contaminated by *M. avium/intracellulare* is often a source.
- Infection may occur from 10 weeks onwards, with lesions seen at slaughter. Lesions can take up to 8 weeks or more to develop.

Clinical signs

M. avian infection has no clinical effect and there is no difference in performance between infected and non-infected pigs.

Diagnosis

TB in living pigs can be diagnosed using the skin tuberculin test but is usually detected in the cervical and intestinal lymph glands at meat inspection. Normal levels are less than 1 %.

Management control and prevention

- If disease is evident at slaughter, consider the above potential sources and eliminate them.

  Remember that avian/intracellulare is an environmental contaminant – the environment is the source.

Vice – (Abnormal Behaviour)

See also Chapter 10 for further information on flank biting and tail biting

Vice (Abnormal Behaviour) in both weaned and growing pigs can be a major problem on some farms causing considerable economic loss. The different types of vice are shown in Fig 9-36.

Why do pigs mutilate each other? From our own experiences, poor environmental conditions and human interactions cause varying degrees of aggravation and this is no different in the pig. If there is a problem on the farm, consider the 3 major contributing factors: management, nutrition and disease.

FIG 9-36: TYPES OF VICE

Piglet	The Weaned Pig	Growers
Facial necrosis	Navel sucking	Tail biting
	Prepuce sucking	Ear necrosis
	Ear sucking	Chewing feet
	Tail biting	Flank biting

Management

Stand for a few minutes and observe pigs that are either tail biting or ear chewing and you will see that there is one overriding feature: the pigs give the impression of being very unhappy. The pig is indicating that the environment provided is far from ideal. The management factors that contribute to this are listed in Fig 9-37.

Study these, identify important ones and make changes until there is a response. Remove offending pigs from the group because once the vice has become an established habit it can be difficult to stop.

Ear biting, ear sucking and penile sucking tend to be problems of the pig from weaning to 30 kg (66 lbs). From 30 kg to finish, flank biting and tail biting are the major vice problems.

Nutrition

Where there is competition for food or poor access, this tends to create aggression in the pen. Observations have shown that there is a greater tendency to tail biting when automatic feeders are used compared to manual systems. With automatic systems there is little empathy between pig and person. Increasing the salt level in the diet to 0.9 % can often produce a reduction in the level of vice. However, make sure there is sufficient water available (Fig 9-38).

FIG 9-37: VICE-PREDISPOSING MANAGEMENT FACTORS

– A change in the diet	– High hydrogen sulphide levels > 10 ppm
– A very humid environment	– Long tails
– Aggressive breeds	– New concrete
– Automatic feeding and little human/pig empathy	– No bedding
– Bad pen designs - badly sited feeders	– Pigs too small for the environment
– High carbon dioxide levels > 3000 ppm	– Shortage of trough space
– Draughts / chilling	– Trauma
– Fluctuating temperatures	– Uncomfortable conditions
– High air speed	– Unhappy pigs
– High ammonia levels > 25 ppm	– Water shortage
– High stocking densities	– Wet pens

Disease

Greasy pig disease or exudative epidermitis is a little-realised but important factor in the development of tail biting, ear and flank chewing. A skin infection or wet eczema starts on the tip of the tail or ears with small areas of serum oozing to the surface. This is often initiated by a combination of feed contaminating the skin and splitting of the skin caused by trauma. *Staphylococcus hyicus* then invades and causes infection. The pig is attracted to the lesion and eventually this leads to vice. This situation is particularly apparent when pigs are first weaned into nurseries or when they are moved into second stage accommodation, particularly if mixing takes place. New concrete has an alkaline surface, and the high pH combined with prolonged pressure to the skin, particularly when the pig is lying on slats, causes sores to develop over the ham or flanks. This can lead to infection and then vice. Greasy pig lesions on the tail are an irritant causing considerable tail movement which becomes attractive to other pigs. Other diseases such as pneumonia can result in disadvantaged pigs being traumatised by others (Fig 9-39).

Treatment

- Determine the antibiotic sensitivity of the *Staphylococcus hyicus* if this is part of the problem and medicate feed for 7 to 10 days. Assess the results of strategic medication.
- Inject traumatised pigs with long-acting preparations of penicillin or OTC amoxycillin or tulathromycin.
- Provide pain management.
- Provide toys and distractions.

FIG 9-38: VICE-PREDISPOSING NUTRITIONAL FACTORS

- Low salt in the diet.
- Inadequate nutrition.
- Rations with small particle sizes.
- Diet changes.
- Poor feed availability.
- Feeding pellets.
- Inadequate water.

FIG 9-39: VICE-RELATED DISEASE FACTORS

- Greasy pig disease.
- Wet eczema.
- New concrete and skin trauma.
- PRRSV skin lesions.
- Colitis.
- Swine pox.
- Skin trauma.
- Pneumonia.
- Parasites.
- PMWS.

Management control and prevention

- Identify and correct the causal factors outlined above.
- Spray pigs with a 1 % skin antiseptic when housing is changed and continue this daily for 2 days.
- Spraying with a heavy industrial scent will help to reduce fighting when pigs are mixed.
- If *Staphylococcus hyicus* infection is part of the problem there will usually be a very good response to in-feed medication with tetracyclines.
- Remove traumatised pigs from the pen to straw-based accommodation immediately.
- Isolate offending pigs.
- Review environmental factors – in particular draughts.
- Increase salt concentrations (NaCl) from the normal 0.5 to 0.9 %.
- Review tail docking and cut tails shorter and more consistently.

Vulval biting can be a problem in gestating/dry sows and this is covered in Chapter 7. Nutrition can also be associated and this is covered in Chapter 10.

Water Deprivation (Salt Poisoning)

See Chapter 14 for further information

Water deprivation is unfortunately common in all ages of pigs and almost without exception is related to water shortage, either caused by inadequate supplies or complete loss. See Chapter 14 for information on this condition.

Yersinia Infection

This bacterium of which 2 species occur in the pig, *Y. pseudotuberculosis* and *Y. enterocolitica,* is found in the intestine. It may cause food poisoning in the human. Infection takes place by mouth and the organism is passed in the faeces for 1 to 10 weeks. It may reside in the tonsils for long periods. It normally causes little or no disease but has been associated with outbreaks of diarrhoea in weaned pigs. *Y. enterocolitica* causes inflammation of the small and large intestines and *Y. pseudotuberculosis* causes small tiny abscesses throughout the carcass. The main significance of the organism relates to cross-reactions that occur in agglutination tests for brucellosis. Pigs that are carrying the organism are likely to react positively. If this is the case it is necessary to determine the point in the rearing system when exposure takes place and break the cycle by management control. There is a response to antibiotics including tetracyclines, synthetic penicillins and fluoroquinolones.

Yersinia enterocolitica is a zoonotic disease and may infect humans, resulting in diarrhoea.

10 SKIN CONDITIONS

10 SKIN CONDITIONS

Structure and Appearance of the Skin

At birth the skin and subcutaneous tissues account for up to 10 % of bodyweight but by the time the animal has matured this has dropped to around 6 %. The boar's skin over the shoulder blade is thickened by a mat of fibrous tissue. This protects the shoulder when fighting.

The structure of the skin consists of three parts: an outer epidermis, which is the scaly surface of the skin, the dermis which is the main thick part and the sub dermis which consists of fat and connective tissue. The clinical appearance of the skin particularly in white breeds can be a useful guide to the health or disease state of the pig. When an examination is carried out the following should be noted.

Colour – In white skinned breeds, this may range from very pale, suggesting anaemia possibly from intestinal haemorrhage or iron deficiency, to red which may be generalised suggesting possible fever or sunburn, or localised or pimple sized suggesting insect bites or mange. Blue/black extremities (ears, feet, tail, snout) may suggest septicaemia e.g. salmonellosis, toxaemia, or circulatory failures.

Eczema – This describes dermatitis, where serum oozes to the surface giving rise to a wet lesion. It is often seen in traumatic lesions to the ears and flanks as a result of vice. It can occur suddenly and can be extremely painful in "dippity pig".

Hair growth – If this is excessive it may be related to low environmental temperatures, poor nutrition, or general ill health resulting from diseases such as pneumonia, swine dysentery or mange. Note that sick pigs may appear "hairy" temporarily when they raise their hair as a reaction to chilling.

Inflammation – Infection and inflammation of the superficial layers is called epidermitis and in the deeper parts, dermatitis. Epidermitis is seen typically in greasy pig disease and dermatitis is associated with bacterial infections such as staphylococci, streptococci and erysipelas. The areas of inflammation may coalesce into large patches or remain as discrete small areas or pimples.

Jaundice – The skin is a slight to moderate yellow colour but these changes are more easily observed in the mucous membranes in the eye. Jaundice may be associated with the blood parasite *Mycoplasma (Eperythrozoon) suis,* leptospirosis, or where there is damage to the liver due to toxins such as aflatoxin, migrating ascarid larvae, or poisons such as warfarin.

Necrosis – When there is restriction of blood supply to an area of the skin the surface tissue dies (called necrosis) leaving a dark area. Such changes are often seen on the teats, tails and knees of piglets as a result of trauma and in skin lesions of erysipelas diamonds, where the causal organisms block the tiny blood vessels supplying small areas of the skin.

Pustules or papules – These are small areas of inflammation usually from 1 to 3 mm in size that have red raised centres that may show evidence of pus, dead black tissue or initially appear as small vesicles. They arise after infection with viruses, streptococci or staphylococci, or allergic reactions to the mange mite.

Vesicles – These are blisters containing clear fluid which are small (< 1 mm) in the case of PRRSV virus infection, or up to 10 mm in pox virus infections. Large confluent vesicles occur around the skin horn junctions and the mouth and tongue in the vesicular diseases such as foot-and-mouth disease, swine vesicular disease or vesicular exanthema in countries where these occur.

How to Recognise Skin Conditions

Skin diseases in the pig can be broadly divided into two groups. Those conditions or specific infections that only infect the skin and have minimal effect on the pig and those that are signs of more generalised disease.

Fig 10-1 lists the conditions that may be observed and the times when they are likely to occur, from birth through to the adult animal and indicates whether or not there is a generalised effect on the pig. Note that there are only six major diseases that have any economic significance; greasy pig disease, mange, necrosis, PDNS, sunburn and the vesicular diseases.

Recognition commences by clinical observations across the herd. The following need to be considered:

- What proportion of pigs are affected?
- What age group is affected (refer to Fig 10-1)?
- Is there a generalised illness associated with the condition and can this be related to a specific disease?
- Has the condition appeared suddenly or is it one that has been present in the herd for some time?
- Have you seen the condition before and can you recognise it? If not it may be advisable to consult your veterinarian or refer to the photographs.
- Do the pigs recover without treatment?

FIG 10-1: SKIN CONDITIONS AND THE AGES AT WHICH THEY MAY FIRST BE SEEN

	Sucking Pig		Growing Pig	Adult Pig	Generalised Effects on the Pig
Approximate Days of Age	0 – 7	8 – 28	29 – 200	200 +	
Abscess	✓	✓	✓	✓	Multiple - condemnation
Anaemia	✓	✓	✓	✓	Poor growth
Aujeszky's disease/pseudorabies virus (AD/PRV)	✓	✓			Illness
Bush foot	✓	✓	✓	✓	Poor growth, pain
Bursitis		✓	✓	✓	None
Carbon monoxide poisoning	✓	✓	✓	✓	Illness/death
Cyanosis (blueing)	✓	✓	✓	✓	Illness
Dippity pig			U	U	Pig very painful
Epitheliogenesis imperfecta (defective skin)	✓	✓			None
Dermatitis/nephropathy syndrome	✓	✓			Illness
Erysipelas		U	✓	✓	Illnes /fever
Flank biting			✓		None
Frostbite	✓	✓	✓	✓	Illness
Granuloma (ulcerated)			✓	✓	None
Greasy pig disease *	✓	✓	✓	U	Severe to none
Greasy skin			✓	✓	None
Haematoma	U	U	✓	✓	None
Haemorrhage	✓	✓	✓	✓	Anaemia
Hyperkeratinization			U	✓	None
Insect bites		✓	✓	✓	None
Jaundice	✓	✓	✓	✓	Illness
Lice	✓	✓	✓	✓	None
Mange *		U	✓	✓	Poor growth
Necrosis *: Teats	✓				None
Mouth	✓	✓			Poor growth
Tail	✓				None
Ears		U	✓		None
Knees	✓				None
Flanks			✓		None
Parakeratosis			✓	U	Poor growth
Photosensitisation			✓	✓	Irritation, outdoor
Pityriasis rosea		✓	✓		None
Porcine dermatitis and nephropothy syndrome (PDNS)*			✓	U	Death
Porcine reproductive and respiratory syndrome virus PRRSV	✓	✓	✓	U	Secondary infection
Porcine stress syndrome			✓	✓	Mortality
Preputial ulcer	U	U	✓	✓	Variable
Pustular dermatitis		✓	✓		None
Ringworm			U	✓	None
Shoulder sores			U	✓	Thin sow in farrowing
Spirochaetosis			✓		None
Swine pox		✓	✓	✓	None
Sunburn *	✓	✓	✓	✓	Irritation
Thrombocytopenic purpura	✓	✓			Mortality
Tail biting			✓		Variable may be lame
Transit erythema			✓	✓	None
Trauma	✓	✓	✓	✓	Variable
Vesicular disease *	U	U	✓	✓	Poor growth

* = Economically important. **U** = Uncommon.

Use the progression pathway shown in Fig 10-3 together with Fig 10-2 which outlines diseases that may be responsible for the symptoms observed.

Identifying the Causes of Skin Conditions

If you have a problem which has skin related clinical signs refer to Fig 10-2, the index or relevant chapter. If you cannot identify the cause consult your veterinarian.

FIG 10-2: OBSERVATIONS AND CAUSES

Dark greasy skin
- Dippity pig
- Eczema
- Greasy pig disease *
- Greasy skin
- Photosensitisation
- Pityriasis rosea

Haemorrhage
- Bacterial or viral infection
- Porcine dermatitis nephropathy syndrome (PDNS)
- Swine fever
- Thrombocytopenic purpura
- Trauma *
- Warfarin poisoning

Jaundice
- Haemolytic anaemia
- Leptospirosis *
- Liver dysfunction
- *Mycoplasma suis*
- Poisons – copper

Necrosis – (Black dead areas/black spots)
- Ear, tail, teat necrosis *
- Erysipelas
- Gangrene
- Greasy pig disease *
- Photosensitisation
- Swine pox
- Vice

Open sores
- Abscesses
- Bursitis
- Dippity pig
- Eczema
- Flank chewing *
- Granuloma
- Photosensitisation

FIG 10-2: OBSERVATIONS AND CAUSES CONT.

Open sores cont.
- Shoulder sores *
- Trauma
- Vice

Reddening (Erythema) and/or blue discoloration (Cyanosis)
- Acute pneumonia
- Actinobacillus pleuropneumonia *
- African swine fever
- Classical swine fever/hog cholera (CSF/HC)
- Dermatitis
- Endocarditis
- Ergot poisoning
- Erysipelas
- Frostbite
- Heart failure *
- *Haemophilus parasuis*
- Mastitis
- PDNS
- Pericarditis
- Photosensitisation
- PRRSV
- Porcine stress syndrome
- Thrombocytopenic purpura *
- Salmonellosis *
- Streptococcal infection
- Sunburn *
- Transit erythema

Small red pimples
- Contact dermatitis
- Insect bites
- Mange *
- PRRSV lesions
- Pustular dermatitis *
- Swine pox

Swellings (under the skin)
- Abscesses *
- Anthrax
- Back muscle necrosis
- Brucellosis (testicles)
- Bursitis *
- Fractured leg
- Haematoma *
- Mastitis
- Oedema (including bowel oedema)
- Ruptured muscles
- Tail biting *
- Tumours

FIG 10-2: OBSERVATIONS AND CAUSES CONT.

Thick and roughened skin
- Callus formation *
- Chronic mange *
- Hyperkeratinization *
- PDNS *
- Parakeratosis
- Pityriasis rosea
- Ringworm

Vesicles
- Aujeszky's disease/pseudorabies (AD/PRV)
- Foot and mouth disease
- PRRSV
- Swine pox
- Swine vesicular disease
- Vesicular exanthema

* = more likely to occur.

HOW TO APPROACH AND RECOGNISE SKIN CONDITIONS

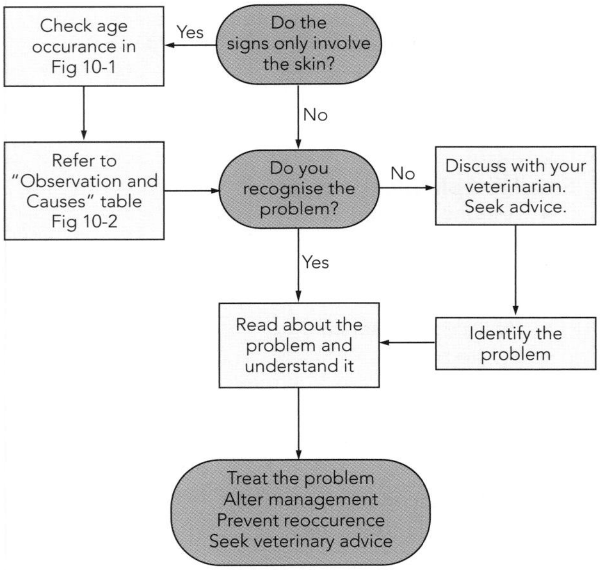

Fig 10-3

Diseases/Pathogens and Disorders Affecting the Skin

Below is a list of the key diseases that affect the skin. These are covered in detail in this chapter.

- Abscesses.
- Anaemia.
- Aujeszky's disease/pseudorabies virus (AD/PRV).
- Bursitis.
- Cyanosis.
- Dippity pig.
- *Epitheliogenesis imperfecta* or defective skin.
- Erythema.
- Erysipelas.
- Flank biting.
- Frosbite.
- Gangrene.
- Granuloma.
- Greasy pig disease (exudative epidermitis).
- Greasy skin.
- Haematoma.
- Haemorrhage.
- Hyperkeratinization.
- Insect bites.
- Jaundice.
- Lice.
- Mange.
- Necrosis of the skin.
- Parakeratosis.
- Photosensitisation.
- Pityriasis rosea.
- Porcine dermotitis and nephropathy syndrome-(PDNS).
- Porcine reproductive and respiratory syndrome virus (PRRSV).
- Preputial ulcers.
- Pustular dermatitis.
- Ringworm.
- Scrotal hemangioma.
- Shoulder sores.
- Sunburn.
- Swine pox.
- Tail biting.
- Thrombocytopenic purpura – bleeding.
- Tumors.
- Ulcerative spirochaetosis (ulcerative granuloma).
- Vesicular diseases.
- Vulval oedema.

Abscesses

Abscesses, as shown in Fig 10-4, are pockets of pus that contain dead cell material and large numbers of bacteria. The bacteria normally enter the body through damage to the skin or via the external orifices. They become walled off from the body tissues, or the bacteria are disseminated by the blood stream to develop abscesses elsewhere in the body. Near the skin surface they may become painful with an inflamed appearance.

Clinical signs

They commonly arise from fighting particularly when sows are grouped at weaning. Initially there is a break in the skin which leaves a scar followed by swellings beneath. Abscesses can also arise as secondary infection to other conditions such as swine pox, PRRSV, pneumonia or tail biting and if they become widespread throughout the body, the result may be emaciation followed by death or condemnation of the carcass at slaughter.

ABSCESS ON THE NECK

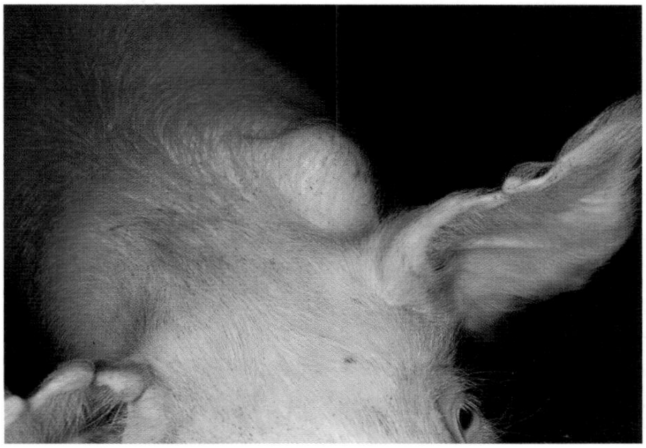

This abscess has resulted following faulty vaccination.

Fig 10-4

Diagnosis

This is based on the clinical signs of abnormal swellings under the skin especially with overlying scars. To confirm the diagnosis, feel and press the swelling to ascertain if the contents are fluid or solid and whether they are beneath the skin or deep seated. To examine the swelling more closely, restrain the pig by a wire noose or by heavy sedation and sample the contents. This is carried out using a 10 ml syringe with an 18 mm 16 gauge needle attached. The needle is inserted at the lowest soft point of the swelling and fluid withdrawn. If it is an abscess a white, yellow or green substance of either a watery or a cheesy consistency will appear.

Similar diseases

Haemorrhage into the tissues from a recently ruptured blood vessel or a haemorrhage of long standing is the only condition likely to be confused with an abscess. In such cases either pure blood or a very thin blood stained liquid will be withdrawn. Such pockets of blood are called haematoma and if they have been present for a long time a clot will have formed, in which case only serum or a clear liquid will be withdrawn.

Treatment

- This is aimed at draining the pus. Sometimes it will occur naturally after the abscess bursts but most require lancing or opening surgically. To do this make an incision approximately 15 to 20 mm long at the lowest point particularly where it is soft and fluctuating. A sharp scalpel blade with only 15 mm exposed is inserted into the abscess in a downward movement to open it up. Carry this out only when the sow is restrained. A quick controlled movement of the blade will cause little pain, far less than trying to infiltrate a local anaesthetic. The pus should be squeezed out and the interior washed using a syringe and sterile saline solution. Such a solution is made by adding 5 grams of salt to 1 litre of previously boiled water. Make a cross cut so that the wound is kept open for at least 3 or 4 days or until all the pus has drained out, otherwise the abscess may reform. See Lancing an Abscess or Haematoma Chapter 15.
- Most of the organisms that cause abscesses in the pig are either penicillin or oxytetracycline sensitive.
- If the area is badly inflamed, squeeze into the hole an antibiotic cream (a cow mastitis tube is ideal) containing penicillin/streptomycin, oxytetracycline, amoxycillin or ampicillin.
- Treatment should be given by intramuscular injection – if the area is inflamed or the sow is ill. Medicines that could be used include:
  - Penicillin/streptomycin daily for 3 to 4 days.
  - Amoxycillin long-acting (LA) every other day.
  - Oxytetracycline (LA) every other day
  - Penicillin (LA) every other day.

Management control and prevention

- Identify projections and sharp objects in the environment. A typical example would be a neck abscess associated with jagged metal on feeders. Long-acting antibiotic injections given at the time of damage will often prevent infection.
- Reduce fighting.
- Prevent Tail Biting. See Vice - Chapter 9.
- Check injection procedures.

Anaemia

See Chapter 14 for further information

Anaemia arises in the sucking piglet due to iron deficiency because the sow's milk is deficient in iron and the piglet has minimal reserves. Iron forms an essential part of haemoglobin in the red blood cells and this is responsible for carrying oxygen. The piglet becomes rapidly anaemic and susceptible to other diseases such as scours, unless iron is supplied either orally or by injection. The skin is very pale particularly if there has been a severe haemorrhage both internal and external and there may also be respiratory distress. Gastric ulcers with internal haemorrhages are a common cause in the growing pig. Ileitis resulting in massive loss of blood into the gut is seen in growing pigs and gilts.

Aujeszky's Disease/Pseudorabies Virus (AD/PRV)

See Chapter 12 for further information

Aujeszky's Disease (AD), also known as Pseudorabies Virus (PRV) is an important disease of the pig and is caused by a herpes virus. The pathogen has been eliminated from a number of countries and regions so is not directly relevant to many pig farmers.

Aujeszky's Disease is not normally regarded as a cause of skin changes, but in acute outbreaks small vesicles of approximately 1 mm in diameter may be seen on the skin around the nose and the mouth. (Fig 10-5) Similar lesions are occasionally seen in PRRSV.

AUJESZKY'S DISEASE

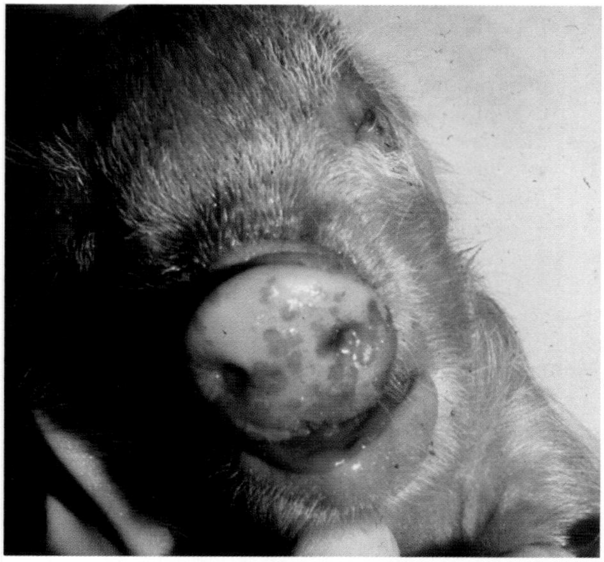

Vesicles on the nose of a 2-day-old piglet.

Fig 10-5

Bursitis

Bursitis (Fig 10-6) is a common condition that arises from constant pressure and trauma to the skin overlying any bony prominence. The periosteum or covering over the bone reacts by creating bone, a swelling develops and the skin likewise responds and becomes thicker, until there is a prominent soft lump. It can commence in the farrowing house, particularly if there are bad floors but it usually starts in the weaner accommodation on slatted floors. As the pig increases in weight there is increased pressure on the leg bones. Swellings develop over the lateral sides of the hocks and elbows and over the points of the hocks. Such swellings are called bursa although strictly speaking they are not. The term should apply to inflammation of bursa that cover tendons. Worn and pitted floor surfaces particularly if sharp aggregate was used in the concrete, can exaggerate the trauma to the extent that the skin is broken and secondary infection develops. If this occurs on wet dirty floors major problems can arise. Where breeding stock is being produced on-farm the management system needs to be adjusted, otherwise rejection rates on breeding gilts will be high.

Wire mesh, woven metal and metal bar floors can produce high levels in weaner pigs in first and second stage housing. Identify the point at which disease first appears and alter the floor surfaces or change the environment.

Clinical signs

These can develop when piglets are 1 to 2 weeks old particularly where farrowing place floors are totally slatted. Metal bars are particularly bad. Most swellings commence on the hind legs below the point of the hock or on the lateral aspects of the elbow. With repeated trauma the lesions increase in size and ultimately fluid appears. This is common in pigs 30 to 70 kg (65 to155 lbs) weight.

Secondary infection with *Mycoplasma hyosynoviae* can occur. Injury may also be seen at the base of the tail over the shoulder blades and the knees.

Treatment

- There is no specific treatment that will reduce the bone reaction. Remove pigs to pens that are well bedded.
- If the swellings have become infected with bacteria inject with either oxytetracycline or ampicillin.
- If *Mycoplasma hyosynoviae* is causing infection use either lincomycin or tiamulin.
- Most lesions do not require treatment.

Management control and prevention

- Move severely affected animals onto deep bedded floors.
- Determine the point at which lesions are occurring and relate to floor surfaces.
- If the problem is arising in flat decks in breeding gilts it may be necessary to change the slats to those covered with plastic. Tri-bar or metal slats and woven mesh are bad surfaces.
- Lime wash which also helps to reduce the floor abrasiveness.

COMMON SITES OF BURSITIS

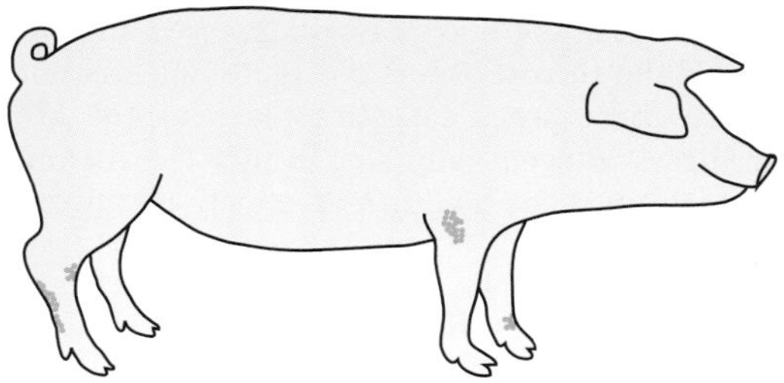

BURSITIS

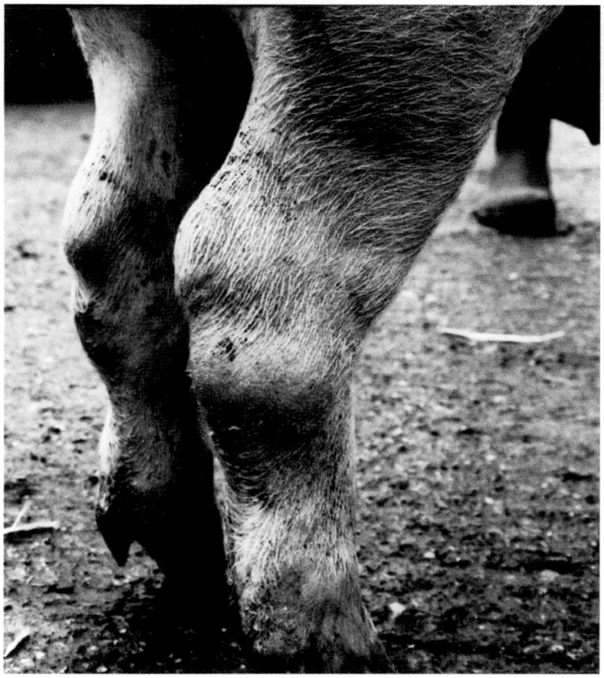

Swellings are evident on both legs below the hocks.

Fig 10-6

Cyanosis

Cyanosis (Fig 10-7) is the name used to describe a blue or red discoloration of the skin which may or may not be localised to small areas, mainly the extremities. It is not a specific skin condition but a symptom of generalised disease.

The colour changes are associated either with poor circulation due to heart disease, toxic conditions, thrombosis of blood vessels or poisoning, particularly by carbon monoxide.

CYANOSIS

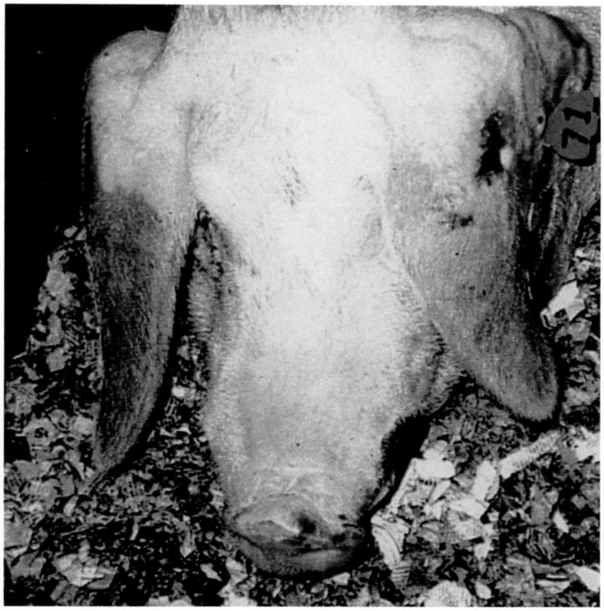

There is a clear line of demarcation on the left ear from the normal skin. This picture is typical of PRRSV or blue ear disease or toxaemias.

Fig 10-7

Dippity Pig

Individual pigs, often pet pigs, are presented in acute pain. The pig will suddenly arch its back, collapse and scream in pain. Examination of the pig reveals patches of wet eczema often in parallel lines across the back. The condition generally heals within a couple of days without treatment. Provide pain relief and clean the affect area of skin. Lincomycin may be helpful if the skin is secondarily affected.

Epitheliogenesis imperfecta or Defective Skin

This congenital condition is where a piglet is born devoid of areas of skin that have failed to develop correctly. (Fig 10-8) If these areas are small they will gradually heal but if they are more than 15 mm in diameter it will be necessary to infiltrate local anaesthetic, loosen the skin and stitch it together. (See Chapter 15: Suturing Skin and Muscle). This will depend on the availability of skin and whether the loss is over the flanks where there is plenty of skin, or over the legs where there is little. In severe cases the piglet should be destroyed.

EPITHELIOGENESIS IMPERFECTA

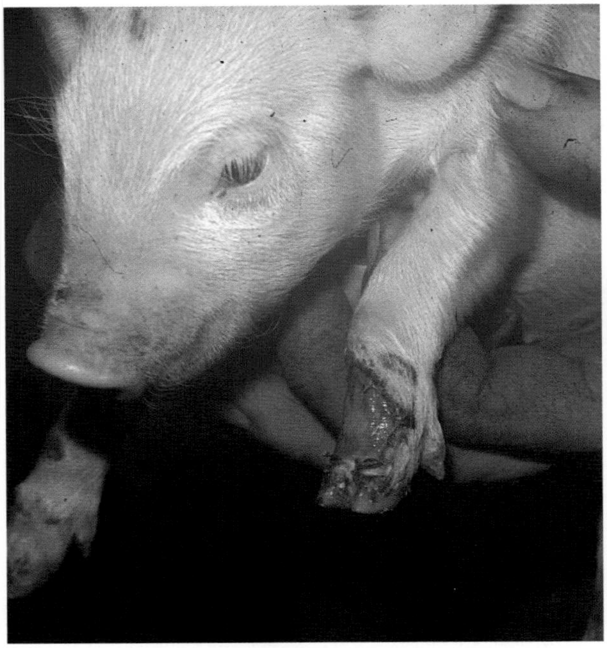

The front leg is a common site.

Fig 10-8

Erythema

This describes a reddening or blue discoloration of the skin, often transient, and commonly seen during transport. The pig's skin is sensitive to irritants such as urine, sawdust and disinfectants. Generalised discoloration particularly the extremities is seen in toxic conditions, bacterial septicaemias and viraemias and where there are circulatory problems. Occasionally when sows are bedded on urine soaked shavings in farrowing houses their complete body turns blue, the sow appearing quite normal. This is usually of no consequence.

Erysipelas

See Chapter 7 for further information

This disease, caused by the tiny bacterium *Erysipelothrix rhusiopathiae (insidiosa),* produces very characteristic skin lesions often described as diamond markings. The organism enters the blood stream causing a septicaemia. In the process it forms small clumps which block the tiny blood vessels supplying the skin. Affected areas appear as raised diamond shaped patches several centimetres across, scattered over the back, flanks and abdomen. Fig 10-9 They are usually pink to purple in colour and in clean white skinned pigs are readily seen but in coloured or dirty pigs they may not be so evident. In such cases run the palm of your hand over the pig's back and sides and they can be felt as raised areas. The affected pigs may or may not be depressed and running a fever. If the thrombosis is complete and treatment is not given these areas can turn black as the affected tissue dies and eventually slough off. The organism is very sensitive to penicillin and effective vaccines are available.

ERYSIPELAS

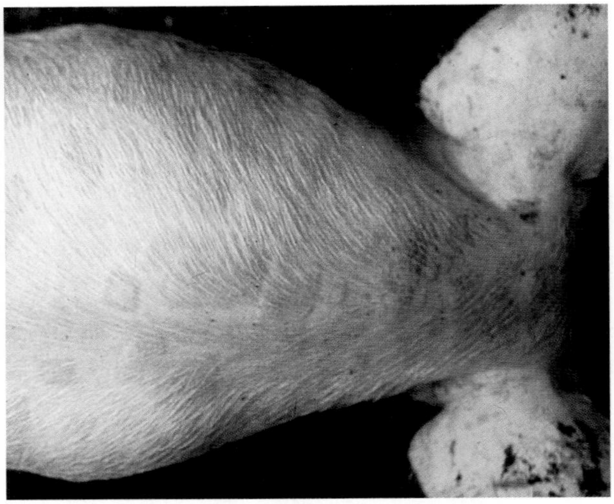

Diamond shaped skin lesions and others on the neck and flank.

Fig 10-9

Flank Biting

See Chapter 9 (Vice - Abnormal behaviour) for further information

This is sometimes seen in intensive pig producing systems but also occasionally in pigs housed in straw yards. It is often associated with ear and tail biting and usually commences as a small dark scab no more than 5 to 10 mm in diameter on the flank of the pig (Fig 10-10).

If the scab is removed a wet eczema or dermatitis is evident from which large numbers of *Staphylococcus hyicus* can be isolated. The condition at this stage is of no consequence until the scab is removed either mechanically or by other pigs that traumatise the area. This rapidly progresses into vice and in extreme cases severe cannibalism. It is important to remove the infected pig into a hospital pen and identify the offending pigs that are responsible for the vice and isolate them too. Provide toys as a distraction.

COMMON SITES OF FLANK BITING

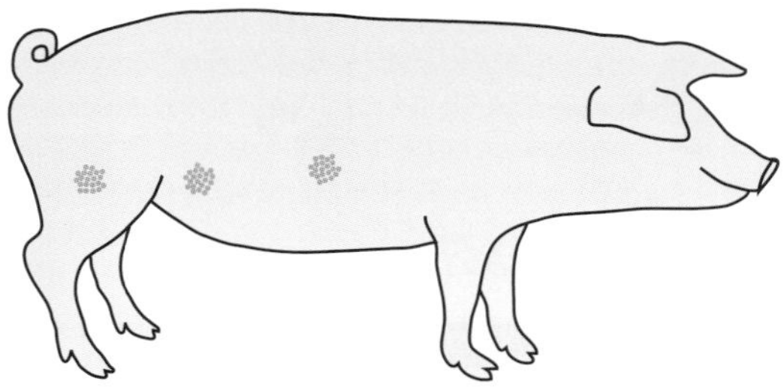

FLANK BITING

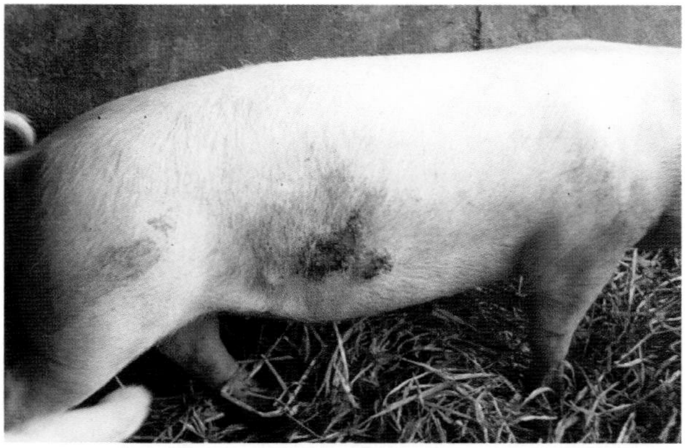

This is a developing lesion with cannibalism just commencing.

Fig 10-10

Frostbite

If pigs are exposed to very cold weather without adequate protection can become frostbitten. The ears and lips of the tails are particularly affected. The major problems occur when the toes become frostbitten and this may require euthanasia.

Gangrene

Gangrene occurs because of a shortage of blood to an area of the skin. This is generally a sequel to other problems. Glässer's disease can lead to severe gangrene of the tips of the ears, tail biting can result in gangrene of the tail, ergot poisoning reduces the blood supply to the peripheral surface and can result in gangrene of the ears, tail and feet.

Granuloma

A granuloma (Fig 10-11) is a large mass of fibrous tissue that has been produced in response to persistent trauma and irritation to the skin and underlying tissues. Granuloma can also arise due to low grade bacterial infections. A typical example would be the large lumps seen in cases of chronic mastitis. The most common sites are over the lateral aspects of the front legs, particularly the knee, hock and elbow joints and on the hind legs over the lateral aspects of the hock and the posterior parts of the legs and feet. Occasionally the granuloma will burst to the skin surface and ulcerate. Animals showing large granuloma, particularly if they are starting to ulcerate, should be slaughtered. It is also possible to amputate large ones and you are advised to discuss this with your veterinarian.

COMMON SITES OF GRANULOMA

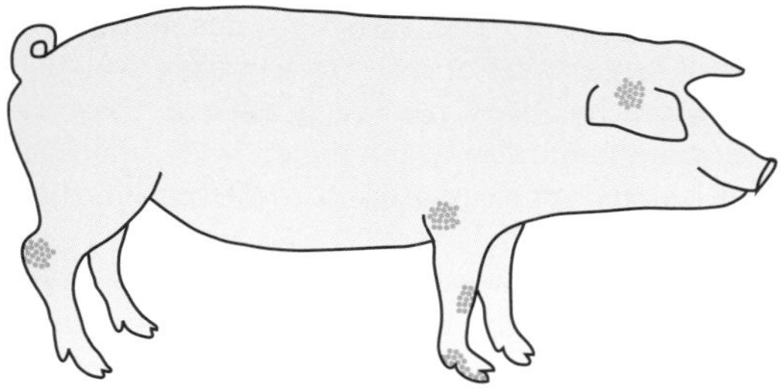

GRANULOMA

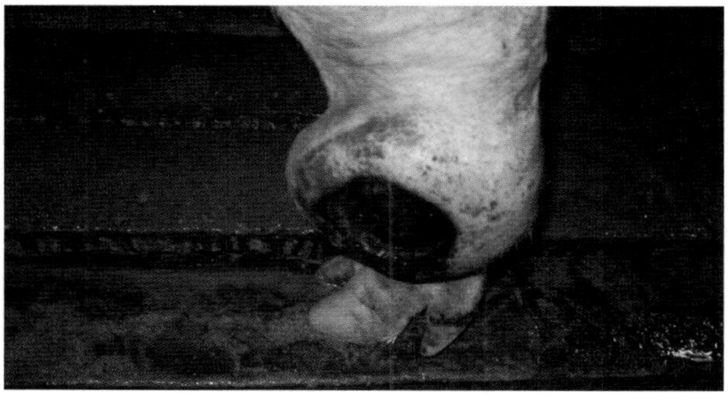

These are common on the legs.

Fig 10-11

Greasy Pig Disease (Exudative Epidermitis)

This is associated by the normal skin bacterium *Staphylococcus hyicus*, which invades abraded skin causing infection. The disease is also called exudative epidermitis which describes the oozing of fluid from the inflamed skin. *Staphylococcus hyicus* produces toxins which are absorbed into the system and damage the liver and kidneys.

In the sucking piglet disease is usually confined to individual animals, but it can be a major problem in new gilt herds and weaned pigs.

It has been shown that during the days immediately preceding farrowing the bacterium multiples profusely in the sows vagina. Piglets are frequently infected during the birth process or soon after.

A mild form of greasy pig may develop as a result of the sharp eye teeth damaging the cheeks during competition for a teat/milk, or when the knees are traumatised when seeking to suck milk. In severe cases where the liver becomes damaged the piglet will die. Often only 50 % of piglets affected during suckling will survive.

Clinical signs

Trauma and subsequent infection of the skin by the bacterium *Staphylococcus hyicus* causes a wet eczema or dermatitis. These usually commence with small, dark, localised areas of infection around the face or on the legs, where the skin has been damaged. The skin along the flanks, belly and between the legs changes to a brown colour to black colour gradually involving the whole of the body. The skin becomes wrinkled with flaking of large areas and it has a greasy feel (Fig 10-12). A more localised picture is seen if the sow has passed some immunity to the piglet, with small circumscribed lesions approximately 5 to 10 mm in diameter that do not spread.

In weaned pigs disease may appear 2 to 3 days after weaning with a slight browning of the skin that progresses to a dark greasy texture and in severe cases the skin turns black. Such cases usually die due to the toxins produce by the staphylococci organisms. In nurseries up to 15 % of the population may be involved. These changes often start around the face and ears or along the abdomen. Mortality can be quite high in the generalised form due the absorption of toxins from the organism and from dehydration. A more localised form (Fig 10-13) is seen in pigs from 5 weeks onwards as small discrete patchy areas of wet inflammation 10 to 30 mm in diameter, often covered over by a black scabs. The organism can also cause eczema of the tail and tips of the ears which leads to vice.

Diagnosis

This is based on the characteristic skin lesions. In an outbreak it is important to culture the organism and carry out an antibiotic sensitivity test. A moist wet area should be identified, the overlying scab removed and a swab rubbed well into the infected area. This should be returned to the laboratory in transport medium to arrive as soon as possible, certainly within 24 hours.

COMMON SITES OF LOCALISED GREASY PIG

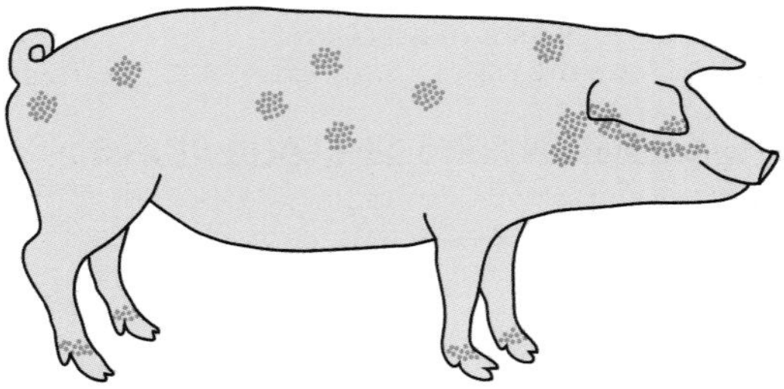

GREASY PIG DISEASE

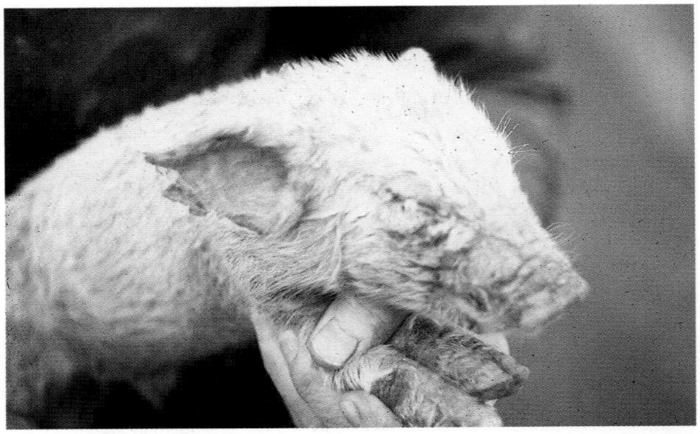

The generalised form.

Fig 10-12

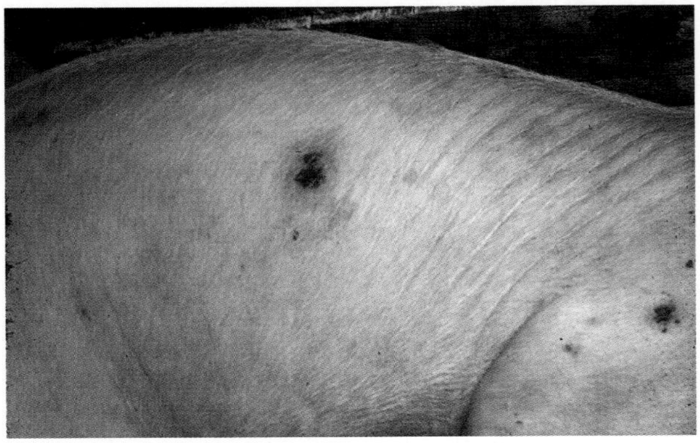

A localised lesion of greasy pig disease.
This may be cannibalised.

Fig 10-13

Treatment

- Determine the antibiotic sensitivity and inject affected piglets daily for 5 days, or on alternate days with a long-acting antibiotic to which the organism is sensitive.
- Antibiotics include: amoxycillin, OTC, ceftiofur, cephalexin, gentamycin, lincomycin or penicillin.
- Topical application of antibiotics used for treating mastitis in dairy cows can also be of use. These can be mixed with mineral oil and sprayed onto the skin or the piglets can be dipped into this solution.
- Piglets become very dehydrated and should be offered electrolytes by mouth.
- Ensure there are no mange problems in the herd. The mange mites damage the skin and allow *Staphylococcus hyicus* to enter.
- Long-acting injections can be given 2 to 3 days before the first signs are likely to appear as a method of prevention. Use either long-acting amoxycillin or OTC if indicated.
- In severe outbreaks an autogenous vaccine can be prepared from the organism and sows injected twice 4 and 2 weeks prior to farrowing to raise immunity in the colostrum. This has proved successful on a number of farms where disease has appeared in both the sucking and weaned pigs.
- If the problem is occurring in gilt litters, cross/split suckling these piglets using older sows at birth for 4 or 5 hours can be of value.
- Ensure the water supply to the pigs is excellent.

Management control and prevention

- Examine the pigs to see where abrasions are taking place. For example, these may be arising from new concrete surfaces or rough metal floors.
- If concrete surfaces are poor, brush these over after cleaning with hydrated lime that contains a phenol disinfectant.
- Check the procedures for removing tails and teeth. Jagged edges of teeth can damage the gums leading to infection around the cheeks particularly when piglets fight for teat access and during mixing after weaning. Consider stopping teeth clipping.
- The skin of the udder is one reservoir of infection. This should be sprayed daily 3 days before and after farrowing with an iodine based skin antiseptic (cow teat dip is ideal).
- If Greasy Pig Disease occurs in the farrowing house, enhance milking ability of the sow.
- Check the feedback programme to enhance antibody levels in the colostrum. Ensure piglet intake of colostrum is adequate.
- Disinfect floors well between farrowings.
- Make sure that sharp needles are used for iron injections and change these regularly between litters.
- If mange is present in the herd treat the sow prior to entering the farrowing house.
- Extremes of humidity and wet pens can encourage the multiplication of the bacteria.
- Metal floors and side panels, in particular woven metal flooring, can cause severe abrasions particularly around the feet and legs. In such cases the first signs of Greasy Pig will be in these areas. Damage to the face by metal feeding troughs can precipitate disease.
- Check the humidity of the weaning accommodation. High levels above 70 % and high temperatures provide an ideal environment for the multiplication of the bacteria on the skin.
- Adopt an all-in/all-out policy in the weaning accommodation. Have the pens bacteriologically checked after they have been washed out and disinfected.
- Review fly control programmes.
- Review water availability.

Greasy Skin

This is a condition that in the initial stages can look like greasy pig disease but is seen only in the growing pig or the sow. The skin, particularly behind the ears, eyes (Fig 10-14) between the elbow and body and the inner parts of the legs, contains a thick brown greasy material. It occasionally may involve the whole of the pig. Unlike greasy pig disease it has little if any generalised effect. It is usually of no consequence.

GREASY SKIN

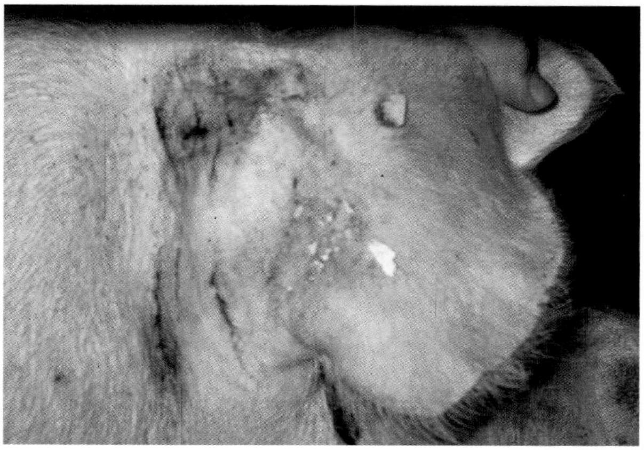

Note the thick brown waxy deposits behind the sow's ear and beneath the eye.

Fig 10-14

Haematoma

A haematoma is a pocket of blood that forms beneath the skin or in muscle tissue and is associated with a ruptured blood vessel (Fig 10-15). It is caused by fighting or external damage that ruptures a small blood vessel. The common sites are the ears and flanks.

Blood from the ruptured vessel causes pressure to build-up in the tissues, once this becomes significant the haemorrhage stops and a clot is formed. In most cases a haematoma will resolve over 2 to 3 weeks once the blood has formed a clot and the serum is reabsorbed. The clot is then slowly removed. Large haematomas on the ears cause considerable discomfort and in such cases the pig should be suitably restrained and a needle entered into the swelling. The removed fluid shows if the blood has clotted.

In the growing pig it may be advisable to lance the ear at the tip leaving a 20 mm open incision to allow drainage. If severe some producers place a band around the base of the ear and allow the ear to slough off.

In some cases the haematoma may be infected and an abscess will develop. This should be dealt with as described under abscess. Haematomas can be confused with abscesses and to differentiate between them it is necessary to use a sterile needle and syringe and sample the fluid contents.

It is not advisable in the sow to lance a haematoma if it has not developed into an abscess. Always sample the fluid first by syringe and needle. If blood is withdrawn the haematoma is of recent origin and if serum is present it is long standing. Leave both alone. Animals with large haematoma of the ear are best culled.

HAEMATOMA

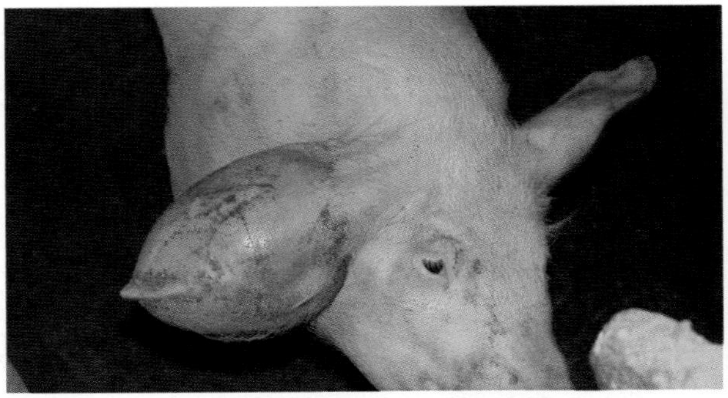

Haematoma of the ear; note the swelling.

Fig 10-15

Haemorrhage

Bleeding into the skin is not uncommon in the pig and is associated with trauma, poisons (warfarin), bacterial or viral infections. See individual conditions for more information.

Hyperkeratinization

This term describes thick layers of surface epithelial cells that become impregnated with black sebaceous material and dust from the environment. It is typically seen in confined sows, particularly over the neck and the back (Fig 10-16). Occasionally, the condition in its extreme form will involve all the upper skin surfaces. It can be confused with mange but there is an easy and simple method of differentiation.

The mange mite burrows into the skin layers causing chronic inflammation and thick scabs. Hyperkeratinization on the other hand only consists of surface debris or scurf which is easily scrubbed away by the hand, leaving a clean smooth normal skin beneath. The condition is unsightly but of no consequence. It has been associated with a shortage of essential fatty acids. 4.5 litres of cod liver oil per 50 sows per week, added to the sow ration will improve the skin appearance. Alternatively essential fatty acids from other sources may be added in the diet.

COMMON SITES OF HYPERKERATINIZATION

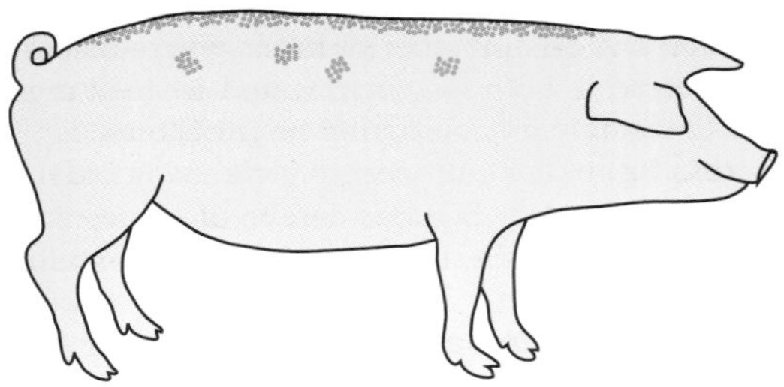

HYPERKERATINIZATION

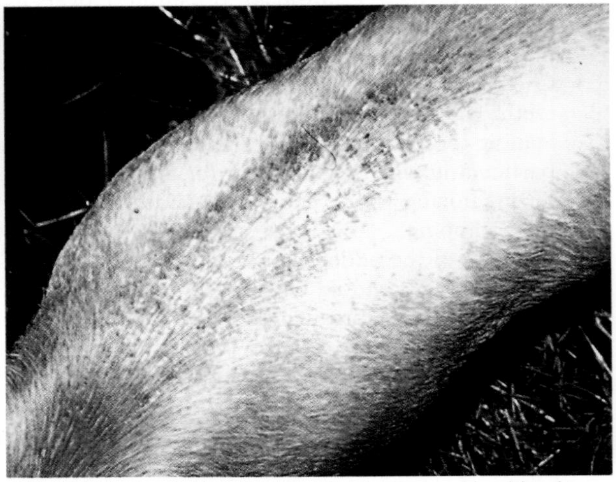

The black dead skin rubs away leaving normal healthy skin.

Fig 10-16

Insect Bites

See Chapter 11 (Flies and Mosquitoes) for further information

Skin damage associated with biting insects is surprisingly common particularly in the summer months and more so where farrowing house floors are slatted. The lesions appear as small red pimples that can look very similar to the allergic form of mange but tend to be localised behind the shoulders and the flanks where biting flies gain access through slatted floors from the slurry. Sometimes it is necessary to take skin scrapings to eliminate mange particularly in herds that are monitored and believed free of the mite.

These can affect the slaughter value of the pig especially if the carcass then needs skinning. With curtain sided buildings beware of stagnent water in the lowered curtains which can act as breeding grounds.

Jaundice

Jaundice is a yellowing of the skin and mucous membranes due to the breakdown of red cells in the blood, the accumulation of the by-products in the liver and the production of a substance called bilirubin. The condition may be seen at any age but is usually confined to individual animals or litter mates.

In sucking pigs it is associated with a haemolytic anaemia where the piglets' red cells become sensitised by antibodies in the colostrum of the sow. (See Purpura later in this chapter). Jaundice will follow the breakdown of red cells caused by the blood borne parasite *M.suis* and be seen in sucking and weaned pigs and occasionally in the sow. Jaundice can also occur in individual pigs under about 3 months of age, when the blood is infected with bacterium, *Leptospira icterohaemorrhagiae*, derived from rats' urine.

It produces a toxin that breaks down red blood cells. Jaundice can also be caused by direct damage to the liver by fungal toxins such as aflatoxin or fumonisin which may be present in-feed components such as peanuts or corn. It can also be caused by coal tar toxicity from eating fragments of clay pigeons, builder's tar or by ingesting high levels of copper in-feed, or by vitamin E and selenium deficiency.

In all of these there are usually other severe clinical signs such as loss of appetite, depression, and respiratory distress. It can also occur (rarely) from heavy ascarid worm infestations blocking the tube from the gall bladder to the intestine. Use the "Observation and Causes" tables to help identify the different causes of jaundice.

Lice

See Chapter 11 for further information

These insects are visible by the naked eye (Fig 10-17). They are approximately 3 mm long and are commonly found behind the ears and elbows. They also congregate between the legs. They suck blood and can be responsible for anaemia and the transmission of blood borne infections.

LICE

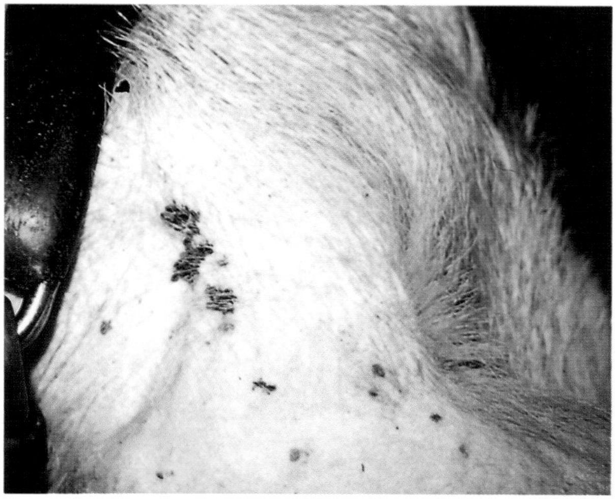

These are easily seen by the eye and are often grouped together behind the ears.

Fig 10-17

Mange

See Chapter 11 for further information

Mange is a parasitic disease of the skin caused by two mites, *Sarcoptes scabiei* var *suis* or *Demodex phylloides,* the latter of which is relatively unimportant in swine.

Sarcoptic mange (sometimes called scabies) is by far the most common in the pig.

The disease is characterised initially by tiny red pimples over the skin, particularly the back and the flanks. In the chronic condition there are thick asbestos like scabs, mainly within the ear but also behind the ears, behind the elbow and on the anterior surface of the hind legs. Often there is slight bleeding evident. Irritation and rubbing are constant findings together with ear shaking (Fig 10-18a and 10-18b).

COMMON SITES OF CHRONIC MANGE

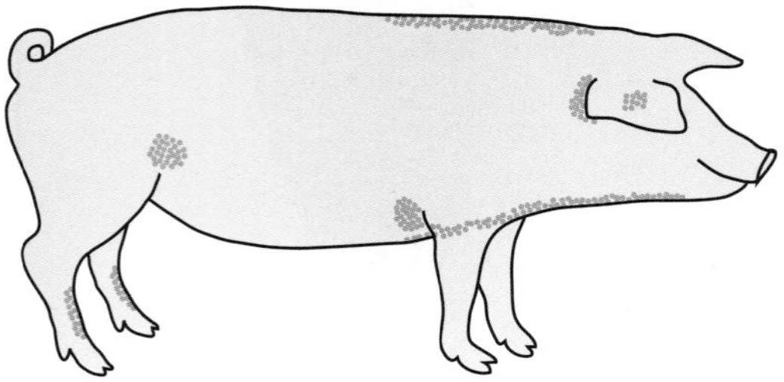

MANGE

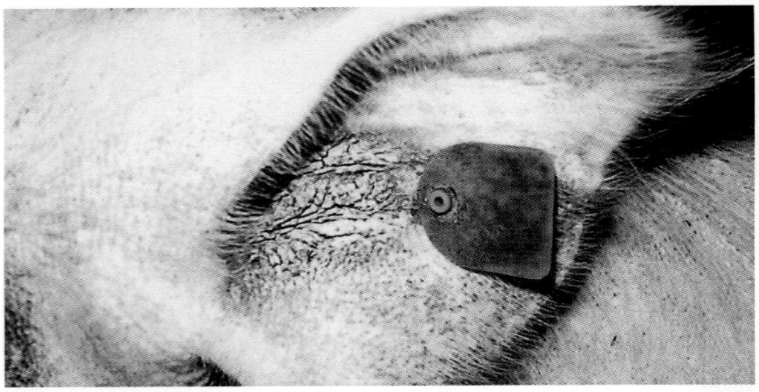

Chronic mange. Note the thick crust in the ears.

Fig 10-18a

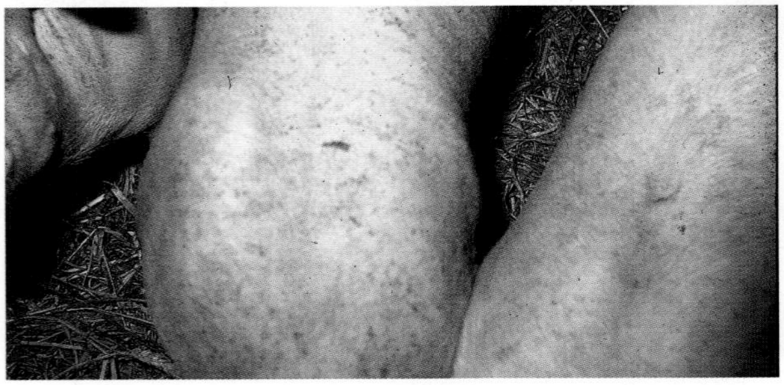

Acute mange. The skin is covered by minute red pimples.

Fig 10-18b

Necrosis of the Skin

Necrosis means that the cells and surrounding tissues have died (Fig 10-19a to d). It arises in one of three ways, from pressure on the skin from the environment, trauma causing necrosis of the teats, knees and the tail, or as a sequel to infectious diseases. It is common on the knees in the sucking pig.

COMMON SITES OF NECROSIS OF THE SKIN

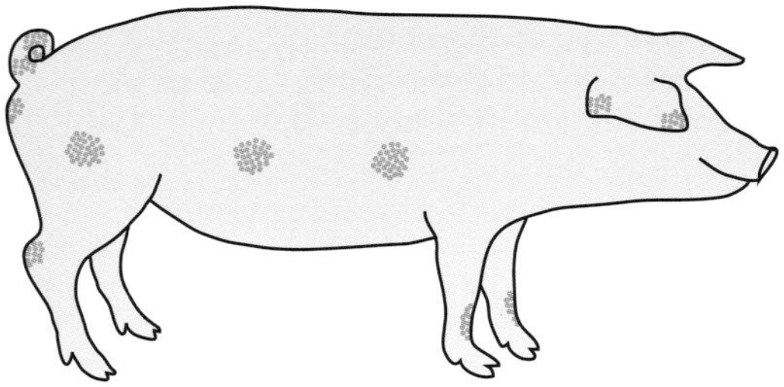

KNEE NECROSIS

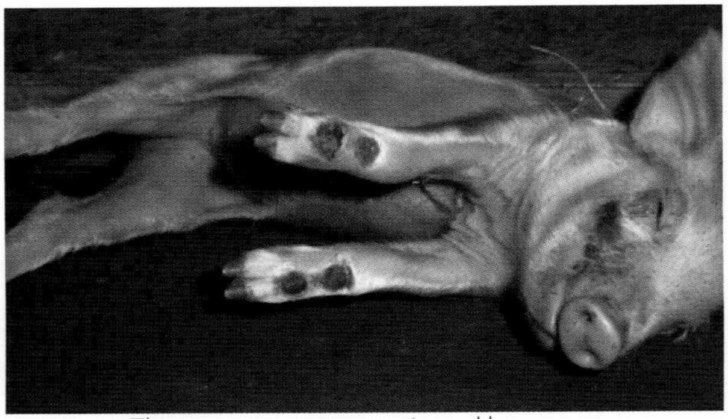

This can cause severe pain and lameness.

Fig 10-19a

TEAT NECROSIS

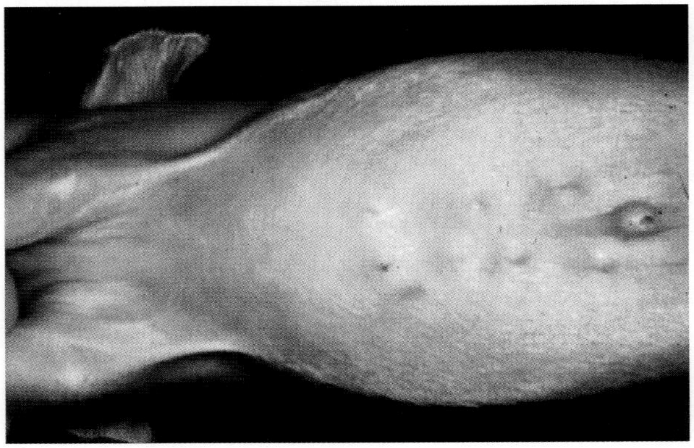

Note the small black damaged teat ends.

Fig 10-19b

TEAT NECROSIS

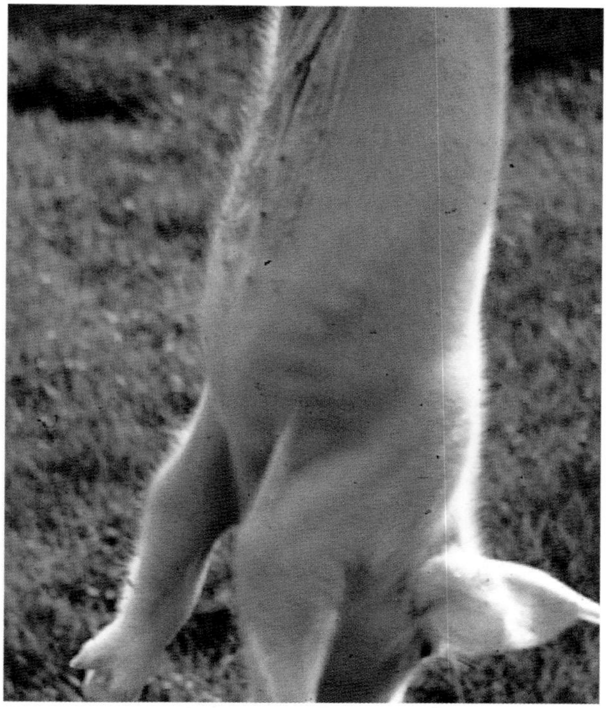

The anterior teats protected by cow gum.

Fig 10-19c

EAR TIP NECROSIS AND BITING

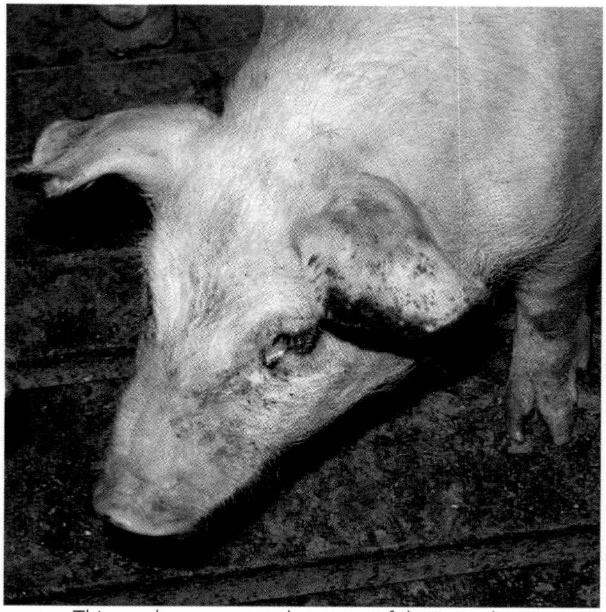

This can be so severe that most of the ear is lost.

Fig 10-19d

Parakeratosis

This is seen in pigs from 5 to 16 weeks of age associated with a deficiency of zinc and/or an excess of calcium which suppresses the availability of zinc in the diet. Up to 50 % of pigs may be affected. The major signs are limited to the skin where gross thickening and roughening occurs over the complete body. It may start initially with small light brown spots or papules on the legs and abdomen and in young pigs it can look like greasy pig disease. Treatment and control involve analysing the levels of calcium in the diet (normal 0.6 and 0.7 %). Zinc oxide or sulphate can be added to the ration at a level of 50 ppm to prevent the disease. The condition is uncommon. It may occur when the nutrients are missed during feed mixing.

Photosensitisation

This occurs in outdoor pigs that have been in contact with substances that make the skin sensitive to ultra violet radiation. These include Alfalfa, Clover, Rape, Lucerne and a Fungus that grows at the base of grass in dry weather, *Pythomyces chartarum*. Certain medicines, in particular tetracyclines and sulphonamides can also have a similar effect following prolonged use. The disease is characterised by a reddening or erythema over the white areas that are exposed to sunlight (Fig 10-20). The affected surfaces are damaged and become coagulated with serum followed by secondary bacterial infection and eventually a thick crust is formed. These changes cause a considerable amount of pain and affected animals should be moved indoors and if necessary given broad spectrum long-acting antibiotic treatment by injection to prevent secondary infections. Amoxycillin could be used. Pregnant animals may abort or absorb the embryos.

PHOTOSENSITISATION

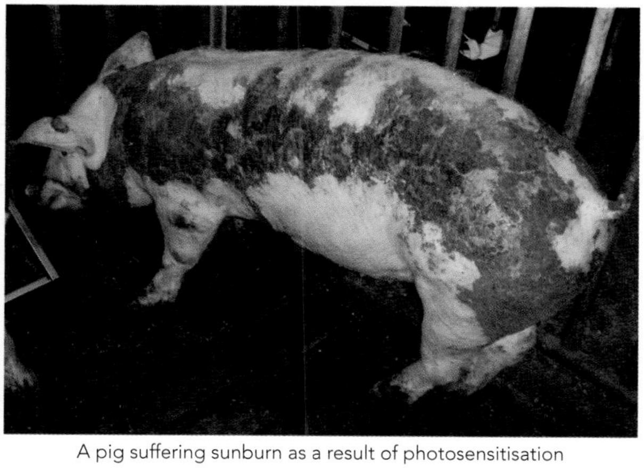

A pig suffering sunburn as a result of photosensitisation

Fig 10-20

Pityriasis Rosea

This is a sporadic condition seen in young pigs from 3 to 16 weeks of age. It is characterised by large coalescing ringworm-like lesions that often start on the abdomen and spread up behind the back legs and ultimately in severe cases involve the whole of the body (Fig 10-21). It is believed to have a hereditary background particular in the Landrace breed. The lesions are characteristic and the condition naturally resolves itself over 6 to 8 weeks. No treatment is required. It is of no consequence apart from being unsightly but may cause customer reactions if you are selling 25 kg (55 lbs) weaners to finishers.

COMMON SITES OF PITYRIASIS

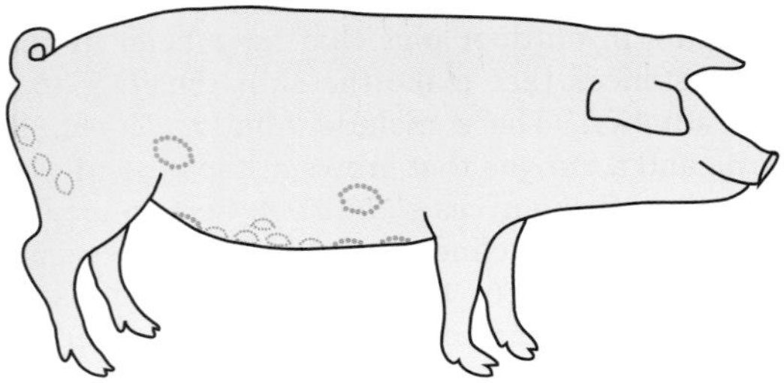

PITYRIASIS ROSEA

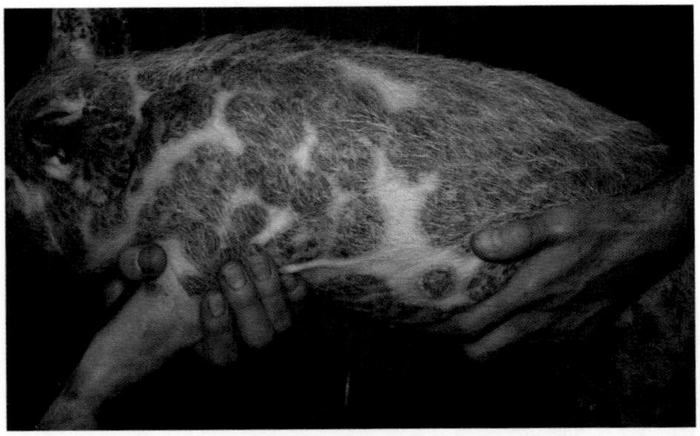

Note the raised ringworm like lesions.

Fig 10-21

Porcine Dermatitis and Nephropathy Syndrome (PDNS)

See Chapter 9 for further information

PDNS occurs mainly in growers and finishers, 12 to 16 weeks of age although can occur sporadically in other age groups. Within a herd, clinical signs may occur in a few pigs and the disease may then go undiagnosed. Alternatively, signs may occur in a bigger proportion of the herd and become economically damaging.

COMMON SITES OF PDNS

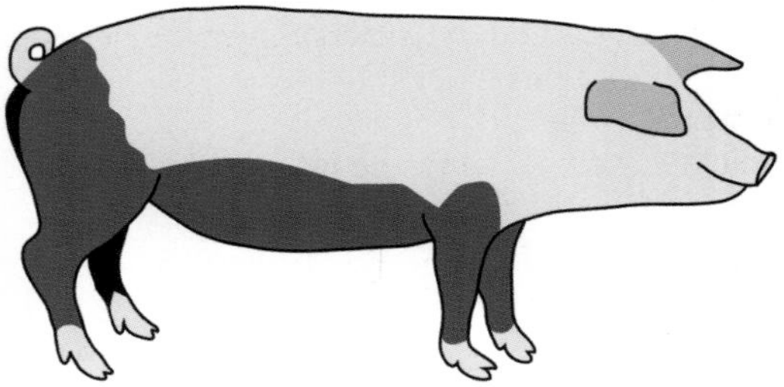

Porcine Reproductive and Respiratory Syndrome Virus (PRRSV)

Occasionally skin lesions are seen in PRRSV that are characterised by small discrete vesicles anywhere on the body but particularly around the nose and the shoulders at points of pressure. The vesicles rupture, become infected and dark coloured and ultimately heal over a 3 week period.

It is not uncommon where PRRSV is active in growing pigs to see a generalised form in 1 to 2 % of pigs (Fig 10-22). The lesions look similar to localised Greasy Pig Disease but close examination will show tiny vesicles covered with black scabs 1 to 10 mm in diameter.

SKIN LESIONS ASSOCIATED WITH PRRSV

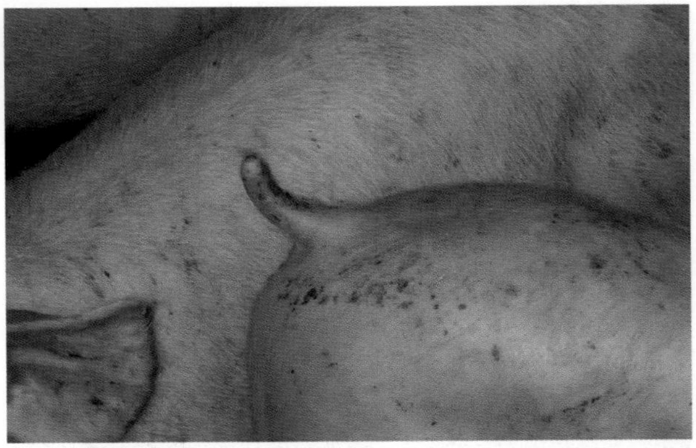

Note the tiny black lesions and the larger ones beneath the hair.

Fig 10-22

Preputial Ulcers
These occur as areas of wet eczema around the skin on the end of the prepuce.

Clinical signs
The ventral part of the prepuce near the opening is swollen, oedematous, red and painful. It is only shown in outdoor boars and is sporadic. The lesions are usually quite obvious and may become extensive.

Diagnosis
The cause of this condition is unknown but the possibility of a virus infection cannot be ruled out. Frost bite may be another cause. Secondary bacterial infection develops.

Treatment
- Isolate the boar until the lesions have healed.
- Apply an antiseptic antihistamine cream.
- Spray with antibiotic.
- If the prepuce is badly infected inject the pig with long-acting amoxycillin.
- Consider culling affected animals.

Pustular Dermatitis
This disease is seen in young growing pigs and occasionally in adults. The skin becomes infected with staphylococci or streptococci bacteria and small circular raised red areas appear (Fig 10-23). These are similar to acne in the human. In some cases the condition can affect large areas of skin. It is usually confined to individual animals and recovery takes place over 2 to 3 weeks. Antibiotic injections will help and lincomycin, amoxycillin or tetracyclines could be used. It can be confused with mange, localised greasy pig disease and pig pox.

PUSTULAR DERMATITIS

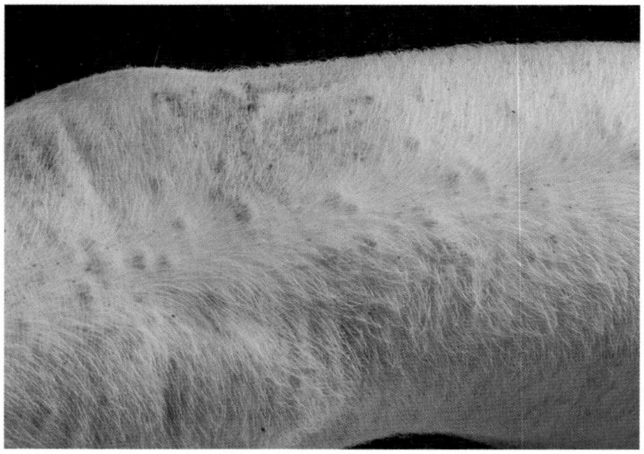

Fig 10-23

Ringworm
This is an uncommon condition in the pig but where it does exist it is of little economic significance. However it is a condition that can be transmitted to the human. It is caused by a dermatophyte fungus. Unlike other animals there is no specific host-adapted species. *Trichophyton* and *Microsporum* species are involved. Infected skin shows gradually increasing circular areas of light to dark brown discoloration behind the ears and on the back and flanks.

Infection can occur in all classes of stock. The fungi enter the skin through abrasions and diagnosis is made by examining scrapings from suspicious areas under the microscope to look for fungal spores. Treatment consists of washing the area with skin disinfectants or fungicides. In cases where infection is widespread, which would be rare, the pig can be treated orally with griseofulvin antibiotic at a level of 10 mg/kg for a period of 7 days.

COMMON SITES OF RINGWORM

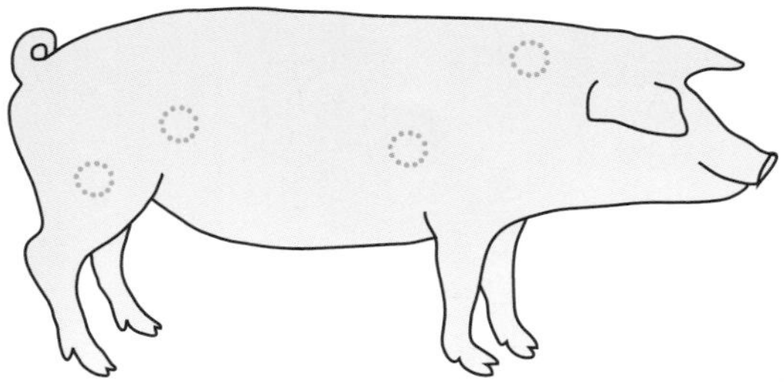

Scrotal Hemangioma
The scrotum of boars can become covered with quite large lumps. These are scrotal hemangioma's and while unsightly are generally irrelevant.

Shoulder Sores
Shoulder sores arise due to constant trauma over the bony prominences on the shoulder blade. Ultimately the skin breaks, there is an erosion and a large sore develops. It is generally associated with totally slatted flooring and individual sows that have a prominent spine to the shoulder blade. It is often first noticed in the farrowing house where floors are slippery and the sow has difficulty in rising, thus constantly bruising her shoulder. Affected sows should be moved onto strawbedding to help the sore heal.

Clinical signs
At the highest point on the spine of the scapula or shoulder blade a reddening of the skin first appears

which gradually forms into an ulcer (Fig 10-24). In severe cases the lesion may extend to 40 to 70 mm (1.5 to 3 inches) in diameter with the development of extensive granulation tissue. Often both sides of the shoulder are affected.

Treatment

- As soon as the condition appears move the sow into a well bedded pen. Feed ad lib for 2 to 4 weeks.
- Cut a hole slightly larger than the sore in a 70 mm (3 inches) square piece of foam or thick carpet and place over the shoulder sore. Hold it in place with contact adhesive. This pad will then protect the sore and allow it to heal.
- Watch for cannibalism by sucking pigs. If this occurs wean the sow.

SHOULDER SORES

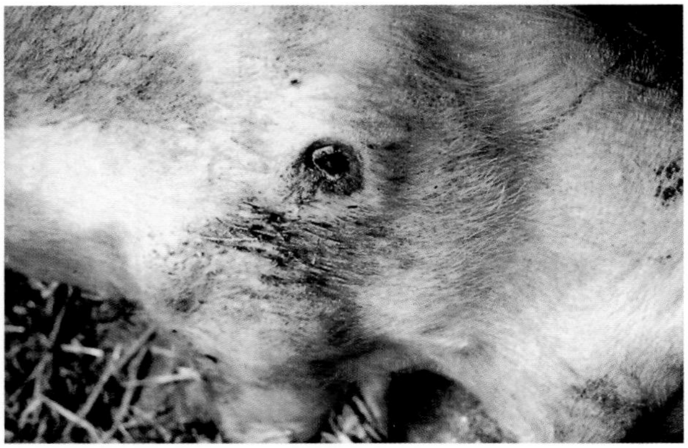

This shoulder sore has become a large ulcerating granuloma.

Fig 10-24

Sunburn

This is common in the white non pigmented breeds, some of which can be highly susceptible to ultra violet radiation. The symptoms are similar to those in the human with rapid reddening of the skin and considerable pain.

In severe cases oedema and oozing of serum may take place with secondary bacterial infection. One major problem with sunburn in outdoor or exposed weaned sows is their refusal to stand for the boar at mating.

Ultra violet radiation can also cause embryo absorption and abortions in pure white breeds. Outdoor pigs can be protected by shades and access to good wallows throughout the year. Ensure that the breeds used have pigmented skins.

Swine Pox

This is a disease caused by the swine pox virus which can survive outside the pig for long periods of time and is resistant to environmental changes. It is a vesicular disease characterised by small circular red areas 10 to 20 mm in diameter that commence with a vesicle containing straw-coloured fluid in the centre. After 2 to 3 days the vesicle ruptures and a scab is formed which gradually turns black. The lesions may be seen on any part of the body but are common along the flanks, abdomens and occasionally the ears. There is no treatment and the condition usually resolves itself spontaneously over a 3 week period. Infection during gestation produces piglets with lesions. The piglets can be quite sick.

It can be spread by lice or mange mites. It can be confused with localised greasy pig disease, pustular dermatitis and the allergic form of mange.

Tail Biting

See Chapter 9 (Vice) for further information

Trauma to the tail and the skin is common under all conditions of management both indoors and outdoors. It is more common however in intensive conditions particularly where pigs are housed on slatted floors or solid ones without bedding.

Fig 10-25 shows a typical case, probably 2 to 3 days old, with considerable infection around the stump. Infection may progress into the spine or be disseminated throughout the pig causing multiple abscesses.

The Pigs environment must be carefully examined to reduce all stress factors. Prevent draughts, check water supply and salt levels.

Provide toys as a distraction.

TAIL BITING

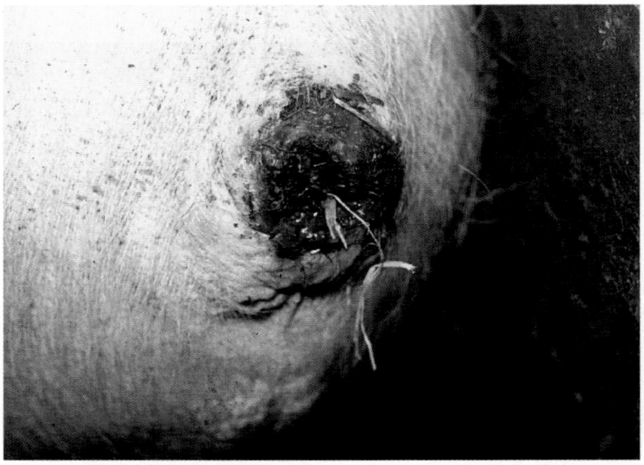

Note the swelling and infection around the tail.

Fig 10-25

Thrombocytopenic Purpura – Bleeding

This is an uncommon condition seen only in young piglets from approximately 3 to 21 days of age, most commonly around 14 days of age. It arises when the sow's colostrum contains antibodies that destroy the piglet's blood platelets (thrombocytes). This occurs because the immune system of the sow during the period of pregnancy has recognised the piglet's platelets as foreign protein and has produced antibodies against them. The formation of these antibodies is also related to the boar that is used.

Clinical signs

These can be sudden and are indicated by good pigs found dead. Look closely at the skin of these and you will see haemorrhages wherever there has been bruising, teeth marks or trauma.

Haemorrhages are evident throughout all body tissues (Fig 10-26). The piglet dies through the failure of normal blood clotting mechanisms. The disease is very sporadic but up to half the litter may be affected. Invariably the pigs die due to internal haemorrhage in one or more organs.

Diagnosis

A typical history of 1 to 4 good piglets dying between 3 to 21 days of age with haemorrhages. Consult your veterinarian.

THROMBOCYTOPENIC PURPURA

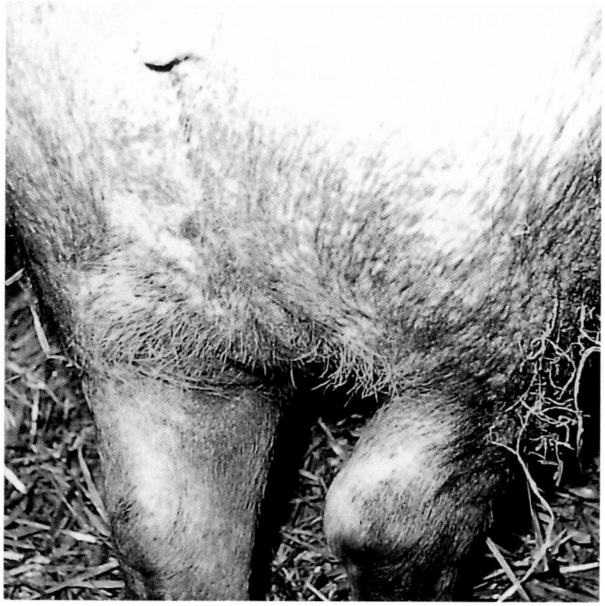

Note the haemorrhages throughout the carcass, skin and lymph nodes.

Fig 10-26

Similar diseases

Thrombocytopenic Purpura has similar signs to acute bacterial septicaemia and needs to be differentiated by laboratory analysis. *Streptococcus suis* and *Actinobacillus suis* should also be considered. The disease can appear similar to both African and classical swine fever particularly with the extensive haemorrhaging throughout the carcass.

Treatment

- There is no known treatment other than good nursing. In the early stages of the disease it is worthwhile cross-fostering litters to remove exposure to any lingering antibodies in the sow's milk.

Management control and prevention

- Where a sow has produced such a litter make sure she is mated with a different boar at the next pregnancy or cull her.
- Another litter should be ok to suckle the sow's milk.

Tumours

The pig can suffer from a variety of skin tumours. These need to be reviewed on an individual basis. They generally occur in the older pig. Full skin biopsy may be required to reach a diagnosis. In pet pigs these tumors may require surgery.

Typical tumors are melanoma, papilloma and squamous cell carcinoma. Squamous cell carcinoma are a particular problem to white pigs in hot sunny climate such as Australia.

Ulcerative Spirochaetosis (Ulcerative Granuloma)

This is caused by a spirochete bacterium called *Borrelia suis* together with secondary infections with other bacteria including streptococci and staphylococci. The disease is seen in pigs from 3 to 10 weeks of age and skin damage is first necessary to allow the organism to enter. It causes considerable irritation in the tissues, severe inflammation and the development of fibrous tissue. Diagnosis requires laboratory examination and isolation of the organism. Control involves improving hygiene, reducing trauma and identifying the areas within the management system where infection first starts and making changes to these. Infected pigs can be treated with either penicillin, tiamulin or lincomycin. It is not a common disease.

Vesicular Diseases

See Chapter 12 for further information

These include foot-and-mouth disease, vesicular stomatitis, swine vesicular disease and vesicular exanthema. All but the last disease are OIE Listed diseases. These conditions are covered in more detail in Chapter 12. These viral infections all produce blisters or vesicles around the snout, the tongue, on the teats and at the hoof skin junctions or coronary bands (Fig 10-27a to c). In most countries they are notifiable and if suspected must be reported to the authorities.

THE COMMON SITES OF VESICULAR DISEASES

VESICULAR DISEASES

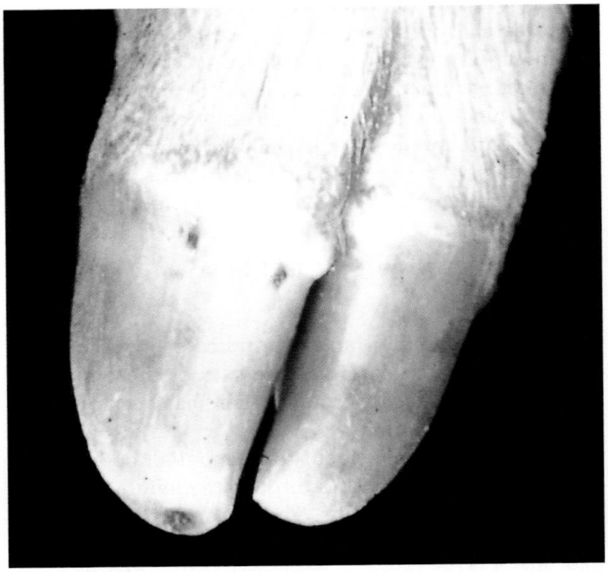

Unruptured vesicles are evident at the hoof skin junction.

Fig 10-27b

VESICULAR DISEASES

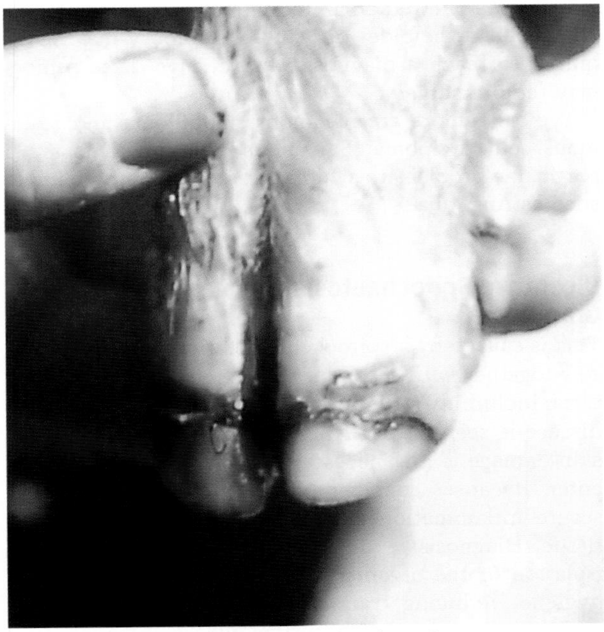

The dark areas on the hoof are due to long standing lesions of SVD.

Fig 10-27a

VESICULAR DISEASES

Ruptured vesicles are evident on the nose.

Fig 10-27c

Vulval Oedema

See Chapter 8 (Udder Oedema) for further information

Fluid accumulating in the vulva at or near farrowing is a common occurrence and in mild cases is a normal physiological process and of no consequence.

Fig 10-28 shows an extreme case extending into the udder, where it can interfere with milk let down and production.

VULVAL OEDEMA

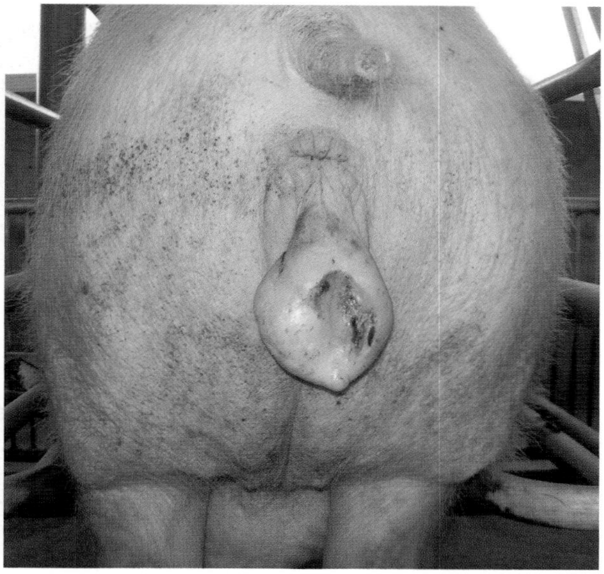

The fluid extends from the vulva between the legs and into the udder.

Fig 10-28

11 PARASITES

Chapter 11

11 PARASITES

Introduction

A parasite is an organism that at some stage must live on or within its host to survive. The relationship is usually a disadvantage for the host but occasionally it may be beneficial in which case it is called a symbiotic relationship (symbiosis).

In its broadest sense this definition would include bacteria and viruses but the term "parasites" as used commonly in veterinary medicine excludes these. Technically, the parasites dealt with in this chapter are all part of the animal kingdom and the cells of their bodies have true distinct nuclei whereas bacteria lack a true distinct nuclei. Viruses are defined differently.

Parasites in the pig are classified into 2 groups, internal (endoparasites) which live inside the body and external (ectoparasites) which live on or in the skin. They are generally host specific but there are exceptions.

Internal Parasites

These must all use nutrients from the host to multiply and survive. They are found in the digestive tract, the kidneys, liver, lungs or the bloodstream. There are 4 main groups in the pig: nematodes (roundworms), thorny-headed worms, tapeworms and protozoa (Fig 11-1).

Location of the different worms are shown in Fig 11-2.

Controlling parasites requires an understanding of their life cycle. Procedures can then be adopted that together with anthelmintics, break this cycle and thus prevent re-infection. There are 2 types of life cycle, a direct one and an indirect one.

FIG 11-1: INTERNAL PARASITES OF THE PIG

Common Name	Scientific Name
Round Worms	**Nematodes**
Kidney worms	*Stephanurus dentatus*
Large roundworms/ascarids	*Ascaris suum*
Lung worms	*Metastrongylus apri* and others
Muscle worms	*Trichinella spiralis*
Nodular worms	*Oesophagostomum dentatum*
Red Stomach worms	*Hyostrongylus rubidus*
Stomach hair worms	*Trichostrongylus axei*
Thick stomach worms (2)	*Ascarops strongylina* and *Physocephalus sexalatus*
Thorny-headed worms	*Macracanthorhynchus hirudinaceus*
Threadworms	*Strongyloides ransomi*
Whipworms	*Trichuris suis*
Tapeworms	**Cestodes**
Pork bladder worms Human tapeworms	*Cysticercus cellulosae* *Taenia solium*
Protozoa	
Balantidium coli	*Balantidium coli*
Coccidiosis	Coccidia, *Isospora*, *Eimeria* species
Cryptosporidiosis	*Cryptosporidium parvum*
Pneumocystis	*Pneumocystis carinii*
Toxoplasmosis	*Toxoplasma gondii*
Other Parasites	
Mycoplasma suis	*Mycoplasma suis*

THE LOCATION OF PARASITES

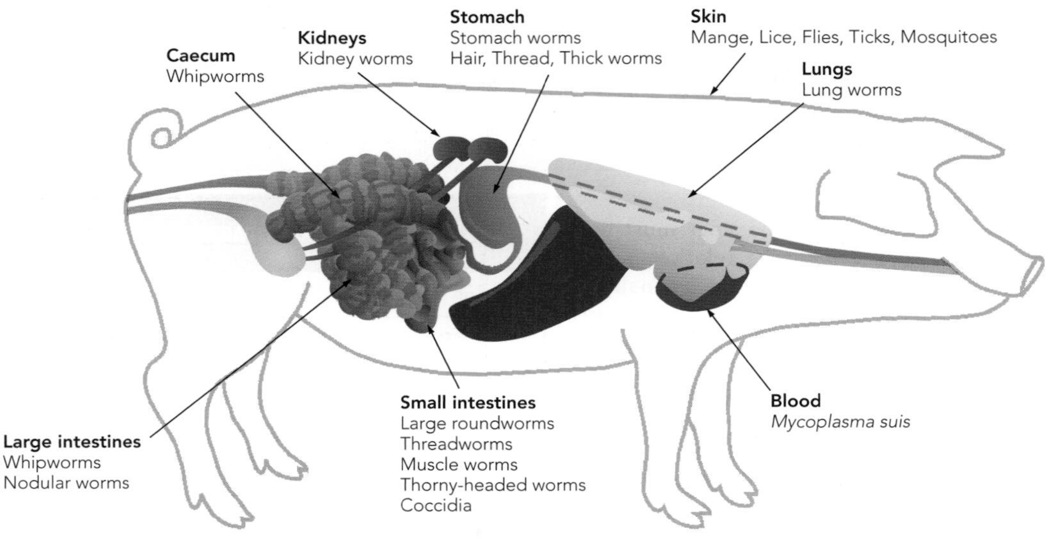

The sites where the different parasites are found.

Fig 11-2

The Direct Life Cycle

This is depicted in Fig 11-3. The adult worm lays its eggs in the intestine and they are passed out in the faeces onto the ground. The eggs then develop through larval stages, but only the last stage can infect a pig and develop into an adult worm. Some larvae (ascarids and lung worm) enter the digestive tract and migrate through the liver to the lungs before they complete their cycle.

The Indirect Life Cycle

This requires an intermediate host as shown in Fig 11-4. It commences as a direct cycle with the eggs leaving the pig with the first stage larvae developing. The egg containing the larva is eaten by a second host such as an earth worm or a beetle, where it undergoes various larval stages before finally becoming infective to the pig. The pig then eats the intermediate host and thus the cycle of reinfection is completed. An indirect cycle always requires another host for development before the larva can infect the pig. Removing or preventing access to the host breaks the cycle of infection.

The length of each cycle is dependent on the temperature and humidity of the environment. Eggs and larvae do not develop in cold conditions and most die in very dry conditions. This survival time outside the pig is important in controlling continuing infections. It also takes a number of days for the larva to develop inside the egg to the infectious stage. If faeces are removed from the environment before this development has been completed then the cycle is broken. The period of time taken for the larva inside the pig to mature to an egg

THE DIRECT CYCLE OF INFECTION

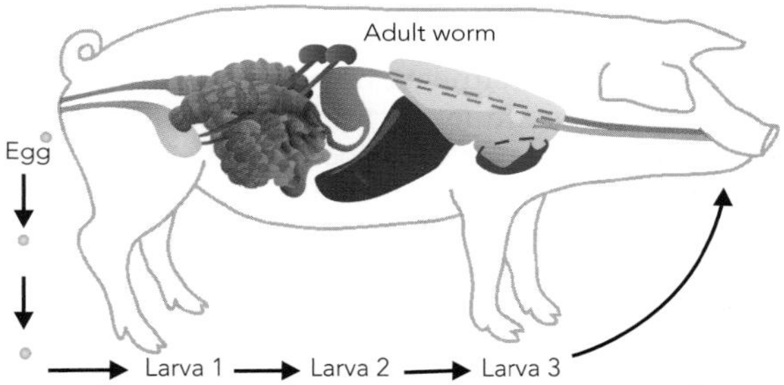

Fig 11-3

THE INDIRECT CYCLE OF INFECTION
THE LUNGWORM

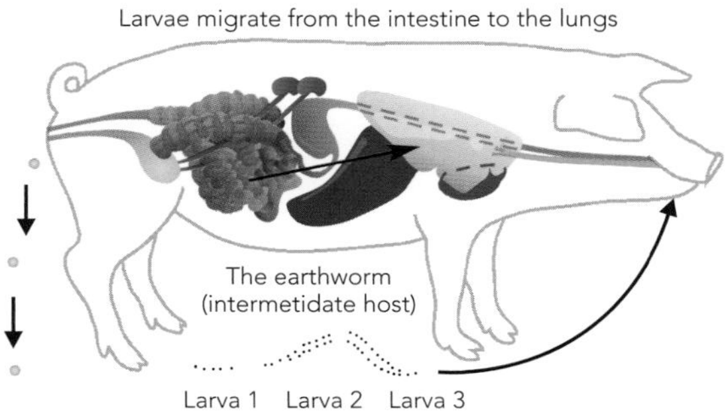

Fig 11-4

laying adult is called the prepatent period. Fig 11-5 shows the prepatent periods and also the time of survival of the eggs and larvae in the environment.

FIG 11-5: THE PREPATENT PERIODS OF THE COMMON INTERNAL PARASITES AND THE TIMES OF SURVIVAL OUTSIDE THE PIG

Parasites	Development Period *	Prepatent Period	Survival Time Outside Pig	Life Cycle Type
Round Worms				
Kidney worm (*Stephanurus dentatus*)	2 – 7 days	> 9 months	1 year	Direct & indirect. Earthworms
Large roundworm (*Ascaris suum*), ascarids	2 – 8 weeks	40 – 60 days	1 – 4 years	Direct
Lung worm (*Metastrongylus apri*)	1 day	21 – 28 days	1 – 2 years	Indirect. Earthworms
Muscle worm (*Trichinella spiralis*)	1 – 2 days	2 months	1 – 11 years	Indirect. Rats
Nodular worm (*Oesophagostomum dentatum*)	5 – 7 days	21 – 40 days	1 year	Direct
Red stomach worm (*Hyostrongylus rubidus*)	5 – 10 days	7 – 60 days	6 months	Direct
Stomach hair worm (*Trichostrongylus axei*)	5 – 7 days	3 weeks	Unknown	Direct
Thick stomach worm (*Ascarops strongylina* and *Physocephalus sexalatus*)	7 – 14 days	30 – 40 days	Unknown	Indirect. Beetles
Thorny-headed worm (*Macracanthorhynchus hirudinaceus*)	3 months	6 months	2 years	Indirect. Beetles
Threadworm (*Strongyloides ransomi*)	1 – 3 days	2 – 10 days	5 years	Direct
Whipworm (*Trichuris suis*)	3 – 4 weeks	35 – 45 days	6 years	Direct
Tapeworms				
Pork bladder worm (*Cysticercus cellulosae*) Human tapeworm (*Taenia solium*)	-	2 months – 2 years	Pig intermediate host	Indirect. People, dogs
Protozoa				
Balantidium coli	1 day	1 – 2 days	2 – 3 months	Direct
Coccidiosis (Coccidia) (*Isospora, Eimeria*)	1 – 3 days	4 – 6 days	2 weeks – 2 months	Direct
Cryptosporidiosis (*Cryptosporidium parvum*)	3 days	3 – 21 days	30 – 50 days	Direct
Toxoplasmosis (*Toxoplasma gondii*)	1 – 2 weeks	Long. Up to 2 years	Long periods	Direct
Other Parasites				
Mycoplasma suis	4 – 7 day	5 – 10 days	Unknown. Probably short	Direct

* Period from egg to larva.

Recognising a Worm Problem

This is carried out by collecting faeces samples from different ages of pigs and examining them for the presence of worm eggs. 25 g samples should be taken from the following animals: 5 lean gestation/dry sows, 5 lean suckling sows, 5 separate samples from weaner faeces at 12 weeks of age and 5 separate samples from finishing pigs at 90 kg (198 lbs). These are then submitted to a laboratory for examinations.

A 2 g portion of each sample is washed through a sieve with saturated salt or zinc sulphate solution and a small amount of this liquid containing worm eggs is flooded into a glass chamber of a known size. The top of the chamber is then examined microscopically as the eggs float to the surface and the numbers of the different eggs are counted (Fig 11-6). The levels per g of faeces are then calculated.

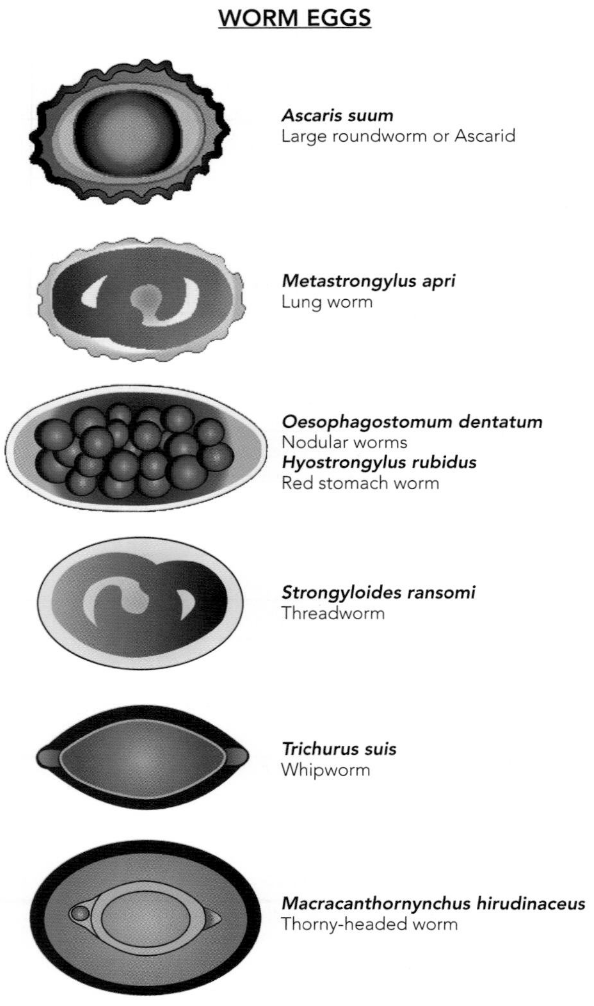

Fig 11-6

Worms of one kind or another are almost always present in commercial pig herds. Low numbers are no problem but large numbers can cause tissue damage with malfunction of the body systems that are damaged and loss of condition. It can be difficult to assess the significance of a parasite burden but the following procedures may be adopted.

Step 1

Assess the body condition, growth rates and clinical symptoms of the group of pigs.

Coughing – consider lung worm but only if the environment could give access to earth worms or beetles. Ascarid larvae as they migrate through the lungs can increase the incidence of pneumonia and coughing.
Wasting – Roundworms, Coccidiosis, kidney worms or *Balantidium coli*.
Blood in the urine – Kidney worms.
Blood in the faeces – Coccidiosis, whipworms.
Anaemic pigs – Stomach worms.

Step 2

Assess the type of environment and the way it could maintain parasites. Look at the ages of the pigs affected.

Step 3

Assess post-mortem and slaughter house information for evidence of parasites in the following organs:

Liver damage/milk spot – Large roundworms, kidney worms.
The kidney – Kidney worms.
The stomach – Stomach worms.
The intestine – Large roundworms, nodular worms.
The large bowel – Nodular worms, whipworms, *Balantidium coli*.
Muscle – Muscle worms.

Step 4

Assess the results of the faecal examination. The egg output each day is variable and the results must be interpreted by assessing all samples, together with the types of worm eggs, their numbers (Fig 11-7) and the clinical picture (Fig 11-8). The output of eggs also varies in the sow with the stage of reproduction, with increased outputs of eggs during lactation. Never make a diagnosis on egg counts alone. Judge their significance by Steps 1, 2 and 3.

FIG 11-7: A GUIDE TO INTERPRETING WORM EGG COUNTS

Worm	Eggs Per Gram of Faeces	
	Low Levels which are Not Significant (No Treatment Necessary)	High Levels which may be Significant * (If more than 50 % of Sample Groups Consider Treatment)
Coccidia	2000 – 5000	> 10,000
Kidney worms	0	any eggs
Large roundworms	< 100	> 300
Lung worms	0	any eggs
Nodular worms	< 300	> 1000
Stomach worms	< 100	> 1000
Threadworms	< 100	> 300
Whipworms	< 100	> 300

* Treatment could be required.

FIG 11-8: AGE OF PIG AND CLINICAL SIGNS

Type of Animal	Common Worms	Symptoms
Sucking piglets	Threadworms	Anaemia Bloody diarrhoea Coughing Mortality Poor growth Vomiting
Sows	Nodular worms Red stomach worms	Poor milk production Variable reproduction
Weaners, Finishers	Ascarids Nodular worms Whipworms	Coughing Diarrhoea Emaciation Liver condemnation Poor daily liveweight gain Poor growth

Management Control and Prevention

"How important are worms?" This is a question often posed with the pressures and perceived necessities to routinely treat. In indoor systems where all-in/all-out procedures are used, internal parasites will not build-up in sufficient levels to require routine treatment.

The objective therefore is to manage the environment to prevent the pig gaining access to faeces after the larva have become infective. It can be seen from Fig 11-5 that in practical terms this is approximately 5 to 7 days. A routine parasite examination of faeces every 6 months will establish the status.

Field experiences over many years have shown that if confined sows have no access to faeces, internal parasites are almost eliminated, with the exception of the large roundworm (*Ascaris suum*). The infective period for this larva is 2 to 8 weeks and generally the longer period. However the egg will survive outside the pig for long periods of time. Provided good hygiene is practised on the farm and faeces and liver surfaces are monitored, it is not necessary to treat.

The danger areas for the build-up of infections are the permanently populated areas such as boar, matingand gilt holding pens. Provided these are cleaned out regularly, herds can be maintained with negligible levels of parasites.

Parasite control in loose-housed sow herds is less predictable but again depends on hygiene, drainage and the regular removal of faeces. Field experiences with the foregoing provisos, also demonstrate it is not usually necessary to treat, but each herd must be assessed individually together with its history. Faeces should be examined every 3 months.

In finisher herds the adoption of multi-site operations or all-in/all-out have virtually negated the necessity for treatment.

In outdoor herds however parasite control is difficult. Twice yearly worming and in-between faeces examination gives good control and an insurance against the build-up of infections.

Because the exposure to parasites in many commercial systems is low the corresponding immunity levels in the pig are also low. This will mean that within any given herd a number of animals will be highly susceptible and therefore the number of larvae required to produce disease is minimal.

Where farms are free of *Ascaris suum* biosecurity protocols need to be enhanced to maintain this status.

Key points in the control of roundworms

- Assess the husbandry system. Is there access to faeces after 4 days?
  - In intensive reared indoor pigs on concrete or slats treatment is unlikely to be necessary.
  - If sow are confined treatment of sows is unlikely to be necessary.
  - If sows are loose-housed – how often are faeces removed?
  - If sows are outdoors – worming is probably necessary.
  - If sows are housed in permanent paddocks worming is essential.
- Are there permanently populated pens on the farm? Could these create reservoirs of infection? For example – gilt pens, boar pens, or grower pens. Check faeces samples.
- Assess the history of the management system. Have parasites been a problem?
- Has the slaughter house been complaining?
- Carry out a faecal screen every 6 months.

- If there is a loose-housed system with faeces remaining for 2 weeks or more a faeces screen should be carried out every 3 months. A worming programme is likely.
- If sows are outdoors use a worming programme every 4 to 6 months.
- Move outdoor sites regularly every 1 to 2 years.
- When pigs are weaned into arcs in the field always move to clean ground.
- Always move farrowing arcs to new ground between farrowings and burn the old bedding.
- Remember worm eggs and larva survive in warm damp wet conditions. Younger animals are more susceptible than older ones. Problems are more likely in summer than winter.
- Use farrowing, weaner, grower and finishing accommodation on an all-in/all-out basis. Wash out using a detergent between batches.
- Assess the body condition of sows.
- Examine livers regularly at the slaughter house.
- If parasites are a problem in growing pigs check the pens.
- Permanently populated grower finisher pens will help to perpetuate ascarid infections.
- When outdoor pigs are moved to fresh paddocks always worm the group of animals 14 days beforehand.

All-in/all-out systems should mean no clinical worm problems.

FIG 11-9: COMMON ANTHELMINTICS

	In-Feed	Injection	Water	Topical
Amitraz				✓
Doramectin		✓		
Febantel	✓			
Fenbendazole	✓		✓	
Flubendazole	✓			
Ivermectin	✓	✓	✓	
Levamisole		✓	✓	
Oxfendazole	✓			
Oxibendazole	✓			
Phosmet				✓
Piperazine	✓			
Thiabendazole	✓			
Thiophanate	✓			

Treatment Programmes for Internal Parasites

- The objective of a routine programme is to maintain infection at negligible levels. It is impossible to maintain a herd completely free of all parasites because the risk of re-introduction is high and always present. The medicines used to treat parasites are called anthelmintics or parasiticides.
- Most are broad acting. Some of the common generic compounds are shown in Fig 11-9.
- Some suggested worming programmes are shown in Fig 11-10.

Consult your veterinarian for further information and trade names.

Fig 11-11 shows the medicines that are active against different parasites and a comparison of approximate cost ratios. If the herd is free from Mange and Lung Worm then less costly anthelmintics can be used. The most effective time to treat outdoor sows and particularly sows remaining in permanent paddocks or yards

FIG 11-10: SUGGESTED WORMING PROGRAMMES

		Criteria	Action
1.	Indoor herd	Confined sows where the herd operates an all-in/all-out system. The faecal screen is negative or low.	No treatment required. Check a faecal screen every 12 months.
2.	Indoor sow herd	Sows loose-housed, faeces removed twice weekly. Faecal screen negative or low.	No need to treat. Monitor growing stock.
3.	Indoor sow herd	Sows loose-housed and faeces removed periodically. The faecal screen is low.	Worm 7 days prior to entering farrowing houses. Faecal screen every 6 months.
4.	Indoor herd	Sows loose-housed, faeces removed occasionally. Faecal screen is high.	Treat sows and boars in-feed every 3 to 6 months. Faecal screen every 3 months.
5.	Boars	Any housing.	Treat every 3 months.
6.	Outdoor sow herd	Carry out a faecal screen every 3 months.	Treat in-feed every 6 months. If the faecal screen is positive treat every 2 months during the summer.
7.	Growing pigs	Milk spot livers.	Treat all the growers in the feed. Develop an all-in/all-out system of pen usage. Wash pens between batches of pigs.

is 7 days prior to farrowing to break the cycle of infection to the piglet. However this may not be easy and the alternate is to treat the complete herd by in-feed medication every 3 to 6 months.

The following methods of treatment can be adopted:-

- Inject sows with either ivermectin or levamisole 7 days prior to farrowing or 2 days before entering the farrowing place.
- Mix the wormer in the feed according to the manufacturer's instructions either:
  – As a single dose given on one day.
  – Continuously for 7 days.
  – Continuously for 14 days.
- Top dress the sow or boars feed individually with the anthelmintic in pellets or use a small amount of the premix.

If treatment is necessary in the herd it is likely that management control procedures are defective. You may be spending money unnecessarily.

FIG 11-11: SOME MEDICINES * AVAILABLE TO TREAT PARASITES OF THE SKIN (ECTOPARASITES) AND INTERNAL PARASITES (ENDOPARASITES)

Active Medicine	Presentation/Dose Levels *	Parasite and Activity (See Fig 11-1 for Alternative Names)																
		Eggs	Kidney worm	Large roundworm	Larvae	Lice	Lung worm	Mange mites	Muscle worm	Nodular worm	Pork bladder worm**	Stomach worm (red)	Stomach worm (thick /hair)	Thorny-headed worm	Threadworm	Ticks	Whipworm	**Withdrawal period (days) ***
Amitraz 12.4 %	Topical liquid concentrate. 40 ml to 10 l water.					✓		✓								✓		7
Amitraz 2 %	Pour on to skin.					✓		✓								✓		7
Doramectin	Injection. 1 ml/33 kg lw. (300 mcg doramectin/kg lw).		✓	✓	✓	✓	✓	✓		✓		✓	✓	✓	✓	✓	✓	28
Febantel	In-feed pellets.	✓		✓	✓		✓			✓		✓			✓		✓	35
Fenbendazole	Pellets for top dressing. In-feed for 1 day.	✓	✓	✓			✓		✓	✓		✓	✓	✓	✓		✓	5
Flubendazole 5 %	Powder. Top dress or in-feed for 10 days.	✓		✓	✓		✓		✓	✓		✓	✓	✓	✓		✓	7
Ivermectin 1 %	1 ml/33 kg. (300 mcg ivermectin/kg liveweight).		✓	✓	✓	✓	✓	✓	✓	✓		✓	✓	✓	✓	✓	✓	28
Ivermectin 0.6 %	Powder in-feed 330 g to 1 kg premix/tonne.		✓	✓	✓	✓	✓	✓	✓	✓		✓	✓	✓	✓	✓	✓	5
Levamisole 7.5 %	Injection.		✓	✓			✓			✓		✓	✓	✓	✓		✓	28
Oxibendazole 2 – 20 %	In-feed for 10 days or pellets for top dressing.	✓		✓					✓	✓		✓	✓	✓	✓		✓	14
Phosmet 20 %	Topical liquid pour onto skin 1 ml/10 kg liveweight.					✓		✓								✓		35
Thiophanate 22.5 %	Powder in-feed for 14 days.	✓		✓	✓					✓		✓	✓	✓	✓		✓	7

* See manufacturer's data sheets for further details.
** Difficult to treat.
Some bendazole compounds may have activity against muscle worm.
lw – liveweight.

Nematodes

Kidney Worms (*Stephanurus dentatus*)

This is an important parasite commonly found in North America.

The life cycle

The adult females form cysts in the kidney fat and pass eggs out into the urine which develops to infective larvae in 2 to 7 days. The cycle can be a direct one through the intake of infected larvae by mouth or penetration through the skin, or indirectly through infected earth worms. The larvae migrate from the intestine throughout the body over a period of 4 to 6 months before they finally arrive at the kidneys to mature. The cycle from egg to adult is a long one (up to a year) and the females lay very large numbers of eggs each day.

Clinical signs

Stephanurus dentatus is found only in warm wet countries because the larva dies out very quickly in cold conditions. The larvae cause severe damage, particularly in the liver, as they migrate throughout the body and they cause loss of appetite and body condition. Blood is often passed out in the urine. There is considerable wasting of muscles.

Diagnosis

This is made at post-mortem by examination of the kidneys. Liver milk spot lesions will also be evident at slaughter, especially in adult pigs. Eggs will be found in the urine.

Treatment

- See Fig 11-10.

Management control and prevention

- Prevent access to infected earthworms.
- As the worm takes so long to develop to an adult it is important to maintain a young breeding herd.
- Keep pens clean and well drained.
- See also management control and prevention discussion earlier in this chapter.

Note: earthworms can live 5 years.

Large Roundworms – Ascarids (*Ascaris suum*)

This is 250 to 400 mm (10 to 15 inches) long and often seen in the faeces of sows and finishing pigs. Female worms are very prolific producing 0.5 to 1 million eggs per day and these will survive outside the pig for many years. They are resistant to drying and freezing but sunlight kills them in a few weeks. It is a common parasite found world-wide and probably the most important one economically. Almost all farms are infected.

The life cycle

This is direct. It takes 2 to 8 weeks for the larva to develop inside the egg and become infective. The eggs after ingestion hatch in the intestine; the larvae migrate through the wall and via the blood enter the liver. They then migrate through the liver to the lungs, finally reaching the trachea where they are coughed up, swallowed and returned to the small intestine to develop into adults. The cycle from egg to egg production is completed within 2 months. The adult worm in the small intestine inhibits the migration of the larvae.

Clinical signs

Large numbers of worms in the intestine absorb food and interfere with digestion. As the larvae migrate through the liver, liver damage (milk spot) results in condemnations at slaughter. The liver lesions heal in 24 days.

Heavy larval migration through the lungs causes coughing and pneumonia and may activate latent respiratory diseases. Both growth rate and feed efficiency may be depressed by up to 10 %.

Diagnosis

This can be difficult and is confirmed by the presence of eggs in the faeces and evidence of liver damage (milk spot) at slaughter.

Treatment

- See Fig 11-10.

Management control and prevention

- Contaminated pens are the most common source of infection hence the adoption of all-in/all-out strategies is important in control.
- See also 'Management Control and Prevention' discussion earlier in this chapter.

Lung Worms (*Metastrongylus apri*)

This slender worm, up to 50 mm (2 inches) in length, is found in the small bronchi (air passages) of the lungs.

The life cycle

This is indirect. The eggs are laid by the adult worm in the bronchi; they are coughed up, swallowed and passed out via the faeces. They are eaten by earth worms in which they develop through 3 larval stages over 10 days to become infective. The cycle is completed by the pig eating the earth worm. Infection therefore only occurs where pigs have access to earth worms, for example in outdoor production. The larvae from the earth worm penetrate the intestine and migrate via the lymph nodes and blood vessels to the lungs undergoing 2 more larval stages in the process. The prepatent period is 3 to 4 weeks.

Clinical signs
These are primarily due to irritation as the larvae migrate through the lungs and the presence of the worms and their eggs in the bronchi. This produces a persistent cough and mild pneumonia. The lung damage can precipitate or enhance other respiratory diseases. Growth rates may be impaired.

Diagnosis
This is determined by the recognition of the characteristic eggs in the faeces but these are not easy to find. The worm and its eggs can be identified at post-mortem examination by cutting the posterior margins of the diaphragmatic lung lobes and extruding them by squeezing. Lung worm infection is unlikely to occur if pigs are reared on concrete.

Treatment
- See Fig 11-10.

Management control and prevention
- This is affected quite simply by preventing access to infected earth worms.
- If pigs are at grass then it is necessary to treat the herd and move onto clean pasture that has not had pigs on before. The larvae can survive inside the earth worm, which can live for 5 years, for a number of years.
- See also 'Management Control and Prevention' discussion earlier in this chapter.

Muscle Worms (*Trichinella spiralis*)
These are very tiny nematode parasites no more than 2 to 4 mm long. They form cysts in the muscles of the pig. When inadequately cooked pork is eaten by humans they cause disease in man.

The life cycle
The female worm is found in the intestine where it produces large numbers of larvae which migrate through the intestinal wall into the blood stream. Some of these larvae eventually form cysts in the muscles and remain viable for many years. For the cycle to develop further the infected cyst must be eaten by another host, either people, rats or other pigs. The pig therefore can act both as an intermediate host and be involved in a direct cycle through tail and ear biting and cannibalism.

Effect on the pig
It has little effect on the pig but it is important from a public health point of view.

Effects on the human
When the larvae are eaten they migrate from the intestine and burrow into muscles particularly those of the diaphragm and jaw. In severe cases death may occur. The adult worm in the intestine can cause vomiting and diarrhoea.

Diagnosis
This is difficult but cysts containing larvae may be found in muscle at meat inspection. A blood test is also available.

Treatment
- There is no practical treatment for the cyst but fenbendazole has an effect.

Management control and prevention
- Control is based on ensuring that pig meat is always well cooked.
- Control of rats is important.
- Control human faeces and prevent contact with pigs.
- See also 'Management Control and Prevention' discussion earlier in this chapter.

Nodular Worms (*Oesophagostomum dentatum*)
This species is found in the large intestine. The adult worms are 7 to 15 mm (1/2 inch) long.

The life cycle
This is direct, the egg developing to the infective larval stage in approximately 1 week. The larvae burrow into the wall of the large intestine where they remain for about 2 weeks and form nodules. They then re-enter the intestine to mature and lay eggs. The life cycle is 40 to 50 days and the adult worms live in the large intestine.

Clinical signs
Heavy infections cause poor growth rate. Damage to the intestine can be severe when the larvae enter and leave the mucosa. Large numbers of nodules interfere with digestion and cause enteritis and colitis with diarrhoea. Affected pigs become pot-bellied.

Diagnosis
This is by identifying the strongyle egg in the faeces and the presence of the worms at post-mortem examination. If eggs are found medicate the pigs and review parasite control.

Treatment

- See Fig 11-10.

Management control and prevention

- See also 'Management Control and Prevention' discussion earlier in this chapter.

Red Stomach Worms (*Hyostrongylus rubidus*)

This red hair like species is less than 10 mm ($^1/_2$ inch) long but can just be seen by the naked eye. It is found world-wide in the stomach and is a common parasite of outdoor pigs particularly sows. It is uncommon in growing pigs.

The life cycle

This is direct. The infective larvae are taken in by the mouth and enter the stomach lining forming nodules. They take about 2 weeks to become adults which then return to the lumen. The complete event from egg to egg takes approximately 3 to 4 weeks. The period from egg to infective larva is from 5 days onwards and therefore if faeces are removed from the pig's environment more frequently than this, the cycle can be broken. Larvae are destroyed by cold and drying.

Clinical signs

The adult worms burrow into the mucous lining of the stomach where they suck blood and cause an inflammation (gastritis). Heavy infections cause anaemia, poor growth rates, loss of condition, thin sows and occasional episodes of diarrhoea.

Diagnosis

This is carried out by identifying the strongyle eggs in the faeces and/or by post-mortem examination. The eggs are similar to those of the nodular and *Trichostrongylus* worms.

Treatment

- See Fig 11-10.

Management control and prevention

- Remove faeces from the pig's environment every 2 to 3 days.
- See 'Management Control and Prevention' discussion earlier in this chapter.

Stomach Hair Worm (*Trichostrongylus axei*)

This is another tiny strongyle worm which as the name implies is found in the stomach. It can only just be seen by the naked eye. It is not a common worm and is considered of low significance.

The life cycle

This is direct and typical of the other strongyles. It takes 5 to 7 days for the larva to become infective and the prepatent period is around 3 weeks under ideal conditions.

Clinical signs

On its own it would be unusual to get high enough numbers to cause problems but the worm is sometimes found in mixed infections with other worms.

Diagnosis

This is by recognising the typical strongyle egg in the faeces.

Treatment

- See Fig 11-10.

Management control and prevention

- See 'Management Control and Prevention' discussion earlier in this chapter.

Thick Stomach Worm (*Ascarops strongylina* and *Physocephalus sexalatus*)

These worms are 10 to 20 mm ($^1/_2$ inch) in length and are found world-wide. They are however relatively uncommon. The life cycle is indirect and involves beetles. Provided faeces are removed regularly, infection cannot occur. They must be present in large numbers to cause problems.

Treatment

- See Fig 11-10.

Management control and prevention

- See also 'Management Control and Prevention' discussion earlier in this chapter.

Thorny-Headed Worm (*Macracanthorhynchus hirudinaceus*)

These measure 100 to 400 mm (5 to 15 inches) in length and are found in the intestine. The heads contain large hooks which hold the worm to the small intestinal wall and they cause nodular lesions. They are found in both temperate and tropical climates.

The life cycle

This is indirect. The larvae are eaten by the grub of the May beetle. The pig must then eat these to complete the cycle. The worms are prolific egg layers but it is an uncommon parasite.

Clinical signs
Large numbers can cause considerable damage to the small intestine and large numbers of nodules are formed. Mild diarrhoea and loss of condition occur.

Treatment
- See Fig 11-10.

Management control and prevention
- See also 'Management Control and Prevention' discussion earlier in this chapter.

Threadworm (*Strongyloides ransomi*)
These worms are very thin and hair like, 3 to 4 mm long and are one of the few species that can also multiply outside the host. *S. ransomi* is more important in warm climates where it is a major parasite of the sucking pig.

The life cycle
Unlike the other round worms of the intestine the threadworm larvae enter the pig by penetrating the skin or mucous membranes of the mouth and are transported by the blood to the lungs, coughed up and swallowed. They then develop to maturity in the small intestine. The infective larvae can also cross the placenta or be excreted by the colostrum and therefore infect piglets within 24 hours of birth. The prepatent period is from 3 to 7 days. Infection is uncommon in good dry farrowing houses.

Clinical signs
Larvae may be found in body tissues of young pigs particularly if infection has taken place via colostrum. Migration causes considerable damage and results in coughing, stiffness, pain, vomiting and bloody diarrhoea particularly from 10 to 14 days of age. Mortality can be high.

Diagnosis
This is carried out by recognising the eggs in fresh faeces or the presence of the worm at post-mortem examination.

Treatment
- See Fig 11-10.

Management control and prevention
- The eggs already contain infective larvae and infection may occur as soon as they leave the sow.
- Control is by good farrowing house hygiene and all-in/all-out procedures to remove the free living forms.
- Treat the sow with broad acting anthelmintics 7 days prior to farrowing before entry into the farrowing house.
- See also 'Management Control and Prevention' discussion earlier in this chapter.

Whipworm (*Trichuris suis*)
This is about 50 to 80 mm (2 to 3 inches) long and shaped like a whip. It can also affect other species including people. It is a common world-wide parasite.

The life cycle
This is direct, the eggs being passed out into the faeces where they become infective within 3 to 4 weeks. They can remain viable for many years outside the pig. After ingestion the larvae hatch out and penetrate the intestinal wall to develop further before moving to the large intestine and caecum where they mature into adults. In dry and hygienic environments this worm is of little significance but in poor conditions it can become a major pathogen.

Clinical signs
Large numbers can cause economic loss with depressed growth rate and feed conversion efficiency. The larvae burrow into the intestinal wall forming nodules, causing irritation, inflammation, haemorrhage and anaemia. Diarrhoea with blood and mucous occur in heavy infections.

Diagnosis
The eggs in the faeces are characteristic. *Trichuris suis* should always be considered when there is diarrhoea with blood. It is important to differentiate this from swine dysentery and colitis.

Treatment
- See Fig 11-10.

Management control and prevention
- See also 'Management Control and Prevention' discussion earlier in this chapter.

Cestodes

Pork Bladder Worm (*Cysticercus cellulosae*) Human Tapeworm (*Taenia solium*)
Cysticercus is the name of the larva or cyst which forms part of the life cycle of the tapeworm *Taenia solium* found in the human. Pigs are the natural intermediate host.

The life cycle

Segments of the tapeworm in people are passed out in the faeces. They contain eggs which are eaten by the pig. The cysticercus which measures 18 mm (1 inch) in diameter develops in the skeletal or cardiac muscles of the pig and the cycle is completed by the human eating inadequately cooked infected pork.

Clinical signs

These are minimal but infected carcasses are condemned at meat inspection.

Diagnosis

Cysticerci are identified at meat inspection.

Treatment

- No highly effective compounds are available for treatment in the pig.

Management control and prevention

- This is achieved by preventing pig access to human faeces, by meat inspection and the burning of infected carcasses.

Protozoa

Protozoa are small single celled organisms that are found in the small and large intestine. There are 4 found in the pig: coccidia, *Balantidium coli*, cryptosporidia and toxoplasma. Of these, coccidia are the only one of importance and then normally only in the young pig. (Fig 11-12).

FIG 11-12: PROTOZOA AND THEIR RELATIVE IMPORTANCE

Protozoa	Pig Age	Infection	Disease	Signs
Balantidium coli	Piglet	Rare	Rare	None
	Weaner	Common	Common	Colitis. Diarrhoea
	Grower	Common	Common	Colitis. Sloppy faeces
	Adult	Common	Rare	None
Coccidia	Piglet	Common	Common	Diarrhoea
	Weaner	Common	Uncommon	Diarrhoea. Poor growth
	Grower	Uncommon	Rare	Sloppy faeces
	Gilt	Common	Uncommon	Loss of weight
	Sow/Boar	Common	Rare	Loss of weight
Cryptosporidia	Piglet	Common	Uncommon	Diarrhoea
	Weaners	Uncommon	Uncommon	Diarrhoea
	Other pigs	Uncommon	Uncommon	Diarrhoea
Toxoplasma	Piglet	Common	Uncommon	Diarrhoea
	Weaners	Common	Uncommon	Diarrhoea
	Other pigs	Rare	Rare	None.

Balantidium coli

This single cell protozoan organism is found in the caecum and large colon as a normal inhabitant. It is debatable whether it is a primary pathogen in pigs and is more likely to be a secondary invader after bacterial or viral infections e.g. *Salmonella*. It is thought however that if abnormal digestion takes place the parasite may multiply to large numbers, causing erosion and mild inflammation of the mucous membrane followed by colitis. Under the microscope it appears as a sphere covered with hair like structures which propel it through the liquid material in the bowel. Once outside the pig the organism rapidly forms a spherical cyst that remains infectious for long periods of time. *B. coli* uses starch from the large bowel as its source of nutrition and certain types of diet or undigested food contribute to its multiplication. The organism can also affect the human causing colitis. Soft liquid faeces that may develop into diarrhoea are seen in pigs from 4 to 12 weeks of age. The cycle of infection is direct.

Clinical signs

These are similar to colitis, sloppy grey faeces and in some pigs there can be considerable loss of condition.

Diagnosis

Post-mortem examinations of affected pigs should be carried out within half an hour of death. Fresh wet scrapings are taken from the lining of the large intestine and examined microscopically.

Treatment

- Sulphonamides and dimetridazole have a moderate effect on *B. coli*.
- Consider those recommended for controlling colitis.
- Change the components of the feed or use a different ration and assess the response.
- Feed meal instead of pellets.

Management control and prevention

- Hygiene is important in preventing a build-up of cysts in the pens.
- Develop all-in/all-out systems.

- Check that other enteric diseases are not present including *E. coli* diarrhoea, salmonellosis, swine dysentery, spirochaetosis, ileitis and non-specific colitis.
- Steam clean pens
- Use bleaches or ammonia based disinfectants to sterilise floors.

Coccidiosis (Coccidia)

See Chapter 11 for further information

Coccidiosis is caused by small parasites called coccidia that live and multiply inside host cells, mainly in the intestinal tract. There are 3 types, *Eimeria, Isospora* and cryptosporidia. Disease, predominately caused by *Isospora suis,* is common and widespread in sucking piglets and occasionally in pigs up to 15 weeks of age. Sows and older animals are generally host to *Eimeria* organisms not *Isospora*.

The life cycle

Tiny-egg like infected structures called oocysts are passed out in the faeces into the environment where they develop (sporulate). This takes place within 12 to 24 hours at temperatures between 25 °C to 35 °C (77 °F to 95 °F). Oocysts can survive outside the pig for many months and are very difficult to kill. They are resistant to most disinfectants. The oocysts are eaten and undergo 3 complex developments in the wall of the small intestine to complete the cycle. It is during this period that damage occurs. The life cycle in the piglet takes 5 to 10 days and disease therefore is not seen before 5 days of age. Sows are not normally the source of the infection, as the major adult coccidia is a different species – *Eimeria*.

Clinical signs

Coccidiosis causes diarrhoea in piglets due to damage caused to the wall of the small intestine. This is followed by secondary bacterial infections. Dehydration is common. The faeces vary in consistency and colour from yellow to grey green, or bloody according to the severity of the condition. Secondary infection by bacteria and viruses can also result in high mortality, although mortality due to coccidiosis on its own is relatively low. Occasionally disease is seen in pigs up to 15 weeks of age and in young floor fed boars and gilts that are housed in permanently populated pens.

Diagnosis

Coccidiosis should be suspected if there is a diarrhoea problem in sucking pigs from 7 to 21 days of age that does not respond particularly well to antibiotics. Diagnosis however is not easy in some outbreaks because identifying oocysts in the faeces of infected pigs can be difficult. In other outbreaks however clear signs are evident at post-mortem examinations. The oocysts do not pass out into the faeces until approximately 3 to 4 days after diarrhoea is seen, by which time the pig may have recovered. Faeces samples for laboratory examination should be taken from semi-recovered pigs rather than pigs with scour. Diagnosis is best made by submitting a live pig to the laboratory for histological examination of the intestinal wall. *Isospora suis* can be differentiated from *Eimeria* spp. by examination of the oocyst in the faeces – *Isospora* have two merozoites whereas *Eimeria* have four.

Treatment

- Treatment is unreliable. To be effective it must be given just prior to the invasion of the intestinal wall. Once clinical signs have appeared the damage is done.
- Adding sulphurmethazine to the water may help.

Management control and prevention

The oocysts rapidly become established in an environment, even in new farms and contaminate the environment via flies, dried faeces, dust and faeces contaminated surfaces. Hygiene and insect control are thus very important.

- Remove sow and piglet faeces daily.
- Improve the hygiene in farrowing houses, in particular farrowing pen floors and prevent the movement of faeces from one pen to another.
- Ensure as far as possible that slurry channels are completely emptied between farrowings.
- Thoroughly wash and disinfect the farrowing houses with substances that are active against oocysts.
- If farrowing place floor surfaces are made of concrete and pitted, brush these over with lime wash and allow it to dry before the next sow comes into farrow. See Chapter 15.
- Keep pens as dry as possible and in particular those areas of the floor where the piglets defecate. An effective method is to cover the wet areas with shavings and remove them daily.
- If creep is fed on the floor stop creep feeding until piglets are at least 21 days old.
- Control flies as described in this chapter.
- In outdoor herds control can be difficult. Always move farrowing arcs to new ground between farrowings and burn bedding.
- Disinfect floor boards in farrowing arcs, if used.
- Wallows can be an ideal focus of infection particularly during lactation. Increase the amount of shade and provide sprays. Provide alternating wallows.
- Position wallows well away from the food.
- Review draughts in the farrowing house.

- Review cross-fostering and colostrum intake protocols to minimise spread.
- Specific disinfectants may be incorporated into the control programme.
- One or two doses of Toltrazuril at a level of 6.25 mg/kg is effective in controlling disease. Follow label perparation instructions closely. A 2 ml dose may be given once at 4, 5 or 6 days of age. If there is no response it is unlikely that coccidiosis is the problem. This treatment can be used as a diagnosis tool. Specifically discuss this method of treatment with your veterinarian who may prepare this for you.
- Elimination is very unlikely.

Cryptosporidiosis (*Cryptosporidium parvum*)

Cryptosporidia are parasites similar to Coccidia. They are small single-cell organisms that infect the cells at the base and top of the finger-like villi in the small intestine. The oocysts are passed out in the faeces and the cycle of infection is direct.

The most common form, *Cryptosporidium parvum,* can survive outside the pig for 6 to 10 weeks. The condition is usually unimportant in pigs unless they are exposed to heavy infections or they are part of a secondary infection following other primary causes of diarrhoea.

The life cycle

This is similar to Coccidia, *Eimeria* and *Isospora* species.

Clinical signs

Often there are none. Where seen, these are associated with villus atrophy in piglets 7 to 21 days old which results in mild malabsorption manifest by diarrhoea. Cryptosporidia also infect humans, rats, mice and other species.

Diagnosis

Oocysts can be detected in the laboratory in stained smears of faecal scour or by histological examinations of the small intestine at post-mortem.

Treatment

- There is no recognised treatment and the condition is not common.

Management control and prevention

- Apply the same criteria as for the control of coccidiosis. If the organism is associated regularly with diarrhoea the management control procedures recommended for coccidiosis should be adopted. Substances that are active against oocysts have an effect against cryptosporidia.

This is an uncommon disease and is usually unimportant in pigs.

Toxoplasmosis (*Toxoplasma gondii*)

This is caused by the protozoa *Toxoplasma gondii* which affects animals and people. The life cycle is indirect. Cats are primary hosts and the only one that sheds infective oocysts in their faeces. Pigs may become infected by ingesting feed or water contaminated with cat faeces, by cannibalism of other infected dead pigs, by ear and tail biting or by eating infected rodents or other uncooked meat. In the pig the organisms form cysts in muscles and other organs where they remain viable for long periods of time. These can develop then into the mature parasites when eaten. Pork therefore is a rare but potential source of human infection.

Clinical signs

Clinical disease in the pig is uncommon. There are usually few signs or none at all. If infection occurs in the first 6 to 8 weeks of pregnancy the organism may cross the placenta and abortion may result but if it is later on piglets may die and become mummified. Occasionally increased stillbirths or premature piglets with tremor and coughing may occur. Piglets may also be born weak and lethargic. Diarrhoea may be seen.

Diagnosis

The presence of the parasite rarely results in clinical disease. In cases of abortion foetal fluids from aborted piglets can be tested for the presence of antibodies. Serology can also be carried out on blood samples from pigs.

Treatment

- Sulphonamides and trimethoprim are effective by in-feed medication but treatment is rarely indicated.

Management control and prevention

- Keep cats out of piggeries.
- Keep cats out of feed and grain stores.
- Control rodents.
- Reduce and prevent cannibalism.

Other Parasites

Mycoplasma (Eperythrozoon) suis

Formally known as Eperythrozoonosis (EPE) but now reclassified, the causal agent of this condition is caused by a small bacterial organism called *Mycoplasma suis (M. suis)*. This blood parasite (a pleomorphic, gram-negative bacteria) attaches itself to the red cells in the blood, sometimes damaging them and causing them to break apart. This causes a reduction in the red blood cell count, reducing haemoglobin (the substance which transports oxygen around the body) thus diminishing the capacity of the blood to carry oxygen. When large numbers of red cells are damaged hemolysis anaemia results and pigs look jaundiced.

A sow may carry *M. suis* and yet remain quite healthy, however, it can cross the placenta and infect pigs in utero causing weak piglets at birth.

Clinical signs

Clinical pictures vary, particularly if there are secondary infections involved. It is useful however, to look at the clinical symptoms in acute and chronic disease.

Acute disease

Affected sows are inappetent with fever 40 to 42 °C (105 to 107 °F) when high numbers of organisms are present in the blood. This clinical picture is seen after farrowing and anaemia is a common symptom. Similar acute infections occur in sows at weaning time together with anoestrus. Primary anaemia and secondary infections may be seen in piglets and weaners.

Chronic disease

Sows become debilitated and pale (anaemia) with jaundice and bleeding into tissues. Infertility signs include delayed returns to oestrus, reduced conception rates and abortion. In weaners, signs may include slow and/or variable growth and poor-doing pigs.

The carrier state

This is common and blood provides a constant source of transmission throughout the herd, particularly when females are vaccinated using common needles.

Diagnosis

This is carried out by making a blood smear on a glass slide, staining it with a special stain (Wright's stain) and looking for the organism under the microscope. The presence of *Mycoplasma suis* in a smear need not necessarily imply disease and there is still controversy over the actual role of this organism and its capacity to cause disease.

Similar diseases

- Actinobacillus pleuropneumonia.
- Chronic respiratory disease complexed with PRRSV and SIV.
- Glässer's disease – *Haemophilus parasuis*.
- Ileitis
- Iron/copper anaemia.
- Leptospirosis (*L. icterohaemorrhagiae* and *L. canicola*).
- Malabsorption and chronic enteritis.
- Pale piglet syndrome – haemorrhages.

Treatment

Consider the following and discuss with your veterinarian:

- If the herd is infected with mange it is important to adopt a control programme.
- Oxytetracycline at a level of 400 g to the tonne for 4 weeks is claimed to have some effect but generally the response to treatment is only moderate.
- Arsanilic acid at a level of 90 g to the tonne has been used to provide a dose of approximately 250 mg per day to each female. This may be illegal in many countries.

Management control and prevention

Mycoplasma suis is spread by blood to blood transmission. In a problem herds it is important to eliminate possible methods of spread including:-

- Vaccination – Wipe the needle between inoculation with cotton wool well dampened with surgical spirit and change every third sow.
- Tagging – Wash the applicators between animals or hold 3 pairs in an antiseptic solution and rotate.
- Tailing, teething piglets – Use in rotation 3 instruments kept in surgical spirits.
- Eliminate lice or mange mites.
- Control biting insects.
- Minimise fighting and vice.

Keeping stress and immunosuppression to a minimum also helps.

Do not let Mycoplasma suis infection get out of control and cause disease. Vaccinate and control external parasites.

External Parasites, Mosquitoes and Flies

There are 5 groups of these (Fig 11-13): ticks, mites, lice, mosquitoes and flies. They can cause considerable skin irritation, sometimes resulting in loss of blood and poor growth.

Some can transmit diseases. For example, the pig louse may carry swine flu viruses or swine pox. Flies can mechanically transmit bacteria and viruses from one pig to another, directly in the case of biting flies or indirectly by contaminating feed. Flies can also transmit infections from one pig farm to another if they are less than 3 km (2 miles) apart. In South East Asia mosquitoes can transmit the deadly Japanese encephalitis virus from pigs to people. In Africa they transmit African swine fever virus from pig to pig.

FIG 11-13: EXTERNAL PARASITES, MOSQUITOES AND FLIES

Common Name	Scientific Name
Flies	*Musca, Drosophila, Stomoxys*
Mange mite Follicle mite	*Sarcoptes scabiei var. suis* *Demodex phylloides*
Lice	*Haematopinus suis*
Mosquitoes	*Anopheles, Culex, Aedes*
Ticks	*Boophilus, Amblyomma, Ixodes*

Mites, lice and ticks are important (Fig 11-14) because of their effect on growth, feed efficiency and spread of disease. They may transmit PRRSV and *Staphylococcus hyicus* and are a nuisance to the pigs and pig attendants.

Flies

As the raising of animals has moved into controlled environments a variety of flies, spiders, cockroaches and other insects have established themselves in these warm places. The most important of these is the house fly of which there are 2 types; the common house fly (*Musca domestica*) and the lesser house fly. Other flies that occasionally cause problems are the blue bottle (*Calliphora*), the stable fly (*Stomoxys calcitrans*) and the fruit fly (*Drosophila*). An understanding of the different life cycles from egg to adult is important in their control.

The life cycle

The common house fly, which is world-wide in distribution, is by far the greatest problem in farrowing and weaner houses. It has a life cycle from egg to egg of 7 to 14 days. The adults lay up to 400 eggs. The fruit fly has a slightly longer life cycle, from 8 to 30 days and is a more prolific egg layer producing up to 900 eggs. The eggs of all species hatch into larvae which develop into

FIG 11-14: EXTERNAL PARASITES, MOSQUITOES AND FLIES: THEIR RELATIVE IMPORTANCE

Parasite	Pig Age	Infection	Disease	Symptoms
Flies	All	Common	Common	Mastitis Greasy pig disease Swine pox PRRSV
Lice	Piglet	No	No	Lice visible.
	Weaner	Uncommon today	Uncommon	Anaemia. Irritation.
	Adult	Uncommon today	Uncommon	Irritation.
Mange	Piglet	Common	Rare	Skin lesions.
	Weaner	Common	Uncommon	Red papules. Thickened skin.
	Grower	Common	Common	Crusts. Rubbing.
	Adult	Common	Common	Crusts. Rubbing. Thickened skin.
Mosquitoes	All	Common	Common	Red bite marks. Slaughter condemnation. Japanese Encephalitis (JEV). PRRSV. Swine pox. Greasy pig disease.
Ticks	Rare indoors. Can occur outdoors in certain areas.			Ticks visible.

pupae and finally adults. The common house fly breeds in slurry, in manure heaps and any damp moist places, particularly if food is present. The fruit fly breeds in damp feed.

At around 20 °C (68 °F) the life cycle of the house fly (*Musca domestica*) can be summarized as:

- Eggs will hatch in 1 day from laying – 1 day;
- Maggots present for 8 days from laying – 9 days;
- Pupae present for 10 days from laying – 19 days;
- Hatched pupa after 19 days.

Use of a fly as a diagnostic tool:

Flies can be a useful diagnostic tool for the freshness and quality of feed. Flies will only colonise stale and mouldy feed which should be removed at the first signs. Using the life cycle above you can estimate the age of the stale feed.

Disease risks

Flies make contact with faeces, skin and discharges from all surfaces of the pig. It follows therefore, that if the number of flies in the environment reaches a high enough level they can become major transmitters of disease organisms, not only within a building, but also between buildings and sometimes between pig herds. Such infections include pathogenic strains of *E. coli, Brachyspira hyodysenteriae* which causes swine dysentery, *Salmonella*, staphylococci, streptococci, rotavirus, PRRSV, TGE and PED. In a laboratory test, a large number of flies from a farrowing house were cultured to determine the bacteria present. *Bacillus* bacteria, moulds, staphylococci, yeasts, streptococci and coliforms were isolated. This illustrates the potential for the dissemination of organisms. Major outbreaks of greasy pig and coccidiosis can be maintained by very high fly populations. When sows are sick with mastitis, flies are attracted to the udder and skin surfaces in great numbers and they can be responsible for enhancing severe outbreaks.

They have also been shown experimentally to transmit *Streptococcus suis* type 2 which causes meningitis and because adult flies can live for up to 4 weeks and travel up to 2.4 km (1.5 miles), transmission between farms becomes a possibility. They can be responsible for piglet diarrhoea persisting in farrowing houses.

If fly populations are allowed to build-up, particularly in farrowing houses, they cause annoyance to stockpeople and distress to sows and piglets. Fly dirt causes heavy contamination of surfaces, particularly around warm areas, lamp surfaces and lights.

Large fly populations on a pig farm can also be a nuisance to nearby communities.

Mosquitoes and flies can cause skin condemnations when severe.

Management control and prevention

Keep the numbers low – This is the most important factor in the control of flies, to prevent the build-up of the population. In countries with warm summers, fly control should commence at the onset of the breeding period and be maintained throughout.

Break the breeding cycle – Flies require a minimum temperature, moisture content and light in order to breed. Eggs hatch best at 35 °C (90 to 100 °F) and multiplication is reduced when the temperature is below 16 °C (63 °F). Moisture is a major requirement and humidity between 25 to 65 % is ideal. It is interesting that the breeding activity is impaired when light levels are reduced.

Identify the breeding grounds – This is an important part of control because if the breeding grounds can be removed, or the conditions for breeding changed, the reproductive cycle will be broken or much reduced. Waste feed that accumulates in and around pens, particularly where there is moisture, provides an ideal environment for flies to lay eggs. Crust on top of slurry becomes major breeding grounds especially if slurry tanks are not completely emptied. Cracks and crevices in walls are an attractive area for flies to breed in, as are solid manure heaps. Farrowing houses harbour large populations that become sources of infection for other houses.

Use all-in/all-out systems – All-in/all-out batching combined with cleaning and washing of the houses between batches of pigs stops the breeding cycle.

Identify resting sites – Where contact insecticides are to be used for control it is important to identify the resting sites of the flies. These are creep lids, lamp tops and walls where it is warm. For the fruit fly this can be a problem, because this species does not move around the building, but lives on roofs, walls and in cracks where there is moist feed at ground level. Residual sprays are of value.

Hygiene – Having identified the breeding grounds, keep them clean by pressure-washing frequently and applying a residual insecticide. Where flies have built up to large numbers it is necessary to completely empty slurry channels and remove the crust material. Bedding used in farrowing houses that contaminates slurry will exaggerate the problem and provide better breeding conditions. Solid muck stored on the farm is a prime breeding site and should be moved well away regularly each week.

Creep feeding – Creep feeds contain high levels of milk products and sugar which provide nutrition and encourage both breeding and feeding. Delay creep feeding until pigs are at least 14 days of age (they probably do not eat much anyway before that age). This will reduce breeding levels.

Monitoring the population – A useful technique is to hang a white card of approximately 150 x 200 mm in size from the roof of each house. This card should be soaked in a sugar solution and dried. Weekly, over a 48 hour period the dots of the fly faeces should be counted. This will give an indication of the build-up over a period of time and predict a population explosion so that intensive prevention and treatment routines can be carried out.

- Use sticky boards to help identify species that are present in the building.
- Manage weed/plant growth around finishing houses. Keep to a minimum.
- Ensure that there is a 2 metre walkway around the outside of each building, to reduce rodent risks and to cut down on fly infestations.

Methods of control

These include electrocution, spray (either daily or residual), paints and baits, larvacides, fly traps and biological predators such as other flies, beetles and wasps. In-feed medication using insecticides that pass through the sow have also been reported. The main chemicals used (Fig 11-15) include pyrethrins, organophosphorus compounds (OPs), lindane and BHC. Fig 11-15 lists some products and their uses. These may or may not be available in your country.

FIG 11-15: SOME COMMONLY USED CHEMICALS FOR FLY CONTROL

Chemical	Application
Azamethipos	Residual spray, bait paint, granules.
Trichlorphon (OP)	Manure heaps.
Neoprene	Larvacide.
Methomyl	Granules, paint-on baits.
Golden malrin	Crystals.
Iodenphenol	Buildings, manure heaps.
Bromophenol	Buildings, yard.
Diazinon (Sheep dip)	Slurry crusts.
Deltamethrin	Sow skin.
Permethrin	Spray, roof/walls.
Fenitrothion	Slurry, faeces heaps, residual larvacide.
Pyrethrins	Knockdown sprays.
Cryomazine	Granules, liquid, larvacide.

The following procedures can be used:

Pulse medication – This involves use of an automatic, battery-powered spray that ejects insecticide into the environment on a periodic basis. This is carried out every 10 to 20 minutes and is a very good method of keeping fly populations at low levels.

Contact baits – These are one of the best methods of controlling flies, because they have immediate access to the insecticide as soon as they hatch. Such surface baits can either be sprayed on the ceiling or painted on the walls, particularly over warm areas where the flies rest. Some products contain a fly attractant. An excellent method of control is to sprinkle small crystals containing the fly attractant and the insecticide over the top of creep lids or other flat surfaces every other day. When selecting the insecticide look at the history of use on that farm and as soon as there is evidence of resistance developing change to an alternate chemical. Make sure the manufacturer's instructions are followed and in particular, that the siting of the insecticide is to best advantage. Alternate contact baits every 3 days with knock down sprays.

Sprays – There are many products available; but if they have to be used daily for control, the battle is lost because large numbers of flies will have built up.

Electrocution – These fly traps are a promising addition to the range of controls.

Sheep dip – This successfully kills maggots and small amounts can be added periodically onto slurry crusts provided there are no contraindications to the product being used as such. Waste oils will have a similar effect.

Larvacides – These are substances such as neoprene which inhibit larval development and stop the cycle. They are sprayed or applied to the breeding grounds such as faeces heaps and slurry.

Biological control – House flies can be controlled using the predator fly, *Orphyra*. This fly is attracted to the areas where the house fly deposits its eggs. Its larva cannibalise the house fly larva and then cannibalise themselves thus the life cycle is a limiting one. This method of control is also effective against the fruit fly. Artificially produced pupae are placed in the contaminated houses in small trays and allowed to pupate over a 10 to 14 day period. It takes 3 to 6 weeks to achieve control and the populations are maintained at a low level by periodic inputs of further pupae. The production of pupae on the farm using purpose built culture chambers and special media is the most cost effective method.

Lice

These are relatively uncommon in herds today, particularly if mange treatment is carried out because this will also destroy the pig louse.

The life cycle

This is direct. The adult female lays 2 to 4 eggs per day over a period of 20 to 30 days. The eggs are attached to the hair by a cement like substance and they hatch out as nymphs 10 to 21 days afterwards. The cycle from adult to adult is approximately 30 days. They are blood sucking and cause a certain amount of irritation but their economic effects are probably relatively low. They are aesthetically however not acceptable and severe infestations can cause anaemia.

Diagnosis

They are easily visible on the skin.

Treatment

- Lice are one of the easier conditions to eliminate from the farm. It is important to treat the whole breeding herd to break the cycle from the sow to the sucking pig. The feeding herd also has to be treated if infested.
- Treat the entire breeding herd in one operation.
- Use either ivermectin, diazinon, lindane or deltamethrin 1 %. Deltamethrin (Coopers spot-on) has long lasting effects and is used as a pour on. Two doses given 10 days apart will totally eradicate lice. All medicines are ineffective against the eggs hence the necessity to treat twice.

Management control and prevention

- Lice are the easiest disease to keep out of the herd. They will only enter on a pig.

Lice are easy to keep out of the herd, but it's important to treat the whole herd to break the cycle.

Mange

Mange is a parasitic disease of the skin caused by one of two mites either *Sarcoptes scabiei* var. *suis* or *Demodex phylloides,* the latter of which is relatively unimportant in swine.

Sarcoptic mange (sometimes called scabies) is by far the most common and important because it is irritant and uncomfortable for the pig, causing it to rub and damage the skin which becomes unsightly. Sarcoptic mange significantly depresses growth rate and feed efficiency. It also increases stress resulting in other diseases becoming more severe. Mange is widespread across countries with up to 60 % of herds affected.

The life cycle

The life cycle of *Sarcoptes scabiei* var *suis* is direct and takes 10 to 15 days to complete. Fig 11-16.

The mite spreads directly from pig to pig, either by close skin contact or contact with recently contaminated surfaces. The boar helps to maintain infection in the herd because he is constantly in direct skin contact with breeding females and he remains a chronic carrier. If pigs are housed in groups there is increased opportunity for spread. The mite dies out quickly away from the pig, under most farm conditions, in around 3 weeks, shorter in the hot summer months. This is an important factor in control. If a herd is free from mange, it is one of the easiest of diseases to keep out because it can only be introduced by carrier pigs. However, once it is introduced it tends to be come permanently endemic unless control measures are taken.

Clinical signs

Acute disease

The common signs are ear shaking and severe rubbing of the skin against the sides of the pen. Approximately 3 to 8 weeks after initial infection the skin becomes sensitised to the mite protein and a severe allergy may develop with very tiny red pimples covering the whole of the skin. These cause intense irritation and rubbing to the point where bleeding may occur. Head shaking is a common symptom and hairs are often rubbed away leaving bare patches. The incubation period to the appearance of clinical signs is approximately 3 weeks although it may be several months before signs are noticed in large pig populations, particularly in grower pigs housed in pens with solid partitions.

Chronic disease

After the acute phase thick asbestos-like lesions develop on the ear, along the sides of the neck, the elbows, the front parts of the hocks and along the top of the neck. (Fig 11-17). The constant itching and scratching will reduce the lifespan of the building.

Diagnosis

Persistent skin irritation with small red spots on the skin developing into asbestos-like thickening suggests the presence of disease. The skins of pigs can also be examined at slaughter for evidence of the small red pimples. Herds with active disease always show a high level of grade 2 or grade 3 lesions (Fig 11-18). Average grade scores at slaughter indicate the degree of infection and its economic significances. Diagnosis is confirmed by demonstrating the presence of the mite. To do this scrapings are taken from suspicious lesions on the skin and particularly inside the ears. A teaspoon is an ideal instrument to scarify material from the interior of the ear. This material can be spread onto a piece of black paper and left for 10 minutes. Mange mites which are rounded in shape and only 0.5 mm in length may be just visible to the naked eye. However to positively identify the mite the scrapings should be submitted to a laboratory for microscopic examination.

Serology

The ELISA test is a very specific test for detecting antibodies to the mange mites with 80 % sensitivity in growing pigs but less in sows. Following an eradication programme the ELISA test can be used to confirm the success, however wait at least 12 weeks and test weaners of at least this age as sow antibody levels may persist for up to a year. The ELISA test is also of value in confirming mange free herds.

LIFE CYCLE OF THE MANGE MITE
(*Sarcoptes scabiei* var *suis*)

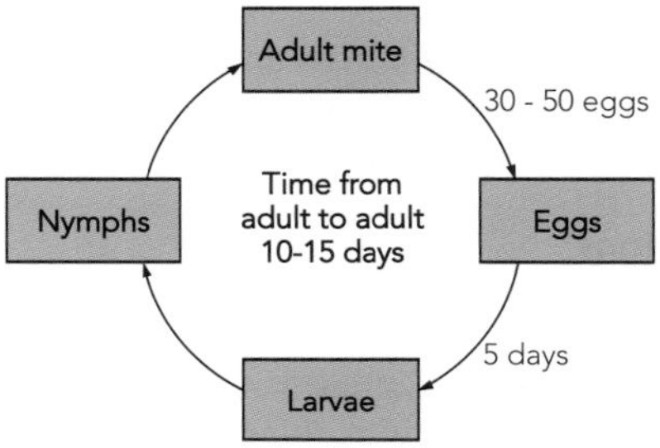

Fig 11-16

COMMON SITES OF CHRONIC MANGE

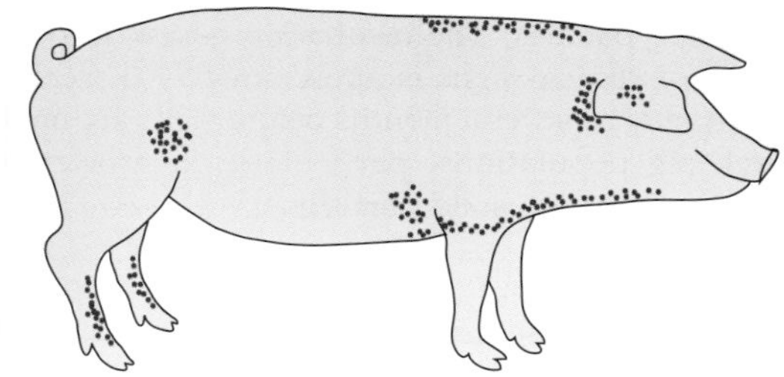

Fig 11-17

Average grade score of pigs examined at slaughter

Less than 0.3 – Very low level, unlikely to be mange. Other factors such as fly bites can cause similar lesions.
0.3 to 0.5 – Disease well controlled no economic effect.
1 to 1.5 – Active disease, considerable irritation and rubbing.
2 or more – Severe disease; expect losses of 0.1 in-feed efficiency and 7 to 10 days extra to slaughter.

Similar diseases

Mange can be confused with the normal dead scurf of the skin (hyperkeratinization – see Chapter 10) that is often seen over the back and the neck but this flaky material can be rubbed away leaving normal skin beneath. The mange scabs on the other hand penetrate the skin surface, are not easily removed and skin damage is evident. Other diseases that might be confused with mange include greasy pig disease, swine pox and sun burn. Occasionally ear scrapings in mange free herds may reveal mite eggs and mites but no clinical disease. This particularly occurs in pigs bedded on old straw that has been contaminated by other animals such as rats or birds with their own host specific mites. These mites, which may be indistinguishable from *Sarcoptes scabiei* var *suis,* do not survive long or usually cause disease in the pig.

Management control, treatment and prevention

Mange is an expensive disease not only because of its economic effects on the pig but also the costs and necessity for repeated treatment. The constant rubbing by the pigs also results in considerable damage to the buildings and equipment. Medicines available are shown in Fig 11-19.

SKIN LESIONS OF MANGE

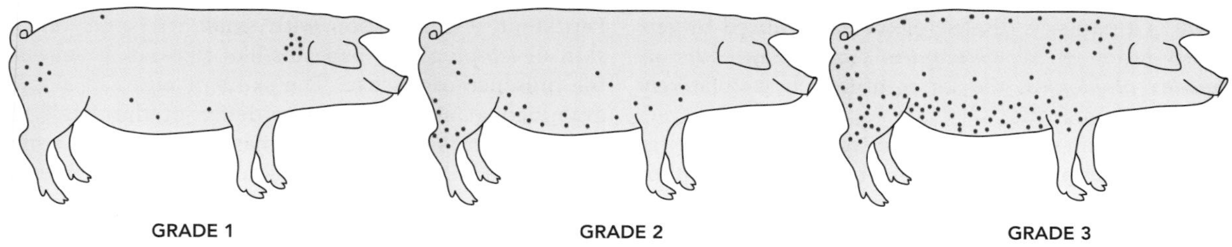

Fig 11-18

FIG 11-19: SOME MEDICINES
*** USED IN THE TREATMENT OF MANGE**

Generic Substance	Method of Application	Comments
Amitraz 0.1 %	Spray, Pour on	Spray 3 times 10 days apart.
Benzyl benzoate	Liquid	Topical application to chronic lesions.
Diazinon 0.05 %	Spray	Spray 3 times 10 days apart.
Doramectin	Injection	Repeat in 10 days.
Ivermectin	Feed, Water, Injection	Feed for 7 – 10 days. Medicate for 5 days. Repeat in 10 days.
Lindane 0.06 %	Spray	(60 days withdrawal). Repeat in 10 – 14 days.
Malathion 0.5 %	Spray	Repeat in 10 – 14 days.
Phosmet 20 %)	Pour on	1 ml/10 kg repeat in 14 days.
Toxaphene 0.5 %	Spray	Repeat in 10 – 14 days.

* For specific uses see the relevant data sheets

There are 3 aspects to consider:

1. Maintain a free herd

If the herd is mange free make sure purchased pigs are also free. Insist upon a veterinary declaration of the pathogen status. Examine incoming pigs carefully during their period of isolation. Take skin scrapings from suspicious lesions. Two injections of ivermectin, doramectin or related compounds 10 days apart usually eliminate the mite but if in doubt keep the pigs out.

2. Control the parasite

This can be can be carried out either by spraying, applying an oily liquid containing phosmet 20 % to the back of the pig, (the medicine is absorbed through the skin) or by in-feed medication or injection. The objective of any control programme however must ultimately be to eliminate the parasite.

Mange control programmes

These are based on the fact that the sows and boars in the breeding herd are the permanent reservoirs of infection, the growers and finishers being constantly removed to slaughter. The aim is to prevent the sucking sow infecting her piglets thereby producing potentially mange free animals. Such grower and finisher pigs are separated from skin contact with the breeding herd.

Control or eradication programmes can be carried out either by injection of ivermectin which have a long period of activity, in-feed medication for 7 to 10 days or by topical application. The first two have given excellent results in the field.

The success of control programmes depends on people and the following should be noted:

- Is there sufficient labour available to carry out the recommended programme?
- Is the correct equipment or methods of feeding sufficient?
- Can the correct dose levels be given?
- Carry out herd inspections to assess the level of disease particularly in sows and boars.
- Maintain the discipline of treating the sow 7 days prior to farrowing. Set up a system of documentation.

Suggested programmes:

PROGRAMME A

Adult stock

- Examine the breeding herd for the presence of chronic lesions. Identify such animals for special treatment. These chronic lesions, found especially in the ears behind the elbow and on the legs, can be difficult to eradicate and they remain a constant source of infection. If they are evident in the ears dress them three times, once every 10 days with either 1 ml of phosmet or spray with 1 % benzyl benzoate. If they are on the skin scrub with amitraz every 10 days.
- Treat the entire breeding herd (gilts, sows and boars) with phosmet 20 % 1 ml to 10 kg weight on one day. Repeat this 10 to 14 days later.
- Repeat this programme every 3 months. Alternatively, treat sows twice yearly but give them one single dose just prior to farrowing.

Weaners

- *Option 1* – Treat pigs on the day of weaning with 0.75 to 1 ml of phosmet using a pump applicator or give an ivermectin injection.
- *Option 2* – Medicate the creep feed with ivermectin for 7 days.
- *Option 3* – Inject pigs at weaning with ivermectin or doramectin.
- Treated pigs should only be moved into **cleaned washed pens** that have been sprayed with a parasiticide such as amitraz and left empty for at least 3 days, preferably 5 to 6.

Growers

If mange is active at the onset in the growing herd medicate with ivermectin in-feed for 7 days.

PROGRAMME B

All breeding adult stock
- Treat the entire sow herd with a single injection of ivermectin or doramectin and repeat every 6 months.
- Treat sows with phosmet 7 days prior to farrowing.

Weaners
- Treat weaners as in programme A.
- Inject boars every 3 months.

Growers
- No treatment should be necessary.

PROGRAMME C

All breeding adult stock
- Medicate feed for 7 days with ivermectin at a level of 100 mcg/kg liveweight.
- Repeat every 6 months. Round worms are also treated and controlled for the same cost.
- Treat sows with phosmet 7 days prior to farrowing.

Weaners
- Treat as in programme A.

Growers
- No treatment should be necessary.

PROGRAMME D

Alternate the phosmet and ivermectin treatments on a 3 monthly basis.

Home-bred gilts
If home-bred gilts are retained for the breeding herd, inject these animals twice with ivermectin 10 days apart.

3. Eradication
If the control programmes outlined above are carried out efficiently for a period of 6 months it becomes possible to eradicate mange mites because there will only be a low population present in the breeding herd. At the point where there is no clinical evidence of mange in the growing pigs or sows a final eradication in the sow herd should be carried out.

FIG 11-20: CONTROL FOLLOWED BY ERADICATION

STEP 1	Only purchase breeding stock from mange free herds. Carry out a control programme for 6 months.
STEP 2	Identify chronically infected animals and cull them.
STEP 3	Examine the growing herd for mange clinically, by ear scrapings and skins at slaughter. All pigs should have been treated.
STEP 4	If Step 3 is clear proceed. If not look to the efficiency of the control programme and delay eradication.
STEP 5	Medicate the sow, boar and gilt rations with ivermectin for 10 days with 1.25 kg of ivermectin premix per tonne (to give 100 mcg of ivermectin per kg liveweight – see manufacturer's instructions). Repeat again in 14 days. Alternatively treat the herd with phosmet followed a further 14 days later with ivermectin in-feed for 10 days. Boars may require extra feed during the treatment periods due to their heavier weights.
STEP 6	Make sure that the hospitalised pigs are treated. They can be a reservoir of infection.
STEP 7	Continue treatment of weaners for 4 weeks after Step 5 has been completed then remove all treatments.

There are two methods:

a) Control first followed by eradication: Fig 11-20
Control requires the production of mange free weaners by treatment postweaning and the movement of pigs to an all-in/all-out system. This is much easier to carry out in the weaner producing herd than the breeder/finisher herd. Where practical segregation of treated weaners from potentially infected ones until their removal is ideal. See Partial Depopulation, Chapter 3.
- Carry out a control programme as described for 6 months.
- Purchase mange free breeding stock during this period.
- Examine the herd for signs of mange particularly chronic symptoms. Test scrape suspicious lesions. Listen for ear shaking in growing pigs. Cull any pigs with chronic lesions.
- Determine the management procedures for eradication i.e. partial depopulation and wean to slaughter depopulation or off site weaning and multi-site production. Partial depopulation is the ideal for breeding finishing herds.
- Proceed as for eradication.

b) Immediate whole herd eradication using partial depopulation: Fig 11-21

Whole herd eradication is expensive but effective. However, if the breeding/grow/finisher herd has problems with other diseases is worth considering.

- Weaned pigs are moved off site over an 8 week period whilst all infected weaners, growers and finishers are sold.
- During the 6 to 8 week period sows and weaners are treated, sows for two 10 day periods with in-feed ivermectin 10 days apart and weaners by in-feed medication for 7 to 10 days postweaning.
- After removal of the infected weaners/growers buildings are cleaned and disinfected prior to reintroduction of the off-site weaners.
- The efficiency of eradication is further improved if early weaning and two or three site production is an option.

Eradication in the combined breeding/finishing herd

- Only proceed when the disease is under control.
- Medicate all weaners and grow/finishing pigs with in-feed ivermectin as per manufacturer's recommendations for 7 to 10 days. Note withdrawal periods.
- Inject all pigs at weaning with ivermectin or use in-feed medication for 14 days. Carry out for an 8 week period.
- Inject all sick pen pigs with ivermectin and continue in-feed medication for 14 days.
- Medicate the entire breeding herd with in-feed ivermectin for 10 days twice 10 days apart.
- Move treated weaners only to totally cleaned and disinfected houses and pens. Spray pens with a topical ascaricide.
- Keep treated weaners separate from the grow/finisher herd. If this is not practical medicate in-feed the grow/finishers again 3 weeks later and continue the weaner medication for a 12 week period. Send all finishers to slaughter.

Note when using in-feed medication programmes, not all animals eat – sick sows, sows in heat, boars may not eat according to their weight and lactating sows have variable feed intake requirements.

Once the programme has been completed wait at least 7 months before success can be claimed. This is determined by the lack of any clinical evidence, the observations of finishing pigs at slaughter and regular scrapings of ears particularly of boars and finishing pigs.

If the eradication has failed the appearance of the allergic skin conditions can be sudden.

FIG 11-21: IMMEDIATE WHOLE HERD ERADICATION

This programme should only be attempted when there are few if any symptoms of disease in the herd.

One month prior to commencement	Examine all pigs in the herd particularly sows and boars for chronic skin lesions. Cull these animals or inject with ivermectin and treat topically with amitraz or other available topical dressing.
STEP 1 – Day 1	Only purchase breeding stock from mange free herds. Medicate all weaners and growers 7 to 50 kg in-feed with ivermectin premix 330 g/tonne of feed for 7 to 10 days (to give 100 mcg of ivermectin per kg liveweight – see manufacturer's instructions).
STEP 2 – Day 1	In planning Step 1 the current recommended withdrawal period should be observed which means no slaughter pigs for the designated period.
STEP 3 – Day 1	Medicate all finisher pigs 50 to 120 kg in-feed with 400 g/tonne of ivermectin premix (to give 100 mcg of ivermectin per kg liveweight – see manufacturer's instructions).
STEP 4 – Day 1	Medicate all sows, gilts and boars in-feed with 1.25 kg of ivermectin premix per tonne (to give 100 mcg of ivermectin per kg liveweight – see manufacturer's instructions). Feed each animal 2.7 kg of feed once per day. Feed lactating sows extra non medicated feed.
STEP 5 Day 1 to 10	Any sick animals that fail to eat during the 10 days inject with ivermectin 1 ml per 33 kg liveweight.
STEP 6 – Day 1	Inject all pigs in the sick pen with ivermectin followed by in-feed medication.
STEP 7 – Day 5	Spray pens sides to pig height with topical dressing.
STEP 8 – Day 7	Wash and spray all boots with topical dressing. Change to new pig clothing.
STEP 9 – Day 10	Stop all medication.
STEP 10 – Day 20	Repeat the programme after the 10 day gap.

Economics

The economics of eradication are sound. The costs of the programmes are approximately the same as the amount of money spent on control over a 6 to 9 month period. The costs however are concentrated into a 4 to 6 week period.

Field experiences show that after eradication sows are less restless with consequent reductions in piglet mortality and increased weights at weaning. The most important effects however are in improvement to feed efficiency by up to 0.1 and in daily liveweight gain by 4 to 10 % depending on severity.

Management control and prevention

Summary

- Treat regularly to prevent a build-up of numbers.
- Disease is less easily spread where sows are housed in individual confinement.
- Treat boars regularly every 2 to 3 months. They are a constant source of infection.
- The mite will only live away from the pig for around 3 weeks.
- The mite is host specific, except that on rare occasions pig attendants develop a localised lesion, and it is easily kept out of the herd.
- Always treat animals twice 10 days apart.
- Leave pens empty for 3 days after infected pigs move out and spray the pen after washing with a mange dressing.

More mange mites
– More disease
– More mange mites.

Demodectic Mange (Follicle Mites)

Demodectic mange is caused by the tiny mite *Demodex phylloides* and is unimportant from a disease aspect in the pig. They live at the base of the hair follicles, around the face and on the abdomen producing small papules. Rarely do they cause clinical disease.

Treatment

The response to treatment is poor but the mite is sensitive to those medicines used for mange control.

Mosquitoes

Mosquitoes and flies can cause skin condemnations when severe. Mosquitoes can transmit PRRSV.

Remove all areas of stagnant water around farm note old tiers. In the summer lowered curtains may create pools of stagnant water allowing mosquitoes to breed next to the pigs. Raise the curtains at least monthly to eliminate this risk. This also helps with mice control. Do not cut the weeds by a finishing building within 7 days of slaughtering the pigs as the slaughterhouse may have problems with skin condemnations.

Do not site the compost pile within 10 metres of the a finishing unit.

Ticks

Ticks are found on most species of animals including humans. They are not generally host specific. They are not commonly found on pigs indoors. They differ in species and incidence from region to region so local knowledge is useful. If they do attach to pigs it is invariably to those kept outdoors. They are not usually themselves of any importance but they can spread a number of diseases, including classical swine fever and African swine fever. With African swine fever specific soft body ticks are an intermediate host.

The life cycle (Figs 11-22 and 23)

This is similar to the mange mite, from the egg to a larva then a nymph and finally the adult. Both the nymph and adult suck blood. Each part of the life cycle may be on one or a number of different species.

THE LIFE CYCLE OF A SINGLE-HOST TICK

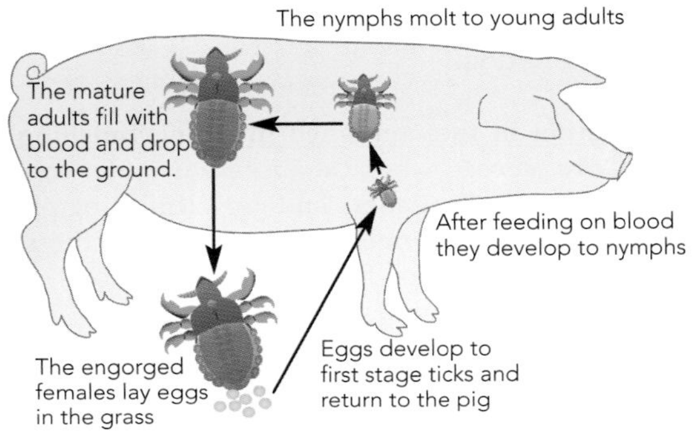

Fig 11-22

THE LIFE CYCLE OF A 2- OR 3-HOST TICK

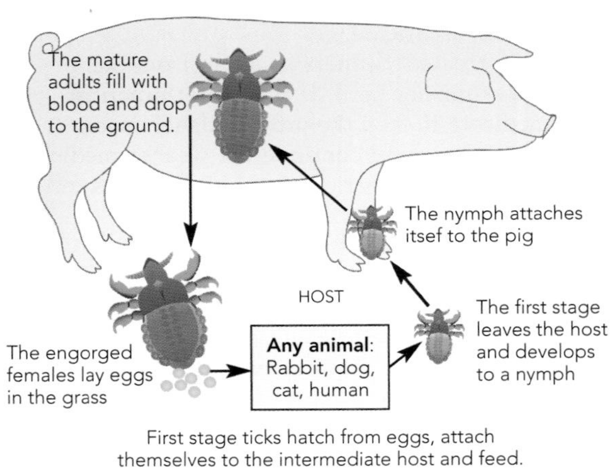

Fig 11-23

Clinical signs and diagnosis
Ticks are easily seen by the naked eye and they are often engorged with blood.

Treatment
- All the medicines used for the treatment of mange are effective.
- Spray contaminated areas in the environment if practical.

Management control and prevention
- Keep pigs away from the infected environments where possible.
- Use the same parasiticide as for mange.

12 OIE AND OTHER DISEASES

Chapter 12

12 OIE AND OTHER DISEASES

Introduction

What are OIE List Diseases?

The OIE list of diseases is the official list of notifiable animal diseases throughout the world. The list consists of transmissible diseases with the following characteristics:

- Have the potential for very serious and rapid spread, irrespective of national borders.
- Have a serious socio-economic or public health consequence.
- Are of major importance in the international trade of animals and animal products.

Generally these diseases do not occur in many regions of the world because they have never been present or because they have been eradicated and are kept out by government control measures. If any of these diseases are suspected they must be reported immediately to the local authorities.

When a new condition is recognised across the planet it is generally added to the OIE list until its impact is understood. This happened when PMWS was first identified, but now it is known to be globally endemic it is no longer on the list.

Free, fringe and enzootic areas

Pig-producing regions of the world can be classified as free, fringe or enzootic for any given disease. These terms are defined as follows:

"Free" describes those regions, countries or continents which are free from the disease.

"Fringe" refers to areas which are generally free but which are always under threat of re-infection from outside (bordering an area where pathogens are common). Fringe areas may suffer sporadic outbreaks which then have to be controlled and stamped out. Such outbreaks may occur several times a year, or several years apart.

"Enzootic" refers to those areas that are permanently infected. The infection is endemic (enzootic) in that population.

Which Countries are Free from Which Diseases?

Fig 12-1 highlights the important diseases and the status of those diseases in numerous countries. It should be noted that the global situation is extremely fluid and subject to rapid change.

Some countries might be free of a disease/pathogen but will practise a vaccination programme.

Countries might be free of a pathogen in commercially farmed pigs, but feral (wild) unfarmed pigs may remain endemically infected.

Also, whereas the information in some countries is fairly reliable, in others it is not and in some there is very little information. The most reliable information is that in the countries listed in North and South America, the Antipodes and the EU. For up-to-date information, see the OIE website.

Terminology

Enzootic (= endemic) disease – This means that the disease is permanently present in a population. In relation to diseases of pigs, the population may be a herd or the herds in a region, a country or a continent. Strictly speaking 'endemic' should only be applied to populations of people (demic is from the Latin demos meaning people/democracy) and 'enzootic' should be used for animals but in practice the two are interchangeable.

Epizootic (= epidemic) disease – A disease which spreads, usually fairly quickly, to a large proportion of the pig population. The term 'epidemic' is often used instead but strictly speaking "epidemic' applies to people not animals.

Exotic disease – A disease which does not occur in the region or country of your pig farm.

Pandemic disease – A disease which is widespread throughout a region or the world.

If you farm in an area free from a pathogen, forget about it.
If not, take care.
Learn to recognise it.

FIG 12-1: DISTRIBUTION OF SELECTED OIE DISEASES IN DIFFERENT PIG REARING-REGIONS 2013

	AD/PRV	ASF	BRUCE	CSF/HC	FMD	JEV	PED[1]	PRCV[1]	PRRSV	SVD	TGE
N. America											
Canada	–	–	+	–	–	–	–	–	+	–	+
Mexico	+	–	+	+	–	–	–	–	+	–	+
USA	– w	–	+	-	–	–	+	+	+	–	+
S. America											
Argentina	+	–	+	–	v	–	–	–	–	–	–
Brazil	+	–	+	–	v	–	–	–	–	–	–
Chile	–	–	–	–	–	–	–	–	–	–	–
Europe											
Austria	+	–	–	+	–	–	+	+	+	–	+
Benelux *	+	–	+	–	–	–	+	+	+	–	+
Bulgaria	+	–	–	+	–	–	+	+	+	–	+
Byelorussia	+	–	–	+	–	–	+	+	+	–	+
Czech Republic	+	–	–	+	–	–	+	+	+	–	+
Denmark	–	–	+	–	–	–	–	+	+	–	–
Finland	–	–	–	–	–	–	–	+	+	–	–
France	+	–	–	–	–	–	+	+	+	–	+
Germany	+	–	–	+	–	–	+	+	+	–	+
Hungary	+	–	–	+	–	–	+	+	+	–	+
Iberia *	+	–	+	–	–	–	+	+	+	–	+
Ireland	+	–	–	–	–	–	–	–	+	–	–
Italy *	+	– +	–	–	–	–	+	+	+	+	+
Latvia	+	–	–	+	–	–	+	+	+	–	+
Lithuania	+	–	–	+	–	–	+	+	+	–	+
Poland	+	–	+	+	–	–	+	+	+	–	+
Russia	+	+	+	+	+	–	+	+	+	–	+
Slovakia	+	–	–	+	–	–	+	+	+	–	+
Sweden	–	–	–	–	–	–	–	+	+	–	–
UK	– (NI+)	–	–	–	–	–	+	+	+	–	+
Ukraine	+	–	–	+	–	–	+	+	+	–	+
Antipodes											
Australia	–	–	–	–	–	–	–	–	–	–	–
New Zealand	–	–	–	–	–	–	–	–	–	–	–
Asia											
China	+	–	+	+	+	+	+	+	+	+	+
India	–	–	+	+	+	+	+	+	+	–	–
Indonesia	+	–	+	+	+	+	+	+	+	–	+
Japan	+	–	–	+	–	+	+	–	+	–	+
Korea	+	–	+	+	–	+	+	+	+	–	+
Philippines	+	–	+	+	–	+	+	+	+	–	–
Thailand	+	–	+	+	+	+	+	+	+	–	+
Africa											
Kenya	–	+	–	–	–	+	–	+	+	–	–
South Africa	–	+	–	–	–	+	–	+	+	–	–
Uganda	–	+	–	–	–	+	–	+	+	–	–
Zimbabwe	–	+	–	–	–	+	–	+	+	–	–

AD/PRV – Aujeszky's disease/pseudorabies virus.
ASF – African swine fever.
BRUCE – Brucellosis (*Brucella suis*).
CSF/HC – Classical swine fever/hog cholera.
FMD – Foot-and-mouth disease.
JEV – Japanese encephalitis virus.
PED – Porcine epidemic diarrhoea.
PRCV – Porcine respiratory coronavirus.
PRRSV – Porcine reproductive and respiratory syndrome virus.
SVD – Swine vesicular disease.
TGE – Transmissible gastroenteritis virus.
– Disease absent.
+ Disease present.
W = wild pigs only.
V = vaccinated.
– (NI+) Northern Ireland positive.
*** Benelux:** Belgium, Netherlands, Luxembourg.
*** Iberia:** Spain and Portugal.
*** Italy:** includes Sardinia and Corsica.
[1] Not OIE diseases, for reference only.

Protecting Your Herd against Serious Infectious Diseases

See Chapter 2 for further information

Vaccinate

If you farm in a country where one or more serious infectious diseases are enzootic or where there is a risk of them occurring, and if vaccines are available and vaccination is allowed, you should vaccinate your herd routinely. This does not completely prevent your herd becoming infected, but it greatly reduces the chances as well as reducing the damage. Of the diseases described in this chapter, effective vaccines are available in enzootic countries against classical swine fever (hog cholera) and Aujeszky's disease (pseudorabies). Less effective or shorter lasting ones are available in some countries against FMD, where the vaccine may be used to slow the spread of the disease.

Biosecurity

Whether you can vaccinate or not, you should protect your investment by applying appropriate biosecurity measures. This is not just to keep out the infectious diseases covered in this chapter, although of course if you are in a high risk area this is crucial, but to keep out some of the more serious diseases covered in other chapters too.

Some viruses, such as foot-and-mouth disease (FMD) virus spread very readily and if they are spreading in your neighbourhood they are difficult to keep out of the herd no matter what precautions you take. This is partly because they are highly infectious so that it only takes the entry of a few viable virus particles to set off a major outbreak. It is also because they are carried on the wind, or on birds or flies which are difficult or impossible to guard against. Fortunately not all infectious agents are like that. Some viruses and bacteria spread less readily. Examples in this chapter are swine vesicular disease (SVD) and classical swine fever (hog cholera). You can keep them out fairly reliably if you take appropriate biosecurity precautions. See Chapter 2 for biosecurity details.

OIE diseases that can infect pigs

- African swine fever.
- Anthrax.
- Aujeszky's disease/pseudorabies virus (AD/PRV).
- Brucellosis (*Brucella suis*).
- Bungowannah virus.
- Classical swine fever/hog cholera (CSF/HC).
- Echinococcosis/hydatidosis (*Echinococcus granulosus*).*
- Foot-and-mouth disease (FMD).
- Japanese encephalitis (JE).
- New World screwworm (*Cochliomyia hominivorax*).*
- Nipah virus (Niv).
- Old World screwworm (*Chrysomya bezziana*).*
- Porcine cysticercosis:
  - Pork bladder worm (*Cysticercus cellulosae*).
  - Human tapeworm (*Taenia solium*).
- Porcine reproductive and respiratory syndrome virus (PRRSV).
- Rabies.
- Rinderpest*.
- Swine vesicular disease.
- Transmissible gastroenteritis virus (TGE).
- Trichinellosis – Muscle worm (*Trichinella spiralis*).
- Vesicular stomatitis (VS).

Other diseases (not covered elsewhere)

- Borna disease.
- Blue eye (BE).
- Getah viruses.
- Tumours (cancer)
- Vesicular exanthema of swine (VES).

*These diseases are capable of infecting pigs, however infection would be very unusual. For information on these and any new pathogens/diseases visit the OIE website.

OIE Diseases

African Swine Fever (ASF)

African swine fever (ASF) is a highly contagious disease that is enzootic in many African countries and the Mediterranean island of Sardinia. The virus has infected Russia and Georgia and is a serious concern to both Eastern Europe and Western Asia. The virus is currently the only member of the Asfarviridae family.

The disease produces a range of clinical signs that are very similar to those of classical swine fever (see Classical Swine Fever).

African swine fever (ASF) resembles classical swine fever (CSF) (hog cholera) so closely that laboratory tests are required to differentiate them. The clinical signs and post-mortem lesions of the two pathogens are almost indistinguishable. ASF is caused by a unique virus which is distinct from that of CSF and which infects domestic and wild pigs and a variety of soft-bodied ticks. The virus is endemic in Africa south of the equator, in warthogs and bush pigs, but the infection in them produces no clinical disease. It circulates between warthogs and the soft-bodied ticks which inhabit their burrows. The ticks transmit it through all stages of their life cycle and perpetuate it. It is also endemic in the domestic pigs of some African countries.

ASF virus is relatively tough and can survive in the environment and in pig carcasses for a long time. Curing and smoking pork products does not destroy it. Its main method of spread from country to country is via waste uncooked pork products fed to pigs. Its spread between herds within a country is by direct and, to a lesser extent, indirect contact between pigs. Indirect contact usually involves contamination from dead pig tissues and secretions.

Importance of ASF

- When virulent strains cross into domestic pigs they cause very serious disease. Virtually all the pigs in the infected herd become ill and the majority die. It is not surprising that all countries regard it as important. There is no vaccine for it and those European, South American and Caribbean countries which have been infected have adopted a slaughter policy to eradicate it. Mild strains of the virus also occur which cause a milder but still serious disease in domestic pig herds.
- It was carried from Africa to Portugal in 1957 and again in 1960. It spread through Portugal and into Spain. Outbreaks then appeared in France, Italy, Belgium, the Netherlands, Malta, Cuba, Brazil, Haiti and the Dominican Republic. It has been eradicated by slaughter from all of these except the Italian island of Sardinia where pigs are kept free-ranging to forage for food. There are no ticks that transmit it there, and although it is a virulent strain which causes severe disease in wild boars on the island they are not a major factor in its spread. The infection spread into Russia in 2006 and remains a primary concern to surrounding countries including the largest pig herd in China.

Should you be concerned about ASF?

Outbreaks occur from time to time in domestic pig herds in Africa even in the large control areas in South Africa where a slaughter policy exists. Herds usually become infected by eating warthog flesh, from tick infestations or from other infected domestic pigs. If your pig farm is in an infected region of Africa it is clearly at risk. To keep it out you should adopt high standards of biosecurity. The central aim of your biosecurity must be to keep your herd well away from live and dead warthogs, their ticks and other domestic pigs.

In indoor herds reasonable biosecurity should prevent contamination. If your pigs run outside, the herd should be double ring fenced. Soft-bodied ticks are not a risk in Italy. While the pathogen is still present in Sardinia, there is a small chance of it appearing in pigs in Western Europe which have been illegally given infected uncooked pork products from Sardinia. Such products would most likely be brought back by holiday-makers ignorant of the risks to domestic pigs and be fed to back-yard pigs or pet pigs. You have a duty to warn anyone you know with back-yard pigs or pet pigs against feeding any household waste.

The main threat to pig herds outside Africa is the introduction of infected pork products in waste food from aeroplanes and ships arriving from the southern half of Africa. This is unlikely to occur in the EU, North America or the Antipodes because of the strict rules governing the disposal of such waste but the rules may not be so strict in some other countries. There is a small theoretical risk of seagulls carrying scraps of such waste from ships to coastal pig farms.

Clinical signs

In the acute form of the disease caused by highly virulent strains, several pigs develop a high fever of 40 to 42 °C (104 to 108 °F) but may not show any other very noticeable signs for a couple of days. They then gradually lose their appetites and become depressed. If they are white-skinned pigs their extremities (nose, ears, tail and lower legs) become cyanotic (blue-purple colour) and discrete haemorrhages appear in the skin, particularly on the ears and flanks. They lie down huddled together, shivering, breathing abnormally, and perhaps coughing and they do not want to get up. If you make them get up they are unsteady on their legs. Within a few days they become comatose and die. Pregnant sows abort. The pathogen spreads through the herd over several days or sometimes more slowly over several weeks and many pigs die. Some may die very soon after they become ill.

Some outbreaks of the disease in African countries, in which the virus is endemic in the domestic pig population, are milder, run a longer course and spread more slowly through a herd with fewer deaths. Some affected pigs become very thin and stop growing and develop signs of pneumonia, skin ulcers, and swollen joints.

Milder strains may also be introduced accidentally into pig herds in countries outside Africa in waste food from ships, aeroplanes or returning travellers. The disease may then go unrecognised for some time.

Diagnosis

Pigs that die early in an outbreak may have no very noticeable lesions but as the disease progresses the lesions then are striking. Bright red haemorrhages in the lymph nodes, kidneys, heart and linings of the body cavities are common findings. There may also be excess haemorrhagic fluid in the body cavities and gelatinous fluid in the lungs. The spleen may be enlarged and darkened, and may crumble on slight pressure.

Before sending any samples, the veterinarian must consult the appropriate veterinary authorities.

The best samples to send are the tonsils, blood, lymph nodes, spleen and in chronic cases, serum for serology.

The tonsils of the pig are very easy to find. Laying the dead pig on its back, cut away the skin and flesh under and between its lower jaw bone and also the tongue. The pair of tonsils are two large red patches each about the size of the end half of your thumb or perhaps slightly bigger. Their surfaces are covered with small pits or depressions.

In South Africa and countries outside Africa it is essential to isolate and identify the virus. Virus may be isolated in primary cultures of pig bone marrow or peripheral blood leucocytes. Infected cells haemadsorb i.e. pig red cells will adhere to them. Virus can also be detected in infected cells by fluorescent antibody tests. ELISA tests are also used to detect antibodies. In doubtful cases samples can be injected into experimental pigs.

Serum antibody titres may be tested in a number of ways. The indirect immunofluorescence (IIF) and the ELISA tests seem to be the most favoured.

Rapid, accurate laboratory diagnosis of swine fever is essential.

Similar diseases

Porcine dermatitis and nephropathy syndrome, which occurs from time to time in most pig rearing areas, can resemble ASF and CSF clinically and at post-mortem examination. Laboratory examination may be necessary to eliminate them from the diagnosis.

Treatment

- There is no treatment.

Management control and prevention

- No effective attenuated or inactivated vaccines have been developed and so none are available.
- Prevention in countries outside Africa has to be on a national basis by restrictions on incoming pigs and pork products, compulsory boiling of waste animal products under licence before feeding to pigs and the application of a slaughter policy when the disease is diagnosed.
- Prevention in Africa is based on measures to keep warthogs and materials contaminated by warthogs away from the herd.

Anthrax (*Bacillus anthracis*)

This is an uncommon disease of pigs in most parts of the world including the EU where it is notifiable. It is caused by the bacterium *Bacillus anthracis* and is characterised by acute illness, fever, respiratory distress and rapid death.

Anthrax should be suspected if a sow is found dead and post-mortem examination shows copious blood tinged-mucus and large haemorrhagic lymph nodes under the skin of the neck and in the abdomen. The post-mortem examination should be discontinued immediately and veterinary help sought.

Diagnosis is then confirmed by taking smears for microscopy and swabs for culture from the affected tissues. (Note that samples are not made from blood as they would be with cattle).

The veterinarian should not fix the smears by flaming because this destroys the bacterial capsule which is a diagnostic feature. Smears are best fixed in Zenker's fluid. Acute infection with *Haemophilus suis* in primary MEW or SPF immunologically naive sows may result in similar signs and lesions to anthrax.

The source of the infection in sows is usually feed containing contaminated feedstuffs from a spore-contaminated area, although in sows kept outdoors in such regions the source may be contaminated soil or other dead animals. The anthrax bacillus is sensitive to penicillin. Care should be taken in handling diseased pigs or carcasses because the disease is communicable to people. Effective vaccines are available in some countries for both pigs and people.

Aujeszky's Disease/Pseudorabies Virus (AD/PRV)

Aujeszky's disease (AD) and pseudorabies virus (PRV) are synonymous (AD/PRV). In Europe the name 'Aujeszky's disease' (AD), after the Hungarian who helped identify the causal agent, is preferred to pseudorabies. This is to avoid confusion with rabies, a name which tends to alarm people. It is totally unrelated to rabies. Nevertheless, pseudorabies virus (PRV) is the name most commonly used in North America and other parts of the world. Another early name was 'mad itch' reflecting the most striking clinical sign in cattle.

Pseudorabies virus (PRV) is primarily an infection of pigs which represent its only known reservoir host. It is sometimes transmitted naturally from pigs to individual cattle, horses, dogs and cats which develop nervous signs and rapidly die. These animals are end hosts and do not usually spread it. It has never been known to cause disease in people.

Should you be concerned about AD/PRV?

There is a wide spectrum of risk and concern over this disease. At one end of the spectrum, if you are involved with pig farming in disease-free areas you can ignore this section, but governments need to keep national biosecurity on high alert.

Herds in Canada, UK, Ireland or Denmark are safe from this disease because it has been eradicated from these countries. However, farms in Canada should be aware that they could become infected from the USA. Whilst the US commercial herd is declared free of the disease, the pathogen is still present in the feral pigs. Also, if you are a British pig farmer you should not be complacent; it could get reintroduced into the UK from other EU countries with the movement of pigs, possibly for slaughter, now that there are no barriers to trade.

The virus could blow over the sea again from Germany to some of the Danish islands as it did before the vaccination of neighbouring German pig herds was financed.

AD/PRV is widespread throughout most regions of Italy, Spain and Portugal, Central and Eastern Europe, South East Asia including China and Central and South America.

For many farmers therefore, this disease is a constant threat and herds should be routinely vaccinated.

Importance of AD/PRV

AD/PRV is an economically damaging viral disease of pigs although not anything like as damaging as swine fever. Some governments (e.g. the Australia, UK, Ireland, Norway, Denmark, Netherlands, Canada, USA, and Chile) take it seriously and have adopted control policies, which may include compulsory vaccination or slaughter and eradication policies.

In a susceptible (unvaccinated) herd large numbers of pigs may be clinically affected and some sows may abort. Mortality in suckled piglets is high but much lower in growing pigs. Unfortunately, it can spread on the wind so standard precautions of farm biosecurity cannot be entirely relied on to keep it out. On the credit side, there is only one serotype of the virus and attenuated vaccines are highly effective.

Clinical signs

The susceptible breeding and sucking herd

The earliest sign is usually a few pregnant sows aborting. If there is a farm dog in close contact with the pigs, it may develop severe acute nervous signs, which are distressing to see and which always progress to death. The farm's cats disappear, presumed dead.

Acute disease

Acute outbreaks of disease occur when virulent strains of the virus first infect an unvaccinated susceptible herd. The virus crosses the uterus and placenta and infects the foetuses. The first clinical signs are abortions, stillbirths and the birth of weak litters which soon die. Abortions may rise to 5 % over about 6 weeks followed by reproductive failure at all stages of the cycle. Embryos are killed and absorbed and sows return to heat. These reproductive problems may occur in up to 20 % of dry/gestating sows. Small vesicles of approximately 1mm in diameter may also be seen on the skin around the nose and the mouth. Similar lesions are occasionally seen in PRRSV. Deaths in suckling sows can occur in this phase.

Carrier state

After the acute phase clinical signs of disease may die out altogether and this is often seen in small herds of less than 100 sows. Sometimes the virus itself may disappear. However it is more likely to persist in a few animals and cause clinical signs sporadically.

Chronic disease

In an unvaccinated herd, when the early acute phase of the disease is over and the herd has developed an immunity clinical signs are sporadic and milder. These signs may be difficult to associate with an infertility problem because of their insidious nature.

Depression of reproductive efficiency across all parameters is a feature of the chronic infection with increased levels of repeats, mummification, stillbirths and piglet mortality. Young carrier females that are stressed shed virus thus maintaining infection throughout the herd. Spread of infection in the breeding herd is low with immunity and infection waning and rising over 1 to 2 year cycles.

Piglets and weaners

In newborn piglets severe and fatal nervous signs develop. This is first evident by rough-haired, listless appearances. They stop sucking and within 24 hours tremble, become uncoordinated, salivate excessively and go into convulsions, rolling their eyes forward and backward. They often emit a high pitched squeal.

Affected sucking piglets may show a variety of clinical signs such as walking in circles, sitting like a dog, or lying on their sides paddling their legs. Some may vomit and others may develop diarrhoea. They die within 24 to 36 hours of the onset of nervous signs. Mortality in piglets may be very high, approaching 100 %. Subsequent litters may be born weak and/or start showing clinical signs as soon as they are born.

The susceptible weaned herd

If there are weaned pigs on the farm, the younger, most recently weaned ones develop early signs similar to, but milder than, the sucking piglets, with fewer numbers affected. Vomiting and diarrhoea are sometimes seen. Affected weaned pigs run a high fever of 41 to 42 °C (106 to 107 °F). Many of those that develop nervous signs and/or severe pneumonia die, but the death rate is usually less than 10 %. Most of the rest recover fully in 5 to 10 days but some may be left stunted and grow poorly.

Some types of the virus may have a tendency to affect one tissue over another – for example a virus that preferentially affects the lung is called pneumotropic; if it affects the nervous system it is referred to as neurotropic.

If the virus is a pneumotropic strain, causing pneumonia as well as nervous signs, older weaned pigs start sneezing, develop a nasal discharge, breathe heavily and start coughing. The virus infection tends to cause flare-ups of mycoplasma (enzootic) pneumonia caused by *Mycoplasma hyopneumoniae* and may trigger off secondary bacterial infections such as *Pasteurella multocida* and *Actinobacillus pleuropneumoniae*.

If sows abort, the farm dog goes into convulsions, and piglets become dull, hairy and start shaking, call your veterinarian urgently. Early vaccination can reduce the losses.

The susceptible grower-finisher herd

The earliest signs here are usually depression, lack of appetite, staring coat (where the pig's hair appears to stand high) and a high fever of 41 to 42 °C (106 to 107 °F). The development of respiratory diseases similar to those described above for older weaners is common. Relatively few pigs develop nervous signs or they may be mild (e.g. muscle tremors). Although all or nearly all of the growers and finishers may be affected, most recover in 5 to 6 days and start to grow well again when they have recovered their appetites. The mortality is usually below 2 %.

Variations on the above clinical picture

These are common partly because of variations in the virulence of different strains of the virus and partly because of the differing immune status of pigs. If the virus is a mild one, a herd may seroconvert to become positive in routine serological tests without the pig attendants being aware of anything seriously wrong. All strains of the virus have an affinity for nerves. Some also tend to have a strong affinity for the respiratory tract, causing severe pneumonia, while others do not.

In modern pig production cattle are not usually reared in close proximity to pigs but on those farms where they are, some of the cattle may become sick, develop a relentless itching and die. Farm cats may also develop nervous signs and die.

Diagnosis

In disease-free regions where controls are in place notification of the Government Veterinarians is required before any tests are taken.

In acute severe outbreaks a strong presumptive diagnosis can be made on the typical clinical signs, particularly those in newborn piglets. The diagnosis is strengthened further if dogs, cats, cattle or horses are affected. On the other hand, in milder outbreaks and in grower-finisher herds where there are no sucking piglets, diagnosis may be difficult or impossible. Furthermore, even when sucking piglets are involved, AD/PRV can easily be confused with porcine reproductive and respiratory syndrome virus (PRRSV) which is caused by an entirely different type of virus. Both diseases cause abortions and the birth of weak and premature litters, and result in high piglet mortality, but with PRRSV there are no nervous signs in piglets and weaners. The disease in grower/finisher pigs can also be confused with SIV which results in similar clinical signs and takes a somewhat similar course.

To confirm the diagnosis laboratory tests are needed. Whole dead piglets should be submitted if possible. If not, tonsils and smears made from the pharyngeal region on microscope slides can be submitted. See Chapter 15 (Swabbing). The laboratory should be able to do a rapid fluorescent antibody test (FAT) and have the results in a matter of hours. This test is fairly reliable in sucking piglets but much less so in grower/finisher pigs and so a negative result may be false and unreliable.

In grower/finisher pigs virus isolation has to be attempted and this delays the result considerably. The best tissues to submit for this are brain, spleen and lung which should be kept chilled on ice during transport. If no dead pigs are available, nasal swabs can be submitted and should be sent in transport medium containing antibiotics. Blood samples from pigs in the early stage of the disease and from recovered pigs can be tested for rising antibodies but this delays the result by about 2 weeks.

Virus isolation is carried out in cell cultures. The virus produces changes in the cells (cytopathic effects) and can be demonstrated by fluorescent antibody tests.

A variety of serological tests have been developed for pseudorabies virus including serum neutralisation tests, ELISAs, and latex agglutination tests. The ELISA is usually the test of choice and yields results fairly quickly, requiring a very small amount of blood. ELISA kits are commercially available and some of them are able to differentiate antibodies produced by gene-deleted vaccine virus from those produced by natural infection from a wild virus.

Post-mortem lesions

AD/PRV damages the pig's nervous system but this damage (i.e. lesions) can only be seen through a microscope, not with the naked eye.

The only gross lesions that can be seen by the naked eye are those in the lungs and respiratory tubes and areas of necrosis on the tonsils and abdominal organs. Necrosis of the tonsil may be seen.

Small, multiple 1-to-3 mm areas of necrosis may be seen in the liver. An untrained pig person would not be able to distinguish AD/PRV lesions from those caused by SIV, PRRSV, *Mycoplasma hyopneumoniae*, *Actinobacillus pleuropneumoniae*, or *Pasteurella multocida*.

Similar diseases

When disease is first introduced into a susceptible breeding herd there are few other diseases except possibly swine fever that would be confused with AD/PRV. Once the disease has become chronic it could be confused with PRRSV and chronic classical swine fever. Laboratory tests would be required to differentiate them.

Treatment

- There is no treatment available specifically against the virus.
- No treatment is effective in sucking pigs.
- Antibiotics will prevent secondary infections particularly of the respiratory system and also reduce bacterial damage.
- Vaccination is the key action to take. As soon as disease is identified all breeding stock should be vaccinated with a gene-deleted vaccine to mitigate the effects and reduce spread of the virus.

Management control and prevention

Vaccination of the herd

In countries which are free of the AD/PRV virus, vaccination is not practised and generally is not allowed.

In enzootic and high risk areas routine vaccination is practised and may be compulsory as part of an eradication scheme.

The development of gene-deleted attenuated-virus marker vaccines made vaccination effective. In these vaccines a small part of the genetic code is removed, rendering the virus non-pathogenic when it multiplies in the pig, thus producing immunity without causing disease. Because the missing code fails to stimulate a full complement of antibodies, vaccinated pigs can be identified from those who have had the disease.

Special commercially available serological test kits can distinguish antibodies which have been stimulated by the marker vaccine from those that have been stimulated by natural infection with a gene-complete wild virus. Pig herds in a control and eradication area can thus be vaccinated to protect them from the worst ravages of virulent AD/PRV virus and at the same time serologically screened for natural infection. The widespread use of gene-deleted marker vaccines in a controlled region has the added advantage of suppressing the spread of wild virus and thus reducing its level in the region.

There is only one main serotype of AD/PRV virus which produces a strong, long-lasting immunity. Vaccinated pigs can become infected but multiplication of the virus in the pigs' tissues is limited and so less is shed into the environment.

Vaccination also prevents the virus from crossing the placenta of pregnant sows to infect the unborn piglets.

Piglets which are suckled by vaccinated sows receive colostral protection which lasts about 6 to 10 weeks. This is the age when the virulent AD/PRV virus would do most damage. During this time the pigs cannot be vaccinated successfully because the maternal antibodies neutralise the vaccine virus before it has had time to stimulate immunity.

National/regional control and eradication programmes

Individual countries and regions are variously trying to suppress or eliminate the AD/PRV virus. The approach used in most of these is blanket vaccination of all 8 to 12 week old commercial slaughter pigs with a relatively cheap inactivated vaccine. Breeding stock is vaccinated with a longer lasting gene-deleted vaccine over a limited period to reduce the level of virus in the pig population. This is combined with routine testing of sow sera using the differential ELISA to monitor for natural infection. The use of meat juice from the slaughterhouse has proven to be very useful for screening. The pathogen may also be made notifiable (i.e. when a veterinarian diagnoses it he has to report it to the authorities who take the appropriate action).

If your pig farm is in a region where AD/PRV is enzootic, vaccinate the herd routinely.

On-farm eradication

First try and identify how it arrived before deciding on the best option. If your herd is infected and the disease is in a stable state there are various methods of eradicating it.

1. Slaughter the herd, clean and disinfect the premises and repopulate with negative breeding stock. This is expensive and is unlikely to be cost effective in most commercial herds.
2. In a weaner production unit, vaccinate all the breeding stock with a gene-deleted vaccine. Later, test all the sows with the differential ELISA and remove those that are positive for wild virus.

Follow this by one of two options:

- **a** Depopulate the weaner accommodation and rear the pigs elsewhere. Then, after all the grower/finishers have been slaughtered and/or the youngest of them reared elsewhere, gradually restock the weaner, grower and finisher accommodation as subsequent sows are weaned; or
- **b** Change to a 3-site or multi-site system with all-in/all out rooms or buildings.

AD/PRV virus can spread from farm to farm on the wind.

On-farm precautions

If your herd is free of AD/PRV virus you should of course endeavour to keep it free but it is not always easy to do so, particularly in a pig-dense area.

Unfortunately, the AD/PRV virus can spread on the wind several km over land and much further over water, so standard precautions of farm biosecurity such as those described earlier in Chapter 2 cannot be entirely relied on to keep it out of your herd.

Also, semen from infected boars is thought to spread it so the use of contaminated semen could in theory infect the herd.

Rats, which are normally thought of as dead end hosts, also become infected and on rare occasions may spread it from farm to farm before they die.

Infected pigs can become long term sub-clinical carriers. Any replacement pigs coming on to the premises should come from known safe sources and be quarantined. These should as a minimum be physically separated from your pigs, for at least a month, preferably 6 weeks and blood tested prior to integration.

AD/PRV virus can be carried on the tonsils of the wild boar and be excreted intermittently. Make sure the farm is wild-boar-proof!

The causal agent is a herpes virus and like the herpes viruses of humans (e.g. the causes of cold sores and chicken pox) the AD/PRV virus can lie latent in the pig's nerve cells. Stress (e.g. the stress of transport) can reactivate it and the pig starts to shed virus again. The main spread between pigs is aerosol and nose-to-nose contact. It is not spread in faeces or urine. Unwashed pig trucks which have carried infected pigs do not seem to be such a major factor in the spread of AD/PRV. Nevertheless, it would be unwise not to adopt safe methods of loading your slaughter pigs.

The AD/PRV virus is not thought to survive long in pig meat and thus feeding of pig meat scraps to neighbouring "back-yard" pigs is unlikely to trigger off an outbreak. However, mink farms have been reported to become infected from being fed pig meat waste products.

If there is a risk of AD/PRV in the area, check on the health status of your replacement stock and segregate them for a month after arrival.

In summary

In regions where farms are exposed to AD/PRV, consider the following:

- AD/PRV is a pathogen you cannot afford to live with.
- Gilts and boars should only be purchased from known free herds and be vaccinated before arrival or in isolation.
- Keep it out of the herd by isolating all purchased breeding stock and blood sampling them before they enter the herd.
- Only buy from AD/PRV-free herds.
- If your herd is at risk, in other words within a 3 km (2 miles) radius of large infected herds, then vaccinate it to prevent disease.
- Vaccination helps to prevent the establishment of the virus.
- If you have AD/PRV, adopt the strategies for eliminating this pathogen as discussed above.

Brucellosis (*Brucella suis*)

This disease, which is notifiable in some countries, is caused by the bacterium *Brucella suis*, of which there are a number of biotypes. It can be spread by venereal infection and the boar is a major source either by direct contact at mating or via artificial insemination. It can be an important source of infection to the pig. In Northern Europe the hare is a natural host and can transmit the disease to outdoor pigs. The disease can be transmitted to people and is serious.

If brucellosis is suspected in your herd, take great care that you and other pig attendants do not get infected. Call in your veterinarian.

Importance of brucellosis

- *Brucella suis* is not highly infectious amongst pigs. It spreads slowly between and within herds and you should be able to keep it out of your herd if you take sensible precautions. But it is a serious pathogen and you must always think of it as such.
- If it gets into your herd, it is difficult to eliminate. It causes long term reproductive losses and can cause disease in humans.

If sows abort at all stages of pregnancy and the farm is in a pig brucellosis region, call in your veterinarian and request an accurate diagnosis.

Clinical signs

In a susceptible breeding herd

The earliest sign is usually a few pregnant sows aborting. No other pigs may appear to be affected. If the early abortions are few in number they may be missed particularly in loose-housed sows and in outdoor herds.

The first signs noticed may be a high return to service and vaginal discharges. These returns are 30-50 days after mating if the disease has been introduced to the herd by an infected boar. The females are infected by the boar at mating. If sows are already pregnant abortions may occur at any stage as a result of infection arising from contaminated rooting materials or vaginal discharges.

In the boar, *B. suis* tends to multiply in the testicles and/or the male accessory reproductive glands and is then shed in the semen for prolonged periods. The reproductive tracts of sows and gilts become infected when served by the boar or inseminated artificially. This results in a large proportion of very early abortions. The infection in the sows' reproductive tracts is not permanent and eventually clears up spontaneously. In contrast, the infection in the boars' reproductive tracts is usually permanent; the damage that it does is irreversible.

It may be noted that some of the boars' testicles are becoming enlarged (orchitis). A few growing pigs and adults, and a slightly greater number of sucking piglets and weaners start to develop partial or total paralysis in their hind quarters. This is the result of infection and damage to the spine. Some pigs may become lame, with swollen joints.

Mortality is likely to be low and the disease may spread slowly but the long term damage to production can be serious.

Diagnosis

The combination of abortions, returns to service, vaginal discharges, swollen boar testicles, lame pigs and young pigs with posterior paralysis is strongly suggestive but laboratory tests have to be done to make a definitive diagnosis. The disease is notifiable in some countries.

Do not open up a dead pig suspected of having brucellosis as the risk to human health is too great. The best samples to submit are aborted piglets, swabs of vaginal discharges, dead pigs, and blood samples from at least 10 sows, preferably from those which have aborted or returned to service.

The most sensitive and accurate laboratory method is to culture and identify the organism on selective media, which is not difficult. The products of abortion are teeming with organisms. The organism can also be cultured from semen, testicles and accessory organs, lymph nodes, fluid from swollen joints, and in the early stages, from blood.

A variety of serological tests can be done on blood samples, including agglutination and complement fixation tests, ELISAs, and card and plate tests. Both false positive and false negative results occur, which is why you need to blood-test at least 10 sows and boars, preferably more. Serological tests are unreliable for diagnosis in an individual pig.

Diagnosis is complicated by false positive responses from the cell wall of *Yersinia enterocolitica* O9, which can make the test positive. This can be very significant for breeding companies exporting animals.

Post-mortem

It would be risky to open a dead pig to look for lesions. In the early stages of the disease the organism is spread throughout the pig's body. Even in the later stages it is fairly widespread. It can infect you through tiny cuts and abrasions on your hands or by being splashed or rubbed into your eye or mouth.

Do not open up a dead pig for post-mortem examination. The risk of catching brucellosis is too great.

Treatment

- Treatment with antibiotics is not very effective and generally should not be attempted. Affected pigs should be destroyed.
- If your herd becomes infected the most reliable method of control is to slaughter the herd, clean up the premises and restock with brucellosis-free pigs. This is also the safest procedure from the pig stockperson and public's standpoint and in the long term is usually the least costly.
  Depending upon the country in which you work, it may be mandatory to do so.
- Other approaches include repeated herd blood tests with removal of positive reactors. This may be effective if only a few pigs are infected but is likely to be unsuccessful if many pigs are positive.

If your herd becomes infected, the best course of action is to slaughter the herd, clean up the premises and restock with Brucella suis-free pigs.

Management control and prevention

Vaccination

Vaccines which have had widespread use in cattle are not effective in pigs. The low incidence of the disease has not made it cost-effective to develop vaccines for pigs and none are available.

National/regional control and eradication programmes

There is a move in a number of countries, including the USA, to gradually eliminate this pathogen, mainly by compulsory slaughter of herds in which the organism is diagnosed combined with various other procedures depending on the particular programme.

Farm precautions

If your herd is free of *Brucella suis* you should endeavour to keep it free by adopting high standards of biosecurity (see Chapter 2).

The main method of spread is by pig-to-pig contact, through venereal transmission during mating, and through on-farm artificial insemination.

Other methods include the eating or rooting of aborted piglets, dead piglets, aborted afterbirths or materials contaminated by vaginal discharges from aborted sows.

Exposure of cut or abraded skin to infected materials may also result in transmission.

Sows shed the organism from the vagina for at least 30 days after aborting, sometimes much longer.

Suckling sows also shed the organism in milk which then infects their piglets.

The organism does not spread on wind and rarely on other vectors so the main defence of your herd depends on preventing exposure to infected pigs and the introduction of brucellosis-free replacement pigs.

Protection of personnel

Brucellosis in people, also called undulant fever, is a serious, long-lasting disease which does not respond well to treatment. Infected people get recurrent attacks of clinical disease over many years. It causes a variety of symptoms including severe headaches, meningitis and bad dreams, severe back-ache, depression and lack of energy and interest, damaged testicles and changes of personality. It is therefore essential that if the organism infects your herd, you take every precaution against people becoming infected.

Main precautions to take:

- Wear protective clothing and take it off and wash your hands before eating.
- Wear rubber gloves when handling affected pigs or infected materials.
- Protect your eyes against splashing infected materials into them.
- Do not touch your eyes or put your fingers or any instruments on or in your mouth when working with affected pigs.
- Protect any bare skin on your arms or face that might have cuts and abrasions.
- Get rid of the pathogen as soon as possible.

If the herd becomes infected, take every precaution against people catching it.

Bungowannah Virus

Bungowannah virus is a recently recognised virus from the *Pestivirus* family, which also includes the viruses that cause classical swine fever, bovine viral diarrhoea and border disease in sheep.

The virus was first identified in an outbreak of mortality in 3 to 4 week old piglets in New South Wales, Australia, in 2003. The infection also resulted in an increase in the birth of stillborn foetuses.

The virus causes a myocarditis of the foetus, which can also be seen in low virulent cases of CSF.

Additionally, similarities in viral proteins between the

Bungowannah virus and CSFV could result in a misdiagnosis of CSF which would create problems in countries that have eradicated CSFV.

However, the genetic make-up of the two viruses is clearly different and genetic analysis could be used to differentiate them in the field.

Classical Swine Fever/Hog Cholera (CSF/HC)

Classical swine fever is caused by virus from the *Pestivirus* family. The pig is the only natural host and in most countries, CSF is a notifiable disease. The viruses survive in frozen carcasses for long periods of time.

Importance of CSF

- CSF is one of the most economically-damaging pandemic viral diseases of pigs in the world. Many governments take it very seriously and adopt strict control policies, which include compulsory vaccination or slaughter and eradication policies.
- In a susceptible (unvaccinated) herd almost all the pigs are affected. It causes generalised disease, including fever, malaise, lack of appetite, diarrhoea, paralysis, abortion, mummification and the birth of shaking piglets. Mortality is high.
- Fortunately, there is only one serotype of the virus and attenuated vaccines are highly effective. Also, it does not spread on the wind or on insects or birds so standard precautions of farm biosecurity should keep it out. However it persists in uncooked and cured meat and these should not be fed to pigs.

Should you be concerned about CSF?

If your pig farm is in the United States, Canada, Chile, Australia, New Zealand, UK or Ireland, which can be regarded as free countries and which are most unlikely to be re-contaminated, you need not be concerned about it unless there is a notified outbreak.

If your pig farm is in Western Europe, most of which can be regarded as a fringe region, (i.e. free from the virus but at risk of re-entry) you should be concerned because although your country is probably free from the virus, there is a real risk that it might get back in from Central Europe or Eastern Europe.

If your pig farm is in Germany, Austria, Italy or Poland, although the domestic pig population is free from the virus, it is endemic in the wild boar population which poses a threat to your herd. The virus is thought to be endemic (enzootic) throughout the rest of Central and Eastern Europe, S.E. Asia and Africa so if you are involved in pig production in any of these areas you may need to vaccinate your pigs routinely.

Clinical signs

If you are involved in pig farming in one of the fringe regions identified above you should be able to recognise the early clinical signs so that you can alert your veterinarian. If the disease is suspected he/she will report it to the authorities. They will carry out laboratory tests and depending on your country's policy, will probably slaughter your herd, a very distressing experience even if you are paid compensation.

The virus that causes CSF varies in virulence. Some strains are highly virulent and cause acute (i.e. rapid) serious disease. Some strains are of low virulence and cause chronic (i.e. long-lasting) disease. Others are intermediate causing sub-acute disease.

Acute disease

Clinical signs usually appear first in a small number of growing pigs which show non-specific signs of depression, sleepiness, and reluctance to get up or to eat. If you get them up they may wander to the feeder but eat very little or nothing and wander away again to lie down. They walk and stand with their heads down and tails limp. Over the following few days these signs get worse and more pigs become affected.

Younger piglets may appear chilled, and may shiver and huddle together.

Initially affected pigs may appear to be constipated but this generally changes to a yellow-grey diarrhoea as the disease progresses. Early on some of the pigs may develop conjunctivitis (inflammation of the eye surface) with thin discharges. This gets worse, and the discharge getting thicker with time until some of the eyelids are completely closed and adhered.

A constant early sign, which persists throughout the disease until just before death, is a high fever, over 42 °C (108 °F). Check the sick pigs' rectal temperatures. If they are all high, suspect CSF.

As the disease progresses the affected pigs become very thin and weak and develop a staggering walk. Initially this is probably through weakness but later it is due to infection of and damage to the spinal nerves. Partial paralysis of the hind end results in a drunken walk and a tendency to fall to a sitting or lying position. Diarrhoea worsens and some pigs vomit a yellowish bile. The pigs' skins go purple, first over the ears and tail, followed by the snout, lower legs, belly and back. Affected pigs die in 10 to 20 days. Some pigs go into convulsions before death.

Sub-acute disease

The early signs in growing pigs are similar but they progress more slowly and are less severe. Affected pigs may be ill for up to 30 days before they die.

If a number of your pigs become drowsy, reluctant to get up or to eat, and hang their heads and tails, take their temperatures. If most are above 42 °C (108 °F), call your veterinarian urgently.

Chronic and aberrant disease and persistent infection

The virus can cross the placenta and infect the piglets in the sow's uterus. Sows that have been inadequately vaccinated that become infected, or sows which become infected with a virus of low virulence, may appear normal but give birth to shaking piglets, many of which die. (Note: there are also other causes of shaking or trembling piglets).

If the virus crosses the placenta before the piglets' immune systems have developed they may be born apparently healthy although possibly weak and may grow on to be persistent carriers without at first showing clinical signs. They shed virus so they are a menace to other pigs. At several weeks or months of age they may develop typical clinical signs but these are likely to be milder, to last longer and to be without the high temperatures.

Virus that infects the piglets in the uterus may cause other effects, namely death, mummification, abortion or the birth of weak piglets, some of which may be deformed. Vaccination of sows during pregnancy with some of the original attenuated virus vaccines has resulted in trans-placental infection of unborn piglets with similar adverse results. The newer attenuated vaccines are claimed to be safer.

Low virulence strains of the virus may also multiply in the reproductive tracts of unvaccinated boars or boars which have been inadequately vaccinated. The vaccine virus itself in some of the older attenuated vaccines was thought to do this, resulting in returns to service and abortions.

Diagnosis

In acute or sub-acute outbreaks a presumptive diagnosis can be made on the typical clinical signs and post-mortem lesions but African swine fever and *Salmonella choleraesuis* infection produce some similar signs and lesions.

The major differential is porcine dermatitis and nephropathy syndrome (PDNS) which may present with all the clinical and pathological findings of CSF.

In chronic or aberrant cases the clinical signs and lesions are less diagnostic and may only raise a suspicion of CSF.

Before sending any samples, the veterinarian must consult the appropriate veterinary authorities.

The best samples to send are the tonsils, blood, lymph nodes, spleen and in chronic cases, serum for serology.

The tonsils of the pig are very easy to find. Cut away the skin and flesh under and between the lower jaw bone including the tongue. The pair of tonsils are two large red patches each about the size of the end half of your thumb or perhaps slightly bigger. See Chapter 15 (Swabbing). Do not freeze them but send them packed with ice.

Rapid accurate laboratory diagnosis is essential.

Virus is present throughout the body. In addition to the tonsils the next best organs to send are the spleen, kidneys and last few inches of the small intestine (before it meets the large intestine).

The laboratory should be able to carry out rapid tests and let you know the diagnosis on the same day they receive the samples.

A quick and fairly accurate test, which has been used for a number of years, is the direct fluorescent antibody test (FAT) carried out on frozen sections of the tonsils or other organs. It can sometimes give false negative results but this can be avoided if a sufficient number of pigs' tonsils are sampled. The CSF virus also cross-reacts in the FAT test with bovine viral diarrhoea virus (BVDV) of cattle and the border disease virus (BDV) of sheep, either of which may sometimes infect pigs. These viruses can cause trans-placental infection of unborn piglets in the sow's uterus, resulting in infertility and piglet problems similar to those of CSF. Such congenital infection may also result in newborn piglets which are shedding virus and thus infecting other pigs.

The false positive results obtained in FAT may cause awkward problems if there is a control or eradication policy in force. Specific antibodies against specific antigens of the CSF virus differentiate it from BVDV and BDV. Fortunately, BVDV and BDV are relatively rare infections in pigs kept separately from cattle and sheep. One source of infection with BVDV, however, may be the feeding of raw milk by-products to pigs. Another source may be the accidental presence of BVDV in live attenuated Aujeszky's disease (pseudorabies) or other vaccines. Recent vaccination (i.e. a few days) of pigs with attenuated CSF vaccine may also cause positive FAT results.

Post-mortem

Normally it requires a trained person to interpret post-mortem examination findings but in the case of CSF an untrained pig person should be able to recognise some of the more marked lesions.

If pigs are laid on their backs after death and opened up for examination, the picture is striking. There are usually many small haemorrhages throughout the body and larger haemorrhages in some organs such as lymph nodes. Some of these may be bright red and filled with blood. Larger haemorrhages may also be present in the lungs and under the skin.

The kidney surfaces are often described as looking like mottled ducks' eggs in that they are covered with variable sized bloody spots.

The spleen may have dark raised areas of dead tissue. Similar areas of dead tissue occur in other organs (e.g. the tonsils) but are more difficult to find.

The lungs may show severe pneumonia, haemorrhage and pleurisy, usually resulting from secondary bacterial infection.

The stomach and gut are usually empty except for scant liquid which may be brightly coloured. Fairly unique lesions to look for are raised, so-called 'button ulcers' on the inner lining of the large intestine near its junction with the small intestine.

Treatment

- There is no treatment.

Management control and prevention

Vaccination

In most national CSF eradication programmes and in countries which are free of the CSF virus, vaccination against CSF is not practised and generally is not allowed.

In enzootic and high risk areas routine vaccination is practised and may be compulsory.

Inactivated vaccines were in common use but they sometimes contained live virus which resulted in infection. Inactivated vaccines have now largely been replaced by live attenuated vaccines, the most recent of which are relatively safe and effective. Pigs develop protective immunity 1 week to 10 days after vaccination and the immunity lasts 2 to 3 years (i.e. the lifetime of many sows and boars). Piglets which are suckled by vaccinated sows receive colostral protection which lasts about 6 to 8 weeks. During this time they cannot be vaccinated successfully because the maternal antibodies neutralise the vaccine virus before it has had time to stimulate an immunity.

There is only one mutationally stable serotype of CSF virus which produces a strong, long-lasting immunity. In a circumscribed region in which the CSF virus is endemic it is usual to blanket-vaccinate all pigs over 2 weeks of age initially. Piglets born to vaccinated sows would be vaccinated over 8 weeks age.

National/regional eradication programmes

A number of countries in which CSF was enzootic have successfully eradicated it, most notably Canada and the USA, the UK and most EU countries on mainland Europe.

Generally, an eradication policy starts with compulsory large scale vaccination over a limited period to reduce the level of virus in the pig population. Vaccination is then stopped and the disease is made notifiable. When CSF is then diagnosed the whole herd and other in-contact animals are slaughtered. Theoretically, it would be useful to carry out national serological testing at the same time (as has been done in eradicating Aujeszky's disease), but vaccination results in almost lifelong positive tests. Also, BVDV and BDV infections cause false positive results.

If CSF/HC is enzootic in your area, vaccinate your herd routinely.

National preventative programmes

Countries which are free of the CSF virus prevent re-infection from outside by controlling the importation of pigs and pig meat products, unless they have been well processed, from regions in which the CSF virus is still present. In addition, swill (waste human food) containing meat products must be sterilised by heating in licensed premises.

The disease is also legally notifiable. If a case does occur the herd is slaughtered, all in-contact pigs are traced and monitored or killed, and a standstill order is placed on pig movements in the area around which the case occurred. Attempts are made to find out the source of the infection. This is likely to be illegally imported pig meat scraps which have been given to pigs without first thoroughly cooking them. Other possibilities are the illegal importation of infected semen, the return of unwashed pig trucks from an infected region and contact with wild boar or unprocessed meat.

On-farm precautions

If you farm in a country where CSF is endemic or where there is a risk of CSF occurring, consider routinely vaccinating your herd if vaccination is allowed. This will greatly reduce the possibility of contamination.

CSF virus does not spread as readily as some other viral infections. Unlike FMD it is not windborne. Thus the conscientious application of simple biosecurity measures should keep it out of the herd.

If CSF is in your country important precautions include reducing visitors to a minimum, taking precautions against contamination from vehicles, and not allowing pig meat products near any pigs.

Any replacement pigs coming on to the premises should come from known safe sources and should be quarantined. In some areas the disease has become very mild and spread can go unrecognised.

Pig buildings should be protected from stray animals, particularly wild pigs and boars.

Echinococcosis/Hydatidosis – *(Echinococcus granulosus)*

This is a new emerging zoonosis that has been identified in a number of Eastern European, Asian and South American countries. It is a dog tapeworm and the pig acts as an intermediary host for the G7 serotype. There are no clinical signs and diagnosis is done at necropsy. Maintain good biosecurity and do not allow dogs onto the farm.

Foot-and-Mouth Disease (FMD)

FMD is called a porcine vesicular disease because one of its most prominent clinical signs is the appearance of vesicles (blisters containing a clear fluid) in the mouth, on the lips and nose, on the teats and around the coronets of the feet just above the hooves.

There are 4 vesicular diseases of pigs which are very difficult to differentiate clinically: FMD, swine vesicular disease (SVD), vesicular exanthema (VES), and vesicular stomatitis (VS). Of these, FMD is the most widespread and important with SVD being of secondary importance in some regions (e.g. the EU). The other 2 have very limited distribution and VES has all but disappeared. Because of the importance of FMD and the difficulty of distinguishing it from the others, all 4 have to be considered in this chapter.

Importance of FMD

- FMD is the most important restraint to international trade in animals and animal products. Consequently, large sums of money have been invested in control and eradication programmes and also into research. As a result, more is known about the FMD virus than about almost any other animal infection.
- It generally produces severe disease in pigs and cattle.
- FMD is so important because it is highly infectious and spreads rapidly throughout animal populations and over long distances on the wind. Hence it is difficult and costly to control. Also because of its damaging and debilitating effect on cattle, a great deal of effort and tax-payers' money has been spent keeping it out of large areas of the world. It would be highly irresponsible to let it back in.
- If you live in an FMD-fringe area that is also free of swine vesicular disease (SVD) you should be aware of what early clinical signs would make you suspicious and what you should do if you suspected them in your herd. If you farm in an endemic area or a fringe area in which SVD is present then you should know a bit more, particularly about the clinical signs in pigs and vaccination regimes.
- If you farm in an FMD-free country that takes sound precautions against its entry, the risk to your herd is negligible unless you farm in western USA where vesicular exanthema may pose a very small risk.

Susceptible animals

Among farm animals, pigs, cattle, sheep, goats and deer are susceptible. In addition, wild and domestic cloven hooved animals such as hedgehogs and rats are also susceptible, as are elephants.

Early clinical signs

In cattle the early clinical signs are much more definitive or suggestive than in pigs. For example in a dairy herd several cows may suddenly show depressed milk yield, go off their feed, run a fever, have a dramatic drop in milk yield, and a little later start salivating profusely, the saliva running from their mouths (slavering). If you see such signs, jump into action: ring the vet. Veterinarians who have to deal with FMD say that if a farmer telephones to say that several cows are salivating profusely they think first "FMD?"

If after cows have started salivating and smacking lips, vesicles are noticed on the lips, on the teats and around the coronets, the areas above hooves – your worst fears are probably true. The probability is that your pig herd has been infected too.

If you keep cattle as well as pigs and some of the cattle go off their food and start salivating profusely, think "FMD?"

In pigs, early signs are lameness and a drop in food consumption, and some pigs will appear depressed and have fevers of about 40.5 °C (105 °F). In piglets, sudden death due to cardiac failure is common. What should make you strongly suspicious is the appearance a little later of vesicles of up to 30 mm diameter, similar to those described above for cattle. They are most plentiful around the coronets but are less plentiful on the nose and lips although this is where you are likely to see them first. They often appear on the teats of recently farrowed sows. By then the sows and some of the other pigs may be dribbling saliva and chomping their jaws. If they are on bedding they may not appear lame but if they are on concrete they probably will.

The early signs of swine vesicular disease (SVD) when it is severe are indistinguishable from FMD so you should suspect it too.

If you farm in Georgia, the Carolinas or Central or South America and it is summer/autumn time, perhaps you should think of vesicular stomatitis. The clinical signs of all 4 diseases are almost indistinguishable. There are a number of other viruses – especially caliciviruses – that may produce vesicles in pigs. However, DNA analysis of tissues will rapidly differentiate these from cases of FMD.

Within 24 hours many of the vesicles will have burst. On the lips and teats they may leave shallow erosions but on the coronets of the feet secondary infection and trauma may convert them into raw, jagged-edged ulcers.

If the pigs are not killed some may lose their complete hooves (so-called "Thimbling"), sows may abort as a result of fever, and in severe outbreaks some may die. Boars may go lame and stop serving sows, so there is an infertility side effect. There may also be an increase in mortality among suckled piglets. This is often the first sign.

In endemic areas where vaccination is carried out routinely the disease is not a serious economic problem in pig herds. In fringe areas, particularly where vaccination is not allowed (e.g. in the EU) it is a serious problem because the herd will almost certainly be slaughtered out and although compensation is likely to be paid, the farm cannot be restocked for at least 6 weeks. It is therefore out of production and in a negative cash flow for a long time.

Diagnosis

Rapid accurate diagnosis is essential.

FMD cannot be distinguished from SVD on clinical grounds, or from VS in western USA, although SVD is often much milder. To differentiate these diseases and confirm the presumptive diagnosis, samples have to be sent to a laboratory capable of making a diagnosis.

The samples sent are blood and pieces of the skin that overlay the blisters, plus vesicular fluid if this is available. Once the samples have been received by the laboratory, diagnosis is fairly rapid.

ELISA tests are used for virus identification and if it is FMD they also indicate what serotype it is. The virus may also be grown in cell culture and the identification confirmed by other tests. PCR may also be used to 'fingerprint" the virus. The gene (genome or RNA) of FMD repeatedly undergoes minor changes as the virus spreads through animal populations, so by identifying the precise sequence in the gene, the laboratory staff are able to make an assumption as to where it may have come from by the most recent isolate with a similar sequence.

Diagnosis of suspected FMD must be rapid and accurate.

Treatment

- There is no treatment. In negative areas, government will decide on the animal's fate.

Management control and prevention

Vaccination (where applicable)

In endemic and high risk areas routine vaccination may be practised, mainly to protect the breeding stock.

Most FMD vaccines are produced in cell suspension cultures and inactivated by ethylenimine derivatives. An adjuvant is added to make them more potent. Oily adjuvants are used in the pig.

Vaccination in pigs is problematical. This is because protection is short-lived, lasting only about 6 months. It is also partly because there are 7 serotypes of FMD and protection against one leaves animals susceptible to the others. Vaccines must be multivalent (several serotypes) in most endemic regions. Some vaccines are now available which can differentiate between vaccinated and wild (field) strains. This may be very important in ring vaccination and control measures in outbreaks.

Serotypes – There are 7 main serotypes: A, O, C, SAT 1, SAT 2, SAT 3 and Asia 1. There are also many strains within serotypes. Careful selection of the strains for incorporation in vaccines is essential to ensure they are effective. In Taiwan there is a pig-specific strain.

Precautions

Countries in free and fringe areas apply strictly enforced national preventative measures against the introduction of infection. The main features of these measures are control over the importation of cloven-hoofed animals and of meat from such animals from counties in which FMD occurs.

The virus does not survive rigor mortis but it can persist in bone marrow and lymph nodes of infected carcasses for several weeks.

If the disease does enter a free or fringe area, a slaughter policy is implemented, with all infected and in-contact animals being slaughtered. A standstill on animal movement is imposed and tracings are carried out to check possible spread of the disease through previous contacts. Ring vaccination may be used around the affected region.

If you farm in a FMD-risk region you should take strict precautions against the contamination of your herd. If you have a cattle or goat herd or a flock of sheep as well as a pig herd you should also adopt preventative measures for them and keep a wary eye for the appearance of typical clinical signs.

Unfortunately, none of the measures described prevent the windborne spread of FMD. Infected pigs can produce huge quantities of infective virus as aerosols. They produce far more aerosol virus than cattle, goats or sheep. In dry weather when there are strong thermals the aerosol virus is rapidly inactivated so the wind does not carry infective aerosols very far. Strong winds, hills and objects such as high buildings and trees create turbulence and disperse the plume of airborne virus as they would a plume of smoke from a bonfire. In humid, overcast weather with a steady light wind blowing over flat countryside, infective virus may survive long enough to infect other herds up to 60 km (36 miles) distant. Over water, given the same climatic conditions, infective virus has been shown to travel up to 300 km (180 miles), so siting your pig herd on an island in a lake is not going to stop it. Windborne infection is impossible to guard against. Even if your pig herd is in closed buildings, the aerosol virus can get in through the ventilation system and you may carry it in from outside on your boots or clothes.

If vaccination is permitted and the pig herd is in a high-risk area you should consider routine vaccination to reduce the susceptibility of your herd.

If a group of pigs go off their feed, and have blisters on their noses and lips and go a bit lame, suspect FMD or SVD. Contact your veterinarian urgently.

Japanese Encephalitis (JE)

Japanese encephalitis (JE) is caused by a virus (a member of the flavivirus group) which is spread by mosquitoes. The pig is the natural amplifier and reservoir host. In other words it serves to multiply up the virus and keep it going. Many other species can be infected, including most domestic animals and many wild animals such as rabbits, mice, birds, bats, snakes and lizards.

Should you be concerned about JE?

You should be if you work in South Asia or visit pig farms there.

Clinical JE is confined to South Asia.

If you live outside South Asia, there is no risk to your herd and no risk to you unless you decide to visit pig farms in the area, in which case get yourself vaccinated twice, preferably 3 times, before you go.

If you intend to visit pig farms in South Asia, get yourself vaccinated.

Importance of JE

- JE is not a very important pathogen of pigs, causing only sporadic reproductive problems.
- Its main importance is its threat to public health.

Japanese encephalitis is transmitted by mosquitoes and causes a serious disease in people.

Clinical signs

The main clinical signs are degeneration of the boar's testicles and infertility, and the birth of abnormal piglets. These include mummified foetuses, dead piglets with subcutaneous oedema (excess clear bodily fluid under the skin), hydrocephalus (water on the brain), and weak piglets, sometimes with nervous signs.

If sows give birth to piglets with a variety of abnormalities, and one or more boars become infertile, and you work in S. Asia, suspect JE.

Diagnosis

This requires laboratory examination of dead stillborn piglets and affected boars' testicles.

Definitive diagnosis depends on the isolation of the virus in tissue culture and demonstration of antibodies in stillborn piglets' serum, usually by an ELISA or PCR.

Similar reproductive problems are caused commonly by parvovirus, Aujeszky's disease, PRRSV, certain strains of SIV, CSF/HC, and some teschovirus.

Post-mortem

There are no specific lesions visible to the naked eye.

Treatment

- There is no effective treatment.

Management control and prevention

- **Vaccination** – In countries in which this disease is endemic, young breeding gilts and boars are vaccinated twice before the mosquito season starts, commonly with an attenuated vaccine (but inactivated vaccines are also available).
- Trying to control mosquitoes in pig herds is a waste of time.

New and Old World Screwworm (*Cochliomyia hominivorax* and *Chrysomya bezziana*)

New and Old world screwworms are parasites that can affect any mammal. As such, pigs may become infected and may be potential carriers. There is no specific treatment, but the authorities should be alerted.

Nipah Virus (NiV)

The Nipah virus was first identified in 1998 during an outbreak among pig farmers in Malaysia. Since then, there have been more than 12 outbreaks, all in South Asia. The disease causes respiratory and occasionally nervous signs in pigs, is zoonotic, and can have a devastating effect on humans.

Nipah Virus is an RNA virus from the paramyxovirus family and is named after the village in Malaysia where it was first identified. Fruit bats of the Pteropodidae family are the natural host of Nipah virus.

Symptoms

Humans

Symptoms may be mild or severe and include:

- Fever, headaches, encephalitis.
- Drowsiness, confusion leading to coma.
- Respiratory failure.
- High mortality of 40 % to 75 % reported.
- A few people have shown no symptoms.
- The incubation period is from 4 to 45 days.

All pigs

- Morbidity is usually high but mortality is low.
- Rapid laboured breathing.
- Very harsh explosive cough.
- In sows clinical signs may be more pronounced, with severe breathing difficulties.
- Convulsions, death.
- Pneumonia.
- Mucopurulent (mucus and pus) discharges from the nose.
- Pigs are infectious during the incubation period, which lasts from 4 to 14 days.

At post-mortem the predominant signs are consolidation of the lungs.

Causes/contributing factors

- Movement of pigs.
- Direct pig-to-pig contact either by mouth, by the respiratory route or by aerosol from urinary excretions.

Diagnosis

This is by serological tests, virus isolation and identification. In infected farms sows show high levels of antibodies and in infected areas antibodies have been widespread in dogs but not in rats.

Management control and prevention

There is no vaccine against Nipah virus. Good disinfection and cleaning of farms (with sodium hypochlorite or other detergents) should minimise infection.

Where an outbreak is suspected, the authorities should be immediately notified and the premises quarantined. Infected pigs should be culled and close attention should be given to the burial or incineration of carcasses to reduce the risk of transmission to people.

Porcine Reproductive and Respiratory Syndrome Virus - (PRRSV)

See Chapter 6 and also Chapters 7, 8, 9 and 10 for information

Porcine reproductive and respiratory syndrome (PRRS) is caused by an RNA virus from the genus *Arterivirus* and causes reproductive failure in breeding stock and respiratory tract illness in young and weaned pigs. PRRSV is on the OIE disease list B.

The main information on PRRSV, including methods of spread, immunity, clinical signs, diagnosis, similar diseases, treatment, management control and prevention, and elimination from a herd are covered in Chapter 6, with additional information in Chapters 7, 8, 9 and 10.

Pork Bladder Worm (Cysticercus cellulosae) Human Tapeworm (*Taenia solium*)

See Chapter 11 for further information

Cysticercus is the name of the larva or cyst which forms part of the life cycle of the tapeworm *Taenia solium* found in the human. Pigs are the natural intermediate host.

Rabies

This is relatively rare in the pig. It occurs when a carnivorous rabid animal such as a racoon, dog, wolf, fox, coyote, jackal or skunk with the furious form of rabies gains access to pigs and manages to bite them. In Argentina certain species of cave-dwelling vampire bats transmit rabies to cattle but they are most unlikely to gain access to pigs.

Importance of rabies

Remember, rabies is a fatal disease, communicable to man.

Clinical signs

After a variable incubation stage the bitten pig suddenly shows depression, develops difficulty in walking and becomes prostrated. Nose twitching, rapid chewing movements, excess salivation and muscular spasms which evolve into tremors follow. This sequence of clinical signs is termed dumb rabies. Death ensues in 3 to 6 days. The furious form of rabies, seen in dogs and foxes, has been reported in pigs but is rare. The pig shows the above signs but in addition becomes aggressive.

Diagnosis
Accurate diagnosis is essential and requires a specialised laboratory.

Treatment
- There is no treatment or method of control suitable for pigs.

Management control and prevention
- Pet pigs in areas where Rabies may be a problem should be vaccinated every 3 years. The authorities, however, may not consider the vaccination effective.
- People are highly susceptible to the disease so unless you have been properly vaccinated, barricade the affected pig away from the others and keep right away from it yourself.
- The saliva of the pig is highly infectious and can exist as an aerosol which you might inhale. Call your veterinarian quickly and let him deal with the situation.
- Beware that other pigs may be incubating the disease. Inspect them all frequently and if any behave suspiciously separate them and keep away from them yourself.
- In regions where rabies is common, farm staff and the farm dogs should be vaccinated.

Rinderpest
Rinderpest virus has been eliminated on the planet. However, pigs could become infected and may be important in the spread of the condition if it ever appeared again.

Swine Vesicular Disease (SVD)
Although the virus which causes swine vesicular disease (SVD) virus is different from that causing foot-and-mouth disease (FMD), it produces a condition in pigs that is clinically indistinguishable from FMD. So if you are concerned about SVD in your pigs, read also the FMD section.

Should you be concerned about SVD?
If your pig farm is in the UK, mainland Europe or S.E. Asia you should be aware of the possibility of it becoming infected with SVD, however low the risk might be. The herd will probably be slaughtered if it gets infected.

If you farm in North or South America, or Australia or New Zealand, the risk of your pig farm becoming infected with SVD or FMD is virtually nil.

Importance of SVD
- Although clinically SVD is similar to FMD it causes little impact on productivity. Often it can be so mild that the pigs do not appear lame particularly if they are on straw bedding. What is more, it is strictly a disease of pigs and does not infect cattle, goats, sheep or other species. Why, then, is it regarded as so important that governments, such as those in the EU, bring in costly slaughter and eradication policies? It is for the very reason first stated, namely, that it is clinically indistinguishable from FMD. They are afraid that if SVD became widespread in the pig populations of FMD-free and fringe areas, pig farmers and pig veterinarians might become accustomed to seeing vesicles on pigs' noses and feet and not report them or even consider FMD.
- One could argue that an expensive slaughter policy is unnecessary; that SVD could be made notifiable (i.e. any pigs with vesicles would have to be reported to the authorities) and that accurate rapid tests (e.g. ELISAs and PCRs) could be available in all diagnostic laboratories, but such an argument would be academic. The fact is that most governments in free and fringe areas would adopt a slaughter and eradication policy if there were a risk of contamination. If you farm in such an area you have to live with such a policy.

Unless you farm in Italy or S. Asia, your herd will be slaughtered if it gets SVD.

Clinical signs
SVD does not infect or affect cattle, sheep, goats or any species other than pigs. So, unlike FMD, if you keep other livestock they will not be affected.

Clinical signs of SVD are much the same as FMD, so read that section.

If there is no slaughter policy and pigs are not killed, some may lose the claws off some of their toes. SVD does not usually cause abortion and boars are not sufficiently lame to stop serving sows. Mortality among all age groups is low.

The pigs recover completely in 2 to 3 weeks but you may see bruises under the claws which gradually move down under the horn as it grows (about 2 mm per week). See Chapter 10: Vesicular diseases.

In fringe areas (e.g. in the EU) it is a serious problem

because the herd will almost certainly be compulsorily slaughtered and although compensation is likely to be paid, the farm cannot be restocked for at least 3 months. This is twice as long as for FMD because the SVD virus is much tougher and survives longer. Furthermore, after that time small numbers of susceptible pigs may have to be introduced onto the farm to act as sentinels. If there is still residual virus in the premises these sentinels are likely to come down with clinical signs and the farm will have to be emptied and disinfected again.

The pig farm is a long time out of production even if the sentinels do not develop clinical signs but if they do it is even longer, with all the financial hardship that this will cause. So-called recrudescences on farms which had been slaughtered out, disinfected and later re-stocked were a major problem in the early days of the SVD eradication programmes in Europe.

Cattle, sheep and goats are not affected by SVD.

Diagnosis

This is the same as for suspected FMD. Read the FMD section. SVD cannot be distinguished from FMD on clinical grounds, so to confirm the presumptive diagnosis samples have to be sent without delay to an appropriate laboratory.

Management control and prevention

- **Vaccination** – There is only one main serotype of SVD and theoretically it should be possible to produce an effective vaccine, but in endemic areas the disease is too mild to warrant it. Vaccination is not allowed in fringe areas because it might mask the disease and go undiagnosed.

Do not let pig-meat products contaminate your pigs or those of your neighbours.

Other precautions

Countries in free and fringe areas apply strictly-enforced national preventative measures against the introduction of SVD. The main features of these measures are control over the importation of pigs and of pig meat products from countries in which SVD occurs. Pig meat products are particularly dangerous because, unlike the FMD virus, SVD virus is tough and survives rigor mortis. FMD is able to survive in bone and lymph nodes. If infected pig meat gets into the food chain there is a risk of uncooked waste being eaten by pigs which could trigger off an outbreak. To prevent this, many countries have brought in cooked-waste feed policies which forbid the feeding of animal products to pigs unless they have been processed through a licensed processing plant.

If the disease does enter a free or fringe area, a slaughter policy is implemented similar to that described for FMD. All infected and in-contact pigs are slaughtered. A standstill on animal movement is usually imposed and tracings are carried out to check possible spread of the disease through previous contacts.

If you farm in the EU, which is an SVD-risk region, you should keep in mind the possibility (small though it may be) of contamination of your herd and you should take simple appropriate precautions.

The greatest risk is from contaminated pig trucks returning from Italy where the disease is still present. The virus survives well in trucks and pigs can become infected by mouth or via skin abrasions. The truck collecting your pigs for slaughter is the obvious danger.

You should not feed pig meat products.

Windborne spread does not occur in SVD so the simple precautions outlined later in this chapter should be effective.

Prevent pig trucks that carry slaughter pigs from contaminating your herd.

Transmissible Gastroenteritis (TGE)

See Chapter 8 for further information

TGE is an economically-damaging highly-infectious disease of pigs that is enzootic in most pig-rearing countries of the world. A few countries are thought to be free of it including Australia, New Zealand, Ireland and Denmark.

If your herd is fully susceptible to TGE and it becomes infected, up to 100 % of piglets less than 2 weeks of age will die. Many older sucking pigs and some weaned pigs will also die. A proportion of recently bred sows will return to service. As a rule of thumb you can predict that the number of young pigs that die will be > 1.5 times the number of breeding females in the herd. The precise number depends on the actions you take and the concurrent diseases present in your herd. If the herd is farrow-to-finish, you will suffer a substantial loss of revenue and negative cash flow for 4 to 5 weeks about 4 to 5 months after the outbreak and another one 4 to 5 months later. This disease is covered in detail in Chapter 8.

TGE is a highly infectious and economically damaging disease.

Trichinellosis (Muscle Worms)

See Chapter 11 for further information

These are very tiny roundworms (*Trichinella spiralis*) no more than 2 to 4 mm long. They form cysts in the muscles of the pig. When inadequately cooked pork is eaten by humans, they cause disease in man.

Vesicular Stomatitis (VS)

This disease occurs mainly in South and Central America, occasionally in the USA and rarely as epidemics extending as far north as Canada and as far south as Argentina. If you farm in any of these areas you should be aware of it. The infection is more widespread, involving wildlife including sea mammals.

It does not occur outside the Americas. The only confirmed reports outside the Americas were France during World War I and South Africa during the Boer War – both associated with horses.

Importance of VS

The VS virus produces a clinical condition in pigs that is clinically indistinguishable from FMD, SVD and VES. Most often, however, infection of pigs is subclinical.

In itself it is not very important in pigs. In severe outbreaks the foot lesions may be painful and make pigs lame and there may be a reduction in growth rate. On the other hand it can be so mild that the pigs do not appear lame or ill.

If you are concerned about VS, read also the sections above on FMD and SVD.

Species affected

In domesticated animals, VS is primarily a disease of horses and cattle, only occasionally causing clinical disease in pigs. A range of wild animals including wild pigs, deer, racoons and sea mammals as well as people can become infected, developing flu-like symptoms.

Vesicular stomatitis is primarily a disease of horses and cattle, only occasionally occurring in pigs.

Clinical signs

The initial signs in affected pigs are drooling saliva and a rise in body temperature to 40 to 41 °C (106 to 107 °F). Thereafter the clinical signs are so closely similar to those of FMD, that they need not be repeated in detail here. Like FMD, the most striking feature is the appearance of vesicles (blisters) up to 30 mm (1 inch) diameter on the nose, lips and teats and around the coronets of the feet, which may make the pigs lame. These burst, leaving erosions and ulcers. Those on the nose and mouth then heal rapidly but those on the feet may become secondarily infected and permanent damage may result. Mortality is usually low and most pigs recover in 1 to 2 weeks.

One difference from FMD is the relatively small proportion of pigs in a herd outbreak that show vesicles, although many may seroconvert as a result of sub-clinical infection. Another difference from FMD is that usually when an outbreak occurs in a pig herd it rarely (if ever) spreads to cattle and horses on the same farm and vice versa.

Occurrence and spread

The Americas can be divided into enzootic and epizootic regions. In the enzootic region the virus is present and cycling among animals and insects all the time. The enzootic region covers Central America, Mexico, the coastal plains of South Eastern USA and the North West of South America. The rest of the USA, Canada and South America constitute epizootic regions.

The virus is not present in these all the time but outbreaks occur seasonally and intermittently, several years often passing between epizootics. Epizootics usually start in the USA in late spring or early summer, often after heavy rain, presumably reflecting the rise in the insect population.

The virus is spread mechanically by a variety of insects and has been isolated from face flies, black flies, eye gnats, sand flies, leaf hoppers and mosquitoes. The virus may multiply in some of these insects and can pass vertically through the ovaries to the offspring. Insects are therefore thought to act as reservoirs, perpetuating the virus in the enzootic regions.

It is thought that in the spread between pigs in epizootic regions, insects get the virus on their mouth parts from feeding on the lesions left after the vesicles have burst, and carry it mechanically to other pigs in the same herd or in neighbouring herds. It is unlikely that they get infected from sucking the pigs' blood. The erosions and ulcers are initially teeming with virus and although the virus is assumed to be carried around the body in the bloodstream, it is at undetectable levels.

The virus can also spread between pigs by direct contact, particularly when pigs are tightly packed together, for example, during transport or when pigs fight after mixing. The virus is thought to be spread in these circumstances by getting into cuts and abrasions.

The virus can also be carried from herd to herd through the movement of pigs but pigs do not appear to become long term sub-clinical carriers.

Diagnosis

VS is notifiable in most epizootic areas, i.e. if you suspect it in the herd, you or your veterinarian have to report it to the authorities.

The clinical signs of VS are similar to those of FMD and SVD, both of which are subject to government slaughter and eradication policy in Canada, the USA, Mexico, Chile, South Brazil, and Argentina. It is therefore crucial to reach a fast, accurate diagnosis. This can only be done by delivering samples to a laboratory equipped for and capable of doing the appropriate tests. The aim is to eliminate the possibility of FMD.

The best samples to submit are vesicular fluid, if available, which has high concentrations of virus and/or vesicular tissue (e.g. the thin superficial skin layer over the vesicle) which also contains virus. If these samples are from pigs or cattle the authorities will probably only allow you to send them to a designated FMD laboratory in case the pathogen is FMD. If they are from a horse then they cannot be FMD (horses do not get FMD, SVD, or VES) and you may be allowed to send them to other laboratories. The possibility of the pathogen being FMD should be eliminated and an accurate identification of the VS virus made. The first of these, namely, elimination of FMD, SVD and VES can probably only be done in the designated FMD laboratories. Other diagnostic laboratories may be able to do the second, namely, identification of the VS virus in samples from horses.

Paired blood samples (i.e. one sample taken during the early stage of the disease and one 10 to 14 days later) may also be taken. The authorities will probably allow these to be tested in non-FMD-designated laboratories. The tests used are generally ELISAs with back-up neutralisation and complement fixation tests. In horses, rising antibody levels to the VS virus have to be demonstrated in the blood samples to be sure that an active VS infection has taken place. This is because in epizootic regions some old horses may have positive antibodies from the last outbreak. Pigs generally do not live so long, so single positive samples would be strongly indicative of active infection.

Unfortunately blood sampling and serology may mean a delay of at least 2 weeks, which is too long.

Management control prevention and treatment

- **Vaccination** – It is possible to produce an effective live attenuated vaccine or an inactivated vaccine but in practice the low incidence of the disease in swine, even in the face of big outbreaks in cattle and horses, makes vaccination uneconomical. Furthermore, in the USA vaccination of pigs against VS is not allowed.

There are numerous serotypes of VS virus but only two (Indiana 1 and New Jersey 1) are known to affect pigs.

National/regional precautions

If your pig farm is in an epizootic area, which most states in the USA are, for example, and VS is confirmed, the herd will not be slaughtered out to get rid of the virus because the pathogen is self-limiting and disappears spontaneously. It is likely, however, that the authorities will quarantine your farm until they deem that the virus has gone.

In the USA, animals in any state in which outbreaks of VS are occurring cannot be legally moved to any other state or to the EU without testing and quarantine.

Countries which are completely free from VS (e.g. countries in Europe) apply national preventative measures against the introduction of VS. The main feature is control over the importation of cattle and pigs from countries in which VS occurs.

In practice it is most unlikely that VS would get into such countries, spread and become established. Pig farmers in these countries should not worry about this pathogen

If the pathogen did enter such a country, which could be through the movement of horses, a standstill on animal movement would probably be imposed and the affected animals would be isolated and might be slaughtered. In fact, it has only been identified once in Europe and that was in American military horses in France in the First World War. It did not spread or persist.

On-farm prevention and pathogen management

As with other infectious pathogens, if you are a pig farmer in an epizootic region be careful about the source of newly introduced pigs into your herd. Isolate them for a month to 6 weeks to ensure that they, or the source from which they have come, are not incubating this pathogen or any other.

VS is not as contagious as, say, FMD or TGE, but if the pathogen breaks out in neighbouring pig herds you should tighten the biosecurity of your own farm.

In the case of VS, the gap in the protective measures is that the virus is spread by insects and, of course, it is difficult to stop their movement.

If the pathogen breaks out in your herd the most urgent thing for you to do is to get an accurate diagnosis. Call your veterinarian immediately.

If possible, make the affected pigs comfortable by providing clean bedding. Also, give them soft food. Put them in clean pens to reduce the likelihood of secondary infections. If secondary infections occur, treat with antibiotics.

Use insect sprays and repellents to reduce the spread in your herd and to your neighbours' herds. If you supply other herds with stock, stop the movement of pigs from your herd for 30 days. (The authorities will probably insist on this anyway).

Other Diseases (Not Covered Elsewhere)

The following diseases, whilst unusual, are known to occur in pigs and are covered here for completeness.

- Borna disease.
- Blue eye (BE).
- Getah viruses.
- Tumours (cancer).
- Vesicular exanthema of swine (VES).

Borna Disease

Borna disease is caused by a virus which damages the nervous system. It has been known in horses in Germany and surrounding countries for over a century. It has also been diagnosed in sheep and more recently in cats. It causes concern from time to time because there is evidence that it may pass to people working with these animals and cause dementia. However, it has never been reported in pigs.

Blue Eye (BE)

This condition is so-called because the whole eye develops a bluish tinge as a result of corneal opacity.

Blue eye (BE) is characterised by clinical signs of nervous derangement, reproductive problems and opacity of the surface of the eyes (corneal opacity). It is caused by a paramyxovirus.

BE was first recognised in central Mexico in 1980 and was later diagnosed in many other regions of Mexico. It has never been diagnosed in any other countries and seems unlikely to spread to them.

In Japan, a paramyxovirus (the *Sendai virus*) can cause nervous disorders and reproductive problems but it is a different virus.

If you are involved with pig farming outside Mexico and your country has strict controls over imports of live pigs, particularly from Mexico, do not be concerned.

But beware! You could confuse Ontario encephalitis (the nervous form of vomiting and wasting disease) which occurs in Europe and North America with BE.

If you farm in Mexico, and piglets suddenly develop convulsions, consider blue eye, Ontario encephalitis or pseudorabies.

Clinical signs

When the condition first breaks out in a susceptible breeding herd the first clinical signs are usually seen in the farrowing house. Piglets are hunched with rough coats for a short time and then suddenly develop acute nervous signs. They have difficulty walking, start to shiver, and sit in abnormal postures. Some lie on their sides, paddle their legs and roll their eyeballs back and forth. The eyes of some piglets water and the eyelids swell and get stuck together. Worst-affected are piglets under 2 weeks of age. Up to 90 % die. Older weaned piglets may develop blueing of the pupils (black centres) of the eyes. Up to 65 % of litters may be affected. The disease disappears slowly and spontaneously after several weeks.

Older pigs of a month or more in age show milder transient signs of loss of appetite, fever, sneezing and coughing. Nervous signs, such as an unsteady walk, are less common.

Some adult pigs also show clinical signs including blueing of the eyes. There is an increased return rate in sows lasting about 6 months. Abortions and stillbirths may occur. Up to 40 % of boars have testicular enlargement, which is usually one-sided, followed by shrinking with a concurrent loss of fertility.

Diagnosis

In acute severe outbreaks a strong presumptive diagnosis can be made based on the nervous signs, corneal opacity, infertility and testicle changes.

For confirmation of the presumptive diagnosis, serological tests have to be done in the laboratory. Virus isolation from brain tissue can also be demonstrated.

BE will not be of concern unless you farm in Mexico.

Post-mortem

Excluding the corneal opacity, there are no lesions indicative of BE that can be seen with the naked eye.

Similar diseases

Diseases with similar clinical signs include Aujeszky's disease (AD), PRRSV, Ontario encephalitis (vomiting and wasting disease), and brucellosis. However, none of them have all the clinical signs of BE.

Treatment

- There is no effective treatment.

Management control and prevention

- **Biosecurity** – This should reduce the chances of your herd being infected. Read the final section of this chapter.
- **On-farm eradication** – Once a herd has been infected and clinical signs have disappeared the virus may disappear. To help this do not introduce replacement stock into the herd for 1 to 2 months.

Getah Viruses

These have been associated with foetal death and abortion in sows in Japan and Korea but few are seen or noted in other countries. They are not of great significance.

Tumours (Cancer)

Tumours can sometimes be seen in the commercial environment, but pigs are often slaughtered before they develop. However they are most often seen in older pet pigs.

Juvenile tumours

Nephroblastoma: – This kidney tumour develops from the renal blastema. Generally the tumour is discovered at slaughter as an incidental finding.

Young adults

Lymphosarcoma – These classically affect young adults – gilt or 1st parity sows. The sow loses weight rapidly and is unresponsive to treatment. At post-mortem, lesions associated with swollen lymph nodes are found scattered throughout the carcass.

Middle-aged adults

Melanoma – These are not infrequently found in pet pigs. They can be removed surgically, or review the growth of the tumour over time. Many of these lumps remain very static over a long period of time without causing any problems or irritation to the pig.

Elderly adults

Reproductive tumours – Tumours of the female reproductive tract are being increasingly reported, particularly associated with the broad ligament and uterus. Ovarian tumours are also seen.

Scrotal papillomatosis – These can be very common in the boar but have little or no significance.

Vesicular Exanthema of Swine (VES)

The virus of vesicular exanthema of swine (VES) is different from those causing foot-and-mouth disease (FMD) and swine vesicular disease (SVD), but it produces a disease in pigs that is clinically indistinguishable from FMD and SVD, both of which are described in detail above. Unlike FMD, it only affects pigs.

You only need be concerned about VES if you have a pig farm on the west coast USA. To everyone else it is no more than a historical curiosity. This account is therefore brief.

Importance of VES

- In fact, it has not occurred since 1959 but the virus is still present in sea mammals and fish so it could occur again and you should not dismiss it completely.
- The main importance of VES is that clinically it resembles FMD and if it occurred again a slaughter policy would be applied.

Where is it?

Viruses that are virtually identical to VES virus are present in marine mammals and fish along the Pacific coast of the USA. It is therefore assumed that the source of VES was waste seafood fed to pigs as garbage or finding its way to pigs from farmed-mink fed seafood.

History

The story of VES is unusual. It was first diagnosed in pigs in Southern California in 1932. Because of its close similarity to FMD all the pigs were destroyed. It kept reappearing in California from time to time, the pigs being slaughtered each time. Then, in 1952, the virus escaped from California in a train-load of infected pork. Garbage-fed pig herds came down with the disease and it spread from them to neighbouring herds until herds in 43 states were affected. It was eventually stamped out in 1956 by a major slaughter policy combined with a ban on feeding uncooked garbage to pigs. The USA was declared free of the disease in 1959.

The only cases outside the USA were in slaughter pigs on a ship from the USA bound for Hawaii in 1947 and in pigs fed uncooked pork scraps from an American military base in Iceland in 1955.

The source of the virus remained a mystery until 1972 when an essentially similar virus was isolated from San Miguel sea lions. When inoculated experimentally into pigs it caused typical signs of VES.

Clinical signs

VES does not affect cattle, sheep, goats or any species other than pigs and sea mammals. So, unlike FMD, if you keep other livestock they will not be affected.

Clinical signs of VES are much the same as FMD so read the section on FMD. Mortality is low but there may be some deaths in sucking piglets. Growing pigs may become debilitated.

Diagnosis

This is the same as for suspected FMD and SVD and requires laboratory tests to identify it.

Management control and prevention

- No vaccines are available.
- The cooked garbage policy in the USA should prevent its reappearance but it is conceivable that

an ignorant or careless person may break the rules.

- If you farm in California do not feed waste seafood to your pigs. Also, do not allow anyone working in your herd to take sandwiches or other food into the pig buildings. Provide a designated eating place for them away from the pigs. This is a good policy anyway.

13 POISONS

13 POISONS

Dose Effect

Most poisons and medicinal products are dependent upon the size of dose for their clinical effects. These progress from:

- No clinical signs.
- A therapeutic effect.
- A toxic effect.
- A lethal effect.

Intake of Poisons

A toxic compound may gain entry to the body through the intestinal tract (via the mouth), the respiratory tract, the eye, the uterus, through the skin or by injection. It is transported via the bloodstream and deposited in various tissues throughout the body. It is important to know which of these are involved so that relevant tests can be carried out.

Detoxification and Excretion

Water soluble poisons are excreted by the kidney in the urine. Other poisons are first detoxified in the liver into water soluble compounds and then excreted in the urine. Some poisons are detoxified into non-water soluble substances which may then be combined with other substances to neutralise them further. Fat soluble poisons are likely to accumulate in the liver. Poisons may also be excreted through the skin or through the milk with possible adverse effects on sucking piglets. Some may be absorbed across the placenta of the pregnant sow and affect the foetus. Tissues used for testing include liver, kidney, blood and stomach contents.

Factors in the Pig that Influence the Effects of a Poison

- The age of the pig.
- Its weight and size. Generally the younger the pig the more severe the effect.
- The nutritional status and feed intake.
- The health status.
- The droplet size of the toxin. The smaller the size the greater the absorption.
- Sex. Gilts and sows, for example, are more susceptible to organophosphorus poisoning.

How to Recognise Poisoning

This is never easy because of the myriad of potentially toxic substances likely to be present on a farm. However always look for the most common and obvious first.

A poison may only affect an individual animal within a group because it was the only one inadvertently exposed to it, for example if a toxic dose of medicine was administered by mistake.

Similarly most or all the pigs in a pen, a number of pens or a complete building may be affected, indicating a much wider exposure. Finally a complete herd may be affected – invariably associated with a common feed, water source or airborne pollution.

Use the process below and consult with your veterinarian to assist with the identification of poisoning.

Step 1- Study the history carefully

- Is the onset rapid – usually within 48 hours? If so, are a number of pigs affected?
- Is only a particular age group affected, for example gilts, sows, sucking piglets? If so, what is common to the group?
- What medicines and management procedures have been applied to the group recently?
- Is a particular area of the farm or are a number of pens affected? Are there any common factors?
- Does the appearance of the condition coincide with the introduction of a new batch of feed or feed ingredients, a change in water or other local change?

Step 2 – Clinical signs

- List the clinical signs (see Fig 13-1 and Fig 13-2).

Common features of poisoning include:

- Rapid onset – (There are however exceptions depending on the dose level and period of exposure).
- A defined group of pigs affected.
- A number of pigs with identical clinical signs.
- Signs not recognisable as a disease.
- Rectal temperatures are usually normal.

Step 3 – Post-mortem examinations and records of mortality

Post-mortem examinations may assist in differentiating between a specific disease and toxic conditions. Samples from tissues are probably required for further laboratory

FIG 13-1: BIOLOGICAL SYSTEMS AND CLINICAL SYMPTOMS ASSOCIATED WITH DIFFERENT TOXINS

Poison	General Malaise	Nervous Signs	Digestive Signs	Circulatory Signs	Repro Signs	Urinary Signs	Skin Signs	Sudden Death	Liver Signs	Resp Signs	Incoord Signs
Algae		✓		✓				✓	✓	✓	
Arsenic: inorganic		✓	✓	✓							
Arsenic: organic		✓	✓								✓
Coal tars		✓			✓		✓	✓			✓
Copper	✓		✓	✓		✓	✓		✓	✓	
Ethylene glycol	✓	✓				✓		✓		✓	✓
Fluorine		✓	✓								✓
Herbicides		✓		✓			✓	✓		✓	
Iron dextran								✓		✓	✓
Insecticides											
Aldrin	✓	✓						✓			
Benzene hexachloride	✓	✓						✓			
Chlorinated hydrocarbons	✓	✓						✓			
OP compounds	✓	✓	✓	✓				✓		✓	
Lead		✓	✓							✓	
Manganese			✓								✓
Medicines – general	✓	✓		✓	✓	✓	✓	✓		✓	
Medicines – specific											
Carbadox	✓	✓	✓				✓	✓			✓
Furazolidone		✓						✓			
Monensin	✓	✓	✓	✓						✓	✓
Olaquindox	✓	✓									✓
Penicillin		✓	✓	✓				✓			
Salinomycin	✓							✓			
Sulphonamides						✓					
Tiamulin	✓						✓	✓			
Melamine	✓					✓	✓	✓			
Mercury		✓	✓			✓				✓	
Metaldehyde		✓								✓	✓
Mycotoxins											
Aflatoxin	✓		✓	✓		✓			✓		
Ergot	✓			✓			✓			✓	✓
Fumonisins	✓	✓			✓		✓		✓	✓	
Gossypol	✓				✓	✓	✓				
Ochratoxin and citrinin	✓		✓			✓			✓		✓
Trichothecenes	✓			✓							
Zearalenone			✓		✓						✓
Nitrates/nitrites	✓		✓	✓				✓		✓	✓
Plants – general	✓	✓	✓	✓		✓				✓	

FIG 13-1: BIOLOGICAL SYSTEMS AND CLINICAL SYMPTOMS ASSOCIATED WITH DIFFERENT TOXINS CONT.

Poison	General Malaise	Nervous Signs	Digestive Signs	Circulatory Signs	Repro Signs	Urinary Signs	Skin Signs	Sudden Death	Liver Signs	Resp Signs	Incoord Signs
Plants – specific											
Bracken								✓		✓	
Cocklebur	✓	✓							✓	✓	
Deadly nightshade		✓						✓			
Oak leaves and green acorns		✓	✓			✓					
Pigweed		✓				✓					✓
Pokeweed	✓		✓			✓					
Sorghum		✓				✓				✓	
Yellow jasmine	✓							✓			✓
Water hemlock		✓						✓			
Plants causing photosensitisation							✓		✓		
Protein			✓								
Selenium	✓					✓		✓		✓	✓
Toxic gases											
Ammonia	✓									✓	
Carbon dioxide		✓									
Carbon monoxide					✓		✓	✓		✓	
Hydrogen sulphide		✓		✓			✓	✓		✓	
methane		✓									
Vitamin A											✓
Vitamin D	✓			✓						✓	
Warfarin				✓			✓	✓		✓	✓
Water deprivation (salt poisoning)	✓	✓						✓			✓

OP - Organophosphorus, **Repro Signs** - Reproductive Signs, **Resp Signs** - Respiratory Signs, **Incoord Signs** - Incoordination Signs

tests. The number of deaths and whether they are sudden or after a short or prolonged illness may characterise certain poisons.

The information assessed from Steps 1 to 3 will raise a suspicion of poisoning.

Step 4 – Identify the possible sources of the poison

- List the chemicals on the farm – sprays, pesticides etc.
- List the medicines on the farm.
- What injections have been given?
- Are rodenticides used and available?
- Are parasiticides used?
- Could any sources of feed be suspected?
- Is there evidence of spoiled or mouldy feed or mould in the feed delivery system?
- Consider water, bedding and other environmental contaminants.
- Are sprays/disinfectants used?
- In outdoor herds consider plants, water and environmental contaminants.
- Do any of the signs fit into Fig 13-1 or Fig 13-2?

Step 5 – Identify the toxin

Use Fig 13-1 and Fig 13-2 to identify the potential toxin or toxins, together with the history and symptoms.

Step 6 – Read about the poison

Refer to the poison in the text and administer treatments in conjunction with veterinarian advice.

Step 7 – Confirm the poison

Refer samples to a lab for confirmation (Fig 13-3). Seek veterinary advice.

Clinical Signs of Different Poisons

A broad outline of how different poisons affect different systems of the pig is given in Fig 13-1. This table, in conjunction with Fig 13-2, which lists typical visual observations and possible poisons, can be used to help identify likely causes.

Specific systems of the body may be affected and develop the following signs:

Circulatory system signs

- Anaemia.
- Cyanosis (blue discoloration of skin).
- Increased respiration.
- Jaundice.
- Haemorrhage.

Digestive system signs

- Abdominal pain.
- Diarrhoea – with or without haemorrhage.
- Rectal prolapse.
- Salivation.
- Vomiting.

General malaise signs

- Generalised sickness and depression.
- Reduced feed intake or complete inappetence.
- Reduced growth.

Locomotor system signs

- Abnormal gait.
- Ataxia.
- Incoordination.
- Lameness.
- Muscle weakness.
- Stiffness.

Nervous system signs

- Blindness.
- Convulsions.
- Excitation.
- Fits.
- Incoordination.
- Spasmodic movements.

Reproductive system signs

- Abnormal oestrus.
- Swollen vulva.
- Abortion.
- Embryo reabsorption.
- Failure of fertilisation.

Respiratory system signs

- Coughing.
- Difficulty breathing.
- Pneumonia (found at post-mortem examination).
- Sneezing, nasal discharge.

Skin signs

- Colour.
- Haemorrhage.
- Irritation.
- Vesicles.

Urinary system signs

- Blood in the urine.
- Cystitis/pyelonephritis (found at post-mortem examination).
- Excess mineral deposits (found at post-mortem examination or in the urine).
- Liver related issues (found at post-mortem examination).
- Pus.

FIG 13-2: OBSERVATIONS AND CAUSES

Anaemia
- Copper
- Warfarin

Abortion/Agalactia
- Aflatoxin
- Ergot toxin

Abdominal Pain
- Arsenic: organic
- Copper
- Deadly nightshade (*Solanum*)
- Melamine
- Nitrates and nitrites
- Oak leaves and green acorns
- Pokeweed (*Phytolacca*)

Blindness
- Arsenic: organic
- Lead
- Mercury
- Water deprivation (salt poisoning)

Colic
- Arsenic: inorganic

Diarrhoea
- Arsenic: inorganic
- Arsenic: organic
- Copper
- Monensin overdose

FIG 13-2: OBSERVATIONS AND CAUSES CONT.

Diarrhoea cont.
- Nitrates and nitrites
- Organophosphorus compounds
- Protein

Diarrhoea – Bloody
- Algae
- Deadly nightshade (*Solanum*)
- Mercury
- Oak leaves and green acorns
- Pokeweed (*Phytolacca*)

General Malaise – Dehydration, Reduced Feed Intake, Poor Growth
- Aflatoxins
- Ammonia
- Arsenic: inorganic
- Carbadox overdose
- Cocklebur (*Xanthium*)
- Copper
- Ergot toxin
- Ethylene glycol
- Fumonisins
- Furazolidone overdose
- Insecticides
- Manganese
- Mercury
- Monensin overdose
- Nitrates and nitrites
- Ochratoxin and citrinin
- Olaquindox overdose
- Pokeweed (*Phytolacca*)
- Salinomycin overdose
- Selenium
- Trichothecenes
- Vitamin D poisoning
- Water deprivation (salt poisoning)
- Yellow jasmine (*Gelsemium*)

Lameness/Stiffness/Incoordination cont.
- Arsenic: organic
- Carbadox overdose
- Ergot toxin
- Iron dextran
- Lead
- Manganese
- Metaldehyde
- Monensin overdose
- Nitrates and nitrites
- Ochratoxin and citrinin
- Olaquindox overdose

FIG 13-2: OBSERVATIONS AND CAUSES CONT.

Lameness/Stiffness/Incoordination cont.
- Pigweed (*Amaranthus*)
- Selenium
- Vitamin A poisoning
- Warfarin
- Water deprivation (salt poisoning)
- Yellow jasmine (*Gelsemium*)

Nervous Signs – Tremors, Convulsions, Prostration
- Algae
- Arsenic: inorganic
- Arsenic: organic
- Carbon dioxide (CO_2)
- Coal tars (clay pigeons)
- Cocklebur (*Xanthium*)
- Deadly nightshade (*Solanum*)
- Ethylene glycol
- Furazolidone overdose
- Herbicides
- Hydrogen sulphide (H_2S)
- Insecticides
- Lead
- Mercury
- Metaldehyde
- Monensin overdose
- Pigweed (*Amaranthus*)
- Sorghum (*Sorghum*)
- Water deprivation (salt poisoning)
- Water hemlock (*Cicuta*)
- Yellow jasmine (*Gelsemium*)

Piglet Malformation/Stillbirths
- Coal tars (clay pigeons)

Prolapses
- Zearalenone

Reproductive Problems
- Carbon monoxide (CO)
- Zearalenone

Respiratory Signs
- Ammonia
- Bracken (*Pteridium aquilinum*)
- Carbon monoxide (CO)
- Cocklebur (*Xanthium*)
- Ergot toxin
- Herbicides
- Hydrogen sulphide (H_2S)
- Metaldehyde

FIG 13-2: OBSERVATIONS AND CAUSES CONT.

Respiratory Signs (cont.)
- Methane gas
- Monensin overdose
- Nitrates and nitrites
- Organophosphorus compounds
- Selenium
- Sorghum (*Sorghum*)
- Vitamin D poisoning

Skin – Discoloured/Irritation/Jaundice/Necrosis
- Aflatoxins
- Algae
- Carbon monoxide (CO)
- Coal tars (clay pigeons)
- Copper
- Ergot toxin
- Fumonisins
- Hydrogen sulphide (H_2S)
- Iron dextran
- Melamine
- Plants causing photosensitisation
- Tiamulin overdose

Sudden Death
- Algae
- Bracken (*Pteridium aquilinum*)
- Carbon monoxide (CO)
- Cocklebur (*Xanthium*)
- Coal tars (Clay pigeons)
- Deadly nightshade (*Solanum*)
- Ethylene glycol
- Herbicides
- Hydrogen sulphide (H_2S)
- Insecticides
- Iron dextran
- Melamine
- Monensin overdose
- Nitrates and nitrites
- Salinomycin overdose
- Selenium
- Warfarin
- Water deprivation (salt poisoning)
- Water hemlock (*Cicuta*)
- Yellow jasmine (*Gelsemium*)

Swollen and Reddening Vulva
- Zearalenone

Urinary – Blood, White Crystals
- Copper
- Melamine
- Mercury

FIG 13-2: OBSERVATIONS AND CAUSES CONT.

Vomiting
- Arsenic: inorganic and organinc
- Cocklebur (*Xanthium*)
- Copper
- Deadly nightshade (*Solanum*)
- Ethylene glycol
- Mercury
- Nitrates and nitrites
- Organophosphorus compounds
- Pokeweed (*Phytolacca*)
- Trichothecenes

FIG 13-3: TOXIC LEVELS OF COMMON POISONS

Poison	Non-Toxic Levels Less Than	Reported Toxic Levels from Mild to Severe Effect
Ammonia	25 ppm	50 – 150 ppm
Arsenic	50 mg/kg lwt.	10 – 200 mg/kg in-feed
Arsenic: inorganic	Liver, 1 ppm Kidney 1 ppm	Liver, Kidney 10 ppm
Arsenic: organic	100 ppm in feed	1000ppm for 2–3 days in feed 400 ppm for 14 days in feed 300 ppm, 21 – 28 days in feed
Carbadox	50 ppm in feed	100 – 300 ppm in feed
Carbon monoxide	30 ppm	60 – 250 ppm
Coal tars	–	10 – 15 g ingested
Copper	200 ppm in feed	200 – 600 ppm in feed Liver > 250 ppm Kidney > 60 ppm 20 mg/kg liveweight
Ergot	0.1 % sclerotia	0.3 % sclerotia
Furazolidone	400 ppm in feed	500 – 1000 ppm in feed
Hydrogen sulphide	50 ppm	80 – 900 ppm
Lead	30mg	600 mg/kg in feed Liver or kidney 4 ppm
Mercury	Kidney < 1 ppm Liver 0.3 ppm Feed 1 ppm	Liver 40 – 70 ppm Kidney 40 – 100 ppm 2 – 13 mg/kg liveweight 1.5 – 10 ppm in feed Single dose of 5 – 15 mg/kg
Metaldehyde		V. small amounts of pellets
Monensin	100 ppm in feed 10 mg/kg lwt.	200 – 400 ppm in feed 20 mg/kg liveweight
Nitrates	150 – 400 ppm	1500 ppm
Nitrites	10 mg/kg lwt.	15 – 20 mg/kg lwt 40 ppm in water
Nutritional poisons	See Chapter 14	
Olaquindox	100 ppm in feed	300 – 500 ppm in feed
Salt	2 % ad lib water	0.5 % in feed - no water
Selenium	0.1 – 0.3 ppm	3 – 10 ppm in feed 12 ppm in liver
Tiamulin	120 ppm in feed	200 ppm in feed
Warfarin		Low levels

ppm = g/tonne = mg/kg   lwt = liveweight

Potential Poisons

Algae

If drinking water becomes heavily contaminated with green and blue toxin-producing algae, acute disease and high mortality can take place. Poisoning is only likely to be seen in pigs outdoors where there is access to ponds used as drinking water (unless of course indoor pigs derive their water from such sources). Large numbers of algae appear in the water during periods of warm sunny weather. Wind blows the organisms to the water's edge where they are ingested by the pig. The algae produce highly toxic substances that cause massive damage to the liver and haemorrhage and/or affect the nervous system causing coma and respiratory failure. Post-mortem examinations show a grossly enlarged liver with haemorrhages.

Clinical signs

The onset of symptoms is usually sudden and within 1 to 2 hours. Pigs die within 24 hours. Pigs show sudden collapse, muscle tremors, convulsions, discoloured skin and bloody diarrhoea.

Diagnosis

A history of pigs having access to water showing evidence of algae blooms suggests poisoning. Clinical signs and post-mortem examinations provide strong diagnostic evidence. A sample of the algae should be sent to a laboratory for identification and information sought on how to control it.

Treatment

- There is no specific treatment.
- Give gastrointestinal absorbents such as charcoal and kaolin as a drench in water or in feed if practicable.
- Pigs should be removed immediately from the source of water and housed away from sunlight in warm surroundings.

Management control and prevention

- Provide an alternative source of water. If this is not possible install a fence to allow access to only a small area of the waterline. Protect this by a muslin barrier to hold back the algae.
- Multi-vitamin injections, particularly the B complex, may help.

Arsenic

Arsenic exists as both inorganic and organic compounds. Inorganic arsenicals are little used today except in a few rodenticides, insecticides and weed killers.

Organic arsenicals are less toxic than the inorganic ones and are used in some countries to control swine dysentery, to treat *Mycoplasma suis* or as growth enhancers. The common chemicals are arsanilic acid, sodium arsanilate, roxarsone (3-nitro – 4-hydroxyphenyl arsonic acid) and carbasone.

Clinical signs

Poisoning with inorganic compounds results in acute illness with severe damage to the intestinal tract. Clinical signs include vomiting, acute diarrhoea and colic, dehydration, convulsions, collapse and death.

The onset of clinical signs in organic arsenical poisoning varies according to the level of intake. At low levels nothing may be seen for 2 to 3 weeks, when mild signs of incoordination (lameness) and possibly blindness occur. With high doses clinical signs may be seen within 2 to 3 days, as incoordination, paralysis of the hind legs and blindness. The blindness is often irreversible. Affected animals continue to grow if they can get to feed and water. Very high doses of organic arsenicals may cause gastroenteritis (diarrhoea, vomiting, abdominal pain).

Diagnosis

This is determined by the history, availability of arsenical compounds and analysis of feed or suspect substances.

Treatment

- There is no effective antidote but if the compounds are removed immediately most pigs recover.

Coal Tars

The distillation of coal tar and crude petroleum products produces a variety of substances including creosols or phenolic compounds, crude creosote and pitch. These substances are used as disinfectants and preservatives and may be eaten by pigs. Pitch is used as a binder in clay pigeons and it only requires 10 to 15 g ingested over a 7-day period to produce mortalities. Tar papers, bitumen on floors and creosote-treated wood are further sources of poisoning. Coal tar preparations are also irritant to the skin.

Clinical signs

The first signs of coal tar poisoning are prostration or coma leading rapidly to sudden death. Pigs may be found dead.

Low levels of intake interfere with the absorption of vitamin A and produce signs of deficiency including piglet malformation and stillbirths. If the skin of a newborn piglet comes into contact with phenolic disinfectants, contact areas, particularly the teats and soles of the feet, may be burnt.

Diagnosis
This includes a history of access to phenolic and other coal tar preparations. Look for evidence of contamination, particularly in outdoor pigs.

Treatment and control
- There is no effective treatment.
- The environment should always be examined for evidence of exposure, particularly where sows are outdoors.
- Injections of vitamin A using multi-vitamin preparations may be of value.

Copper
Copper sulphate is added routinely in some countries to grower rations as a growth enhancer at levels between 50 to 175 ppm. Levels above 200 ppm may interfere with normal growth rate, particularly if the levels of zinc and iron in the ration are low. If the levels of zinc and iron are normal a level of 500 ppm becomes toxic. Plants such as subterranean clover may also produce a mineral imbalance in outdoor pigs by increasing the retention of copper.

Clinical signs
These are usually gradual in onset unless there has been a massive intake of copper. In acute cases there is severe gastroenteritis, abdominal pain, diarrhoea and jaundice, an enlarged liver and blood in the urine. In less acute cases there is anaemia and reduced growth rate.

Diagnosis
This is based on history, clinical signs and post-mortem findings. Laboratory tests for copper levels of more than 250 ppm in the liver and 60 ppm in the kidney confirm the diagnosis.

Treatment
- The response to treatment is poor.
- Calcium versonate by injection can be of help.
- Seek veterinary advice.

Electrocution
While an unusual poisoning, if a group of pigs are found dead suddenly, consider hydrogen sulphide or electrocution. Take extreme caution when entering the room. Whilst electrocution is an uncommon cause of death, many farm electricity supplies may be in a poor state of repair and electrocution is a possibility. Stray electricity, especially through the water supply, may be a cause of reluctance to drink and eat. Watching the behaviour of the pig or pigs when using the drinker or feeders may raise the alarm.

Ethylene Glycol
This is a substance commonly used in antifreeze liquids in engine coolants. Ethylene glycol is very toxic and only 4 to 5 ml per kg bodyweight are required to produce signs.

Clinical signs
In acute poisoning pigs vomit, become depressed, lose locomotor function, develop kidney failure and die. Death may occur within 12 to 24 hours and is often the only sign observed.

The tissues throughout the body become acid. In less acute cases pigs suffer depression, gastroenteritis, abdominal distension, difficult breathing and nervous signs.

Diagnosis
This is based on a history of access to antifreeze, clinical signs and post-mortem lesions.

Treatment and control
There is no treatment although some success has been reported using 5 % sodium bicarbonate intravenously to reduce the acid state of the blood. Seek veterinary advice.

Fluorine
Fluorine poisoning is uncommon in pigs but may occur on heavily contaminated pastures. In acute cases of poisoning, diarrhoea, lameness and nervous signs may be seen. In chronic cases, however, lameness may be the only symptom.

Diagnosis
This is carried out by assessing fluorine levels in bone which should be less than 600 ppm.

Herbicides
Poisonings associated with herbicides are much more common in cattle and sheep than in pigs. The important ones however include arsenic, borax and pentachlorophenols (PCPs). The latter can be absorbed through intact skin or mucous membranes to which they are highly irritant.

PCPs are used widely as preservatives, insecticides, fungicides and molluscicides but generally they are inactivated in soils when used correctly. Contact with treated wood used in feeding troughs can cause salivation and severe irritation to the mouth.

Clinical signs
These include respiratory distress and nervous signs such as tremors, muscle weakness and convulsions. Herbicide poisonings usually result in death.

Diagnosis
This can be difficult but a careful examination of the products on the farm and access to them will often assist.

Treatment
- If nervous signs are severe use a sedative. Give electrolytes.

Iron Dextran
Iron injections as iron dextran are given by intramuscular or subcutaneous injections to piglets between 1 and 7 days of age. This is to correct the development of iron deficiency that leads to anaemia. A dose level of 100 to 200 mg per piglet is required. If sows or gilts become deficient in vitamin E and/or selenium during pregnancy, piglets are also born deficient. Under such circumstances the enzymes which metabolise the iron cannot function and the iron then becomes toxic.

Clinical signs
The piglet becomes acutely lame, a dark swelling occurs at the site of the injection (usually the thigh) and about 50 % of the litter or more die within hours of the injection.

Diagnosis
The association between injections and symptoms is usually clear and the clinical signs are typical. The cut muscle surface where the iron has been injected loses all its structure and appears almost like wet fish muscle. This is due to necrosis (death) of the tissues.

Treatment
- The initial reaction often is to blame the iron. It is true that inferior quality iron dextrans (usually the cheap ones) are more likely to be associated with severe disease than better quality ones but the primary problem is vitamin E/selenium deficiency.
- As soon as the condition is recognised all the piglets that have had iron within the last 2 days should be injected with vitamin E/selenium according to the manufacturer's recommendations.
- All females due to farrow within 7 days should be injected with vitamin E/selenium.
- Sows within the last month of pregnancy should be injected with vitamin E, 2 weeks before farrowing.

Management control and prevention
- Excessive oxidation of fats in the sow feed is the most common cause of low vitamin E status. Oxidation results from poor storage of cereals or corn and particularly in situations where the bottoms of feeds bins are not cleaned out and moisture has gained access.
- As a precaution add an extra 150 g/tonne of vitamin E to the sow feed for the next 2 months.
- Check the sources and storage facilities of all feed grains.

Insecticides
Insecticides by virtue of their actions are toxic or lethal poisons and it is important that the manufacturer's instructions are followed when using them. Furthermore most of these chemicals are absorbed through the skin and therefore become potentially hazardous to the operator. Take note of the detailed recommendations for use, particularly when handling the concentrate. Poisoning in pigs from insecticides may arise from incorrect application, pigs eating contaminated feed or accidental exposure. Insecticides are available as 4 different chemical types; carbamate, chlorinated hydrocarbons, pyrethrins and organophosphorus compounds (OPs).

All insecticides act on the nervous system and the general symptoms are similar. These include general malaise, muscle tremors, hyperactivity and ultimately convulsions and death. OPs also cause vomiting, diarrhoea and respiratory distress.

Carbamates
The most common of these are carbaryl and methomyl which are often used to destroy worms and other insects. They act by blocking the transmission of nervous impulses at nerve muscle junctions and act in a similar manner to organophosphorus compounds. The antidote is atropine sulphate by injection and veterinary advice should be sought.

Chlorinated Hydrocarbons
These include aldrin, benzene hexachloride (BHC), lindane, chlordane, dieldrin, methoxychlor and toxaphene. They are highly effective against insects, but due to their side effects and toxicity to humans a number of them are now prohibited. BHC is still used in large animals and dogs and lindane is the best known proprietary name used in pigs. Most chlorinated hydrocarbons are restricted to use on crops.

Clinical signs
All chlorinated hydrocarbons stimulate the central nervous system and signs include twitching of the muscles of the face, generalised muscle trembling and shivering, followed by fits, coma and death. Some pigs may stand with their heads pressed against the wall, continually licking and chewing; others show loss of leg

function (ataxia). Symptoms appear within 24 hours. The chlorinated hydrocarbons are absorbed by the body fat.

Diagnosis

Whilst the history assists in indicating the type of poison, a laboratory analysis of the brain, kidney, liver and fat tissues is required for confirmation. Aujeszky's disease (pseudorabies virus) and water deprivation (salt poisoning) can give similar signs.

Treatment and control

- There are no known antidotes but severe nervous signs can be controlled by barbiturates.
- Chlorinated hydrocarbons are excreted slowly from the body over a long period of time and suspected carcasses should be destroyed and not used as food.

Pyrethrins

Pyrethrins are naturally occurring substances found in the chrysanthemum plant and have a powerful knock down effect on insects. Synthetic compounds are manufactured that have similar effects. Toxicity is relatively low and it is unlikely to be seen unless massive doses are either absorbed through the skin or taken in by mouth. All compounds have an effect on the central nervous system causing excitation, convulsions, coma and death. Marked muscle trembling and paralysis may also be seen. Most pigs die from respiratory failure.

Treatment and control

- Pigs usually recover naturally and none is required. Activated charcoal or attapulgite given by stomach tube will reduce absorption.

Organophosphorus Compounds (OPs)

These groups of medicines are widely used as insecticides and pesticides. OPs and similar carbonate insecticides affect the chemical transmitters that control the nerve endings in muscles. The result is over-activity causing muscles to go into a continual spasm.

Clinical signs

The continual and excessive stimulation of the muscles in acute poisoning causes excessive salivation, the passing of faeces and urine and a very stiff awkward gait. This may be followed by vomiting, diarrhoea and muscle tremors of the face and body. Death can be fairly rapid, within 1 to 4 hours, and is generally caused by respiratory failure.

Diagnosis

A history of exposure to OPs gives rise to a suspicion and together with the clinical signs point to a diagnosis. The insecticides will be identified in the stomach or in the suspect material such as feed, but detection is unlikely in body tissues.

Treatment and control

- The antidote is atropine sulphate at 0.5 mg/kg bodyweight by intramuscular injection. The response is usually seen within 3 to 5 minutes. Consult your veterinarian immediately.
- Move affected pigs to well-bedded hospital pens.
- Give activated charcoal or attapulgite by mouth if possible to absorb the insecticide.

Lead

Compared to ruminants pigs are not easily poisoned by lead and as a result lead poisoning in pigs is rare. A level of at least 30 mg per day is required before any signs are seen.

Clinical signs

Lead acts on the central nervous system causing pain, grinding of the teeth, blindness, severe muscle twitching and incoordination. Kidney tissue is a good source for sampling and levels of more than 3 ppm indicate poisoning.

Treatment

- Calcium EDTA 1 % solutions are used for treatments at a level of 110 mg/kg.

Manganese

Normal levels in the diet range from 3 to 10 ppm. At 60 ppm feed intake and daily gain are affected. Levels above 3000 ppm may produce clinical signs including a slight stiffness and abnormal gait. Poisoning would be rare.

Medicines

Sometimes reputable medicinal products, which normally at correct dosages may be highly beneficial, cause a toxic or allergic reaction. This may be because they are given to the wrong species by accident, e.g. through cross contamination of feed, by administration of too high a dose, or by being given with another medicine with which they are incompatible. They may cause a hypersensitivity reaction in an individual. If the medicament causing the problem is in the feed, analysis of feed samples should confirm the tentative diagnosis. If the feed is suspected, random samples should be retained in case a liability dispute arises (see Chapter 15 - Sampling Feeds).

Carbadox

This medicine is used in the treatment and prevention of swine dysentery at 50 ppm. Poisoning may commence at 100 ppm in a mild form and acute disease at 300 ppm or more, usually because mistakes have been made in mixing it in feed. At the lower level of 200 ppm there is reduced feed intake and growth rate, hard faeces and excessive hair growth. Incoordination, posterior paralysis and death occur after 5 to 10 days of continuous consumption. There is no antidote. At post-mortem examination lesions may be found in the kidneys and adrenal glands.

Furazolidone

Furazolidone is a member of a group of medicines called the nitrofurans. This is banned in some countries.

At levels above 500 ppm inappetence and mild nervous signs are seen, progressing to ataxia, fits, coma and death. The medicine is fairly quickly excreted from the system and pigs return to normal 4 to 5 days after withdrawal. Newborn piglets become depressed, hypothermic and lay on their sides, paddling and frothing at the mouth. These effects are most likely to occur if the medicine is given immediately at birth rather than 6 to 7 hours after colostrum intake. Low birth weight piglets (under 800 g) are most susceptible to being poisoned.

Monensin

There is a narrow gap between therapeutic and toxic levels. The therapeutic level in pigs is 10 mg/kg or 100 g to the tonne. Clinical signs are seen when levels are twice this and heavy mortality occurs when they reach 400 g/tonne. Most poisonings occur in pigs either through the accidental addition of monensin or where the calculated inclusion level has been incorrect.

Clinical signs

Clinical signs of poisoning usually occur within 12 hours of intake and include heavy and difficult breathing, frothing around the mouth, loss of use of the hind legs and generalised muscle weakness. Diarrhoea may also be seen. The toxic effects of monensin are exaggerated if either of the antibiotics salinomycin at 60 ppm or tiamulin at 100 ppm are also contained in the feed. Some affected pigs, particularly those that have lost the use of their hind legs, may take 2 to 3 weeks to recover but many of these die.

Diagnosis

This is based on the clinical signs and post-mortem lesions, both of which are characteristic, together with the history of the use of monensin. The lesions observed at post-mortem examination include pale areas in the muscles of the diaphragm, thighs, lower shoulders, back and ribs, and heart necrosis (with equal distribution on each side). The bladder may contain red-brown urine. Feed samples should be examined for the presence of monensin. A test for this is fairly quick and straightforward. Contamination of pig feed is not uncommon because in some countries the medicine is used widely as a growth promoter in cattle and if it occurs in the presence of tiamulin or salinomycin can become toxic.

Treatment

- There is no antidote for monensin poisoning.

Olaquindox

This is used to control postweaning diarrhoea. The margins between therapeutic and toxic doses are narrow. The normal level is up to 100 ppm but 300 ppm or more will give rise to inappetence and incoordination with progressive loss of limb function as the dose rate is increased.

Penicillin

Penicillin (and in particular the combination of procaine penicillin and benzathine penicillin) occasionally causes vomiting in individuals or small groups of pigs when given by intramuscular injection. This is particularly marked in piglets of less than 10 days of age where an acute sensitivity reaction sometimes occurs. Within 3 to 4 minutes of injection piglets collapse and lay on their side paddling and frothing at the mouth. Fortunately this episode passes off within 30 minutes with most pigs returning to normal. Long-acting injections of penicillin given to weaners will also cause a proportion of pigs to vomit but this is of no consequence.

Salinomycin

It should not be used simultaneously with therapeutic levels of tiamulin in feed, water or injection and there must be at least 7 days between the last exposure to salinomycin and the commencement of medication at treatment level with tiamulin. Studies have shown that at the preventative level of 30 to 40 ppm no interaction occurs between tiamulin and salinomycin (see monensin). Poisoning is associated with a marked drop in feed intake and growth rate and sudden death.

There is no antidote.

Sulphonamides

High levels of sulphonamides form crystals in the kidneys and cause damage but poisoning is uncommon because when high levels are inadvertently mixed in feed the pigs will often reject it. Some sulphonamides and in particular sulphadimidine are excreted in the faeces and urine and are recycled back into the pigs again. If recycling occurs during the withdrawal period tissue residues will appear at slaughter. In consequence some countries have either banned or restricted its use. Other sulphonamides used for treatment do not recycle to the same extent.

Tiamulin

This antibiotic is used extensively to treat swine dysentery and enzootic pneumonia. Levels of up to 120 ppm have no toxic effects on the pig but above 200 ppm there have been reports that the skin may react with contaminated faeces or urine, with marked red discoloration. After prolonged use bleeding into the muscles may occur along with increased mortality. The maximum recommended level of the use of tiamulin in the UK is 100 ppm and in other countries 220 ppm.

Tiamulin should not be used in conjunction with salinomycin or monensin because it enhances the toxicity of the monensin or salinomycin (see monensin and salinomycin). There is no antidote.

Cold medicine temperature shock

Medicines taken straight from the refrigerator, including vaccines, may evoke a shock reaction within 10 minutes of administration, particularly in small pigs. Ensure that the medicines are warmed up before use. Note some medicines – sulphonamides, for example – may precipitate out in the cold and create problems due to inappropriate dosing levels.

Melamine, Cyanuric Acid and Ammeline

Melamine is widely used in plastics, adhesives, countertops, dishware, whiteboards. In China, melamine was found to have been added to watered-down milk products to boost the apparent protein content.

If fed to pigs (milk supplement) they appear inappetent, pale and yellow and take to increased drinking. Progressive wasting followed by death due to renal failure is seen. At post-mortem the kidneys are yellow and are populated with urinary calculi.

There is no treatment, but drinking plenty of fresh water helps to flush the contaminants from the kidney.

Mercury

Mercury exists in 2 forms, organic and inorganic. Organic compounds are used as fungicides to treat seed grains prior to sowing. Poisoning occurs if pigs are fed corn or cereals that have been treated or contaminated with such a fungicide. Inorganic mercury as mercury chloride is used as a disinfectant and in some paints, batteries and thermometers.

Clinical signs

Mercury is a cumulative poison so clinical signs vary depending upon the duration of intake as well as the dose and the type of mercury. As a poison, mercury acts primarily on the gastrointestinal tract causing vomiting and bloody diarrhoea. Dead (necrotic) pieces of tissue may be seen in the faeces. Damage also occurs to the kidney leading to signs of uraemia. Pigs stop eating and lose body condition. White crystalline deposits may be seen in the urine. Nervous signs maybe seen including ataxia, blindness, wandering, partial paralysis, coma and death.

Diagnosis

The clinical signs and post-mortem lesions may suggest mercury poisoning which may be confirmed by finding the source of poisoning and by laboratory analysis of mercury levels in the kidney and liver (normally less than 1 ppm).

Treatment

- This is carried out by a combination of sodium thiosulphate 20 % solutions given intravenously at 1 ml / 5 kg bodyweight together with dimercaprol by intramuscular injection at a level of 3 mg/kg bodyweight. Seek veterinary advice.
- Feed pigs with either milk or egg protein, which act as an absorbent.

Metaldehyde

Metaldehyde is widely used to kill slugs and other garden pests and it is presented either as a meal with bran or pelleted. Clinical signs start within 1 hour of ingestion and affect the nervous system, with severe tremors, incoordination, convulsions, high temperatures and death due to respiratory failure. If pigs eat large amounts, it is necessary to anaesthetise them using barbiturates for a period of 6 to 12 hours whilst the medicine is excreted. There is no specific antidote.

Mycotoxins

Under certain conditions fungi multiply on cereals, corn, cotton seed and other food materials, sometimes producing chemicals called mycotoxins (Fig 13-4). They require adequate moisture, oxygen and carbohydrates to multiply and temperatures from 10 °C to 25 °C (50 °F to 77 °F). Multiplication may still take place, however, outside these ranges, and crops that are already diseased are more likely to succumb to fungal infection. The presence of fungi including recognised toxic species however does not necessarily mean that the toxins are present. Each requires precisely the right substrate and environmental conditions to produce toxins. The common fungi-causing disease (mycotoxicosis) in the pig include species of *Fusarium, Aspergillus* and *Penicillium,* but because of the variable requirements for growth and toxin production particular species tend to predominate in certain geographical areas.

Toxins are not destroyed by heating but modern treatments used in the processing of animal feeds such as temperature and pressure may reduce the actual fungal load.

Factors that may increase the likelihood of mycotoxins in feed

- Cereals left over after screening.
- Damaged or broken grains (hail damage).
- Storage of moist grain.
- Storage in warm, damp conditions.
- Damaged, leaking feed bins.
- Fluctuating environmental temperatures.
- Fungal growth in liquid feeding systems.

Diagnosis of mycotoxicosis

This can often be frustrating because although the clinical signs may be suggestive they are rarely diagnostic. It may be impossible to detect the toxin in the feed because of patchy distribution and/or the samples taken are toxin-free. Also the effect of the toxin may have been delayed and the feed containing the toxin has been consumed. Alternatively the laboratory may be testing for the wrong toxin.

It is possible to test the serum for the presence of the toxin.

Effects of mycotoxins

The specific effects of the various toxins on the pig are shown in Fig 13-4. *Fusarium* species require high levels of moisture and relative humidity (> 88 %) for multiplication and toxin production whereas *Aspergillus* and *Penicillium* multiply at lower levels.

Aflatoxins and some of the ochratoxins are immunosuppressive and can enhance effects of generalised disease.

Methods of preventing mycotoxicosis

Wherever a fungal toxin is suspected consider the following actions:

- Immediately replace the feed or cereal sources with alternative ones.
- Examine carefully the meal or pellets for evidence of fungi.
- Empty all feed bins out and examine for bridging of feed or presence of mouldy feeds. If feed bins are contaminated, empty them and treat with a

FIG 13-4: A GUIDE TO MYCOTOXIN LEVELS IN FEED: MILD TO SEVERE DISEASE

Fungus	Toxins	No Clinical Effect	Toxic Level	Clinical Signs
Aspergillus sp.	Aflatoxins	< 0.1 ppm	0.3 - 2 ppm	Poor growth Liver damage Jaundice Immunosuppression
Aspergillus sp. and *Penicillium* sp.	Ochratoxin & citrinin	<0.1 ppm	0.2 - 4 ppm	Reduced growth Thirst Kidney damage
Fusarium sp.	T2 DAS DON (Vomitoxin)	< 0.5 ppm	> 1 ppm	Reduced feed intake Immunosuppression Vomiting
	Zearalenone (F2 toxin)	< 0.05 ppm	1 - 30 ppm	Infertility Anoestrus Rectal prolapse Pseudopregnancy
			> 30 ppm	Early embryo mortality Delayed repeat matings
	Fumonisin	< 10 ppm	> 20 ppm	Reduced feed intake Respiratory symptoms Fluid in lungs Abortion
Ergot	Ergotoxin	< 0.05 %	0.1 - 1.0% ergot bodies by weight (sclerotia)	Reduced feed intake. Gangrene of the extremities. Agalactia due to mammary gland failure.

ppm – parts per million.
sp. – species - each of these fungi have several species, only some of which are toxic.
Note these toxic concentrations may be reduced when more than one toxin is present at the same time.

non-toxic fungicide.

- Examine all automatic equipment and in particular feed hoppers and automatic dispensers for evidence of mouldy feed.
- If wet, moist grain is stored, mould inhibitors such as propionic acid, calcium propionate or sorbic acid will prevent growth.
- Once mycotoxins have developed in feed there are no methods that can destroy them. However their effects can be mitigated by regrinding the feed and mixing with an alternate source at a ratio of 1:10 and feeding to growing stock on a test basis first.
- A feed sample should be sent to a laboratory for examination (see Chapter 15 - Sampling feeds).
- Check other sources of poisoning, including straw and wet bedding materials.
- Production of distillers' grains can concentrate toxins which are present in the grain.

Aflatoxins

These are produced from the fungus *Aspergillus* and are probably the most common and important of the mycotoxins in pigs. The *Aspergillus* species commonly concerned grows in maize, soya beans and peanuts. The main species involved are *A. flavus* and *A. parasiticus* and in countries where maize and peanuts are not fed aflatoxicosis is uncommon. The effect of the toxins is dependent upon the dose and the age of the pig; the younger the animal and the larger the dose the greater the effect. It is more common for pigs to be exposed to low levels of aflatoxins for long periods of time and the effects therefore tend to be sub-acute rather than acute. Feed levels up to 0.2 ppm produce clinical signs and levels above 0.3 ppm may be fatal.

Clinical signs

These include reduced growth and feed efficiency and at the upper levels the effects of liver damage and failure include jaundice. Aflatoxins do not have a direct effect on reproductive efficiency but abortion and agalactia may occur. They are immunosuppressive and therefore increase the severity of concurrent diseases, such as PRRSV, SIV and mycoplasma (enzootic) pneumonia. There are different types of aflatoxins designated B_1, B_2 and G_1, G_2 and the effects of the toxins are enhanced where poor quality, low protein diets are fed. If levels in the feed are low, it can take 4 to 6 weeks for symptoms to appear.

Diagnosis

Wherever there is poor growth in a herd associated with poor nutrition and chronic infectious diseases the possibility of toxins in the feed should be considered and tested for. Post-mortem lesions include jaundice, anaemia, fluid in the abdomen, poor clotting of the blood and liver haemorrhage. Histological examinations of the liver may help to confirm the diagnosis. Follow the steps outlined at the beginning of the chapter and also see "Diagnosis of mycotoxicosis".

Treatment

- There is no specific treatment. Remove the suspect source until it has been tested.
- Raising the vitamin and protein levels in the feed by 10 % for 2 to 3 weeks could be beneficial, as could raising the lysine content of the diet to growing pigs by 0.2 % for 4 weeks.

Ergot Toxins

These are produced from the fungus ergot (*Claviceps purpurea*) that affects wheat, oats, ryegrass and other grasses by entering the seed and developing into a dark elongated body called a sclerotium. This contains toxic alkaloids, one of which is ergometrine which has the effect of contracting small blood vessel walls, thus restricting the blood supply particularly to the mammary gland and the body extremities.

Clinical signs

Levels of more than 0.1 % sclerotium in the ration will produce clinical signs. These usually occur over a period of weeks and are associated with poor growth rates, increased respiration and general depression. The most sensitive blood vessels are those found in the mammary glands of maturing pregnant gilts. The restricted blood supply then causes agalactia, which is unresponsive to oxytocin in lactating animals and gives rise to increased piglet mortality. Lameness is also common due to necrosis and sloughing of the hooves. Tail and ear necrosis are also common.

Diagnosis

There are 2 methods: the examination of food to identify black/brown sclerotium bodies which can be seen with the naked eye and the laboratory identification of the alkaloids.

Treatment

- There is no specific antidote.
- Remove the affected feed.
- Treat areas of gangrene with antibiotics.
- Some pigs may have to be destroyed on humane grounds.

Fumonisins

These are produced by the fungus *Fusarium moniliforme* growing in maize and the toxins cause excess fluid to leak out into lung tissue (pulmonary oedema). This is an uncommon disease called porcine pulmonary oedema syndrome. Toxins also have a mild effect on the liver resulting in jaundice and orange-yellow coloured lesions evident at post-mortem examination. It usually takes 5 to 10 days of continual exposure for the signs of acute respiratory distress, cyanosis (blue colour) of the skin, jaundice, high morbidity and mortality to appear.

Gossypol

Gossypol is a natural phenol often found in cottonseed meal. The toxic level is more than 100 ppm. The most affected pig is often the biggest in the pen, as it has consumed the most food. Affected pigs may be found dead, but more often are depressed and may have signs of pneumonia and perhaps red urine. Post-mortem may demonstrate pulmonary oedema. Cottonseed is often used in cattle feed and poisoning may occur as a result of cross contamination.

Ochratoxin and Citrinin

Other aspergillus species, *Aspergillus ochraceus* and *Penicillium viridicatum,* produce toxins called ochratoxin and citrinin. These species are ubiquitous in northern climates and are found in oats, barley, wheat and maize. Levels of toxin of more than 1 ppm in the diet cause mild clinical signs particularly if spread over long periods of time. If the levels of intake exceed 1 ppm mortality may occur.

Clinical signs

The main signs include reduced growth and feed efficiency. Liver damage occurs but the main effect is on the kidneys resulting in increased water intake.

In young growing pigs oedema (fluid between the tissues) may occur with generalised stiffness. In acute poisoning mortality can be high. Gastric ulceration is also a consistent finding and in herds with high levels, the possibility of these toxins being present in feed should be considered.

Diagnosis

The clinical signs and post-mortem findings are indicative of ochratoxin and citrinin poisoning and this may be confirmed by identifying the toxins in the feed or in kidney tissue at slaughter. The toxins can also be identified in serum. The kidney may exhibit white areas of nephritis.

Treatment

- There is no specific antidote.
- Remove the suspected sources.
- Increase the vitamin levels in feed.
- Recovery is slow.

Trichothecenes

Trichothecenes are produced by other species of the fungus *Fusarium* mainly *F. graminearum,* and *F. sporotrichioides* growing in wheat and maize and the common toxins are called T-2, DAS and DON (vomitoxin). Their significance is that they are immunosuppressive and also affect the production of blood cells by the bone marrow.

Clinical signs

Levels of 1 ppm or more reduce feed intake and at high levels (10 to 20 ppm) they cause vomiting and inappetence but at such levels pigs refuse to eat the food.

Diagnosis

This is based on clinical symptoms, a history of sudden vomiting within 10 to 15 minutes of feeding and inappetence. In such cases feed should be changed and samples tested.

Treatment

- None is required – the toxins are rapidly excreted.
- Change the feed source.

Zearalenone

This toxin, called F2, is produced by a strain of *Fusarium graminearum* which appears in maize. It is an oestrogenic toxin and it is produced in high moisture environments in maize growing areas well before harvest. Rectal and vaginal prolapses are common symptoms in the young growing stock.

Clinical signs

The most striking clinical feature is the swollen red vulva of immature gilts. The other signs are dependent on the levels present in the feed and the stage of pregnancy. The following may be used as guidelines to the symptoms that may be observed.

Boars – Semen may be affected with feed levels above 30 ppm, but not fertility. At higher levels poor libido, oedema of the prepuce and loss of hair may occur.

Gilts (pre-puberty - 1 to 6 months of age) – 1 to 5 ppm in feed causes swelling and reddening of the vulva and enlargement of the teats and mammary glands. Rectal and vagina prolapses also occur in the young growing stock.

Gilts (mature) – 1 to 3 ppm will cause variable lengths of the oestrus cycle due to retained corpora lutea, infertility and pseudopregnancy) with oestrus being delayed until 63 days or more after mating.

Sows – Levels of 5 to 10 ppm can cause anoestrus, which may also be associated with pseudopregnancy due to the retention of corpus luteum. Note the F2 toxin mimics the 1st and 2nd embryonic signals produced at 10 and 14 to 17 days – thus the sow believes she is pregnant. F2 toxin will not normally cause abortion, however. If sows are exposed during the period of implantation litter size may be reduced. In lactation piglets may develop enlarged vulva and teats, although, to a degree, this is normal.

Effects on pregnancy – Low levels of 3 to 5 ppm do not appear to affect the mid part of pregnancy, but in the latter stages piglet growth in utero is depressed, with weak splay legged piglets born. Some of these may have enlarged vulvas.

Effects on lactation – 3 to 5 ppm has no effect on lactation but the weaning to service interval may be extended.

Diagnosis

The clinical signs are distinctive. Rations that are suspected of contamination should be examined both for the presence of zearalenone and also other oestrogen like substances. Removal of the suspect feed will reduce the symptoms within 3 to 4 weeks.

Treatment

- None is required provided the toxin source is removed.
- Sows that are in deep anoestrus may respond to injections of prostaglandins.

Nitrates and Nitrites

If these substances are absorbed from the small intestine in sufficient quantities, nitrite reacts with haemoglobin in the blood to form met-haemoglobin which reduces its oxygen carrying capacity. Poisoning occurs when nitrates and ammonia in slurry and straw are converted to nitrites by bacteria, or if the pig drinks drainage water that has become heavily contaminated. Nitrate itself is only slightly toxic, until it is converted to nitrite. Silage effluent is a particularly heavy source of nitrites.

Clinical signs

These are related to a shortage of oxygen in the bloodstream. There is a marked increase in the respiratory rate with animals showing a staggering gait and general weakness. The mucous membranes become dark red, almost blue. There may be sufficient irritation of the digestive tract to cause gastroenteritis. Clinical signs of poisoning will develop when 10 to 20 mg of nitrites per kg liveweight are eaten. Mortality becomes very high above the latter level.

Diagnosis

Sudden prostration of groups of pigs or a pen of pigs that have had access to potential sources of nitrates and nitrites, including whey and milk by-products, must always raise a suspicion. Post-mortem examinations show that the blood and musculature are a dark brown colour due to the formation of met-haemoglobin. This is a diagnostic feature.

Treatment

- Intravenous injections of 10 mg/kg methylene blue. Consult your veterinarian.

Plants

Poisoning from plants eaten by pigs kept outdoors is rare but the following species may cause problems from time to time.

Bracken (*Pteridium aquilinum*)

Bracken produces a toxin that destroys the vitamin thiamine. Pigs are normally resistant to bracken poisoning and it needs in excess of 6 weeks of constant exposure, particularly to fresh bracken shoots and the rhizomes.

Clinical signs

Sudden death is often the only sign, although prior to this respiratory distress may be seen associated with oedema or fluid in the lungs.

Diagnosis

This is based on history, clinical signs if seen and enzyme tests carried out on serum.

Treatment and control

- If sows are likely to have access to large amounts of bracken, injections of thiamine should be given every 2 weeks. Where clinical signs are seen the entire group is at risk and should be injected with thiamine.

Cocklebur (*Xanthium*)

This is found in waste places and the edges of ponds and rivers. Poisoning occurs when pigs eat the two-leaf seedling. The mature plant is unpalatable. The toxins cause depression, vomiting, weakness, rapid breathing, convulsions and lowered body temperature. Death occurs within a few hours. Lesions seen at post-mortem examination include ascites (liquid in abdominal cavity) and liver congestion and necrosis. Treat with mineral or sunflower oil by mouth.

Deadly Nightshade (*Solanum*)

These plants are found in waste areas and hay fields. The berries are very poisonous and will produce an acute haemorrhagic gastroenteritis with considerable salivation, trembling and paralysis, leading to coma and finally death. Poisoning in pigs is rare. Treatment requires the use of pilocarpine. Consult your veterinarian.

Oak Leaves and Green Acorns
Whilst ruminants are more susceptible to acorn poisoning, outdoor pigs may be affected occasionally. Young oak leaves or green acorns are the major sources and signs are seen 2 to 3 days after ingestion. These include abdominal pain and constipation followed by haemorrhagic diarrhoea. The kidneys may also be affected. Pigs traditionally are reared on mature acorns in some countries. There is no treatment but remove the pigs from the sources immediately.

Pigweed (*Amaranthus*)
This is a characteristic, clearly defined disease associated with the accumulation of oedema (fluid) around the kidneys. The signs start with trembling and an uncoordinated gait and pigs characteristically lie on their bellies. Acute kidney damage associated with oxalic acid build-up can occur within 2 to 3 days which in severe cases is followed by coma and death. The characteristic changes of fluid around the kidneys and history of access to the plant are diagnostic. There is no treatment and pigs should be removed immediately from the source.

Pokeweed (*Phytolacca*)
This plant produces oxalic acid which causes acute abdominal pain, vomiting, haemorrhagic diarrhoea and haemorrhage from the kidneys. There is no treatment.

Red Clover (Slaframine)
Red clover (*Trifolium pratense*) can easily be ingested by outdoor pigs. The clover is susceptible to a fungus called *Rhizoctonia leguminocola* (black patch disease) especially in cool, wet springs and autumns. Other legumes (white clover, alsike, alfalfa) can sometimes be infected. The fungus produces a toxic substance called slaframine that stimulates the salivary glands causing profuse slobbering. Signs include excessive drooling and tearing of the eyes. Diarrhoea, mild bloat, and frequent urination may also be seen.

Diagnosis is tentatively based on the signs and the presence of "black patch" on the forages. Removing the source of contamination resolves the issue.

Red clover (*Trifolium pratense*) contains a photosensitization agent and in rare cases may cause skin problems if consumed by outdoor pigs.

Sorghum (*Sorghum*)
These coarse grasses contain hydrocyanic acid and intake causes difficult breathing, convulsions and a very bright red coloured or chocolate brown blood associated with the formation of nitrites. Treatment is as for nitrites.

Yellow Jasmine (*Gelsemium*)
This plant is found in woodlands and can cause poisoning. The onset is sudden and acute with incoordination, weakness, coma and death, usually within 48 hours. There is no specific treatment. Sedate to control convulsions.

Water Hemlock (*Cicuta*)
This is found in open, moist or wet ground and it is highly toxic. Death can occur within 30 to 50 minutes preceeded by violent muscular spasms, dilated pupils, convulsions and coma. There is no specific treatment. Use sedatives to control convulsions.

Plants Causing Photosensitisation – Enhanced Sensitivity to Sunlight
The sap of at least 30 wild and cultivated plants including parsnip tops, parsley, celery tops and giant hogweed contains substances (furocoumarins) which, in contact with bare skin, enhance the skin's sensitivity to the ultra violet rays in direct sunlight.

Clinical signs
The affected skin becomes red and inflamed and then develops blisters and extensive angry looking skin erosions. Similar signs may occur following the ingestion of plants such as buckwheat, white and alsike clover and St. John's wort. The toxic principle may either be absorbed unchanged into the bloodstream and be deposited in the skin or it may damage the liver with subsequent photosensitisation.

These types of photosensitisation tend to occur in herbivores and people but not in pigs although there are reports of photosensitisation in pigs in central Europe from eating buckwheat. Celery tops sometimes have the same effect but this is due to fungi growing on them.

Sows and boars may be most obviously affected on their snouts and ears. The relatively hairless udders of lactating sows, which have most contact, may also be badly affected and become too painful for them to allow their piglets to suck.

Boars sometimes become affected on their undersides which make them reluctant to mate.

Diagnosis
This is based on the appearance of typical skin blisters and erosions and the availability of plants containing the toxic substances.

Similar diseases
The early blisters and erosion on the nose could be mistaken for foot-and-mouth-disease, swine vesicular disease or vesicular stomatitis but the history of exposure to plants containing furocoumarins and the progress

of the disease over a day or two would allow ready differentiation.

Treatment
- There is no antidote to these photosensitising plant substances.
- Affected pigs should be removed from the plants and from direct sunlight or at least provided with shade, until the skins heal.

Management control and prevention
- Give antibiotic cover to prevent secondary bacterial infections.
- Remove offending plants.
- Move pigs to different pastures.
- Inject pigs with B vitamins.

Protein
Mistakes may occur in feed manufacture and feed ingredients may be added in excess. If the soya inclusion is doubled, excess protein (30 %) may be present in the feed. The pigs will generally refuse the feed. Those that have eaten the feed – often the biggest pigs – may demonstrate diarrhoea and intestinal pain.

Selenium
See Chapter 14 – Vitamin E/Selenium for further information
Selenium is a highly toxic mineral but it is required in very minute amounts for normal bodily functions. Selenium poisoning is rare and usually occurs when selenium supplement has been wrongly mixed into the ration. Levels above 3 ppm in the diet have a clinical effect on the pig and when they reach 10 ppm severe clinical signs develop. The toxic dose of selenium by injection is approximately 0.8 mg/kg. Problems are more likely to occur with a deficiency producing muscle myopathies and mulberry heart disease.

Clinical signs
Pigs become anorexic with loss of hair and separation of hooves at the coronary band. Paralysis of front and hind legs is common. As the disease progresses there is liver and kidney failure and the pigs become toxic.

Diagnosis
The clinical picture of selenium toxicity is characteristic and almost diagnostic. Confirmation is by the identification of abnormal levels in feed, and in the liver and the kidneys of affected pigs. Levels above 3 ppm are diagnostic.

Treatment
- There is no specific treatment.

Toxic (Slurry) Gases (Fig 13-5)

Ammonia (NH_3)
Ammonia is the most common poison in the pig's environment. The concentrations of the various gases found in piggeries are expressed as parts per million (ppm). Generally ammonia levels are less than 5 ppm in well run pig houses. The human respiratory tract can detect levels at around 10 ppm. Levels of 50 to 100 ppm affect performance, particularly daily gain, which may be reduced by up to 10 % during prolonged periods of exposure. At levels of 50 ppm and above the clearance of bacteria from the lungs is also impaired and therefore the animal is more prone to respiratory disease.

Clinical signs
These include increased coughing and respiratory rates, irritation of the mucosa lining the respiratory tract and an increased incidence of pneumonia. Pigs are restless, uncomfortable and may show increased levels of vice, such as tail biting, ear biting and flank chewing.

Diagnosis
This is based on the assessment of air quality by the pig person and observed effects on the pig. Ammonia levels can be measured easily using glass sampling tubes. These are thin glass tubes containing a chemical to which syringes are attached. By drawing air through the chemical, a colour change takes place which indicates the ppm of the gas being tested for. (See Chapter 15 – Sampling the air for toxic levels of gases).

Treatment
- Increase ventilation rates.

Management control and prevention
- Where levels reach above 25 ppm ventilation must be improved and other measures taken. This concentration requires concern because of human health and safety issues.
- Increase the drainage of urine from the house and remove solid faeces daily. If a slurry system is used, remove slurry regularly and prevent a crust developing on the surface. Proprietary products that include an extract of the yucca plant are effective in reducing ammonia production in slurry.
- When designing houses keep the surface area and depth of the slurry to a minimum.
- Slurry systems that are flushed every 20 minutes reduce ammonia levels.

- Do not store slurry under slats. Empty frequently to holding areas. Make alterations to the diet to reduce nitrogen excretion.

FIG 13-5: A SUMMARY OF TOXIC (SLURRY) GASES

Gas	Acceptable Levels in Piggeries	Toxic Levels	Clinical Effects in People	Common On-Farm Problems and Dangers
Ammonia	< 25 ppm	> 50 ppm	Difficult breathing Irritation	Agitation of the slurry pits when emptying.
Carbon dioxide	< 2000 ppm	Uncommon	Difficult breathing Headaches Drowsiness	Power failure. Ventilation failure.
Carbon monoxide	< 30 ppm	> 60 ppm	Difficult breathing Asphyxiation Drowsiness	Faulty heaters. Deep slurry pits.
Hydrogen sulphide	< 5 ppm	> 80 ppm	Irritation Sickness Headaches Unconsciousness Death	Release of gas after slurry pits are agitated.
Methane	Non-toxic unless 80 % or more in air.	> 80 %	Irritation Explosive atmosphere > 1 %	Rare.

Carbon Dioxide (CO_2)

This gas is a normal component of air in levels up to 300 ppm. It is only toxic in very high levels > 3000 ppm and rarely causes problems in piggeries. However if CO_2 levels are high then there are significant risks that the levels of other more toxic gases will be critical. Increased levels cause pigs discomfort and restlessness which may lead to vice and cannibalism.
Carbon dioxide concentrations can be easily measured by gas sampling tubes.

Carbon Monoxide (CO)

Carbon monoxide poisoning in pigs occurs where faulty gas heaters are used in farrowing houses and ventilation is poor resulting in an increase in stillbirths. Also, the decomposition of faeces, particularly in slatted floored finishing houses produces high levels. 50 ppm is not uncommon and suggests inadequate ventilation. Levels of 250 ppm, however, interfere with the uptake of oxygen by the haemoglobin in the pig's blood, and act similarly in people. Instead of oxyhaemoglobin, carboxyhaemoglobin is formed and the oxygen carrying capacity of the blood is reduced. This markedly increases the number of stillborn piglets. Levels may rise to 50 %. Carbon monoxide affects the blood of unborn piglets more quickly than that of the sow.

Clinical signs

Sows may show dark coloration of the mucous membranes and stillborn piglets are often bright red due to the formation of the carboxyhaemoglobin in the blood. Such levels of carbon monoxide can have a similar effect on people so care should be taken in investigating the housing area.

Diagnosis

If gas heaters are used in farrowing houses and a sudden rise in stillbirths occurs, inadequate combustion must immediately be suspected together with poor ventilation. Carbon monoxide levels can be measured using glass sampling tubes.

Treatment

- Improve the ventilation immediately, and identify and remove the source of production.
- Administer oxygen but this may be impractical.

Hydrogen Sulphide (H_2S)

This is one of the important toxic gases found in piggeries and it is potentially lethal. The gas is produced by anaerobic bacterial decomposition of organic matter, particularly that found in faeces and slurry. The greatest source of H_2S and indeed the greatest potential for disasters comes from slurry that is held in pits beneath slatted finishing houses. Pockets of gas may become trapped and when the slurry is agitated and removed from the house there is a risk of acute and fatal poisoning not only to pigs but also to people. There is a low concentration of H_2S in most pig houses, usually less than 50 ppm. H_2S concentrations can be detected by the human nose within the range of 0.05 to 200 ppm. When levels get above the upper limit the sensitivity of the nose to detect the gas decreases significantly and the situation becomes potentially dangerous. The gas is toxic when you cannot smell it.

Clinical signs

These are dependent on the levels in the air and the following sequence of events can be considered in both pigs and humans:

- Less than 50 ppm – little effect but detected by people.
- 200 to 250 ppm – slight breathing difficulties, distress and irritation of the eyes, nose and back of the throat. The environment is becoming dangerous.
- 250 to 400 ppm – breathing starts to become very difficult with muscular spasms and disorientation.
- 400 to 600 ppm – pigs become totally

disorientated and some start to relapse into a coma.
- 600 to 1000 ppm – the pig develops fits and convulsions and becomes extremely short of oxygen. Its skin goes blue (cyanosed) and death ensues.

Methane
Methane gas is highly explosive when levels reach 1 % or more. This is the main risk on farms. As a poison the gas is largely inert and would only cause problems to pigs if the levels were so high as to displace oxygen. This would take place above a concentration of 80 % and the signs would be acute respiratory embarrassment.

Vitamin A Poisoning
Excess vitamin A either in the diet or by injection, can cause pathological changes in the pig. If oil based injections are given to excess at weaning time high levels may interfere with the growth of the embryo and foetus.

In older stock exposure to more than 25,000 iu/kg in the feed causes growth plate changes.

Clinical signs
- Extreme cases of OCD.
- Bent legs at the growth plate.
- Shortened bones.
- Dipped pasterns in breeding stock.
- Lameness.
- Piglets born with growth plate changes.

Diagnosis
A history of feeding high levels of vitamin A to both sows and growing pigs and shortened bent legs in piglets and weaned pigs would suggest poisoning.

Treatment
- The changes in the bones are irreversible.
- Reduce the vitamin A levels in the diet.

Vitamin D Poisoning
An excess of vitamin D would be unusual but where it occurs either by faulty additions to feed or by long acting injections clinical symptoms would be seen.

These would include poor growth rates, and because calcium may be deposited in the lungs and heart, signs of respiratory problems. Provided the case has not progressed removal of the excess from the diet will allow a response in 3 to 4 weeks.

Warfarin
Warfarin is one of a group of chemicals (rodenticides) that are used to control rat and mice populations. Poisoning by these substances is common in pigs due to accidental exposure. A single dose of warfarin of 3 mg/kg is fatal. Doses of less than 0.06 mg per day for 7 to 10 days will produce poisoning. Warfarin acts by preventing blood clotting.

Clinical signs
Severe haemorrhage is present throughout the carcass of the dead pig and in the skin, particularly where trauma has occurred. Lameness with haemorrhage in the joints and anaemia are consistent features.

Diagnosis
This is based on the history of the use of warfarin and the typical post-mortem lesions of widespread haemorrhage throughout the carcass.

Treatment
- Give injections of vitamin K, 5 mg/kg onDay 1 and then 2 mg/kg daily for 5 days.

Water Deprivation (Salt Poisoning)
Water deprivation (salt poisoning) is the most common poisoning to be seen in pigs. It arises where there is a shortage or complete lack of water and the normal salt in the diet then becomes toxic. The normal levels of salt in the ration vary between 0.4 and 0.6 % and even at these levels water deprivation can result in toxicity after 48 hours. The higher the level of salt in the diet the shorter is the period of water deprivation before signs are seen. However in the presence of ad lib water the pig can tolerate up to 2 % or more of salt in the diet. The first signs are inappetence and whenever this occurs in a pen of apparently healthy pigs or an individual always check the water supply first.

Clinical signs
These appear within 24 to 36 hours of water deprivation. Pigs become inappetent, wander aimlessly, are blind and stand with their heads pushed into the wall of the pen. Recurrent fits are common. These start with a characteristic twitching of the nose, the head then goes back and the pigs fall over. They eventually lie on their side with convulsive leg movements, froth at the mouth and become comatose and die.

Diagnosis
This is based on clinical signs and a history of water deprivation. Diagnosis can be confirmed by histological examination of the brain in which the lesions are diagnostic.

Treatment

- The response to treatment is poor particularly if pigs have developed fits. Rehydration of the pig is important and this can be achieved by dripping water through a flutter valve (see Chapter 15) into the rectum or allowing water to drip onto the tongue from a hose pipe. An alternate technique is to inject sterile water at body temperature into the abdominal cavity. This technique requires veterinary advice and direction.
- Clinical signs may become more acute once rehydration starts due to the fluids flowing to the brain, causing it to swell.

If you suspect poisoning, always seek veterinary advice.

14 NUTRITION AND HEALTH

14 NUTRITION AND HEALTH

Introduction

There are complete books dedicated to pig nutrition and this chapter is not intended to cover detailed nutritional aspects of pig production. The aim of this chapter is to review the role of nutrition as it affects health and disease on the farm. A major role for the stockperson is to judge the interaction between the pig, its age and/or productive cycle against the quality, content and intake of feed. The role of management in this respect has an important influence not only on the levels of disease in the herd but also on whether the pig maximises its biological potential.

The essential nutrients include water, protein and amino acids, energy, essential fatty acids, vitamins and minerals. A guide to the normal requirements is shown in Fig 14-1 both by weight of pig and ration type.

If you feel you have a nutrition-related problem, study the Observations and Causes list (Fig 14-2). First identify the problem by symptoms and this will suggest potential nutrient deficiencies or problem areas. You would be advised at this time to first check your water supply; if that is in good condition consult with your feed supplier because a knowledge of the composition of the diet will then assist in determining more specific areas, for example insufficient energy or lysine for the particular age group of pigs in that environment.

FIG 14-1: A GUIDE TO BASIC NUTRIENT REQUIREMENTS AT DIFFERENT STAGES OF LIFE
FINAL DIETARY LEVELS

	Liveweight kg (lb) Ad libitum Feeding							
Type of Ration	**3 – 5 (6 – 10) Creep**	**5 – 10 (10 – 22) Piglet**	**10 – 20 (22 – 45) Weaner**	**20 – 50 (45 – 110) Grower**	**50 – 120 (110 – 260) Finisher**	**Dry/ Gestating Sow**	**Lactating Sow**	**Boar**
Crude protein %	25 – 27	23 – 25	19 – 23	17 – 19	13 – 17	13.5 – 13.8	17 – 18	14
Crude fibre %	1 – 3	1 – 3	2 – 4	2 – 5	3 – 4	4 – 5	2 – 5	4
MJ NE/kg	11	11	10	9.9 – 10	9.6 – 9.7	9.4 – 9.5	9.6 – 9.8	9.4
MJ DE/kg	8 – 15	15 – 16	14.5 – 15	13.8 – 14.1	13.5 – 13.8	13.3 – 13.8	14.2 – 14.8	14
Essential Fatty acid (linoleic) mg/kg		100	Unknown, not considered necessary			300	300	
Amino acids requirements (as a % of total lysine)								
Lysine	100	100	100	100	100	100	100	100
Arginine	42	42	42	40	34	89	59	89
Histidine	32	32	32	32	32	30	42	30
Isoleucine	55	55	55	55	55	57	59	57
Leucine	100	100	100	100	100	94	121	94
Methionine	28	28	28	29	29	27	28	27
Methionine and cysteine	58	58	58	58	60	70	51	70
Phenylalanine	60	60	60	60	60	58	59	58
Phenylalanine and tyrosine	95	95	95	94	94	100	100	100
Threonine	62	62	62	63	64	76	66	74
Tryptophan	17	17	17	16	16	18	20	18
Valine	65	65	65	65	65	68	89	68

FIG 14-1: A GUIDE TO BASIC NUTRIENT REQUIREMENTS AT DIFFERENT STAGES OF LIFE FINAL DIETARY LEVELS CONT.

	Liveweight kg (lb) Ad libitum Feeding							
Type of Ration	**3 – 5 (6 – 10) Creep**	**5 – 10 (10 – 22) Piglet**	**10 – 20 (22 – 45) Weaner**	**20 – 50 (45 – 110) Grower**	**50 – 120 (110 – 260) Finisher**	**Dry/ Gestating Sow**	**Lactating Sow**	**Boar**
Minerals								
Calcium g/kg	9	8	8	7	6	8 – 11	9 – 12	7
Phosphorus g/kg (digestible)	6	5	4	3	2.5	6 – 8	7 – 9	8
Sodium chloride g/kg	2.5	2.0 – 2.5	1.5 – 2.0	1.0 – 1.5	1.0	5	5	5
Magnesium mg/kg	10 – 40	10 – 40	10 – 40	10 – 40	10 – 40	10 – 40	10 – 40	10 – 40
Iron mg/kg	100 – 200	100 – 200	80 – 160	60 – 160	50 – 100	80 – 160	80 – 160	80 – 200
Zinc mg/kg	100 – 200	100 – 200	80 – 160	80 – 160	60 – 120	75 – 150	75 – 150	100 – 200
Manganese mg/kg	30 – 60	30 - 60	20 - 40	30	20	200	200	200
Copper mg/kg	6 – 175	6 – 175	6 – 175	5 – 100	5 – 100	5 – 15	5 – 15	5 – 20
Iodine mg/kg	0.3 – 1.0	0.3 – 1.0	0.31 – 0.75	0.3 – 0.5	0.1 – 0.3	0.3 – 0.4	0.3 – 0.4	0.15 – 0.5
Selenium mg/kg	0.2 – 0.3	0.2 – 0.3	0.3	0.2	0.2	0.1	0.1	0.15 – 0.25
Vitamins and fatty acids which may need to be added to the diet								
Vitamin A iu/kg	1500 – 2000	1200 – 2000	1000 – 2000	800 – 22000	2000 – 8000	4000 –10000	4000 –12000	12000
Vitamin C mg/kg	0 – 200	0 – 200	–	–	–	0 – 250	0 – 250	0 – 400
Vitamin D3 iu/kg	200 – 2000	200 – 2000	200 – 2000	200 – 2000	200 – 2000	200 – 2000	200 – 2000	200 – 2000
Vitamin E iu/kg	100 – 250	100 – 200	50 – 150	15 –100	50 – 100	50 –100	50 –125	50 – 125
Vitamin K mg/kg	0.5 – 10	0.5 – 7	0.5 – 7	0.5 – 7	0.5 – 7	0.5 – 7	0.5 – 7	0.5 - 7
Vitamin B2 – riboflavin mg/kg	8 – 15	8 – 15	6 – 12	4 – 10	4 – 8	8 – 10	8 – 10	8 – 10
Vitamin B3 – niacin mg/kg	40 – 80	40 – 60	30 – 40	20 – 30	20 – 30	20 – 30	20 – 40	20 – 40
D-Pantothenic acid B5 *mg/kg	25 – 50	15 – 40	15 – 40	15 – 40	15 – 40	20 – 30	20 – 30	20 – 30
Vitamin B12 mg/kg	0.04 – 0.07	0.04 – 0.06	0.04	0.04	0.04	0.04	0.04	0.04
Choline mg/kg	800	600	400	300	300	500	800	800
Vitamin B1 mg/kg	5.0 – 5.5	4.0 – 5.0	3.0 – 4.0	2.0 – 3.0	1.0 – 2.0	2	2.5	2
Vitamin B6 mg/kg	0 – 8	0 – 8	0 – 6	0 – 4	0 – 3	0 – 5	0 – 5	0 – 5
Biotin mg/kg	0.2 – 0.4	0.2 – 0.4	0.3 – 0.3	0.2	0.2	0.2 – 0.5	0.2 – 0.8	0.2 – 0.8
Folic acid mg/kg	2 – 3	2 – 3	2 – 3	1.5 – 2	1 – 2	2 – 5	2 – 5	2 – 5

mg/kg = g/tonne = ppm   mg/kg x 0.0001 = %   g/kg x 0.1 = %
CP = Crude Protein **NE** = Net Energy **DE** = Digestible Energy **MJ** = Megajoules
iu = International Units
* If using Ractopamine provide at least 1% lysine.

FIG 14.2: OBSERVATIONS AND CAUSES

Abortion
- Energy deficiency.
- Iron deficiency.

Anaemia
- Copper deficiency.
- Cyanocobalamin B_{12} deficiency.
- Feed: no food, small particle size (gastric ulcers).
- Iron deficiency.
- Protein deficiency.
- Vitamin E – Selenium deficiency.
- Vitamin K deficiency.
- Water deprivation (causing gastric ulcers).

Bone fracture, malformed bones, lameness
- Biotin deficiency.
- Calcium deficiency.*
- Choline deficiency.
- Magnesium deficiency.
- Manganese deficiency.
- Nicotinamide deficiency.
- Pantothenic acid deficiency.
- Phosphorus.*
- Vitamins* – A, D_3, E deficiency.

Colitis (diarrhoea)
- Copper deficiency.
- Excess protein.
- High levels of wheat.
- Increased levels of potassium and magnesium.
- Iron deficiency.
- Iron excess in water.
- Magnesium excess.
- Nicotinamide deficiency.
- Pantothenic acid deficiency.
- Potassium excess.
- Selenium excess.
- Sodium chloride deficiency.
- Sodium chloride excess in water.
- Sulphate excess in water.
- Vitamin E deficiency.
- Zinc deficiency.

Haemorrhage
- Copper excess.
- Vitamin K deficiency.

FIG 14.2: OBSERVATIONS AND CAUSES CONT.

Nervous symptoms, incoordination, lameness
- Biotin deficiency.
- Calcium.
- Copper deficiency.
- Magnesium deficiency.
- Manganese deficiency.
- Pantothenic acid deficiency.
- Potassium deficiency.
- Phosphorus.
- Selenium excess.
- Sodium chloride excess.
- Vitamins* – A, B_6, D_3,* E.
- Water deprivation (salt poisoning).*

Poor growth, poor appetite (Postweaning ill-thrift)
- All aspects of nutrition.*
- Amino acids* – lysine, arginine, histidine, isoleucine, methionine, cystine, threonine, tryptophan.
- Calcium deficiency.
- Choline deficiency.
- Copper deficiency.
- Energy deficiency.*
- Feeds of poor digestibility.*
- Food deficiency – poor feed space etc.
- Folic acid deficiency.
- Iron deficiency.*
- Magnesium deficiency.
- Nicotinamide deficiency.
- Pantothenic acid deficiency.
- Phosphorus deficiency.
- Potassium deficiency.
- Protein deficiency.*
- Sodium chloride deficiency.*
- Thiamine deficiency
- Vitamin deficiency * A, B_6, B_{12}, * D_3, choline, riboflavin, pantothenic acid niacin.
- Water deprivation (salt poisoning).*
- Zinc.

Poor litter size
- Choline deficiency.
- Energy deficiency.*
- Folic acid deficiency.
- Lysine deficiency.*
- Other essential amino acids deficiency.
- Protein deficiency.
- Vitamin E deficiency.*
- Water deprivation.*

FIG 14-2: OBSERVATIONS AND CAUSES CONT

Reproductive failure
- Choline deficiency.
- Energy deficiency.*
- Feed deficiency (in gilts) – poor feed space etc.
- Folic acid deficiency.
- Iodine deficiency.
- Lysine deficiency.*
- Manganese deficiency.
- Other essential amino acids deficiency.
- Protein deficiency.
- Riboflavin deficiency.
- Vitamin deficiency – A, B_2, B_{12},* E.
- Water deprivation (salt poisoning).*

Respiratory diseases
- Energy deficiency.
- Protein deficiency.
- Vitamin E deficiency.
- Water deprivation (salt poisoning).*

Skin changes
- Copper excess.
- Iron deficiency.*
- Essential fatty acids deficiency* – linoleic acid.
- Nicotinamide deficiency.
- Potassium deficiency.
- Riboflavin deficiency.
- Salt deficiency.
- Water deprivation (salt poisoning).
- Zinc deficiency.

Sudden death
- Copper excess.
- Selenium.*
- Thiamine (B_1) deficiency.
- Vitamin E deficiency.*
- Water deprivation (salt poisoning).

* Likely to occur. Others uncommon or rare.

Water

Water is the forgotten nutrient and the ready availability of clean fresh water is essential. Insufficient attention is given to this on many farms and should be your first point of investigation related to every health issue. It is useful to consider the role that water plays in the normal metabolic functions of the pig.

- It helps to maintain and control body temperature, through both the intake and during exhalation when the heat is dissipated from the pig. It is lost in 3 ways, either by respiration, in the urine or in the faeces.
- An imbalance between water intake and loss results in dehydration and increased concentration of urine. Clinical signs include very dry faeces, hollow eyes and a dehydrated skin.
- It is responsible for transporting food and waste products throughout the body. Waste products are eliminated via water through the kidneys.
- Hormones are transported around the body through the bloodstream.
- Water regulates the acid/alkali balance in the body through the controls exerted by the kidneys.
- Water is used in protein synthesis. The digestive process will not function without it. Any restriction of water therefore will affect the above vital functions.

The farm should monitor water use via the internet every 15 minutes. If this is not possible then monitor at least daily. Water usage can provide critical information on the health of the pigs predicting feed intake and temperature requirements. An outbreak of SIV will result in a 10 % drop in water use even before the first sneeze! Water intake is a major driver of feed intake and thus daily liveweight gain.

The piglet

Within 6 hours of birth, water should be made available in a shallow dish or a trough because fluid intake is so vital at an early age. An efficient dish used for both creep and water in the farrowing pen is shown in Fig 14-3. It is interesting to note how many piglets within 24 hours will drink small amounts of water when given the opportunity. Nipple drinkers are not a very attractive method of presenting the water to the piglet. Water consumption by piglets during lactation is also influenced by the farrowing house temperature, and at 28 °C (82 °F) in a warm creep area water requirements will increase dramatically. The provision of water to the piglet in the first week causes no harm and is more likely to be of benefit, particularly if diarrhoea occurs on the farm, as the pig will already be accustomed to water. For pigs from 1 to 3 weeks of age clean water is best presented in an open type drinker rather than a nipple drinker. The water in dishes and drinkers must be clean and fresh.

The weaned pig

The pig experiences dramatic changes at weaning by the sudden move from a liquid to a solid diet. The conditioned reflex, calling the pigs to suckle regularly, is also lost. Dehydration associated with poor water intake and marked villus atrophy is a common occurrence within the first 7 days of weaning. Ensure the flow rate is at least 0.7 litres per minute from nipple drinkers. It is advisable to offer water in small open drinkers or water bowls daily for the first 5 to 7 days postweaning. The loss of milk at weaning time and villus atrophy reduce the availability of liquid to the pig for the first 48 hours. Check the salt concentration in the weaner's water supply.

The sow

The changes in water intake from pregnancy to lactation are considerable. Sows that have a lower water intake during lactation generally rear poorer litters and it is important therefore to encourage the sow to drink the moment she enters the farrowing quarters. This is best carried out by giving 4.5 litres of water twice daily into the feed trough until 2 to 3 days post farrowing. The water flow through a nipple drinker for the lactating sow should approximate 1.5 to 2 litres per minute. Water intake in the dry/gestating sow varies from 9 to 18 litres per day and in lactation from 18 to 40+ litres per day.

Guidelines for water requirements, water flow rates, drinker heights and drinker to pig ratios are given in Figs 14-4 and 14-5.

Water supply examination

The water supply needs to be examined from the entry point into the room through to the drinker. In some countries, the UK for example, water from the mains supply cannot be used directly because of the risk of backflow contamination, and a non-return valve, often in the form of a header tank, is used to prevent backflow. If present, examine the header tank hygiene and use. Is the size suitable for the number of pigs in the room? The header tank can be valuable for possible water medication routines. There should be a lid, acting as a light seal to reduce algae growth and to reduce aerial contamination of the water.

While walking around the room, examine as many drinkers as possible. In the farrowing house, all sow drinkers should be examined as a matter of course twice a day. Water is that important to lactation feed intake.

A CREEP DISH FOR EITHER FEED OR WATER

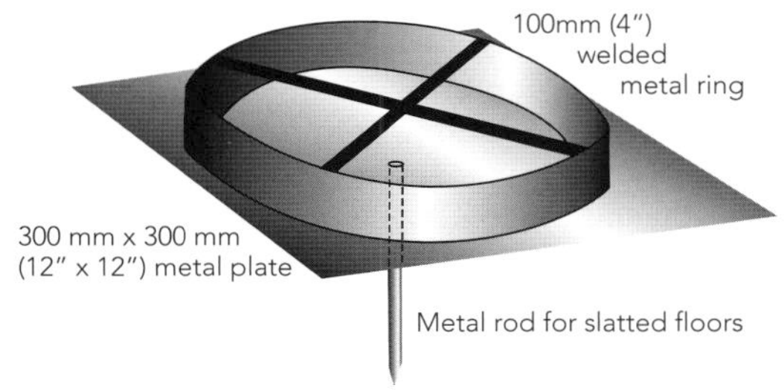

Fig 14-3

FIG 14-4: WATER REQUIREMENTS

Weight of Pig (kg)	Daily Requirement (Litres)	Minimum Flow Rate through Nipple Drinkers (LItres/min)	Guideline Heights for Nipple Drinkers (mm)
Newly weaned	1.0 – 1.5	0.5	100 – 130
Up to 20 kgs	1.5 – 2.0	0.7 – 1.0	130 – 300
20 kgs – 40 kgs	2.0 – 5.0	1.0 – 1.5	300 – 460
Finishing pigs up to 100 kgs	5.0 – 10	1.2 – 1.5	460 – 610
Sow and gilts pre-service and in pig	9 – 18	2.0	610 – 760
Sows and gilts in lactation	40 +	2.0	610 – 760
Boars	5.0 – 12	2.0	760 – 910

Nose/bite drinkers for lactating sows should to be set at 760 to 910 mm

FIG 14-5: DRINKER TO PIG RATIOS

Type	Rationed Feeding	Ad lib Feeding
Nipple	1 : 10	1 : 10
Bite	1 : 10	1 : 10
Bowl	1 : 20	1 : 30
Double sided trough (per 300mm)	Up to 15 kg weight – 35 pigs per trough. 15 kg to 35 kg weight – 30 pigs per trough. Over 35 kg weight – 25 pigs per trough.	

Examine the drinker for:

1. Any leaking drinker – These are costly and a major cause of wasted water.
2. Cleanliness – a dirty looking drinker is generally not working properly.
3. Location – can the pig gain access to the drinker properly?
4. Height and angle – is it appropriate for the pigs using the drinker? Classic mistakes occur when drinkers are not lowered in between batches of pigs.
5. Flow of water – this may need to be measured using a 250 ml cup and a stop watch.
6. Temperature of the water – measured by an infra-red gun on the collected water above.
7. The colour and possible taste of the water. Water high in iron is reddish and if high in sulphate can be foul-tasting.
8. Pressure of water from the drinkers.

Any drinker that fails to supply adequate water should be examined in more detail. This is likely to involve removing the drinker from its down-pipe bracket. When the drinker is removed, check that the flow of water is adequate from the down-pipe. If the flow is poor, examine the pipe in more detail. Partially blocked pipes from lime and other sediments, including biofilms, can reduce the internal bore.

Continue the examination by dismantling the drinker, in particular look for blocked and dirty filters, drinker settings and damaged seals.

Once the drinkers are fixed or replaced, reattach the drinker and ensure that the water supply is restored to the pigs.

Water quality

The variables in water quality include organisms, the physical characteristics and the mineral content. Water can become contaminated with pathogenic and non-pathogenic bacteria and viruses. The presence of coliform bacteria (i.e. *E. coli* and related bacteria) is an indication of faecal contamination and a potential source of disease (Fig 14-6).

The chemical quality of water can be assessed by determining the total dissolved solids, the pH (the alkalinity or the acidity), the iron content and the presence of nitrates or nitrites. Further testing would include levels of sulphates, magnesium, chloride, potassium, calcium, sodium and manganese. The total solids in water represent the amount of matter that is actually dissolved. If this level is less than 1000 ppm it is of no significance but once it reaches over 2000 ppm it becomes unfit for pigs. Generally if the total solid content is low it is usually of good quality and the water is safe to drink. The pH level of good water varies between 6.5 and 8.

Hardness of water is dependent on the levels of calcium and magnesium present but these have no effect on animal health. Hardness does however result in the accumulation of scale, causing pipes to gradually block and the flow rate to drop unnoticed. This is a common problem on farms that have metal pipes of at least 4 years' standing.

Iron can cause problems in water, with brown coloured staining. Certain types of bacteria can grow and cause blockage of pipes.

High levels of nitrates and nitrites can interfere with the use of vitamin A by the pig and they may be responsible for high still-birth rates.

FIG 14-6: WATER QUALITY GUIDELINES FOR PIGS

Nutrient	ppm (parts per million) Less than
Calcium	1000
Chloride	400
Copper	5
Fluoride	2 – 3
Hardness Calcium carbonate	< 60 Soft > 200 Hard
Iron	0.5
Lead	0.1
Magnesium	400
Manganese	0.1
Mercury	0.003
Nitrites	10
Nitrates	50
Phosphorus	7.8
Potassium	3
Sodium	150
Selenium	0.05
Solids dissolved	1000
Sulphate	1000
Zinc	40
Total viable bacterial counts (TVC) per ml 37 °C (99 °F) 22 °C (72 °F)	Low but more important no fluctuation between samples. Target $< 2 \times 10^2$ $> 1 \times 10^4$ poor
Coliforms/100ml	Zero

A summary of the effects of high mineral levels in water

Sodium and chloride

- If this is above 500 ppm then a brackish taste may develop.
- High levels of sodium chloride (salt) affect palatability and can adversely affect pig productivity and performance.
- Sodium sulphate is a laxative and mildly irritant (causes diarrhoea).

Calcium and magnesium

- There are no effects on animal health unless there are high levels of the sulphates which result in the accumulation of scale (as $Mg(OH)_2$ and $CaCO_3$) and over a period of time the diameter of pipes is reduced with the poor flow rates.

Iron and copper

- High levels of copper have a catalytic effect on the oxidation of iron and if the iron levels are high precipitation of iron occurs when water is pumped, resulting in problems with the delivery system.
- Iron also supports the growth of certain types of bacteria causing foul odours and blocked water systems.
- High levels of iron in the water reduce lactation feed intake.

Sulphate

- High levels of sulphate in association with magnesium and sodium can cause diarrhoea.
- High sulphate may cause palatability problems.
- High sulphate may reduce lactation feed intake.

Manganese

- High levels promote oxidation leading to a reddish tinge in the water.

Nitrates/nitrites

- Nitrites can change the structure of the haemoglobin in blood rendering it incapable of transporting oxygen. If levels are high the blood is a dark colour due to lowered levels of oxygen.
- Extremely high levels of nitrates/nitrites in water impair the utilisation of vitamin A in pigs and a reduction in performance – such levels however are very rarely found under practical conditions but levels can be sufficiently high to increase stillbirths.

Minerals and Vitamins

Fig 14-7 lists the clinical signs associated with deficiencies or excesses of minerals and Fig 14-8 of nutrients and vitamins. Details of individual nutrients are given in subsequent pages.

One of the most common problems on farms can be the failure of the diet to satisfy the amino acid and energy requirements of the pig.

Amino Acids

These are substances which when linked together in different combinations form different proteins. There are approximately 22 amino acids and whilst the pig can synthesise the majority of these, there are a number it cannot and these are described as essential for normal health and metabolic processes.

The essential amino acids are:

- Arginine.
- Isoleucine.
- Histidine.
- Leucine.
- Lysine.
- Methionine plus cystine.
- Phenylalanine plus tyrosine.
- Threonine.
- Tryptophan.
- Valine.

Field experiences constantly reinforce the importance of good quality proteins and amino acid availability particularly during periods of stress, management change and when the immune system is challenged.

Critical time periods are in the first 14 days post-weaning and the period from 6 to 12 weeks of age when maternal antibodies are declining, especially in endemically infected environments where the pigs are exposed to mycoplasma (enzootic) pneumonia, PRRSV and App. During these periods of challenge you are advised to feed or continue feeding the higher quality diet. The advent of segregated early weaning, which removes many pathogens and environmental contaminants that are normally exposed to the pig, has increased the nutritional requirements necessary to satisfy the increased growth. This is particularly true of lysine and energy.

The quality of the protein in the pigs' diet is a reflection of the amount and the availability of these essential amino acids. High quality protein contains all of the essential amino acids at acceptable levels; poor quality protein is deficient in one or more. When proteins enter the intestinal tract they are broken down into the separate amino acids which are absorbed into the bloodstream and transported around the body. These amino acids are then built into different types of proteins to

FIG 14-7: MINERALS. CLINICAL SIGNS OF DEFICIENCIES AND EXCESSES		
Mineral	**Signs of Deficiency**	**Signs of Excess**
Calcium	Agalactia. Depressed milk yield. * Fractures. * Hypocalcaemia. Osteomalacia. * Osteoporosis. * Posterior paralysis in sows. * Rickets.	Changes in bone formation. Reduced strength of bone. If zinc is low (parakeratosis) more than 1 % may cause problems. May bind up other minerals - Zn and Cu. High calcium in the diet can result in raised pH in the stomach allowing more pathogens to cause intestinal problems.
Copper	Anaemia. Leg weakness. Loose faeces if suddenly withdrawn. Poor growth.	* Jaundice 200 to 600 g/tonne. Haemorrhage. Death.
Iodine	Enlarged thyroid glands. Reproductive failure. Weak hairless pigs at birth.	Rare > 800 mg/kg.
Iron	* Anaemia. * Increased respiration. More prone to piglet diseases. Poor growth. Pale skin.	Death in piglets already deficient in vitamin E. Muscle degeneration > 5000 mg/kg.
Magnesium	Abnormal gait. Infertility Rare. Poor growth. Weak joints.	Loose faeces > 0.5 % in diet.
Manganese	Infertility Rare. Lameness. Poor growth. Weak piglets.	Inappetence > 2000 ppm.
Phosphorus (note effect of phytases in reducing phosphate requirements).	Poor growth. * Rickets. Phosphorus and calcium need to be in balance. Soft bones.	Changes in bone formation. Posterior paralysis in sows. Phosphorus and calcium need to be in balance.
Potassium	Anorexia (rare). Heart malfunction. Incoordination. Poor growth.	Loose faeces > 1.2 % in diet.
Salt (sodium chloride)	Low water intake – salt poisoning. Poor growth and feed efficiency. Unthriftiness. Diarrhoea. Poor feed intake in lactation.	* Common. Any level if water is short. Death > 2 - 8 % if water short. Fits. Incoordination. Thirst.
Selenium	* Mulberry heart disease. Muscle changes. Sudden death.	Diarrhoea. Feet deformity. Lameness. Respiratory distress. Sudden death 5 – 10 g/tonne
Water	* All systems affected. Failure to thrive. Nervous convulsions (water deprivation). Predisposition to disease.	Colic.
Zinc	* Dry thick skin (parakeratosis). Poor appetite.	Reduced feed intake > 3000 g/tonne. Up to 2500 g/tonne in diet none.

* Likely to occur. Others uncommon or rare.

FIG 14-8: NUTRIENTS AND VITAMINS. CLINICAL SIGNS OF DEFICIENCIES AND EXCESSES

Nutrient	Signs of Deficiency	Signs of Excess
Amino acids	A predisposition to disease. Poor growth.	Digestive disturbances.
Biotin	Infertility. Anoestrus. * Lameness. * Poor hoof quality.	
Choline	Poor litter size. Poor growth.	Unknown. Unlikely.
Cyanocobalamin (B12)	Poor growth. Infertility. Anaemia.	Unknown. Unlikely.
Energy *	* Infertility. Loss in weight. Poor fat deposition. Predisposition to: * Cystitis pyelonephritis. * Postweaning enteritis. * Respiratory disease. * Villus atrophy and malabsorption. * Thin sow syndrome.	Deposition of excess fat.
Fat and fatty acids * (Linoleic)	Dry skins in sows and piglets. Loss of weight in lactation. Poor growth.	Colitis. Digestive disturbances. Loose faeces.
Folic acid	Anaemia. Poor litter size. Poor growth.	Unknown. Unlikely.
Nicotinamide (Niacin)	Diarrhoea. Dermatitis. Poor growth. Paralysis.	Unknown. Unlikely.
Pantothenic acid (B5)	Poor appetite and growth. Goose-stepping gait. Diarrhoea.	Unknown. Unlikely.
Protein *	Lean tissue gain reduced. Poor growth. More prone to disease.	Diarrhoea.
Pyridoxine (B6)	Poor growth.	Unknown.
Riboflavin (B2)	Infertility. Weak piglets.	Unknown. Unlikely.
Thiamine (B1)	Poor appetite and growth. Sudden death.	Unknown. Unlikely.
Vitamin A	Rare but reports of: Infertility. Incoordination. Poor bone growth. Poor sight. Congenital defects, born blind.	* Epiphyseal plate changes. * Increased incidence of OCD. * Increased requirements for vitamin E. Joint pain. * Leg weakness. * Mulberry heart disease.
Vitamin D	Fractures. Lameness. Rickets. Rubbery bones or osteomalacia. Swollen joints.	Calcification of soft tissues.

* Likely to occur. Others uncommon or rare.

FIG 14-8: NUTRIENTS AND VITAMINS. CLINICAL SIGNS OF DEFICIENCIES AND EXCESSES CONT.		
Nutrient	**Signs of Deficiency**	**Signs of Excess**
Vitamin E * (Mulberry heart disease)	Agalactia. Discoloration of fat. * Gastric ulcers. * Liver, heart and muscle changes. Oedema disease. Porcine stress syndrome. * Predisposition to: App. E. coli diarrhoea. Respiratory disease. Swine dysentery. * Reduced immune responses. Sudden death. Udder oedema.	
Vitamin K	Enhances warfarin poisoning. Poor blood clotting.	Unknown. Unlikely.

* Likely to occur. Others uncommon or rare.

satisfy the many diverse requirements of the body. It can be seen therefore, that where there is a deficiency of one or more essential amino acids in the diet, the metabolic functions of the pig are compromised leading to biological inefficiency and possibly disease. The major roles of the amino acids are in the production of muscle protein, digestive enzymes, haemoglobin in the blood, gamma globulins (antibodies), milk protein and hormone metabolism. Since the proteins used in pig diets are of variable quality some of the essential amino acids may be deficient. These are called the limiting ones and in most cases lysine is the most likely, followed by methionine and both are often added to diets routinely. If the diet is deficient in one or more of these essential amino acids then protein synthesis will only continue to the level associated with the first limiting amino acid. The amounts of each amino acid required in the diet are expressed as a percentage of the total lysine requirement (Fig 14-1).

Enteric diseases such as *Escherichia coli* (*E. coli*) enteritis in the sucking pig, transmissible gastroenteritis, colitis and swine dysentery, which severely damage the lining of the intestine and its capacity to absorb nutrients can have a profound effect on the absorption of amino acids and exacerbate the effects of the disease. It is important when dealing with such diseases to ensure that the diet has a high level of amino acids during the recovery period.

Energy

Energy in the diet is measured either by kilocalories (kcal) as used in the USA and Canada or megajoules (MJ) as used in Europe. In some countries the megacalorie (Mcal) is used. 1 Mcal = 1000 kcal. To convert kilocalories to megajoules, multiply by 4.184.

The most common nutritional deficiency in pigs is that of energy and the amount available in the diet is usually measured either as, net energy (NE), digestible energy (DE) or metabolisable energy (ME). (ME = 0.96DE).

Net energy is the energy that is retained by the animal for maintenance and productive purposes. This more accurately describes the animal's energy requirements. This is more difficult to calculate for each feedstuff. But if it is used, it will reduce the feed costs as it is the more accurate predictor of the animal's energy requirements.

Digestible energy is the amount of energy which is present in the feed and readily digested and absorbed from the intestine into the body. It is basically the total energy in the feed minus the energy that remains in the faeces. This is therefore relatively easy to calculate.

The pig could not operate without adequate sources of energy because it is the fuel that supports maintenance and drives the whole of the metabolic processes resulting in the production of meat and milk. Fig 14-9 shows the key factors that contribute to an energy-deficient state. If there is a poor growth problem in your herd you would be advised to check this list and identify those factors that are likely to have a bearing on your problem and then refer to the index or other chapters relating to the problem area.

Energy requirements are determined by the weight of the pig, its growth rate, the amount required for maintenance and its stage in the reproductive cycle. The requirements for energy in the lactating sow should not be underestimated. Fig 14-10 shows the critical periods

FIG 14-9:
KEY FACTORS THAT CREATE AN ENERGY DEFICIENCY

- A low energy diet.
- A low protein diet.
- An incorrect diet for the age of the pig.
- Poor feed intake – especially postweaning.
- Poor palatability.
- Floor feeding.
- Poor access to the feed.
- Faulty feed hoppers.
- Toxic gases affecting palatability.
- Shortage of water.
- Distance to the feeder.
- Distance to the drinker.
- High stocking density.
- Wet pens.
- Draughts.
- Poor insulation of floors and buildings.
- Changing / low environmental temperatures.
- Faulty heaters.
- The movement of pigs from one house to another.
- The mixing of pigs.
- Movement from a solid to a slatted floor.
- Wet bedding making movement difficult.
- A change from dry to wet feeding or vice versa.
- The genetic make-up of the pig – the more lean the pig the more susceptible it is to the environment.
- Exposure to infection and disease.
- The requirement of body tissues.
- Excessive use by body tissues and body products.
- Parasites.
- Reproduction and pregnancy.
- Growth of the foetus.
- Lactation.

USE THIS AS A CHECKLIST IN YOUR HERD

FIG 14-10: THE ENERGY CRISES

The Critical Time Period	Diseases or Conditions to which a Deficiency of Energy may Contribute
From birth to Day 3	*E. coli* enteritis. Hypoglycaemia. Joint infections.
Day 3 to weaning	*E. coli* enteritis. Glässer's disease. Immunosuppression. Septicaemia.
The weaned pig to 7 days postweaning	*E. coli* enteritis. Glässer's disease. Malabsorption. Poor growth. Starvation – postweaning ill-thrift. Streptococcal meningitis. Streptococcal septicaemia.
Growing pig 5 - 14 weeks old	Actinobacillus pleuropneumonia. Colitis. Mycoplasma (enzootic) pneumonia. Immunosuppression. Mycoplasma arthritis. *Pasteurella* pneumonia. Poor metabolism and growth. Porcine reproductive and respiratory syndrome. Progressive atrophic rhinitis.
Finishing pig	Aujeszky's disease (pseudorabies). Enteric disease. Immunosuppression. Pneumonia. Poor metabolism and growth.
Dry/ gestating sow or gilt	Abortion in late pregnancy. Anoestrus. Cystitis pyelonephritis. Embryo resorption. Litter size. Pseudopregnancy.
Lactating sow	Poor subsequent litter size and fertility. Extended wean to service intervals.

of energy demand in the pig through its life and diseases that may be associated if an energy deficiency arises. Daily feed intake and the energy level per kg of feed are crucial factors which help the pig to maintain a positive (anabolic) energy state instead of a negative (catabolic) one. The intake of energy is also essential to maximise reproduction performance (see Chapter 5). The relationship between feed intake and energy used by the pig on the one hand and that lost to the environment is a complex one. The survival of the piglet in the first 2 to 3 days of life is highly dependent on a regular supply of energy and if the sow's nutrition is inadequate leading to poor quality milk, the susceptibility to disease and piglet mortality rises. Diarrhoea in the neonatal period and up to 14 days of age is often precipitated by intermittent periods of catabolism associated with low environmental temperatures. The introduction of creep feed at an inopportune time can, through indigestion and abnormal fermentation, initiate a sequence of events leading to scour or increased susceptibility to rotavirus, PRRSV, joint infections or Glässer's disease. If you have an intransigent scour problem on your farm and you are feeding creep to sucking piglets, try stopping it; it may reduce the problems.

In the newly weaned pig the quality and availability of carbohydrates and other energy sources are vital if a healthy rapid growing pig is to be produced. Within 12 to 24 hours after weaning most pigs become energy deficient for a short period, which affects the degree of villus atrophy and the rate of their regeneration. The immune system also does not respond efficiently and the results are more disease or a greater incidence. The important changes in management and feeding practices are considered in detail in Chapter 9.

The significance of maintaining a positive energy balance in the pig and its role in the precipitation of disease is often not appreciated on the farm. Aspects of this are dealt with in the relevant chapters and specific diseases. Conditions are however highlighted here from field experiences, so that you can assess them in relation to problems on your farm.

Do not feed creep to piglets if you have a pre-weaning scour problem.

Common Diseases and Conditions Associated with Nutrition

Figs 14-7 and 14-8 highlight signs of deficiencies and excesses documented in the literature. In practice today, using modern well formulated premixes, problems with most of the individual nutrients are uncommon or rare. Problems usually only occur due to gross mismanagement in the preparation of the premixes, prolonged storage under adverse conditions, the failure to add nutrients over a prolonged period of time or incorrect mixing.

The roles of energy and protein in disease generally have already been highlighted but there are a number of specific conditions that are initiated by faulty feeding in one form or another, as listed below.

- Abortion and seasonal infertility.
- Anaemia.
- Colitis.
- Diarrhoea.
- Fractures.
- Gastric ulcers.
- Lameness.
  - Leg weakness.
  - Osteochondrosis.
  - Osteodystrophy.
  - Osteomalacia.
  - Osteoporosis.
  - Rickets.
- Mulberry heart disease.
- Mycotoxicosis.
- Postweaning ill-thrift.
- Prolapse of the rectum.
- Reproduction.
- Respiratory disease.
- Torsion of the stomach/intestines.
- Udder oedema and mastitis.
- Water deprivation (Salt poisoning).

Abortion and Seasonal Infertility

See Chapter 5 for information on non-infectious causes of abortion and seasonal infertility. See chapter 6 for infectious causes of abortion.

Anaemia

Anaemia is a condition associated with either a reduction in the number of red cells in the blood, in the amount of haemoglobin they contain, or in the volume of the red cells themselves.

It can arise in one of 3 ways:

1. Loss of blood through haemorrhage. Typical examples would be gastric ulceration, trauma to the vulva or a ruptured liver.
2. Lack of haemoglobin due to dietary insufficiencies, particularly iron and copper.
3. Reduced numbers of red cells. These are produced in the bone marrow and any disease, infection or toxic state affecting it may result in anaemia.

The common causes of anaemia are shown below in Fig 14-11 but iron deficiency and anaemia in piglets are the most common and important. From a purely nutritional aspect anaemia in growing and adult animals is uncommon but disease will often give rise to a secondary anaemia. Anaemia is also common, secondary to specific diseases such as actinobacillus pleuropneumonia or Glässer's disease.

Clinical signs

The pig is pale and becomes breathless on exertion, sometimes (but not always) with a slight check in growth. The mucous membranes of the eyes are pale. The colour of the skin may take on a slight yellow or jaundiced appearance. In severe cases breathing is rapid, particularly with exercise, and there may be a predisposition to scour. Postweaning diarrhoea may be more common in piglets with iron anaemia. Haemorrhage

FIG 14-11: THE COMMON CAUSES OF ANAEMIA

- *Mycoplasma suis* in all ages.
- Fungal toxins.
- Gastric ulcers.
- Iron deficiency in piglets.
- Mange (*Sarcoptes scabiei* var *suis*).
- Mycotoxins.
- Ileitis.
- Stomach worm (*Hyostrongylus rubidus*).
- Whipworm (*Trichuris suis*).
- Umbilical haemorrhage.
- Vaginal haematoma.
- Vulval biting.
- Warfarin poisoning.

may be obvious to the exterior or it can occur through bleeding into the tissues or the gut. Warfarin poisoning causes signs of anaemia due to severe bleeding into the tissues.

Diagnosis

Anaemia can be diagnosed on clinical grounds, the lack of any supplemental iron and by examining a sample of blood. This is tested for the red cell volume and the haemoglobin levels. Normal levels in a young pig are 9 to 15 g/100ml; if levels are less than 8 g/100 ml the piglet is becoming anaemic. A stained blood smear will also confirm the shape and size of the red cells and whether there are any bacteria present. Specific cell types are involved in the different anaemias. Iron dextran toxicity associated with vitamin E deficiency may give rise to anaemia in piglets.

Treatment

- The intestine can absorb only small amounts of iron daily which may not be enough to reverse the anaemia quickly. Nevertheless iron and copper levels in the feed should be checked. It is also helpful to give anaemic pigs an injection of iron dextran (300 to 500 mg) depending on the age of the pig.
- In severe cases electrolytes can be given either by injection or by mouth. It is possible to give electrolytes par rectum.
- Specific treatment will depend on the cause. Refer to the specific conditions in the index.

Management control and prevention

Piglets

- Inject piglets at 3 to 5 days of age with 200 mg of iron dextran.
- The sites of injection should be the muscles of the neck. See Chapter 15.
- Iron is best given from 3 to 5 days of age and not at birth. A 2 ml dose at birth causes considerable trauma to the muscles. A single dose should be sufficient.
- Iron can also be given orally but this method is time consuming and the pig must be treated on 2 or 3 occasions at 7, 10 and 15 days of age.
- Oral pastes available ad lib have been used but the uptake within any litter is variable and a few piglets remain anaemic.

General

- Monitor iron levels in-feed and supplement all pig diets with iron where necessary.
- Carry out regular worming programmes and/or check faeces samples for parasites every 3 to 6 months.

Colitis

Colitis means inflammation of the large bowel and it is very common in some countries in growing pigs. It is characterised by sloppy "cow pat" type faeces, with no blood and little if any mucus but the condition may progress to severe diarrhoea. Affected pigs are usually 6 to 12 weeks of age, and in any one group up to 50 % of the population may be affected. It is not seen in adult or sucking pigs. A number of organisms have been implicated but spirochetes and in particular *Brachyspira pilosicoli*, an organism distinct from a similar one that causes swine dysentery, is thought to be important. Colitis may also be associated with *E. coli*, coccidiosis, *Clostridium perfringens* type A, *Brachyspira hyodysenteriae* and *Salmonella* infections. However dietary factors also precipitate disease and pelleted feed is much more likely to be associated with the disease than meal. If the incriminating pellets are ground back to meal colitis still results, demonstrating an effect of the pelleting process. Certain components in the feed are also implicated including poor quality oils and carbohydrates. Specific ones have not been identified but may include anti trypsin factors and oligosaccharides.

Clinical signs

These usually appear in rapidly growing pigs from 8 to 14 weeks old, fed ad lib on high density diets. The early signs are sloppy faeces but with pigs appearing clinically normal. As the disease and its severity progress, very watery diarrhoea, with dehydration, loss of condition and poor growth become evident in the pigs. During the affected period daily gain and food conversion can be severely affected, with feed conversion worsening by up to 0.2.

Diagnosis

This is based on clinical signs and the elimination of other causes of diarrhoea, in particular swine dysentery or ileitis. Faecal examinations in the laboratory are necessary to assist with diagnosis together with post-mortem examinations and laboratory tests on a typical untreated pig. It is possible that ileitis may be involved in the clinical syndrome and if the herd has a severe problem examination of the terminal parts of the small intestine in pigs at slaughter would be advised together with PCR tests on faeces.

Treatment

- Antibiotic therapy is not always successful because it depends on the presence of primary or secondary bacteria, but the following medicines have given responses on problem farms, using in-feed medication.
  - Dimetridazole – 200 g/tonne if available (banned in many countries).

- Lincomycin – 110 g/tonne.
- Monensin – 100 g/tonne if available.
- Oxytetracycline – 400 g/tonne.
- Salinomycin 60 g/tonne.
- Tiamulin – 100 g/tonne.
- Tylosin – 100 g/tonne.
- Valnemulin – 25 to 75 g/tonne.

- For the individual pig daily injections of either tiamulin, lincomycin, tylosin or oxytetracycline may be beneficial.
- Weaned pigs are often fed zinc for the first 2 weeks postweaning to prevent *E. coli* enteritis. Colitis may develop in the 2 to 3 weeks following its removal from the diet. The response to continuing zinc oxide in the feed at 2 to 3 kg per tonne should be considered.

Management control and prevention

- Check the drinking water for bacterial contamination.
- Check for draughts and other stress factors.
- Check protein concentration in the feed – mistakes in mixing can result in colitis diarrhoea.
- Disease is seen in pigs that are growing well on ad lib feeding and often it is associated with a change of diet.
- It is more common with diets high in energy and protein: 9.9 MJ NE/kg (14.5 MJ DE/kg) 21 % protein. It is experienced using all types of diets but particularly those that have been pelleted rather than fed as a meal. It is thought that the pelleting process may have an effect on fats in the diet and thereby initiate digestive disturbances in the large bowel.
- It is common when fat sprayed diets are fed; try diets without.
- Mortality is low but morbidity can be high, ranging from 5 to 50 %. Adopt all-in/all-out management of pens.
- The same diet can be used on 2 separate farms and disease only appears on one, suggesting inherent causes.
- The presence of certain types of bacteria in the large bowel such as *Brachyspira pilosicoli* obviously plays a part in the disease. Use preventative medication in-feed to control *Brachyspira pilosicoli*.
- It is uncommon on home milled cereal based diets.
- The response to treatment can be variable.
- Control consists of assessing the above key factors, which would include changing the diet formulation (less added fats), changing the source of feed, acidifying the diet and improving pen hygiene. A change from pellets to meal feeding is usually effective. The following anti-colitis diet fed as a meal has been effective:
  - Wheat 50 %.
  - Barley 11 %.
  - Full fat soya 15 %.
  - Fish meal 7.5 %.
  - Hypro soya 6.5 %.
  - Sharps (wheat by-product) 5 %.
  - Skim milk 2.5 %.
  - Vitamin lysine mineral supplement 2.5 %.
  - Analysis: Protein 24 %, 9.7 MJ NE/kg (14.6 MJ DE/kg), Lysine 1.35%.

Diarrhoea

See Chapter 8 for further information

Whilst diarrhoea is generally caused by disease it may also be initiated by the diet. Dietary association with diarrhoea is often related to the immediate postweaning period. A number of changes occur to the intestine of the pig at weaning and these are covered in detail in Chapter 9. Briefly, only minimal amounts of solid food are eaten during the suckling period and very little before 10 days of age. The sow's milk provides all the nutrients the piglet requires. At weaning the piglet has its main source of food removed (milk) and needs to survive on a solid diet. Thus it is important that efforts are made to assist this transition. There are a number of actions that can be taken both before and after weaning to achieve this.

Key points to maximising feed intake after weaning

- Pigs at weaning time will eat a warm gruel better than a solid food.
- Gruel feeding reduces the degree of villus atrophy and dehydration.
- Pigs need to be encouraged to feed in the first 3 days postweaning because the maternal discipline of suckling every hour is lost.
- Manually provide creep feed for the first 72 hours in long non-metal troughs 5 to 6 times a day (recently washed metal troughs have unattractive smells). This will encourage the pigs to eat whilst avoiding over-eating. Ensure that there is sufficient space for all the weaners to eat at the same time – this requires 100 mm per weaned pig (50mm if the trough is 2-sided). Talk to the weaners when you are feeding them.
- Only provide sufficient food to be eaten in 10 minutes.
- By experiment place the feeders in the most attractive part of the pen, but not in the sleeping area.
- A small pellet or crumb will increase intake. Pellet size should be 2 mm or less.

- If you are still teeth-clipping, examine the piglet's mouths at weaning time to ensure there has been no damage to the gums during teeth removal. Pigs with sore infected gums will not eat. Ideally teeth-clipping should be stopped. There is no requirement if lactation is adequate.
- Use a highly palatable diet possibly with added milk.
- Provide a draught free sleeping area.
- Ensure water supply is adequate, easy and close to the feeder. Gruel feed provided for the first couple of days postweaning will enhance feed intake further. There is a 10-day advantage in finishing if pigs eat well in the first week postweaning.

The pig must balance its energy requirements during the first 72 hours post-weaning. If not, disease is likely to occur (Fig 14-12).

Creep feeding/options

The term "creep feed" here means the pre-starter diet offered to piglets before and just after weaning until they can be changed to a cheaper starter diet. The name is such because with loose-housed sows the pre-starter has to be placed in a "creep" where the sows cannot get to it.

There are a number of options for managing the feeding of creep to piglets:

- No creep given pre-weaning.
- Different creeps given pre and postweaning.
- Mixed creeps given postweaning.
- A high density diet used pre-weaning and a low one postweaning.
- A low density diet used pre-weaning and a high one postweaning.
- Restricted feed for varying periods of time.
- Choice feeding.
- By trial and error to determine the best methods that produce a healthy rapid-growing weaner.

MAINTAINING A POSITIVE ENERGY STATE: THE BALANCE

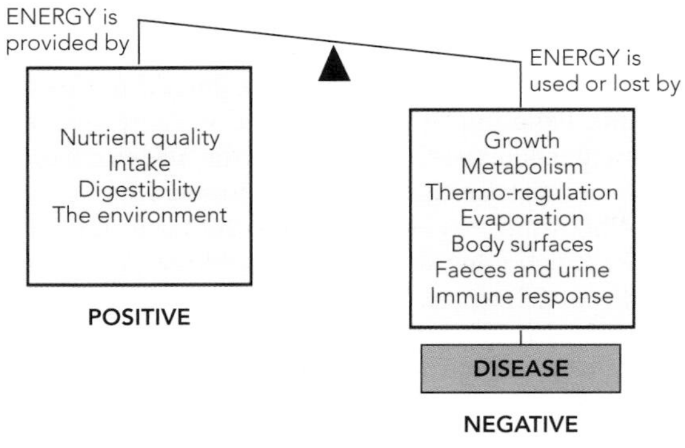

Fig 14-12

On most farms the best method is to offer very small quantities of fresh creep feed several times a day for the last 7 to 10 days before weaning and to continue this for 1 to 3 days after weaning while gradually changing to starter rations.

For 20-day weaning creep feed should be avoided or minimised. For 27-day weaning creep feeding has to be exceptional to achieve expected growth rates.

Nutritional components of a good creep diet

Whilst it is not the purpose of this book to discuss nutrition in detail, it is important to understand the effects on growth rate (daily liveweight gain – lwg) of a simple diet compared to a complex one (Fig 14-13).

A complex diet could consist of the following: Cooked cereals 38 %, maize oil 11 %, milk products 45 %, glucose and sugars 5 % plus minerals and vitamins, providing MJ NE/kg 11 (MJ DE/kg 16.4) energy, protein levels of 21 to 23 %, lysine at 1.3 to 1.4 % and 20 % oil content.

POSTWEANING GROWTH

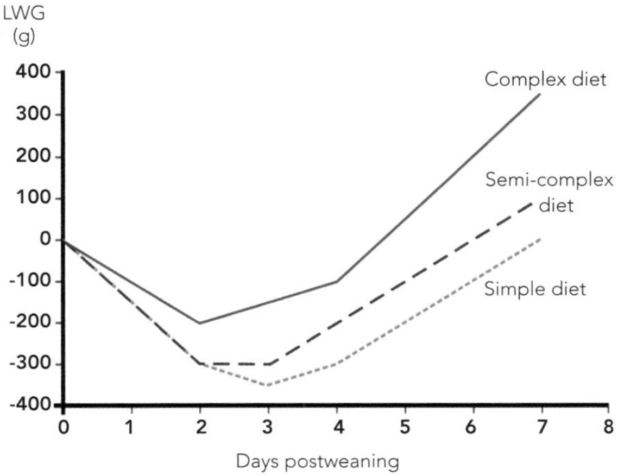

Complex diet – Cooked cereals and skim milk.
Semi complex and simple diets – Wheat, barley, soya.

Fig 14-13

Fractures

Bone fractures are not uncommon in growing pigs and are usually the end result of trauma and fighting, although spontaneous ones occur in bone disease such as osteomalacia, associated with calcium phosphorus and vitamins A and D, and osteochondrosis (OCD). Gilts may fracture their legs immediately postweaning due to lowered bone calcium levels as a result of excessive milk production during lactation. Bone fractures may also occur in gilts when naturally mated with a large, heavy boar.

Clinical signs

The onset is invariably sudden, the animal being unable to rise on its own without difficulty. A significant feature is the reluctance to place any weight on the affected leg. The muscles and tissues over the fracture site are often swollen and painful and the pig is very reluctant to move unless on 3 legs. An examination is best carried out when the pig is lying down. Crepitus or the rubbing together of the 2 broken ends of the bone can often be felt. Fractures of the spinal vertebra are common in the first litter female particularly during lactation and in the immediate postweaning period. The pig usually adopts a dog sitting position and exhibits severe pain on movement. Such animals should be destroyed.

Diagnosis

This is based upon the history, symptoms and palpation to detect crepitus.

Similar diseases

These include acute laminitis, arthritis, muscle tearing, bush foot and mycoplasma arthritis.

Treatment

- Prescribe pain relief to alleviate the suffering.
- The affected animal should be euthanised on the farm as soon as possible.

Management control and prevention

- If fractures are a recurring problem it is necessary to check that there are no diseases such as osteomalacia, osteoporosis or leg weakness (OCD).
- Check the calcium phosphorus and vitamin D levels in the feed.
- Check management procedures during the affected period.

Gastric Ulcers

Erosion and ulceration of the lining of the stomach is a common condition in all pigs. It occurs around the area where the oesophagus enters the stomach (called the *pars oesophagea*). In the early stages of the disease the pars becomes roughened and gradually changes as the surface becomes eroded until it is actively ulcerated. Intermittent haemorrhage may then take place leading to anaemia, or massive haemorrhage may occur resulting in death. The incidence in sows is usually less than 5 % but in growing pigs up to 60 % may show lesions at slaughter.

The causes of gastric ulceration are simple but complicated by multifactorial issues.

These be categorised as follows:

- Nutritional factors.
- Physical properties of the feed.
- Managerial related.
- No food for more that 24 hours.
- Infectious in origin.
- Miscellaneous factors.

The key element of gastric ulceration is a result of the pig not eating. However, once the ulcer starts to form, healing is prevented by the following contributory factors:

Nutritional factors:

- Low protein diets.
- Low fibre diets. (The introduction of straw reduces the incidence.)
- High energy diets.
- High levels of wheat in excess of 55 %.
- Deficiencies of vitamin E or selenium.
- Diets containing high levels of iron, copper or calcium.
- Diets low in zinc.
- Diets with high levels of unsaturated fats.
- Diets based on whey and skimmed milk.

Physical aspects of the diet that increase the incidence:

The size of feed particle is critical. The more the meal is ground the smaller the particle size and the higher the incidence of ulcers. Particle size should not be below 500 μm. Pelleting to feed does not mitigate against too small a particle size, and in fact the pelleting process itself can increase the incidence of ulcers.

- Where there is a problem on the farm, have the feed examined to assess the varying percentages of particle sizes. This is carried out by sifting the meal through a series of 12 to 14 tiller screens and weighing the residual amounts remaining in each screen. Particle size is also affected by the type and moisture content of the cereals that are being used, the condition of the hammer mills and the screen and the rate of flow through the grinding system. The smaller the particle size, the greater the incidence.

- Sometimes there can be problems in changing from pellets to meal and a compromise is to feed alternately.
- If the feed is home-produced and is a meal, then it is necessary to check the size and quality of the screen that is being used.
- Using cereals with a high moisture content.
- Rolling cereals as distinct from grinding them will often produce a dramatic drop in the incidence but the penalties of feed use have to be taken into consideration.

If ulcers are a problem, consider changing from pellets to coarse ground meal.

Management factors that increase the incidence:

- Irregular feeding patterns and shortage of feeder space.
- Periods of starvation.
- Increased stocking densities and movement of pigs. Look carefully at the environment in the pens and in finishing houses. Are there undue stresses or aggressions?
- Poor management of sow houses.
- Transportation.
- Poor availability of food or water.

Miscellaneous factors:

- Stress associated with fluctuating environmental temperatures.
- Adverse environmental conditions that create an unhappy environment.
- Psychological stress resulting from bad or harsh stockmanship.
- The condition is more common in castrates and boars than in gilts but the reason for this is unknown. Difference in feeding patterns may be a factor.
- Split sexing may help.
- Breed. More common in certain genotypes particularly those that have low back fat measurements and a capacity for rapid lean tissue growth.

Infectious causes:

- There is a clear relationship between outbreaks of pneumonia and the incidence of gastric ulceration.
- Ulceration may occur following bacterial septicaemias such as those associated with erysipelas and swine fever.
- In the breeding sow gastric ulceration is usually confined to the individual animal and is often secondary to a specific disease.

If ulcers are a problem try increasing the screen size to 3.5 mm.

Clinical signs

These depend on the severity of the condition. In its most acute form healthy animals are found dead.

The most striking sign in these cases is the paleness of the carcase due to internal haemorrhage.

In the less acute form the affected pig is pale, and weak, and may show breathlessness, grinding of the teeth due to stomach pain and vomiting. The passing of dark faeces containing digested blood is often a persistent symptom. Usually the temperature is normal. When the condition becomes chronic the pig has an intermittent appetite and may lose weight. The faeces vary from normal to dark coloured depending on the presence or absence of blood. Feed intake, feed efficiency and daily gain can be affected. Nutritional deficiencies particularly of minerals and vitamins can increase the incidence of gastric ulceration.

Diagnosis

Ulceration should always be considered in sows or pigs which are pale lose body condition and develop a variable appetite particularly if the faeces are black and tarry.

A sample of faeces should be examined for the presence of blood and to eliminate parasites. Although the disease is usually confined to individuals or less than 5 % of sows, occasionally it can become a herd problem. In such cases poor body condition is widespread through the breeding herd.

Whenever black tarry faeces are seen gastric ulceration should be suspected. In the growing herd an examination of stomachs at slaughter should be carried out. Be aware that ulcers can develop quickly and may appear in the lairage immediately prior to slaughter.

Similar diseases

Haemorrhage from the bowel can also arise from the intestine in cases of ileitis but usually this is confined to young gilts and growing pigs.

Anaemia in pigs can also be associated with *Mycoplasma suis*, the stomach worm *Hyostrongylus rubidus*, chronic mange and ileitis.

Treatment

- Move the affected animal from its existing housing into a loose bedded peaceful environment.
- Feed a weaner type diet containing highly digestible materials.
- Inject multi vitamins and in particular vitamin E together with 0.5 to 1 g of iron intramuscularly and repeat on a weekly basis.
- Add an extra 100 g vitamin E/tonne to the diet for 2 months and assess the results.
- Cull affected pigs.

Management control and prevention

- Consider the above factors and their relevance to your situation. Make alterations and adjustments accordingly.

Lameness

There are numerous causes of lameness in the pig. It is the second most common cause of sows being culled with most cases occurring from weaning through to the point of farrowing (see Chapter 7). Lameness is also prevalent in growing pigs with levels ranging from 1 to 5 % (see Chapter 9).

The causes of lameness can be both infectious and non-infectious, with nutrition playing a role. This section reviews the nutritional aspects associated with lameness in the pig.

Calcium and Phosphorus

With modern dietary formulations actual deficiencies arising due to defective diet would be unusual. Problems however occur due to faulty storage, the incorrect application of the feed or interactions that reduce the availability to the pig. The latter can result from intestinal disease, metabolic failures or adverse interactions between nutrition, the pig, management and the environment.

Calcium contributes to a wide variety of functions in the body including blood clotting, muscle and nervous activity, hormone production and milk production, to name but a few. Its major role however, together with phosphorus, is in the formation of bone where on a dry matter basis it forms approximately 38 % of the structure and phosphorus 20 %.

Bone is a very strong and dynamic structure with minerals constantly being removed and replaced. The intestines control the rates of absorption both into the body and skeleton and these are necessary to maintain equilibrium between demand and excretion. The ratio of calcium to phosphorus in the diet is also an important factor in this equation and this should not rise above 2:1. When it does, the absorption of calcium may be impaired; likewise if the ratio drops below 1:1. The ideal is approximately 1.25:1 to 1.50:1. Vitamin D_3 is also required in calcium metabolism together with controlling hormones produced by the parathyroid gland.

The level of phytase in the diet has an effect on the amount of phosphorus that is available to the pig. Phytase is an enzyme that helps release phosphorus trapped in grains and oil seeds enabling it to be readily digested by the pig.

Excess calcium in the diet may have a negative impact on the availability of other minerals.

Osteochondrosis or Leg Weakness

See Chapter 7 for further information

Degenerative changes in the joints and cartilage are generally described under the term leg weakness or osteochondrosis (OCD) resulting in lameness. These changes involve erosion of the articular cartilage and alterations to the normal patterns of growth in the growth plates at the ends of the long bones.

The use of both vitamins and minerals to try and prevent the condition have been singularly disappointing and it is doubtful if specific nutrient factors are involved. This condition is covered in Chapter 7.

Osteodystrophy

This is a general term to describe specific diseases that arise whenever there is a failure of bone structure and metabolism due to faulty nutrition. Such diseases include osteoporosis (OP), osteomalacia (OM), rickets, periostitis – describing disease of the periosteum – and osteomyelitis, disease of the centre or medullary cavity of the bone.

Osteomalacia (OM)

Osteomalacia (OM) and osteoporosis (OP) are becoming more common in commercial production systems, particularly in the first litter gilt where the skeleton is still growing and there are heavy demands on calcium for milk production. Both diseases have very similar signs and x-ray analysis is required to differentiate.

Osteomalacia (OM) is a condition responsible for the downer sow syndrome – where sows cannot get up. Fractures of the long bones at the mid shaft and fractures of the lumbar vertebrae are common, with the sow becoming paraplegic. The condition is due to inadequate levels of calcium, phosphorus and vitamin D in the ration. Sometimes sows cannot absorb sufficient micro-nutrients in spite of there being adequate levels in the diet. OM is also associated with immature skeletons, an imbalance of calcium and phosphorus and vitamin D and/or a failure of the sow to consume adequate feed and satisfy her nutritional requirements. Large amounts of calcium and phosphorus are excreted into milk from the bones resulting in weaker, less dense bone which predisposes to fractures. Bone mass is also lost due to lack of exercise during confinement in the farrowing place.

Clinical signs

The condition is common in first litter animals and up to 30 % of such animals may be affected. The history is one of sudden acute lameness often with the animal completely off its legs. The lameness is usually precipitated when the sow is moved from the farrowing place, during mixing or when the boar mounts at mating. Other symptoms include a stiff gait, difficulty in rising, discomfort in the hind legs and a dog sitting position.

Diagnosis

This is based on history, clinical signs and examinations.

Treatment

- In the early stages move the sow to well bedded loose-housing.
- If there are no fractures inject with calcium and vitamin D3.
- If there is a problem in first litter females inject them with vitamin D3 after farrowing and 7 days later.
- Supplement the diet with dicalcium bone phosphate, 30 g day of sterilised bone flour.
- If there are fractured bones affected, sows should be culled or destroyed.

Management control and prevention

Once OM has developed treatment is of minimal effect although injections of calcium phosphorus and vitamin D may help. If your herd has a problem consider the following:

- Feed a high dense diet in lactation 9.9 MJ NE/kg (14.5 MJ DE/kg) and18 % protein.
- Check the levels of calcium and phosphorus (minimum 0.9 % and 0.75 %). In first litter animals it may be necessary to raise the levels to 1.2 % and 1 %.
- Give up to 100,000 iu vitamin D3 by injection 10 days before farrowing. Inject pregnant animals with 50,000 iu vitamin D3 3 weeks prior to farrowing. Repeat again in the second week after farrowing.
- Top-dress the feed in lactation with calcium/phosphorus. Feed a good lactation diet during suckling and consider top-dressing the diet daily with 20 g of dicalcium bone phosphate.
- Bone ratios of calcium/phosphorus in affected sows are often 3:1 (normal < 2:1). Check the ratio of calcium:phosphorus in bone ash. The normal ratio is approximately 2:1 or less. In problem sows this is often 3:1 or more.
- Only mate gilts from 220 days onwards and if the disease is a persistent problem in a particular genotype, change the source.
- Use a good lactation diet and feed through to 28 days post-mating.
- Provide non slip floors in farrowing places. Check that floor surfaces are not slippery.
- Wean first litter females singly and use AI, or a light weight boar.
- Provide exercise to the pregnant gilt. Increase exercise during pregnancy if possible.
- Check parasite levels to ensure no dietary insufficiencies arise.
- The problem is less common in outdoor herds.
- Outbreaks are often more apparent in new gilt herds. Selection of animals for good conformation is essential.
- Investigate the growth rates and nutrition and feeding in the gilt.
- Maximise feed intake to appetite during lactation.
- Review the diet of replacement gilts between 60 and 120 kg bodyweight.

Osteoporosis (OP)

See Chapter 8 for further information

Osteoporosis (OP) and Osteomalacia (OM) are becoming more common in commercial production systems particularly in the first litter gilt where the skeleton is still growing and there are heavy demands on calcium for milk production. Both diseases have very similar signs and laboratory analysis is required to differentiate.

Bones affected with OP are quite normal in their structure but they become thinner particularly in the dense parts and shafts of the long bones. As a result they become more prone to fracture. OP can arise due to a shortage of calcium in the diet and imbalance of calcium and phosphorus, poor or inadequate absorption from the diet, heavy losses during lactation and where there is a lack of exercise.

Clinical signs

These are most common in the first litter female and occasionally after the second litter.

The onset may be gradual with the pig having difficulty rising and showing pain or sudden lameness associated with complete fracture of the long bones. Spinal fractures occur in some animals and they often remain in a dog sitting position. Most pigs are affected in late lactation or shortly after weaning, associated with the onset of oestrus and the trauma that results from other animals, or the weight of the boar at service.

Diagnosis

This is based on clinical signs, a history in lactating and newly weaned sows and evidence of fractures of the long bones. If the herd has a problem it is necessary to examine the bones of an affected animal by x-ray to differentiate between OP and OM.

Similar diseases

These include:

- Leg weakness or osteochondrosis.
- Spinal fractures.
- Torn muscles at their insertions into the bones.
- Mycoplasma arthritis – *Mycoplasma hyosynoviae* infection.
- Osteomalacia (OM).

Treatment

- In the early stages move the sow to well bedded loose-housing.
- If there are no fractures inject with calcium and vitamin D3.
- If there is a problem in first litter females inject them with vitamin D3 after farrowing and 7 days later.
- Review the diet of replacement gilts of 60 to 120 kg bodyweight.
- Supplement the diet with dicalcium bone phosphate, 30 g day of sterilised bone flour.
- In cases of bone fracture the sow is best destroyed on humane grounds.

Management control and prevention

- Outbreaks are often more apparent in new gilt herds. Selection of animals for good conformation is essential.
- Investigate the growth rates, nutrition and feeding in the gilt.
- Increase exercise during pregnancy if possible.
- Check the levels of calcium and phosphorus in the diet. They should be 10 to 12 g/kg of calcium and 8 to 10 g/kg of phosphorus.
- Only mate gilts from 220 days onwards and if the disease is a persistent problem in a particular genotype, change the source.
- Check that floor surfaces are not slippery.
- The problem is less common in outdoor herds.
- Feed a good lactation diet during suckling and consider top dressing the diet daily with 20 g of dicalcium bone phosphate.
- Inject pregnant animals with 50,000 iu vitamin D3 3 weeks prior to farrowing. Repeat again in the second week after farrowing.
- Maximise feed intake to appetite during lactation.
- Check the ratio of calcium:phosphorus in bone ash. The normal ratio is approximately 2:1 or less. In problem sows this is often 3:1 or more.

Rickets

This arises in a similar way to OM except it occurs in young growing animals, again as a failure of mineralisation of bone and growth plate cartilage. Phosphorus and vitamin D deficiencies are the common cause, but the condition today is rare.

Where it occurs, young animals are often housed in dark surroundings and fed starch feed waste with no mineral vitamin supplements. It usually takes 6 to 8 weeks before symptoms become evident, by which time the disease has progressed to become irreversible.

The symptoms are similar to OM but because the growth plate and cartilage does not develop to bone the joints swell with stiffness and pain is evident. Bone fractures are also common. In the few cases treated the response to injections of vitamin D3 has been very poor with pigs being totally uneconomical.

Vitamin A

The classical descriptions of vitamin A deficiency are described in Fig 14-8. However, most (if not all) rations are well fortified with the vitamin, indeed in many cases to excess, and problems associated with overconsumption are more likely to be experienced, but may not well recognised.

Two problems arise in the field. The first is where high levels of vitamin A (up to 15,000 to 18,000 iu/kg) are fed. This has been shown to lower the vitamin E status of the pig and therefore make it potentially more susceptible to mulberry heart disease. This depression of vitamin E may also reduce antibody production and thereby increase susceptibility to disease. This scenario has coincided with outbreaks of respiratory disease in the field, and lowering the vitamin A levels to around 8,000 iu/kg and raising the vitamin E by 50 to 100 iu/kg had been undertaken with improvements.

Piglets born from sows fed high levels of vitamin A may produce piglets with a low vitamin E status and this can be a fruitful line of investigation where iron dextran problems persist in sucking pigs.

The second problem arises when excessively high levels of vitamin A – 25,000 to 30,000 iu/kg are fed. At these levels the growth plates of the foetus become affected with classical signs of leg weakness and grossly shortened and bent bones in pigs as young as 3 weeks of age.

Fig 14-14 shows a typical affected pig from a herd where such high levels produced widespread changes in the fine-boned Landrace breed but were not so evident in the Pietrain. The condition disappeared 4 months after the levels in the sow rations were dropped to 12,000 iu/kg.

Further evidence for the effects of vitamin A on growth plates was also illustrated in a severe outbreak of leg weakness in weaners and growers fed rations

EXCESS OF VITAMIN A

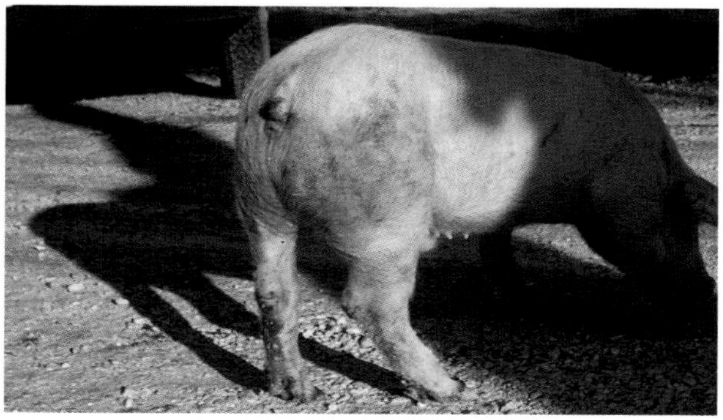

Vitamin A poisoning. Note the bending and shortening of bones (90 kg pigs).

Fig 14-14

containing high lysine 1.5 % and high vitamin A levels of 18,000 iu/kg. The pigs were housed on very smooth slats, the surfaces of which sloped to the edges. This resulted in the claws remaining in the gaps for long periods and by the time the pigs were 16 weeks of age the claws were completely crossed over due to a combination of pressure and weakened growth plates. When levels were reduced to 10,000 iu/kg and the lysine levels dropped to 1.1 % and the slat surfaces roughened, the problem gradually disappeared.

Prolapse of the Rectum

See Chapter 9 for further information

This is a widespread condition occurring in good growing pigs from 8 to 20 weeks of age. The onset is sudden. The size of the prolapse varies from 10 to 80mm and if small it will often revert to the rectum spontaneously. In most cases however the prolapse remains to the exterior and is often cannibalised by other pigs in the pen, as evident from blood on the noses of the offending pigs and on the flanks of others. The fundamental cause is an increase in abdominal pressure which forces the rectum to the exterior.

The following nutritional factors may be considered as causal or contributory when adopting control measures.

- Ad lib feeding – Feeding pigs to appetite results in continual heavy gut fill and indigestion. There is then a tendency for abnormal fermentation in the large bowel because undigested components of the feed arrive in greater amounts.
- High density diets and in particular lysine levels increase growth rates, and outbreaks may often subside either by a change to restricted feeding or using a lower energy/lysine diet.
- Wet feeding systems, especially if there is any aggression at feeding time.
- Water shortage – This can lead to constipation.
- Diets high in starch may predispose to prolapse – Try adding 2 to 4 % grass meal to the diet.
- The presence of mycotoxins in-feed – If there is a problem make sure that the bins have been well cleaned out. Examine the cereal sources.
- Change of diet – By studying the timing of the problem it is sometimes possible to identify rectal prolapses not only with a change of diet but also a change of housing.

This condition is covered in more detail in Chapter 9.

Reproduction and Nutrition

It is beyond the scope of this book to discuss nutrition in detail, but you should keep in mind that the quality of the feeds and the way in which they are fed are important in the management of not only disease but also reproduction. The boar, gilt, lactating sow and dry/gestating sow all have different nutritional requirements, each of which may vary with genotype. It is therefore difficult to be prescriptive about nutritional requirements other than give best practice guidelines, which this chapter aims to do.

There are a number of important factors that need to be considered to ensure the diet does not impact on the reproductive performance of the herd. These aspects are covered in other areas of the book as follows:

1. Nutrition and feeding the gilt and boar: See Chapter 5
2. Nutrition and feeding the gestation/dry period: See Chapter 7
3. Nutrition and feeding the lactating sow: See Chapter 8

Respiratory Diseases

See Chapter 9 for further information

These are a major problem on many farms today and they usually appear at predictable times that coincide with an increased demand for energy together with a change in the environment. Of all the factors highlighted in Fig 14-9 the following 3 will always be important if there is a disease problem:

1. An incorrect diet during and for 7 to 10 days after pig movement that creates a negative energy state.
2. A change in environmental temperatures associated with movement that creates a negative energy state.
3. Exposure to infection and the development of disease at the same time.

A typical example would be the development of respiratory disease caused by actinobacillus pleuropneumonia.

Pigs carrying these organisms will often develop acute pneumonia when they are moved from solid concrete floors onto slats. This increases the lower critical temperature for the pig and therefore more energy is required. A change in feed or system of feeding, together with a reduction in stocking density, further exaggerates the problem. The immune system does not respond efficiently, the metabolism of the pig alters and acute disease results. Energy cannot be considered alone to the exclusion of other nutrients and the following other factors need to be considered:

- Identify the time when disease first becomes evident.
- Note when nutritional changes take place. When pigs move from one house to another there is often a drop in the nutrient density of the diet. This also coincides with a drop in intake for the first 2 to 7 days.
- This reduced feed intake and change can result in catabolism and the pig dropping below the LCT.
- Maintain high energy diets at critical times. Do not change the feed for at least 5 days when pigs move from one house to another.
- Provide adequate water at all times. Look at the type of nipple drinkers, their availability and accessibility to the pig, particularly when it moves from one house to another. Monitor the flow and use of water. Record daily water intake.
- Look at the methods of feeding. Are there any features of design or feeder placement that inhibit access to the feed?
- Is there sufficient hopper space?
- Make sure there are adequate levels of vitamin E in the diet. Add an extra 50 to 100 iu to the tonne if there is a disease problem.

Torsion of the Stomach and Intestines (Twisted Gut)

This is one of the most common causes of sudden death in the growing pig and it is usually one of the best pigs in the group. It can also affect other pigs.

The design/layout of the pig's gut makes it particularly susceptible to becoming twisted. The intestinal tract in the pig is suspended from a common point by a membrane call the mesentery. This arrangement is unstable and makes the intestines liable to rotate and finally twist. A build-up of excessive gas in the gut is usually a precondition. Sudden movement can then cause the gut to twist. The gas build-up can be the result of overeating, fermenting feed and high density diets. Where pigs have colitis they can be more prone to gas production. The inclusion of whey in feeding rations may lead to bloat and torsions.

In older animals, the condition can manifest a gastric dilation and torsion, whereby the stomach becomes over inflated with gas, causing severe bloating. Signs are similar.

Clinical signs

There are usually none because the pig is found dead. The carcass is fresh, the pig is very pale and the abdomen is very distended. If a pig is seen early on in the condition it will be in severe pain but once the gut tissue dies the pain recedes and the pig will die peacefully.

Diagnosis

A post-mortem examination is necessary to confirm diagnosis. This shows the small and large intestines heavily congested and full of blood. The gut will be twisted, sometimes over 360 degrees.

Management control and prevention

- Deaths are usually sporadic although they can be of significance where, for example, whey is being fed and bloat occurs.
- Over-feeding and abnormal fermentation of the contents of both the small and large intestine result in gas formation, increased pressure and a twist.
- Mortality in weaned and growing pigs should normally be less than 5 % and up to a third of this may be caused by torsion. If it reaches 1 % or more the dietary components should be examined closely to see if there are starch based ingredients that might cause excessive fermentation. If torsions are consistent causes of death collect information about each one, including weight, age, sex, house, stocking density, environmental temperatures and feed changes. A study of this may give guidance as to contributing causes.
- Increasing fibre intake (barley straw or bran) seems to reduce the risk of excessive fermentation and gas production in the sow around farrowing.
- More regular feeding (2 to 3 times a day) can also reduce the risk of fermentation.
- Watch out for changes in feeding times, particularly at weekends and holidays. A consistent feeding routine is essential.

Abdominal Catastrophe

This is a common cause of sudden death in growing and adult pigs. Pigs are simply found dead and bloated. Typically the condition is characterised by a torsion/twist of one or more of the non-intestinal abdominal contents. A twist in the abdomen is most commonly associated with the intestinal tract at the mesentery root or associated with a mesenteric tear. This is known as torsion of the intestines or twisted gut, as discussed above.

However, occasionally the spleen, a liver lobe or lung lobe can twist (this may or may not include the various elements of the intestines) causing an "abdominal catastrophe".

Udder Oedema and Failure of Milk Let Down

This presents itself as a failure of milk let down associated with excess fluid in the mammary tissues and is a condition seen in both gilts and sows. It is characterised by a clinically normal animal with no fever or loss of appetite. The distinguishing features are a firmness of all the glands, discomfort on high pressure but no actual pain. The oedema or fluid can be both in the skin and deep in the udder tissue. The pressure produced in the glands once farrowing has ceased prevents a good milk flow and there is a reduction in both the quantity and quality of the colostrum which means a lowered immune status of the piglet. Severe oedema, particularly in the rear glands, may result in poor accessibility of teats at sucking time. Such glands often dry off. When piglets eventually find the teat they will not thrive but waste away.

Clinical signs

Usually there is a history on the farm of poor milking amongst all ages and 1 or 2 pigs per litter having to be fostered at around 5 to 7 days of age due to poor growth. Scouring problems can sometimes be related back to udder oedema and a poor intake of colostrum. Palpitation of the udder shows fluid either just beneath the skin or deep in the gland and often extending between the legs towards the vulva. The vulva is also often involved.

Diagnosis

This is based on the demonstration of oedema of the udder, by appearance and palpation and the appearance of the litter. Oedema and congestion can lead to mastitis.

Treatment

- Recognise the condition early and medicate.
- Treat the sow with small doses of oxytocin (5 to 10 iu) every 4 to 6 hours on 4 occasions. If the intra vulval route is used oxytocin doses can be halved.
- Supplement the piglets with rice and artificial milk and make water available in dishes.
- Give a preventative injection of long-acting antibiotic, either penicillin, amoxycillin or OTC if there are any signs of mastitis.

Management control and prevention

Whilst udder oedema usually occurs in individual animals it can become a problem at a herd level. Some genetic lines are potentially susceptible. If this is the case the following actions should be considered:

- Look at feed levels and the development of the udder 7 to 10 days pre-farrowing. Excessive tissue growth can be associated with high feed intake, particularly high energy levels. Feed can be dropped to 1.5 kg a day of a standard dry/gestation feed.
- Maintain sows on the same bulk level of feed pre-farrowing and from the time of entering the farrowing house to 2 days post-farrowing.
- Use the same ration pre-farrowing until 2 days post-farrowing and then change to a lactation ration.
- Low water intake 2 to 3 days before farrowing can also predispose. Add water to dry feed.
- Constipation can be a predisposing factor. Bacterial toxins become absorbed from the gut and interfere with the circulation in the udder tissue. In some cases a response will be obtained by feeding increased levels of fibre before farrowing to increase bulk and reduce constipation.
- Alternatively the levels of feed can be reduced but it is important to increase the fibre content with bran or another available fibre source. A typical example would be 2 kg of breeding diet with 0.5 to 0.75 kg of bran. This is better fed wet in the trough to improve palatability.
- A change from straw yards to farrowing houses is associated with a marked reduction in fibre intake. In such cases give the sow straw for the first 3 to 4 days pre-farrowing.

Water Deprivation (Salt Poisoning)

Water deprivation is unfortunately common in all ages of pigs and almost without exception is related to water shortage either caused by inadequate supplies or complete loss. The exception would be where the salt concentration in the water exceeds 3000 ppm.

Where the pig fails to ingest sufficient water it becomes dehydrated and the normal levels of salt in the diet (0.4 to 0.6 %) become concentrated and toxic as a result of osmotic changes in the brain. Whilst most animals are affected, the pig seems to be particularly susceptible to this condition.

Where pigs are dehydrated, often symptoms may not be seen until the water supply is restored. The high concentration of salt in the brain draws in water, causing it to swell and resulting in a pressure build-up on the brain which causes a number of the clinical signs.

Clinical signs

The very early stages of disease are always preceded by inappetance and whenever a sow or groups of pigs are not eating always check the water supply first. The first signs are often pigs trying to drink from nipple drinkers unsuccessfully. Nervous changes are the major signs and in more advanced cases involve fits, with animals wandering around apparently blind. Often the pig walks up to a wall, stands and presses its head against it in a characteristic position. Pigs may also take up a dog sitting position with their head held upwards and backwards. One symptom strongly suggestive of salt poisoning is nose twitching just before a convulsion starts. Pigs may die no more than a few hours after showing clinical sings. Signs develop within 24 to 48 hours.

Diagnosis

This is based upon the clinical signs and lack of water. Examination of the brain histologically at post-mortem confirms the disease, with a characteristic infiltration of eosinophils.

Similar diseases

Aujeszky's disease (pseudorabies), classical and African swine fever, streptococcal meningitis, oedema disease and Glässer's disease all produce nervous signs. The condition might also be confused with middle ear infection but this only affects one individual rather than a group of pigs.

Treatment

- The response to treatment is poor but involves rehydrating the animal. At a practical level this can be achieved by dripping water into the mouth of the pig through a hose pipe or alternatively via a flutter valve into the rectum where it is absorbed (see chapter 15: Flutter Valve).
- Rehydration should be slow, as the high concentration of salt in the brain draws in water, causing it to swell. This results in a pressure build-up on the brain and ultimately death in many cases.
- Discuss the possibility of administering sterile water into the abdomen with your veterinarian.
- Corticosteroids may also help.

Management control and prevention

- It must be a daily routine to check that all sources of water are adequate, free flowing and available, especially in the farrowing house as onset can be much quicker.

Common Conditions Associated with Minerals and Vitamins

Biotin Deficiency

Biotin supports the normal function of the reproductive tract, nervous system and thyroid and adrenal glands.

It is often thought that because biotin is present in most nutrient sources used for pigs and is produced by organisms in the gut that a deficiency is unlikely. However field experience indicates that biotin-deficient conditions can and do occur in breeding herds worldwide.

Reports from studies and field observations have highlighted the following associations:

- Diarrhoea.
- Dermatitis.
- Excessive hair loss.
- Extended weaning to mating intervals.
- Haemorrhages on the solar surfaces of the feet.
- Lameness and laminitis.
- Poor litter size.
- Poor food conversion.
- Reduced growth rates.
- Transverse cracks in hooves.

Pig farmers experiencing poor reproductive performance in their sow herds, associated with excessive loss of hair and severe foot lesions, should evaluate the biotin content of their sow diet and consider supplemental biotin.

Clinical signs

Widespread lameness will be a constant feature of biotin deficiency, particularly in sows. Detailed examinations should be carried out on at least 15 to 20 affected animals and the nature of the changes in the hooves documented. Examinations are best made when sows and gilts are at rest. The hooves will be soft over the walls and the soles will show slight evidence of haemorrhage. Dark transverse cracks will be seen on the hoof walls. Assess trauma from poor floor surfaces as a cause.

Diagnosis

This is based on the clinical picture and the fact that the herd or a group of animals will be affected. The onset is usually gradual and this will distinguish the lameness from foot-and-mouth disease. Chronic lesions of swine vesicular disease could be confused with biotin deficiency. Levels in the ration can be determined but firm recommendations are not available. It would appear 100 to 200 mcg/kg is adequate.

Treatment

- Where a herd shows widespread lesions add up to 500 to 1000 mcg/kg to the diet. Any response will be slow; up to 9 months. Supplementing higher levels of 2000 to 3000 mcg/kg may reduce this timeframe.

Management control and prevention

- Add biotin to the diet as a routine.

Choline

Choline forms part of the chemical substances that enable electrical pulses to pass between nerve endings. It is produced from the amino acid methionine. Signs of deficiency include poor growth and reduced litter size. A deficiency would be unusual.

Copper

This, like iron, is necessary to allow red blood cells to form normally and deficiencies can lead to anaemia. Copper is also important in enzyme systems as well as acting to suppress bacterial growth.

Copper is often added to the diet at levels of up to 175 ppm for pigs to 16 weeks of age and 100 ppm for pigs over 16 weeks of age (EU legal requirements) to enhance growth rates. There have been occasions in the field where it has been inadvertently and suddenly removed from creep diets which has resulted in diarrhoea, bacterial enteritis and poor growth probably associated with the sudden multiplication of pathogenic bacteria. Fortunately, nutritional deficiencies in pigs are very rare.

Cyanocobalamin (Vitamin B12)

A deficiency of this vitamin is very unlikely because it is produced by bacteria in the gut and access to faeces provides a continual source. Synthetic sources are however usually added to diets.

Folic Acid

A deficiency in folic acid results in anaemia, poor weight gain and loss of hair colour. While the pig is capable of producing the vitamin, addition may help maintenance and improvements in litter size. The addition of 1 g/tonne may have a beneficial effect if litter sizes are poor and nutrition appears to be involved.

Iodine

Iodine is necessary for the production of the hormone thyroxin by the thyroid gland. This gland regulates the rate of body metabolism and if there is a shortage of iodine in the diet a condition called goitre arises. This denotes an increase in the size of the thyroid gland. Substances called glucosinolates found in winter-sown rape seeds are sometimes present in the diet and can prevent the gland using iodine.

Canola meal, an improved rape seed with low levels of glucosinolates, has enabled them to be used safely in pig diets. An iodine deficiency should be considered where large numbers of litters are born with piglets weak and hairless. With modern diets this would be rare. Iodised salt containing 0.008 % of iodine provides sufficient iodine in the diet.

Iron

See Chapter 14: Anaemia for further information

Iron forms part of the haemoglobin molecule that transports the oxygen around the body. Where there is an iron deficiency the oxygen carrying capacity of the blood is greatly reduced, causing anaemia. Anaemia due to a primary deficiency of iron is unusual in pigs over 6 weeks of age unless they have not been administered iron in the first 1 to 2 weeks of life and have had no access to a supplemented creep diet. It is usual to administer iron to the piglet by intramuscular or subcutaneous injections of 200 mg of iron dextran between 3 and 5 days of age. Consider iron deficiency in intractable postweaning diarrhoea and increased stillborn numbers. Monitor iron levels in feed and supplement all pig diets with iron where necessary.

Magnesium

This mineral is also found in bone but its main role is in the composition of many of the enzymes in the body. Most feed ingredients contain sufficient magnesium and it is not normally necessary to supplement the diet. Reported signs of deficiency include an abnormal gait as a result of lack of enzymes in the nervous system, incoordination and weak knees and hock joints with loss of tendon tension. If an excess of calcined magnesite is added to supplement rations, diarrhoea may result.

Manganese

This is important in enzyme production and the development of bone. It is also required for normal reproductive function. Deficiencies are reported to be associated with lameness, irregular oestrus, delayed sexual maturity and weak pigs at birth.

The levels required in the diet are small at around 4 g/tonne and problems likely to be experienced in the field would be rare.

Pantothenic Acid (Vitamin B5)

This vitamin is an important enzyme in energy metabolism and one which swine diets are deficient. A crystalline salt called d-calcium pantothenate is included in vitamin premixes.

Deficiencies are associated with a high stepping gait particularly of the hind legs which is described as goose-stepping, with increasing incoordination and finally posterior paralysis. Loss of hair and diarrhoea are also seen. The condition is occasionally seen in swill-fed units particularly where waste bakery foods are used. The feed can be supplemented with 5 to 10 g/tonne of calcium pantothenate. Where modern diets are fed, the condition is rare.

Potassium

Potassium is an important mineral because of its role in maintaining water and acid balance in the body. It is also involved in maintaining a normal heart rate and in the transport of materials in and out of the cells. Clinical signs of deficiency include emaciation, reduced appetite and incoordination. Excess potassium in the diet may precipitate colitis. A deficiency of potassium is uncommon because nutrient sources contain adequate amounts.

Riboflavin (Vitamin B_2)

Riboflavin is a water-soluble vitamin that is involved with carbohydrate, protein, and fat metabolism. Most diets are supplemented with this vitamin because the level in basic cereal ingredients is very low. Whey, however, is a good source. A deficiency would be rare but anoestrus, reproductive failure, hair loss and vomiting have been reported.

Sodium and Chloride

Although these are 2 individual components in the diet it is normal practice to fulfil the need with common salt. A high level of sodium particularly in water causes scouring and poor growth rates. A low level of chloride in the ration can depress the growth rate and feed intake. Normal levels in the diet would be between 0.4 to 0.6 % of salt, although provided there is ad lib drinking water, levels of up to 2 % or more are well tolerated by the pig and are often used at these levels to reduce tail biting and vice.

Thiamine (Vitamin B_1)

Most cereals are good sources of thiamine and it is not normal to supplement the diet but if it is considered necessary levels of 1 to 2 mg/kg is required. The signs of deficiency are reported to include poor appetite and daily gain, often vomiting, low body temperatures and poor heart rates. Sudden death may also occur.

Vitamin A

See Lameness earlier in this chapter.

Vitamin B_3 (Niacin/Nicotinic Acid)

This vitamin is involved in enzyme systems associated with metabolism of carbohydrates, fats, and protein. A deficiency results in reduced weight gains, poor appetite, often a very dry skin and diarrhoea. The skin may turn a yellow colour with dermatitis and loss of hair. Posterior paralysis may also occur. At post-mortem examination necrotic lesions may be seen in the intestines.

The amino acid tryptophan acts as a precursor and if this amino acid is deficient there is a greater risk of niacin deficiency. If deficiencies are suspected nicotinic acid can be added at 15 mg/kg. Alternatively the levels of tryptophan could be increased. Deficiencies in the feed are rare.

Vitamin E/Selenium (Mulberry Heart Disease)

Problems associated with either the lack of availability of vitamin E and selenium, or absolute deficiencies in either can cause major problems on some farms. These can arise due to high levels of polyunsaturated fats in diets, which are used as sources of energy. High levels of copper, vitamin A or mycotoxins in the feed can have a similar effect.

The active principles of vitamin E are called tocopherols and they are widespread in feedstuffs including vegetable oils, cereals and green plants. Tocopherols are used in pig rations as dl-alpha-tocopherol acetate, measured in international units. The international unit (iu)1 iu of vitamin E is defined as 1 mg of a standard preparation of a specific tocopherol acetate.

Vitamin E is necessary for the optimum function and metabolism of the nervous, muscular, circulatory and immune systems, and the latter highlights its importance in maintaining the health of the pig.

Its function is to prevent the breakdown of oxygen at a cellular level (oxidation) when toxic products including hydrogen peroxide and hydroxyl radicals are produced. These oxidising agents are powerful tissue poisons.

The function of vitamin E in the pig

- To increase the efficiency of the immune system. Adequate levels must be available at critical times particularly as maternal antibody is dropping and pigs are being challenged by infectious agents. This highlights the importance of both diet quality and levels of energy lysine and vitamin E at these times.

- It acts as a tissue antioxidant. Heart muscle is particularly sensitive to oxidising agents.
- It helps to maintain the integral structure of muscles in the digestive and reproductive systems.
- It is involved in the synthesis of certain amino acids and vitamin C.
- It has a close relationship with selenium metabolism.

Selenium is an essential nutrient in its own right and part of an enzyme called glutathione peroxidase which also acts as an antioxidant and thus has a complementary role to vitamin E. The less selenium in the diet the greater is the requirement for vitamin E.

The recommended requirements to give a maximum boost to the immune system range from 75 to 220 iu/kg. According to age of the pig and diet; this is in the first stage creep 220, the second stage 150, the grower 100, the finisher 60 and sow 50 iu/kg. These levels are probably higher than those necessary for maximum growth, which may be 50 % less.

Polyunsaturated fatty acids (PUFAs) cause considerable oxidation at tissue levels and when added to diets 3 iu of vitamin E should be added for each g of PUFA.

Moulds will destroy vitamin E in-feed.

Vitamin E and selenium-related diseases

Gastric ulcers – These are often stress oriented and the incidence increases where vitamin E levels are low.

Hepatosis dietetica (HD) – A condition where there is necrosis or death of liver cells.

Muscular or nutritional dystrophy (MD) (also called a myopathy) – This results from a degeneration of muscle fibres whether they be skeletal, smooth or cardiac. Oedema or fluid is often produced around the tissues and muscles (PSE) as a result.

Mulberry heart disease (MHD) (also called a myopathy) – A specific disease of heart muscle and a common cause of sudden death.

Reproduction disorders – Vitamin E is involved in sperm production and ovarian function. The actual role of vitamin E on the farm is difficult to clarify.

Clinical signs

These vary according to the system affected. HD, MD and MHD are usually associated with sudden deaths in rapid growing pigs without any prior clinical signs; usually the best pigs in the pen are affected and they range from 15 to 30 kg (33 to 66 lbs) in weight. Diets being fed often contain high levels of fats and yet in many cases vitamin E levels appear within normal ranges. Post-mortem symptoms are characteristic and include:

- Large amounts of fluid around the heart and lungs.
- Haemorrhagic and pale areas in heart muscle.
- Fluid in the abdomen with pieces of fibrin.
- Pale muscle areas (necrosis) particularly in the lumbar muscles and hind muscles of the leg which contain excesses amounts of fluid.
- If the liver is involved it is enlarged and mottled with areas of haemorrhage interspersed with pale areas. With liver rupture the abdomen may be filled with blood.

Diagnosis

Histological examinations of the liver, heart or skeletal muscle will confirm diagnosis and this is the most accurate method. Serum samples should be taken from pigs at risk and tested for levels of vitamin E. Normal levels are variable from pig to pig; however they should be more than 1.8 mg/litre. The availability of selenium can be assessed by measuring the levels of glutathione peroxidase in the serum. If levels are less than 0.025 μg/ml in the serum or 0.1 mg/kg in the liver a deficiency should be suspected and rations checked.

If MD is the major change stiffness and muscle trembling may be seen. If back muscle necrosis is involved sudden acute lameness occurs, particularly in gilts, especially outdoors, when they are moved into paddocks for the first time. This sudden exercise precipitates disease in association with the porcine stress syndrome (PSS). Stress-related problems include gastric ulcers and where lesions occur in more than 20 % of pigs at slaughter, the addition of 50 iu/kg of vitamin E should be assessed.

The role of vitamin E and selenium in reproductive performance is more difficult to quantify. Improvements have been noted in herds with persistent cases of agalactia and udder oedema by raising the levels in the lactating diet to 100 iu/kg.

Similar diseases

These include:

- Actinobacillus pleuropneumonia.
- Glässer's disease.
- Oedema disease.
- Streptococcal septicaemias.

Specific diseases associated with deficiencies, or lack of availability include:

- Actinobacillus pleuropneumonia.
- *E. coli* diarrhoea.
- Oedema disease.
- PRRSV.
- Swine dysentery.
- Those diseases that occur during periods of immunosuppression.

Treatment

- Where a population is at risk inject all the pigs with vitamin E/selenium. A rate of 70 iu vitamin E and 1.5 mg selenium per 50 kg is adequate. Also seek veterinary advice.
- Where there are sudden deaths in piglets following iron injections, inject sows 14 days prior to farrowing with vitamin E/selenium.
- Water soluble preparations are sometimes available as alternatives.
- Multi-vitamins that include vitamin E and/or selenium may be used. Refer to the recommended treatment levels on the bottle label.
- Move individual sick pigs to hospital pens for treatment.
- Increase vitamin E levels in creep and growing rations to around 200 iu/kg.

Management control and prevention

- If problems persist change to another diet with less added fats.
- Check the levels of PUFAs in the diet.
- Check the levels of vitamin E and selenium in the diet.
- Check the levels of vitamin A. If more than 10,000 iu/kg this may be increasing the requirement for vitamin E.
- Check for high copper levels – do pigs have access to copper pipes etc?
- Reduce stocking densities if pigs are overcrowded.
- Check there are no parasite burdens.
- Review grain/feed storage. Grains/feed stored with high moisture content in high temperatures may have fungal growth and thus low levels of vitamin E.
- Do not breed from animals that carry the stress gene.
- Rapid growth may be a contributing factor. In newly established herds the pigs may grow extremely quickly and have an increased requirement for vitamin E which the diet is not satisfying.

Vitamin K

This vitamin is necessary to maintain normal blood clotting mechanisms and a deficiency causes haemorrhages throughout the tissues. Low levels of vitamin K have been implicated in navel bleeding in newborn piglets although the response to the addition of 2 g/tonne of vitamin K is usually disappointing. Sulphadimidine and warfarin act as vitamin K antagonists (inhibitors). Oral and injectable preparations of vitamin K are available as menadione bisulphate. Treatment can be given by intramuscular injection at 2.5 mg/kg liveweight (see Warfarin Poisoning).

Zinc

Most of the enzymes in the pig require zinc for their normal structure and function. It forms an essential part of insulin. Pigs deficient in zinc show poor growth, poor appetite and a skin thickening called parakeratosis. Excessive calcium in the diet also reduces the availability of zinc and leads to parakeratosis but the condition in practice is uncommon. Generally, the diet should include between 50 to 100 g/tonne. In the period from weaning to 21 days postweaning the inclusion of zinc oxide BP (80 % zinc) at a level of 3.1 kg/tonne in creep diets has been found to have a profound effect on preventing diarrhoea associated with *E. coli*. Such an inclusion provides 2.5 g/kg of elemental zinc and at this level most pathogenic *E. coli* are inactivated. Occasionally the withdrawal of the zinc 2 to 3 weeks after weaning can result in diarrhoea and in such cases it may be necessary to continue its use for a further 2 to 3 weeks. No adverse effects at these levels have been recorded but a reduction in appetite has been suggested.

Where phytase is included in the diet the zinc levels can be reduced. Discuss with your nutritionist.

Non Nutritional Supplements

A variety of chemical and non-chemical compounds and organisms are added to diets to treat or prevent disease or to enhance growth.

They include:

- Antibiotics and antibacterial compounds (see Chapter 4, Fig 4-17 – Therapeutic medicines that may be available for in-feed use).
- Parasiticides (see Chapter 4).
- Growth/health enhancers (see Chapter 4).
- Probiotics and prebiotics.

The addition of such substances in most countries is strictly controlled by law and the actual medicines available and the dose levels allowed also vary.

Types of products that aim to enhance growth

There are a number of different products that claim to enhance growth in pigs (Fig 14-15).

FIG 14-15: GROWTH ENHANCERS

Feed Additives	Efficiency
Copper sulphate	+++
Enzymes	+++
Fermentation	+
Immunoglobulins	++
Lactose	++
Mineral clays	?
Organic acids	+
Probiotics	+
Probiotics (fermentation)	++
Zinc oxide	++++

++++ Most efficient

Acids

Acids act by either lowering the pH of the intestinal contents and/or by having an antibacterial effect.

The former include acids such as lactic, citric, fumaric and malic acids. Those having an antibacterial effect include acetic, propionic and formic acids. These can be benificial postweaning.

Antibiotics

The use of antibiotics for promoting growth is now widely banned, mainly due to concerns over antibiotic resistance. Where still allowed, approved products are becoming more limited and producers need to adapt their production methods to eliminate the use of these products.

Betaine

This is a chemical which can be added to the diet of older lactating sows to enhance feed intake. This can be extremely useful in tropical hot climates. Betaine may also assist the older sow to have larger litters when fed during gestation.

Enzymes

These are many and varied and found in the digestive system of all mammals. Pigs have difficulty in breaking down the complex cell structures, particularly in wheat. Enzymes assist in the breakdown of fibre and also enhance the efficiency of digestion of vegetable proteins, peas, beans and soya beans. Each enzyme is specific in its action.

The digestion of dietary fibre is enhanced by carbohydrates such as beta glucanase or xylanase which act upon fibres such as arabinoxylans and beta-glucans. Such fibres contain anti nutritional factors, that are also found in proteins. New enzymes have appeared on the market (proteases) to assist in the breakdown of such proteins and results have shown reductions in digestive disturbances, particularly in weaned pigs.

Fermentation

Liquid diets provide the opportunity to create acid conditions by fermentation through the inclusion of certain micro-organisms such as lactobacillus species prior to feeding. The establishment of the latter in the intestine, particularly in the weaned pig, helps to prevent the establishment of pathogenic bacteria. Fermentation also increases the digestibility of the diet with increased efficiency of use.

Metallic Substances

The use of copper as a growth promoter is well documented. Zinc oxide has proved to be very efficient in preventing postweaning diarrhoea at levels of 2500 ppm of elemental zinc.

Mineral Clays

These are substances that have the ability to absorb toxins from the digestive systems which may allow pathogenic bacteria to proliferate. Attapulgite is a typical example. Their value has not been established.

Nutraceuticals

These are substances originating from different plants and whilst many claims are made for their effectiveness they are yet to be substantiated with confidence. Fatty acids and plants with high levels of vitamin E claim to increase the efficiency of the immune system. Claims are also made for garlic, ginseng, oregano and extract of cinnamon, aniseed, rosemary, peppermint and propolis extracted from honey.

Pre- and Probiotics

These are live bacteria or micro-organisms that are either mixed in the feed or administered individually by mouth to produce a beneficial effect on the organisms in the gut.

Under normal conditions the intestines of the pig contain a complex of 400 or more organisms (the microflora) which constantly protect against disease. In addition, specific bacteria are able to protect against specific pathogens.

Probiotics are believed to act in the following ways:

- Neutralise toxins in the intestinal tract.
- Prevent the adhesion of pathogens to the mucosal surface by competition.
- Stimulate local immune defences.
- Reduce the numbers of pathogens by competition.

The common organisms used include lactobacilli, yeasts, streptococci, enterococci and non-pathogenic *E. coli*. Lactobacilli are the most common and at least 7have been identified as having a beneficial effect. They produce large amounts of lactic acid which creates an unfavourable environment for certain bacteria. In the piglet this creates high acid conditions in the stomach, helping to prevent the establishment of pathogenic *E. coli*.

Probiotics have been available for use in pig feeds for a number of years and yet the results of their use are inconsistent and in the field often remain unconvincing.

Oral preparations given to piglets and their response to *E. coli* diarrhoea can be difficult to assess on the farm because their use often starts with ongoing diarrhoea problems, most of which go away with better management and hygiene. A typical example was their use in a 1000 sow operation where 30 % of all litters scoured. This was associated with a continual pig flow policy in the farrowing houses. This was changed to all-in/all-out and at the same time probiotics were given to all piglets at birth. The problem gradually disappeared. Three months later due to costs, the probiotic use ceased. The scour problem did not return.

Probiotics appear to have little effect on piglet diarrhoea after 5 days of age.

Lactobacilli have been used for growth enhancement but consistency of results in the field is poor and it is difficult to see large numbers of well controlled trials that give sufficient confidence for their continual use. Field experiences in the immediate postweaning period have in a few herds shown a beneficial effect but in most an economic return could not be justified.

Ractopamine Hydrochloride

This is a Beta-adrenergic agonist (Beta-agonist) which is added primarily to the finishing diet. The benefit of ractopamine is to increase the amount of lean deposited by the finishing pig, which would normally be depositing fat. This substantially improves the rate of growth and the efficiency of lean growth. Pigs on ractopamine need to have a diet enhanced with lysine (1%) in order to allow the extra lean to be manufactured. Use of the product is banned in many countries.

532 Managing Pig Health

15 SURGICAL, MANIPULATIVE AND PRACTICAL PROCEDURES

15 SURGICAL, MANIPULATIVE AND PRACTICAL PROCEDURES

Introduction

This chapter aims to provide a basic understanding of the various surgical and practical procedures regularly carried out on the pig farm.

However the regulations which govern those procedures that can be carried out by the pig farmer and those that can only be carried out by the veterinarian differ from country to country. Suggestions are made where veterinary help and advice should be sought but you should ascertain what you are allowed or not allowed to do in your country.

Some of the following procedures may require initial instruction from your veterinarian. If you are uncomfortable carrying out any of these procedures, consult your veterinarian.

Basic Blood Sampling Methods

Materials Required

- Syringes: 2 ml, 10 ml, 30 ml.
- Needles: 20 g 16 mm (5/8") and 25 mm (1"), 18 g 38 mm (1½"), 16 g 100 mm (4").
- Vacutainers: 7 to 10 ml (Fig 15-1).
- Surgical spirit.
- Cotton wool.
- Labels.
- Wire/rope noose.
- Ear protectors.
- Two people.

Use Fig 15-1 to select the method of sampling that best fits your needs.

FIG 15-1: SELECT THE METHOD OF SAMPLING

Site	Method	Age of Pig	Comments
Ear Vein	Collect using a 20 g needle and syringe.	Sow Gilt Boar	Only small samples. Sample contaminated. Haemorrhage sometimes follows, apply pressure.
Jugular Vein	Syringe or vacutainer. Vaccutainer.	All ages Weaners	Method of choice for 10 – 30 ml. Best option for weaners.
Anterior Vena Cava	Vacutainer or syringe.	Piglets to adults	Good choice for large samples.

Ear Vein

The veins are raised by pressure at the base of the ear. The skin is cleaned with cotton wool and surgical spirit. Use a syringe and 20 g 16 mm (5/8") needle. Store the blood in a glass vacutainer tube (Fig 15-2).

Jugular Vein

This is the method of choice in pigs from 30 to 200 kg (65 lbs to 440 lbs).

The position of the blood vessels is shown in Fig 15-3 (transverse section of the neck). Use ear protectors for this procedure.

Restrain the pig with a wire noose or rope (see procedure for restraining the pig). The external jugular vein lies 25 to 40 mm below the skin. Raise the head of the pig to define the jugular groove and direct the needle slightly to the mid line 120 mm above the point of the shoulder. Sixty pigs per hour can be bled by this method. Once you have mastered the technique it is easy but you may need instruction from your veterinarian initially. The secret is to snare and hold the pig in the correct position to define the jugular groove (Fig 15-4). Keep the pig on its feet and straight. The head has to be held taut.

Anterior Vena Cava

This is used to bleed pigs from 2 to 30 kg weight and for adult pigs. The small pig is restrained on its back (see Fig 15-5) with the front legs held back and the chin pressed downwards. The right hand inlet of the pig between the first rib and breast bone is determined. A vacutainer or syringe can be used.

Needle sizes:

Pigs up to	10 kg	20 g 25 mm (1")
	45 kg	18 g 38 mm (1½")
	100 kg	18 g 50 mm (2")
	Sows	16 g 100 mm (4")

Older pigs and adult stock are bled in a standing position. The needle is inserted alongside the front of the breast bone directed slightly inwards towards the spine, upwards and at a slight angle backwards. Take care not to swing the needle, to prevent tearing of the blood vessel, because you might cause haemorrhage and even death.

Insert it straight in and slowly withdraw it in the same line with a light negative pressure on the syringe. If you have missed the blood vessel insert it again at a slightly different angle. You will need instruction from your veterinarian initially.

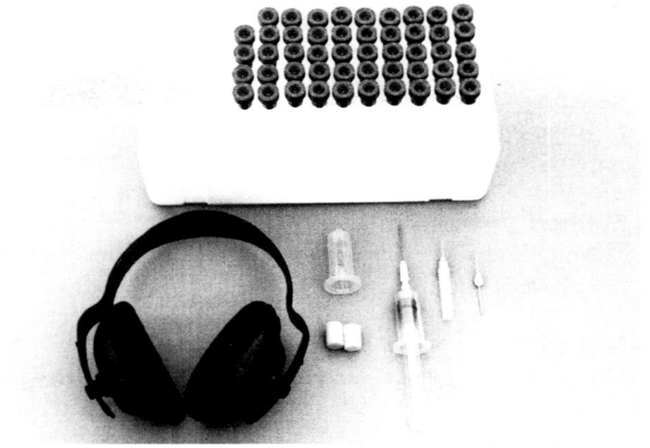

A blood sampling kit including vacutainers. The glass tubes are under vacuum and draw blood when pierced by the double ended needle. Ear protectors and plugs are also shown.

Fig 15-2

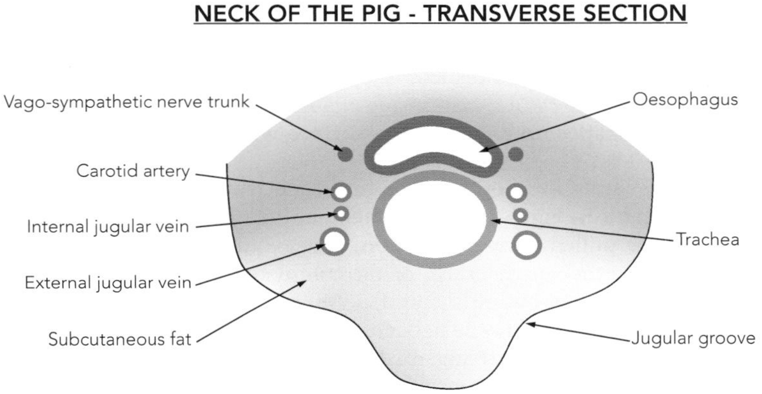

Fig 15-3

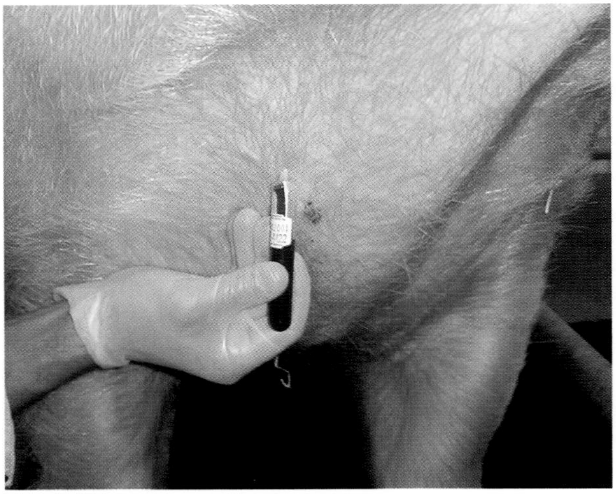

Pig snared for bleeding showing the jugular furrow and positions for bleeding.

Fig 15-4

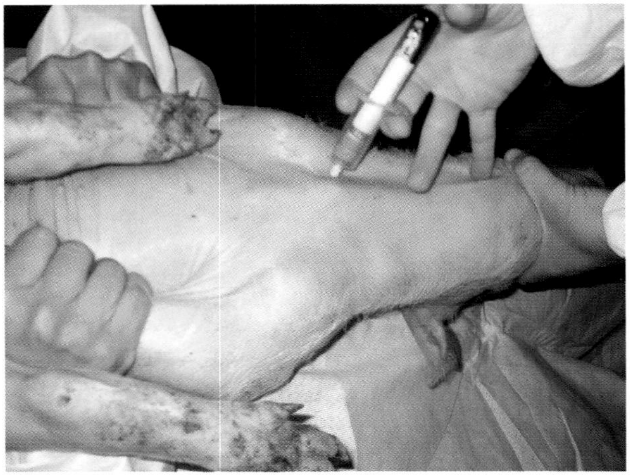

Bleeding from the anterior vena cava in pigs up to 30 kg weight. The centre of the triangle indicates the point of entry on the right side of the pig as shown.

Fig 15-5

Castration of the Normal Pig (Surgical)

Materials Required

- A bucket of warm water, antiseptic solution and cotton wool.
- A surgical blade and handle.
- A marker spray.
- Two people or one if a stand is used.
- Ear protectors.

Reason

In some countries entire boars are accepted at slaughter but in others castration is required. Where possible, it is advisable not to castrate.

Method

This method covers surgical removal of both testicles and should ideally be conducted before the piglet is a week old.

Procedure

1. Only commence castration after the farrowing pen has been cleaned out and the pen or creep area has been bedded with suitable dry bedding materials if applicable.
2. Hold the litter in a clean, dry area or a box with no floor and covered with bedding.
3. Remove the females.
4. One person holds the piglet between the legs with the testicles presented as shown, or a stand may be used. Study Fig 15-6 showing the anatomy.
   The operator must not be involved in catching pigs and at all times must keep their hands clean. Scrub and wash hands before commencing.
5. The skin over the testicle is wiped clean with cotton wool and antiseptic solution.
6. Each testicle is raised to the surface with the thumb, first and second fingers (Fig 15-7).
7. A separate incision is made into each testicle through the skin, and the tunic (point a in Fig 5-16) and attachment of the tunic (b) is broken. The testicle is either pulled away by traction, endeavouring to break the blood vessels (c) and cutting the vas deferens, or the complete spermatic cord (blood vessels and vas deferens) are cut at (c and d).
8. After both testicles have been removed it is important to raise the skin incisions to make sure no strings of tissue are left behind. If so, cut cleanly away (see Fig 15-8).
9. Pigs over 7 days of age should only be castrated under local anaesthetic on welfare grounds. Inject ½ ml of anaesthetic into each testicle and under the skin. Castration should commence 5 minutes later. This may require a veterinarian.
10. Once complete, return the pig to the pen.

Outdoor sow herds – Hygiene and sow aggravation are problems and it is wise to avoid routine castration of outdoor piglets by finding a castration free outlet for the young pigs.

The use of chemical castration (vaccination) using a licensed product is a useful alternative to surgical castration.

ANATOMY OF THE NORMAL AND RUPTURED PIG

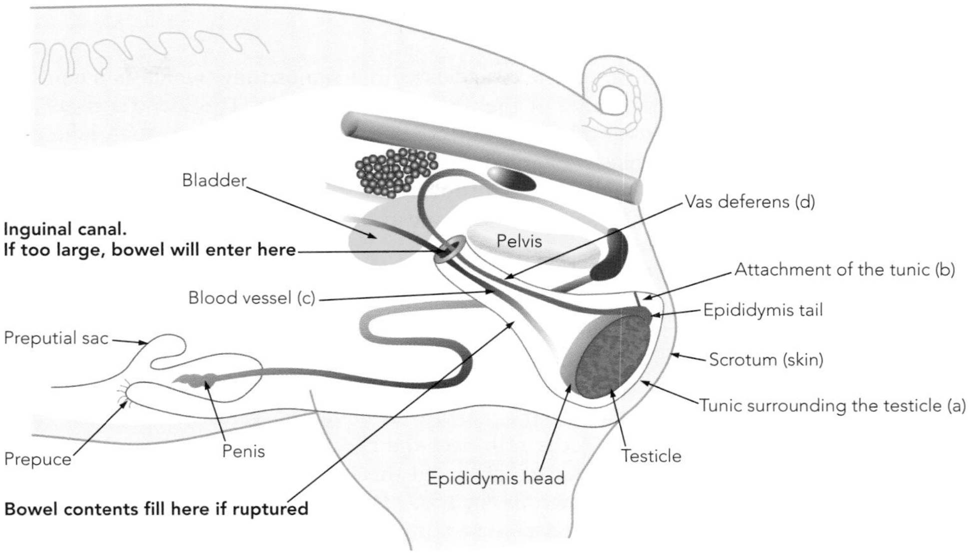

Fig 15-6

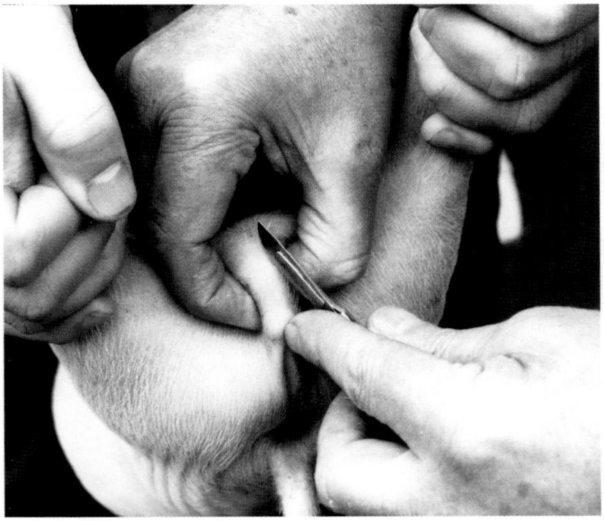

Castration. Squeezing the testicle ready for castration.
Fig 15-7

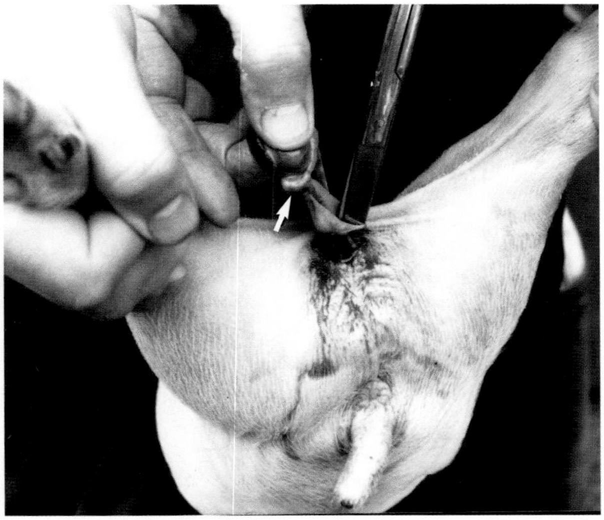

Castration. Showing the ligament attachment and the tail of the epididymis (arrow). The scalpel is placed in the tunic.
Fig 15-8

Castration of the Ruptured Pig

Materials Required

As for normal castration but include:

- Azaperone, acetylpromazine injection or other available sedative.
- Local anaesthetic.
- Syringe 21 g 16 mm ($^5/_8$") needle.
- Half curved triangular suture needles 38 mm ($1^1/_2$") and nylon thread or catgut.
- Scissors and tissue forceps.
- Penicillin cream or CTC powder.
- Two people.

Reason

The testicle is enclosed in a sac or tunic which is a continuation of the peritoneum or lining of the abdomen (Fig 15-6). The blood vessels and vas deferens leave the abdominal cavity to reach the testicles via a small hole called the inguinal canal. If this is enlarged, bowel and abdominal contents enter the sac to produce a hernia, usually called a rupture but they are contained within the tunic. If a pig is castrated in the normal way and the tunic opened the bowel contents are exposed.

Procedure

1 Identify the pig and prepare for the operation as for normal castration. It is best carried out at 3 to 4 weeks of age under the guidance of your veterinarian.
2 Sedate the pig with the sedative, inject ½ ml of local anaesthetic into each testicle and skin where the incision is to be made and wait for 5 minutes.
3 Hold the pig with its belly uppermost, identify the rupture and raise the testicle to the skin over the inguinal canal.
4 Very carefully cut through the skin and loose tissues down to the tunic as illustrated in Fig 15-7. **DO NOT CUT INTO THE TUNIC** otherwise this will open the sack and abdominal contents will spill out. If this happens, euthanise the piglet.
5 Twist the tunic and sac so that the ruptured contents are squeezed back into the abdomen (Fig 15-9).
6 With the testicle held, pass a curved suture needle and nylon suture through the twisted cord and sac as near to the body as possible. Wind round twice and tie.
7 Cut off the sac on the testicle side of the suture to remove the testicle.
8 Place penicillin over the suture.
9 Stitch the skin together with mattress sutures (see procedure for suturing skin and muscle).
10 Return the pig to the pen. The sutures may be removed after 10 days or left in.

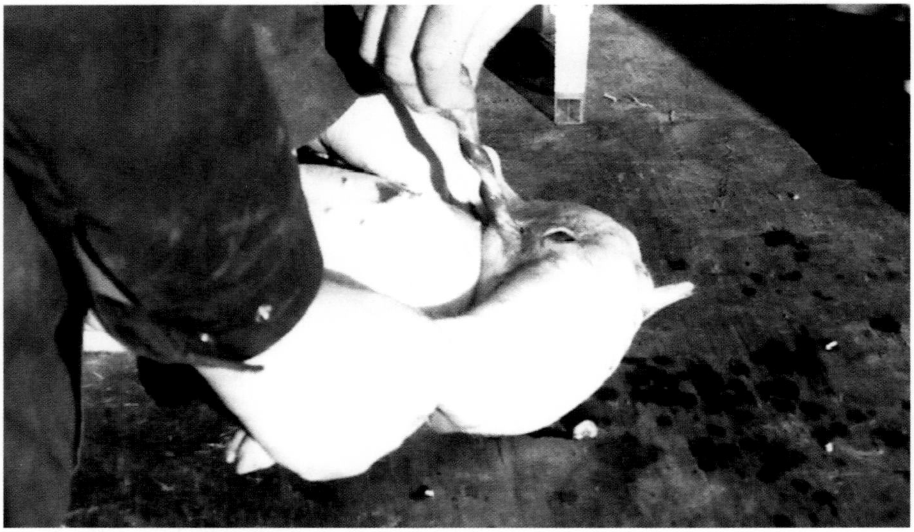

This shows the testicle held inside the tunic, and by twisting, the rupture is squeezed back into the abdomen.

Fig 15-9

Cleaning and Disinfection of Buildings

Materials Required

- A good supply of water.
- A pressure washer/steam cleaner with up to 1000 psi.
- Waterproof protective clothing.
- Eye protectors.
- Hand protectors.
- Detergent, disinfectants: either phenol and iodine or chlorine based or organic acids detergent steriliser. Non toxic disinfectants for terminal aerosol use.
- One person.

See also in this chapter procedures for water – cleaning and sterilising a system, fumigation of houses using formaldehyde vapour, and lime washing concrete floors

Reason

There are a number of important reasons for the disinfection of buildings and other areas of the farm, including; reducing the numbers of organisms and risk of disease, preventing the spread of pathogens, maximising growth and performance, and creating pleasant working conditions.

Procedure

1. Ensure the last set of pigs have eaten all the feed. If not, remove all feed from the house.
2. Remove all dirt and faeces and empty all slurry channels and tanks.
3. Disconnect all moveable equipment, feeders, lamps etc. and open all inaccessible areas e.g. channels.
4. Isolate the electricity supply.
5. Brush down and sweep out the house, including fan boxes.
6. Soak the complete building, roof to floor with a farm detergent or water, and leave for 24 hours if possible.
7. Soak all moveable equipment and clean down.
8. Drain and flush out the water system, bowls, nipples, water tanks etc. and fill with a detergent steriliser. Leave for two hours, empty and then refill with water.
9. Pressure wash the complete building using hot water or a steam cleaner. Use 500 psi but take care not to damage concrete surfaces.
10. Visually check the building.
11. Disinfect the complete house including all equipment and surrounds using the pressure washer at 200 psi or spray.
12. Place a disinfectant foot bath outside the house and use prior to entry. Use an iodophor disinfectant which is brown when active and yellow when no longer active.

Do not restock the house until dry (a minimum of 48 hours). If you have to restock the house before this then use a space heater to dry out the surfaces.

De-Tusking a Boar

Materials Required

- Azaperone, acetylpromazine injection or other available sedative – in large boars a light anaesthetic administered by the veterinarian may be required as well.
- Wire or rope noose for physical restraint.
- A piece of wood to hold the mouth open.
- A pair of shears or hoof cutters or embryotomy wire and handles (Fig 15-10).
- Two people.
- Ear protectors.

Reason

Both the upper and lower tusks of the boar become large, sharp and dangerous with increasing age. On safety grounds it is advisable that all boars have the exposed portions removed at around 6 months of age. The remaining stumps will not then be capable of causing such severe damage.

Method

Remove the teeth either using shears or hoof cutters or use embryotomy wire.

Carry out under the guidance of your veterinarian.

Procedure

1. Heavily sedate the boar in his pen as per sedative instructions.
2. Wait until boar is sedated.
3. Place a noose made of strong material, such as nylon calving rope, around the upper jaw and fasten the other end to a post or bar.
4. In young boars use hoof cutters. In old boars remove the tusks with embryotomy wire. This is abraded stainless steel wire with handles attached. It is placed around the tooth which is removed with a sawing action at gum level. Do not leave any sharp edges.
5. During sedation the penis may protrude. Cover with a damp cloth or replace. Do not carry out this operation with other pigs in the same pen.

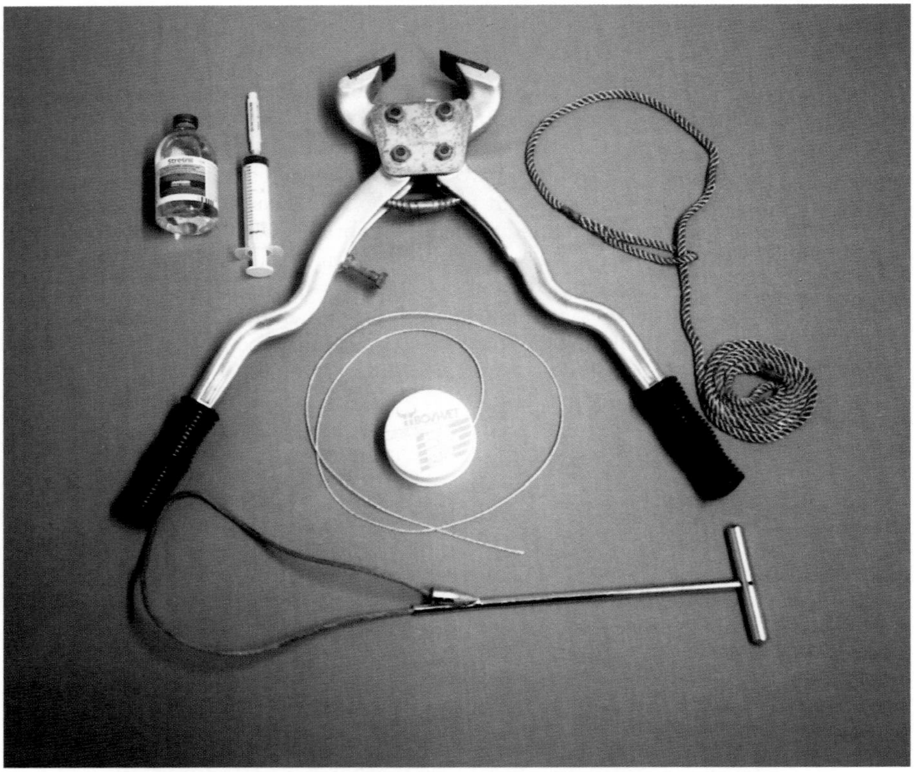

Equipment required for de-tusking a boar, showing clippers and embryotomy wire.

Fig 15-10

Docking (Tail Clipping) Piglets

Materials Required

- One of the following: teeth clippers, scissors (Fig 15-11), scalpel blade, gas cauteriser (Fig 15-12) or a burdizzo instrument turkey debeaker. A sharp surgical knife or scissors are recommended here because the cut is clean with a minimum of damaged tissue. A good blood clot is formed.
- A spray or suitable marker.
- A small container with water and a mild antiseptic. Iodine cow teat dip antiseptics are ideal.
- Ear protectors.
- Best carried out with two people but can be accomplished by one.

Reason

To prevent tail biting. Until the reasons for tail biting are fully understood and preventative methods can be applied with 100 % success, removing most of the tail at birth can reduce welfare problems in later life. Tail biting causes infection, the development of abscesses in the spine, severe pain and carcass condemnations at slaughter.

Procedure

1. Wash all instruments with hot soapy water and disinfect.
2. Never remove tails at birth. Always wait until the piglet has consumed colostrum (>24 hours after birth). Use designated instruments for this procedure. Do not use the same instrument used for teeth-clipping for example.
   The tail is best removed after the piglet has consumed colostrum, from 12 to 72 hours after birth. The ideal time to dock a piglet's tail is on Day 3 when iron injections are given.
3. Hold the piglets in the creep area or confine to a bottomless box, well bedded with either shavings or sawdust.
4. The operation is best carried out with two people but can be accomplished by one. Hold the tail into the scissors and remove with a quick cut. Leave only 16 mm of tail length. Do not be tempted to leave half the tail. This is important because tail biting will occur at this length. Alternatively remove the tail by cautery or apply a burdizzo. A turkey debeaker is extremely useful in tail docking piglets.
5. Mark the piglet and place in a clean creep area. Any bleeding should cease in ½ minute.
6. Place the instrument in the disinfectant when not in use.
7. Examine all piglets five minutes later to check that none are bleeding.
8. If bleeding does occur apply a tourniquet to the tail using a piece of string. Leave this on for 15 minutes, then remove.
9. Wash all instruments with hot water and then disinfectant.

NOTE: Ensure all tail stubs are the same length. Do not let different stockpeople create groups with different tail stub lengths.

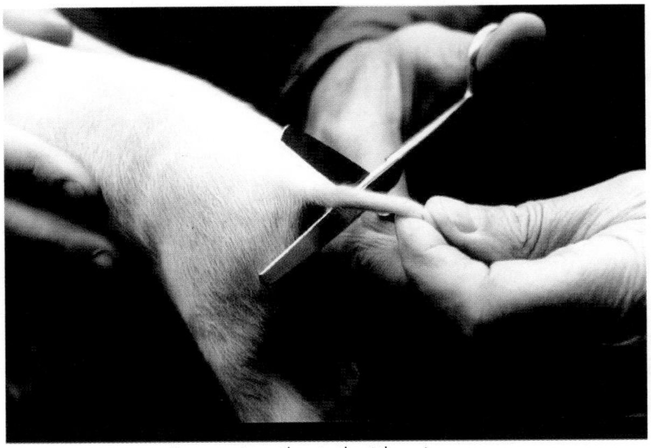

Removing the tail with scissors.

Fig 15-11

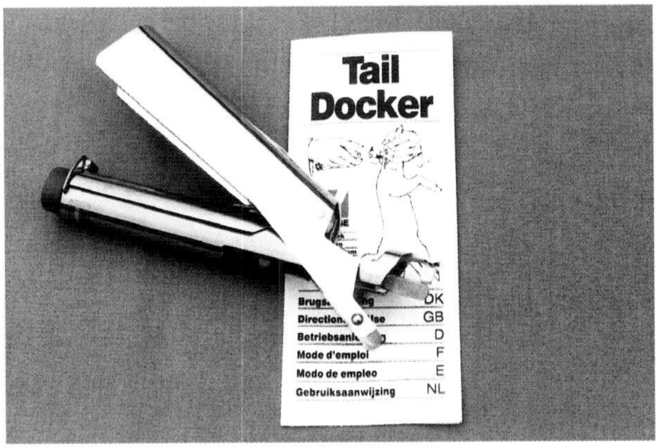

A gas cautery.

Fig 15-12

Epididectomy

Materials Required

As for normal castration plus a pair of surgical scissors and tissue forceps.

Reason

Epididectomy is a simple method of rendering a boar sterile but not impotent (i.e. he still wants to mate) by removing the tail of the epididymis of each testicle (see Fig 15-6). The epididymis is that part of the testicle that stores the mature sperm prior to its ejaculation. Epididectomy achieves the same objectives as vasectomy (details later in this chapter) but it is much easier and simpler to perform (Fig 15-13).

Procedure

These procedures should be carried out by or under the guidance of your veterinarian.

1. Proceed as for normal castration when piglets are 10 to 21 days old but use local anaesthetic. The procedure can be done in older growing pigs.
   DO NOT CUT INTO THE TESTICLE ITSELF.
2. Lift one testicle up and identify the large blue swelling of the tail of the epididymis.
3. Hold the tail with the tissue forceps and cut it away making sure not to damage the testicle. The arrow on Fig 15-8 points to the tail of the epididymis and the attachment of the tunic.
4. Let the testicle return into the scrotum making sure that no tissue protrudes between the skin.
5. Apply antibiotic cream into the wound and suture the skin.
6. Castrate the other testes.

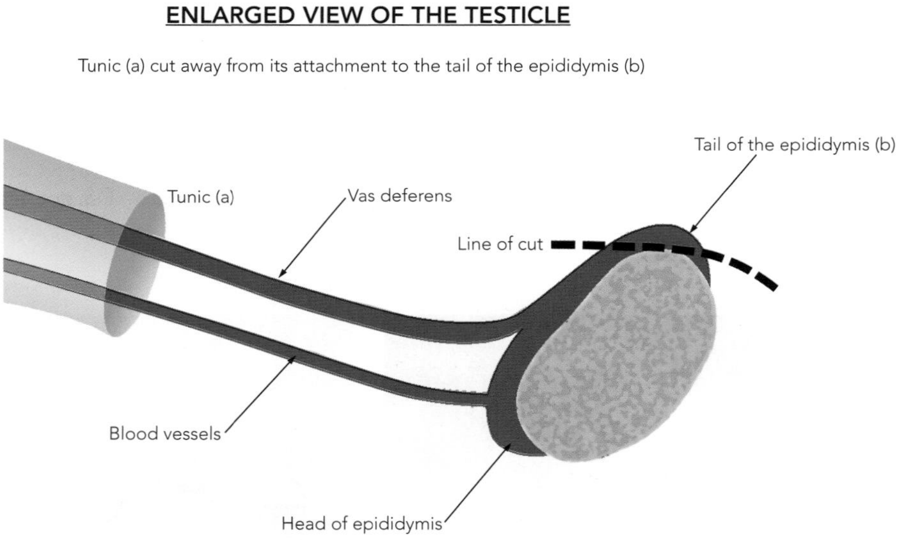

Enlarged view of the testicle showing the line of incision for epididectomy at the tail of the epididymis.

Fig 15-13

Feedback

Materials Required

Sources:

- Sow faeces.
- Weaner faeces. This is a good source of enteric infections.
- Piglet scour.
- Respiratory feedback. Use a rope given to weaners. They chew on the rope and transfer saliva to the rope. The rope is then given to gilts.
- The feedback of animal protein (placenta, stillborn or mummified pigs etc) is illegal in some countries.

Reason

To use potentially infected material to stimulate immunity prior to expected disease challenge. It is of value in TGE, PED, PCV2, PRRSV, PPV and rotavirus diseases. It is also used to stimulate immunity to *E coli* in gilts, but the effectiveness varies.

Procedure

1. Identify the pathogen for which feedback is to be used. Decide with the advice of your veterinarian the advisability of feedback.
2. Collect 1 to 2 kg of mixed weaner faeces and mix with water in a bucket. Alternatively soak piglet scour in shavings, sawdust or meal and mix with water. Piglet scour can be collected using newspaper.
3. Expose the target group via the drinking water or feed.
4. Determine the number of occasions when feedback is to be practised, usually twice weekly.
5. Where legally allowed, in outbreaks of TGE/PED liquidize the small intestine from piglets that have died, to harvest the virus. Expose to all pigs immediately to spread the disease and produce immunity across the herd. However avoid sows 4 weeks from farrowing.

NOTE: For further information on this topic see Chapter 3: Feedback.

Flutter Valve and Its Use

Materials Required

- A flutter valve.
- 500 ml (1 pint) glass bottles and cord to suspend them.
- Saline solution 0.5 % or water warmed to body temperature (Fig15-14).
- Two people.

Reason

To drip water into the rectum in cases of water deprivation (salt poisoning) and streptococcal meningitis. Can also be used to treat dehydration in boars.

If pigs are constipated this can be relieved by enema.

Procedure

1. Raise the bottle to a height and suspend it by a cord above the pig so that water slowly drips.
2. Insert the end of the rubber tube 75 mm in the rectum.
3. Repeat every 2 to 3 hours.

Flutter valve. The end of the tube is inserted into the rectum and the bottle suspended above.

Fig 15-14

Fumigation of Houses Using Formaldehyde Vapour

Materials Required

- Face mask (dampened).
- Goggles.
- Rubber gloves/hand protectors.
- Electric aerosol or heat generator.
- Formalin 40 % solution or powder for use in generators.
- Potassium permanganate crystals.
- Empty metal containers – 2 litres capacity.
- Two people – one as an observer from a safety aspect.

Reason

This is often used to disinfect the complete environment. Formalin is a toxic substance and care is necessary when it is used. The gas can be produced either by aerosol using an electric generator, from powder heated in an electric pan or by mixing formalin with potassium permanganate.

Procedure

1. Clean and pressure-wash the house (see procedure for cleaning and disinfection of buildings) and fumigate whilst the building is still wet.
2. Make sure the house is sealed and gas cannot escape to other buildings or pigs. If this is not possible do not use this method.
3. Calculate the approximate cubic capacity of the building.
   Use 500 ml formalin per 28 m^3 with 200 g potassium permanganate crystals. Place metal containers evenly along the centre of the house and add the crystals.
4. Add the formalin to each container starting furthest away from the door.
5. Leave the building closed for 12 hours.
   Place a warning sign on the door: **FUMIGATION IN PROGRESS.**
6. Open the building and ventilate for 24 hours before use.

NOTE: The use of formalin vapour is banned in some countries and you should be aware that it is toxic and dangerous. Great care should be taken when using it. An alternate is to fog the building with a product which is non-toxic and a safer and equally effective option.

Hysterectomy – Emergency or Planned

Materials Required for an Emergency Hysterectomy

- Two people.
- A captive bolt gun, cartridges and pithing rod.
- A nylon rope or wire snare.
- A sharp knife and a 150 mm nail or pair of scissors.
- Navel clips.
- A bucket of warm water.
- A box to hold the piglets.
- Two large towels or absorbent paper to dry the piglets with.
- All people should wear gloves.

Reasons

A hysterectomy is the removal of the complete pregnant uterus from the sow at or near term. It is used routinely in organisations such as breeding companies, research institutes or Specific Pathogen Free (SPF) associations to produce piglets which are free from specified pathogens (SPF piglets), or alternatively as a remedy when a sow cannot farrow. She is destroyed but the litter is saved.

Procedure for an Emergeny Hysterectomy

These procedures should be carried out by or under the guidance of your veterinarian.

1. Move the sow to a clean area that can be easily washed down.
2. Restrain the sow. (See procedure for restraining the pig).
3. Shoot and pith. Pithing involves passing the flexible rod through the hole made by the captive bolt in the skull to the back of the brain and down the spinal cord to destroy it. If this is not possible the sow should be bled by cutting her throat. The hysterectomy **MUST NOT BE CARRIED OUT** until the sow has been pithed or bled.
4. With the sow on her side open up the abdomen at the mid line between the udder tissue. Make a 600 mm incision. Pull both horns of the uterus containing the piglets completely out. Open up the uterus by holding each piglet and tear the wall with scissors to prevent damage to the piglets.
5. Remove the mucus from the piglet's mouth to allow it to breathe.
6. Massage the umbilical cord back towards the piglet's abdomen, squeeze the blood it contains towards the abdomen of the piglet, apply an umbilical clip and cut the cord away.
   DO NOT carry this out until the piglet is breathing well and fully active or pale piglets will result.
7. Foster the litter to a newly farrowed sow (see Chapter 8 – Fostering Piglets).

Materials Required for a Planned Hysterectomy to Produce SPF Piglets

At the slaughter end:

- A pig-free area to hold and slaughter the sow.
- A captive bolt gun, cartridges and pithing rod, knife and wire noose.
- A bucket of warm water, disinfectant and cloth.
- One slaughter person, one operator, two bath carriers.
- A two-handled bath tub holding 25 litres of warm water and disinfectant.

Procedure for a Planned Hysterectomy to Produce Specific Pathogen Free (SPF) Piglets

This method involves removing the intact uterus from the sow to a clean area at least 50 m away before the piglets are removed. Alternatively the uterus can be removed and passed through a disinfectant trap into a sealed room where the piglets are then removed. The operation is carried out on either Day 112 or 113 of pregnancy. Day 113 is best. As a precaution the sow could be injected with 300 mg of progesterone to delay farrowing.

1. Move the sow from the farm to isolation premises 4 to 9 weeks beforehand. Carry out any testing procedures to confirm the disease status.
2. Move the sow to the pig-free slaughter area on Day 113 of pregnancy in cleaned and disinfected transport. (In some breeds Day 112 or 114 may be more appropriate; discuss with your vet). Assess the pregnancy state by examination of the udder. If the sow has started farrowing, abandon the operation.
3. Fill the bath with warm water and disinfectant sufficient to cover the complete uterus.
4. Restrain, shoot and pith the sow. If pithing is not successful bleed out the sow. This operation should not be carried out by the operator (Fig 15-15a – h).
5. Hold the sow on her back and wash the skin over the udder with disinfectant.
6. Make a careful incision into the abdomen at the sternum sufficient to pass the hand into the abdomen.
7. Place the knife with the blade uppermost into the abdomen and cut open up to

Hysterectomy – Emergency or Planned cont.

At the clean area:

- A pig-free area to receive the uterus and piglets.
- Three reception personnel who have had no contact with pigs for 48 hours and are wearing clean non-pig-contact clothing.
- A disinfected reception table.
- Warm water.
- Disinfectant.
- Sterile towels.
- Scissors and a reception box containing two hot water bottles.
- The reception box should be approximately 0.6 m^2, insulated with a lid and hold a false bottom, beneath which two hot water bottles can be placed. Shavings make suitable bedding.
- Navel clips and tubular bandage.
- Electricians' nylon ties which can be sterilised may be used instead of clamps.
- A pre-planned reception farm with one to three newly farrowed sows.
- The sows can be synchronised to farrow with prostaglandin injections.

the pelvis, taking care not to cut the uterus or the intestines.

8 Hold the disinfectant bath to the edge of the abdomen and pass the uterus and piglets into it. Each horn of the uterus separates itself from the ovaries.
9 Identify the cervix and cut through its centre. The uterus must be completely covered by the warm disinfectant.
10 Carry the bath 50 m away to the reception point. Piglets can survive for up to 4 minutes after the sow has been shot.

The Reception Point

1 The uterus is placed onto the perforated metal reception table and the disinfectant allowed to drain away. The transporting personnel immediately return to the slaughter point.
2 Each piglet is grasped and the uterus torn open with blunt scissors.
3 Mucus is cleaned away from each piglet's mouth and they are left to breathe normally and until they have become active.
4 The blood in the umbilical cord of each piglet is squeezed towards its abdomen and a clamp applied.
5 The cord is cut, the piglet is wiped dry with the towels and tubular gauze is placed over the body to hold the clamp.
6 The hot water bottles are placed beneath the false bottom, then the piglets are placed in the reception box prior to transportation to a newly farrowed sow on the recipient farm.

NOTES:

- Do not remove piglets until a minimum of $1^1/_2$ minutes after slaughter.
- Piglets will survive for up to 4 minutes after slaughter.
- Sometimes pithing is difficult. Shoot the sow again.
- If pithing cannot be carried out bleed the sow by cutting completely across the throat.
- Remove the uterus only when bleeding has ceased. In this case the piglets must be removed within $1^1/_2$ minutes.
- If a piglet breaks from the uterus it must be discarded.
- Do not feed the piglets before they suckle the foster sow. Pre-feeding will close down the intestines to the absorption of colostral antibodies.
- Piglets can survive well in a warm, dark insulated box for several hours before sucking the sow.
- Farrowing may be delayed for up to 24 hours by injecting progesterone, 300 mg, on Day 113 of pregnancy.
- The carcass must be condemned.

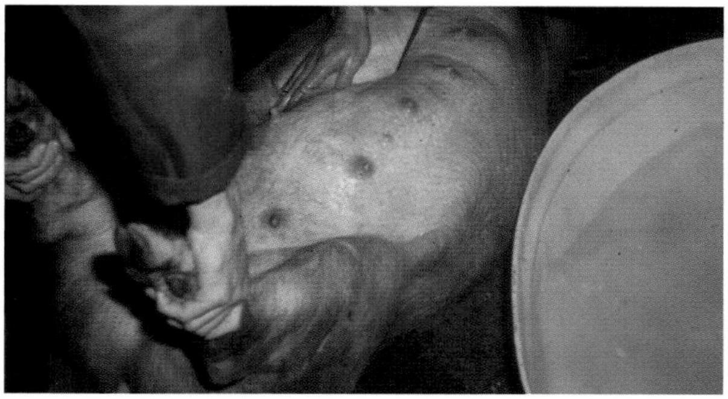

Opening up the sow.
Fig 15-15a

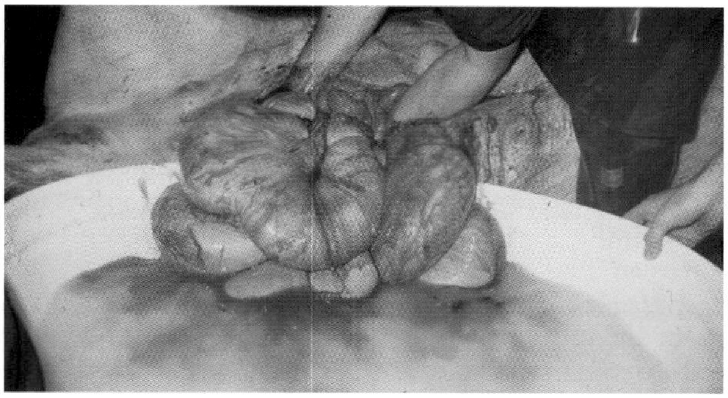

Passing the uterus into the bath.
Fig 15-15b

Transporting the bath to the reception table.
Fig 15-15c

The reception table.
Fig 15-15d

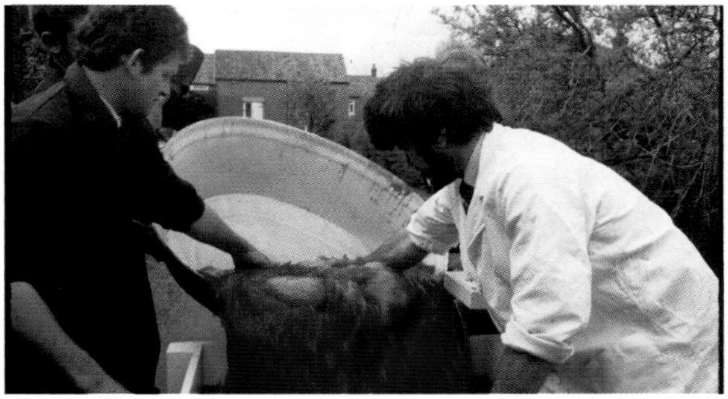

Emptying the uterus onto the table.
Fig 15-15e

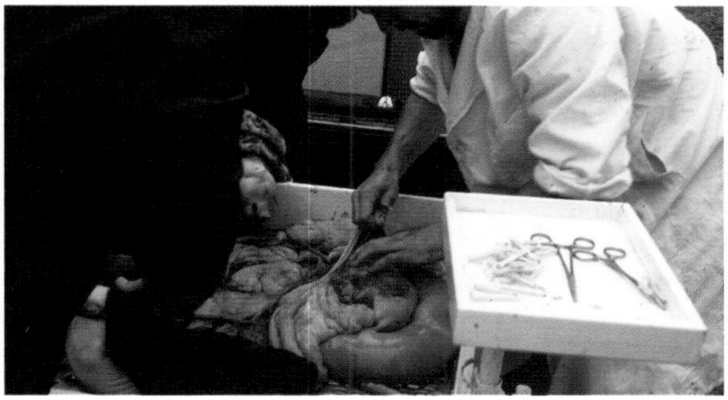

Opening the uterus.
Fig 15-15f

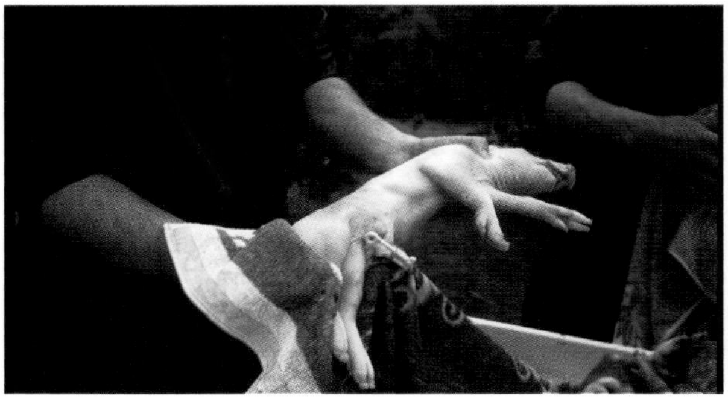

Clamping the navel.
Fig 15-15g

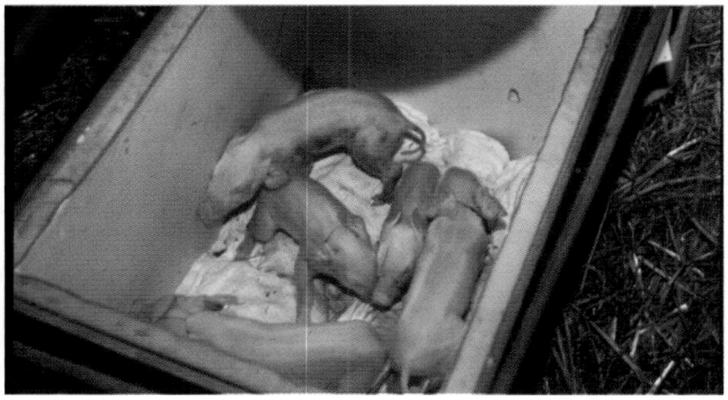

The reception box.
Fig 15-15h

Identification – Tattooing, Slap Marking, Tagging, Transponders, Implants, Ear Notching

Materials Required
- One person.

Tattooing
- Green ink.
- Tattooing applicators for the different dies.
- Four sets of each size of dies numbered 0 to 9.
- A clean divided box, easily cleaned, to hold dies.
- A container to soak and clean dies.
- Disposable gloves.
- Toothbrush.
- Recording equipment.

Slap marking
- Slap marker with unique farm tattoo plate.
- Black ink and ink pad.
- Methods of restraint.

Tagging
- Tags and applicators.
- Tissues and surgical spirit.

Electronic ID
- Electronic collars
- Reading equipment

Radio transponders
- Implanting equipment (for implanted radio transponders).
- Transponders
- Reading equipment.

Ear notching
- Ear notching pliers

Reason
These are the methods used for the permanent identification of pigs. They may be required for recording and management purposes, and for the identification of pigs under treatments and of carcasses at slaughter.

Methods

Tattooing/Slap Marking
Dies or spiked numbers or letters are used.
Three sizes are needed

1. 8 mm ($^5/_{16}$") for piglets for 3 to 21 days.
2. 10 mm ($^3/_8$") for pigs from 10 to 90 kg (20 lbs to 200 lbs).
3. 16 mm ($^5/_8$") for sows and slap marking.

Two sites are used. The ear or on the neck behind the ear for use on the farm, and over the shoulder or back (slap marking) on pigs destined for slaughter. Green tattoo ink shows best and is retained longer in white-skinned breeds for permanent marking on the farm. Use black for pigs at slaughter. Ensure the equipment is in good order and all pins are present and not bent.

Tagging
There are many types of tags on the market. They are used only in the ear. Positioning and hygiene are important. Infection after tagging can in some herds be a major problem. Make sure tags are stored in a clean, dust-proof container. Pigs can also be allergic to certain types of plastic.
If there is a problem, consider the following:

- Make sure instruments are clean and disinfected and tags are not contaminated.
- Wipe the skin clean with antiseptic (e.g. iodine solution) before application.
- Change the age when pigs are tagged.
- Carry out bacteriological examinations to determine the cause.
- Change the type of tag.
- Assess the cleanliness of the environment – wet, dirty pens predispose.

It is usual to place a tag in each ear. Use tags where possible that are numbered on both sides.

Electronic Identification
These are electronic collars attached to the pigs that can be read by or respond to electronic impulses. They are used in automatic feeder systems to monitor feed intake. They can also be used to store data and be identified electronically at a distance from the pig.

Radio Transponders
These are transistors enclosed in very small non-reactive injectable implants that are placed beneath the skin at the base of the ear. Alternatively they can be placed in an ear tag. They communicate with receivers which can detect the signal remotely (up to 60 cm away). The data can be retrieved and stored/analysed on a computer. The initial set up can be costly, but the investment is rewarded with accurate management information and identification at slaughter. Transponders are removed and destroyed at slaughter.

Ear Notching

This involves creating notches in the pigs' ears in a specific pattern/system to individually identify each pig.

Timing

Piglets up to 21 days old	– Are best tattooed at 3 to 5 days of age to coincide with iron injections.
Pigs at weaning	– Transponders.
Gilts at selection	– Either tattoo or tag.
Sows	– At any time with tattoos, tags or transponders.

Procedures

Tattooing

Prepare the equipment and hold in a dust-free, clean box. Avoid external contamination.

1. Suitably restrain the pig.
2. Clean the skin with surgical spirit.
3. Place ink on the dies with the toothbrush and apply the tattoo to the ear. Rub the ink into the skin with the brush.
4. Repeat for each pig that needs identifying.
5. On completion of tattooing soak the dies and applicator in warm water and detergent. Clean, then soak in a mild disinfectant for 15 minutes and then dry.
6. Tattooing may be done up to 30 kg (70lbs) of weight.

Slapmarking

Prepare the equipment and hold in a dust-free, clean box. Avoid external contamination.

1. Identify the pigs and area of application.
2. Press the ink pad onto the marker and slap the dies onto the skin either over the shoulder or the back near the tail. Restraint is not necessary for slaughter pigs, but will be necessary when marking breeding stock.
3. On completion of tattooing, soak the dies and applicator in warm water and detergent. Clean, then soak in a mild disinfectant for 15 minutes and then dry.

Tagging/Transponders

1. Suitably restrain the pig.
2. Clean the skin with surgical spirit.
3. For tagging, apply the tag to the centre of the ear, taking care to avoid the veins.
4. For implants, place the implant into the loose skin at the base of the ear.

Ear Notching

This procedure is best carried out at birth using a pair of notching pliers. Different methods are used but some breeding organisations have a specific method. One ear, usually the right ear, is used to identify the litter and the other to identify the individual pig. Pigs may be notched from birth to 2 weeks of age.

Each ear is divided into areas and permanent numbers allotted to each. The numbers required are then identified and the notches made accordingly. There are different systems for notching pigs' ears; identify the system that best suits your circumstances. Discuss with your vet.

1. Clean and disinfect the pliers.
2. Clean the ear with surgical spirit. Clean the pliers between litters.
3. Tattooing is more welfare-friendly and is replacing this method, but for small numbers it is still useful.

Injecting Piglets with Iron

Materials Required

- One person.
- Marker spray.
- Iron dextran (100 mg or 200 mg per ml).
- Syringes 16 mm (5/8") 21 g needles or an automatic syringe delivering 1 or 2 ml.
- Use sterile or disposable ones.
- A bottle of surgical spirit and cotton wool or a surgical spirit sprayer (in outdoor herds).
- A platform to hold the materials.
- In outdoor herds carry the above in a small container that can be washed and disinfected between sessions.

Reason

Piglets are born with minimal reserves of iron and sows' milk contains insufficient iron to satisfy their needs. They will become anaemic by 10 days of age unless they have access to iron orally or are injected with an iron preparation.

Method

Iron can be given by mouth as a ferrous sulphate paste on Days 4, 10 and 15. This is not a common method and it is time-consuming. The best procedure is to give 150 to 200 mg of iron dextran either by subcutaneous or intramuscular injection. Injections are available that contain either 100 mg or 200 mg per ml.

If the piglets are to be weaned at less than 3 weeks and are provided with creep feed containing iron, the smaller dose is sufficient. Outdoor herds on certain soil types may not require an iron injection.

Timing

Iron injections are best given between 3 and 5 days of age. Do not administer on Day 1 because the injection will cause considerable stress to the piglet.

It is good management practice to give iron injections at the same time as tails are removed and the piglets are tattooed.

Procedure

1. Collect the litter together in a bottomless box that is well bedded or hold in a clean creep area. This is a one-person operation.
   In outdoor sow herds, routine iron injections are given when the sows are feeding. It is best done inside the hut on clean straw and with a second person on watch outside the entrance with a pig board in case the sow returns.
2. Fill the syringe with iron solution. Use a new syringe and needle.
3. Wipe the needle clean with surgical spirit and cotton wool every third injection or in outdoor sow herds spray it with a surgical spray.
   In indoor units use a new needle after every three litters, or if it becomes damaged. For outdoor units use a new needle for every litter.
4. The site of the injection can be intramuscular into the neck muscle behind the ear or subcutaneous under the skin in the inguinal fold. Follow the manufacturer's recommendations.
5. Roll the skin over the muscle with the thumb and introduce the needle at a 45° angle. Inject either 1 or 2 ml (200 mg). 1 ml is preferred.
6. Roll the skin back and apply pressure to the injection site.
7. Mark the piglet.

Lancing an Abscess or a Haematoma

Materials Required

- Wire or rope noose
- Azaperone, acetylpromazine injection or other available sedative.
- A 10 ml syringe with a 38 mm (1½") 18 g needle.
- Scalpel blade and handle.
- Water and antiseptic.
- Saline solution (1 % salt).
- Two people.

Reasons

A swelling that appears just under the surface of the skin is usually either an abscess or a haematoma. Haematoma commonly occur on the ears and soft tissues overlying the shoulders and sides. They are pockets of blood formed after a blood vessel has ruptured. If a swelling has appeared suddenly it is probably a haematoma and is often best left to repair on its own. If the swelling appears gradually it is possibly an abscess, in which case the pus must be drained. Abscesses are common due to faulty vaccination procedures (Fig 15-16).

Procedure

1. Restrain the pig.
2. Examine the swelling.
3. Determine the softest area of the swelling and pass the needle into the mass. Withdraw fluid.
4. If the fluid is fresh blood – proceed no further. It is a recent haematoma.
5. If the fluid is clear or blood tinged it is a haematoma that has ceased bleeding and is healing. Generally it should be left alone.
6. However, if a large haematoma over the ear is causing pain open it up on the inside top of the ear and treat as for an abscess.
7. If the fluid is pus it is an abscess and requires opening and draining. Usually there will be evidence of skin damage and scar tissue.
8. If treatment is required, move the pig to a clean, disinfected pen.
9. For an abscess, make an incision 30 to 40 mm long at the lowest point of the abscess and squeeze out the contents. Restraint and sedation may be needed depending upon the size of pig.
10. Make a cross-cut to help keep the hole open. If the abscess is large, flush it out with large volumes of water from a hose for example.
11. Flush out with the syringe and saline and keep the incision open with a cleaned gloved finger, or a pair of scissors. Repeat daily for 3 to 4 days.
12. Inject the pig on alternate days with either OTC, amoxycillin or penicillin (long-acting).

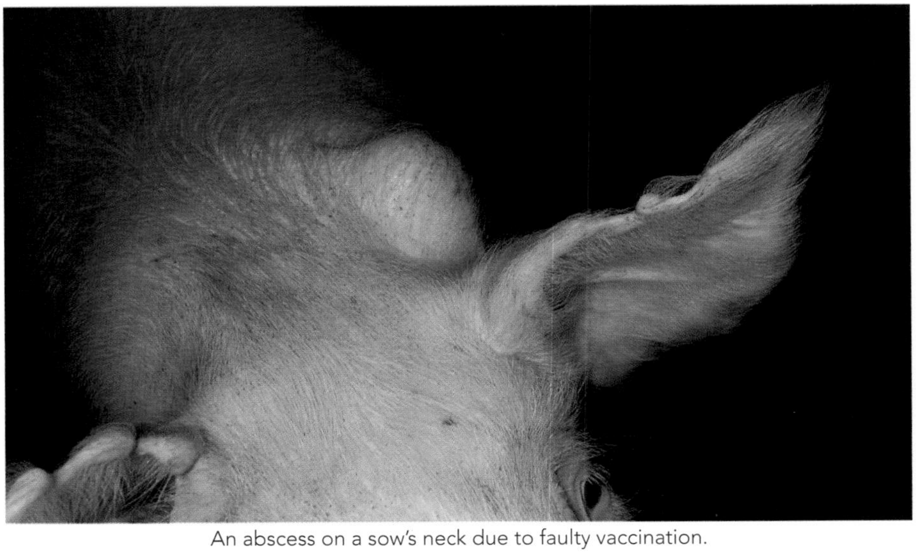

An abscess on a sow's neck due to faulty vaccination.

Fig 15-16

Libido Checking – Training Boars to Use a Stool

Materials Required

- A mounting stool or dummy or a gilt on heat.
- A dry pen at least 9.3 m^2. Good light.
- One person.

Reason

This may be carried out to check a boar's libido before sale, in preparation for semen collection on the farm, or at an AI stud.

Procedure

1. Establish a good rapport and empathy with the boars. Be quiet and patient.
2. Only test boars from 6 months of age onwards.
3. Young boars may be housed in groups in warm, draught-free accommodation but ideally in pairs. They must have visual contact with other animals.
4. Test the boar in his pen, if possible in his familiar surroundings.
5. Place the stool in the pen each day early in the morning when it is cool.
6. Allow the boar to make contact. Add saliva from other boars to the surface of the stool or use after another boar. Alternatively use urine from a sow in oestrus. The boar may take a few days before he will mount.
7. If there is no interest within 10 to 15 minutes, remove the dummy and try again the following day.
8. See procedure for semen collection and artificial insemination on the farm for more information.
9. If the boar has poor libido, enhance using a natural prostaglandin (2 to 5 ml) 20 minutes before exposing to the dummy sow.

Local Anaesthesia

Materials Required

- 2 ml and 10 ml syringes.
- 21 g, 20 g, 18 g needles 0.8 to 1.1 mm, 25 to 38 mm (1 to 1½") long.
- Local anaesthetic.
- Cotton wool.
- Water and antiseptic.
- One person.

Reason

Local anaesthesia in the pig is used to facilitate minor surgical techniques that include: suturing skin and muscle, castration of older pigs, replacing prolapses and inserting retaining sutures.

Procedure

Depending upon what procedure you intend to follow, you may need the help or guidance of your veterinarian.

1. Clean the tissues to be anaesthetised with cotton wool or gauze soaked in antiseptic.
2. Fill the syringe with anaesthetic and place the needle 5 mm in front of the tissues to be anaesthetised.
3. Squeeze anaesthetic ahead of the needle point as the needle is slowly pushed forward.
4. As a guide use 0.5 ml for each 25 mm of length.
5. Wait five minutes. Test the tissues by lightly pricking the area with a needle to assess the completeness of anaesthesia.

Lime Washing Concrete Floors

Materials Required
- Builders' lime ($CaCO_3$) or dehydrated lime powder (calcium carbonate). (Concrete cement can also be used).
- Water.
- A large bucket and stirrer.
- A pair of eye protectors or goggles.
- A soft brush and handle.
- A phenolic disinfectant.
- One person.

Reason
Over a period of time concrete surfaces wear away and become pitted, making it difficult to clean and disinfect them adequately. Bacteria and coccidia become embedded in the concrete and rise to the surface as the floor dries out after cleaning. Covering the surfaces with lime wash reduces exposure to organisms; it has a disinfectant action and the surface becomes less abrasive.

Procedure
1. Wear eye protectors at all times. Lime wash can burn.
2. Place lime in the bucket and add water in a ratio of approximately 1:1 to form a thick consistency, sufficient to brush onto the concrete surfaces.
3. Add 30 ml of a phenol-based disinfectant to 5 litres of water.
4. Clean and wash the concrete surface.
5. Brush off surplus water.
6. Brush the lime wash over the surface.
7. Allow the surface to dry for at least 48 hours.
   DO NOT ALLOW SKIN CONTACT WITH WET MATERIAL.

Mating the Sow with the Boar

Materials Required
- A large pen for mating, minimum size 3 m x 3 m with no projections and each side at least 2.4 m.
- A dry, non slip or well bedded floor.
- Good lighting.
- Plastic disposable gloves.
- A movement board for protection.
- One person.

Reason
It is important to make sure the sow is at the correct time of the oestrus period before mating takes place. See Chapter 5, Fig 5-24 under Group 3 Losses.

Procedure
1. Identify the sow or gilt that is in oestrus by ear number or tattoo.
2. Examine the animal for lameness, and the vulva for abnormal discharges and malformations.
3. Select a suitable boar that has not been used within 24 hours.
4. Move the sow to the boar pen or mating pen.
5. Observe the behaviour of boar and sow and check that the sow is standing firmly.
6. Allow the boar to mount. If there is no dissension by the sow, allow mating to commence.
7. As the penis spirals out guide it, if necessary, into the vulva by holding the hand as a funnel against the vulva.
   DO NOT HANDLE THE PREPUCE. It is a heavily contaminated area.
8. Watch for the penis to lock into the cervix. The boar will stop thrusting.
9. Observe below the boar's tail for pulsation of the urethra as insemination takes place.
10. Observe that there is no leakage of semen.
11. Mating will take from 5 to 15 minutes. Observe during this period.
12. Remove the sow to a quiet area and do not disturb for 2 hours.
13. Record the mating.

Outdoor sow herds – Some adopt individual mating so the same principles apply. Where group mating is used it is important to ensure that the boar to sow ratio is correct for the system (e.g. dynamic service system) and that all boars are actually serving and serving correctly. Guard against overweight boars overfeeding.

Mating the Sow by AI with a Cervical Catheter

Materials Required

- Small pen or ideally a simple sow crate to constrain the sow.
- A dry, non slip or well bedded floor.
- Good lighting.
- Plastic disposable gloves.
- Semen.
- Catheter and bag.
- Breeding belt.
- One person.

Procedure

Preparation for service

1. Remove the boar from sight/sound or the sow for at least 1 hour prior to breeding.
2. Parade a boar in front of the female outside the pen. If the signs of oestrus are demonstrated check that the sow will stand still to back pressure (Fig 15-17a).
3. Only inseminate females which demonstrate the "standing still reflex".
4. Place breeding belt onto the sow's back
5. Place gloves on both hands.
6. Wipe vulva clean with a dry tissue. If initially, clean do not wipe with tissue.
7. Select semen pack by checking it is from the required boar.
8. Mix settled semen by gently rotating flat pack then place mixed semen in overall top pocket.
9. Take a new catheter. Do not use any catheter which is not clean. You do not need to lubricate a foam tipped catheter. If you do, avoid applying the lubricant to the hole.

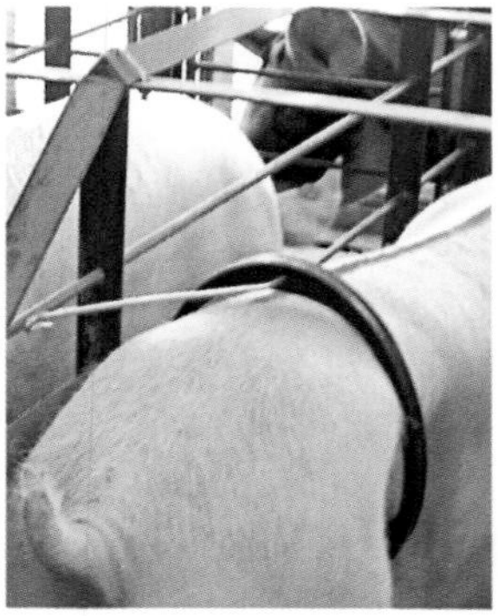

Sow in oestrus standing with breeding belt. Note the boar in front of the sow to stimulate oestrus.

Fig 15-17a

Insemination procedure

1. Open the semen pack.
2. Place the semen pack on the catheter.
3. Part vulva lips and gently insert catheter at a 45° angle into the vagina, forwards and upwards to prevent the catheter entering bladder (Fig 15-17b).
4. Push the catheter straight forward until the ridges of the cervix are felt. Do not twist. Do not rotate a foam tipped catheter.
5. Raise the semen pack to prime the catheter. Leave the catheter horizontal.
6. Attach the semen pack to the breeding belt.
7. Using your free hands, rub the female's flanks, and apply back pressure during insemination to stimulate and maintain the standing still reflex (Fig 15-7c).
8. Allow the female to draw the semen (Fig 15-17d).
9. When the semen pack is empty slowly remove catheter by rotating clockwise.
10. Record service and score quality of service.
11. If this is the first insemination, re-inseminate 24 hours later if the female passes the standing still reflex.

NOTE: Using this technique 3 sows can be mated at the same time. Take at least 7 minutes to complete the mating procedure.

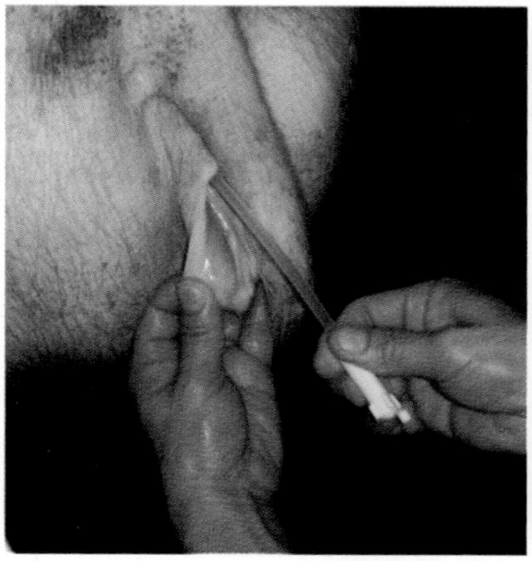

Entering the catheter into the vagina.

Fig 15-17b

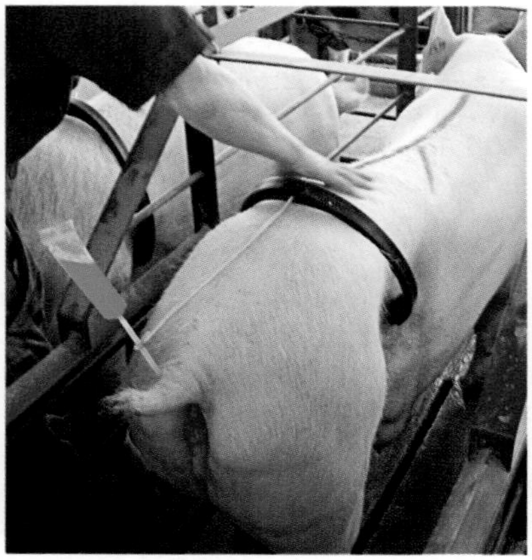

Insemination using a semen pack.

Fig 15-17c

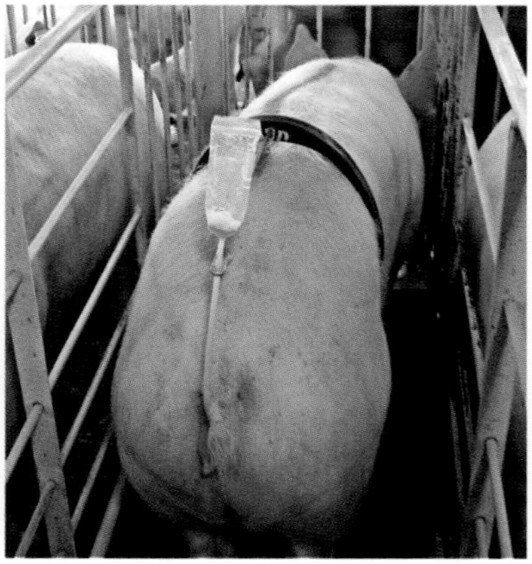

Allow the sow to draw the semen.

Fig 15-17d

Mating the Sow by AI with a Deep Uterine Catheter

Materials Required

- Small pen or ideally a simple sow crate to constrain the sow.
- A dry, non slip or well bedded floor.
- Good lighting.
- Plastic disposable gloves.
- Semen.
- Catheter and bag.
- One person.

Procedure

Preparation for service

1. Remove the boar from sight/sound of the sow for at least 1 hour prior to breeding.
2. Parade a boar in front of the female outside the pen. If the signs of oestrus are demonstrated check that the sow will stand still to back pressure.
3. Only inseminate females which demonstrate the "standing still reflex".
4. Place gloves on both hands.
5. Wipe vulva clean with a dry tissue. If the vulva is clean do not wipe with tissue.
6. Take a new catheter. Do not use any catheter which is not clean. You do not need to lubricate a foam tipped catheter. If you do, avoid applying the lubricant to the hole.
7. Part vulva lips and gently insert catheter into vagina forwards and upward to prevent catheter entering bladder.
8. Push the catheter straight forward until the ridges of the cervix are felt. Lock the catheter into the cervix. Do not insert the inner catheter.
9. Move the boar onto the next sow.
10. Wait about 30 to 90 seconds.
11. Inset the inner catheter through the cervix and into the uterine body.
12. No particular force should be required.

Insemination procedure

1. Select semen bottle by checking it is from the required boar.
2. The boar is not required to be present during the actual insemination. In fact his presence can cause problems of excitement.
3. Mix settled semen by gently rotating bottle (top over bottom 6 times) then place mixed semen in overall top pocket.
4. Take semen from pocket, gently rotate (top over bottom once), cut top 1 cm off spout.
5. Insert semen bottle firmly into catheter. Apply gently pressure to the bottle to prime the catheter. Keep bottle at an upwards angle of about 30°.
6. Apply gentle pressure on the bottle to overcome vacuum pressure. Continue administration of the semen. It may be necessary to remove bottle half way through to allow air back into bottle to release vacuum pressure.
7. When bottle is empty remove and seal catheter.
8. After 5 minutes remove the inner catheter from the uterus.
9. Remove the whole catheter from the sow with a rotating motion.
10. Record service and score quality of service.
11. If this is the first insemination, re-inseminate 24 hours later if the female passes the "standing still reflex".

Pregnancy Diagnosis

Materials Required

- Real time ultrasound scanner.
- Coupling liquid.
- Marker spray and note book.
- One person.

Reason

An early diagnosis of pregnancy indicates the efficiency of reproductive performance and also any impending management or disease problems.

Procedure

1. The instrument passes high frequency sound waves from the probe into the abdomen of the pregnant sow. The machine produces an image of the uterus and developing piglet (Fig 5-18). Carry out routine pregnancy checks at 4 weeks and repeat again at 5 weeks. If there is a fertility problem in the herd, check again at 6 weeks.
2. Visually check sows at 10 weeks.
3. Sows that give a negative or doubtful reading should be moved next to a boar and observed for oestrus.

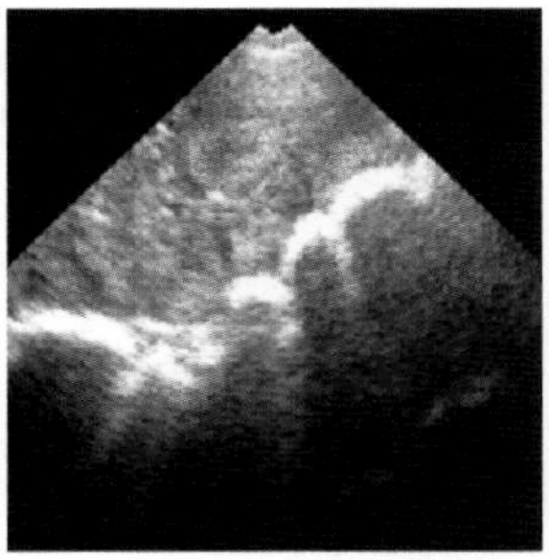

Non-pregnant, loops of gut showing.

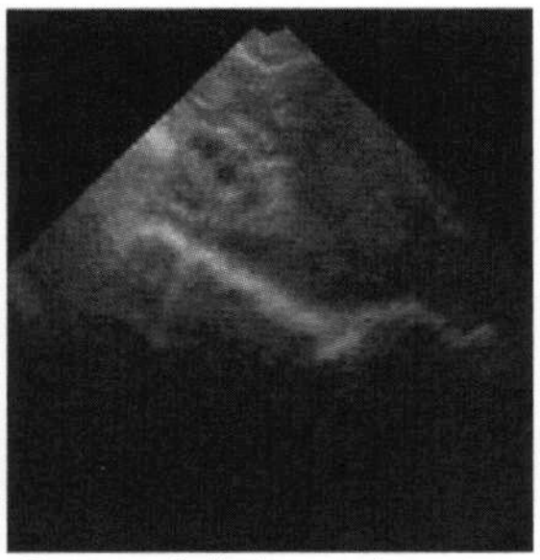

Ovarian follicles and uterine outline.

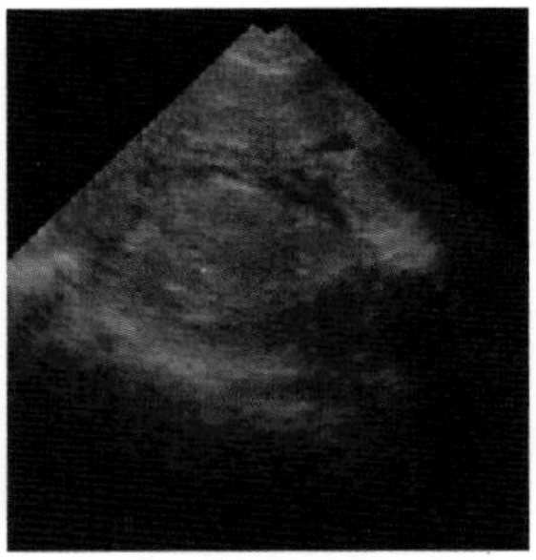

Sow in oestrus.

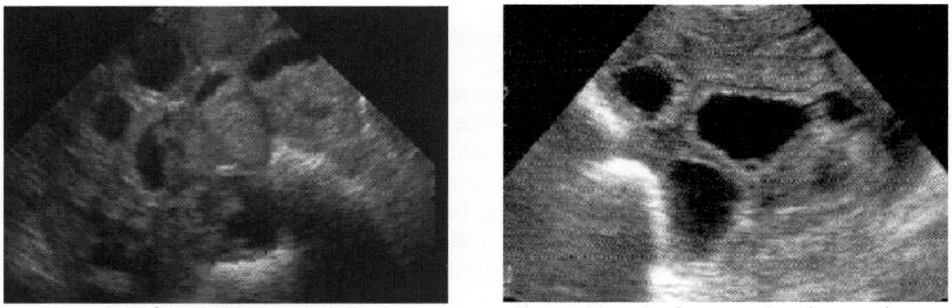

Pregnancy foetal vesicles at 28 days. It is essential to visualise the foetus within the vesicle.

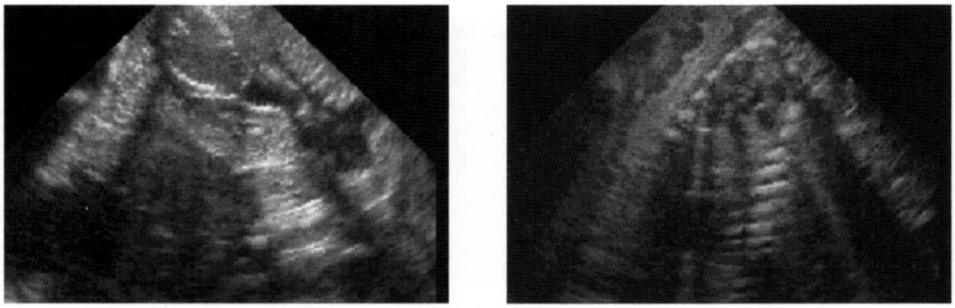

Foetal skeleton.

Fig 15-18

Penis Examination

Materials Required

- An empty clean pen to house the boar.
- Sedative.
- Gauze and saline solution.
- Suture materials and local anaesthetic.
- Surgical scissors, tissue forceps and artery forceps.
- Two people.

Reason

Examination of the penis is necessary for surgical purposes, anatomical faults or where the mating procedures are suspect.

Procedure

1. Sedate the boar as per the sedative instructions. At this level the penis will protrude on its own.
2. Wait, and do not disturb the boar, for 15 minutes.
3. Pull out the penis, gripping it with a piece of gauze and examine.
4. Carry out any procedures as necessary.
5. Return the penis to the prepuce as far as it will go and cover the remainder with wet gauze.
6. Leave the boar to recover and until the penis is fully retracted.

Prolapse of the Rectum

Materials Required

- A method of restraint or sedative. See procedure restraining the pig.
- Local anaesthetic, syringe and 21 g 25 mm (1") needles.
- Curved suture needles 25 to 50 mm (1 to 2") in size.
- Nylon or cat gut suture for pigs up to 100 kg (220 lbs).
- Suture tape for sows.
- Scissors, tissue forceps.
- 12 mm diameter corrugated plastic tubing.
- Obstetrical lubricant.
- Warm water, antiseptic, cotton wool.
- Two people.

Reason

This is a common condition in growing pigs and occasionally it occurs in breeding females.

Methods

There are four options for treatment, but you may need instructions from your veterinarian.

1. Replace and retain with a suture. This is the preferred method.
2. Place a piece of corrugated tube into the prolapse and apply a tourniquet to allow the prolapse to eventually slough off. Used in growing pigs.
3. Suture and then amputate.
4. Place the pig in a hospital pen and allow the prolapsed tissue to slough away. Not a preferred option.

Procedures

You may need guidance or instruction from your veterinarian.

Replacing the Prolapse

1. If the pig is in a pen with others remove it as soon as possible.
2. Sedate or restrain the pig by placing the pig's head down in a round container or hold over a wall.
3. Clean the area around the prolapse and remove all faeces.
4. Infiltrate around the rectum tissue with local anaesthetic. Usually only necessary in pigs over 60 kg (130lbs).
5. Select the nylon or synthetic suture material and insert around the rectum using either a mattress suture or a purse string suture. See Fig 15-19 and 15-20.
6. Cover the prolapse with obstetrical lubricant and using fingers of both hands squeeze it into the rectum. Replace the prolapse even if torn – the pig will have a greater chance of survival.
7. Tighten the suture with a double tie to prevent it slipping, and leave a finger width for the passage of faeces.
8. The prolapse in sows may be swollen and full of fluid. Steady gentle pressure may then be required for 10 to 15 minutes to reduce it before it can be returned to the

A MATTRESS SUTURE

Suitable for retaining a prolapsed rectum in pigs up to 120 kg weight

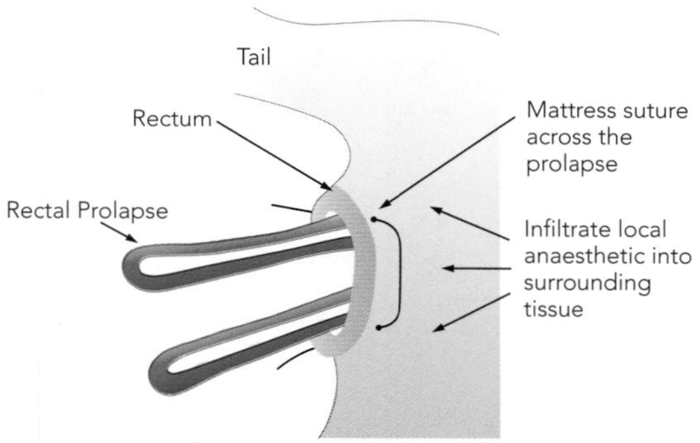

A mattress suture ready for tying.

Fig 15-19

A PURSE STRING SUTURE

Ideal for a Sow

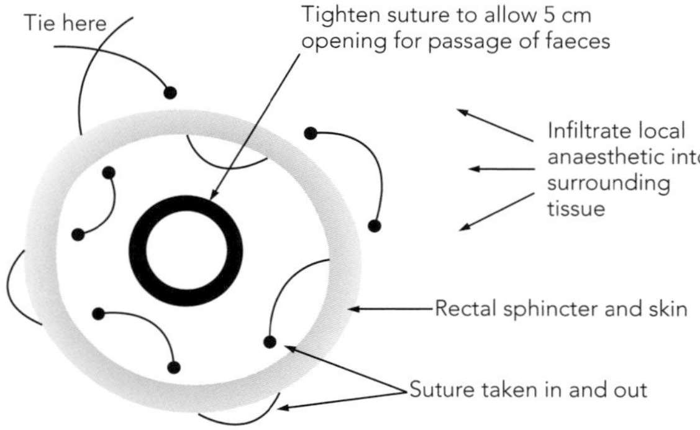

Prolapse replaced, place a finger in the rectum to leave an opening as tying occurs.

Fig 15-20

rectum. If possible the prolapse should be pushed well inside the abdomen after it has been replaced. Tighten the purse string suture allowing two fingers' width. Tie leaving a bow so that it can be slackened if necessary.

9 Inject the pig intramuscularly with long-acting penicillin.
10 Mark the pig and monitor its progress. The suture may be removed after 7 days or left in situ.

Using Corrugated Tubing

1 Proceed as in 1, 2 and 3 previously.
2 Insert a 75 mm x 18 to 20 mm diameter piece of corrugated plastic tubing into the prolapse and place suture material or an elastic band around the prolapse next to the skin. Use electrical conduit tubing.
3 Tie tightly to restrict blood supply and eventually the prolapse will drop off.
4 Inject the pig intramuscularly with long-acting penicillin. Use 2 injections 3 days apart.

Suture and Amputate

If the prolapse cannot be replaced amputation is an option. Proceed under the guidance of your veterinarian.

1 Proceed as for steps 1 to 4 (replacing the prolapse).
2 Place two fingers into the prolapse, just past the anal ring. Insert the half curved needle and suture material 6 mm from the skin into the prolapse, through to the two fingers, and out again. Place the interrupted sutures completely around the prolapse, each one overlapping the other (Fig 15-21).
3 Once all the interlocking sutures are in place and tied the prolapsed tissue is cut off and the stump is slipped into the rectum.
4 The pig should be injected intramuscularly every other day with long-acting penicillin on three occasions.
5 Cull the pig as soon as possible. The majority of rectal prolapses will eventually progress to a rectal stricture (see Chapter 9).

Fig 15-22 and 15-23 show typical examples of rectal prolapses. Pigs suffering rectal prolapses should be made comfortable and culled at the first opportunity.

AMPUTATION OF A PROLAPSE USING INTERUPTED SUTURES

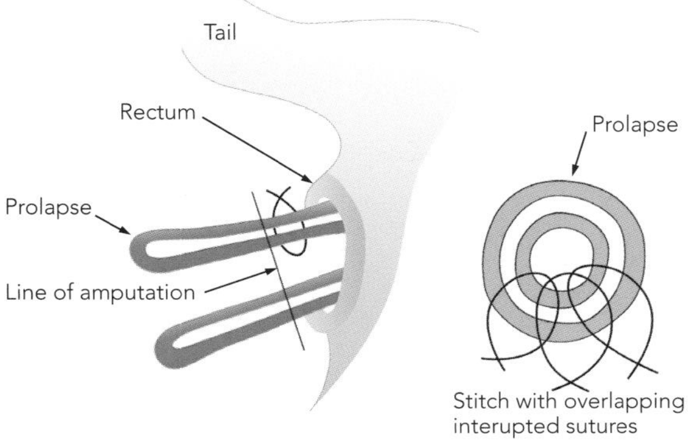

Amputate the prolapse along the line shown and stich up using interrupted sutures.

Fig 15-21

EXAMPLES OF RECTAL PROLAPSES

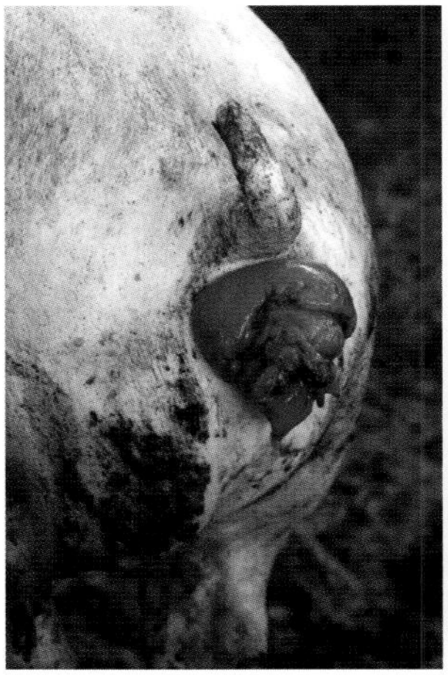

Fig 15-22

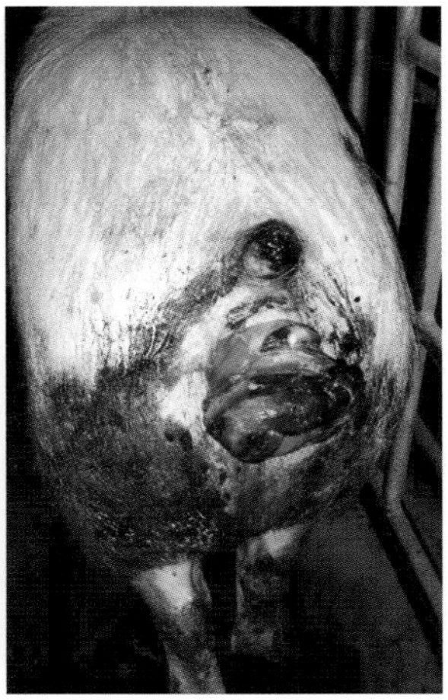

Fig 15-23

Prolapse of the Cervix

Materials Required

- Two people.
- Sedative and/or physical restraint.
- Local anaesthetic 10 ml syringe and 25 mm (1") 21 g needles.
- Large curved needles 38 mm (1½") and 100 mm (4").
- 75 mm curved suture needle.
- Suture tape, scissors.
- Warm water and antiseptic.
- Obstetrical lubricant.

Reason

This occurs when the tissues that support the cervix (neck of the uterus) are weak or fail. With advancing pregnancy and increased abdominal pressure the tissues are squeezed out through the vulva.

Procedure

1. As soon as the prolapse is noted and if abdominal pressure is causing the prolapse, move the sow into loose-housing.
2. If the prolapse ceases no further action is required.
3. Restrain the sow and infiltrate local anaesthetic under the skin and tissues down each side of the lips of the vulva. Replace the prolapse. Apply a tape mattress suture across the vulva as shown in Fig 15-24. Leave a 230 mm overlap of tape and tie in a bow (not so tightly that blood supply is cut off) so that it can then be loosened if farrowing is imminent. Use a 75 mm curved suture needle.
4. Tie the tape to hold the lips of the vulva together and prevent a further prolapse. Untie the tape at point of farrowing.
5. Before the sow is moved into the farrowing crate, build a wooden floor in the crate that slopes downwards to the front. Start at least 150 mm high at the back. Fill in the side of the floor with straw. This slope will hold the prolapse in during farrowing.
6. After the sow has finished farrowing and passed afterbirth, tie the tape sutures again to leave a small aperture.
7. Cull the sow after weaning.

TAPE MATTRESS SUTURE ACROSS THE VULVA

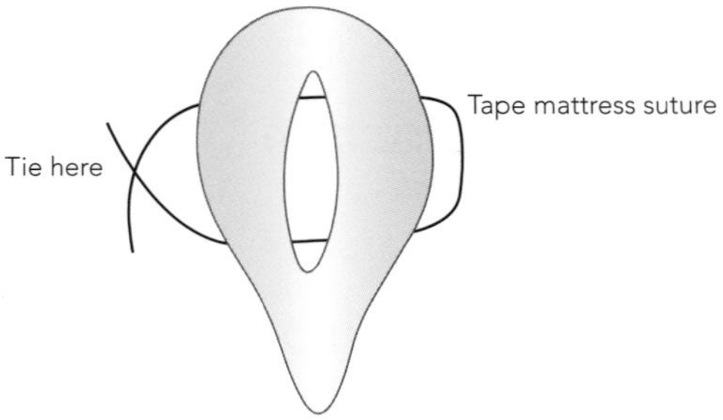

Fig 15-24

Prolapse of the Uterus (Womb)

Materials Required

- Two people.
- Sedative or physical restraint.
- Local anaesthetic 10 ml syringe and 25 mm (1") 21 g needles.
- A small, clean empty bottle.
- Large curved needles 38 mm (1½") and 100 mm (4").
- Suture tape, scissors, warm water and disinfectant.

Reason

This occurs within 2 to 3 hours of farrowing. The uterus turns partially or completely inside out and protrudes as a large mass from the vulva (Fig 15-25 and 15-26).

Methods

Options: euthanise the sow, replace the prolapse, or amputate it (should be carried out only by a veterinarian). In most cases euthanasia is the best option.

Procedure

1 If the uterus has prolapsed completely and the sow is pale and anaemic, shooting is the best course of action. Prior to this give the sow 1 ml 10 iu of oxytocin to contract the uterus and let down colostrum for the piglets, for 5 to 10 minutes. This will ensure a better rate of survival of the piglets for fostering.
2 If the uterus is not completely prolapsed, replacement may be possible provided internal haemorrhage has not occurred.
3 Remove the sow carefully to an open pen, holding the prolapse together in a clean towel or sheet during movement and taking care not to tear it.
4 Sedate the sow and wait 20 minutes.
Keep the prolapse covered with a clean towel or sheet.
5 Clean the surface of the uterus with obstetrical lubricant and place a clean towel beneath it.
6 If possible raise the sow by the hind legs to create a negative pressure in the abdomen and place a bale of straw or suitable support beneath the pelvis to elevate the prolapse.
7 Invert and replace the prolapse. This is not easy and sometimes the full bladder is within the prolapse. It may be necessary to drain it with a large needle.
8 It is essential to return both horns completely to their original position by inserting a clean arm, covered with lubricant and mild antiseptic full length into the replaced uterus. A small bottle held in the hand and covered in obstetrical lubricant will sometimes help lengthen the arm. If the uterine horns are not completely returned to their normal position they will prolapse again.
9 Place two pessaries into each horn.
10 Inject 5 iu of oxytocin. Give a long-acting penicillin injection every other day for three injections. This technique has a poor success rate.
11 Amputation – An operation for your veterinarian. It has a low success rate.

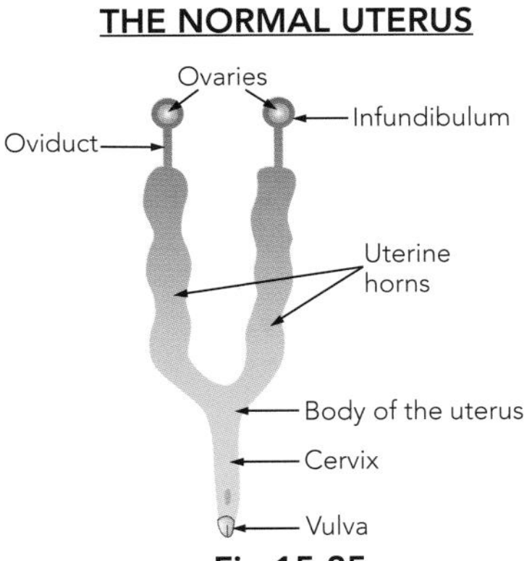

Fig 15-25

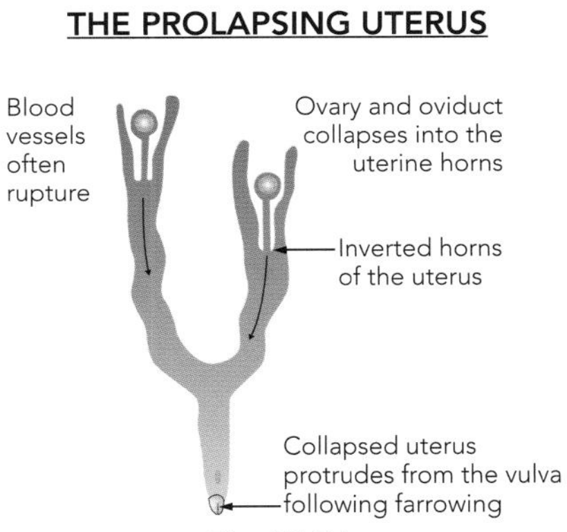

Fig 15-26

Prostaglandin Injections to Initiate Farrowing

Materials Required

- Prostaglandin injection as prescribed by the veterinarian.
- 2 ml syringe.
- Either 38 mm (1½") 18 g (1.2 mm) needle or 21 g (0.8 mm needle).
- Records of sows mating and farrowing dates.
- One person.

Reason

To make the sow farrow at a more predictable time so that management procedures can be adopted to improve piglet survival.

Procedure

1. Calculate the average length of pregnancy in the herd.
2. Identify the farrowing date for each sow.
3. Examine the sow's udder. Is it developed as expected for the farrowing date? The posterior glands always develop to maximum size last. If in doubt delay the injection.
4. The sow can be injected from 112 days of pregnancy onwards but it is best at Day 113. Give by intramuscular injection.
5. The sow will farrow 14 to 30 hours later.
6. Dispose of the needle and syringe immediately.
7. Read the precautions for use in Chapter 4: Prostaglandins and Chapter 8: Controlled Farrowings.

Health and Safety Issues

Ensure that prostaglandins are not used by women of childbearing age or asthmatics. Read the label carefully and follow health and safety instructions.

Prostaglandin and Oxytocin Injections (Intravulval)

Materials Required

- A 21 gauge (⅝") needle (piglet iron injection needle).
- Prostaglandin induce farrowing.
- Instead of a 2ml dose use 1 ml 18 hours before expected farrowing date – and no sooner than Day 114 of gestation.
- Oxytocin – use 2.5 iu (this is 0.25 ml if the concentration is 10 iu/ml) to assist farrowing.
- Oxytocin – use 5 iu (this is 0.5ml if the concentration is 10 iu/ml) to assist milk let down and to assist cleaning the uterus (ecbolic).

Reason

There is a more direct communication between the blood in the vulva and the uterus and ovary.

It is possible to capitalize on this when administering reproductive hormones and reduce the dose required to achieve the desired response. By injecting intravulvally (Fig 15-27) the dose of some reproductive hormones can be halved.

Method

This method covers the injection of prostaglandin/oxytocin into the vulva of a sow or gilt to stimulate farrowing.

Procedure

1. Massage the side wall of the vulva concentrating on the area where the vulva meets the normal skin of the back of the sow.
2. Inject just above this junction – on the vulva side, keeping the needle perpendicular to the skin.
3. Inject quickly.
4. Remove the needle and syringe.
5. The injection may sting the sow/gilt who may even jump up.
6. Relieve the pain response by vigorous rubbing of the area and the pain response will quickly subside.

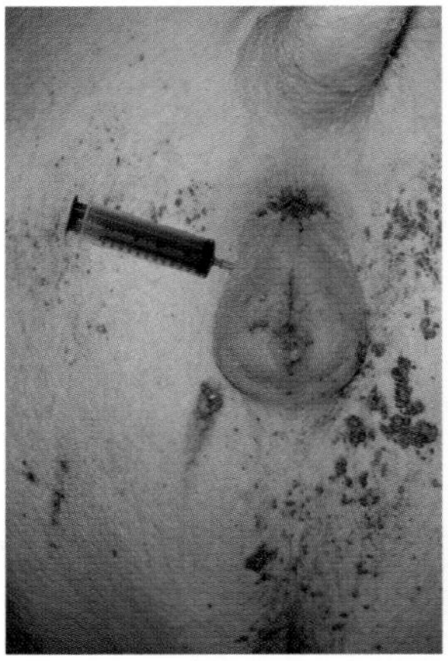

Intravulval injection.

Fig 15-27

Recording Temperature, Humidity and Air Movement

Materials Required
- One person.
- A wind vane anemometer.
- Smoke generation kit.
- Thermohygrometer.
- Infrared thermometer.

Reason
The temperature and humidity, and speed and direction of air flow in a house determine the quality of the environment for the pig.

Measuring Temperature and Humidity
At a basic level, a max-min thermometer can be used to measure temperature. Place at pig level in a protected area e.g. behind a feeder. Read at regular intervals.

Humidity can be recorded using a hygrometer. Alternatively, the thermohygrometer shown in Fig 15-28 is a simple and easy method of recording both temperature and humidity at the same time. Infrared thermometers are useful for monitoring the temperature an efficiency of heat mats.

Alternatively, electronic recorders can be used to record at any interval over a period of days or weeks. Using a computer programme the results can be compiled graphically and analysed.

See Chapter 3 for ideal temperature zones. In cold weather maintain a relative humidity between 60 to 70 %. Below 60 % and in dusty conditions there is a greater predisposition to respiratory disease.

Measuring Air Movement
Air movement can be assessed in a number of ways:
1. Observation of the pigs and their lying patterns. Ideally provide a window in all doors and enter rooms quietly to observe pigs before they are disturbed.
2. The use of non-toxic white vapours that are formed from compounds when they make contact with the air (Fig 15-29).
3. A wind vane anemometer (Fig 15-30).
4. Electronic recorders are also an option as for temperature and humidity.

The comfort zone for pigs at floor level is <0.2 m/s or very little air movement.
At 0.2 m/s the pig may experience little or no heat loss.
At 0.5 m/s the pig may experience a drop of 3 °C (5.4 °F).
At 1.0 m/s the pig may experience a drop of around 6 °C (10 °F).

Medicine and AI storage fridges
Temperature recording should also be done in the medicine fridge and the AI store. Ideally, all medicine stores and AI stores should be attached to the alarm system.

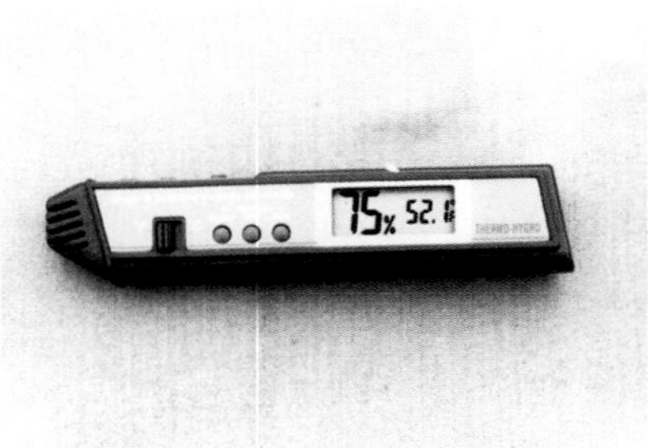

An thermohygrometer provides instant measurements of both temperature and relative humidity.
Fig 15-28

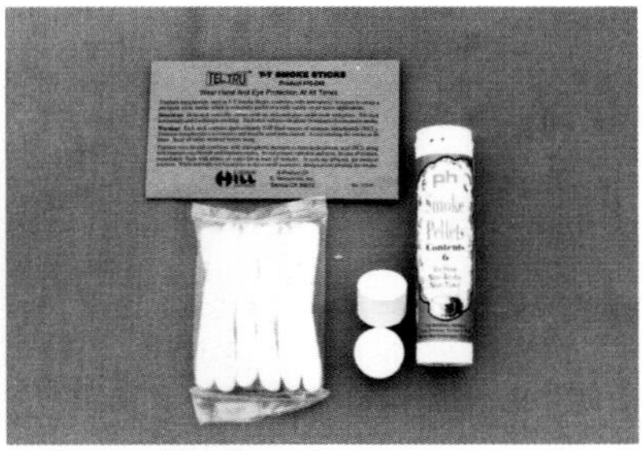

Two types of smoke generators. The sticks release vapour on contact with air. The pellets are started with a match.
Fig 15-29

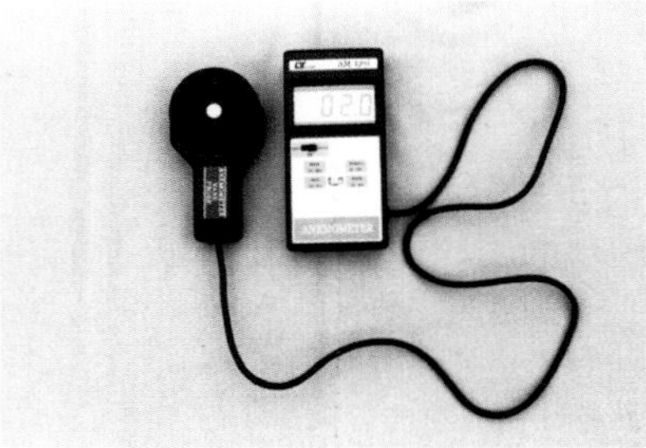

An anemometer. The vane is held in the air and the rate of flow is recorded.
Fig 15-30

Restraining the Pig

Materials Required

- One person.
- Azaperone, acetylpromazine injection or other available sedative.
- 10 ml syringe 25 to 38 mm (1 to 1½") x 16 g needles.
- Nylon rope with loops on each end. Cow calving ropes are ideal, or use a wire noose (Fig 15-31a and b).
- A post or wall ring to attach the rope to.
- A metal or plastic drum.
- Wear ear protectors.

Reasons

These include:

- Close examination of the animal.
- Lancing abscesses.
- Surgical procedures, suturing, replacing prolapses etc.
- Blood sampling.
- Prior to anaesthesia.
- De-tusking boars.
- Euthanasia.

Methods

There are three methods of restraint: physical restraint (e.g. with a noose), sedation by injection or a combination of both.

Procedure

1. Decide whether to use a rope or wire noose. The rope is better if a sow or boar is to be restrained for a period of time.
2. Stand by the side of the pig's shoulder. Hold the wire or rope in two hands and pull into the mouth behind the tusks of the upper jaw.
3. Tighten the noose with an upward movement and pull forwards to restrain the pig. The pig will generally pull backwards.
4. If a rope is used only one person is needed because the rope can be fastened to a post, bar or ring. This is ideal for restraining a sow in a farrowing place. If a wire noose is used it is usually attached to a handle which has to be held.
5. To remove the noose keep the tension above the pig's's nose and then release downwards, pushing the noose towards the pig to allow the pig to shake it off.
6. Physical restraint can also be applied on smaller pigs <30 kg (<65 lbs) by holding the pig upside down in a plastic or metal drum with the hind legs held. Alternatively the pig can be held over the top of a wall. Both methods are useful for replacing rectal prolapses in small pigs.
   For some procedures, e.g. blood sampling from the anterior vena cava, small pigs <30 kg (<65 lbs) can be held on their backs in a trough or on a straw bale.

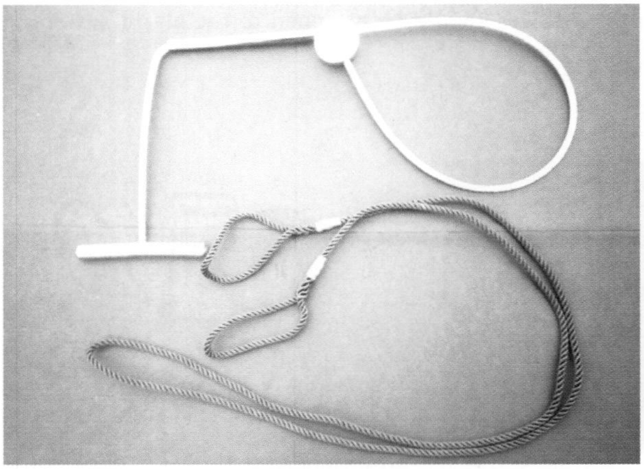

Restraint. A double-looped nylon rope.
A wire noose.
Fig 15-31a

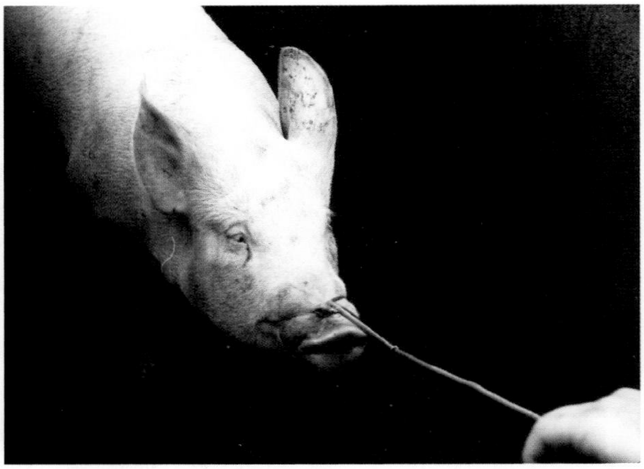

Restraint. Using a wire noose. Note the noose is behind the upper tusks.
Fig 15-31b

Sampling Air for Dust Levels

Materials
- One person.
- A commercial sampling kit (Fig 15-32).
- Pre-weighed filters.
- Access to an analytical balance.

Reason
Air contains 5 elements that may affect both pig and human health. They are total dust (mg/m³), respirable dust (mg/m³), endotoxins (mcg/m³), toxic gases (ppm) and bacteria and fungi (cfu/m³).

Methods
Both total and respirable dust levels are measured using pumps that draw known volumes of air through different sized filters over given periods of time. The filters are weighed at the onset and on completion of the test period and from this information the levels of exposure can be calculated.

Procedure
1 Determine the time of monitoring, usually measured over an 8-hour working period.
2 The pump is calibrated according to whether total or respirable dust is being measured.
3 Load the sampling head with a pre-weighed filter.
4 Attach the harness and pump behind the back.
5 Clip the sampling head at shoulder height.
6 Wear for the monitoring period during work.
7 The filter is removed and weighed by a specialist laboratory.
8 **Calculations**:
The weight of dust is measured and the personal exposure determined. This is often expressed as mg/m³ of air over an 8-hour time-weighted average.
9 **Interpretation**:
There are no universally accepted levels and many countries have their own standards. Measurements can also vary according to the type of equipment and methods used. Suggested guidelines would be:
Total inhalable dust – 10mg/m³;
Respirable dust – 5mg/m³ dust particles between 1.6 and 3 microns.

Carbon dioxide
8 hour exposure	5,000 ppm
15 minute exposure	15,000 ppm

Ammonia
8 hour exposure	25 ppm
15 minute exposure	35 ppm

Hydrogen sulphide
8 hour exposure	10 ppm
15 minute exposure	15 ppm

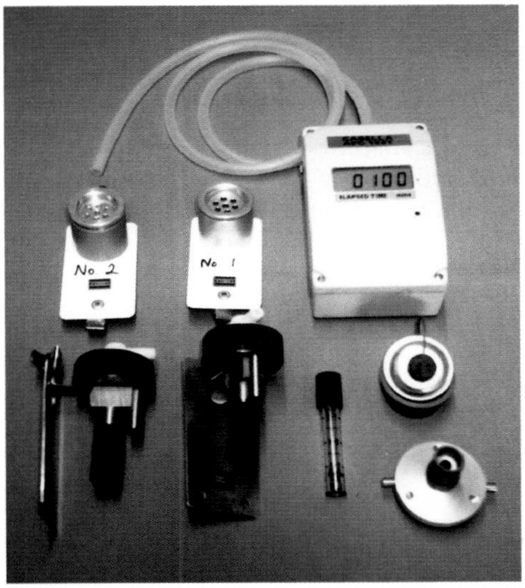

This shows a battery operated vacuum pump and pre-weighed filters.
Fig 15-32

Sampling Feeds for Laboratory Testing

Materials Required

- A new paper bag. A paper bag is preferred over plastic as it prevents the feed sample from sweating and possibly causing a false mycotoxin concentration.
- An indelible marker pen.
- Labels.
- A witness (independent if possible) to the sampling.

Reason

To collect representative samples for despatch to a laboratory for testing. The collection must be made in such a way that the results become meaningful both from a diagnostic and a legal standpoint.

Procedure

1. Check with the laboratory to find out if there are any special requirements necessary relative to the problem.
2. Identify the material to be sampled by date received, batch number, invoice and feed bin if appropriate.
3. If a liquid take three separate samples of at least 500 ml. Always use sterile glass containers. If in doubt, be advised by the laboratory.
4. If sampling feed, take 8 separate samples of 1 kg from different parts of the material and bulk together. Mix these well.
5. Separate into 4 x 2 kg samples.
6. Seal each one. Identify as 1 to 4 and label with full details. Sign and date each bag.
7. Hold samples in a refrigerator at 4 °C (34 °F) until despatch. Deep freeze if more than 3 to 4 days.
8. Send one or two samples to separate laboratories. Send one to the supplier and retain one.
9. Send a full history of the problem, potential poisons suspected or tests required.

Sampling Milk from a Mammary Gland with Mastitis

Materials Required

- One person.
- Swabs in transport medium.
- A sterile container.
- Cotton wool.
- Surgical spirit.
- Oxytocin.
- Needle and syringe.

Reason

In herd outbreaks of mastitis it is often necessary to identify the organism involved and its antibiotic sensitivity.

Procedure

1. Identify the affected gland.
2. Inject the sow intramuscularly with 0.5 ml iu of oxytocin to release milk and wait 2 to 4 minutes.
3. Clean the teat end with cotton wool soaked in surgical spirit and then wipe dry. Wash your hands thoroughly or wear disposable gloves.
4. Squirt milk either onto the swab or into the sterile container. Place the swab into the transport medium.
5. Return the sample to a laboratory as soon as possible but within 24 hours.
6. Note that getting a milk sample from a gland with mastitis can be difficult as milk from an adjacent normal gland is more likely to be expressed.

Sampling the Air for Toxic Levels of Gases

Materials Required

- One person.
- Air sampling device (Fig 15-33).
  These consist of an aspirating pump which draws a known volume of air through glass tubes containing reagents that are specific for the different gases. The tubes required for testing gases in the buildings should include carbon monoxide, carbon dioxide, hydrogen sulphide, methane and ammonia.

Reasons

To assess the efficiency of the environment and ventilation systems for health and safety purposes. To maximise the living environment/welfare of the pig.

Procedure

1. Read the instructions for each gas carefully and attach the glass tube to the aspirator, after removing the end of the tube.
2. Assess the environment to be tested to ensure there is no hazard to human health. Always have two people in the house.
3. Sample the air at pig level by drawing back the suction handle. Repeat the test again at human level.
4. Wait the appropriate time and record the changes.
5. Calculate, from the instructions, the level of the gas.

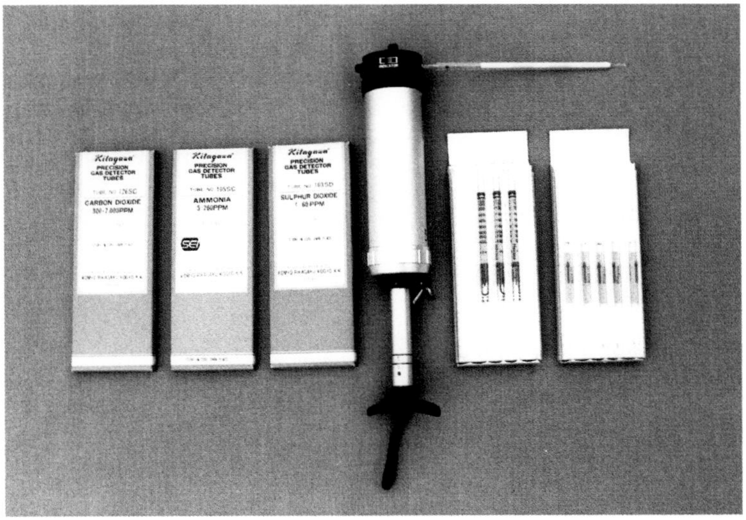

Sampling air for toxic levels of gases. Air is drawn through the glass tubes and colour changes measured.

Fig 15-33

Semen Collection on the Farm

Materials Required
Semen Collection

- One person.
- A separate clean collecting area approx. 3.5 m x 3 m containing a stool or sow dummy. Collection can also be carried out in the boar pen.
- Place posts 0.9 m away from the wall, 0.7 m high and 0.3 m apart to provide a means of escape.
- An environmental temperature of 21 to 22 °C (70 to 72 °F) in the collecting area.
- A dry, non-slip floor,
- A boar protection board.
- A large polystyrene cup for semen collection and a 150 mm diameter plastic funnel.
- Fine muslin cloth or sterile gauze or milk filters.
- Soft disposable polythene gloves (non-toxic to sperm).
- A marker pen and elastic band.
- Items as described under semen preparation.

Reasons

The objective is to collect a sample of fresh semen for examination, or fresh semen samples.

Semen Preparation

- A clean processing area held at 30 °C (86 °F) as shown in Fig 15-34, with an area to change into clean clothing. All bench surfaces must be smooth and easily cleaned.
- Use as much disposable equipment as possible. This reduces the risk of contamination.
- Purified and distilled or de-ionised water.
- Microscope glass slides and disposable pipettes.
- Water bath or incubator.
- Ready made packs of semen diluent.
- Weight scales.
- 1.5, 2.5 or 5 litre semen bags.
- Semen bag holder and dispenser.
- Semen packs.
- Disposable catheters.
- Two thermometers.
- Tissue papers.
- Polystyrene boxes for storage.

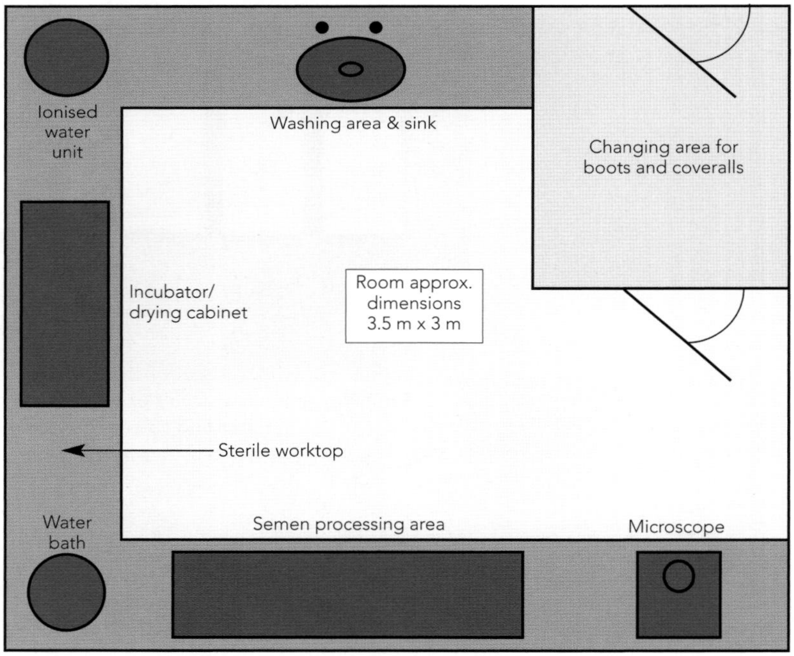

A simple on the farm AI laboratory.

Fig 15-34

Collection Procedures

Initially you will probably need guidance from your veterinarian.

1 Place the gauze over the polystyrene collecting vessel and retain by an elastic band.
2 Prepare fresh semen diluent by adding powder to water, check the temperature in the water bath and maintain at 35 °C (95 °F) for 2 hours before use.
3 Introduce the boar to the pen, squeeze out fluids from the prepuce and clean the area with tissues. Allow the boar to mount the stool.
4 Put on plastic gloves and clean the prepuce area again with tissues.
5 Use a clean, gloved hand for collection. Wait for the penis to emerge and grasp it firmly around the spiral to lock into the hand. Relax the hand slightly as semen flows. See Fig 15-35.
6 Commence collecting the ejaculate into the gauze-covered cup. Avoid any preputial fluids entering the semen. A tissue held in hand will prevent this, or bend the penis during collection.
7 Continue collection (5 to 15 minutes) until all fluid has ceased and the penis twitches, and then remove the sample immediately to the processing area. Approximately 150 to 300 ml will be collected.
8 If a boar is not trained to a dummy sow and a sample is only required to check that the semen is fertile allow the boar to serve a female. Divert his penis away from the vulva and collect into a prepared cup.
9 Return the boar to his pen, thank him and feed him. Make sure to clean and wash the pen after use. Hygiene is very important.

Processing

1 Prepare the sample in the clean processing area where the above materials are kept. Maintain a room temperature of 17 °C (63 °F).
2 Determine volume of the ejaculate. This will vary between 100 to 500 ml. Examine its colour. It should be creamy white with no evidence of blood. A weigh scale can be very useful. 1 ml of semen weighs 1 g.
3 Determine the number of sows to be inseminated on the farm and the amount using 100 ml doses.
4 Check the temperature of the semen and the diluent. They should be within 1 to 2 °C. Dilute the semen to a minimum of 1 in 4 and a maximum of 1 in 10 and place in warmed disposable collapsible packs.
5 Place 2 drops of semen on a warm glass slide and check for wave motion under the microscope.
6 Semen can be stored for up to 5 days at 17 °C (63 °F).

Collecting semen on the farm.

Fig 15-35

Slaughter – Humane Destruction

Materials Required

- A captive-bolt pistol and blank cartridges or a shotgun.
- A wire snare or rope (Fig 15-37).
- A sharp knife.
- A flexible rod to destroy the spinal cord.
- A bottle of barbiturate, 20 ml syringe 38 to 50 mm ($1\frac{1}{2}$ to 2") 18 g needles.
- Two people.
- An area to hold the equipment.
- Ear protectors.

Methods

This can be carried out in one of four ways:

1. Stunning with a captive-bolt pistol. Ideal and recommended for pigs from 6 kg (15 lbs) to adults.
2. Shooting with a shot gun; 4.10 or 12 bore. Suitable for pigs from 100 kg (220 lbs) to adults.

If methods 1 or 2 are used on pigs over 50 kg (110 lbs) weight they should either be bled or the spinal cord destroyed by pithing after stunning.

3. A sharp blow to the head. Suitable for pigs up to 6 kg (15 lbs) of weight.
4. Injecting a strong solution of barbiturate into the heart. Suitable for pigs up to 15 kg (35 lbs) weight. In most countries barbiturates can normally only be obtained and used by veterinarians. Their advice should be sought.

Procedures

Using a Captive-Bolt Gun

1. Instruction must have been given in the use and maintenance of the gun from a licensed slaughterer or a veterinarian. A firearms licence may be required.
2. Check the gun to see that it is assembled correctly. Load with a blank cartridge in the slaughter area. Cock the gun and apply the safety catch.
3. Place the loaded gun on the platform with the other materials. Always point the gun away from the body and other people. Have a second cartridge ready in case it is required.
4. Move the pig to the slaughter area. Make sure the floor is not slippery.
5. Snare the pig and identify the point that bisects lines drawn from the base of each ear to each eye as illustrated (Fig 15-36).
6. Load gun. At all times gun safety must be practised.
7. The person holding the snare stands well forward and the gun is pressed firmly to the head of the pig (Fig 15-38).
8. Remove the safety catch.
9. Place the finger on the trigger at this point.
10. Shoot the pig and either bleed or pith.
11. To bleed either cut the throat transversely or stick the knife deep into the inlet in the chest on the right hand side of the pig to sever the main arteries and veins. Do not enter the chest.
12. To pith, pass the rod into the hole made in the skull by the captive bolt and push to the back of the brain and into the spinal cord to destroy it.
13. Dismantle the gun, clean and oil and return to its secure housing.

NOTE: Some old sows and boars have very thick skulls and a normal cartridge may not be sufficient. Always have heavy duty ones available. This problem does not arise if a shotgun is used.

Using a Shotgun

This is an effective method for pigs from 30 kg (70 lbs) weight up to boars. A 4.10 bore can be used up to 150 kg (330 lbs) and a 12 bore for sows and boars.

1. Remove the pig to an outside area. Do not shoot indoors.
2. Restrain the pig using a wire noose or nylon rope but with sufficient length to allow the person to stand behind the operator using the gun. Alternatively place feed on the floor and wait until the head is still and presented without restraint.
3. Once the pig is restrained, stand in front of the head.

4 Load the gun and close the barrel.
5 Aim and hold the nozzle 0.5 m from the head. Do not place the barrel of the shotgun against the forehead.
6 Release the safety catch and shoot.

Care in the Use of a Shotgun

1 In many countries a certificate/licence is necessary to both hold and use a gun.
2 Always keep equipment and cartridges locked in a secure place.
3 If you have not used firearms before seek expert instruction. Discuss procedures with your veterinarian.
4 Check the gun is unloaded at the outset and leave the breech open. Always point away from people.
5 Select the correct and most suitable area and restrain the pig as appropriate. Make sure that personnel involved are always behind you.
6 Only load the gun after No. 3 above.
7 If the gun fails to fire wait five seconds before opening it to replace the cartridge. Have spare cartridges available in case it is necessary to shoot again.
8 As soon as the gun has been used, open the breech and remove the cartridges. Dismantle, clean and return to its safe custody.

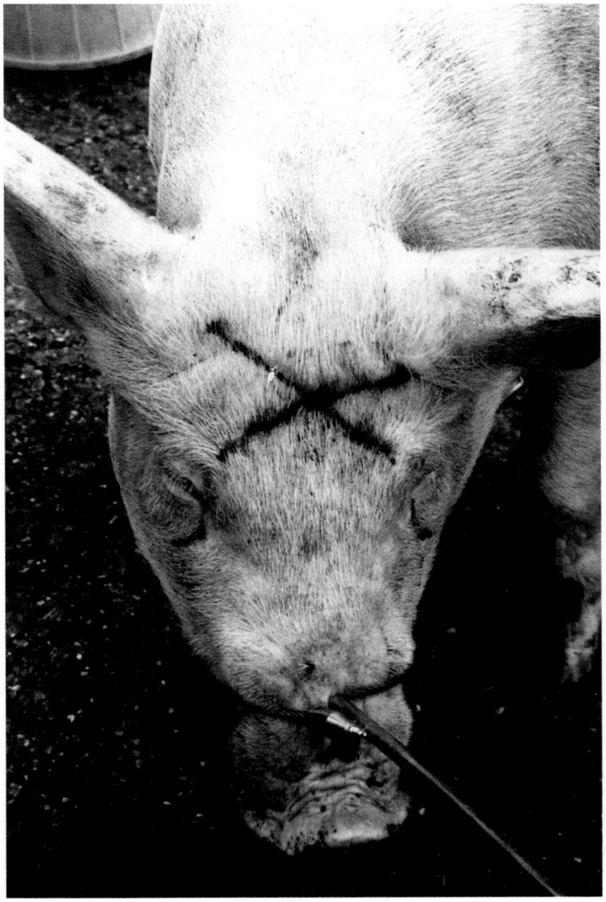

The position for shooting.
Fig 15-36

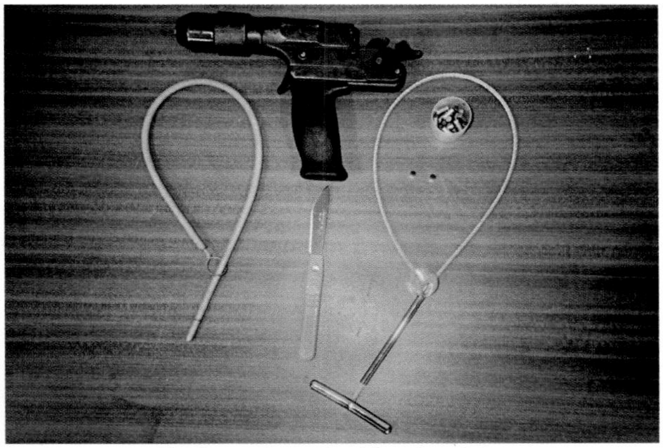

Equipment required for slaughter.
Note the pithing rod on the left of the picture.
Fig 15-37

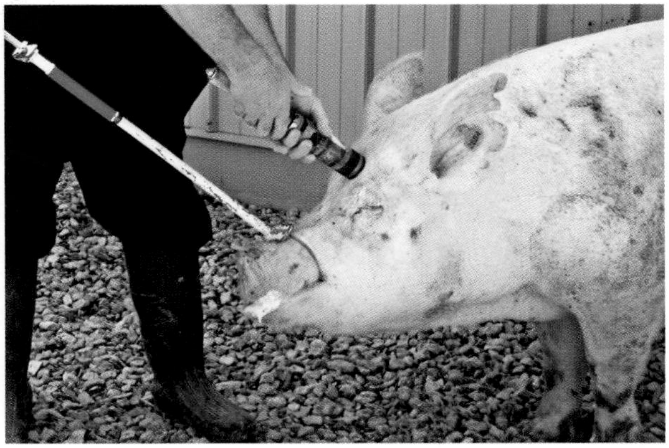

The angle of the gun.
Fig 15-38

Stomach Tube – How to Use One

Materials Required

- One person.
- A plastic stomach tube 4 mm and syringe (Fig 39a).
- Sows' colostrum milked into a container at farrowing, or
- Irradiated cows' colostrum with added porcine immuno globulins or 20 % dextrose solution.

Reason

This procedure is usually carried out at birth to give the piglet colostrum, either the natural product or an alternative source. Dextrose sugar solution can also be given to provide an instant source of energy.

Procedure

1. Fill the syringe and tube with the liquid.
2. Lubricate the tube with either colostrum or liquid paraffin.
3. Identify the piglet and place the little finger at the back of the tongue. Only carry on with the procedure if there is a suckling reflex.
4. Hold the pig under the arm with the hand over the head and the third finger placed in the angle of the jaw.
5. Introduce the tube over the back of the tongue and gently push in feeling for a swallowing reflex. Accidental introduction of the tube into the trachea (windpipe) instead of the oesophagus rarely occurs. If it does the piglet will cough violently. Withdraw it and try again (Fig 15-39b).
6. Gauge the distance from the mouth to the stomach. Resistance will be felt after the tube enters the stomach.
7. Introduce no more than 20 mls at any one time. Repeat the procedure in 2 hours if thought necessary.

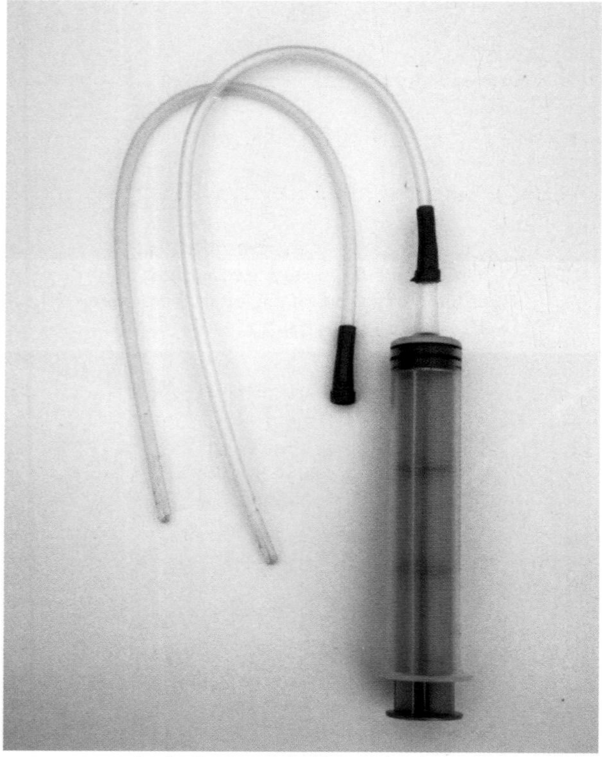

A plastic stomach tube and syringe.

Fig 15-39a

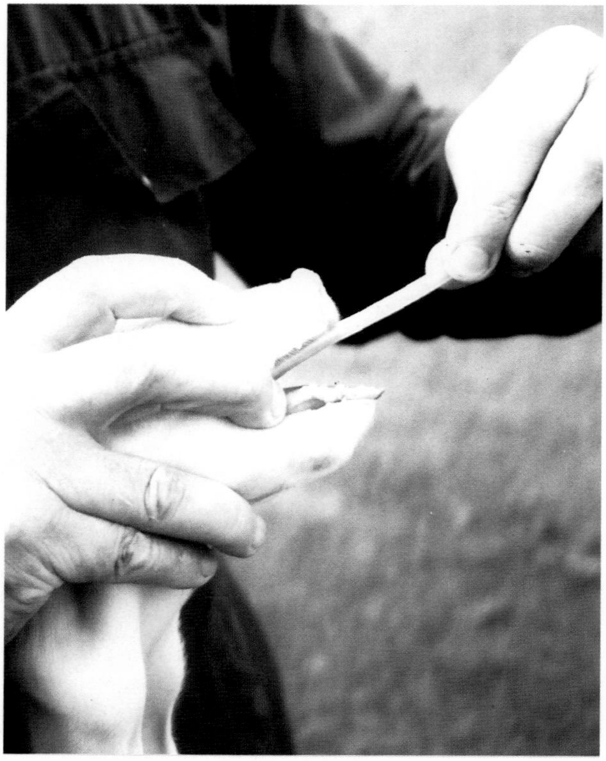

Passing the tube.

Fig 15-39b

Suturing Skin and Muscle

Materials Required (Fig 15-40a-c)

- Two people.
- Azaperone, acetylpromazine injection or other available sedative.
- Local anaesthetic.
- 2 or 10 ml syringe and either 25 mm (1") 21 g needle or 38 mm (1½") 18 g needles.
- Curved suture needles 25 to 50 mm (1 to 2") in size.
- Scissors, forceps, scalpel blades and handle.
- Suture materials, see below.
- Suture tape.
- Warm water, cotton wool, mild antiseptic.
- A table to hold the instruments.

Reason

The common causes of skin and muscle damage arise from trauma in the case of piglets and fighting in older animals.

Suture Materials

These are classified by diameter, using numbers. Two methods are in use.

The imperial system	4/0	3/0	2/0	0	1	2	3	4
The metric system	1.5	3	3.5	4	5	6	7	8
	Small			→		bigger		

Uses: (metric)

1.5, 3	Suitable for piglets' skin
3.5, 4	Suitable for pigs from 30 to 100 kg (70 to 220 lbs) weight
5, 6, 7, 8	Suitable for the skin of sows and boars

Suture materials consist of: synthetic non-absorbed ones, usually made from nylon or cotton tape, and absorbed ones such as catgut.

Non-absorbed materials are used for suturing the skin and rectal and vaginal prolapses. Catgut is used in muscle or subcutaneous tissues. Tape is used to retain a prolapse of the cervix prior to farrowing. The vulva can be easily opened and tied again.

Skin sutures should be removed after 14 days but catgut remains in the tissue and is gradually absorbed. Do not leave any suture material in pigs intended for human consumption.

Procedure

1. Prepare the suture materials and equipment and decide where the operation will take place.
2. Sedate the pig, as per instructions, if over 5 kg (10 lbs) weight or if it cannot be restrained manually.
3. With larger animals it may be necessary to restrain the pig by using a wire/rope noose over the upper jaw.
4. Infiltrate local anaesthetic beneath the skin and wait five minutes.
5. Shave the skin over the wound and clean with warm water and cotton wool. Make sure all hair is removed from out of the wound.
6. Pick up the skin edges with the forceps and loosen them from the underlying tissues with the scalpel blade.
7. Decide the type of suture to be used. There are three methods (ask your veterinarian for advice):
   - **i** Interrupted sutures (Fig 15-41) – Used where there is plenty of loose skin and no tension. Ideal for small piglets. Place the sutures 6 mm apart.
   - **ii** Mattress suture (Fig 15-42) – This is used where there is tension on the skin. The suture turns the edges of the skin outwards. It is best used in pigs over 10 kg (20 lbs) weight. It is important to free the skin from underlying tissues so that it will turn outwards.
   - **iii** Continuous suture (Fig 15-43) – This is used for suturing subcutaneous tissue and muscle and also for the skin. If used in tissue, use catgut, which dissolves away.
8. Clean the suture line after completion and apply an antibiotic cream or inject the pig with long-acting penicillin.
9. Leave the pig to recover and then return it to its pen.
10. Remove the sutures after 10 days.

Suturing Skin and Muscle cont.

Outdoor herds – Suturing may be difficult but when sows or boars suffer skin injuries e.g. after fighting it is wise to administer penicillin by injection and in hot weather to watch out for fly strike. The maggots may get into the damaged skin under the healing skin. If so use ivermectin injections.

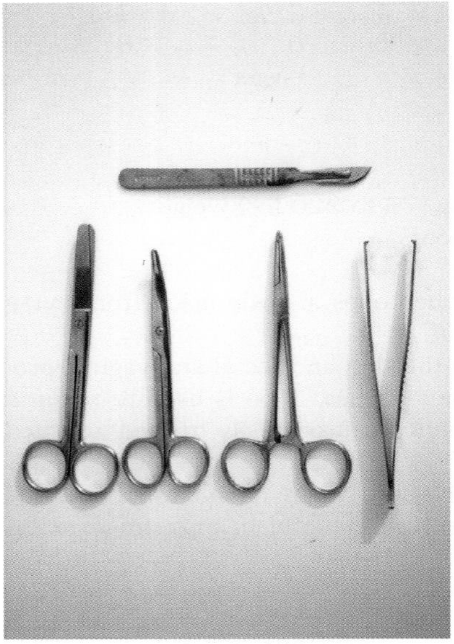

Materials required for suturing. Note the tooth or tissue forceps on the left.

Fig 15-40a

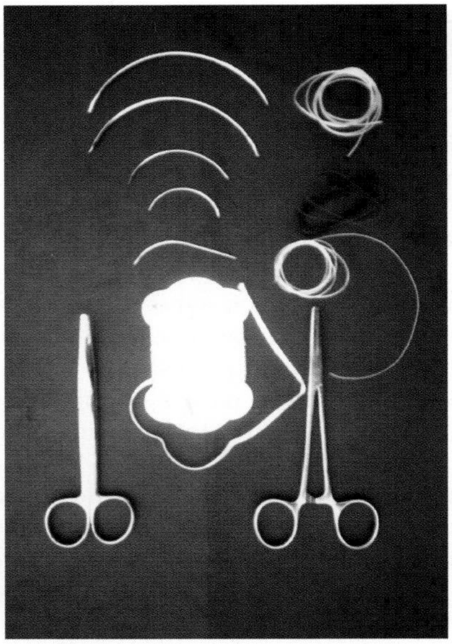

Suture materials. The tape is used for suturing the vulva.

Fig 15-40b

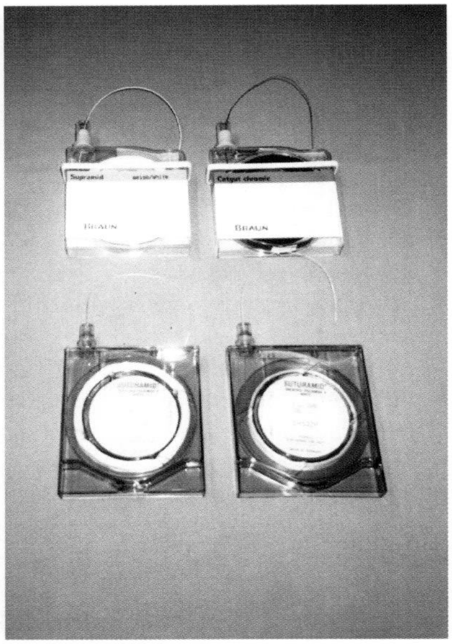

Suture holders and contents.

Fig 15-40c

INTERRUPTED SUTURES

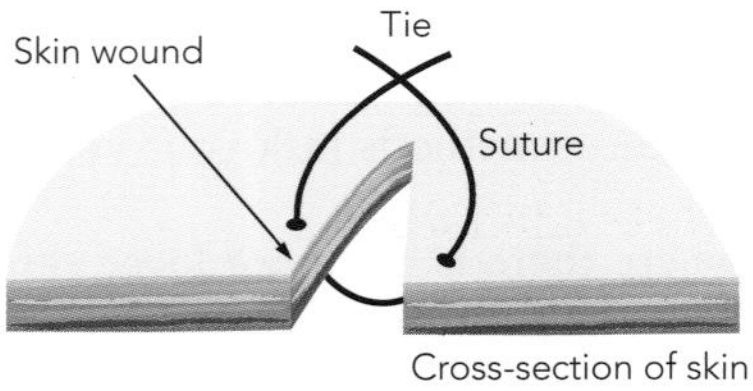

Fig 15-41

MATTRESS SUTURE

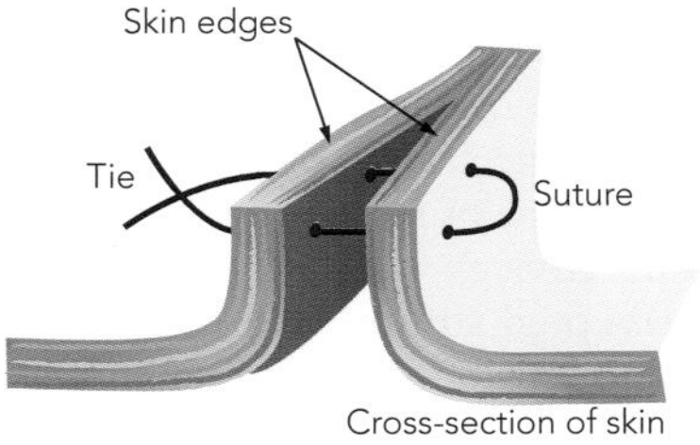

Fig 15-42

CONTINUOUS SUTURE

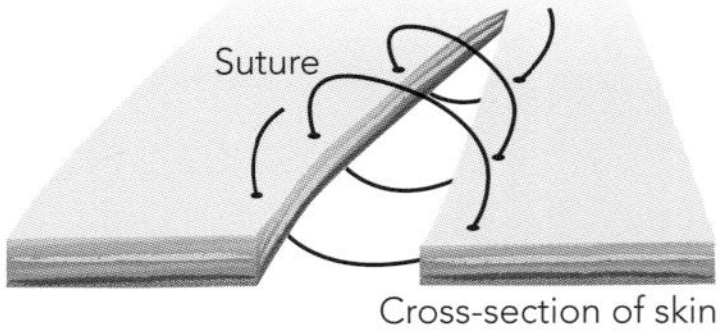

Fig 15-43

Swabbing the Nose and Tonsils

Materials Required (Fig 15-44)

- Two people.
- Wire noose for restraint.
- Mouth gag, small and large.
- Wire or flexible nasal swabs in transport media.
- A polystyrene box for posting swabs to the laboratory.
- An ice pack for use in the box.
- Identification material i.e. tags.
- Torch.
- Ear protectors.

Reason

If a pig herd is affected by progressive atrophic rhinitis it will be carrying toxigenic strains of *Pasteurella multocida*. These organisms are found in the nose and tonsils.

Procedure

1. Clean the nose with cotton wool and remove the swab from its container.
2. Restrain the pig. Pigs less than 10 kg (20 lbs) in weight can be restrained manually, otherwise use a noose.
3. Place the hand over the pig's mouth and nose and rest the other hand on the wrist (Fig 15-45).
4. Insert the swab 25 to 70mm into each nostril to the back of the nose and replace in the protective cover with the end dipped in the transport medium.
5. If the tonsils are to be swabbed hold the pig's head by the ears or restrain by a noose. The operator then opens the mouth using a gag (Fig 15-46).
6. Restrain the pig with a snare if over 15 kg (35 lbs). Open the mouth with the gag (Fig 15-47), the tongue beneath the bar. Depress the tongue and identify the tonsils using the torch. They are rough areas on each side of the soft palate at the back of the throat (Fig 15-48).
7. Rub the swab over each tonsil and replace in the swab case in the liquid. Label with the pig number.
8. Deliver the swabs together with the ice pack to the laboratory within 24 hours.

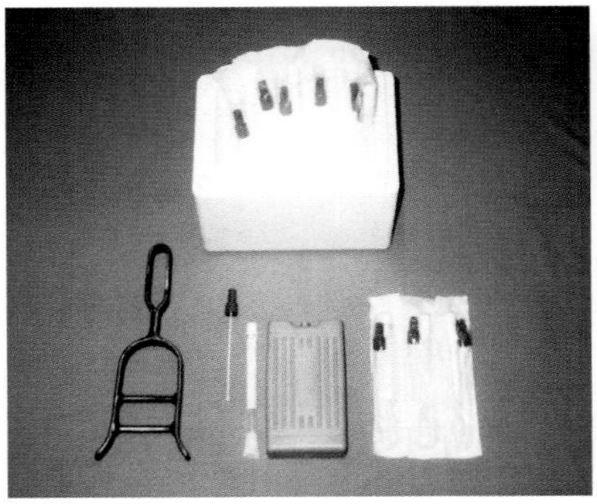
Materials required for swabbing.
Fig 15-44

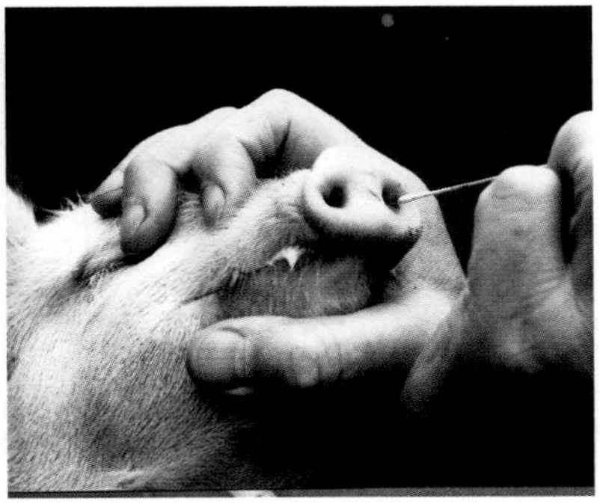
Swabbing the nose.
Fig 15-45

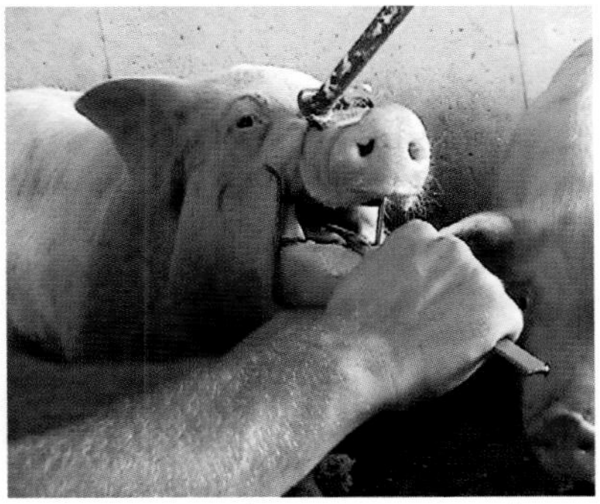
Using a mouth gag to examine the tonsils.
Fig 15-46

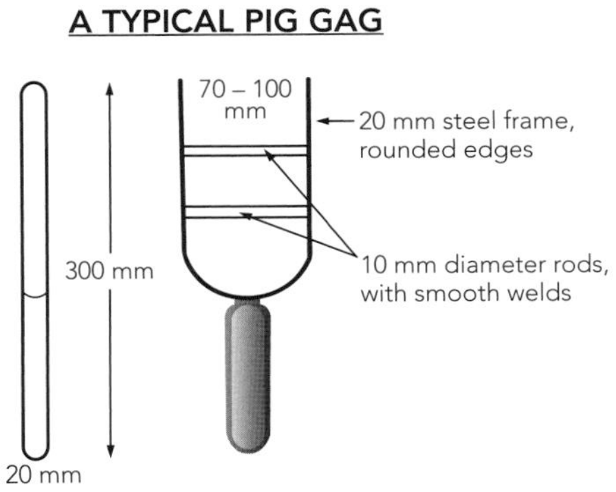

Fig 15-47

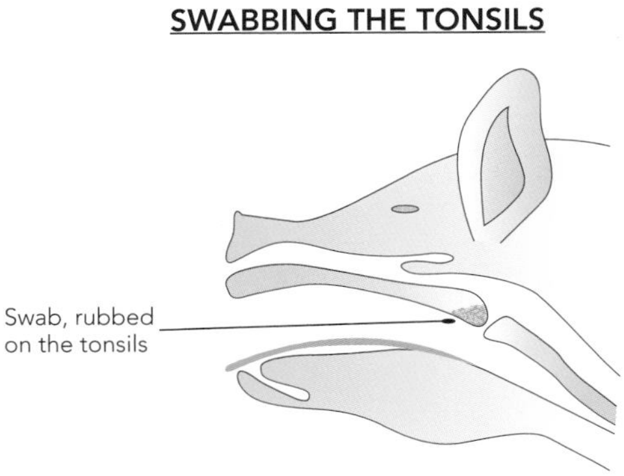

Fig 15-48

Syringes and Needles and Their Use

Materials Required

- Needles – These are measured by diameter in both mm and standard wire gauge (g) (Fig-49a and 49b). Sizes are given in mm and the table below gives comparisons of measurements.

Reason

Syringes can be bought already sterilised and intended for disposal after use or they may be designed for repeated usage after re-sterilisation. Disposable materials should always be used where possible.

Length		gauge (g)	mm diameter	uses
16 mm	(⅝")	23, 21	0.6 to 0.8	Piglets
25 mm	(1")	21, 20, 18	0.7 to 1.2	Piglets, weaners
35 mm	(1½")	18, 16	1.4	Weaners, finishers, sows
50 mm	(2")	18, 16	1.4	Sows
100 mm	(4")	18, 16	1.4	Sows, boars (blood sampling)

Similar sizes may be used for both intramuscular and subcutaneous injections.

- Syringes – multi-dose automatic syringes for vaccinating large numbers. Disposable 2 ml, 10 ml, 20 ml, 30 ml, 50ml.
- Cotton wool.
- Surgical spirit.
- A small dish to hold needles in.

Procedure

1. Select the appropriate syringe size for the volume to be administered.
2. Select the needle by size and length. Always use disposable ones with a protective plastic cap attached to protect needle and operator.
3. Select the bottle to be used and wipe the rubber cap clean with surgical spirit and cotton wool. Shake well and invert.
4. Draw an equivalent amount of air into the syringe, place the needle through the rubber cap into the liquid, expel the air and draw liquid out into the syringe. This is not necessary with a collapsible bottle.
5. Remove the bottle. Place the protective cap over the needle after use.
6. Use a fresh needle after it has been used 5 times.
7. If a number of animals are to be injected keep a 25 mm (1") 16 g needle in the top of the bottle and re-attach the syringe to it each time for filling.
8. Where large numbers of animals are being injected in one period with an antibiotic or preserved inactive vaccine, wipe the needle before each use with cotton wool soaked in surgical spirit. Change the needle every 5 animals and hold other needles soaking in the container of surgical spirit.
   Shake out each needle before use.
   Do not use surgical spirit or disinfectant to store needles or wipe them if you are injecting sensitive substances such as live attenuated vaccines.
9. Always keep part-used bottles in a fridge, do not re-use them too often or store them too long and follow the manufacturer's instructions. Most vaccines need to be used entirely on the day of opening or disposed of. Record the date the bottle was first opened with a marker pen on the side of the bottle.

Self-Inoculation

If self-inoculation occurs take the following actions:

- Immediately inform someone on the farm.
- Read the literature and safety sheets that should be present on the farm.
- Are there any immediate actions which should be taken?
- If it is an oil-based vaccine seek medical advice immediately.
- Ring your veterinarian, doctor or supplier for advice.

Disposal

- Special containers should be used for the disposal of needles. These consist of special plastic boxes with a non-returnable flap. Some have facilities for cutting the needlepoint prior to entry into the box. The total contents are finally incinerated.
- Syringes should be placed in separate containers for incineration.
- Your veterinarian will have facilities for disposal.
- Dispose of all out-of-date medicines appropriately.
- Follow any broken needle policy.

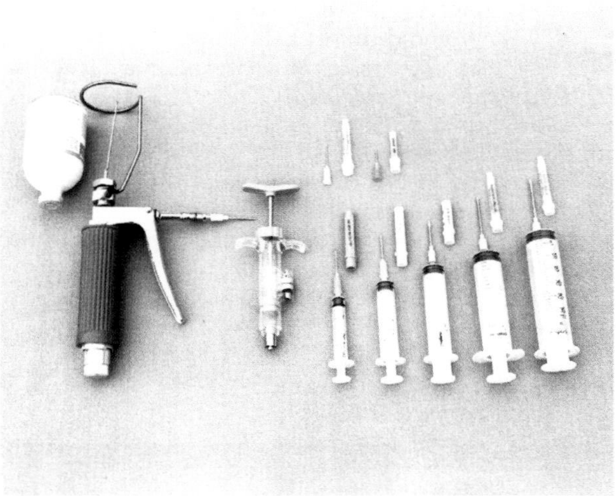

This shows 2 – 3 – 10 – 20 and 30 ml disposable syringes and re-useable multidose ones.

Fig 15-49a

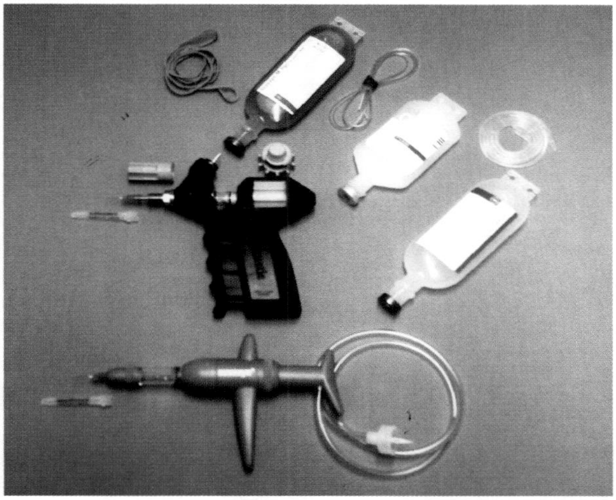

Multidose syringe and flat packs.

Fig 15-49b

Teeth-Clipping

Materials Required

- One person.
- A sharp pair of clippers, either stainless steel or piano wire cutters.
- A small container with mild skin antiseptic or cow teat dip (iodine based). Hold the clippers in the container when not in use.
- A toothbrush to clean the instruments.
- A marker spray.
- Ear protectors.

Reason

Ideally it is best to avoid having to teeth-clip. Teeth clipping is generally unnecessary if the milk supply from the sow is optimized resulting the in sow being able to provide adequate milk for all the piglets.

Piglets are born with needle-sharp canine (eye) teeth and incisors at the corners of the upper and lower jaws. If there is a shortage of milk piglets will fight over teats, traumatising both the sow's teats and the faces of other piglets. Damage to the skin can lead to greasy pig disease. The sow may become reluctant to suckle.

Timing

Teeth should not be clipped until at least 6 hours after birth when all the piglets have taken in adequate amounts of colostrum. Teeth removal at birth before suckling will predispose to joint infection.

Procedure

1. Wash the clippers in hot, soapy water and disinfect before use.
2. Close the blades and hold to the light to check for any damage. If evident, use a new instrument. Always have a spare pair available.
3. Hold the piglets in the creep area or in a bottomless box. Bed the creep area with wood shavings or other suitable bedding.
4. Hold each piglet with the third finger placed in the angle of the jaw and the fourth finger across the trachea to suppress squealing.
5. Place the clippers parallel to the jaw bone. Do not point the clippers into the gum. This will cause damage and infection (Fig 15-50a-c).
6. Make sure that no sharp points of teeth are left. These will predispose to skin damage around the face and the development of greasy pig disease.
7. Mark the piglet.
8. Clean and dry the clippers on completion. Store correctly.

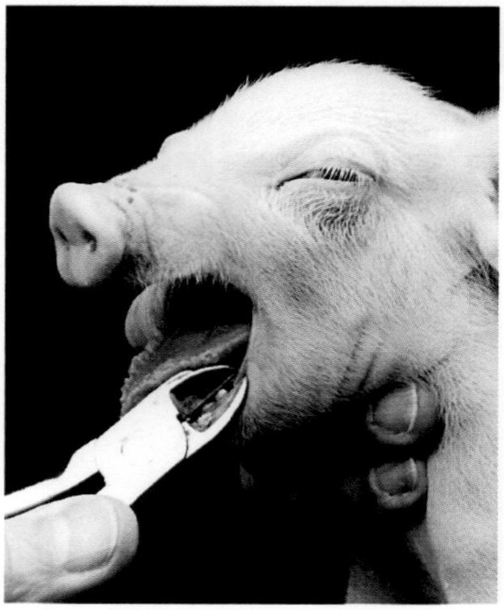

Teeth clipping. Note the holding position and angle of the clippers parallel to the gum.

Fig 15-50a

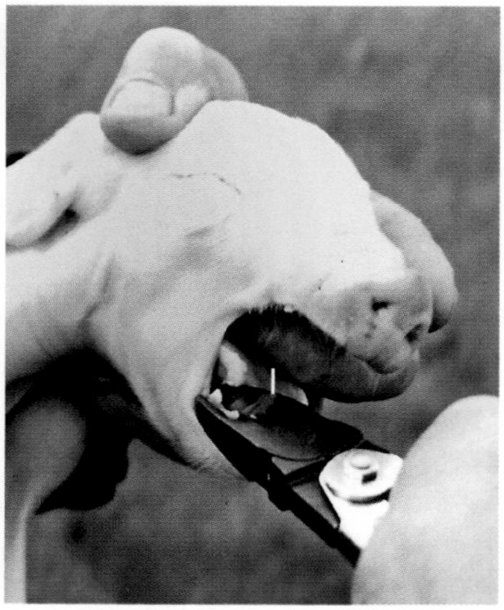

Teeth clipping. Note the wrong angle of the clippers pointing into the gum.

Fig 15-50b

Infected gums due to faulty clipping of teeth.

Fig 15-50c

Temperature Recording from the Rectum

Materials Required
- One person.
- Two clinical thermometers (one spare), mercury bulb type held in a metal casing.
- Surgical spirit placed in the casing to sterilise the thermometer. Always use your own on-farm thermometer.
- As part of the farm's biosecurity the farm should provide the vet with a rectal thermometer.
- Wire noose (may be required occasionally).
- Cotton wool.

Reason
The temperature of the pig indicates that it is either normal, hypothermic, shocked, toxic or fevered.

Temperature °F	°C	Clinical state
99	37.2	Approaching death
99.5	37.5	Approaching death
100	37.7	Toxic or hypothermic
100.5	38	Toxic or hypothermic
101	38.3	Toxic or hypothermic
101.5	38.6	Normal
102	38.9	Normal
102.5	39.2	Normal
103	39.4	Fevered
103.5+	39.7+	Fevered
109	42.8	Approaching death

Procedure
1. If the pig is in a stallwhen becomes traumatised, make it stand. Remove the tail gate if it interferes with the procedure.
2. If the pig is loose-housed it may be necessary to restrain it, or hold the tail and thermometer together.
3. Shake the mercury down the glass tube of the thermometer towards the bulb and examine to see that it records below 36 °C (97 °F). Wipe clean with cotton wool.
4. Lift the tail of the pig, place the thermometer 50 mm into the rectum and angle gently into the lining or mucous membrane.
5. Record for 30 seconds, remove and read the temperature.

Udder – Methods of Examination

Materials Required

- One person.

Reason

Changes in the udder can be divided into three broad areas: poor development, congestion or oedema and mastitis. It is important to differentiate between these to determine cause and subsequent action.

Procedure

1. Examine the sow lying down or standing. In outdoor herds this is best carried out when the sows are feeding and particularly when they are drying off post-weaning.
2. Assess the visual appearance of the glands.
   Are any enlarged? This suggests mastitis or congestion.
   Is there evidence of trauma over the skin and teats? This suggests mastitis.
   Is the skin overlying the glands discoloured and red? This suggests mastitis and toxaemia or fever.
   Is there generalised discoloration of the skin? This suggests toxaemia or fever.
3. Palpate each gland separately.
4. To do this place the teat in the palm of the hand and firmly hold the gland with the fingers and thumb grasping it all.
5. Find a totally soft normal gland and apply pressure until the sow shows signs of discomfort.
6. Apply slightly less pressure to the remaining glands. Any reaction by the sow will indicate discomfort and pain and possibly early mastitis.
7. Press the thumb or first finger deep into the udder tissue, hold for 2 seconds and then feel for and look for an impression left in the tissue. This will indicate oedema.
8. If all glands are normal with no pain and poor udder development, suspect agalactia, possibly due to water shortage.

Umbilical Cord – Applying a Clamp

Materials Required

- Two people.
- Plastic self-fastening umbilical clips – those used for babies. Electricians' nylon wire binders are also useful.
- Scissors or scalpel blade and handle. Use ones that can be reused.

Reason

Persistent haemorrhage from the umbilicus at birth occurs in certain herds at a high frequency and occasionally those using prostaglandins. Piglets die from anaemia. Clamps should be used on all hysterectomy-derived piglets and on farms where piglets are confined immediately after birth if navels become damaged.

Procedure

1. This is a two-person operation ideally, but with practice one person can become proficient.
   Hold the piglet by the hind legs with the umbilicus facing the operator.
2. Place the clip across the cord approximately 6 mm below the skin. Do not fasten too near the skin because this can damage the umbilicus and may cause umbilical ruptures (Fig 15-51). Small cable ties are an alternative option.
3. Squeeze the clip to close it.
4. Cut the cord below the clip.
5. It may be necessary to place a piece of adhesive tape around the body to protect the clip. This is essential if hysterectomy-derived piglets are being transported.

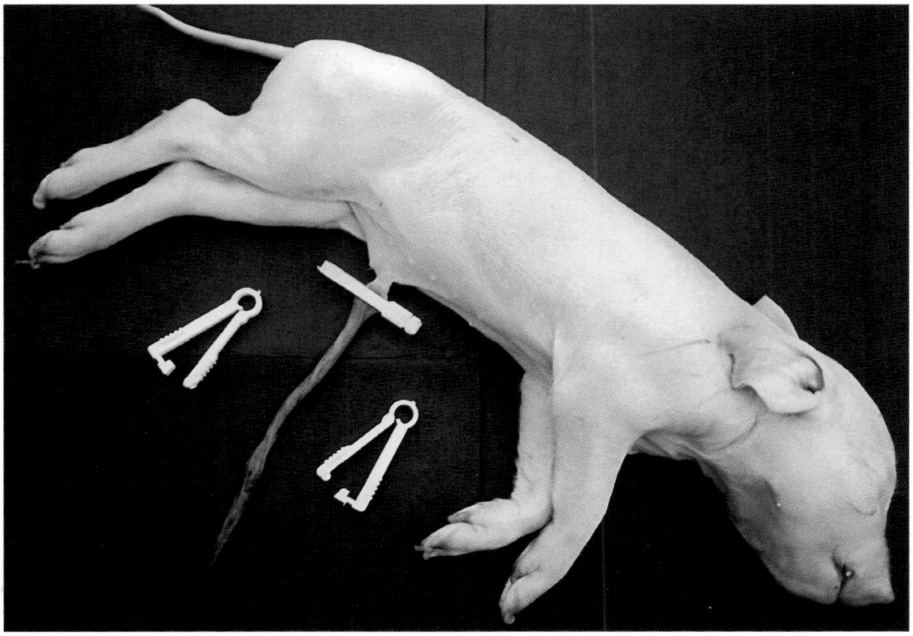

Umbilical clamp. Note the position on the cord.

Fig 15-51

Vasectomy

Materials Required

- A surgical kit containing scissors, scalpel blade, artery forceps and suture materials.
- Sedative and barbiturate anaesthetic.
- Warm water and antiseptic.
- Cotton wool.
- A bale of straw or low table.
- A clean, well-lit pen.
- Syringes 16 mm (5/8") 23 gauge needles.
- Two people plus the veterinarian.
- Penicillin cream and long-acting injection.

Reason

A vasectomy consists of cutting and removing a piece of the vas deferens (sometimes called the spermatic cord) to make the boar sterile but still capable of a normal but aspermic mating. This is a procedure that will only be carried out by the veterinarian, but see also the procedure for epididectomy.

Method

This involves a general anaesthetic although it is possible to perform the operation under sedation and local anaesthetic or epidural anaesthetic. It can be carried out at any time but is best from 50 to 100 kg (110 to 220 lbs) weight.

Procedure

1. Identify the boar by ear tag or tattoo. Starve him for 12 hours and sedate per sedative instructions. Wait for 20 minutes.
2. Restrain with a noose and inject anaesthetic using pentobarbitone into the ear vein. Anaesthetise until the eye reflex is just lost – and no more.
3. Place the boar on his back supported by the people at the head and the tail. The operator should not handle the pig but scrub up in preparation for the operation.
4. Wash and clean the area between the hind legs and over the testicles with warm water containing anaesthetic. Dry and finally soak in surgical spirit.
5. Identify the skin area above and just to the rear of the inguinal canal. Make a 50 mm incision through the skin on one side.
6. Using the two forefingers blunt, dissect through the subcutaneous tissue until the complete cord and blood vessels are seen. It helps if the testicle is moved to highlight the structure. Pass a finger round the complete cord and pull upwards to the surface. Hold in position by passing forceps beneath.
7. Identify the vas deferens, a small white tube beneath the clear sac or the tunica.
8. Make a small incision into the tunica to expose the vas, raise it and clamp artery forceps to it approximately 40 mm apart. Remove this portion.
9. Tie off each end of the vas and fold back and re-tie.
10. Return the vas into the sac and replace the complete cord.
11. Suture the skin with mattress sutures.
12. Repeat the operation on the other side.
13. Inject the boar with long-acting penicillin. Ideally tattoo the ear with a V or number for permanent identification.
14. Place the two removed pieces of the vas in liquid preservative for histological examination to confirm its structure. It can also be identified by squeezing fluid from the tubes onto a glass slide and examining each drop for the presence of sperm.
15. Do not use the boar for 3 weeks and then only after a semen sample has given clearance. Record the date of the operation and the boar number on the bottle containing the vas.

Vulval Haematoma - Treating a Haematoma

Materials Required

- Two people.
- A noose for restraint or sedative.
- A 1 metre length of strong cord or bandage.
- Warm water, cotton wool, and mild antiseptic.
- Scissors, suture material or tape and needles.
- Local anaesthetic syringe and 21 g 25 mm (1") needles.
- Tissue forceps, sterile artery forceps.

Reason

Vulval haematomas are common in gilts, occasionally in second-litter females and rarely in sows. If they rupture, continual haemorrhage may be fatal (Fig 15-52).

Procedure

1. Sedate or restrain the sow.
2. Clean the vulva.
3. Infiltrate local anaesthetic into the skin anterior and to the front of the haematoma.
4. **Option 1** – Place the cord or bandage between the lips of the vulva, continue behind the haematoma and tie as a tourniquet. It may be necessary to keep tightening this. Observe for 2 to 3 minutes to see if the bleeding stops.
   Option 2 – If haemorrhage continues place a mattress suture behind the haematoma and tighten.
   Option 3 – If haemorrhage continues open up the haematoma and identify the bleeding vessels. Clamp with the sterile artery forceps and tie off the bleeding points. Seek veterinary advice.
5. Place clean bedding behind the sow so that good observation can take place. If the sow is on slats use a paper bag.
6. Pad the back of the crate or alter it to prevent further trauma and crushing of the vulva.
7. Inject the sow with 10 to 15 mls of penicillin.
8. Watch the sow every 15 minutes over 2 to 3 hours to ensure no further bleeding occurs.

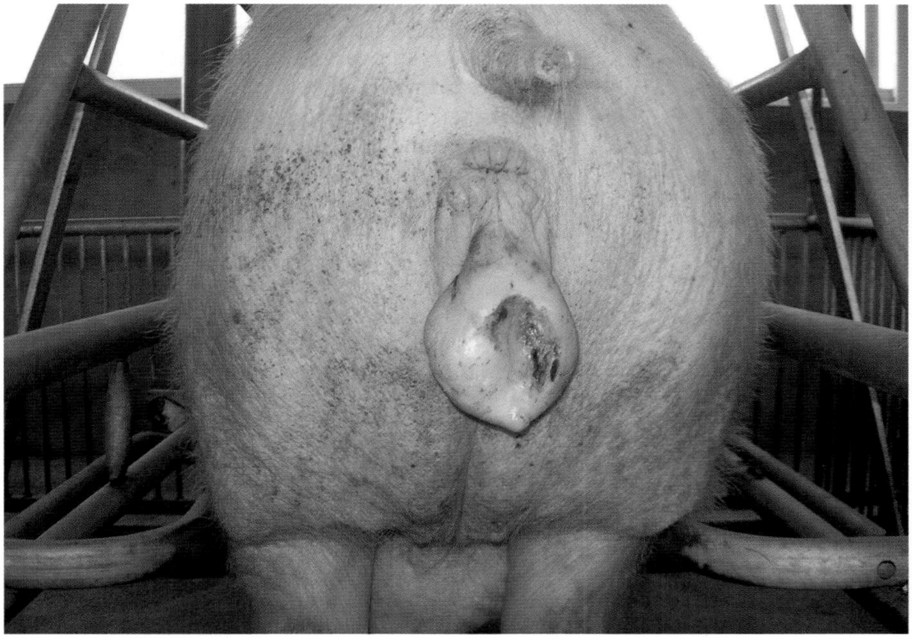

Vulva oedema which can become a haematoma if it becomes traumatised.
Gilts can bleed to death as a result.
Fig 15-52

Water – Cleaning and Sterilising a System

Materials Required

- Sterilising detergent.
- Spanner.
- Pressure washer or piped source of water.

Reason

Any water system with or without a header tank may become contaminated with bacteria, dirt, rodents or potential poisonous substances such as fungal or bacterial toxins or warfarin. Regular cleaning is essential and the frequency is dependent upon the degree of contamination.
Farrowing houses – clean and sterilise between batches.
Flat decks and nurseries – clean and sterilise between batches.
Water bowl and troughs – inspect and clean daily.

Procedure

1. Turn the water off or fasten the float up in the header tank.
2. Remove the last nipple drinker or bowl on the line. Periodically remove all nipples, clean and replace.
3. Drain out the system. Inspect and clean out the header tanks.
4. If water bowls are used open up, clean and wash out.
5. Open up the water supply and allow to flush through. Use a pressure washer if necessary.
6. Add a detergent steriliser or chlorine based dairy steriliser to the header tank and allow to flush through. Hold this water in the system for 30 minutes.
7. Drain the system, refill and check all nipples are flowing.
8. Periodic check list:
   Assess the number of pigs per drinker.
   Check the rate of water flow and:
   - The colour and quality.
   - The build-up of deposits in the pipes.
   - The pressure from the holding tank.
   - When the system was last cleaned.
   - The covers over the holding tank.
   - The presence of vermin in the holding tank.
   - If there is a mastitis or scour problem carry out bacteriological examinations of the water in the pipes, tanks, and drinkers. Sterile containers and swabs can be used.

16 WELFARE AND HEALTH

This Chapter was updated with assistance from FAI Farms
who develop 3E sustainable food and farming.
FAI Farms are based in Oxford, UK.
www.faifarms.co.uk

Chapter 16

16 WELFARE AND HEALTH

Guidelines to Good Welfare Practices

The reasons for considering animal welfare may have an ethical basis driven by consumer concern, but increasingly science is demonstrating that contented animals receiving proper husbandry care are healthy animals that produce safer, better quality food. Importantly, research is finding ways to improve animal welfare that do not necessarily increase overall production costs. Better husbandry techniques, increasing product quality and reducing losses from disease or injury can all contribute to offsetting production costs.

These guidelines are based upon the requirements of EU legislation and on several countries' industry guidelines, in particular Australia, New Zealand, Canada and the United States. These laws and guidelines are summarised below and give a lead to future developments across the world. Good health is of primary importance in supporting the welfare of farmed pigs and many of the factors associated with health are discussed at the beginning of the relevant chapters in this book, in particular the management and control procedures specific to particular diseases. The reader will be referred to these throughout.

Different countries have different attitudes towards standards of welfare and the points raised here should be considered in relation to the accepted standards and legal requirements in your country. Wherever possible, you should aspire to the highest standards of welfare on your pig farm.

EU welfare regulations

The UK and other EU Country standards are derived from European Council Directive 2008/120/EC which lays down minimum standards for the welfare of pigs. This directive acts as a minimum baseline for pig welfare legislation at individual country level throughout the European Union.

Australian welfare regulations

Legal requirements for livestock welfare in Australia are covered by the Australian Animal Welfare Standards and Guidelines. Pig welfare is specifically covered in the Model Code of Practice for the Welfare of Animals: Pigs – Third Edition (2008) and covers a range of topics similar to those in the EU.

New Zealand's welfare regulations

Welfare guidelines for pigs come under the Animal Welfare (Pigs) Code of Welfare (2010) that sets out the standards and recommendations concerning all aspects of pig care. Owners and persons responsible for the animals' care must achieve the minimum standards in order to fulfil their obligations under the Animal Welfare Act (1999). The codes are reviewed every 10 years.

Canadian welfare guidelines

At the time of publication, pig welfare guidelines in Canada were governed by the Code of Practice for the Care and Handling of Pigs (1993). This was in the process of being updated by the National Farm Animal Care Council with completion stated for summer 2013. Priority issues cover pain control, methods of euthanasia, space requirements and social management of sows. (See the Animal Care at Work documentation.)

US welfare guidelines

Pig welfare guidelines for US producers are covered in the Swine Care Handbook (2003) published by the National Pork Board. Swine welfare is directed by the Pork Board's Animal Welfare Committee.

Welfare Auditing

Several countries have attempted various auditing programmes. While superficially beneficial, there are often many 'assured' farms which do not meet higher standards. Political and social pressures have been blamed for this failure to achieve standards.

The UK Real Welfare scheme initially looked at finishing pigs, concentrating on a statistically significant number of pigs and then assessing the number of pigs in this group which should have been hospitalized, those which were lame and those with tail lesions and body marks. This was then compared to a national median value. Farms which were in the worst 25 % of assessed farms were then given advice on how to improve the situation.

The US has a similar scheme (SWAP – Swine Welfare Assurance Program), which for many years looked at the whole farm, assessing not only body lesions but also pig behaviour in terms of flight times.

All these schemes provide excellent teaching opportunities for veterinarians and other advisers. However, the time and cost of some of the schemes can be prohibitive and the rewards in terms of measurable improved pig welfare, cost of production and profits often make them difficult to justify. Unfortunately, the driver for some of these schemes is supermarkets looking for a marketing edge and by the time the industry has changed, the supermarkets' focus has also changed and the promised welfare payment "bonus" does not materialise.

We have to enhance the welfare and wellbeing of the pigs we farm and this chapter discusses these issues from a welfare-focused perspective. It is also understood that economics, farming practicalities and the fact that customers consider pig meat a cheap source of nutrients - which is why pork is the world's favourite food – all need to be taken into account at the farm level.

The Five Freedoms

The five freedoms represent the basic requirements of animals and provide a starting point for the provision of good welfare for your own pig herd:

1. **Freedom from hunger and thirst** - Ready access to fresh water and a balanced ration which maintains full health and vigour.
2. **Freedom from discomfort** - Provision of a suitable environment and a comfortable resting area.
3. **Freedom from pain, injury and disease** - Prevention, and prompt diagnosis and treatment when injuries or disease occur.
4. **Freedom to express normal behaviour** - Provision of sufficient and appropriate space, interest and the company of other pigs.
5. **Freedom from fear and distress** - Sympathetic stockmanship, stable environmental conditions and freedom from aggression by other pigs.

Inappropriate provision of these freedoms can lead to stress as the animal fails to cope physically and mentally within the restrictions of its environment. Stress has been shown to increase susceptibility to disease and leads to reduced growth rates in piglets and lower reproductive fitness in sows. There is, therefore, a clear incentive to ensure the welfare of the animals to encourage productivity.

Of these five conditions, freedom to express normal behaviour is the most difficult to provide on intensive units and welfare is inevitably compromised. Expression of normal behaviour requires sufficient space and environmental stimuli to enable the pig to move freely and carry out innate behaviours including foraging, rooting, grazing, wallowing and social interaction with other pigs. Different housing systems enable provision of this freedom to varying degrees.

Where a suitable husbandry system is in place, the approach and ability of the stockperson can have the greatest influence on provision of the five freedoms and therefore good welfare of the pigs. The importance of practising good management is discussed throughout the chapter. Other aspects which are not spelt out by the five freedoms but which may be implied by them are:

- The provision of a caring and knowledgeable management team.
- The provision of light during the hours of daylight.
- The avoidance of unnecessary mutilation.
- The provision of emergency arrangements to cover disasters such as fire, the breakdown of mechanical services and the disruption of supplies.
- The use of appropriate transport and handling.
- Humane slaughter.

Good welfare conditions generally result in less disease, better production and potentially greater profits.

Welfare Recommendations for All Pigs

There are fundamental factors that affect the welfare of all pigs throughout their life stages. These encompass both good management and appropriate environmental design.

Management of Welfare

Quality of management

As already stated, quality of management is crucial. People should be selected for their standard of stockmanship which includes a caring nature and ability to establish empathy with their pigs (see Chapter 3). Well-motivated, capable staff must have the correct knowledge and skills for providing appropriate welfare to the animals, particularly in handling, preventing ill health or distress to the animals, and providing appropriate care where necessary. The farm manager must ensure that staff are fully trained and that all required tasks are performed and guidelines are adhered to. This can be supported by the use of a health and welfare plan that is updated on a regular basis or when changes are made to the farm's husbandry practice. Attention to detail and the monitoring of daily routines are vital.

Education and understanding

There should be an ongoing process of education as part of the management system. Young people entering the farm should be trained by more experienced stock people, in particular on how to recognise normal pig behaviour and be able to identify deviations from this. To do this requires a recognised training programme on the farm that is continually reinforced. Special training must be provided to staff carrying out tasks that compromise pig welfare and cause pain to the animal (e.g. teeth-clipping/grinding, tail docking). The farm manager is responsible for ensuring that personnel are trained and highly competent in these procedures.

Poor welfare of the growing pig leads to inefficient growth and feed conversion.

Observation and timely response

Each day a detailed examination of all the pigs on the farm should be carried out. Compromised and sick animals should be identified and procedures adopted to ensure that they are comfortable and not victims of aggression from other pigs. This should involve movement in to a designated hospital pen (see below). Whilst identifying and moving a compromised animal may seem simple, nevertheless it is most important. A typical example would be the failure to identify a tail-bitten pig. The consequences of this are considerable, not only for the welfare of the pig but also because excessive tail damage invariably results in abscesses of the spine and a condemned carcass. The procedures for carrying out and documenting clinical observations on the farm have been considered in Chapter 3. It is strongly recommended that these formats are used daily by each person responsible for a section or part of the farm and for the supervision of trainees. In cases where pigs do not respond to treatment or the cause is unknown veterinary expertise should be sought.

To ensure the health and wellbeing of the pigs, daily checks should also include inspection of the equipment and the immediate environment surrounding the animals. Areas requiring cleaning, adjusting or fixing should be dealt with promptly. Importantly, thorough observations and a timely response require that adequate time and sufficient staff are made available for the tasks.

Special care of compromised, sick or injured animals

Compromised, sick or injured animals have poor welfare and must be dealt with immediately by administering the appropriate treatment and where appropriate carefully moving them to a designated hospital pen (see Chapter 3). If the cause of illness or correct treatment is unknown, or if the animal fails to respond to provided treatment, contact your farm vet for advice without delay. The hospital pen should be kept clean and dry and provide an appropriate supply of water and feed. Compromised animals should be attended to and checked regularly. Animals that do not respond to treatment (7 and 14 day rule, see Chapter 3) should be euthanised using a humane method as quickly as possible so that suffering is not prolonged. Lame and sick animals should not be transported except for travel for veterinary assistance.

Handling

The quality of human interaction can have a significant impact on the welfare of the pigs and the stockperson should encourage positive interactions that minimize stress at all times. Handling and moving of pigs should be done in a calm and gentle manner. Pigs should be moved in small groups and allowed to move at their own pace. Using well lit, clear routes with solid sides and curved corners will encourage the pigs to move in the required direction (Fig 16-1). Ideally the route should allow two pigs (which can see each other) to walk side by side within their own passageway. Avoid slippery floors and steep ramps (must be less than 20°) as these can cause injury, and minimize stressful incidents such as shadows, floor changes and sudden or loud noises. Applying excessive pressure or force when handling or moving pigs may cause pain and stress and reduces wellbeing. Electric goads should not be used in any circumstances.

Tethering of pigs should only be used for restraining animals undergoing examination/treatment or used as a temporary measure of restraint for moving a boar, for example. Continuous tethering of animals compromises the five freedoms and should only be used when other housing and management alternatives have failed. In this circumstance, ensure that the tether does not cause any injury or discomfort to the pig and that it is of adequate length to provide the animal with freedom of movement to fulfil its behavioural needs and reach all provisions. Tethers must be regularly inspected and adjusted or replaced wherever necessary. They are banned in many countries.

Mixing pigs

Pigs are highly social animals that form complex social hierarchies within their group. Introducing new pigs into a group will cause varying degrees of fighting and trauma as the individuals re-establish their rank within the unit. Mixing events compromise the welfare of the animals as stress levels increase within the group and individuals suffer pain and distress associated with injury. The adverse effects on growth often go unrecognised but growth rates may be reduced. Much of the aggression within an established group is caused by competition for space and resources and so an appropriate management technique is essential to reduce rivalry for food and water. Where possible, keep pigs in the same groups throughout the rearing period, and introduce new members of the group slowly whilst providing appropriate shelters where victims of aggression can hide. The stockperson should closely monitor the levels of fighting within a group and intervene where necessary to remove very aggressive individuals or injured/deprived victims. Additional factors that may reduce fighting within mixed groups include reducing group size so that hierarchy is quickly established, increasing space (reducing stocking density) and providing obstacles or barriers to enable victims to avoid dominant animals, minimising external stressors (e.g. temperature fluctuations, feed disruptions) and providing appropriate foraging material to prevent boredom and satisfy behavioural instincts. Reducing the light intensity for 24 hours after mixing and using boar pheromones may help reduce fighting. Where practical, place pigs together in the evening, provide, with some feed and turn out the lights. Generally, pigs that wake up together will settle better. Injectable sedatives are available that may help reduce aggression, although this is not economical or suitable for regular, general use.

Unwanted behaviour

Unwanted behaviours include tail biting and ear and navel sucking in weaned pigs; tail biting, ear biting, chewing feet and flank biting in growers; and bar biting and sham chewing in adult pigs. Unwanted behaviours may result from misdirected instinctive behaviours such as ear and navel sucking in weaned pigs, or as a result of lack of environmental stimuli to satisfy the animals' behavioural needs. In all cases, stress is an important factor in triggering unwanted behaviour and can arise as a result of any number of environmental and physiological disturbances such as temperature changes, group mixing, nutritional imbalances or disease. Competition for resources can contribute to aggressive behaviour and stress within groups.

The financial impacts of unwanted behaviours can be considerable. Studies have shown that the majority of all carcass condemnations are the result of infectious conditions. Associated with these are tail biting lesions and hind-limb bursitis. Although tail docking is used to reduce the impact of this behaviour, this practice in itself can cause long-term discomfort for the animal and, in many cases, does not stop tail biting. Removing tails does not address the root causes of this unwanted behaviour and can mask poor husbandry systems. Every effort should be made to ensure the environment satisfies the pigs' physiological and behavioural needs. Considering the genetics of the pigs and selecting against stress and aggression within breeding programmes is also important.

Mutilations

Mutilations include nose-ringing of sows kept outdoors, tattooing, ear tagging or notching, tail docking, teeth clipping or grinding, castration and boar tusk removal. Legislation in some countries accepts the use of certain procedures to counteract severe abnormal behavioural problems arising in some husbandry systems. These

A LOADING RAMP DESIGN

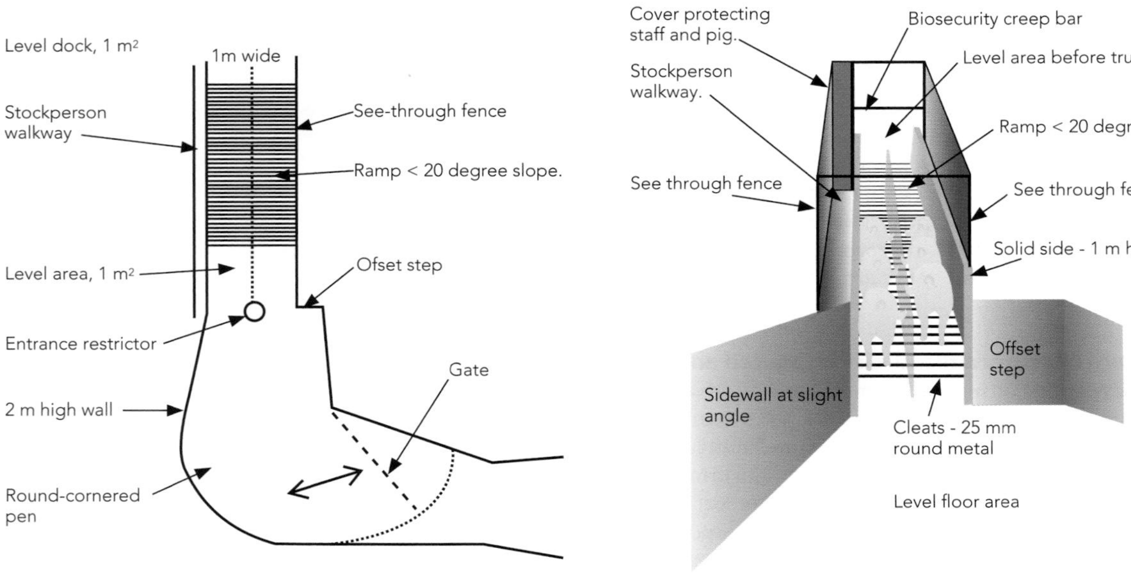

Fig 16-1

procedures cause pain and distress and are detrimental to the welfare of the animal and should therefore be avoided wherever possible in favour of other management options. For example, nose-ringing can be avoided by altering management strategies of outdoor pig housing, for example using increased field rotations, improving the provision of bedding and forage material, providing a sacrificial rooting area and burying low density feeds like root vegetables. Keeping the nutritional status of the sow optimum may also help reduce foraging damage to the field.

Following consultation with your farm vet, any procedures that you deem necessary on your pig farm should be carried out by well-trained operators, at the correct age, and in a manner that minimises pain and distress caused to the animal (see Chapter 15). In the EU, castration and tail docking without anaesthetic must be carried out within 7 days of birth. Following this, a trained veterinarian and the use of anaesthesia and prolonged analgesia are required to carry out these procedures. All equipment should be well-maintained and kept clean and disinfected. Castration has virtually ceased in many countries. In regions where male pigs are grown to slaughter weights beyond the age of sexual maturity, it should be noted that the management and housing requirements for rearing entire males differ from those of castrated males, and should address the increased prospects of heightened aggression and sexual behaviour in entire males to ensure that staff safety and pig welfare is not compromised. Immunocastration and genetic selection of males for later sexual development can be used to reduce boar taint levels.

Both teeth-clipping and grinding are likely to cause pain to piglets. If absolutely necessary, teeth-clipping should be carried out before 7 days of age and on as few animals as possible. However, when lactation is adequate, there should be no requirement for either teeth-clipping or grinding. A plentiful milk supply can be achieved through proper management of the sow before and during lactation. When lactation is sufficient, teeth-clipping has a limited benefit on mammary injuries on the sow and piglet performance. Sows with large litter sizes, mastitis and young gilts will be at particular risk of low milk yields and piglet competition for milk.

Breeding

Using suitable pig breeds and breeding methods can greatly impact upon the welfare of the animals and the European Council Directive 98/58/EC specifies that "natural or artificial breeding or breeding procedures which cause or are likely to cause suffering or injury to any of the animals concerned must not be practised". Balanced breeding strategies that incorporate multiple traits, in addition to fertility, litter size and weaning weight, are crucial and ultimately affect the sustainable production of a pig enterprise. Other factors include temperament, disease resilience, good mothering ability, low levels of aggression and suitability to the climate/production system.

Housing Systems

Restrictive or controlled housing systems provide fewer opportunities for animals to adapt to unfavourable conditions, for example wallowing or huddling to regulate body temperature. Changes to housing systems should not be made without thorough consideration of the effects upon the animals' wellbeing. All housing systems should aim to optimise the following factors.

Stocking densities

It is important to work within recognised and accepted legislation/standards for various ages and weights of pigs. This is to ensure good welfare, optimum growth rates, and a low level of disease and mortality. The total square metres of accommodation required for growing and finishing pigs needs to be determined from the batch mating programme, farrowing rates, litter sizes and pigs weaned. This is needed to ensure adequate capacity to meet legislative requirements/guidelines and to maximise production. See Chapter 3: Stocking Densities for more information on this topic.

Floor and wall surfaces

The floors should be free from projections, well maintained, easily cleaned and, if solid, good drainage towards the exterior of the pen should be provided. Slats should be well maintained and of a suitable size for the age of the pig to prevent trauma and disease (see Chapter 3). The use of bedding has many benefits to welfare including standing, lying and thermal comfort, minimising injuries, and meeting behavioural needs such as foraging and nesting. Bedding must be kept clean and dry, especially in lying areas. All solid floors should be well insulated. The walls and partitions should be constructed and maintained so that there are no sharp edges or protrusions that may cause injury or distress. All surfaces should be capable of being cleaned and disinfected and be kept free of debris. Loose-housed service areas should not have slatted or slippery flooring.

Hospital pens

There must be adequate provision to cope with the number of compromised or sick and injured pigs that have to be moved from their normal environment during periods of illness and treatment. Aspects of this are discussed in detail in Chapter 3. Any ill pig that cannot fend for itself must be moved immediately to a hospital pen e.g. tail-bitten, severely lame, acutely ill etc. The failure to provide adequate well-managed hospital pens is a cause of serious economic loss on many farms. The experiences on one large farm are worthy of consideration. No ill pig was ever treated in its pen but moved to one of a series of small hospital pens. The owner often remarked that many of the hospital pen pigs reached slaughter weight before their contemporaries - a rather sobering thought.

Electrical installations

These must not be accessible to pigs and should be correctly earthed with trip-out switches. They should be protected against damage and contamination by water during cleaning and disinfection.

Automatic equipment

This should be thoroughly inspected on a daily basis. Where breakdowns occur provision should be made to ensure that the welfare of the pigs is not compromised.

Food

Food should be presented in a manner that allows all pigs to eat without distress or fear. Feed hoppers or dispensers should be examined daily to ensure any automatic delivery systems are functional and are working efficiently. Pigs should be fed at least once a day. If fed only once a day all the pigs in a group should have access to food at the same time. It is strongly recommended that, in addition to concentrate feed, all pigs are provided with a manipulable, high fibre substrate that enables foraging behavioural instincts and gut satiety.

Water

All pigs over 2 weeks of age must have permanent access to a sufficient quantity of fresh water (see Chapter 3). There must be a sufficient number of accessible drinkers per pen for every pig to drink readily, including the pigs at the bottom of the pecking order. Every drinker should be checked daily to ensure adequate flow rate. In the farrowing house the sow's drinker should be checked twice a day. If a header tank is used it should have a lid to prevent contamination by debris, dust and vermin. This also helps to prevent pipework blockages. If necessary, pipes should be insulated to prevent freezing. An inadequate flow rate from partially blocked drinkers is one of the most common management faults found in commercial piggeries. As well as being a welfare issue water shortage leads to poor productivity and a predisposition to disease. This is particularly serious when single nipple drinkers with poor flow rates are incorporated into farrowing pens. This results in low milk yield, loss of appetite and sow condition, poor piglet growth and an extended wean-to-oestrus period. It is a wise precaution to have a single tap system on the water line over the sow's trough which can be turned on twice a day to let the sow have a good drink. The trough must of course be watertight to prevent wet floors, a common cause of predisposition to mastitis and piglet diarrhoea. A single space hopper with an integral nipple drinker should not be the only source of water to a pen of growing pigs.

The importance of feed and water in disease is grossly underrated and experiences over the years in the investigation and control of disease outbreaks constantly reinforce this. Access, availability and quality are the three important components.

Management for housing systems

Housing systems for pigs from birth to weaning and for lactating and weaned sows should be used on a batched all-in/all-out basis, keeping pigs of similar age within a common environment. There should be provision for the cleaning and disinfection of each section between each batch of pigs. This is a major component in disease control and hence good welfare. The adoption of this principle on a number of breeding-finishing farms has, together with other management methods, resulted in significant improvements in feed efficiency (up to 0.4) and improved daily liveweight gains (up to 120g). Segregating one age group of pigs from another is now the most important aspect of disease control. These procedures and principles are discussed in detail in Chapter 3.

Environmental Design

Environmental enrichment

Performing natural behaviours such as rooting, exploring and chewing are essential for pig welfare and good production. Pigs have been shown to spend up to 75 % of their day engaged in these activities. The importance of this provision is such that EU legislation requires that all pigs have access to manipulable substrate at all times, such as straw, hay, wood, sawdust, mushroom compost, peat, or a mixture of these that does not harm the health of the pigs. Toys and play objects are less attractive as they are easily damaged and quickly lose their novelty factor. Providing appropriate manipulable material provides the opportunity for foraging as well as a comfortable bedding material and a fibrous component to the diet to achieve gut satiety. In some housing systems appropriate bedding material can also help reduce waste emissions from manure. Providing appropriate environmental enrichment is especially important for reducing unwanted behaviours such as ear and tail biting. Exposing weaned pigs to a more complex environment is also thought to help prepare the animals for future stresses of handling and transport.

Temperature
Pigs are highly susceptible to temperature changes and should be maintained within their thermo-neutral zones at their comfort level (see Chapter 3). The temperature will depend on liveweight, group size, airflow and feed intake. Thermal comfort can be assessed by the lying patterns of the group, as pigs will either huddle together or move away from one another. It is important to ensure the building is sufficiently insulated to prevent uncontrolled heat loss or gain. Pigs are particularly prone to heat stress and where animals cannot regulate their own temperature through natural wallowing behaviours, cooling methods must be provided. These may be in the form of air vents flowing across part of the house, water sprays/mists or by wetting designated sections of the floor.

Past experiences have shown that the failure to maintain a stable comfortable temperature is one of the most important trigger factors in the development of disease. This is particularly so in postweaning diarrhoea and respiratory disease (see Chapter 9). Avoid rapid fluctuations in temperature as this stressor may trigger unwanted behaviours such as tail biting.

Humidity
Relative humidity should be maintained between 50 and 75 %.

Ventilation
Provision of fresh air is essential to remove noxious gases and help regulate air temperature. Ventilation should be maintained at a controlled level so the pigs remain within their thermo-neutral zone (see Chapter 3), so that moisture is removed and to ensure toxic gases are adequately removed. However, a light draught (e.g. an increase in air movement from 0.2 m/sec to 0.5 m/sec) can lift the lower critical temperature (LCT) by 4 °C (7 °F) or more. Mechanically ventilated systems must have an alarm to identify when a failure occurs and have provision for alternative means of ventilation under these circumstances. Alarm systems should be tested at least every 7 days, and a backup system should be available when electricity supplies fail.

Light
Pigs should not be kept in darkness during daylight hours, but should have sufficient light of at least 40 lux to satisfy their behavioural and physiological needs. A pig's natural environment is woodland and they are adapted to the light intensity of this environment. The period of light should be at least that of natural light. Sufficient light should also be available to allow the pigs to be thoroughly inspected at any time. Adequate light has an important role in maximising reproductive efficiency, as discussed in Chapter 5. Breeding requires a timed lighting programme of 16 hours on and 8 off.

Noise
General noise levels should ideally be kept below 35 decibels. Noise levels above 85 decibels should be avoided in that part of any building where pigs are kept. This can sometimes be difficult, particularly when handling pigs, for instance during bleeding. Veterinarians and assistant staff should always wear ear protection during bleeding.

Fire and emergencies
Suitable provision must be made to enable stockpeople to vacate a pig building in the case of fire. While it is honourable to save as many pigs as possible, it is essential that the stockpeople are safe. Note pigs also have the unfortunate habit of running back into a burning building.

Summary of Factors Essential for Good Welfare
A suggested quick guideline to ensure the welfare of pigs is:

- Caring and considerate stockpeople to look after them.
- A good clean water supply.
- A draught-free sleeping area.
- Wholesome food.
- Housing and flooring that is non-traumatic.
- Appropriate environmental enrichment.
- Disease prevention, medication and care while sick.
- Good genetics and breeding.

Welfare and Housing
There are many different types of housing available for dry/gestating sows, lactating sows and weaning, growing and finishing pigs. Fig 16-2 shows the different types of sow housing and Fig 16-3 the feeding systems, both manual and mechanical, that are often used. The following is a résumé of these different systems to illustrate the welfare and disease problems that may arise.

FIG 16-2: TYPES OF HOUSING SYSTEM
Dry/Gestating Sow Housing
Tethers (banned in many countries)
Confinement stalls (restricted in many countries and banned in a few)
Cubicles/free access stalls
Straw yards
Slatted/concrete yards
Yards and individual feeders
Paddocks and huts (outdoor)

FIG 16-3: TYPES OF FEEDING SYSTEM
Dry/Gestating Sow Feeding Systems
Individual/mechanical
Trickle feeding
Manual into a trough or onto the floor
Dump or drop feeders - floor feeding
Spin feeders onto the floor
Ad-lib feeding
Electronic sow feeders (ESF)
Wet feeding

It is not however intended to provide details of housing designs but only sufficient information to highlight the interactions between the structure, management, welfare and disease.

Housing Systems and Welfare Failures

Each production system has its own benefits and problems. Fig 16-4 highlights the welfare related observations and problems that are likely to be seen in the different gestating sow housing systems. The majority of these problems can be overcome with good quality housing and high levels of husbandry.

FIG 16-4: OBSERVATIONS AND PROBLEMS WITH GESTATION SOW HOUSING

	Confinement	Cubicles	Loose yards	ESF	Outdoors
Abortion					
High air flow.	✓	✓			✓
Lack of boar contact.	✓	✓	✓	✓	✓
Lameness.	✓	✓	✓	✓	✓
Low feed levels.	✓	✓	✓	✓	✓
Low intensity of light.	✓	✓		✓	
Low temperatures.	✓	✓	✓	✓	✓
Poor building insulation.	✓	✓		✓	
Poor hygiene.	✓			✓	
Trauma.		✓	✓	✓	✓
Lack of bedding.	✓	✓	✓	✓	✓
Poor body condition.			✓	✓	✓
Wet bedding.		✓	✓	✓	
Cold wet conditions.					✓
Draughts.	✓				
Heat stress.	✓				✓
Parasites.					✓
Season.					✓
Sunlight.					✓
Abscess					
Arthritis.	✓	✓	✓	✓	✓
Bad slats.	✓	✓	✓	✓	
Bullying.		✓	✓	✓	✓
Bursitis from trauma.	✓	✓	✓	✓	✓
Bush foot from trauma.	✓	✓	✓	✓	✓
Environmental trauma.	✓	✓	✓	✓	✓
Fighting.		✓	✓	✓	✓
Poor concrete.		✓	✓	✓	
Stalls too small.	✓				
Trauma.			✓	✓	✓
Aggression					
Establishment of pecking order.		✓	✓	✓	✓
Uneven group mixing of sows.		✓		✓	✓
Adding new sows.			✓	✓	✓
High stocking density.			✓	✓	

FIG 16-4: OBSERVATIONS AND PROBLEMS WITH GESTATION SOW HOUSING CONT.

	Confinement	Cubicles	Loose yards	ESF	Outdoors
Mixing.		✓	✓	✓	✓
Group systems.				✓	✓
Shape of yard.				✓	
Anaemia					
Gastric ulceration.	✓		✓	✓	
Haemorrhage - vulva.		✓	✓	✓	✓
Arthritis					
Bullying.		✓	✓	✓	✓
Bush foot - poor concrete.		✓			✓
Fighting.		✓	✓	✓	✓
Leg weakness (OCD).	✓	✓	✓	✓	✓
Slippery floors.	✓	✓	✓	✓	
Stalls too small.	✓				
Worn slats.	✓	✓	✓	✓	
Damage at oestrus.		✓	✓	✓	✓
Bursitis					
Fighting.		✓	✓	✓	✓
Bullying.		✓	✓	✓	✓
Shortage of bedding.		✓	✓	✓	✓
Lameness, bush foot, foot rot etc.					
Bad slats.	✓	✓			
Concrete kerbs.		✓			
Poor concrete surfaces.	✓	✓	✓	✓	
Cystitis pyelonephritis					
Frozen pipes.	✓	✓	✓	✓	
Leg weakness.	✓	✓			
Poor access to water.	✓	✓	✓	✓	
Slippery concrete surfaces.		✓	✓	✓	
Sows reluctant to drink.	✓	✓	✓	✓	
Water shortage.	✓	✓	✓	✓	
Infrequent urination.	✓				
Lameness.	✓	✓	✓	✓	
Poor access to water.	✓		✓	✓	
Clostridial infections					
No vaccination.					✓
Erysipelas					
Straw systems.		✓	✓	✓	✓
Wet pens.	✓	✓	✓	✓	✓
Dirty bedding.		✓	✓		
Ease of spread from clinically ill animals.			✓	✓	✓
Fear and distress					
From bullying.		✓	✓	✓	✓
Mixing sows.		✓	✓	✓	✓
Fractures					
Leg weakness.	✓		✓		
Slippery floors.		✓	✓	✓	

FIG 16-4: OBSERVATIONS AND PROBLEMS WITH GESTATION SOW HOUSING CONT.

	Confinement	Cubicles	Loose yards	ESF	Outdoors
Haematoma					
Vulval biting.		✓	✓	✓	✓
Fighting.		✓	✓	✓	✓
Hygroma					
Pressure sores.	✓	✓			
Trauma.	✓	✓	✓		
Poor concrete surfaces.	✓	✓	✓	✓	
Infertility					
High return rate and "not in pig" from aggression.		✓		✓	✓
Unwanted behaviours.	✓				✓
Stress induces embryo losses.			✓	✓	✓
Stress after feeding.			✓	✓	✓
Stress at feeding.				✓	✓
Fighting.		✓	✓	✓	✓
Hot weather.	✓	✓	✓	✓	✓
Lack of adequate wallows.					✓
Lame boars.					✓
Overuse of boars.					✓
Shortage of shade.					✓
Trauma to the boar's penis.					✓
Haemorrhage from the penis.					✓
Associated with sandy soils.					✓
Loss of sow identification					
Lost transponder tags/collars.				✓	✓
Thin sow.				✓	✓
Stockperson education - understanding.				✓	✓
Poor design and maintenance.				✓	✓
Leptospirosis					
Associated with contaminated water.					✓
Rodents exposure.	✓	✓	✓	✓	✓
Mastitis					
No bedding.	✓	✓	✓	✓	
Poor concrete.	✓	✓	✓	✓	
Poor hygiene.	✓	✓	✓	✓	
Trauma.	✓	✓	✓	✓	✓
Wet pens.		✓	✓	✓	✓
Bad draining.	✓	✓	✓	✓	
Solid gates in stalls.	✓				
Udder contamination.	✓		✓	✓	✓
Permanently bedded yards.			✓	✓	
Overgrown, distorted claws					
Lack of exercise.	✓				
Leg weakness factors.	✓	✓			
Slats wrong way.	✓	✓			
Gaps too wide.	✓	✓			
Smooth surface.	✓	✓			

FIG 16-4: OBSERVATIONS AND PROBLEMS WITH GESTATION SOW HOUSING CONT.

	Confinement	Cubicles	Loose yards	ESF	Outdoors
Parasites					
Continuous access to faeces.		✓	✓		✓
Inability to remove faeces adequately.		✓			✓
Wet dirty area.			✓	✓	
No worming programme.					✓
Old pastures.					✓
Prolapsed rectum or vagina					
Badly designed tail gates.	✓	✓			
Slippery floors.	✓	✓			
Short standing area and a deep step.	✓				
Sloping floor.	✓				
Abdominal pressure.			✓		
Nutrition.			✓		
Slow farrowing, lack of muscle tone					
Old sows.	✓	✓			
Stereotypic behaviour					
Boredom.	✓				
Lack of appropriate forage.	✓				
Sunburn					
Lack of shelter.					✓
No wallows.					✓
White breeds.					✓
Thin sow syndrome					
Disease.	✓	✓	✓	✓	✓
Draughts.	✓	✓			
Low feed intake.	✓	✓		✓	✓
Low house temperatures.	✓	✓		✓	
Parasite burdens.	✓	✓	✓	✓	✓
Poor environments.	✓	✓	✓		
Poor nutrition.	✓	✓		✓	✓
Bullying.		✓	✓	✓	✓
Parasites.			✓		
Variable feed intake.			✓	✓	✓
Variable litter size					
Embryo reabsorption/stress.				✓	✓
Vulval biting					
Occurs towards the latter end of pregnancy.		✓	✓	✓	✓
Feeding methods.			✓	✓	✓
Pen design.			✓	✓	
Stocking density.			✓	✓	
Vulval discharge					
Dirty weaning pens.		✓			✓
Dirty wet boar pens.	✓	✓			✓
Wet defecating areas.		✓	✓	✓	
Feeder design.				✓	
Poor pen/yard design.				✓	
Queuing for feed.				✓	

Welfare of Pigs in Indoor Housing Systems

Sow Stalls and Confinement

Sow stalls were developed as a means to minimising the aggression that develops between large sow groups and increase sow stocking density. In addition, sow stalls are thought to simplify management (ease of feeding, waste disposal and handling) and reduce labour costs. They also provide safety for the stockpeople handling the pigs.

Confinement systems, such as sow stalls, do not provide for all of the five freedoms. As with any system, when combined with poor husbandry, sow stalls can result in sows displaying signs of poor mental and physical health and unwanted behaviours such as bar biting and sham-chewing.

During the service period holding sows and gilts in stalls may improve reproductive efficiency. Stalls may be used as appropriate during veterinary treatment.

Confinement by tethers is banned in the EU and in some other countries.

Where confinement is still practised the following welfare issues should be considered.

Design requirements

Size – Width 0.5 to 0.68 m. The smaller width of stall is used for gilts to prevent them turning around and defecating in the trough at the front, increasing the risk of infections.

– Length 2.28 m. If the rear part of the stall is slatted, gilts housed in long stalls will tend to defecate on the solid concrete area, leading to sore legs and other leg problems. Longer lengths may be needed for large sows.

Rear gate – Ideally this should be made of vertical bars so that faeces and urine can spill away from the back of the slats. Care is required, however, to make sure that the bars do not cause pressure sores. Where solid rear gates are used, their bottom edges should be raised 80 mm above the slats so that faeces cannot build up, contaminate the vulva and lead to vaginitis and endometritis (but not so high they cause pressure sores). The height of the rear gate should be sufficient to prevent the sow from sitting on top of it. Outbreaks of rectal prolapse and abortion have been associated with this.

Floors – These may be part slatted or totally solid. Totally slatted floors should not be used as they are uncomfortable for the sow and create difficulties when the sow stands, increasing the incidence of leg weakness. On part-slatted floors the solid area should be approximately 1 m at the front and the remainder slatted. Ideally, slatted flooring should not cover more than 25 % of the floor area. The slats should run parallel to the sow to provide a better, less traumatic surface for the feet. If sows are housed on solid floors there should be a 1:20 fall in the last 300 mm of the floor. This allows drainage of urine and easy removal of faeces. If the whole length of the floor slopes to the back it may predispose to vaginal prolapse, and in gilts and second litter females create a predisposition to osteochondrosis or leg weakness. All the accessible concrete edges should be round and not sharp and the surfaces of the slats flat and not sloping to the gaps.

Bedding – If bedding is used, the faeces- and urine-soiled material must be removed from behind the sow daily.

Water – Fresh water should be made available to the sow in a trough or drinker at the front at all times. If this is supplied manually, it may be necessary to provide extra in hot weather. Ensure that drinkers do not leak as wet floors can lead to skin sores. A shortage of water will predispose to cystitis, pyelonephritis and increased sow mortality.

Cooling can be provided by individual drip cooling/ spray systems.

Feeding – Sows may be fed once or twice daily, preferably once a day to prevent agitation and anticipation in sows awaiting their second feed. From a welfare point of view, automatic dispensers held in front of each sow are best so that all animals can be fed at the same time. This will reduce the risk of torsion of the intestines, trauma due to excitation and stress-associated reproductive failure. Fibrous, manipulable material such as straw should be provided at all times to enable foraging and feeding behaviours and to aid gut satiety. Feeding in the afternoon releases the morning for health and heat checking.

Group size – There are no constraints on the numbers of animals held in any one building.

Temperature requirements – Ideally 18 to 20 °C (64 to 68 °F). Drip cooling systems and evaporative cool cells can be extremely useful in hot climates.

Management, welfare and health

If you keep a sow in a stall or tether her, she is totally confined and has no control over her environment. It is therefore important that the stall is long enough (some modern-day sows grow longer than previous breed types), the floor is comfortable to lie on, she has sufficient feed to satisfy her appetite, the environmental temperature remains consistent day and night and there are no draughts. You can tell whether you are achieving these if you quietly open the door to the dry/ gestation sow house when no staff have been present for a period; over 95 % of the sows should be lying down on their sides.

The majority of sows lie down most of the time and will only rise to drink, urinate and defecate, at feeding time, or during heat checking. If urination and water intake is inadequate this can lead to problems such as cystitis and pyelonephritis. If they are only fed one main meal a day, the ration should contain sufficient fibre to satisfy appetite and keep the faeces soft.

All animals should be examined daily in a standing position to detect any signs of lameness or leg weakness. Additionally feet, legs, shoulders and hips should be examined for any sores, swellings or granulomas. If used, neck or girth tethers should be checked regularly, and adjusted as necessary. A monthly assessment of body condition and body score should be carried out and the skin should be examined for evidence of lice or mange. A frequent examination should be carried out of the slats, to check that they are not worn and causing trauma to feet and legs. All metal work and rear gates should be examined regularly to check there are no parts likely to cause trauma. A maximum and minimum thermometer should be checked daily to ensure that the sows are being maintained within their comfort zone. Adequate light has an important role in maximising reproductive efficiency, as discussed in Chapter 5. In artificially lit buildings, light provided should be at least 40 lux with a minimum 8 hours dark period.

If sows are weaned into stalls, both feed and water should be available at all times and the floor must be kept very clean and well drained. If straw is used, keep surfaces clean and well bedded to prevent udder contamination and the development of mastitis. A combination of the above factors can often be responsible for embryo reabsorption, abortion and heavy culling rates, resulting in persistent low farrowing rates. Ensure that all cooling systems are working adequately.

Cubicles and Free Access Stalls

When correctly managed, cubicles and free access stalls enable a higher level of welfare compared to sow stalls and tethers. In cubicles and free access stalls a greater range of movement is available to the individual sow and division of dunging and lying areas support her natural inclination to separate the two. Provision of space to turn around and lie down freely enables a greater level of exercise and comfort, improving bone strength and muscle tone. However, in order to establish a dominance hierarchy, sows usually exhibit aggression, which can be particularly marked in a confined space. Division of sows into stable family groups and provision of adequate resources can help minimise aggression.

Design requirements

Size – Width 0.68 m. Length 2.28 m to the front of the trough. The sow lies in its own stall but it has free access behind to a defecating passage and it can move around. Some designs have self-opening and closing tail gates and are often referred to as free-access stalls.

The width of the dunging passage will be determined by the width of each cubicle and whether they are in groups of 3 or 4. In a group of 4 the passage widths would be approximately 2.74 m. The design should allow the pigs to turn around and the dimension of any stall or pen used for holding individual pigs should be such that the internal area is not less than the square of the length of the pig, and no internal side is less than 75 % of the length of the pig; the length of the pig in each case being measured from the tip of its snout to the base of its tail while it is standing with its back straight.

Floors – These should consist of solid concrete, well-insulated lying areas and not-too-smooth defecating areas. The cubicle divisions may be barred or solid; preferably the former.

Bedding – Sufficient material should be provided to allow physical and thermal comfort and enable rooting and foraging behaviours. If the dunging area is solid, manure should be removed at least 3 times per week. This area should also be well drained.

Water – This must only be provided in the dunging area either by a nipple drinker, bowl or trough. If water is placed in the feeding troughs, sows tend to defecate and urinate in the lying areas resulting in lameness, sores on the feet and legs and a predisposition to leg weakness.

Feeding – Feed is put into the trough once daily. Feed levels required may be higher than in stalls to satisfy the sow's environmental requirements. Fibrous, manipulable material such as straw should be provided at all time to enable foraging and feeding behaviours and to aid gut satiety.

Group size – These are usually 3 or 4 sows but there can be up to 10 in free access stalls depending on the size of herd. Sows of similar size and age should be grouped together. Gilts and sows that have lost condition should be housed separately to prevent bullying.

Temperature requirements – Because of the low stocking density and mating system well-insulated buildings are advised. In temperate climates such housing is best ventilated naturally. In very cold climates heating will be required.

Management, welfare and health

The total lying and defecating area, per sow, ranges from 2.25 to 2.3 m^2. In such confined spaces, however, sows should not be mixed together for the first time but rather in large open pens for 24 hours, because severe trauma fractures and lameness may result from fighting. Not only can they damage each other physically and cause fear and distress, thus denying two of the five freedoms, but also because the aggression occurs in the early weeks of pregnancy it can cause pregnancy failure and raise the rate of regular and irregular returns to oestrus. The sows that return can be mated again but at some point they then have to be mixed into another group. If this is done in the first 6 weeks of pregnancy aggression may result in more returns and a reduction in litter size. It is best to move sows to this type of accommodation in the second half of pregnancy (after Day 35 of gestation) when pregnancy is more secure and sows are more docile. Dividing sows into smaller family groups within the unit and avoiding group mixing enables the sows to establish a social rank and limits further aggression. In addition, limiting competition for resources, such as feed, and providing adequate forage material prevents boredom and reduces hunger, which can help limit fighting within the group. The lying areas should be raised above the defecating passage but not by a kerb, since this increases the incidence of foot and leg damage and mastitis. Furthermore, with a kerb, faeces and urine tend to be retained in the lying bed. If the cubicles are covered, observation and management become difficult. If free access stalls close automatically behind the sow there is no need to confine them during feeding. However with the cubicles it may be advisable to close the gate at feeding time to prevent aggression and bullying. Sows should be fed and examined from a central feed passage.

Group Sow Housing

In a well-managed system, group housing can fulfill the five freedoms of good welfare for the sow. Where adequate space is provided, behavioural needs including exercise, socialising and division of lying and dunging areas is possible. Good, fibrous, bedding material provides a comfortable substrate for lying and thermal control as well as foraging and feeding. Aggression between sows can be avoided with appropriate management (see below). However, in poorly managed systems, group housing of sows can create terrible welfare conditions with pigs forced to sleep wet in their own faeces.

Design requirements

Pens – Allow 2.7 to 2.8 m² per sow for lying and defecating. Do not house groups of sows in long narrow pens. These will increase the incidence of vulva biting.
Yards – Allow 3.4 m² per sow. At the recommended stocking density there is a marked reduction in fighting episodes when sows are mixed.
Floors and bedding – Solid concrete floors bedded with straw or other suitable materials are best. From a purely welfare perspective, totally slatted floors are not recommended due to high levels of lameness, arthritis and the foot problems that occur. Solid defecating areas should be well drained and the manure removed 3 times weekly.
Water – This should be provided ideally by a self-levelling small trough or 1 bite drinker per 15 sows.
Feeding – Electronic sow feeder systems (ESF) are common. Here the sow is individually fed a maximum daily amount when it enters the feeder station in response to a transponder placed in the ear or around the neck. One feeder station will accommodate 50 sows. Trickle feeder systems have been developed whereby the feed is dispersed in small amounts at a time, sufficient to keep the sow feeding continuously. The troughs are divided by short divisions to separate each animal during feeding. They are suitable for groups of up to 12 sows. With automated feeds regular equipment checks are needed to ensure no sow goes unfed. Floor feeding can be provided by feeders suspended above the lying area and the feed dropped to the floor (dump feeders) or alternatively a single feeder that spins the feed out across the lying area. Wet feeding systems are also being used, successfully feeding the sows from troughs placed down the sides of large straw yards or along the front of the pens. Where group feeding systems are used, feed must be widely distributed so that all sows have access.
Group size – Housing sows in groups of up to 30 enables the establishment of a stable social hierarchy that, with plentiful resources, limits aggression and fighting. Some farmers keep much larger groups of between 100 to 250 sows in a dynamic group using electronic sow feeders. These large groups require a complex housing system and very able management to ensure the wellbeing of all the sows. The stockperson must ensure that all sows have equal access to resources through provision of adequate space, bedding areas, feed stations and drinkers. Careful monitoring of the group is needed to ensure dominant individuals are not restricting access of subordinate sows through blocking passageways or lying areas. Where sows are bullied or excluded from feed and clean resting areas, loss of condition can lead to reduced fertility and sow performance.
Temperature requirement – This depends on the amount of bedding and floor insulation and ranges from 15 to 20 °C (59 to 64 °F).

Management, welfare and health

Welfare problems and disease can arise due to the aggressive nature of sows in group-housed systems. Furthermore this aggression creates problems of varied feed intake and predisposes the individual under-nourished sow to disease. Dividing the sows into smaller, stable, family groups is ideal. High stocking densities and mixing between groups should be avoided. In addition, provision of plentiful resources such as forage material is essential to prevent boredom and competition between sows. It is a requirement of EU legislation that appropriate fibrous substrate is made available at all times. Adequate space and shelters can help weaker sows escape bullying and stockpersons should closely monitor any sows that are injured or failing to feed. Removal of aggressive sows may be necessary.

The ideal management system should allow newly weaned sows to mix in a large pen with some boar contact for the first few days. Sows are then served naturally or artificially inseminated. Following service sows are confined to a stall for a short period before returning to the group. In the UK the sow can only be confined in a stall for 2 hours. Ideally, sows should be maintained in the same group throughout the whole of pregnancy. In some countries outside the EU, the sows are weaned into stalls, mated in boar pens, and kept in stalls or cubicles for the first half of pregnancy before being group-housed. Pregnant animals mixed together between 2 to 21 days post-service are at risk of higher embryo mortality and more variable litter size. Where sows are separated from the group for only a few hours, problems of aggression are minimised.

Wet-feeding, electronic sow feeding (ESF) and trickle-feeding systems give a more even feed intake compared to group feeding systems. If aggression is a problem, particularly when sows are mixed, a series of soft rubber mats (1.2 m x 1.2 m) suspended vertically, 500 mm above the floor, over the lying area in large pens provides a means of separating one sow from another when a fighting episode commences.

Keeping parity 1 and 2 gilts and sows separate from the older sows helps to reduce lameness problems and aggression, particularly when the older sows exhibit oestrus.

Sows fed with electronic feeders have the advantage of individual feeding but aggressive behavioural patterns can develop, particularly when they are waiting to enter the feeders. This can cause considerable stress, particularly to animals that have just joined a large group. Severe bullying with trauma and skin damage can take place and poor hygienic conditions and wet areas often develop around the feeder stations.

ESF systems require careful design and skilled, experienced stockmanship. It is best to use feeders with a separate front or side exit door so that sows do not back out and be confronted by sows which are eager to feed. Some sows tend to lie in the feeder, thus blocking others. If the design of the layout is not well thought out and the stockperson not properly trained there may be an unacceptably high level of vulva biting and badly scratched and bruised skin.

If you decide to construct an ESF system you should first visit and study successfully established units and get competent training in ESF management. Training gilts to ESF systems is necessary.

If possible the stockperson working the system should be involved in the selecting of the equipment and the design of the yards. He should be enthusiastic, committed to making the system work and should understand the computer technology. He should be competent to undertake everyday maintenance. In areas of frequent power failure, a standby generator will be necessary.

Identification of ESF-fed sows by ear tags needs careful management as the ear tags can come out, resulting in the sow being unable to obtain feed.

Yards and Individual Feeders

Well-managed yards also fulfil the five freedoms of good welfare for the sow and can further limit problems of aggression between sows as a result of small group sizes and provision of individual free-access sow stalls.

Design requirements

The yards may be of variable size housing from a few sows to large groups. They resemble a cubicle system in that sows are fed in individual stalls.

Size – The lying area: 1.4 m^2. Defecating area: 1 m^2. Total: 2.4 m^2 per sow.

Floors – The lying area is insulated concrete with minimum bedding or deeply bedded with straw. Slats in the defecating area are not common and tend to result in high levels of lameness.

Water – Trough drinker or nipple drinker placed in the dunging area.

Feeding – Feeder stalls should be 0.6 m wide and 2 m long. Appropriate fibrous foraging substrate should be provided at all times.

Temperature requirements – Temperatures are well maintained usually because the lying areas consist of small pens with solid sides and a well-insulated roof.

Management, welfare and disease – Due to individual feeding many of the problems associated with group housing are avoided. There are still problems from mixing, particularly at weaning, but bullied animals should be able to escape into stalls.

It is important to ensure that a minimum of 16 hours of light, with 500 lux, are available in the lying/sleeping areas.

Welfare of Lactating Sows and Sucking Piglets

The majority of sows in indoor units are farrowed in farrowing places (generally a crate) of various designs, although a small but increasing number are kept in loose-housed, free-farrowing pens and yarded free-farrowing systems. Outdoor sows are farrowed in arcs.

Given the genetics of the modern pig, whatever system the health team selects, the design should be such as to accommodate 12 weaned piglets grown to a weight of 100 kg per farrowing place.

The farrowing crate was developed to limit piglet mortality caused by the sow crushing or cannibalising her piglets and to protect the stockperson. Many factors have been implicated in predisposing sows to piglet crushing including the size of the sow and her litter, mothering ability, availability of space and material for nesting, and sow stress during farrowing and lactation. The farrowing crate has been used to reduce piglet mortality where these factors have increased the risk of piglet crushing, and to simplify sow management. However, the farrowing crate heavily restricts the sow's ability to undertake natural behaviours and regulate her own environment and therefore greatly compromises her welfare.

Sows are generally placed in farrowing places 1 week prior to farrowing. EU legislation requires that, where appropriate, manure disposal must be in place and bedding material must be provided. Bedding materials such as shredded paper can be used to boost sow welfare and will not interfere with slurry systems.

The health and welfare of suckling piglets is essential for a profitable pig business. Piglet mortality has multiple causes, some of which are discussed in Chapter 8. Maintaining a strict hygiene practice is essential. Large litter size can increase the chances of weaker piglets that fail to thrive and also encourage fighting and injury caused by teat competition. Piglets should be provided with a heated safety area with adequate bedding, forage and creep feed. Appropriate space allowance and provision of environmental enrichment can help reduce piglet fighting and minimise injuries. Mutilations (see Recommendations for all Pigs) should be avoided and where necessary carried out in line with your country's legislation and veterinary guidelines.

Depending on the country, piglets may be weaned between 1 to 4 weeks.

In the EU, weaning must not take place before 28 days, except where piglets are orphaned or requiring special attention. EU requirements are that piglets may be weaned up to 7 days before Day 28 (i.e. at 21 days) provided the weaned pigs are housed in specialist accommodation.

The design specification of the nursery for early weaned piglets is higher than those weaned older. Piglets that are weaned later often achieve faster growth rates and require simpler management and less veterinary intervention than early-weaned litters.

Design requirements

The farrowing pen – There are a wide variety of designs but they should be approximately 2.4 m in length and wide enough to enable the sow to stand up and lie down with ease. In some designs the pen width is only 1.68 m in which case the pen has to be longer, with a forward creep. The creep lying area should be about 0.6 m^2 and be either in front of or beside the head of the sow.

The farrowing place should be a minimum length of 2.3 m preferably with a moveable rear bar or gate for shorter gilts. The width between the bottom bars of the farrowing place should be 730 mm to allow for ease of suckling and at a height 230 mm from the floor.

Modifications in design are largely aimed at preventing the sow crushing the piglets as she stands up and lies down or moves suddenly. They include angled projections on the bottom bars, or bottom bars which move in when the sow stands and move slowly out when she lies down again. Cold air fans which automatically blow across the floor when she stands can be used to encourage the piglets to retreat to their warm creep.

EU regulations allow a female pig to be confined in a farrowing place for the period between 7 days before the predicted day of farrowing and the day on which the weaning of her piglets (including any fostered by her) is complete.

Tail gate – This should be designed to prevent trauma to the vulva (parallel retaining bars at right angles to the farrowing place are likely to cause this) and yet allow the sow room to farrow. A good design consists of two "D" shaped bars that project into the farrowing place and hold the sow from the tail gate itself.

Floor – This may be either totally perforated using metal or plastic slats or part solid, the front half of the crate floor being insulated concrete. Alternatively the complete floor may be insulated concrete. The floor may be raised slightly to provide good drainage if slats are used with bedding. Make sure that the area where the front feet make contact with the floor is not slippery otherwise the sow has great difficulty standing (causing trauma to the piglets) and this increases levels of leg weakness and shoulder sores.

To meet EU regulations part of the total floor where the piglets are, large enough to allow the animals to rest together at the same time, must be solid or covered with a mat or be littered with straw or any other suitable material.

Water – Provide water using a bite-type nipple drinker delivering 2 litres per minute minimum, located at the most forward point of the farrowing place at head height. Alternatively, this can also be provided through a nipple/button drinker delivering a minimum of 2 litres per minute that projects into a trough either at head height or at floor level, with an over-flow to drain away surplus water.

Bite drinkers can also be used. Some farrowing places also have a tap to fill the trough after feeding. All farrowing pigs should have access to clean water at all times. Sucking piglets drink little water during the first 2 weeks, but a small drinker should be attached to the farrowing place or pen wall for piglet access. Water should be made available in small dishes for the first 48 hour after farrowing.

To maximise feed intakes during lactation, water could also be added at feed times.

Feed – This is fed in the trough to appetite throughout lactation. Fibrous foraging material should be provided wherever possible.
Group size – The numbers of farrowing pens in any one house or section should be restricted to the numbers of sows that farrow in any 1 week, up to a maximum of 12.
Bedding – Bedding should be made available to the sow prior to farrowing as well as in the creep area for the first 10 to 14 days after farrowing.
Temperature – Creep area – This should be capable of being maintained at between 28 to 34 °C (82 to 93 °F). The farrowing house should be maintained at 20 °C (68 °F) during farrowing and between 16 to 18 °C (61 to 64 °F) during lactation although some houses are operated much lower.
Free-farrowing pens – These systems share a number of common features designed to provide the sow with the space and resources to express her natural behaviours, without affecting piglet mortality levels and productivity, including: a nest area with angled walls, rounded corners and other structures to reduce piglet crushing; an adjacent heated creep area for the piglets; separate lying, feeding and dunging areas and adjustable walls/penning area that allow the sow to be confined if necessary (e.g. during farrowing or treatment).
Loose-housed pens – Loose-housed pens consist of a common housing area for feed, drink, dunging and exercise, surrounded by individual farrowing arcs that provide secluded nests for the sows. A barrier across the entrance prevents the piglets leaving the arc in the initial days post-farrowing. The system provides extensive freedom of movement for the sow and greater freedom for expression of natural behaviours, including nest building, foraging and exercise. The family group arrangement throughout farrowing preserves stable group hierarchies and enables continued social interaction between sows and later piglets, in preparation for weaning. In addition, the simple design means existing buildings are easily adapted. Piglet mortality can be high, however.
Yarded free-farrowing systems – A number of variations exist around a central design that encompasses a weather-proof farrowing arc surrounded by an enclosed outdoor area for feeding, drinking and loafing. The farrowing arcs are deep-bedded and often have angled walls to reduce piglet crushing when the sow lies down, and a barrier across the entrance prevents the piglets leaving the arc in the initial days post-farrowing. The outdoor area enables the sow to exercise, regulate temperature and control attention levels from the piglets, as well as retaining visual and nasal contact with other sows in surrounding pens.
Outdoor farrowing arcs – Where the environment and climate are suitable, outdoor sows are farrowed in arcs. See Welfare of Pigs in Outdoor Housing Systems for details.

It is essential in all systems to provide the lactating sow with toys or suitable manipulable materials.

Management, welfare and health

Alternative free and loose-housed farrowing systems enable the sow to move more freely and interact with her offspring. These systems may achieve comparative levels of piglet mortality where good pen design and management practices are put in place, including provision of sufficient bedding material and a safety area for the piglets. Selection of breeds for strong maternal traits and good performance in these systems as well as trained staff is important.

Welfare of Newly Weaned Sows

The process of weaning is stressful for the sow and all additional strains should be minimized.

Methods of housing the newly weaned sow include:

- Sows housed individually in sow stalls or tethers from weaning.
- Sows housed in groups of 2 or 3 in spacious pens.
- Sows mixed into 1 group at weaning.
- Sows grouped from the day of weaning together with a boar and then removed to confinement during the mating period. Sows then regrouped with the boar within 48 hours after mating.
- Dynamic group system.

Design requirements

Breeding stalls – See the design requirements for sow stalls. The back of the stall should be designed with breeding in mind. The rear of the pen should allow the stockperson easy access to the vulva so that the AI catheter can be easily inserted without the need for entering the pen.
Loose-housing – The design for loose-housed sows should consider the following:
Size – In loose-housed weaning accommodation allow approximately 3.4 m² of floor area per sow.
Floors – Keep these as dry as possible at all times and ensure that they are comfortable and well drained. Ideally use straw bedding but if this is not available use other materials. Replace bedding daily. Soiled wet bedding can cause mastitis, vaginal infections and infertility.
Water – Site nipple drinkers or troughs in a well-drained area.
Feed – Feed to appetite from Day 1 after weaning through to the day of mating, preferably from an ad lib feeder. If your sows are group-housed allow a minimum of two separate feeders spaced well apart to reduce

stress at feeding time and prevent anoestrus developing in disadvantaged sows.
Group size – Ideally 6 to 15 sows.
Temperature – For the first 4 to 5 days postweaning this should approximate that being used in the farrowing houses at the time of weaning; around 20 °C (68 °F).

Management, welfare and health
The weaning to mating period is a critical time for the group-housed sow because there can be a considerable amount of disturbance and aggression. Good non-slip floor surfaces and smooth walls are essential to prevent trauma and injury. Where practicable, sows should be re-grouped into their original family groups or by size. Particular care should be taken with the weaned first litter gilt and thin sows to ensure adequate food intake and reduce any fighting and aggression.

Welfare of Weaned and Growing Pigs
Weaned and growing pigs are usually group-housed in pens of 5 to several hundred pigs per pen. Problems stemming from high stocking densities, group mixing and provision of poor environmental enrichment have the greatest impact on grower pig welfare. Mixing unfamiliar pigs leads to aggressive behaviour, and serious injuries can result from fighting. High stocking densities and lack of environmental enrichment prevent weaker pigs from escaping aggressive individuals. Lack of environmental enrichment and manipulable material to occupy the pigs' behavioural needs can exacerbate aggressive tendencies and encourage additional unwanted behaviours, e.g. tail biting. Other factors that increase these behaviours may include lack of appropriate nutrients and external stressors such as poor ventilation and temperature control. Genetic selection is important, for example research has shown that selection for lean meat correlates with increased tail biting activity.

Flooring in rearing systems is usually fully or part slatted or full concrete to enable ease of management. Bedding is rarely provided as it interrupts slurry removal. These flooring conditions frequently cause injuries to the feet and legs, increasing the incidence of lameness and bursitis.

Design requirements
Size – This is dependent on the numbers of pigs per pen, their weight, the type of floor surface, the bedding used (if any) and the shape of the pen. Refer to stocking density requirements discussed earlier in this chapter. The space must be sufficient to allow all pigs to sleep, feed and exercise without restriction.
Floors – These can be totally solid, insulated and drained towards the defecating area, part solid and part slatted, totally slatted or straw-based. If the quality of the slats is good and they have a round edge, a totally slatted area provides the best performance and welfare conditions provided that the temperature, air flow and humidity are controlled within the pigs' requirements. Bedding should be provided but for this to be satisfactory it must be available, clean, dry and deep enough to provide warmth and comfort.
Water – This can be provided either through nipple drinkers, automatic bowls or troughs. The water requirements for pigs are provided earlier in this chapter. Water should be available at all times.
Feed – This can be presented to the pig by a wet feeding system 3 to 4 times daily, restricted feeding or ad libitum through feed hoppers. Wet feeding systems have several advantages. Within a group all pigs tend to drink at about the same speed whereas there is a wide variation between pigs in the amounts of dry feed they can eat.
Group size – Housing weaned pigs in groups of up to 30 enables the establishment of a stable social hierarchy that, with plentiful resources and suitable stocking density, limits aggression and fighting. Some farmers keep much larger groups of several hundred pigs. These large groups require a complex housing system and very able management to ensure the wellbeing of all the pigs. The stockperson must ensure that all pigs have equal access to resources through provision of adequate space, bedding areas, feed stations and drinkers. Good housing design and careful monitoring of the group is needed to ensure dominant individuals are not restricting access of subordinate pigs through blocking passageways or lying areas. Where pigs are bullied or excluded from feed and clean resting areas, reduced growth rates and disease can result. Where entire males are group-reared to sexual maturity, same sex groups must be avoided as excessive fighting can occur as gilts come into oestrus.
Temperature – A guide to requirements is given in Chapter 3 but always adjust to maintain the pigs in a comfortable lying pattern.

Management, welfare and health
Ensuring good health and welfare for growing pigs is essential to achieve good growth rates, carcass quality and disease control. If the pigs' requirements are not satisfied, disease problems are created particularly if a catabolic state develops. Housing should be managed on an all-in/all-out basis with a total depopulation, cleaning and disinfection between groups. This is important for both respiratory and enteric disease control. Reduced feed availability, lower temperatures or poor quality food will also lead to stress and an increased predisposition to enteric and/or respiratory disease as well as behavioural abnormalities. See Chapter 9 for more details on managing health and disease in weaned and growing pigs.

Wherever possible, group mixing should be minimized and group stocking densities should be kept as low as possible with appropriate space and obstacles for weaker individuals to seek shelter from aggressive behaviour. Where groups must be split to avoid overcrowding, creating smaller subgroups that avoid mixing is best. The EU directive requires that manipulable material is provided to occupy the pig's behavioural interests. A fibrous, rootable substrate such as straw, hay, wood, sawdust, mushroom compost, peat, or a mixture of these that does not harm the health of the pigs, enables foraging behaviours and supports comfort during lying and standing. Provision of environmental enrichment and appropriate forage material can also help reduce aggressive behaviour.

Welfare of Pigs in Outdoor Housing Systems

Well-managed outdoor systems can provide high levels of welfare for the sow as a result of the increased space allowance and a greater level of autonomy conferred to the sow, including exercise, foraging and temperature control. Production levels are however generally lower with outdoor production. Optimal pigs weaned per sow being around the 10 mark against a target of 12 indoors.

The general welfare requirements to maintain healthy outdoor pigs are similar to those required for indoor production, but are often more difficult to control. For example, a greater management input is required during extremes of weather, for example the provision of water during periods of freezing. During wet periods extra bedding is essential in both the dry/gestating sow and farrowing huts. Huts should be sited on free-draining land and on higher ground during periods of snow and heavy rainfall. Adequate shade and wallows must be provided in the summer. Using breeds that can cope with the climate is essential.

Design requirements

Site – It is vital that the ground used is light, free draining and flat. Ideally the soil should be sandy but chalk and gravel are also satisfactory. The site should be low lying in a valley rather than on a hill, be in an area of low wind and have good vehicular access. The huts for dry/gestating sows should be approximately 2.5 to 3 m, hold 4 to 6 sows, and face away from the prevailing wind.

Water – This should be available in troughs which should be no more than a short walking distance for suckling sows. They should have a minimum depth of 6 cm of water.

Feed – This should be presented on the ground in the form of cobs/large pellets dropped along a line. Feeding lines should be at least 3 m apart and changed frequently to prevent competition and aggression between sows. Feed may need adjusting in response to changes in the weather.

Stocking rate – There should be 15 to 20 sows per hectare (6 to 8 sows per acre). In less suitable sites, the stocking density should be reduced.

Bedding – Farrowing huts should be placed on the higher parts of the land and bedded twice weekly with chopped straw. They should also be checked daily for draughts and wet floors.

Temperature – Outdoor pig breeding can only be carried out efficiently in temperate climates and ideally all the huts should be insulated. However due to the small cubic capacity of dry/gestating sow huts and the heat given off by groups of sows there are usually less problems here in maintaining adequate temperatures.

Wallows – There should be two in each paddock near the fence line so that they can be used alternately to allow drying for disease control. Not only do they help to cool sows in hot weather but they also allow sows to cake their skins with mud to prevent sunburn.

Shade – This must be provided to lessen the effects of strong direct sunlight.

Fencing – If electric fencing is used it must be properly installed and maintained to give mild discomfort when contacted by the pigs, not to inflict pain. New pigs must be introduced to electric fencing using a training paddock. It is essential that the electric fencing is insulated around the water troughs and is away from the feeder or feeding areas.

Management, welfare and health

Often, welfare problems in outdoor production are related to climate and therefore involve the group as a whole, although it should be remembered most welfare problems are a result of poor management (people). Feed intake and quality of feed are important in relation to disease, the maintenance of body condition and reproductive performance. The latter can be a particular issue in the late summer where farrowing rates can fall significantly. It is very important to feed sows well in summer and autumn to provide good fat cover for the winter. Therefore it is to be expected that sow feed costs will be higher than in controlled environments. Routine control of external and internal parasites is essential in maintaining the health and welfare of the group. Arcs should be rotated with a suitable rest period and all bedding removed before new groups of sows are brought into the paddock.

During farrowing and suckling, ensure the arc contains deep, clean bedding for nesting and thermal comfort. Arcs should be well maintained as farrowing sows, piglets and newly weaned pigs are particularly susceptible to draughts, and damp conditions can encourage disease. A barrier across the entrance will prevent piglets leaving the arc in the initial days post-farrowing. In warmer climates, adjustable sides on the arcs may be needed to increase ventilation and prevent heat-stress or the sow leaving her litter to seek a cooler environment. In periods of inclement weather piglets suffer due to cold and wet conditions and increased trauma by the sow. Foxes, crows and other large birds can kill and cannibalise piglets. Three-strand electric fences will generally deter foxes but the fencing needs to be well maintained. The identification and treatment of sick piglets can be difficult and careful monitoring is required. It should be remembered that sows can be particularly dangerous, especially with their piglets. Pigs kept outdoors can cause pasture damage and soil erosion due to over-foraging. Trees are particularly likely to suffer from pig damage. Appropriate techniques to limit damage include frequent pasture rotations, low stocking densities and adequate diet. Nose ringing should be avoided as it causes persistent pain to the pig and prevents an important natural behaviour.

Welfare of Boars

Boars are commonly housed individually or, in more extensive systems, may be kept in small, familiar groups. As with all animals, high levels of stress not only compromise the wellbeing of the boar but also leave boars more susceptible to disease. Reducing stress through appropriate husbandry and housing is important to maintain the welfare and productivity of boars.

Individual Boar Pens

Design requirements

Size – Within the EU the minimum floor space is 6 m^2 per boar. This increases to 10 m^2 in pens that are also used for natural service. Elsewhere in the world boars are generally housed in stalls.

The majority of boars are used for heat checking purposes or on an AI station.

Floors – These should consist of solid concrete, well-insulated lying areas and separate dunging and feeding areas. The use of slats is not ideal as there could be a higher incidence of lameness. Deep bedding is preferable and should be kept dry and clean. This is particularly important if the pen is also used for natural service of sows in order to avoid slipping and associated injury during service. A layer of coarse sand also provides grip, reducing potential injury from slips during service.

Walls – Walls are high enough to prevent boars from escaping into adjacent pens. The cubicle divisions may be barred or solid, preferably the former to enable visual or nasal contact of other pigs.

Water – Trough drinker or nipple drinker should be provided in the dunging area.

Feeding – Feeder stalls should be at least 0.6 m wide and 2 m long. Appropriate fibrous foraging substrate should be provided at all times.

Temperature requirements – Temperatures should be well maintained as fluctuations and thermal discomfort can lead to temporary reduced fertility. As boars are generally housed alone appropriate bedding can help improve the thermal environment for the boars.

Management, welfare and disease

Boars are commonly housed individually so as to avoid aggressive interactions. Boars must not be visually isolated as this can affect their behaviour. Boars must therefore have visual access to other pigs from their pen and a minimum of 16 hours of light at 200 lux, or natural daylight should be provided. Stress associated with confinement or inadequate housing, similarly to sows, can manifest itself in unwanted behaviours.

Boars may also be kept in small familiar groups; this practice is generally found in more extensive systems. Where boars are kept in groups, these should be maintained throughout the life of the boar as individuals will display extreme aggression to unfamiliar boars. In group-kept boars, space should be available for subordinate boars to escape and victim boars should be identified and removed from the group as soon as possible. High levels of stress render boars more susceptible to disease and lower performance. Reducing stress by appropriate husbandry and housing is thus paramount to maintain the welfare and productivity of boars. It is essential to provide the boar with toys or suitable manipulable materials.

Boars can be managed in pairs or even in groups of 3. This can be used successfully for AI stations. In outdoor production such systems can be used for boar accommodation. However, it should be noted that 1 boar can become dominant and attempt to do all the mating, with subsequent reduction in farrowing rates and litter size.

17 HEALTH AND SAFETY

17 HEALTH AND SAFETY

Introduction

There are three reasons why a chapter on health and safety should appear in a book dealing with managing pig health.

The first is perhaps obvious: some of the environmental hazards commonly associated with pig production, such as dust and slurry gases, can affect the health of both people and pigs. The need to assess and control these environmental factors is therefore a health and safety issue as well as being an integral part of disease control.

The second is often overlooked: a positive pro-active approach to health and safety, just as with health control, can help you maximise production, reduce costs and increase profitability.

The third is the most important: good management practice is now as applicable to health and safety as it is to other business activities - such as finance, personnel management and the control of pig diseases.

Over the last thirty years health and safety legislation has been changing in many countries. In the past employers could achieve compliance simply by following specific directives. However, these directives were often retrospective, reactive and prescriptive. They were introduced as a result of accidents that had already occurred in an attempt to prevent them from happening again. They placed the same constraints on all employers regardless of individual circumstances or the levels of risk involved.

Now, in many countries, the emphasis of health and safety legislation is on self-regulation. This requires employers to set their own safety standards, within minimal legal requirements, and to define and implement the controls necessary to achieve them. The most effective way for the employer to approach this is to carry out **Risk Assessments** and national regulations in many countries may make it a legal obligation to do so.

To achieve full compliance however, employers have to show that they have addressed all aspects of health and safety that apply to their operations and that they are doing everything they should. The legal term "due diligence" may be used to describe this.

This view is now so widely accepted that in many countries it has (or is likely to) become a legal requirement for employers to implement a 'Health and Safety Management System'. An example of such legislation in the UK is:

"The Management of Health and Safety at Work Regulations 1999", the original version of which came into force in the UK in January 1993, in compliance with a European Council Directive.

This chapter explains methods of risk assessment and health and safety management that are straightforward and appropriate for pig farms, large or small.

Study the outline given (Fig 17-1) so that an appreciation of the structure can be understood. Steps 3, 6 and 8 are the key stages.

DEVELOPING A HEALTH AND SAFETY SYSTEM FOR YOUR FARM

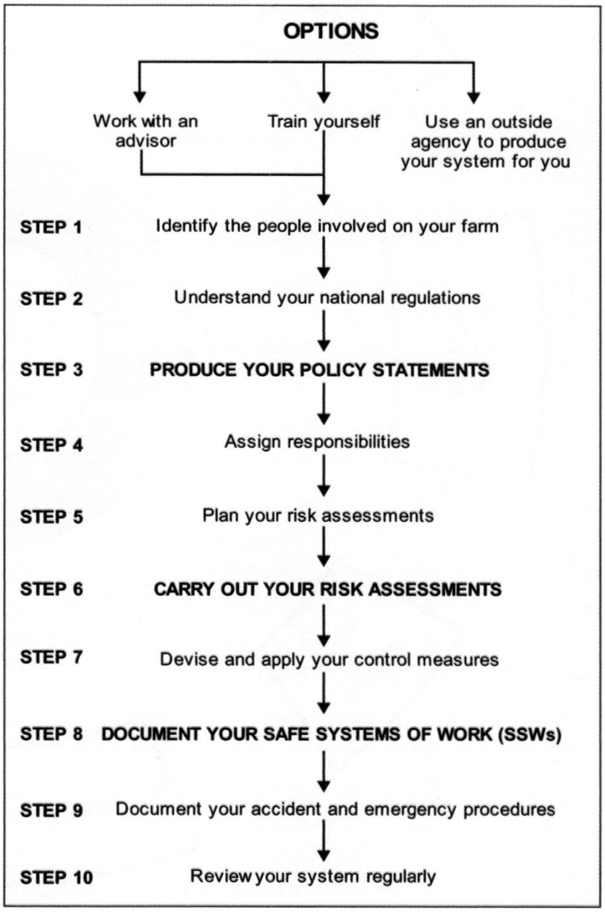
OPTIONS
Work with an advisor
Train yourself
Use an outside agency to produce your system for you
STEP 1 Identify the people involved on your farm
STEP 2 Understand your national regulations
STEP 3 PRODUCE YOUR POLICY STATEMENTS
STEP 4 Assign responsibilities
STEP 5 Plan your risk assessments
STEP 6 CARRY OUT YOUR RISK ASSESSMENTS
STEP 7 Devise and apply your control measures
STEP 8 DOCUMENT YOUR SAFE SYSTEMS OF WORK (SSWs)
STEP 9 Document your accident and emergency procedures
STEP 10 Review your system regularly

Fig 17-1

The Management of Health and Safety

Several books have been written on the subjects of health and safety management and risk assessment and there are many different ways to approach them. Pig farmers are free to adopt or devise any system that suits them - provided that the end results comply with their national legislation and their own requirements.

Taking a pro-active, structured approach to health and safety is key.

The following approach, developed by Beeford Laboratories Ltd., is to consider risk assessment as one key component of health and safety management and to devise methods whereby all of the components can be combined in an overall working system.

The resulting management system shown in Fig 17-2 revolves around the development and implementation of Safe Systems of Work (SSW). These incorporate both the work instructions (how to do a job safely) and the management procedures whereby you can ensure and demonstrate that:

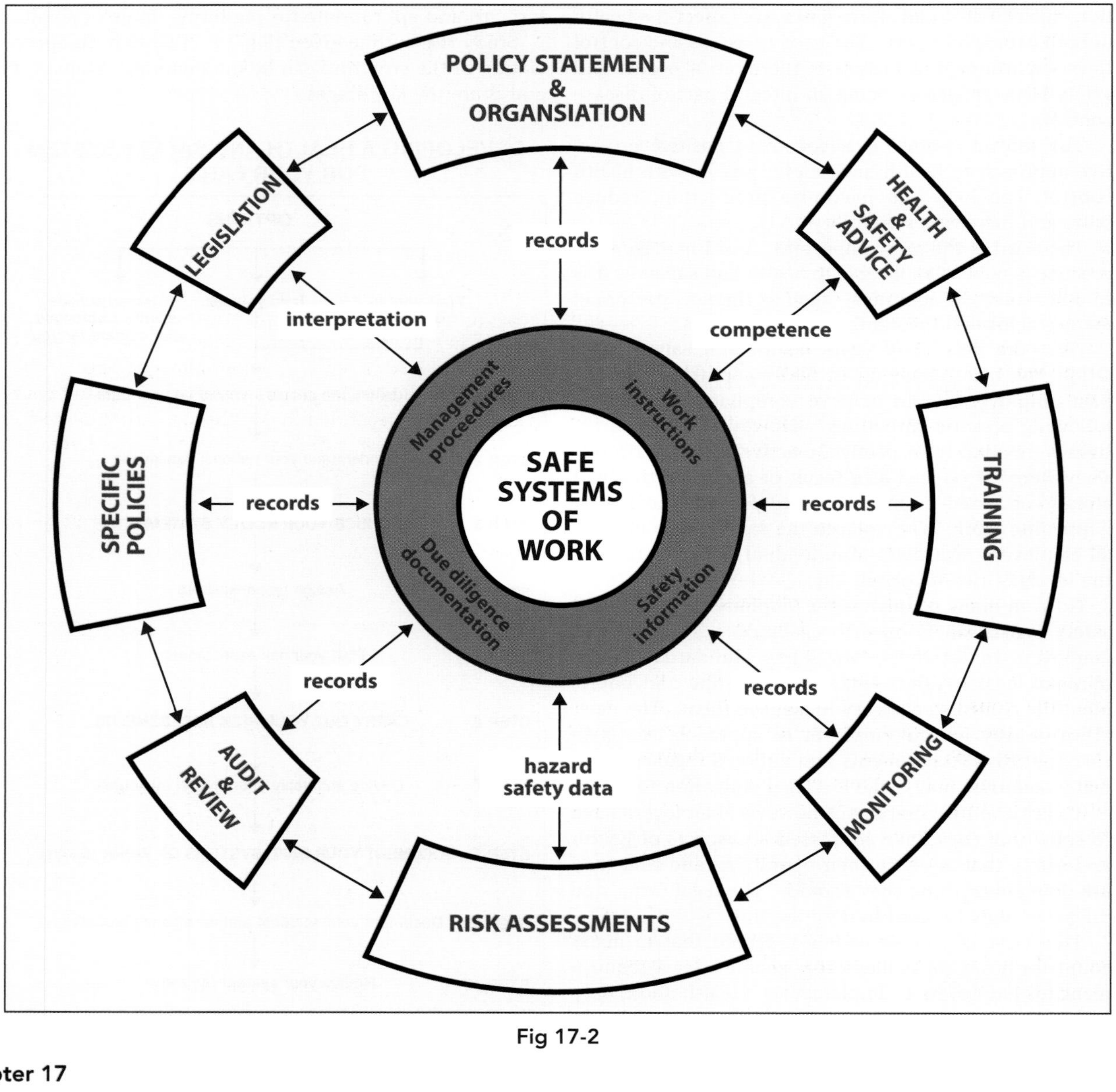

Fig 17-2

- The relevant Regulations, Approved Codes of Practice and Industry Standards are being addressed appropriately.
- The work areas and activities have been, and continue to be, fully assessed.
- The methods of work are safe and without risk, so far as is reasonably practicable.
- The persons carrying out the work are aware of any associated hazards and have received sufficient information, instruction and training to enable them to carry out their work safely.
- The methods of work are adhered to.
- The necessary personal and/or environmental monitoring is carried out.
- The necessary building, machinery and equipment safety checks and routine services are carried out.
- The necessary occupational health checks are carried out.
- The necessary records are kept.
- The necessary safety audits and reviews are carried out.

The SSW are the hub of the working system and you should have them for every aspect of health and safety that applies to your operation. The way in which you should document them will depend on the size and type of your farming operation. The aspects of health and safety applicable to you as a pig producer and appropriate ways to compile and present SSW are discussed later.

Fig 17-2 illustrates the components of health and safety management. Clearly a considerable amount of health and safety knowledge is required in order to develop a workable system and regulations may state that employers must appoint competent persons to assist in their development.

Large pig organisations may have the necessary expertise in house to fulfil these obligations but the majority of pig farmers will not. It is possible to gain the necessary expertise yourself or to send a member of your staff on a health and safety management course, but first consider the cost benefits of this approach. As with other management operations requiring expert knowledge, such as accounting, building design, veterinary medicine and legal matters, it is usually more cost effective to seek outside advice or to appoint a professional agency. Choosing the right partner can bring savings in your time and in insurance premiums. (See the next section on Cost Benefits).

You should appoint a safety professional with a proven track record and considerable first hand experience of pig farming operations. The person should be competent to:

- Advise on the hazards and safety risks associated with pig production systems.
- Advise on the regulations, industry guidelines and standards in your country - and how to observe them.
- Devise safe working systems that are applicable to your situation.
- Draw up policy statements and organisation arrangements that are workable.
- Carry out risk assessments and report the findings in a way that will satisfy you, your employees, and health and safety inspectors.
- Advise on maintenance and safety of machinery and buildings.
- Advise on personal protective clothing and equipment.
- Advise on and help with appropriate training.
- Organise personal and/or environmental monitoring of hazards such as dust, noise and slurry gases and advise on methods to reduce them or control exposure to them.

Good health and safety management depends on effective risk assessment.

The Cost Benefits

Few pig farmers are fully aware of the cost implications of poor health and safety management or of the benefits that can be associated with good management.

Agriculture is a dangerous industry. In many countries it ranks second only to the construction industry for the highest numbers of deaths, accidents and working days lost each year. In the UK and the USA it is top of the list.

The figures that are officially reported are only the tip of the iceberg (Fig 17-3).

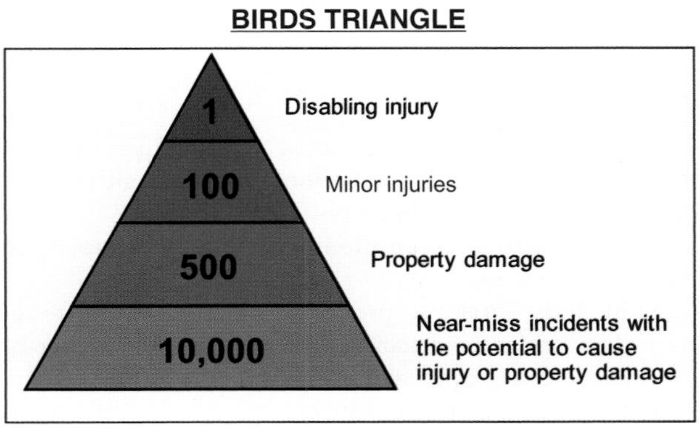

Fig 17-3

Accidents can and very often do seriously harm a business. In addition to the suffering caused to those individuals (and their families) directly affected (e.g.

personal injury, disability, loss of employment) there can be a number of hidden, uninsured costs which include:

- Plant and building damage.
- Tool and equipment damage.
- Legal costs.
- Spending on emergency supplies.
- Loss of expertise/experience.
- Pig disease problems due to loss of management expertise.
- Pig production problems due to loss of management expertise.
- Overtime and temporary labour requirements.
- Investigation time.
- Supervisors' time.
- Clerical paperwork.
- Fines.

The responsibility for health and safety is YOURS.

The investment needed to generate extra profits to cover accidental losses could be much greater than the investment needed to minimise those losses. The principles of loss control should therefore be applied to health and safety, just as they are to other aspects of business.

In addition to controlling the losses described, an effective management system can also result in:

- Lower insurance premiums.
- Proof of due diligence, reducing the likelihood of civil claims.
- Higher efficiency and productivity.
- Less pig disease.
- Higher staff health, morale and motivation.
- Better utilisation of plant and equipment.
- Better legislative compliance.
- Better cost control.
- A better image for your organisation.

Many insurance companies now offer reduced premiums to businesses that can demonstrate a positive commitment to health and safety.

Further reductions in premiums may be achieved in one of three ways:

1. By individual pig producers demonstrating a reduction in their accident rates and the number of successful claims against them.
2. By groups of pig producers or organisations with several farms implementing the same management system so that combined accident and claim statistics could be used.
3. By individual pig producers adopting and implementing a well-recognised health and safety management system that was already being used by a number of other businesses, whose combined accident and claim statistics already showed significant, measurable results.

A poor health and safety record leads to increased insurance premiums.

How to Develop a Health and Safety Management System for Your Farm

When planning your health and safety management system, and before writing policy statements, conducting risk assessments and developing safe systems of work, you will need to be fully aware of two things:

1. The categories of people that you have a duty to protect.
2. The health and safety regulations that apply to you and your farming operations in your country.

The remainder of this chapter deals with setting up the safe systems of work.

Step 1 - Identify the People Involved on Your Farm

All employers have a duty to protect the following categories of people:

- Employees
- Employees with known disabilities
- Trainees/students/temporary workers
- Visitors and the general public
- Contractors
- Trespassers

The points to address for each of these categories are as follows:

Employees

Before asking an employee to carry out a task:

- You must take account of their capabilities.
- This should include consideration of the employee's age, sex, physical strength, size, experience and competence, in relation to the tasks that you will expect them to perform.
- You must provide them with adequate training:
  - when they are first appointed;
  - as and when there are changes in procedures, systems of work etc.;
  - on a routine refresher basis.
- You must provide them with information regarding:
  - the risks to health and safety associated with their jobs;
  - the health and safety control measures in place;

- their obligations regarding health and safety;
- the safe systems of work to be observed;
- the accident, fire and emergency procedures;
- the results of any monitoring and health surveillance.

Employees with known disabilities

In addition to the above:

- You must ensure that employees are not asked to perform tasks or handle substances that may, for them, present particular risks. Examples of this would be asking a pregnant female worker or a known asthmatic to administer prostaglandin to sows. Other examples of disabilities that may be relevant to pig farm workers include dermatitis, a history of back or joint weakness, hypersensitivity to certain antibiotics and respiratory problems.
- You may be obliged to implement a staff health surveillance scheme. This will depend on the regulations that apply to you (seek advice). Even when it is not a specific requirement, it is good practice to implement a health surveillance system anyway. This should include pre-employment health checks (questionnaires) and ongoing records of employee absence, sickness and complaints or requests to do with health and safety. This can be used to identify existing disabilities and also to spot any health effects associated with particular tasks and individuals. Respiratory disease would be a typical example.

Trainees/students/temporary workers

Because of their inexperience and unfamiliarity with your work environment trainees are a particularly high risk group.

- You must provide the same degree of protection as for full employees (see above).
- You must stipulate the special provisions made for the trainees regarding your systems of work etc.
- You must be able to demonstrate, on record, the point at which trainees no longer require special provision (i.e. are deemed to be competent) for each system of work.

Visitors and the general public

You have a duty to ensure that your premises and activities do not present a risk to visitors or the general public.

- You should not leave visitors unaccompanied or allow them access to hazardous areas unnecessarily.
- You should ensure that appropriate directions, instructions and safety notices are displayed.
- You must ensure that areas such as changing rooms, toilets, showers and offices that may be used by visitors are kept as clean and tidy as practicable.
- You must ensure that visitors are provided with sufficient safety information, for example regarding dust, noise, fumes, electrical safety and fire precautions.
- You must ensure that sufficient protective clothing is available for use by visitors and that it is kept clean and in good order.
- You should have a plan of the premises available for emergency services (e.g. in the event of fire) which shows the positions of hazards - such as high voltages, gas cylinders, dangerous chemicals, slurry tank covers, dangerous animals (e.g. boars), etc. How many farms have this available?
- You must consider your impact on the general public with regard to environmental hazards such as dust, noise and fumes and also physical hazards such as concealed farm exits, heavy works traffic at certain times and mud on the road.

Contractors

Although to some extent you share health and safety responsibility with contractors it will remain primarily with you as the pig farmer.

You have a duty to ensure that the contractor is not exposed to hazards as a result of your acts or omissions and that your employees are not exposed to hazards as a result of the acts or omissions of the contractor.

The points regarding visitors will also apply to contractors. In addition:

- You must ensure that all contractors are competent and that they and their equipment comply with statutory provisions.
- You must ensure that contractors are fully informed of the hazards and risks to health and safety which they may be exposed to on your farm.
- You must inform contractors of the measures that you have taken and of those that they must take in order to ensure compliance with legislative requirements and your safe systems of work.

A practical way to ensure that the above points are covered is to use a "contractors' declaration" record form. This can be a standard form that specifies what you require of contractors (i.e. that they comply with the appropriate statutory provisions and levels of competence) and has a section for you to give details of the hazards present at the work site; the accident, fire and emergency procedures; location of the nearest first aid box; the name of their contact on the farm; etc. The contractor and the farm safety manager should both sign the form and each retain a copy. The record can then be kept as proof that you complied with your obligations - should such proof be needed at any time.

Trespassers

The law regarding pig farmers' duties towards trespassers is different in each country but it is generally the case that unless you are certain that trespassing will never occur you must take steps to offer some protection against risks which you know to be present. In most cases the repair of any unsafe or damaged items such as loose hand rails and missing slurry tank covers and the provision of warning signs and measures to deter entry may suffice.

Step 2 - Understand Your National Regulations

These may include specific regulations, approved codes of practice (ACOPs), industry standards, local authority by-laws, duties of care, etc.

Each country has its own health and safety regulations and how they apply to pig farmers can depend on a number of variables. It is therefore not appropriate here to cite specific regulations and ACOPs etc. However, Fig 17-4 lists the topics that may be applicable to your pig farm and that are likely to be covered by regulations in most countries. The list is by no means complete and you should use it as a guide to compile your own comprehensive register. For your management system to be effective you must be aware of the regulations etc., know how they apply to you, and know what you must do to comply with them. You may need outside help with this.

Policy Statements and Policy Organisation

Step 3 - Produce Your Policy Statements

The need for you to write down a policy statement and to make it available to all employees may or may not be compulsory but it is good practice to do so.

In its most basic form your policy statement can simply be that you:

'undertake to do all that is reasonably practicable to ensure the health, safety and welfare of all employees, contractors and visitors'.

However, it should ideally also show your commitment to health and safety in that you have addressed all of your obligations under the prevailing legislation and have made your employees aware of theirs.

Examples of a policy statement and employee duties are shown in Figs 17-5 and 17-6 respectively.

In addition to the general policy statement you will need to adopt and possibly document policies on specific aspects of health and safety. To further define these policies and the associated SSW you will need to carry out risk assessments. Once defined these specific policies should be coupled with, or cross referenced to, the SSW.

FIG 17-4: TOPICS TO WHICH REGULATIONS, ACOPs AND INDUSTRY STANDARDS ARE LIKELY TO APPLY

Health and Safety Topic	Regulation/ACOP/Duty of Care/Industry Standard/Guidance
Abrasive wheels	
Accident/fire/emergency	
Asbestos (roofing & insulation)	
Confined spaces	
Construction	
Contractors	
Electrical safety	
First aid	
Hazardous substances	
Ladders	
Legionella (in water and air con. systems)	
Lead (e.g. in plumbing, paints, fuels)	
Liability insurance	
Lift trucks	
Management of health and safety	
Manual handling	
New employees/trainees	
Noise	
Occupational health (health surveillance, working hours, rest rooms)	
Overhead power lines	
Permits to work	
Personal hygiene (toilets, washing facilities)	
Personal protective equipment	
Pesticides	
Pregnant women	
Pressure systems	
PTO shafts	
Roof work	
Safety signs	
Stairways	
Tractor safety	
Vehicles	
Veterinary medicines	
Waste disposal	
Work at height	
Work environment (heat, lighting, ventilation)	
Work equipment (machines, hand tools)	
Working alone	
Young people/children	

FIG 17-5: THE MANAGEMENT OF HEALTH & SAFETY AT WORK

Statement of Policy

In compliance with the Health and Safety at Work Legislation it is our policy to do all that is reasonably practicable to ensure the health, safety and welfare of all employees, contractors and visitors.

To this end we will adopt policies on all matters of Health and Safety that are compatible with the provisions of all relevant Health and Safety Acts, Regulations, ACOPs, Industry Standards and Duty of Care.

All personnel will receive appropriate information and training to ensure that they:

- are aware of the hazards at their workplace;
- are familiar with all relevant Safety Rules, Procedures & Safe Systems of Work;
- know where and how to access all necessary safety data;
- know where to find and how to use first aid and fire fighting equipment;
- are familiar with the procedures for reporting accidents and for reporting or raising other health and safety issues.

We also undertake to ensure that:

- staff will be supervised until fully trained and assessed as competent;
- machinery, equipment and safety devices are regularly maintained and inspected and are safe and suitable to use;
- the workplace is safe and suitable in terms of comfort, space, heating , lighting, ventilation, cleanliness and freedom from unnecessary hazards;
- working practices are regularly reviewed to improve health and safety;
- accidents and incidents are investigated and appropriate actions taken to prevent recurrence;
- individual members of staff will not be expected to perform tasks that may present risks to them specifically due to their age, sex or health status;
- an ongoing health surveillance scheme is in operation and that the records are regularly reviewed to check for possible links between working practice and ill health and that any such links are fully investigated and appropriate action taken.

Signed

On behalf of

FIG 17-6: GENERAL DUTIES OF EMPLOYEES

It is the duty of every employee:

- To take reasonable care for the health and safety of himself or herself and of other persons who may be affected by his or her acts or omissions at work.
- To co-operate with their employer and Safety Manager so far as is necessary to enable them to comply with any duties or requirements imposed on them under any of the relevant statutory provisions.
- Not to intentionally or recklessly interfere with, or misuse, anything provided in the interest of health, safety or welfare.

Detailed in .... are the Rules, Procedures and Safe Systems of Work which are to be followed by ALL personnel.

All employees must sign the declaration to indicate that they have read and understand them.

The four key points regarding our Health and Safety Policy are listed below.

1. All staff will receive formal instruction and training in the relevant health and safety aspects pertaining to them. Written records of this training will be retained by the employer and by the individual staff members.
2. Staff must not perform any tasks for which they have not received formal instruction or training detailing the health and safety procedures involved.
3. Staff must not perform any task or handle any equipment or hazardous substance if they know that the required health and safety procedures cannot be complied with.
4. When authorisation, training, advice or health and safety information/equipment is needed for specific purposes, staff must approach the Safety Manager or his/her deputy.

Your own Health and Safety policy should be:

IF IN DOUBT ... DON'T DO IT

Step 4 - Assign Responsibilities for Health and Safety

It is at this stage that you must decide who is to be responsible for what, giving due consideration to competence, training needs, available time, etc. The titles (e.g. Safety Manager) and names of the designated persons should be stated on the policy and if the designated person is an employee you may also wish to include their health and safety duties in their job descriptions (see Fig 17-7).

The policy statements should be prepared by the employer in consultation with the designated people. Individuals must be fully aware of their duties and must have the chance to voice their opinions. The whole system will be more likely to work if the employer is seen to be taking it seriously and if it is not considered to be just a paper exercise. It will almost certainly fail if people do not know what is expected of them or they do not feel competent to discharge their responsibilities or if they think that aspects of the policy are impractical or irrelevant.

FIG 17-7: POLICY - ORGANISATION

Our policy with respect to all matters concerned with health, safety and welfare is formulated and carried out by:

The SAFETY MANAGER ....
who undertakes to:

- maintain an up-to-date knowledge of all relevant Health & Safety legislation;
- ensure the implementation of the policy throughout;
- organise the staff training and instruction in methods of safety;
- keep staff aware of the problems of safety and of their responsibility for the safety of themselves and of those around them;
- advise those responsible for the design and construction of new buildings and the modification of existing buildings on all matters which affect safety, and to ensure that the necessary safety features are included in all designs;
- ensure the provision of first aid materials, occupational health supervision and monitoring, where appropriate;
- compile and review data concerning accidents, incidents and staff health issues and ensure that all are recorded, reported and investigated in accordance with the Safe Systems of Work and legislative procedures.

The AREA SAFETY OFFICERS ....

....
who undertake to:

- disseminate information on safety matters within their areas;
- advise on and check procedures to ensure the safety of work within areas;
- ensure that new employees are fully aware of the safety policy and standards, and all area safety arrangements and procedures;
- ensure that all employees and the Safety Manager are made aware of any special or new hazards about to be introduced into areas;
- ensure that the first aid boxes and the safety and fire equipment are checked regularly and that any deficiencies or faults are made good;
- ensure that all accidents, incidents and staff health issues are reported promptly to the Safety Manager in accordance with procedures.

The FIRE OFFICER ....
who undertakes to:

- supervise, or arrange for, the routine inspection and maintenance of fire fighting equipment, fire escapes, fire detection/alarm systems and any fixed systems e.g. sprinklers;
- undertake the posting of appropriate warning signs and notices;
- provide instruction in the use of emergency fire fighting equipment;
- ensure that appropriate fire drills are carried out;
- ensure that the proper liaison is maintained with the Local Authority Fire Protection Officer and that all architects' drawings are submitted for approval;
- provide advice on any topic related to the fire precautions;
- maintain liaison with outside inspectors e.g. fire officers and insurance company assessors and ensure that the appropriate licences and fire certificates are obtained and their conditions observed;
- maintain a record of all fires reported and of the steps taken to ascertain their cause.

The FIRST AIDER ....
who undertakes to:

- maintain an up-to-date knowledge of current accepted first aid practices;
- carry out or witness all initial first aid treatments in the event of accidents and/or act as first call when medical advice is sought by a member of staff;
- maintain an up-to-date knowledge of the approved first aid box contents as recommended by the HSE and report any deficiencies or discrepancies to the Area Safety Officers;
- check the contents of first aid boxes and eye wash bottles and to re-stock or renew as necessary;
- ensure that all accidents are reported promptly in accordance with specified work procedures.

Risk Assessments

Step 5 - Plan Your Risk Assessments

Risk assessment is now a legal requirement of health and safety regulations in many countries and it has become the subject of many conferences, workshops, articles and even regular journals. Only the basic requirements are covered here but hopefully enough to enable you to carry out assessments of the major aspects of your pig farm. You should check the specific requirements of your regulations, particularly with regard to the extent of the assessment required and the need to appoint a 'competent person'.

Risk assessment is a key stage towards the development of SSW and the implementation of a health and safety management system. To satisfy the legal requirements it must involve the following five stages.

1 Careful consideration of what in your pig farm could cause harm to people.
2 Careful consideration of which people might be affected.
3 An evaluation of whether you are taking enough precautions or should do more to prevent or reduce the likelihood of harm.
4 A record (depending on the regulations), and informing your employees, of your findings.
5 A periodic review and revision (from time to time or when significant changes take place).

It should also involve an evaluation of the key aspects of your management systems and the measures necessary to control them. For example, if an assessment determines that there is a potential risk of injury and/or damage associated with a specific procedure and that it can only be adequately controlled by strict adherence to a specific working practice, then that risk assessment should not only identify the need for a detailed written work instruction (how to do the job safely) but also for appropriate staff training, instruction and information, an inspection/monitoring regime, suitable records, specific responsibility for the process and an assessment review date. All of these controls (management and work process) should then be incorporated into the appropriate SSW.

Before conducting risk assessments you will need to be familiar with the following:

- Hazards and safety data.
- Risk rating.
- Acceptable risks.
- Record keeping requirements.

Hazards and safety data

A hazard is anything that has the potential to cause harm.

Hazards are categorised (for simplicity) as follows:

Physical – e.g. noise (too much), light (too little), temperature (too hot/cold), fire, electricity, gravity.
Chemical – e.g. dust, drugs, cleaning agents, gases.
Biological – e.g. bacteria, fungi, viruses, parasites.
Mechanical – e.g. machines, equipment, tools, the pigs themselves.
Ergonomic – e.g. design, layout, inadequate space to work comfortably/safely.
Psychosocial – e.g. stress caused by workload, working alone, concerns about personal safety.

You will see from Fig 17-8 that the hand-feeding of pigs may be more than just a dusty job!

Identifying the hazards present on your farm is the first step in conducting risk assessments. Not many farmers have the expertise to identify all the hazards, which is partly due to their familiarity with their own pig farms. The fresh eye of an independent specialist or that of a friend with experience of risk assessment and pig production can make a useful contribution at this stage.

The checklist in Fig 17-9 gives some examples of common hazards on pig farms. It is a memory jogger to refer to when you are conducting risk assessments. The list is not exhaustive and you should add other hazards known to be present on your farm.

FIG 17-8: THE FOLLOWING EXAMPLE ILLUSTRATES THE DIVERSE NATURE OF HAZARDS
Hand Feeding of Pigs

Hazard	Potential	Possible Harm
Lifting a sack of feed.	Muscular/joint stress.	Back injury/joint injury/rupture.
Work from a gantry.	Fall to ground.	Head, back, limb injury.
Dust from feed.	Respiratory sensitiser.	Bronchitis/respiratory distress.
Penicillin in feed.	Hypersensitivity.	Breathing difficulty/collapse.
Noise from squealing pigs.	Sound energy (loud noise).	Hearing damage.
Rotating auger.	Entanglement.	Physical injury.
Hectic work schedule.	Mental stress.	Hypertension/depression/suicide.

FIG 17-9: HAZARD CHECK LIST

The list below describes hazards that are commonly found on pig farms. It is not exhaustive. It is intended as a memory jogger for you to use when conducting risk assessments. Make a note of those hazards which apply to your farm.

No.	Hazard	Potential Harm	Safety Data
1	Adhesives.	May harm skin, eyes, lungs.	Product safety data sheet.
2	Ammonia (slurry gas).	Respiratory problems. Watering eyes.	Chemical safety data sheet.
3	Asbestos (roofing/insulation).	Respiratory problems. Carcinogenic.	Official H&S information.
4	Carbon monoxide (exhaust gas).	Toxic/fatal by inhalation.	Chemical safety data sheet.
5	Compressed gases.	Explosion. Harmful by inhalation.	Product safety data sheets.
6	Conveyor belts.	Entanglement. Physical injury.	Official and supplier information.
7	Damaged electrical cables.	Electrocution, burns, death.	Electrical Wiring Regulations.
8	Debris/Obstructions.	Slip, trip injury. Hinder escape.	-
9	Diesel.	Skin and respiratory problems.	Product safety data sheet.
10	Disinfectants.	May harm skin, eyes, lungs, body.	Product safety data sheets.
11	Drugs.	May have serious health effects.	Product safety data sheets.
12	Faeces, urine, tissue, body fluids.	Zoonotic diseases.	Zoonoses safety data.
13	Dust.	Respiratory problems.	Official and industry information.
14	Faulty wiring/earth.	Electrocution, burns, death.	Electrical Wiring Regulations
15	Flame/Arc.	Ignition source; burns; Arc Eye.	-
16	Flammable gases.	Explosion; Burns.	Product safety data sheets.
17	Flying debris.	Puncture, cuts, impact, eye damage.	-
18	Fragile roof covering.	Falls.	-
19	Gantries/scaffold/stairs.	Falls, manual handling problems.	-
20	Hand tools (knives/hammers).	Cuts, impact injury, trapping injury.	-
21	Heat/cold.	Exposure, hypothermia, heat stroke.	-
22	Heating oil.	Burns. Harm eyes, skin, lungs, body.	Product safety data sheet.
23	Herbicides.	Toxic. Harmful by ingestion/skin contact	Product safety data sheets.
24	Hydraulic equipment.	Entrapment. Crushing.	Manufacturer's information.
25	Hydrogen sulphide (slurry gas).	Toxic by inhalation (fast acting).	Chemical safety data sheet.
26	Hypodermic needles.	Puncture. Effects of drugs/body fluid.	-
27	Insecticides.	Toxic or (very) harmful.	Product safety data sheets.
28	Insecurity.	Stress effects. Depression.	-
29	Lack of oxygen.	Asphyxia. Danger of death.	-
30	Machinery (general).	Cuts, entrapment, entanglement.	Manufacturer's information.
31	Marker sprays.	Harmful by inhalation; skin contact	Product safety data sheets.
32	Methane (slurry tank gas).	Highly flammable (burns). Explosion.	Chemical safety data sheet.
33	Moving vehicles.	Cuts, crushing, entrapment.	-
34	Noise (high level/repetitive).	Hearing damage/Loss of concentration.	Official H&S information.
35	Overhead power lines.	Electrocution, burns, death.	-
36	Paints.	May harm skin, eyes, lungs, body.	Product safety data sheets.
37	Paraffin.	Burns. Skin & respiratory effects.	Product safety data sheet.
38	Pesticides (crops).	Many are highly toxic. All harmful.	Product safety data sheets.
39	Petrol.	Burns. Skin & respiratory effects.	Product safety data sheet.
40	Pigs (or other animals).	Physical injury. Zoonotic diseases.	Specific zoonoses safety data.
41	Poor lighting.	Eye strain. Cannot see other hazards.	-
42	PTO shafts.	Entanglement.	-
43	Restricted access/egress/space.	Can't avoid hazards. Hinders escape.	-
44	Rodenticides.	Toxic/harmful by mouth, skin, lungs.	Product safety data sheets.
45	Rotating parts (fan blades/auger).	Cuts, amputation, entanglement.	Manufacturer's information.
46	Solvents.	Harmful by mouth, skin, lungs, eyes.	Product safety data sheets.
47	Steam.	Burns.	-
48	Trailing cables.	Trips.	-
49	Uneven/slippery floors.	Slips, trips.	-
50	Unstable stacks (bales/stores).	Injury, crushing by falling objects.	-
51	Used engine oils.	Carcinogenic.	-
52	Vermin (rats, mice & their urine).	Zoonotic disease (esp. Weil's disease).	Zoonoses safety data.
53	Welding fumes.	Respiratory problems.	-
54	Wet plugs/sockets.	Electrocution, burns, death.	-
55	Work - pressure, harassment, abuse.	Stress effects. Depression.	-

The fourth column on the check list is headed Safety Data. The person conducting the risk assessments must have knowledge of the harm that each hazard can cause. For some hazards you may need to refer to additional information. For example, you should obtain safety data for all of the drugs and chemicals used on your farm. In some countries it is a legal obligation to have such data available on site so that the information can be used in assessments and referred to for first aid advice in the event of exposure. The safety data that may be required for other hazards include such things as plant and machinery operating instructions and specifications that detail the machine operating performance limits, maximum load capacities, service requirements, safety test parameters, personal protective equipment required, and operator training requirements.

You will also need to know the legislative requirements regarding personal exposure limits for dust, noise and the various slurry gases, the required lighting levels for work areas and store rooms (the absence of light can constitute a hazard) - and anything else that would contribute to the assessment of the risks associated with the hazards identified.

Zoonoses safety data

Zoonoses are those diseases/infections that can be directly transmitted from animals to man. Therefore, in addition to being a welfare/production/management issue, individual or groups of pigs exhibiting any signs of disease/infection should be considered a potential risk to human health and should be accurately diagnosed and appropriately treated as soon as possible. It is also important to appreciate that healthy pigs can serve as reservoirs of zoonotic organisms (bacteria, fungi, parasites and viruses) and that strict adherence to good hygiene practices is required to minimise risk.

Information on the zoonotic organisms associated with pigs is given in Fig 17-10. Each one is a specific hazard with different properties and potential to cause harm. They should be evaluated individually.

FIG 17-10: SOME ZOONOSES ASSOCIATED WITH PIGS

Source	Organism/Disease	Symptoms In People
Most common or most serious to human health		
A, P	*Brucella suis*	Recurrent fever, weakness, sweating, headache, backache, joint pain.
R	Influenza - some serotypes	Typical influenza symptoms - pneumonia fever and malaise.
I	Japanese encephalitis	(S.E. Asia only.) Viral encephalitis and fever.
A, U, P	*Leptospira* (*pomona* mainly)	Fever, influenza-like symptoms, headache, muscle ache.
F, D, P	*Salmonella typhimurium* and others.	Watery diarrhoea, dehydration, abdominal pain and fever.
Uncommon in people (from pigs) may be common in pigs		
S, F, U, B	Bacteria, various, mainly faecal	Contaminated wounds becoming septic and possibly necrotic.
S	*Erysipelothrix* - human erysipeloid	Local skin lesions, usually on hand. Mainly fish and meat handlers.
S	Ringworm	Typical superficial circular skin lesions.
S	*Streptococcus suis* type 2, 4, 14	Fever, arthritis, sometimes meningitis and deafness.
Rare in people or unimportant		
D	Anthrax	Skin lesions (black scab), pneumonia, diarrhoea with blood.
F	Ascariasis	Asthma, coughing, pneumonia, possible abdominal pain.
A	Chlamydia	Rarely isolated from pigs. No reports of pig-to-human transmission.
F	Clostridia	Tetanus, gas gangrene, diarrhoea. Rarely from pigs.
F	*Cryptosporidium*	Diarrhoea, vomiting, headache, abdominal pain (immuno-suppressed people).
A, F, M	*Listeria*	Fever, headache, vomiting, PREGNANT WOMEN ABORT (pigs unlikely source).
F, S	Swine vesicular disease	Vesicles on skin and fever.
D	*Taenia* and *Trichinella*	In meat. Not direct from pigs. Muscle cysts.
A	Toxoplasma	Theoretically possible but no cases from pigs.
D, P	Tuberculosis (avian)	In meat. Not direct from pigs. Immuno-suppressed people.

A = Abortion, still birth, discharge
B = Bites and scratches
D = Dead animals and carcasses
F = Faeces and dirty bedding
I = Biting insects, mosquitoes
M = Mastitis
P = Doing a post-mortem examination
R = Respiratory disease - inhalation of aerosols
S = Skin contamination, wounds, abscess
U = Urine and soiled bedding

To help you put each in perspective the more common or serious ones are dealt with here. Fortunately there are relatively few serious zoonotic organisms associated with pigs and some are limited to particular regions.

Salmonella - The most common source of food poisoning world-wide. Mainly *Salmonella typhimurium* but also from time to time other serovars including *Salmonella enteritidis* and *Salmonella derby*. The host-adapted serovar in pigs, *Salmonella choleraesuis*, rarely causes food poisoning in people but when it does it is serious. Food poisoning derived from pig meat products results from unhygienic handling and inadequate cooking and is not directly related to what you do on the farm unless you are selling pig products through a farm shop. From the viewpoint of health and safety salmonella are likely to be present on most pig farms but the levels are so low that they do not normally present a hazard to farm workers. If clinical cases of salmonellosis occur in your pigs then the levels rise and become high enough to cause human disease. Everyone on the farm must then take hygienic precautions, particularly before eating.

Campylobacter - Also a common source of food poisoning. Fortunately the majority of campylobacter found in pigs are not the same types that cause human disease (although some are) and pig products are not usually implicated. Nevertheless, you should be aware that they might be present in your herd and you should insist on strict hygiene when eating and drinking.

Provide an eating room for your pig workers and insist on cleanliness and hygiene when they are using it.

Leptospira - Some of the leptospira serovars in pigs, particularly *Leptospira pomona*, can cause a serious generalised infection in man, with fever and meningitis, called "Swine herds' disease" or "Weil's disease" (caused by *L. icterohaemorrhagiae*). In pigs the main clinical signs are abortion, stillbirths and infertility. Fortunately *L. pomona* is now relatively rare in Northern Europe and North America and pig-derived disease in humans is uncommon but you should still be aware of it. The main source of infection to people is pigs' urine or vaginal discharges contaminating cuts (e.g. on the hands) or mucous membranes (e.g. the eyes or lips). Cuts should be covered and pigs' urine kept away from the face. The disease is amenable to prompt antibiotic therapy so if a person working on the farm comes down with fever, depression and headache, call a doctor quickly.

Wear protective clothing and rubber gloves when doing post-mortem examinations.

Brucella suis - This is now rare in Northern Europe and North America but is more common in some other parts of the world. The main clinical effects are abortions, swollen testes and joint problems, particularly of the spine. It causes a very serious disease in people and if it is suspected the greatest care should be taken not to contaminate cuts in the skin or mucous membranes (such as the mouth or eye) when dealing with aborted materials, vaginal discharges or cutting up a dead pig.

If your farm is in a region where *Leptospira pomona* or *Brucella suis* are prevalent then it would be a wise precaution to blood test your herd routinely to check that it is not infected and to get prompt veterinary diagnosis and advice if suspicious clinical signs occur.

If a farm worker develops an acute skin lesion, or an acute fever with a headache, or severe diarrhoea, call a doctor promptly.

Streptococcus suis serotypes 2, 4 and 14 - These are widespread in pig populations around the world and sometimes cause fever, joint problems, meningitis and deafness in people. Most people who become affected are people who handle meat. Cases in farm workers are rare. The number of people affected is tiny relative to the exposure rate that must take place so infection cannot be regarded as a high risk. Nevertheless, protective clothing, particularly rubber gloves, should be worn when carrying out post-mortem examinations and a doctor should be sought promptly if anyone should develop fever, dizziness and a headache. The disease is amenable to antibiotic therapy if treated early.

Influenza - Some strains of the influenza virus can cross-infect between pigs and people, sometimes causing clinical disease in both.

Japanese encephalitis - This is the most serious zoonotic infection of pigs. It is spread from pigs to people by mosquitoes. It is confined to herds in South East Asia and does not occur in Europe or the Americas. People working with pigs in endemic areas should be vaccinated.

Disease in people caused by other potential zoonoses listed in Fig 17-10 is extremely rare and they warrant only brief mention here.

Anthrax - This is rare in pigs outside anthrax incubated areas. In such areas the disease is well understood and controlled.

Erysipelothrix rhusiopathiae - This is widespread in pigs everywhere causing the common pig disease erysipelas. The organism can (rarely) cause skin lesions in people, usually on the hands but mainly in fish and meat handlers. (Note that the disease "erysipelas" in people is unrelated and is caused by a streptococcus).

Clostridium perfringens type A - This is common in pig faeces but poses little direct threat to pig workers. Food poisoning results from poor food preparation.

General safety precautions

If a pig is exhibiting clinical signs of disease or it is in contact with diseased pigs and may be infected take the following precautions (Fig 17-11):

1. Cover cuts and abrasions with waterproof dressings.
2. Wear gloves and coveralls.
3. Disinfect gloves and coveralls after use.
4. Always wash your hands after handling the animal or its products even if gloves were worn.

FIG 17-11: PIGS ZOONOSES SAFETY PRECAUTIONS

Source	Precautions
Abortion, stillbirth and vaginal discharge	Take care not to get vaginal discharges in cuts or near your mouth, nose or eyes. Pregnant women must not work with farrowing sows.
Diarrhoea	Take special care with animals suffering from diarrhoea not to get faeces near your mouth.
Urine	Take care not to inhale or swallow urine or splash it in your eye.
Skin disease	Avoid skin contact.
Respiratory disease	Take care not to get the pigs' nasal discharges or sputum near your lips or nose e.g. if it has been rubbed onto your clothing or boots. Wear a face mask.
Dead pigs	Carcasses should be removed and disposed of promptly. Disinfect the area occupied by the carcass.
Bites	Take care with unrestrained pigs. Disinfect bites quickly. See a doctor if you think the wound is infected.

All staff involved with handling pigs or their excreta should be vaccinated against tetanus.

Risk rating

Determine whether the level of risk associated with each of the hazards identified is high, moderate or low. For risk assessments to be of greatest value you should adopt a risk rating and priority system.

Risk has two components: the likelihood of a harmful event occurring; and the degree of damage if it should occur (i.e. the consequences). If these two contributory factors are given numerical values, say from 1 to 3, the risk rating is the product of these values multiplied together (Fig 17-12).

FIG 17-12: DETERMINING RISK

Likelihood of event (i.e. exposure to a hazard sufficient to cause harm)	X	Consequences (i.e. severity of harm)
1 = Highly likely		1 = Severe (death/major injury)
2 = Moderately likely		2 = Serious (off work more than 3 days)
3 = Unlikely		3 = Slight (off work less than 3 days)

From this it can be seen that the risk ratings will be 1, 2, 3, 4, 6 or 9 with **1** representing very high risk (highly likely to happen, with severe consequences) and **9** representing negligible risk (unlikely to happen, slight consequences if it did). These risk rating values can be used to set priorities for actions with **1** requiring immediate decisive action to eliminate or reduce the risk (i.e. very high priority) and **9** requiring little or no action (i.e. very low priority).

Using a risk rating/priority system in your assessment will make it less likely that you ignore what you consider to be trivial risks. They may indeed turn out to be so but to show due diligence it is better that your assessment is recorded.

In order to determine the likelihood of a harmful event you have to make informed judgements based on:

- What hazards are present.
- What harm could they cause.
- Under what conditions they would be present in a form to cause harm and who would be affected (number of people, sex, health status).
- What precautions are already in place to prevent exposure.

The task of moving sacks of feed can be used to illustrate this point:

It is highly likely that a back injury will result if a slightly built person with no manual handling training has to carry a full sack of feed several times a day along a slippery walkway, up narrow steps to an unstable gantry and then raise the sack above the shoulders to tip it into the auger.

It is unlikely that a back injury will result if a strongly built person, trained in manual handling techniques, has to slide a sack of feed from a pallet, across a stable gantry then open the top of the sack and let the contents empty into the auger.

You should also consider your past accident/incident record and those of the pig farming industry in general.

If you judge an event to be unlikely and yet such events have occurred on your farm or are common in the industry you should think again.

Statistics show that within any organisation there are likely to be many near-miss incidents. These are accidents waiting to happen and many organisations now accept that risk assessments must also take account of these near-misses.

Once you have decided on the likelihood of an event you can consider the consequences if that event should occur. When doing this you will need to take into account the number of people that could be affected and whether certain individuals would be more seriously affected than others because of an existing disability or medical condition. For example, the consequences of damage to an eye as a result, say, of using a grinding tool would be far greater for a person who is poorly sighted in one eye than it would be for a person with two good eyes.

In addition to the harm done to a person, consequential loss should also be included. For example if the person at risk is a key employee whose absence from work even for a short period would affect production, the risk rating should be higher than it would otherwise be.

You can see from these examples that this form of risk rating is not an exact science and you may need to modify it in order to give yourself more flexibility. For example, with certain hazards you may want to multiply in a third component by splitting up "consequences". Thus you could have three columns i.e.

- Likelihood of event
- Degree of harm to people
- Consequential loss/cost

You would then have a wider range of scores, namely, 1 (the worse risk) 2, 3, 4, 8, 12, 18, 27 (the lowest risk).

It is important to recognise that this method of risk rating is not intended to give an objective quantitative measure of risk but rather to create a list of priorities and to provide a record that shows you have carefully considered likelihood and consequences.

Acceptable risks

Although the ideal situation would be to eliminate all risks or reduce them to a trivial level this is obviously not possible. Pig farming is a physical occupation involving manual handling, potentially dangerous machines, potentially hostile environments and large, strong unpredictable animals. There will always be some risks associated with it. With the emphasis now on self-regulation, the onus is on you if you are the employer to determine what the acceptable levels of risk are in your situation and to be able to justify them.

This is summed up by your legal obligation (and policy) *"to do all that is reasonably practicable to ensure the health, safety and welfare of all employees, contractors and visitors"*. The implications of the phrase *"... do all that is reasonably practicable ..."* are significant here.

Most employers appreciate that they are entitled to take cost and inconvenience into account when considering what actions would be required to reduce a risk, unless the risk is high. Perhaps less well appreciated are other criteria that must also be taken into account.

The term "reasonably practicable" applies to both legislative and industry standards which means that you have to ensure that the levels of risk in your pig farm are at least within any specified legal limits and also that they are not significantly higher than those on other similar pig farms.

If other pig farmers are achieving fewer accidents and incidents than you, you cannot claim that it is not reasonably practicable for you to reduce risks also.

Therefore, before you can finally determine the appropriate control measures you have to find out what the legislative and industry standards are. This type of information may be available in Government publications and from agricultural information bureaux or farmers' trade associations/unions. You may have to compare notes with other pig producers and/or take advice from a consultant. You have an over-riding duty to exceed legislative and industry standards if it is reasonably practicable for you to do so.

The following three examples will hopefully put these points into context:

Example 1. Noise is a hazard that many pig producers consider 'goes with the business' and is therefore to be expected and accepted. Furthermore, because the effects of noise can take years to develop many farmers are not aware that the levels of noise on their farms may pose a significant risk to health. Consequently, the majority of farmers do not have a policy to reduce personal exposure to noise. The result, as surveys have shown, is that 25 % of indoor pig workers have a noticeable hearing loss by the age of 30 and 50 % have significant hearing loss by the age of 50.

Given these statistics it would be difficult for you to claim that it is not necessary or reasonably practicable for you to reduce noise exposure levels on your farm, particularly since control measures (using ear plugs at noisy times) would be simple and cheap to implement.

You probably do not have this option anyway because in many countries it is a legal requirement for employers to assess noise at work and if they think that there may be a risk to hearing, to measure the level of personal exposure to noise and if necessary apply controls to ensure that it is kept below legally defined limits. In this case you must control the risk. Cost and inconvenience only have a bearing on your choice of control measures.

Example 2. Back injuries are common in the pig industry and a certain level of risk is to be expected. But what level? Some countries have regulations stipulating that all manual handling activities must be assessed and that if risks are identified appropriate controls must be implemented. But they do not specify the levels of risk to be achieved other than that they must be as low as is reasonably practicable.

In this situation you must look to industry standards and your own records to provide a baseline for assessment. The pig industry may have set standards for, or national statistics may show, the average number of days lost per employee per year due to back injuries. If your record is worse than the standards or the national average, your employees and the enforcement agencies would be entitled to expect you to reduce the manual handling risks on your farm.

Example 3. Dust is another hazard associated with intensive pig production and not an easy one to cope with. The high levels present in many farms constitute a health and safety issue. More than 60 % of all pig farmers suffer from dust-induced respiratory problems during their working lives and in many cases these continue after they retire. The effects can be debilitating.

Research has attempted to determine the safe levels of exposure to dust and practical ways to control it. However, the complex make-up of piggery dust and the lack of long term clinical data make this a difficult task. Consequently, although occupational exposure limits (OELs) are specified in many countries they not only vary widely between countries but also refer to dust in general rather than to piggery dust specifically. Piggery dust has many potentially harmful components such as animal and feed proteins, bacteria and viruses, fungal and bacterial spores, mycotoxins and endotoxins. The reality is that nobody knows the safe exposure levels for piggery dust, which has implications for you as an employer. Although you may be able to achieve and demonstrate (by dust monitoring) levels of exposure to dust that are below the specified OELs (and any industry standards that apply), if any of your existing or past employees exhibit signs of respiratory problems (persistent cough, chest tightness, wheezing, shortness of breath, constantly runny nose, or sore throat) that are thought to be associated with the dust on your farm, you have to implement additional controls to further reduce the levels of personal exposure. Alternatively you must be able to justify why it is not reasonably practicable for you to do so.

All three of these examples highlight the need for you to keep abreast of the legislation, pig industry standards and general health and safety information. They also underline the importance of pre-employment and ongoing health checks.

In the event of a criminal or civil claim being made against you, ignorance of the law is no defence.

Step 6 - Carry out Your Risk Assessments

The risk assessment procedures described here involve systematic appraisals of all of the areas in and around the pig farm and the procedures that take place in them.

Before conducting the assessments it will be necessary to compile the following information.

A list of the areas to be assessed. For example:

- Barns.
- Drying sheds.
- Feed bagging rooms.
- Feed milling rooms.
- Feed stores.
- Fields.
- Grain stores.
- Hospital pens.
- Loading bays.
- Muck heaps.
- Offices.
- Pig houses.
- Rest rooms.
- Silos.
- Tractor sheds.
- Workshops.
- Yards.

It is important not to ignore any areas. A field may look harmless enough but several farmers are killed each year when raised tractor shovels touch overhead power lines.

A sketch map of the site is useful to help list the areas in a logical sequence in preparation for a walk-round inspection. It can also be used to mark safety features and hazards such as the positions of first aid boxes and fire extinguishers, fuel stores and power lines etc.

A list of the procedures to be assessed. For example:

- Feed/grain silo management.
- Feed milling/mixing/bagging.
- Grain handling/drying.
- Hand feeding.
- Hazardous chemicals - handling/storage/use.
- Maintenance/repairs.
- Pest control - rodents/vermin/flies/mites.
- Pressure washing.
- Sick pigs - handling/treatment/carcass disposal.
- Slurry channel drainage.
- Straw handling.
- Use of mechanical scraper.
- Veterinary medicines - handling/storage/use.
- Weighing pigs.

All hazards on the farm need to be assessed. The lists of areas and procedures should therefore be as comprehensive as possible. However, there will be some hazards associated with general items and activities that cannot be described as areas or specific procedures (e.g. equipment and machinery, animal handling). You must ensure that these are also assessed and that the relevant points are addressed and covered in the appropriate SSW.

Human factors

You should compile a list of all employees and others who may be affected by the hazards present. Include contractors and maintenance personnel who may be on-site occasionally or at specific times. Alongside each name (or category of people) include any information that will help you to evaluate the risks to their health and their contribution toward reducing or increasing risks to themselves and others. For example:

- Levels of training.
- Competence.
- Responsibilities.
- Specific skills.
- Disabilities.
- Health status.
- Health record.
- Accident/incident record.
- Attitude to health and safety.

Management practices

Make a note of any existing health and safety systems. For example:

- Accident/incident records and reviews.
- Health and safety policy statements.
- Monitoring of hazards (dust, noise, fumes).
- Occupational health surveillance.
- Organisation and health and safety responsibilities.
- Previous assessments (for hazardous substances, manual handling, etc.).
- Safety checks/audits.
- Work instructions.

You should also give an indication of the current standards of health and safety on the farm and what priority is given to improving them.

These lists can now be used to provide information for the assessments.

Although it may not be a legal requirement for you to record the assessments, it is your duty to conduct them and to inform employees of the findings. It is therefore sensible to record them so that you can prove your due diligence if necessary.

Area assessments

For each area to be assessed you should compile a record sheet with 12 columns. For the first 10 column headings, the information to be considered and the points to record are as follows:

1. **What is the identity of the area?** Simply record the name and basic description of the area, e.g.: Farrowing House 1. 10 farrowing places, slatted floors at rear, slurry channel under slats. Mechanical ventilation.
2. **What procedures are carried out here?** Consult the list of procedures to be assessed and note those that apply to this area. It may be useful to number all of the procedures and simply record the relevant numbers here.
3. **What hazards are present?** See hazards and safety data. Carefully consider all categories of hazards. Use the hazard check list.
   List all the hazards (significant and trivial) associated with the area, i.e. those that are always, usually or sometimes present here.
   Include the hazards associated with the procedures that take place here.
   Include any hazardous substances that are stored here, even if they are used elsewhere. They may seem to present no risk but breakage, spillage or fire could result in personal exposure to them.
4. **What harm could they cause and how?** Refer to the hazard check list and Safety Data. Specify the nature of the harm and how it could occur e.g. a chemical could be strongly irritant if splashed into eyes. A wobbly gantry could cause serious injury if a person fell from it.
5. **To whom could they cause it?** List the people who work in the area regularly or occasionally. Specify the length of time that they are present in the area. Refer to the human factors for consideration and record any significant points.
6. **What precautions are currently taken?** See control options. List the physical and personal control measures and the management practices.
   If risk assessments have already been carried out for particular aspects of health and safety you do not need to repeat them but can refer to them, e.g.: Noise assessment carried out. Results recorded. Assessment is that ear plugs must be, and are, worn by all personnel present when hand-feeding of the pigs takes place.
7. **What is the likelihood of harm occurring?** See risk rating. In the cases of dust, noise and slurry gases you will probably need to carry out personal monitoring to determine the levels of these hazards in order to assess likelihood.

8 **What would be the degree of the harm?** See risk rating. You will possibly need to refer to the hazard Safety Data and occupational health information to be able to judge this.
9 **What would be the consequential loss/cost?** See risk rating and cost benefits.
10 **What is the level of risk (priority for actions)?** See risk rating.

You have now identified and assessed the risks. The remaining two columns on the record sheet are for you to specify what you will do in order to eliminate or reduce those risks, and when you will do it by

Procedure assessments

For each procedure to be assessed you should compile a record sheet of 12 columns with headings similar to those above except for questions 1, 2, 3 and 6. The information to be considered here and the points to record for each procedure are as for the area assessments with the exception of:

1 **What is the identity of the procedure?** Simply record the name of the procedure e.g. slurry channel drainage.
If there is a written work instruction for the procedure you must refer to it here.
If no work instruction exists you should include brief details of the procedure here.
2 **Where does this procedure take place?** Consult the list of areas to be assessed and note those where this procedure is performed. It may be useful to number all of the areas and simply record the relevant numbers here.
3 **What hazards are present?** See hazards and safety data. Carefully consider all categories of hazards. Use the hazard check list. List all the hazards (significant and trivial) associated with the procedure.
4 **What precautions are currently taken?** See control options. List the physical and personal control measures and the management practices.
Specify whether any written work instructions are available for this procedure, and if so refer to them.

If risk assessments have already been carried out for particular aspects of health and safety you do not need to repeat them but can refer to them, e.g.:
Hazardous substances assessment carried out. Results recorded. The assessment record and the work instructions specify the control measures.

They must be, and are always, implemented by persons performing this procedure.

Self-assessment by employees

No matter how thorough you are with your assessments, there will always be some situations that you cannot fully address until they arise and you cannot always be there when they do. Therefore, if you want the system to work you must also be able to rely on your employees to look out for the well-being of themselves and the wellbeing of the business as a whole. To achieve this you will need to demonstrate a positive, pro-active commitment to health and safety and encourage all employees to conduct their own risk assessments before carrying out any new activities. If they think that there is a risk to themselves or to others they should not continue with the activity but should inform you of their concerns so that you can fully assess the situation. This self-assessment need not involve them having to write anything down but should be a simple 'stop and think' exercise.

For example:

1 What am I about to do? Where will I be doing it? How long will it take?
2 Do I know what hazards are present or will be produced?
3 Do I know what harm they could cause - and how?
4 Are sufficient safety precautions in place?
5 Is it safe for me (and those around me) if I continue?

If they cannot answer yes to all of the points 2 to 5, they should not continue.

Step 7 - Devise and Apply your Control Measures

Control options

You must now decide how to eliminate or reduce the risks identified by your assessments, and legislative and industry standards are likely to have a bearing on your selection of the appropriate control measures. There are a number of ways to control risks but legislation may, and often does, require you to adopt a hierarchical approach when deciding which control option to use. In other words you must consider the best (safest) option first and if that is not reasonably practicable you should consider the next best - and so on. Once again, if other pig producers are controlling specific risks by using control measures higher up the hierarchical scale than you are, you should review your situation.

Some aspects of health and safety are governed by regulations that state precisely what control measures must be applied. You must be aware of any that apply to you (see Step 2) and must obviously comply.

For other aspects of health and safety several regulations may apply which have their own specific hierarchy of control measures to be considered. It is not possible to list all of the variations here but the general principles are explained below. Control measures are divided into two categories: physical controls, which are directed against the hazard itself; and personal controls, which apply to the people at risk of exposure to the hazard.

Physical controls

These are usually the most effective and should be considered first.

In order of priority they are:

- Remove.
- Replace.
- Restrict.
- Reduce.

Remove the hazard completely by eliminating the task or making design or organisational changes. This is the most effective control measure. For example:

- Mechanise an activity to eliminate manual handling.
- Re-design a loading bay to avoid having to reverse vehicles.
- Repair floors to eliminate the trip hazard.
- Replace a flammable solvent-based paint with non-flammable water-based paint.
- Replace an electric pressure-washer with a diesel-powered one to eliminate the risk of electric shock.
- Replace a diesel-powered pressure washer with an electric one to eliminate hazardous exhaust fumes.

This last example is to illustrate the point that eliminating one hazard may mean that a different one will have to be assessed.

Replace the hazard with a proceedure that is less hazardous. This is different from removal (and less effective) as the same type of hazard is still present but the likelihood of an event or its consequences are reduced. For example:

- Replace a disinfectant that is highly caustic to skin and eyes with one that is only a mild irritant.
- Replace a highly flammable solvent-based paint with one that is less flammable.
- Replace an aggressive sow with one that is docile.

Restrict access to the hazard by physical means. For example:

- Put guards on dangerous parts of machinery.
- Put barriers around hazards such as machinery and slurry pit openings.
- Put guard rails on raised gantries.
- Fit an automatic lock-out so that the auger cannot be accessed until it has stopped rotating.

Reduce the level of the hazard or the duration of exposure to it by physical means. For example:

- Use dust-extracted straw for bedding to reduce dust levels.
- Fit a residual current device to electrical equipment.
- Store sacks of feed closer to the points of use to reduce the amount of carrying involved.
- Fit baffles or box-in noisy machinery (feed milling and bagging plant) to reduce the level of noise.

Personal controls

These will be required in all cases in addition to the physical controls.

Such personal controls may include:

- Training.
- Instructions.
- Information.
- Supervision.
- Permits to work.
- Protective devices.
- Personal protective equipment.

Training, instructions and information must be provided for all employees by law in most countries but many employers are unsure of the distinction between the three terms.

Training can be considered to have been sufficient once a person is judged to be competent to perform a task safely without supervision.

Instructions may be written or verbal and usually consists of a list of the steps to take to complete a task (i.e. work instructions) or a statement that a specific control must be implemented (e.g. switch off the power before opening the cover of the auger).

Information must be given regarding the hazards and risks that the employees may be exposed to in an area of the pig farm or when performing particular tasks. For example the nature of any hazardous chemicals involved (caustic, irritant, toxic, etc.), how the chemicals can cause harm (by ingestion, inhalation, skin contact, etc.), what level of risk is associated with the chemicals in particular areas or situations, how to prevent or minimise exposure to them and what to do in the event of accidental exposure.

Employees must also be provided with information regarding accident, first aid, fire and emergency procedures.

When assessing risks and evaluating control measures, the existing standards of employee training, instruction and information must be considered.

Supervision - of all new employees and of existing employees undertaking new tasks must be carried out and must be sufficient for the employer to determine whether the procedures are always performed safely, or whether additional training, instructions, information and supervision are needed.

Permits to work - are documents that must be obtained by an employee or contractor before certain works can be carried out. The permit must state that the necessary controls have been implemented and that it is safe to proceed with the work. In many countries it is a legal obligation to operate a permit-to-work system for certain activities. You must check how your regulations apply to you.

Even if it is not a legal requirement it is sensible to operate your own permit-to-work systems for high-risk activities. These need not necessarily be written documents but could include, for example, a stipulation in the work instructions that the employee must obtain permission before starting a given job so that somebody is aware that the activity is about to take place and can implement the appropriate controls or simply be standing by e.g. fumigation using formalin. Other situations where such a permit system may be required are high voltage electrical work and entry into confined spaces e.g. slurry pits and feed silos.

Protective devices - can be used by a person to reduce the likelihood of a harmful event. An example is the use of a pig board.

Personal protective equipment (PPE) - includes such items as ear defenders, hard hats, dust masks, respirators, eye shields, gloves, boots, coveralls, waterproof aprons and washing facilities.

Coveralls, boots and gloves are obviously essential items for pig farmers and should be used routinely, which they generally are. Other items of PPE are often used inappropriately, incorrectly or insufficiently.

Each situation needs to be carefully assessed. The important points to consider are:

- Legislation usually demands that PPE must be the last control option and should only be used if exposure to hazards cannot be controlled by any other reasonably practicable means.
- PPE can itself be a hazard. Dust masks for example make breathing laboured and can put unnecessary strain on the heart and lungs. Ear plugs may mean that audible safety warnings such as fire alarms are less effective.
- The item of PPE used must be appropriate for the level and type of hazard. Nuisance dust masks afford little protection. A mask that is adequate for general piggery dust is unlikely to be adequate for dust that could contain antibiotics (medicated feed). Different types of mask will be required for aqueous mists (e.g. from pressure washers), slurry gases and welding fumes - all common hazards. Ear plugs and ear defenders come in grades appropriate for the level and type of noise.
- PPE can be used as a stop gap to reduce a known risk until it is reasonably practicable to implement a more effective control.
- Only use PPE that has official approval (all PPE in Europe must be CE marked).
- PPE can lead to a false sense of security. Employees must be instructed to continue to take due care.
- PPE is ineffective and possibly hazardous if badly fitted. Dust masks and ear plugs must be close fitting.
- Oversize gloves can result in clumsy handling of hazards. Oversize boots are a potential trip hazard.
- PPE is ineffective and possibly hazardous if damaged, dirty or worn out. Employees must be instructed in the use and care of PPE.

Once you have considered the control options you can complete the last two columns on both your assessment record sheets, i.e.:

11 **What measures will be implemented to control the risk?**
Specify the physical and personal controls you think are appropriate.

12 **What management procedures are needed to ensure the controls are maintained?**
This section should specify, what if anything, is needed in terms of a staff training programme, spot checks. machinery, service contracts, health surveillance schemes, assigning responsibilities, safety audit and review systems, etc.

Records

Record keeping requirements can arise in relation to a wide range of health and safety issues. In many cases specific risk assessments must by law be documented but a number of other records should also be kept.

They are an additional control measure in that they serve as a reminder that things need to be done, they facilitate auditing and reviews, and they demonstrate your due diligence.

You therefore need to decide what records are applicable to your situation.

Some examples of the types of records that may be needed are given below. This list is by no means complete and you should compile your own up-to-date version.

1. Organisation/administration - appointment of competent persons; policy changes.
2. Training - official health and safety qualifications; in-house training (re: SSW, risk assessment procedures); assessment of competence for tasks; specific health and safety instruction courses e.g. pesticides, tractors, forklift trucks, etc.
3. Employment and occupational health records - pre-employment and on-going health surveillance; permits to work e.g. confined spaces.
4. Monitoring survey results - dust, noise, slurry gases.
5. Personal protective equipment (PPE) - PPE issue notes; examination records; test certificates.
6. Fire and emergency - alarm tests; fire drills; fire extinguisher/sprinkler system checks; staff training.
7. Work equipment - machinery and equipment inventories; routine safety checks; certification; guard checks.
8. Buildings and services - inspection of electrical installation; water treatments; construction and design safety plans; assembly/removal by competent persons e.g. asbestos work; contractors' declarations.
9. Risk assessments - hazardous substances, noise, manual handling; management of health and safety.
10. Accidents and incidents - accidents and incident reports (official and in-house); first aid treatments; follow-up actions.

Safe Systems of Work (SSW)

All of the components of your risk assessments and management procedures can now be combined into a series of working documents - the Safe Systems of Work (SSW).

Step 8 - Document Your SSW

The assessment process should have identified the things that need to be done. The management procedures should ensure that they are done. This combined health and safety working system is therefore designed to ensure your due diligence. In the event of an accident, if you cannot prove due diligence you will have little or no defence against legal action or civil claims.

The key points that should be incorporated into the SSW are as follows:

The regulations

Each SSW should include a note of the Specific Regulations, ACOPs, Industry Standards, etc. that apply to the particular aspect of health and safety covered (e.g. noise) and a brief description of what you are obliged to do to comply.

Policy statement

Your policy on the particular aspect of health and safety should be stated, along with who is responsible for ensuring that the policy is implemented.

Risk assessment

The significant findings of the risk assessments with regard to the specific topic should be noted and attention should be drawn to the detailed assessment records that are on file and who is required to read them. This will only apply to those employees who work in areas or carry out procedures that were assessed as presenting risks. The assessment records will mention any specific controls but the SSW should specify whether any general control measures are required (such as the provision of safety notices), when they will be implemented (priority rating), and what interim measures must be adopted. It should also specify any control measures already in place that must be complied with.

Instructions

The work instructions for specific procedures are not usually included on the same page but are referred to elsewhere - such as in a separate file of work instructions, on the assessment records or as laminated sheets that are on permanent display at the points of use. The work instructions specify the actual steps to take to perform a task safely. General safety points are usually included on the SSW and specify the codes of practice that all employees must adhere to in order to ensure that risks remain adequately controlled.

Records

The SSW should specify what records are to be kept e.g. personal monitoring records, work equipment inventories and inspection records, staff training in the correct use of PPE, etc. and who should keep them, how, when and where.

Hazard and safety data information

Together with the information regarding legal obligations, responsibilities and employee duties, the nature of the hazard (e.g. noise) and how it can cause harm (i.e. damage hearing, affect concentration) should also be described on the SSW. It should also contain references to any other information or safety data that is relevant or which may be needed for assessment purposes (e.g. machinery and equipment literature that gives details of noise levels under different conditions).

The overall aim of this type of management system is to have a series of single documents, each one of which provides all of the key information about one aspect of health and safety. Everyone must be familiar with all of the SSW.

Each SSW should clearly state where individual employees can find other information of relevance to them, for example specific work instructions, risk assessment records and safety data. In this way employees only have to read what directly applies to themselves.

Each SSW should also state when, how and by whom the necessary ongoing safety checks, reassessments and audits will be carried out and the information updated and reviewed. These details should also be noted in a diary or on a year planner so that the people responsible can be reminded.

This type of system ensures that all aspects of health and safety have been appropriately addressed and enables you to demonstrate that they have been. It also makes safety inspections and audits very simple for you, external safety consultants and inspectors.

The UK example SSW for noise (Fig 17-14) shows the type of information to be included. It is in a format that can be used for most health and safety topics and easily adapted to suit large or small organisations.

You should have SSW for all aspects of health and safety that directly impact on you.

A list of farm topics for which you should have SSW is given in Fig 17-13. This list is not exhaustive and you will need to compile your own.

Step 9 - Document your Accident, First Aid, Fire and Emergency Procedures

Most countries have their own regulations concerning these, including the assessment, staff instruction, recording and reporting duties of employers.

In effect, safe systems of work are also needed for these topics but because of their importance and to comply with the regulations, the information needs to be presented in a different format.

FIG 17-13: FARM TOPICS YOU SHOULD HAVE SSW FOR

Abrasive wheels.	Personal hygiene.
Accident prevention.	Personal monitoring.
Animal handling.	Personal protective equipment.
Asbestos (roofing & insulation).	Personal safety and security.
Cleaning.	Pesticides.
Compressed gas cylinders.	Post-mortems.
Confined spaces.	Pregnant women.
Construction.	Pressure systems.
Contractors.	PTO shafts.
Dust.	Public access.
Electrical safety.	Roof work.
Environmental monitoring.	Safety signs.
Fire prevention.	Slurry gases.
Handling of animals.	Stairways.
Hazardous chemicals.	Tractor safety.
Hostile environments.	Training.
Ladders.	Vehicles.
Lift trucks.	Veterinary medicines.
Manual handling.	Visitors.
New employees/trainees.	Waste disposal.
Noise.	Work at height.
Occupational health.	Work environment.
Overhead power lines.	Work equipment.
Pathological specimens sent by post.	Working alone.
Permits to work.	Young people/children.

Accidents

Legislation sets out an employer's legal obligations for reporting accidents and dangerous occurrences to the Enforcing Authorities. In the UK this is covered by the **Reporting of Injuries, Diseases and Dangerous Occurrences Regulations 1995 (RIDDOR) Amended 2012.** As defined in your policy statement, it is a key function of the Safety Manager to ensure that these obligations are fulfilled.

It is also important to compile your own accident and incident statistics and to review these periodically to determine what steps, if any, should be taken to prevent recurrence in the future. These reviews should also be coupled with health surveillance and be incorporated into risk assessments so that any control measures needed can be identified and given the necessary priority.

You should document and prominently display "The procedures to be followed in the event of an accident at work". The information should include:

- Accident reporting procedures i.e. how to inform the Safety Manager and the enforcement authority and how to record all accidents, even minor ones, in the accident book.

FIG 17-14: EXAMPLE SAFE SYSTEM OF WORK

for

NOISE

The Management of Health and Safety at Work Regulations 1999
The Control of Noise at Work Regulations 2005

— Prolonged exposure to loud noise can damage hearing (hearing loss, tinnitus, etc.) but the effects are often not apparent until much later in life. Continual exposure to "nuisance" noise can cause stress and/or affect concentration. This may result in risks to individuals and also to those affected by their actions.

— To comply with the above regulations we have a duty to assess the levels of noise to which staff are exposed, and to take steps to reduce or control the noise or exposure if necessary.

— Any areas and procedures on the farm that have been identified as presenting a possible risk to health from noise will be fully assessed.
  - Any noise monitoring required will be carried out by..
  - The results of any noise monitoring surveys will be recorded and filed in ....
  - The results of the assessments will be recorded and made known to all relevant staff.
  - Any remedial actions considered necessary will be implemented as soon as is practicable.
  - Assessments will be reviewed annually, or as necessary, by ....

— The noise contributed by machinery and equipment must be included in the assessments (worn machinery may become noisier and new machinery must meet noise specifications).
  - For further information regarding machinery assessors must see the Work Equipment SSW.

Listed below are some specific areas and procedures associated with the farm which can expose staff to high or nuisance noise levels.

If any of them apply to you, you must read:
- the specific **Area or Procedure Assessment** record (filed in ....)
- the relevant procedure **Work Instructions** (in the Work Instructions Manual and on display).

AREAS	**PROCEDURES**
Feed Mill	Feed Mill Operation
....	Hand Feeding of Pigs
....	....
....	....

IN GENERAL - all staff must adhere to the following Codes of Practice:

1. If normal conversation is difficult to hear across a distance of two metres, for significant periods of time, the background noise levels may be above acceptable limits and may present a risk of damage to your hearing.
   - If in doubt consult the Area Safety Officer. Ear protection will be made available if needed.
2. Repetitive background noise (maintenance work etc.) can result in loss of concentration and is therefore a safety hazard.
   - Individual staff and particularly Area Safety Officers should assess the potential risks in each situation and implement control measures (turn off the noise if possible, stop maintenance work for set periods, use ear protection) as necessary.
3. The pigs can create high noise levels at certain times (feeding, weighing, bleeding).
   - Staff exposed to these conditions for significant periods should assess the situation and use ear plugs or defenders as necessary. Seek advice if you are unsure.
4. All situations will be assessed at least annually, but if you consider any area, equipment or procedure on the farm to be 'noisy' you must inform the Area Safety Officer so that an assessment can be made sooner and the appropriate controls implemented.
5. Some areas of the farm present a definite risk to hearing and have been designated as **"Hearing Protection Zones"**. Safety signs are displayed on the doors to these areas.
   - **All persons MUST wear hearing protection in these areas - at all times or as instructed.**
   - Ear defenders are available in these areas. Extra ear plugs are kept in ....
6. Before using ear protection you must be familiar with the SSW for Personal Protective Equipment.

- How to summon first aid assistance.
- The location of the first aid boxes.
- The telephone numbers of local doctors, hospitals and emergency services.

First aid

You may need to seek advice regarding what you are obliged to provide in terms of first aid facilities. Some authorities stipulate that there must be a qualified first aider on-site (no matter how few people you employ) and there may be specific requirements concerning the number and location of first aid boxes, eye wash stations, signs, posters and rest room facilities.

Fire

You must obviously ensure that your fire precautions and procedures are adequate in terms of:

- Building construction.
- Access and egress.
- Emergency exits.
- Emergency lighting.
- Fire fighting appliances (sufficient, appropriate and maintained).
- Fire drills and alarm checks.
- Storage of flammable/combustible materials (gas cylinders, fuel oils, wood and paper waste).
- Identification of fire/explosion risk areas and instructions for isolating power, fuel, gas etc.
- Evacuation procedures and responsibility for roll calls.
- Employee training in procedures and general fire safety practices.

UK legislation **"The Regulatory Reform (Fire Safety) Order 2005"** requires all employers to conduct a 'Fire Risk Assessment' of their work premises. This shares the same approach as health and safety risk assessments and can be carried out either as part of an overall risk assessment or, ideally, as a separate exercise. Based on the findings of the assessment, employers must ensure that adequate and appropriate fire safety measures are in place to minimise the risk of injury or loss of life in the event of a fire. Similar legislation is in effect in many other countries (including all those in the EU). It is wise to check the legislation in your own country and to obtain guidance from your Local Authority fire service on the matters listed above. Once you have established appropriate facilities, systems and procedures you should document them and have the notice prominently displayed. As mentioned earlier, it is also useful to have a plan of your premises which clearly shows the sites of fire hazards (fuel stores etc.), fire extinguishers and escape routes.

Your fire procedures can also include details of the arrangements for evacuation or protection of the pigs but it must be made clear that personal safety must be the primary concern of all employees.

Emergencies

Similar procedures to those for accidents and fire are required for emergencies, to cover such events as gas leaks, explosions, pressure vessel rupture, building collapse and chemical leaks.

Step 10 - Review Your System Periodically

Once your health and safety working system has been developed and implemented you will need to review it periodically. The more comprehensive your written SSW and records are, the easier this will be. Reviews serve three important purposes:

1. **To comply and keep up-to-date with legislation**
   Many health and safety regulations stipulate that risk assessments, monitoring results and equipment checks etc. must be reviewed at specific intervals and whenever you make any significant changes. You will need to be aware of those that apply to you. Changes in the legislation or hazard safety data may also mean that assessments will have to be amended. For example, the significance of endotoxins in piggery dust is being investigated and the findings may result in lower occupational exposure limits being set.
   In general you should review your risk assessment records at least annually and as and when you introduce changes such as new areas, procedures, staff or hazards. Once your initial assessments have been revised to take account of any new control measures, most records will hopefully not require further amendments, but you should still document the fact that you have checked them and that they are still valid.
   If you have introduced any changes you must re-assess the relevant areas and procedures, and revise the existing records or compile new ones. For example, if you have an outbreak of salmonella in pigs in the finisher/grower house, you may need to introduce additional controls to minimise the risk of employees contracting or transmitting the disease. The new controls must be documented on the assessment record and brought to the attention of the employees.
   You must also review your risk assessments if circumstances suggest that problems are occurring. For example, you may have assessed the risks associated with dust in the pig houses to be low (providing your specified control measures are implemented) but find that employees still suffer from, or begin to suffer from, respiratory problems. This may be due to an incorrect assessment

originally, or because a change has gone unnoticed e.g. a different type of feed or bedding, perhaps. Whatever the reason, you will need to investigate and re-assess.

2 **To monitor and improve the system**

As with all management practices it is necessary to check whether the system is working effectively, or if it should be improved.

For example, your SSW may specify that something is to be inspected by somebody at given times and that the results are to be recorded. If your review of the system shows that the inspection is not being carried out or that the results are not being recorded you will need to revise the system. It may be that the person is not aware of their responsibility to perform the task, does not appreciate the importance of it, or does not have enough time to do it.

The solution may be to leave the SSW unchanged but to re-train the individual concerned, emphasise the importance of the task and/or make more time available for them to do it.

Alternatively the responsibility for the task may have to be assigned to somebody else and the SSW should be re-written accordingly.

3 **To demonstrate your commitment to health and safety**

If you want the system to work and for everyone to play their part you must convince all concerned that you will continually check it and make changes as necessary. Formal reviews will achieve this. If, however, it is seen as simply a one-off paper exercise with no follow-up, it will not be taken seriously and the time and effort spent developing the system will have been wasted.

If you think all of this is too much (and for the small business it invariably is) use the services of a competent organisation or consultant that specialises in health and safety in pig farms in your country. It will save you considerable time and effort and allow you to concentrate on what you are good at - pig production. Beeford Laboratories Ltd. can provide this service both in the UK and internationally. Beeford Laboratories can be contect via Garth Veterinary Group. See www.garthvet.co.uk for contact details.

APPENDIX
QUICK REFERENCES AND USEFUL INFORMATION

Further Reading and Information

Useful Websites

ThePigSite: www.thepigsite.com
OIE website: www.oie.int.
Iowa vet website: vetmed.iastate.edu/departments/vdpam.
fai: www.faifarms.co.uk.
3tres3: www.3tres3.com.

ThePigSite.com
The home of premium international pig news insight, analysis and features

Books/Magazines

Asian Pork Magazine: www.asian-agribiz.com.
Atlas of Topographical Anatomy of the Swine, Prof. Peter Popesko.
Diseases of Swine, Various.
Feed Efficiency in Swine, Patience, John F.
Feed International Magazine: www.wattagnet.com.
Garth Pig Stockmanship Standards. John Carr.
Handbook of Pig Medicine, Jackson and Cockcroft.
International Pig Topics, UK: www.positiveaction.info.
Manual of Pig Production in the Tropics, CAB International, Oxfordshire, UK.
Mechanistic Modelling in Pig and Poultry Production, C. Fisher, R. M. Gous, Trevor Morris.
Modern Pig Production Technology: A Practical Guide to Profit, John Gadd.
National Hog Farmer, USA: www.nationalhogfarmer.com.
Nutrient Requirements of Swine, National Research Council, USA.
Nutritional and Physiological Functions of Amino Acids in Pigs, Blachier, Francois, Wu, Guoyao, Yin, Yulong.
Optimum Vitamin Nutrition, Various.
Pathology of the Pig: A Diagnostic Guide, L. D Sims and J. R. W Glastonbury.
Pigs: A Handbook of Breeds of the World, Valerie Porter.
Pig Diseases, David Taylor.
Pigs: Keeping a Small-scale Herd for Pleasure and Profit, Arie Mcfarlen.
Pig International: www.wattagnet.com.
Pig Production: Biological Principles and Applications, John McGlone, Wilson G. Pond.
Pig Production: What the Textbooks Don't Tell You, John Gadd.
Pig Progress Magazine: www.pigprogress.net.
Pig World Magazine: www.pig-world.co.uk.
Pork Magazine: www.porknetwork.com.
Potbellied Pig Behavior and Training, Priscilla Valentine and Steve Valentine.
Recognising and Treating Pig Diseases, Muirhead and Alexander.
Recognising and Treating Pig Infertility, Muirhead and Alexander.
Small-scale Outdoor Pig Breeding, Wendy Scudamore.
Strategies to improve the health and production of weaner pigs, Danka Oljaca Halas, John R. Pluske.
Sustainable Swine Nutrition, Chiba, Lee I.
Swine in the Laboratory, Swindle, M. Michael
Swine Science, Palmer J. Holden and M. E. Ensminger.
US Pork Center of Excellence: www.usporkcenter.org.
– *Swine Nutrition Guide*: www.usporkcenter.org/Projects/506/NationalSwineNutritionGuide.aspx.
Whittemore's Science and Practice of Pig Production, Colin T. Whittemore, Ilias Kyriazakis.

Organisations/Conferences/Proceedings

Asian Pig Veterinary Society (APVS).
Allen D Leman Swine Conference.
Asociación Nacional de Porcinocultura Científica (anaporc): www.anaporc.com.
American Association of Swine Veterinarians (AASV): www.aasv.org.
Banff Pork Seminar: www.banffpork.ca.
Bpex: www.bpex.org.uk.
International Pig Veterinary Society (IPVS): www.ipvs.org.
National Pig Association, UK: www.npa-uk.org.uk
NSW DPI Pork Production: www.dpi.nsw.gov.au/agriculture/livestock/pigs.
Pig Veterinary Society (PVS): www.pvs.org.uk.
Swine Disease Conference for Swine Practitioners: www.iucs.iastate.edu/mnet/swinedisease/home.html

Equipment, Materials Chemicals and Medicines You May Require on the Farm for Maintaining Pig Health

EQUIPMENT AND MATERIALS	
AI catheters disposable.	Nasal swabs.
AI cattle catheters disposable.	Navel clips.
AI equipment.	Needles 20 g 16 mm and 25 mm, 18 g 38 mm, 18 g 50 mm, 16 g 100 mm.
Antiseptic for skin use.	Paper tissues.
Bandages – tubular, elastoplast, white and adhesive tape.	Paper towels.
Capture bolt pistol or 12 bore shot gun.	Pithing rod.
Corrugated plastic tubing for treating a prolapsed rectum.	Plastic bags.
Cotton wool.	Post-mortem knife.
Detailing equipment – knife, scissors, burdizzo or gas.	Pressure washer.
Detergent.	Rectal thermometers.
Disinfectant for aerosol use.	Rubber gloves.
Disinfectant for surfaces.	Scissors – 1 pair tissue forceps, 2 pairs artery forceps.
Disinfectant or foot baths.	Snare or wire noose.
Disposable gloves.	Surgical blades (handle).
Pregnancy tester and coupling liquid.	Surgical spirit.
Dust masks.	Suture materials.
Ear muffs.	Suture needles.
Embryotomy wire.	Suture tape.
Eye protectors.	Swabs in transport medium.
First aid box.	Syringes 2 – 5 – 10 – 20 – 30 ml.
Flutter valve.	Tattoo equipment – slap marker, tattoo pastes, tags, ink pad, ear notchers.
Hoof cutters.	Teeth clippers.
Indelible marker pen.	Temperature electronic recorder.
KY jelly.	Thermometer – maximum/minimum.
Labels self adhesive.	Tissues.
Liquidiser.	Vacutainers.
Marker sprays, wax crayons.	Water steriliser.
Mouth gag.	

CHEMICALS	
Aerosol disinfectant	Obstetrical fluid
Dehydrated lime powder	Organophosphorus
Detergent	Potassium permanganate
Fenitrothion	Pyrethrins
Fly control	Skin disinfectant
Foot bath disinfectants	Surface disinfectants
Formaldehyde 40 %	Surgical spirit
KY or AI lubricants	Water steriliser
Methomyl	

Medicines
(As advised by your veterinarian)

The active names of products are given. Product names, availability and licencing vary by country/region. Ask your veterinarian for advice.

Trade Names in your Country

The table on pages 644 - 645 provides a guide to medicines for use in the pig. This tables lists the chemical name of the medicines. Use the space provided to write in the Trade names of the medicines available in your country.

ANTIBIOTICS AND ANTIBACTERIAL SUBSTANCES	
Injectable	**Oral and/or In-feed**
Amoxycillin. Ampicillin. Apramycin. Baquiloprim. Ceftiofur. Cephalexin. Enrofloxacin. Erythromycin. Framycetin. Gentamycin. Griseofulvin. Lincomycin. Neomycin. Oxytetracycline. Penicillin. Spectinomycin. Streptomycin. Sulphadiazine. Sulphadimidine. Tiamulin. Trimethoprim/Sulpha. Tylosin.	Chlortetracycline. Enrofloxacin. Griseofulvin. Lincomycin spectinomycin. Lincomycin. Neomycin. Nitrofurans. Oxytetracycline. Phenoxymethyl penicillin. Sulphonamides. Tilmycosin. Trimethoprim sulphonamides. Tylosin.

OTHER MEDICINES	
ACTION ON THE UTERUS	
Antibiotic pessaries. Buscopan – pain relief. Monzaldon assists farrowings.	Oxytocin – contraction. Pituitary extract – contraction. Prostaglandins – induction of farrowing.
ANTI-INFLAMMATORY INJECTIONS	
Betamethasone. Dexamethasone.	Flunixin. Phenylbutazone.
ANTHELMINTICS	
Doramectin. Fenbental. Fenbenazole. Flubendazole. Ivermectin.	Levamisole. Oxibendazole. Piperazine. Thiophanate.
COCCIDIOSTATS	
Amprolium. Dimetridazole. Monensin.	Salinomycin. Sulphonamides. Toltrazuril.
HORMONES	
Altrenogest. Gonadotrophin. Oestradiol benzoate.	Oxytocin Testosterone.
MISCELLANEOUS SUBSTANCES	
Liquid paraffin.	
NUTRITION AND METABOLISM	
Calcium boroglucinate 40 %. Dextrose. Electrolytes. Glucose 40 %. Iron dextran.	Multivitamins. Respirot – oral respiratory stimulant. Vitamin E/selenium. Vitamin minerals.
PARASITICIDES – TOPICAL USE	
Amitraz. Deltamethrin.	Diazinon. Phosmet.
SEDATIVES	
Acetyl promazine – injection. Azaperone – injection.	Primidone – tablets.

GUIDE TO MEDICINES FOR USE IN THE PIG						
Medicine	**Trade Name to Complete**	**Suggested Dose**	**Use**	**Injection**	**Oral**	**Feed**
Acetaminophen		30 mg per kg	Other		Y	Y
Altrenogest		20 mg per gilt or sow	Reproductive		Y	Y
Amoxycillin		20 mg per kg	Antibiotic water/oral	Y	Y	Y
Apramycin		12 mg per kg	Antibiotic	Y	Y	Y
Atropine		0.05 mg per kg	Other	Y		
Azaperone		2 mg per kg	Other	Y		
Calcium boroglucinate		0.1 mg per kg	Other	Y		
Carbetocin		0.2 – 0.3 mg per kg	Reproductive	Y		
Carbodox		50 ppm	Antibiotic			Y
Cefquinome		2 mg per kg	Antibiotic	Y		
Ceftiofur		3 mg per kg	Antibiotic injection	Y		
Chlortetracycline		40 mg per kg	Antibiotic feed/oral	Y	Y	Y
Colistin		50,000 iu per kg	Antibiotic	Y		
Danofloxacin		1.25 mg per kg	Antibiotic	Y		
Dexamethazone		0.06 mg per kg	Other	Y		
Diazepam		10 mg per kg	Other	Y		Y
Dihydrostreptomycin		10 mg per kg	Antibiotic	Y		
Dormectin		0.3 mg per kg	Parasiticide	Y		
Enrofloxacin		2.5 mg per kg	Antibiotic injection	Y	Y	
Erythromycin		22 mg per kg	Antibiotic	Y		
Fenbendazole		5 mg per kg	Parasiticide		Y	Y
Florfenicol		15 mg per kg	Antibiotic injection	Y	Y	Y
Flubendazole		5 mg per kg	Parasiticide		Y	Y
Flunixin		2.2 mg per kg	Other	Y		
Framycetin		5 mg per kg	Antibiotic	Y	Y	
Gentamycin		5 mg per kg	Antibiotic		Y	
GnRF vaccine		Vaccine one dose per pig	Reproductive	Y		
Gonadotrophin serum		400 iu per pig	Reproductive	Y		
Gonadotrophin chorionic		200 iu per pig	Reproductive	Y		
Iron		150 mg per kg	Other	Y	Y	
Ivermectin		0.3 mg per kg	Parasiticide	Y	Y	Y
Ketamine		15 mg per kg	Other	Y		

GUIDE TO MEDICINES FOR USE IN THE PIG CONT.						
Medicine	**Trade Names to Complete**	**Suggested Dose**	**Use**	**Injection**	**Oral**	**Feed**
Ketaprofen		3 mg per kg	Other	Y	Y	
Lincomycin		10 mg per kg	Antibiotic injection	Y	Y	Y
Marbofloxacin		8 mg per kg	Other	Y		
Meloxicam		0.4 mg per kg	Other	Y		
Metronidazole		25 mg per kg	Antibiotic		Y	Y
Monensin		100g/tonne feed	Antibiotic			Y
Neomycin		5 mg per kg	Antibiotic	Y	Y	Y
Oxytetracycline		40 mg per kg	Antibiotic injection	Y	Y	Y
Oxytocin		0.25 iu per dose	Reproductive	Y		
Peforelin		see data sheet	Reproductive	Y		
Penicillin		10 mg per kg	Antibiotic injection	Y	Y	Y
Pentobarbitol		30 mg per kg	Other	Y		
$PGF_{2\alpha}$		10 mg per pig	Reproductive	Y		
Phenylbutazone		8 mg per kg	Other		Y	
Phosmet topical		0.1 ml per kg	Pasasiticide topical		Topical	
Salicylate		35 mg per kg	Other		Y	Y
Sulfadiazine		12.5 mg per kg	Antibiotic	Y	Y	Y
Telazol		4 mg per kg	Other	Y		
Tiamulin		10 mg per kg	Antibiotic injection	Y	Y	Y
Tildipirosin		4 mg per kg	Antibiotic	Y		
Tilmicosin		20 mg per kg	Antibiotic		Y	Y
Toltrazuril		20 mg per kg	Antibiotic		Y	
Trimethoprim		2.5 mg per kg	Antibiotic	Y	Y	Y
Tulathromycin		2.5 mg per kg	Antibiotic injection	Y	Y	Y
Tylosin phosphate		5 mg per kg	Antibiotic feed/oral	Y	Y	Y
Tylvalosin		4.25 mg per kg	Antibiotic		Y	Y
Valneumulin		4 mg per kg	Antibiotic			Y
Vetrabutine		2 mg per kg	Other	Y		
Xylazine		4.4 mg per kg	Other	Y		
Zinc Oxide		2500 ppm	Antibiotic feed/oral		Y	

Note the dose rate suggested is for information only.
Many medicines can be used over a range of dose rates consult with your veterinarian.
New products are constantly being developed.
Not all of these products may be legal with your country.

Physiological Data

BLOOD SAMPLING REQUIREMENTS FOR SEROLOGICAL TESTS			
Test	Sample Required	Minimum Sample Required	Anticoagulant Required
Bilirubin	Serum or plasma	1 ml	Heparin
Calcium	Serum or plasma	1 ml	
Chloride	Serum or plasma	1 ml	
Copper	Serum or plasma	1 ml	
Cortisone (cortisol)	Plasma	2 ml	Heparin or EDTA
Creatinine	Serum or plasma	1 ml	
Glucose	Blood	0.2 ml	Oxalate-fluoride
Haematological tests (PCV, Hb, RBC, WBC)	Blood	3 ml	EDTA
Inorganic phosphate	Serum or plasma	1 ml	Oxalate-fluoride or heparin
Iron	Serum or plasma	1 ml	
Lead	Blood	7 – 10 ml	Any lead-free container
Magnesium	Serum or plasma	1 ml	
Platelets	Blood	1 ml	EDTA (not a vacutainer)
Potassium	Serum or plasma	1 ml	
Protein (total, albumin or globulin)	Serum or plasma	5 ml per test	
Selenium	Blood	2 ml	Any
Enzymes	Serum or plasma	5 ml	For the (GSH-Px-test use whole blood – heparin)
Sodium	Serum or plasma	1 ml	
Urea	Serum or plasma	1 ml	
Zinc	Plasma	1 ml	Any one, but avoid contact with rubber materials

HAEMATOLOGY SI (STANDARD INTERNATIONAL)			
		Units	Normal Range
Red Cells	Numbers	x 10^{12}/l	6. 0 – 9.0
	Haemoglobin	g/dl	11.0 – 17.0
	Packed cell volume	l/l	0.37 – 0.5
White Cells	Leukocytes	x 10^{9}/l	10 – 23
	Lymphocytes	x 10^{9}/l	9.8 – 11.4
	Monocytes	x 10^{9}/l	1.0 – 2.5
	Eosinophils	x 10^{9}/l	0.8 – 1.0
	Basophils	x 10^{9}/l	0.1 – 0.2
Platelets		x 10^{9}/l	100 – 900

SEMEN	
Volume	50 – 400 ml - mean 250 ml
Colour	White cloudy
Consistency	Clear/cloudy/gel
Motility	Active some wave motion
Sperm density	10^{6} /100ml
pH	7.3 – 7.8

URINE	
Volume l/day	2 – 6
Specific gravity	1.020 – 1.040
pH	6 – 8
Bilirubin	None
Blood	None
Glucose	None
Protein	None

TEMPERATURE, RESPIRATION, PULSE RATES

Age/Weight	Rectal Temperature °C	Rectal Temperature °F	Respiratory Rate (per minute)	Pulse Rate (per minute)
At birth	39.0	102.0	40 – 50	200 – 250
During suckling	39.2	102.5	30 – 40	80 – 110
At weaning	39.3	102.7	25 – 40	80 – 100
25 – 45 kg	39.0	102.5	30 – 40	80 – 90
45 – 90 kg	38.8	101.8	30 – 40	75 – 85
Pregnant sow	38.6	101.6	15 – 20	70 – 80
During farrowing	39.0 – 40.0	102.0 – 104.0	40 – 50	80 – 100
During lactation	39.1	102.5	20 – 30	70 – 80
Boar	38.6	101.5	15 – 20	70 – 80

BIOCHEMISTRY

Normal Ranges		
		SI Units
Total Protein	g/l	60 – 80
Albumin	g/l	15 – 40
Globulin	g/l	25 – 50
Bilirubin	mcmol/l	0.9 – 5.0
Calcium	mmol/l	2.0 – 3.8
Copper	mcmol/l	10 – 30
Glucose	mmol/l	3.5 – 7.5
Inorganic phosphate	mmol/l	1.4 – 3.2
Iron	mcmol/l	17 – 26
Lead	mcmol/l	0.5 – 2.4
Magnesium	mmol/l	0.8 – 1.2
Potassium	mmol/l	4.0 – 5.5
Selenium	mcmol/l	> 1.5
Vitamin E	mcmol/l	> 2.3

ENZYME TESTS IN SERUM OR PLASMA INTERNATIONAL UNITS (I.U.)

Enzyme	Normal Range
ALT (GPT)	10 – 18
AP	10 – 50
AST (GOT)	10 – 50
CK	< 200
GSH - P x (iu/ml red cells)	> 10
Gamma GT	8 – 50
LD (LDH)	40 – 700
SDH	< 5.0

SAMPLING HERDS TO DETECT EVIDENCE OF INFECTION 95% CONFIDENCE LIMIT

Herd Size	Sample Size to detect > 1+VE Sensitivity (= % = +ve) = Prevalence							
	0.1	0.2	0.5	1	2	5	10	20
50	50	50	50	50	48	35	22	12
100	100	100	100	95	78	45	25	13
150	150	150	148	130	95	49	26	13
200	200	200	190	155	105	51	27	14
300	300	300	260	189	117	54	28	14
500	500	475	349	225	129	56	28	14
750	750	648	412	246	135	57	28	14
1000	950	777	450	258	138	57	29	14
1500	1279	947	493	271	142	58	29	14
2000	1553	1054	517	277	143	58	29	14
5000	2253	1294	564	290	147	59	29	14

99% CONFIDENCE LIMIT

Herd Size	Sample Size to detect > 1+VE Sensitivity (= % = +ve) = Prevalence							
	0.1	0.2	0.5	1	2	5	10	20
50	50	50	50	50	50	42	29	17
100	100	100	100	99	90	59	36	19
150	150	150	150	143	117	68	38	20
200	200	200	198	180	136	73	40	20
300	300	300	286	235	160	78	41	20
500	500	495	420	300	183	83	42	21
750	750	715	530	343	197	85	43	21
1000	950	890	601	368	204	86	43	21
1500	1430	1177	687	395	212	88	44	21
2000	1800	1367	737	410	216	88	44	21
5000	3008	1844	840	438	223	89	44	21

Units of Measurement used in Biological Science

There are two methods used; – the imperial system, also called the foot-pound-second and the metric system, based upon the centimetre-gram-second system. The metric system was modernised in 1960 and an international system of units (SI) agreed. This is now used throughout the world.

Most pig producing countries use the metric system but in some, part metric and part imperial are also used, for example in the USA, who also have US units for some measurements.

This book uses the metric system throughout in the text but for ease of reference, some tables include imperial measurements.

METRIC UNITS AND RELATIVE VALUES

Weight	Volume	Length	Relationship
Gram (g)	Litre (l)	Metre (m)	1
Kilogram (kg)	Kilolitre (kl)	Kilometre (km)	1,000
Decigram (dg)	Decilitre (dl)	Decimetre (dm)	1-10
Centigram (cg)	Centilitre (cl)	Centimetre (cm)	1/100
Milligram (mg)	Millilitre (ml)	Millimetre (mm)	1/1,000
Microgram (mcg)	Microlitre (mcl)	Micrometre (mcm)	1/1,000,000
Nanogram (ng)	Nanolitre (nl)	Nanometre (nm)	1/1,000,000,000
Picogram (pg)	Picolitre (pl)	Picometre (pm)	1/1,000,000,000,000

( ) = abbreviations

HOW TO CONVERT UNITS OF MEASUREMENTS

Imperial to Metric			Metric to Imperial		
To convert	**into**	**multiply by**	**To convert**	**into**	**multiply by**
Length			**Length**		
inches	millimetres	25.4	millimetres	inches	0.0394
inches	centimetres	2.54	centimetres	inches	0.0397
feet	metres	0.3048	metres	feet	3.2808
yards	metres	0.9144	metres	yards	1.0936
miles	kilometres	1.6093	kilometres	miles	0.6214
Area			**Area**		
square inches	square centimetres	6.4516	square centimetres	square inches	0.155
square feet	square metres	0.093	square metres	square feet	10.764
square yards	square metres	0.836	square metres	square yards	1.196
acres	hectares	0.405	hectares	acres	2.471
square miles	square kilometres	1.6093	square kilometres	square miles	0.386
Volume			**Volume**		
cubic inches	cubic centimetres	16.387	cubic centimetres	cubic inches	0.061
cubic feet	cubic metres	0.0283	cubic metres	cubic feet	35.315
cubic yards	cubic metres	0.7646	cubic metres	cubic yards	1.308
fluid ounces	millilitres	28.41	millilitres	fluid ounces	0.0352
pints	litres	0.568	litres	pints	1.760
gallons	litres	4.55	litres	gallons	0.220
Weight			**Weight**		
ounces	grams	28.35	grams	ounces	0.0352
pounds	kilograms	0.45359	kilograms	pounds	2.2046
tons	kilograms	1016	kilograms	tons	0.000984
tons	tonnes	1.016	tonnes	tons	0.9842

Quick Conversion Tables

LENGTH – INCHES – MILLIMETRES – CENTIMETRES

Approximate 1 in = 25 mm (25.4 mm) 1 mm = 1/32 in (0.039 in)
= 2.5 cm (2.54cm) 1 cm = 2/5 in (0.393 in)
(Exact values)

Inches	←→	Millimetres	Inches	←→	Centimetres
	¼	6.4		¼	0.6
	½	12.7		½	1.3
	¾	19.0		¾	1.9
0.04	**1**	25.4	0.39	**1**	2.5
0.08	**2**	50.8	0.79	**2**	5.1
0.12	**3**	76.2	1.18	**3**	7.6
0.16	**4**	101.6	1.57	**4**	10.2
0.20	**5**	127.0	1.97	**5**	12.7
0.24	**6**	152.4	2.36	**6**	15.2
0.28	**7**	177.8	2.76	**7**	17.8
0.31	**8**	203.2	3.15	**8**	20.3
0.35	**9**	228.6	3.54	**9**	22.9
0.39	**10**	254.0	3.94	**10**	25.4
0.43	**11**	279.4	4.34	**11**	27.9
0.47	**12**	304.8	4.72	**12**	30.5

LENGTH – FEET – YARDS – METRES

Approx. 1 ft = 0.3 m (0.3048) 1 m = 3 ft 3/8 in (3.2808 ft)
1 yd = 0.9 m (0.9144) 1 m = 1 yd 3 in (1.0936 yds)
(Exact values)

Feet	←→	Metres	Yard	←→	Metres
3.3	**1**	0.3	1.1	**1**	0.9
6.6	**2**	0.6	2.2	**2**	1.8
9.8	**3**	0.9	3.3	**3**	2.7
13.1	**4**	1.2	4.4	**4**	3.7
16.4	**5**	1.5	5.5	**5**	4.6
19.7	**6**	1.8	6.6	**6**	5.5
23.0	**7**	2.1	7.7	**7**	6.4
26.2	**8**	2.4	8.7	**8**	7.3
29.5	**9**	2.7	9.8	**9**	8.2
32.8	**10**	3.0	10.9	**10**	9.1
49.2	**15**	4.6	16.4	**15**	13.7
65.6	**20**	6.1	21.9	**20**	18.3
82.0	**25**	7.6	27.3	**25**	22.9
98.4	**30**	9.1	32.8	**30**	27.4
114.8	**35**	10.7	38.3	**35**	32.0

LENGTH – MILES – KILOMETRES

1 mile = 1.6 km (1.6093 km)
1 km = 5/8 mile (0.6214 mile)
(Exact values)

Miles	←→	Kilometres
0.6	**1**	1.6
1.2	**2**	3.2
1.9	**3**	4.8
2.5	**4**	6.4
3.1	**5**	8.0
3.7	**6**	9.7
4.3	**7**	11.3
5.0	**8**	12.9
5.6	**9**	14.5
6.2	**10**	16.1
12.4	**20**	32.2
18.6	**30**	48.3
24.9	**40**	64.4
31.1	**50**	80.5

AREA – SQUARE FEET – SQUARE METRES		
1 sq ft = 0.1 m² (0.0929 m²) 1 m² = 103/4 ft² (10.764 ft²)		
Square Feet	⟷	Square Metres
10.8	**1**	0.09
21.5	**2**	0.19
32.3	**3**	0.28
43.1	**4**	0.37
53.8	**5**	0.46
64.6	**6**	0.56
75.3	**7**	0.65
86.1	**8**	0.74
96.9	**9**	0.84
107.6	**10**	0.93
118.4	**11**	1.02
129.2	**12**	1.11
139.9	**13**	1.21
150.7	**14**	1.30
161.5	**15**	1.39
172.2	**16**	1.49
183.0	**17**	1.58
193.8	**18**	1.67
204.5	**19**	1.77
215.3	**20**	1.86

WEIGHT – POUNDS – KILOGRAMS		
1 pound – 0.5 kg (0.454 kgs) 1 kg = 2¼ lbs (2.205 lbs)		
Pounds	⟷	Kilograms
2.2	**1**	0.45
4.4	**2**	0.91
6.6	**3**	1.36
8.8	**4**	1.81
11.0	**5**	2.27
13.2	**6**	2.72
15.4	**7**	3.18
17.6	**8**	3.63
19.8	**9**	4.08
22.2	**10**	4.54
24.3	**11**	4.99
26.5	**12**	5.44
28.7	**13**	5.90
30.9	**14**	6.35
33.1	**15**	6.80
35.3	**16**	7.26
37.5	**17**	7.71
39.7	**18**	8.16
41.9	**19**	8.62
44.1	**20**	9.07

VOLUME – IMPERIAL/METRIC

1 imperial pint = 0.5 litre (0.568 litres)
1 litre = 1¾pints (1.7598 imp pints) = ¼ gallon (0.22 gallon)
1 gallon (8 pints) = 4.5 litres (4.546 litres)

Imp. Pints	↔	Litres	Imp. Gallons	↔	Litres
1.76	**1**	0.57	0.22	**1**	4.55
3.52	**2**	1.14	0.44	**2**	9.09
5.28	**3**	1.70	0.66	**3**	13.64
7.04	**4**	2.27	0.88	**4**	18.18
8.80	**5**	2.84	1.10	**5**	22.73
10.56	**6**	3.41	1.32	**6**	27.28
12.32	**7**	3.98	1.54	**7**	31.82
14.08	**8**	4.55	1.76	**8**	36.37

Note:
One US Pint = 0.833 Imperial Pints
One US Gallon = 0.833 Imperial Gallon
One US Gallon = 3.785 Litres

TEMPERATURE CONVERSION °C – °F

°F = (°C x 1.8) + 32 °C = (°F – 32) √1.8

°C	↔	°F	°C	↔	°F	°C	↔	°F
-32.8	**-27**	-16.6	-16.6	**1**	33.8	-1.7	**29**	84.2
-32.2	**-26**	-14.8	-16.7	**2**	35.6	-1.1	**30**	86.0
-31.7	**-25**	-13.0	-16.1	**3**	37.4	-0.6	**31**	87.8
-31.1	**-24**	-11.2	-15.6	**4**	39.2	0.0	**32**	89.6
-30.6	**-23**	-9.4	-15.0	**5**	41.0	0.6	**33**	91.4
-30.0	**-22**	-7.6	-14.4	**6**	42.8	1.1	**34**	93.2
-29.4	**-21**	-5.8	-13.9	**7**	44.6	1.7	**35**	95.0
-28.9	**-20**	-4.0	-13.3	**8**	46.4	2.2	**36**	96.8
-28.3	**-19**	-2.2	-12.8	**9**	48.2	2.8	**37**	98.6
-27.8	**-18**	-0.4	-12.2	**10**	50.0	3.3	**38**	100.4
-27.2	**-17**	1.4	-11.7	**11**	51.8	3.9	**39**	102.2
-26.7	**-16**	3.2	-11.1	**12**	53.6	4.4	**40**	104.0
-26.1	**-15**	5.0	-10.6	**13**	55.4	5.0	**41**	105.8
-25.6	**-14**	6.8	-10.0	**14**	57.2	5.5	**42**	107.6
-25.0	**-13**	8.6	-9.4	**15**	59.0	6.1	**43**	109.4
-24.4	**-12**	10.4	-8.9	**16**	60.8	6.7	**44**	111.2
-23.9	**-11**	12.2	-8.3	**17**	62.6	7.2	**45**	113.0
-23.3	**-10**	14.0	-7.8	**18**	64.4	7.8	**46**	114.8
-22.8	**-9**	15.8	-7.2	**19**	66.2	8.3	**47**	116.6
-22.2	**-8**	17.6	-6.7	**20**	68.0	8.9	**48**	118.4
-21.7	**-7**	19.4	-6.1	**21**	69.8	9.4	**49**	120.2
-21.1	**-6**	21.2	-5.5	**22**	71.6	10.0	**50**	122.0
-20.6	**-5**	23.0	-5.0	**23**	73.4	10.6	**51**	123.8
-20.0	**-4**	24.8	-4.4	**24**	75.2	11.1	**52**	125.6
-19.4	**-3**	26.6	-3.9	**25**	77.0	11.7	**53**	127.4
-18.9	**-2**	28.4	-3.3	**26**	78.8	12.2	**54**	129.2
-18.3	**-1**	30.2	-2.8	**27**	80.6	12.8	**55**	131.0
-17.8	**0**	32.0	-2.2	**28**	82.4	13.3	**56**	132.8

Growth

The Possible Effects of Feed Changes on the Growing Pig

Increasing the lysine intake by 1 g per day:-
DLWG (daily live weight gain) is increased by up to 10 g per day.
Feed efficiency is improved by up to 0.05.
Fat depths (P2) are reduced by up to 0.1.

Increasing energy intake by 1 MJ/DE per day:-
DLWG is increased by up to 16 g per day.
Feed efficiency is improved by up to 0.06.
Fat depths (P2) increase by up to 0.6 mm.

Increasing the liveweight at slaughter by 1 kg:-
DLWG is increased by 2.5 g per day.
Feed efficiency deteriorates by up to 0.01.
Fat depths (P2) increase by up to 0.2 mm.

Increasing feed intake by 100 g per day:-
DLWG is increased by 30 g per day.
Feed efficiency is virtually unchanged.
Fat depths (P2) increase by up to 0.6 mm

Calculating Days to Slaughter (Growth Rate Unknown)

Information Required	**Example**
Weaning age in weeks	= 3
Average sold per week over the last 3 months	= 90
Average number of weaners, growers and finishers over past 3 months	= 1900
$\frac{\text{month end figures}}{3}$	$\frac{(1800 + 1900 + 2000)}{3} = 1900$

Growth rate : –

$$\frac{\text{Average No. pigs on farm}}{\text{Nos sold/week}} + \text{weaning age} \times 7 = \text{days to slaughter} \qquad \frac{1900}{90} + 3 \times 7 = 168 \text{ days}$$

If the sale weight is 100kg

$$\text{Daily live weight gain} = \frac{100}{168} = 595\text{g from birth to slaughter}$$

$$\frac{100}{147} = 680\text{g from weaning to slaughter}$$

EFFECT OF VARIABLE GROWTH RATE ON DAYS TO SLAUGHTER

Growth Rate per Day (g)	Weight Gain (kg)												
	30	35	40	45	50	55	60	65	70	75	80	85	90
	Days to slaughter												
450	67	77	88	100	111	122	133	144	155	166	177	188	200
500	60	70	80	90	100	110	120	130	140	150	160	170	180
550	54	63	72	81	90	100	109	118	127	136	145	154	163
600	50	58	66	75	83	91	100	108	116	125	133	141	150
650	46	53	61	69	76	84	92	100	107	115	123	130	138
700	43	50	57	64	71	78	85	92	100	107	114	121	128
750	40	46	53	60	66	73	80	86	93	100	106	113	120
800	37	43	50	56	62	69	75	81	87	93	100	106	112
850	35	41	47	52	58	64	70	76	82	88	94	100	105
900	33	38	44	50	55	61	66	72	77	83	88	94	100
950	31	36	42	47	52	57	63	68	73	78	84	89	94

Mating to Farrowing Dates:

MATING TO FARROWING DATE INDICATOR			
Date of Service	Expected Date of Farrowing	Date of Service	Expected Date of Farrowing
Jan 1	Apr 25	Jul 5	Oct 27
6	30	10	Nov 1
11	May 5	15	6
16	10	20	11
21	15	25	16
26	20	30	21
31	25	Aug 4	26
Feb 5	30	9	Dec 1
10	Jun 4	14	6
15	9	19	11
20	14	24	16
25	19	29	21
Mar 2	24	Sep 3	26
7	29	8	31
12	Jul 4	13	Jan 5
17	9	18	10
22	14	23	15
27	19	28	20
Apr 1	24	Oct 3	25
6	29	8	30
11	Aug 3	13	Feb 4
16	8	18	9
21	13	23	14
26	18	28	19
May 1	23	Nov 2	24
6	28	7	Mar 1
11	Sep 2	12	6
16	7	17	11
21	12	22	16
26	17	27	21
31	22	Dec 2	26
Jun 5	27	7	31
10	Oct 2	12	Apr 5
15	7	17	10
20	12	22	15
25	17	27	20
30	22		

Appendix

INDEX

Index

Your Notes

Your Notes

Your Notes

Your Notes

Your Notes

Your Notes

ABBREVIATIONS USED IN THE BOOK

ACOP Approved codes of practice.

AD Aujeszky's disease, synonymous with pseudorabies.

AI Artificial insemination.

AML Acute myeloid leukaemia.

App Actinobacillus pleuropneumonia.

AR Atrophic rhinitis.

ASF African swine fever.

AVM-GSL Authorised veterinary medicine – general sales list.

BDV Border disease virus.

BHC Benziene hexachloride.

BE Blue eye disease.

BVDV Bovine viral diarrhoea virus.

CFT Compliment fixation test. A laboratory test for detecting antibody.

CL Corpus luteum.

CRD Chronic respiratory disease.

CSF Classical swine fever, also called just swine fever.

CT Congenital tremor.

CTC Chlortetracycline. A broad spectrum antibiotic.

DFDM Dark firm dry muscle.

DHHS Defined high health status.

DNA Deoxyribonucleic acid - genetic material.

ED Epidemic diarrhoea. A virus disease also called porcine epidemic diarrhoea (PED).

ELISA A laboratory test to detect antibodies.

EMCV Encephalomyocarditis virus.

EP Mycoplasma (enzootic) pneumonia.

Epe Eperythrozoonosis (old phraseology). A blood disease caused by the bacterium *Mycoplasma suis*.

ESF Electronic sow feeding.

FAT Fluorescent antibody test used for diagnosing disease.

FCE Feed conversion efficiency.

FMD Foot and mouth disease.

FSH Follicle stimulating hormone.

GSL General sales list. Relates to the availability of medicines.

HC Hog cholera, synonymous with classical swine fever.

HD Hepatosis dietetiea.

HEV Haemagglutinating encephalitis virus. Also called vomiting and wasting disease and ontario encephalitis.

HI Haemagglutinating inhibition test.

Hps *Haemophilus parasuis*.

HSE Health and Safety Executive.

IFA Indirect fluorescent antibody test.

IM Intramuscular.

IMPA Immunoperoxidase monolayer assay.

IU International units.

IV Intravenous.

JE Japanese encephalitis.

LH Luteinising hormone. The hormone that causes eggs in the ovaries to be released.

MD Muscular dystrophy.

MEW Medicated early weaning.

MH Malignant hyperthermia or PSS.

MHD Mulberry heart disease.

MJ Megajoules a unit of energy in feed.

MMA Mastitis metritis agalactia.

MMEW Modified medicated early weaning.

MRL Maximum residue limit. Defines the maximum legal amount of drug residue in a tissue.

NE Necrotic enteritis.

NE Net energy.

NFA-VPS Non-food animal medicine – veterinarian, pharmacist, Suitably Qualified Person.

NIP Not in pig.

NPD Non productive days.

OCD Osteochondrosis.

OD Oedema disease.

OEL Occupation exposure limit.

OIE Office International des Epizooties

OPs Organophosphorus compounds.

OTC Oxytetracycline. A broad spectrum antibiotic.

P2 A point over the last rib 65mm from the spine at which fat depth measurements are made.

PAR Progressive atrophic rhinitis.

PCMV Porcine cytomegalovirus.

PCPs Pentachlorophenols.

PCR Polymerase chain reaction. A diagnostic test used for identifying viruses.

PCV Packed cell volume. A measure of the volume of red and white cells in the blood.

PCV2 Porcine circovirus 2.

PCVD Porcine circovirus diseases.

PD Pregnancy diagnosis.

PDNS Porcine dermatitis and nephropathy syndrome.

PE Porcine enteropathy.

PED Porcine epidemic diarrhoea.

Pen V Phenoxymethyl penicillin an oral form, of penicillin.

PFTS Peri-weaning failure to thrive syndrome.

PHE Porcine haemorrhagic enteropathy.

PIA Porcine intestinal adenomatosis.

PMCV Porcine myocarditis virus.

PMt Toxin producing strains of *Pasteurella multocida*.

PMWS Postweaning multisystemic wasting syndrome.

POM Prescription only medicine.

POM-V Prescription-only medicine – veterinarian.

POM-VPS Prescription-only medicine – veterinarian, pharmacist, Suitably Qualified Person.

PPE Personal protective equipment.

PPV Porcine parvovirus.

PRCV Porcine respiratory coronavirus.

PRDC Porcine respiratory disease complex.

PRRSV Porcine reproductive and respiratory syndrome virus.

PRV Pseudorabies, synonymous with Aujeszky's disease.

PSE Pale soft exudate.

PSS Porcine stress syndrome.

PTO Power take off.

QACs Quaternary ammonium compounds.

RNA Ribonucleic acid.

RI Regional ileitis.

RIDDOR Reporting of injuries diseases dangerous occurrences regulations.

SDC Segregated disease control.

SD Swine dysentery.

SEW Segregated early weaning.

SI Standard International units.

SIV Swine influenza virus.

SM Streptococcal meningitis.

SMEDI Stillbirths, Mummification's, Embryo, Death, Infertility.

SN Serum neutralisation test.

SSW Safety system of work.

SVD Swine vesicular disease.

SW Segregated Weaning.

TGE Transmissible gastroenteritis.

TMS Trimethoprim sulpha. Synthetic antibacterial compounds.

TPs Touch preparations.

VES Vesicular exanthema of swine.

VS Vesicular stomatitis

VWD Veterinary written directive or prescription for medications.